Distributed Computing

Principles, Algorithms, and Systems

Distributed computing deals with all forms of computing, information access, and information exchange across multiple processing platforms connected by computer networks. Design of distributed computing systems is a complex task. It requires a solid understanding of the design issues and an in-depth understanding of the theoretical and practical aspects of their solutions. This comprehensive textbook covers the fundamental principles and models underlying the theory, algorithms, and systems aspects of distributed computing.

Broad and detailed coverage of the theory is balanced with practical systems-related problems such as mutual exclusion, deadlock detection, authentication, and failure recovery. Algorithms are carefully selected, lucidly presented, and described without complex proofs. Simple explanations and illustrations are used to elucidate the algorithms. Emerging topics of significant impact, such as peer-to-peer networks and network security, are also covered.

With state-of-the-art algorithms, numerous illustrations, examples, and homework problems, this textbook is invaluable for advanced undergraduate and graduate students of electrical and computer engineering and computer science. Practitioners in data networking and sensor networks will also find this a valuable resource.

Ajay D. Kshemkalyani is a Professor in the Department of Computer Science, at the University of Illinois at Chicago. He was awarded his Ph.D. in Computer and Information Science in 1991 from The Ohio State University. Before moving to academia, he spent several years working on computer networks at IBM Research Triangle Park. In 1999, he received the National Science Foundation's CAREER Award. He is a Senior Member of the IEEE, and his principal areas of research include distributed computing, algorithms, computer networks, and concurrent systems. He currently serves on the editorial board of *Computer Networks*.

Mukesh Singhal is Full Professor and Gartner Group Endowed Chair in Network Engineering in the Department of Computer Science at the University of Kentucky. He was awarded his Ph.D. in Computer Science in 1986 from the University of Maryland, College Park. In 2003, he received the IEEE

Technical Achievement Award, and currently serves on the editorial boards for the *IEEE Transactions on Parallel and Distributed Systems* and the *IEEE Transactions on Computers*. He is a Fellow of the IEEE, and his principal areas of research include distributed systems, computer networks, wireless and mobile computing systems, performance evaluation, and computer security.

Distributed Computing

Principles, Algorithms, and Systems

Ajay D. Kshemkalyani
University of Illinois at Chicago, Chicago

and

Mukesh Singhal
University of Kentucky, Lexington

CAMBRIDGE UNIVERSITY PRESS
Cambridge, New York, Melbourne, Madrid, Cape Town,
Singapore, São Paulo, Delhi, Tokyo, Mexico City

Cambridge University Press
The Edinburgh Building, Cambridge CB2 8RU, UK

Published in the United States of America by Cambridge University Press, New York

www.cambridge.org
Information on this title: www.cambridge.org/9780521189842

First published 2008
First paperback edition (with corrections) 2011

A catalogue record for this publication is available from the British Library

ISBN 978-0-521-87634-6 Hardback
ISBN 978-0-521-18984-2 Paperback

To my father Shri Digambar and
my mother Shrimati Vimala.

Ajay D. Kshemkalyani

To my mother Chandra Prabha Singhal,
my father Brij Mohan Singhal, and my
daughters Meenakshi, Malvika,
and Priyanka.

Mukesh Singhal

Contents

Preface

Background

The field of distributed computing covers all aspects of computing and information access across multiple processing elements connected by any form of communication network, whether local or wide-area in the coverage. Since the advent of the Internet in the 1970s, there has been a steady growth of new applications requiring distributed processing. This has been enabled by advances in networking and hardware technology, the falling cost of hardware, and greater end-user awareness. These factors have contributed to making distributed computing a cost-effective, high-performance, and fault-tolerant reality. Around the turn of the millenium, there was an explosive growth in the expansion and efficiency of the Internet, which was matched by increased access to networked resources through the World Wide Web, all across the world. Coupled with an equally dramatic growth in the wireless and mobile networking areas, and the plummeting prices of bandwidth and storage devices, we are witnessing a rapid spurt in distributed applications and an accompanying interest in the field of distributed computing in universities, governments organizations, and private institutions.

Advances in hardware technology have suddenly made sensor networking a reality, and embedded and sensor networks are rapidly becoming an integral part of everyone's life – from the home network with the interconnected gadgets to the automobile communicating by GPS (global positioning system), to the fully networked office with RFID monitoring. In the emerging global village, distributed computing will be the centerpiece of all computing and information access sub-disciplines within computer science. Clearly, this is a very important field. Moreover, this evolving field is characterized by a diverse range of challenges for which the solutions need to have foundations on solid principles.

The field of distributed computing is very important, and there is a huge demand for a good comprehensive book. This book comprehensively covers all important topics in great depth, combining this with a clarity of explanation and ease of understanding. The book will be particularly valuable to the academic community and the computer industry at large. Writing such a comprehensive book has been a Herculean task and there is a deep sense of satisfaction in knowing that we were able complete it and perform this service to the community.

Description, approach, and features

The book will focus on the fundamental principles and models underlying all aspects of distributed computing. It will address the principles underlying the theory, algorithms, and systems aspects of distributed computing. The manner of presentation of the algorithms is very clear, explaining the main ideas and the intuition with figures and simple explanations rather than getting entangled in intimidating notations and lengthy and hard-to-follow rigorous proofs of the algorithms. The selection of chapter themes is broad and comprehensive, and the book covers all important topics in depth. The selection of algorithms within each chapter has been done carefully to elucidate new and important techniques of algorithm design. Although the book focuses on foundational aspects and algorithms for distributed computing, it thoroughly addresses all practical systems-like problems (e.g., mutual exclusion, deadlock detection, termination detection, failure recovery, authentication, global state and time, etc.) by presenting the theory behind and algorithms for such problems. The book is written keeping in mind the impact of emerging topics such as *peer-to-peer computing* and *network security* on the foundational aspects of distributed computing.

Each chapter contains figures, examples, exercises, a summary, and references.

Readership

This book is aimed as a textbook for the following:

- Graduate students and Senior level undergraduate students in computer science and computer engineering.
- Graduate students in electrical engineering and mathematics. As wireless networks, peer-to-peer networks, and mobile computing continue to grow in importance, an increasing number of students from electrical engineering departments will also find this book necessary.
- Practitioners, systems designers/programmers, and consultants in industry and research laboratories will find the book a very useful reference because it contains state-of-the-art algorithms and principles to address various design issues in distributed systems, as well as the latest references.

Hard and soft prerequisites for the use of this book include the following:

- An undergraduate course in algorithms is required.
- Undergraduate courses in operating systems and computer networks would be useful.
- A reasonable familiarity with programming.

We have aimed for a very comprehensive book that will act as a single source for distributed computing models and algorithms. The book has both depth and breadth of coverage of topics, and is characterized by clear and easy explanations. None of the existing textbooks on distributed computing provides all of these features.

Acknowledgements

This book grew from the notes used in the graduate courses on distributed computing at the Ohio State University, the University of Illinois at Chicago, and at the University of Kentucky. We would like to thank the graduate students at these schools for their contributions to the book in many ways.

The book is based on the published research results of numerous researchers in the field. We have made all efforts to present the material in our own words and have given credit to the original sources of information. We would like to thank all the researchers whose work has been reported in this book. Finally, we would like to thank the staff of Cambridge University Press for providing us with excellent support in the publication of this book.

Access to resources

The following websites will be maintained for the book. Any errors and comments should be sent to ajayk@cs.uic.edu or singhal@cs.uky.edu. Further information about the book can be obtained from the authors' web pages:

- www.cs.uic.edu/~ajayk/DCS-Book
- www.cs.uky.edu/~singhal/DCS-Book.

CHAPTER

1 Introduction

1.1 Definition

A distributed system is a collection of independent entities that cooperate to solve a problem that cannot be individually solved. Distributed systems have been in existence since the start of the universe. From a school of fish to a flock of birds and entire ecosystems of microorganisms, there is communication among mobile intelligent agents in nature. With the widespread proliferation of the Internet and the emerging global village, the notion of distributed computing systems as a useful and widely deployed tool is becoming a reality. For computing systems, a distributed system has been characterized in one of several ways:

- You know you are using one when the crash of a computer you have never heard of prevents you from doing work [23].
- A collection of computers that do not share common memory or a common physical clock, that communicate by a messages passing over a communication network, and where each computer has its own memory and runs its own operating system. Typically the computers are semi-autonomous and are loosely coupled while they cooperate to address a problem collectively [29].
- A collection of independent computers that appears to the users of the system as a single coherent computer [33].
- A term that describes a wide range of computers, from weakly coupled systems such as wide-area networks, to strongly coupled systems such as local area networks, to very strongly coupled systems such as multiprocessor systems [19].

A distributed system can be characterized as a collection of mostly autonomous processors communicating over a communication network and having the following features:

- **No common physical clock** This is an important assumption because it introduces the element of "distribution" in the system and gives rise to the inherent asynchrony amongst the processors.

- **No shared memory** This is a key feature that requires message-passing for communication. This feature implies the absence of the common physical clock.

  It may be noted that a distributed system may still provide the abstraction of a common address space via the distributed shared memory abstraction. Several aspects of shared memory multiprocessor systems have also been studied in the distributed computing literature.
- **Geographical separation** The geographically wider apart that the processors are, the more representative is the system of a distributed system. However, it is not necessary for the processors to be on a wide-area network (WAN). Recently, the network/cluster of workstations (NOW/COW) configuration connecting processors on a LAN is also being increasingly regarded as a small distributed system. This NOW configuration is becoming popular because of the low-cost high-speed off-the-shelf processors now available. The Google search engine is based on the NOW architecture.
- **Autonomy and heterogeneity** The processors are "loosely coupled" in that they have different speeds and each can be running a different operating system. They are usually not part of a dedicated system, but cooperate with one another by offering services or solving a problem jointly.

1.2 Relation to computer system components

A typical distributed system is shown in Figure 1.1. Each computer has a memory-processing unit and the computers are connected by a communication network. Figure 1.2 shows the relationships of the software components that run on each of the computers and use the local operating system and network protocol stack for functioning. The distributed software is also termed as *middleware*. A *distributed execution* is the execution of processes across the distributed system to collaboratively achieve a common goal. An execution is also sometimes termed a *computation* or a *run*.

The distributed system uses a layered architecture to break down the complexity of system design. The middleware is the distributed software that

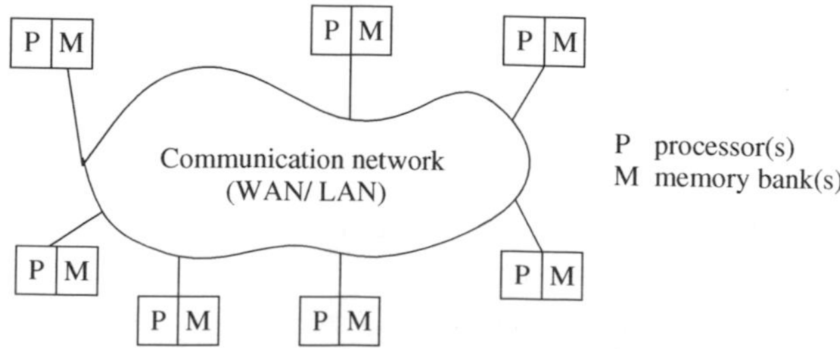

Figure 1.1 A distributed system connects processors by a communication network.

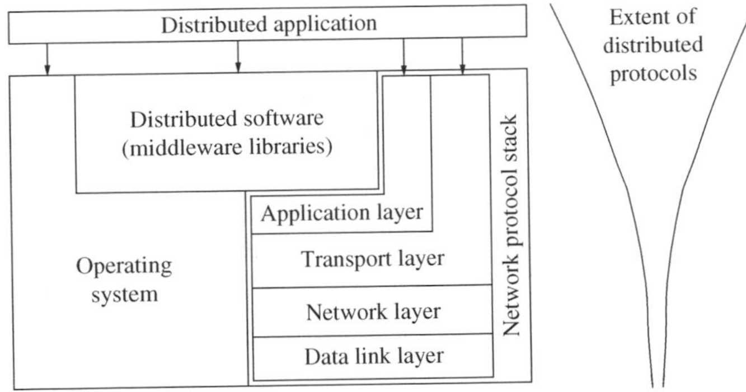

Figure 1.2 Interaction of the software components at each processor.

drives the distributed system, while providing transparency of heterogeneity at the platform level [24]. Figure 1.2 schematically shows the interaction of this software with these system components at each processor. Here we assume that the middleware layer does not contain the traditional application layer functions of the network protocol stack, such as *http*, *mail*, *ftp*, and *telnet*. Various primitives and calls to functions defined in various libraries of the middleware layer are embedded in the user program code. There exist several libraries to choose from to invoke primitives for the more common functions – such as reliable and ordered multicasting – of the middleware layer. There are several standards such as Object Management Group's (OMG) common object request broker architecture (CORBA) [36], and the remote procedure call (RPC) mechanism [1,11]. The RPC mechanism conceptually works like a local procedure call, with the difference that the procedure code may reside on a remote machine, and the RPC software sends a message across the network to invoke the remote procedure. It then awaits a reply, after which the procedure call completes from the perspective of the program that invoked it. Currently deployed commercial versions of middleware often use CORBA, DCOM (distributed component object model), Java, and RMI (remote method invocation) [7] technologies. The message-passing interface (MPI) [20,30] developed in the research community is an example of an interface for various communication functions.

1.3 Motivation

The motivation for using a distributed system is some or all of the following requirements:

1. **Inherently distributed computations** In many applications such as money transfer in banking, or reaching consensus among parties that are geographically distant, the computation is inherently distributed.
2. **Resource sharing** Resources such as peripherals, complete data sets in databases, special libraries, as well as data (variable/files) cannot be

fully replicated at all the sites because it is often neither practical nor cost-effective. Further, they cannot be placed at a single site because access to that site might prove to be a bottleneck. Therefore, such resources are typically distributed across the system. For example, distributed databases such as DB2 partition the data sets across several servers, in addition to replicating them at a few sites for rapid access as well as reliability.

3. **Access to geographically remote data and resources** In many scenarios, the data cannot be replicated at every site participating in the distributed execution because it may be too large or too sensitive to be replicated. For example, payroll data within a multinational corporation is both too large and too sensitive to be replicated at every branch office/site. It is therefore stored at a central server which can be queried by branch offices. Similarly, special resources such as supercomputers exist only in certain locations, and to access such supercomputers, users need to log in remotely.

   Advances in the design of resource-constrained mobile devices as well as in the wireless technology with which these devices communicate have given further impetus to the importance of distributed protocols and middleware.

4. **Enhanced reliability** A distributed system has the inherent potential to provide increased reliability because of the possibility of replicating resources and executions, as well as the reality that geographically distributed resources are not likely to crash/malfunction at the same time under normal circumstances. Reliability entails several aspects:
   - availability, i.e., the resource should be accessible at all times;
   - integrity, i.e., the value/state of the resource should be correct, in the face of concurrent access from multiple processors, as per the semantics expected by the application;
   - fault-tolerance, i.e., the ability to recover from system failures, where such failures may be defined to occur in one of many failure models, which we will study in Chapters 5 and 14.

5. **Increased performance/cost ratio** By resource sharing and accessing geographically remote data and resources, the performance/cost ratio is increased. Although higher throughput has not necessarily been the main objective behind using a distributed system, nevertheless, any task can be partitioned across the various computers in the distributed system. Such a configuration provides a better performance/cost ratio than using special parallel machines. This is particularly true of the NOW configuration.

In addition to meeting the above requirements, a distributed system also offers the following advantages:

6. **Scalability** As the processors are usually connected by a wide-area network, adding more processors does not pose a direct bottleneck for the communication network.

7. **Modularity and incremental expandability** Heterogeneous processors may be easily added into the system without affecting the performance, as long as those processors are running the same middleware algorithms. Similarly, existing processors may be easily replaced by other processors.

1.4 Relation to parallel multiprocessor/multicomputer systems

The characteristics of a distributed system were identified above. A typical distributed system would look as shown in Figure 1.1. However, how does one classify a system that meets some but not all of the characteristics? Is the system still a distributed system, or does it become a parallel multiprocessor system? To better answer these questions, we first examine the architecture of parallel systems, and then examine some well-known taxonomies for multiprocessor/multicomputer systems.

1.4.1 Characteristics of parallel systems

A parallel system may be broadly classified as belonging to one of three types:

1. A *multiprocessor system* is a parallel system in which the multiple processors have *direct access to shared memory* which forms a common address space. The architecture is shown in Figure 1.3(a). Such processors usually do not have a common clock.

   A multiprocessor system *usually* corresponds to a uniform memory access (UMA) architecture in which the access latency, i.e., waiting time, to complete an access to any memory location from any processor is the same. The processors are in very close physical proximity and are connected by an interconnection network. Interprocess communication across processors is traditionally through read and write operations on the shared memory, although the use of message-passing primitives such as those provided by

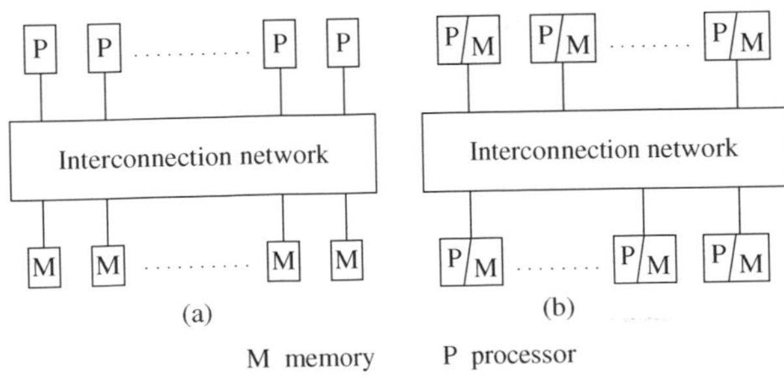

Figure 1.3 Two standard architectures for parallel systems. (a) Uniform memory access (UMA) multiprocessor system. (b) Non-uniform memory access (NUMA) multiprocessor. In both architectures, the processors may locally cache data from memory.

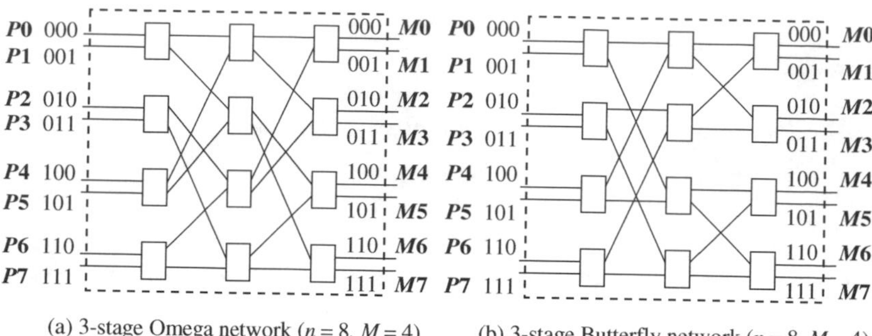

Figure 1.4 Interconnection networks for shared memory multiprocessor systems. (a) Omega network [4] for $n = 8$ processors *P0–P7* and memory banks *M0–M7*. (b) Butterfly network [10] for $n = 8$ processors *P0–P7* and memory banks *M0–M7*.

the MPI, is also possible (using emulation on the shared memory). All the processors usually run the same operating system, and both the hardware and software are very tightly coupled.

The processors are usually of the same type, and are housed within the same box/container with a shared memory. The interconnection network to access the memory may be a bus, although for greater efficiency, it is usually a *multistage switch* with a symmetric and regular design.

Figure 1.4 shows two popular interconnection networks – the Omega network [4] and the Butterfly network [10], each of which is a multi-stage network formed of 2×2 switching elements. Each 2×2 switch allows data on either of the two input wires to be switched to the upper or the lower output wire. In a single step, however, only one data unit can be sent on an output wire. So if the data from both the input wires is to be routed to the same output wire in a single step, there is a collision. Various techniques such as buffering or more elaborate interconnection designs can address collisions.

Each 2×2 switch is represented as a rectangle in the figure. Furthermore, a n-input and n-output network uses $log\, n$ stages and $log\, n$ bits for addressing. Routing in the 2×2 switch at stage k uses only the kth bit, and hence can be done at clock speed in hardware. The multi-stage networks can be constructed recursively, and the interconnection pattern between any two stages can be expressed using an iterative or a recursive generating function. Besides the Omega and Butterfly (banyan) networks, other examples of multistage interconnection networks are the Clos [9] and the shuffle-exchange networks [37]. Each of these has very interesting mathematical properties that allow rich connectivity between the processor bank and memory bank.

Omega interconnection function The Omega network which connects n processors to n memory units has $(n/2)log_2\, n$ switching elements of size 2×2 arranged in $log_2\, n$ stages. Between each pair of adjacent stages of the Omega network, a link exists between output i of a stage and the input j to the next stage according to the following *perfect shuffle* pattern which is a left-rotation operation on the binary representation of i to get j. The iterative generation function is as follows:

$$j = \begin{cases} 2i, & \text{for } 0 \leq i \leq n/2-1, \\ 2i+1-n, & \text{for } n/2 \leq i \leq n-1. \end{cases} \quad (1.1)$$

Consider any stage of switches. Informally, the upper (lower) input lines for each switch come in sequential order from the upper (lower) half of the switches in the earlier stage.

With respect to the Omega network in Figure 1.4(a), $n = 8$. Hence, for any stage, for the outputs i, where $0 \leq i \leq 3$, the output i is connected to input $2i$ of the next stage. For $4 \leq i \leq 7$, the output i of any stage is connected to input $2i+1-n$ of the next stage.

Omega routing function The routing function from input line i to output line j considers only j and the stage number s, where $s \in [0, log_2 n - 1]$. In a stage s switch, if the $s+1$th MSB (most significant bit) of j is 0, the data is routed to the upper output wire, otherwise it is routed to the lower output wire.

Butterfly interconnection function Unlike the Omega network, the generation of the interconnection pattern between a pair of adjacent stages depends not only on n but also on the stage number s. The recursive expression is as follows. Let there be $M = n/2$ switches per stage, and let a switch be denoted by the tuple $\langle x, s\rangle$, where $x \in [0, M-1]$ and stage $s \in [0, log_2 n - 1]$.

The two outgoing edges from any switch $\langle x, s\rangle$ are as follows. There is an edge from switch $\langle x, s\rangle$ to switch $\langle y, s+1\rangle$ if (i) $x = y$ or (ii) x XOR y has exactly one 1 bit, which is in the $(s+1)$th MSB. For stage s, apply the rule above for $M/2^s$ switches.

Whether the two incoming connections go to the upper or the lower input port is not important because of the routing function, given below.

Example Consider the Butterfly network in Figure 1.4(b), $n = 8$ and $M = 4$. There are three stages, $s = 0, 1, 2$, and the interconnection pattern is defined between $s = 0$ and $s = 1$ and between $s = 1$ and $s = 2$. The switch number x varies from 0 to 3 in each stage, i.e., x is a 2-bit string. (Note that unlike the Omega network formulation using input and output lines given above, this formulation uses switch numbers. Exercise 1.5 asks you to prove a formulation of the Omega interconnection pattern using switch numbers instead of input and output port numbers.)

Consider the first stage interconnection ($s = 0$) of a butterfly of size M, and hence having $log_2 2M$ stages. For stage $s = 0$, as per rule (i), the first output line from switch 00 goes to the input line of switch 00 of stage $s = 1$. As per rule (ii), the second output line of switch 00 goes to input line of switch 10 of stage $s = 1$. Similarly, $x = 01$ has one output line go to an input line of switch 11 in stage $s = 1$. The other connections in this stage can be determined similarly. For stage $s = 1$ connecting to stage $s = 2$, we apply the rules considering only $M/2^1 = M/2$ switches, i.e., we build two butterflies of size $M/2$ – the "upper half" and the "lower half" switches. The recursion terminates for $M/2^s = 1$, when there is a single switch.

Butterfly routing function In a stage s switch, if the $s+1$th MSB of j is 0, the data is routed to the upper output wire, otherwise it is routed to the lower output wire.

Observe that for the Butterfly and the Omega networks, the paths from the different inputs to any one output form a spanning tree. This implies that collisions will occur when data is destined to the same output line. However, the advantage is that data can be combined at the switches if the application semantics (e.g., summation of numbers) are known.

2. A *multicomputer parallel system* is a parallel system in which the multiple processors *do not have direct access to shared memory*. The memory of the multiple processors may or may not form a common address space. Such computers usually do not have a common clock. The architecture is shown in Figure 1.3(b).

   The processors are in close physical proximity and are usually very tightly coupled (homogenous hardware and software), and connected by an interconnection network. The processors communicate either via a common address space or via message-passing. A multicomputer system that has a common address space *usually* corresponds to a non-uniform memory access (NUMA) architecture in which the latency to access various shared memory locations from the different processors varies.

   Examples of parallel multicomputers are: the NYU Ultracomputer and the Sequent shared memory machines, the CM* Connection machine and processors configured in regular and symmetrical topologies such as an array or mesh, ring, torus, cube, and hypercube (message-passing machines). The regular and symmetrical topologies have interesting mathematical properties that enable very easy routing and provide many rich features such as alternate routing.

   Figure 1.5(a) shows a wrap-around 4×4 mesh. For a $k \times k$ mesh which will contain k^2 processors, the maximum path length between any two processors is $2(k/2-1)$. Routing can be done along the Manhattan grid. Figure 1.5(b) shows a four-dimensional hypercube. A k-dimensional hypercube has 2^k processor-and-memory units [13,21]. Each such unit is a node in the hypercube, and has a unique k-bit label. Each of the k dimensions is associated with a bit position in the label. The labels of any two adjacent nodes are identical except for the bit position corresponding to the dimension in which the two nodes differ. Thus, the processors are labelled such that the shortest path between any two processors is the *Hamming distance* (defined as the number of bit positions in which the two equal sized bit strings differ) between the processor labels. This is clearly bounded by k.

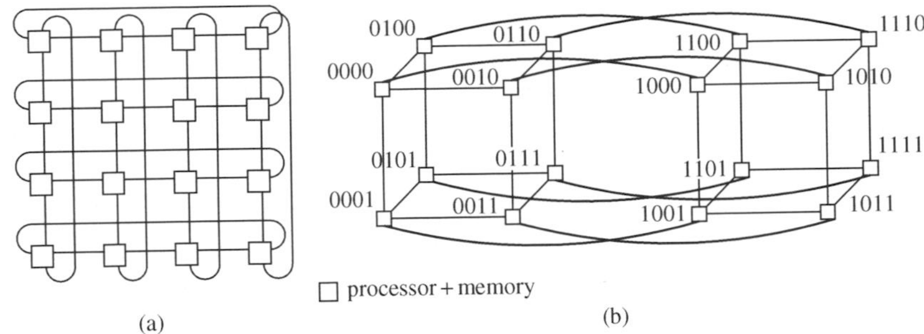

Figure 1.5 Some popular topologies for multicomputer shared-memory machines. (a) Wrap-around 2D-mesh, also known as torus. (b) Hypercube of dimension 4.

Example Nodes 0101 and 1100 have a Hamming distance of 2. The shortest path between them has length 2.

Routing in the hypercube is done hop-by-hop. At any hop, the message can be sent along any dimension corresponding to the bit position in which the current node's address and the destination address differ. The 4D hypercube shown in the figure is formed by connecting the corresponding edges of two 3D hypercubes (corresponding to the left and right "cubes" in the figure) along the fourth dimension; the labels of the 4D hypercube are formed by prepending a "0" to the labels of the left 3D hypercube and prepending a "1" to the labels of the right 3D hypercube. This can be extended to construct hypercubes of higher dimensions. Observe that there are multiple routes between any pair of nodes, which provides fault-tolerance as well as a congestion control mechanism. The hypercube and its variant topologies have very interesting mathematical properties with implications for routing and fault-tolerance.

3. *Array processors* belong to a class of parallel computers that are physically co-located, are very tightly coupled, and have a common system clock (but may not share memory and communicate by passing data using messages). Array processors and systolic arrays that perform tightly synchronized processing and data exchange in lock-step for applications such as DSP and image processing belong to this category. These applications usually involve a large number of iterations on the data. This class of parallel systems has a very niche market.

The distinction between UMA multiprocessors on the one hand, and NUMA and message-passing multicomputers on the other, is important because the algorithm design and data and task partitioning among the processors must account for the variable and unpredictable latencies in accessing memory/communication [22]. As compared to UMA systems and array processors, NUMA and message-passing multicomputer systems are less suitable when the degree of granularity of accessing shared data and communication is very fine.

The primary and most efficacious use of parallel systems is for obtaining a higher throughput by dividing the computational workload among the processors. The tasks that are most amenable to higher speedups on parallel systems are those that can be partitioned into subtasks very nicely, involving much number-crunching and relatively little communication for synchronization. Once the task has been decomposed, the processors perform large vector, array, and matrix computations that are common in scientific applications. Searching through large state spaces can be performed with significant speedup on parallel machines. While such parallel machines were an object of much theoretical and systems research in the 1980s and early 1990s, they have not proved to be economically viable for two related reasons. First, the overall market for the applications that can potentially attain high speedups is relatively small. Second, due to economy of scale and the high processing power offered by relatively inexpensive off-the-shelf networked PCs, specialized parallel machines are not cost-effective to manufacture. They additionally require special compiler and other system support for maximum throughput.

1.4.2 Flynn's taxonomy

Flynn [14] identified four processing modes, based on whether the processors execute the same or different instruction streams at the same time, and whether or not the processors processed the same (identical) data at the same time. It is instructive to examine this classification to understand the range of options used for configuring systems:

- **Single instruction stream, single data stream (SISD)**
  This mode corresponds to the conventional processing in the von Neumann paradigm with a single CPU, and a single memory unit connected by a system bus.
- **Single instruction stream, multiple data stream (SIMD)**
  This mode corresponds to the processing by multiple homogenous processors which execute in lock-step on different data items. Applications that involve operations on large arrays and matrices, such as scientific applications, can best exploit systems that provide the SIMD mode of operation because the data sets can be partitioned easily.

  Several of the earliest parallel computers, such as Illiac-IV, MPP, CM2, and MasPar MP-1 were SIMD machines. Vector processors, array processors' and systolic arrays also belong to the SIMD class of processing. Recent SIMD architectures include co-processing units such as the MMX units in Intel processors (e.g., Pentium with the streaming SIMD extensions (SSE) options) and DSP chips such as the Sharc [22].
- **Multiple instruction stream, single data stream (MISD)**
  This mode corresponds to the execution of different operations in parallel on the same data. This is a specialized mode of operation with limited but niche applications, e.g., visualization.

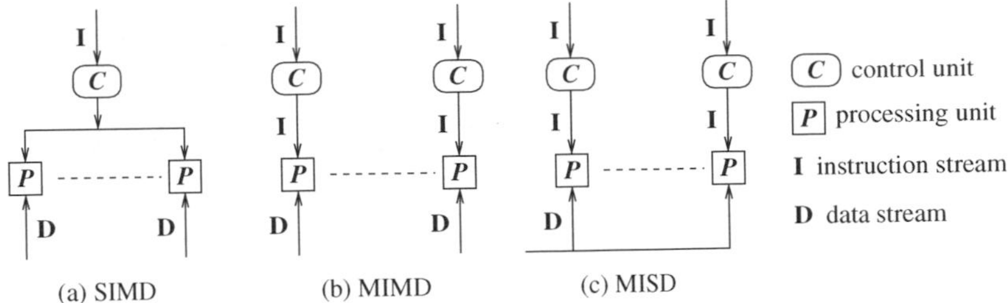

Figure 1.6 Flynn's taxonomy of SIMD, MIMD, and MISD architectures for multiprocessor/multicomputer systems.

- **Multiple instruction stream, multiple data stream (MIMD)**
  In this mode, the various processors execute different code on different data. This is the mode of operation in distributed systems as well as in the vast majority of parallel systems. There is no common clock among the system processors. Sun Ultra servers, multicomputer PCs, and IBM SP machines are examples of machines that execute in MIMD mode.

SIMD, MISD, and MIMD architectures are illustrated in Figure 1.6. MIMD architectures are most general and allow much flexibility in partitioning code and data to be processed among the processors. MIMD architectures also include the classically understood mode of execution in distributed systems.

1.4.3 Coupling, parallelism, concurrency, and granularity

Coupling

The degree of coupling among a set of modules, whether hardware or software, is measured in terms of the interdependency and binding and/or homogeneity among the modules. When the degree of coupling is high (low), the modules are said to be tightly (loosely) coupled. SIMD and MISD architectures generally tend to be tightly coupled because of the common clocking of the shared instruction stream or the shared data stream. Here we briefly examine various MIMD architectures in terms of coupling:

- Tightly coupled multiprocessors (with UMA shared memory). These may be either switch-based (e.g., NYU Ultracomputer, RP3) or bus-based (e.g., Sequent, Encore).
- Tightly coupled multiprocessors (with NUMA shared memory or that communicate by message passing). Examples are the SGI Origin 2000 and the Sun Ultra HPC servers (that communicate via NUMA shared memory), and the hypercube and the torus (that communicate by message passing).
- Loosely coupled multicomputers (without shared memory) physically co-located. These may be bus-based (e.g., NOW connected by a LAN or Myrinet card) or using a more general communication network, and the processors may be heterogenous. In such systems, processors neither share

memory nor have a common clock, and hence may be classified as distributed systems – however, the processors are very close to one another, which is characteristic of a parallel system. As the communication latency may be significantly lower than in wide-area distributed systems, the solution approaches to various problems may be different for such systems than for wide-area distributed systems.

- Loosely coupled multicomputers (without shared memory and without common clock) that are physically remote. These correspond to the conventional notion of distributed systems.

Parallelism or speedup of a program on a specific system

This is a measure of the relative speedup of a specific program, on a given machine. The speedup depends on the number of processors and the mapping of the code to the processors. It is expressed as the ratio of the time $T(1)$ with a single processor, to the time $T(n)$ with n processors.

Parallelism within a parallel/distributed program

This is an aggregate measure of the percentage of time that all the processors are executing CPU instructions productively, as opposed to waiting for communication (either via shared memory or message-passing) operations to complete. The term is traditionally used to characterize parallel programs. If the aggregate measure is a function of only the code, then the parallelism is independent of the architecture. Otherwise, this definition degenerates to the definition of parallelism in the previous section.

Concurrency of a program

This is a broader term that means roughly the same as parallelism of a program, but is used in the context of distributed programs. The *parallelism/concurrency* in a parallel/distributed program can be measured by the ratio of the number of local (non-communication and non-shared memory access) operations to the total number of operations, including the communication or shared memory access operations.

Granularity of a program

The ratio of the amount of computation to the amount of communication within the parallel/distributed program is termed as *granularity*. If the degree of parallelism is coarse-grained (fine-grained), there are relatively many more (fewer) productive CPU instruction executions, compared to the number of times the processors communicate either via shared memory or message-passing and wait to get synchronized with the other processors. Programs with fine-grained parallelism are best suited for tightly coupled systems. These typically include SIMD and MISD architectures, tightly coupled MIMD multiprocessors (that have shared memory), and loosely coupled multicomputers (without shared memory) that are physically colocated. If programs with fine-grained parallelism were run over loosely coupled multiprocessors that are physically remote, the latency delays for the frequent communication over the WAN would significantly degrade the overall throughput. As a corollary, it follows that on such loosely coupled multicomputers, programs with a coarse-grained communication/message-passing granularity will incur substantially less overhead.

Figure 1.2 showed the relationships between the local operating system, the middleware implementing the distributed software, and the network protocol stack. Before moving on, we identify various classes of multiprocessor/multicomputer operating systems:

- The operating system running on loosely coupled processors (i.e., heterogenous and/or geographically distant processors), which are themselves running loosely coupled software (i.e., software that is heterogenous), is classified as a *network operating system*. In this case, the application cannot run any significant distributed function that is not provided by the application layer of the network protocol stacks on the various processors.
- The operating system running on loosely coupled processors, which are running tightly coupled software (i.e., the middleware software on the processors is homogenous), is classified as a *distributed operating system*.
- The operating system running on tightly coupled processors, which are themselves running tightly coupled software, is classified as a *multiprocessor operating system*. Such a parallel system can run sophisticated algorithms contained in the tightly coupled software.

1.5 Message-passing systems versus shared memory systems

Shared memory systems are those in which there is a (common) shared address space throughout the system. Communication among processors takes place via shared data variables, and control variables for synchronization among the processors. Semaphores and monitors that were originally designed for shared memory uniprocessors and multiprocessors are examples of how synchronization can be achieved in shared memory systems. All multicomputer (NUMA as well as message-passing) systems that do not have a shared address space provided by the underlying architecture and hardware necessarily communicate by message passing. Conceptually, programmers find it easier to program using shared memory than by message passing. For this and several other reasons that we examine later, the abstraction called *shared memory* is sometimes provided to simulate a shared address space. For a distributed system, this abstraction is called *distributed shared memory*. Implementing this abstraction has a certain cost but it simplifies the task of the application programmer. There also exists a well-known folklore result that communication via message-passing can be simulated by communication via shared memory and vice-versa. Therefore, the two paradigms are equivalent.

1.5.1 Emulating message-passing on a shared memory system (*MP* → *SM*)

The shared address space can be partitioned into disjoint parts, one part being assigned to each processor. "Send" and "receive" operations can be implemented by writing to and reading from the destination/sender processor's address space, respectively. Specifically, a separate location can be reserved as the mailbox for each ordered pair of processes. A P_i–P_j message-passing can be emulated by a write by P_i to the mailbox and then a read by P_j from the mailbox. In the simplest case, these mailboxes can be assumed to have unbounded size. The write and read operations need to be controlled using synchronization primitives to inform the receiver/sender after the data has been sent/received.

1.5.2 Emulating shared memory on a message-passing system (*SM* → *MP*)

This involves the use of "send" and "receive" operations for "write" and "read" operations. Each shared location can be modeled as a separate process; "write" to a shared location is emulated by sending an update message to the corresponding owner process; a "read" to a shared location is emulated by sending a query message to the owner process. As accessing another processor's memory requires send and receive operations, this emulation is expensive. Although emulating shared memory might seem to be more attractive from a programmer's perspective, it must be remembered that in a distributed system, it is only an abstraction. Thus, the latencies involved in read and write operations may be high even when using shared memory emulation because the read and write operations are implemented by using network-wide communication under the covers.

An application can of course use a combination of shared memory and message-passing. In a MIMD message-passing multicomputer system, each "processor" may be a tightly coupled multiprocessor system with shared memory. Within the multiprocessor system, the processors communicate via shared memory. Between two computers, the communication is by message passing. As message-passing systems are more common and more suited for wide-area distributed systems, we will consider message-passing systems more extensively than we consider shared memory systems.

1.6 Primitives for distributed communication

1.6.1 Blocking/non-blocking, synchronous/asynchronous primitives

Message send and message receive communication primitives are denoted *Send()* and *Receive()*, respectively. A *Send* primitive has at least two parameters – the destination, and the buffer in the user space, containing the data to be sent. Similarly, a *Receive* primitive has at least two parameters – the source from which the data is to be received (this could be a wildcard), and the user buffer into which the data is to be received.

There are two ways of sending data when the *Send* primitive is invoked – the buffered option and the unbuffered option. The *buffered option* which is the standard option copies the data from the user buffer to the kernel buffer. The data later gets copied from the kernel buffer onto the network. In the *unbuffered option*, the data gets copied directly from the user buffer onto the network. For the *Receive* primitive, the buffered option is usually required because the data may already have arrived when the primitive is invoked, and needs a storage place in the kernel.

The following are some definitions of blocking/non-blocking and synchronous/asynchronous primitives [12]:

- **Synchronous primitives** A *Send* or a *Receive* primitive is *synchronous* if both the *Send()* and *Receive()* handshake with each other. The processing for the *Send* primitive completes only after the invoking processor learns that the other corresponding *Receive* primitive has also been invoked and that the receive operation has been completed. The processing for the *Receive* primitive completes when the data to be received is copied into the receiver's user buffer.
- **Asynchronous primitives** A *Send* primitive is said to be *asynchronous* if control returns back to the invoking process after the data item to be sent has been copied out of the user-specified buffer.
  It does not make sense to define asynchronous *Receive* primitives.
- **Blocking primitives** A primitive is *blocking* if control returns to the invoking process after the processing for the primitive (whether in synchronous or asynchronous mode) completes.
- **Non-blocking primitives** A primitive is *non-blocking* if control returns back to the invoking process immediately after invocation, even though the operation has not completed. For a non-blocking *Send*, control returns to the process even before the data is copied out of the user buffer. For a non-blocking *Receive*, control returns to the process even before the data may have arrived from the sender.

  For non-blocking primitives, a return parameter on the primitive call returns a system-generated *handle* which can be later used to check the status of completion of the call. The process can check for the completion of the call in two ways. First, it can keep checking (in a loop or periodically) if the handle has been flagged or *posted*. Second, it can issue a *Wait* with a list of handles as parameters. The *Wait* call usually blocks until one of the parameter handles is posted. Presumably after issuing the primitive in non-blocking mode, the process has done whatever actions it could and now needs to know the status of completion of the call, therefore using a blocking *Wait()* call is usual programming practice. The code for a non-blocking *Send* would look as shown in Figure 1.7.

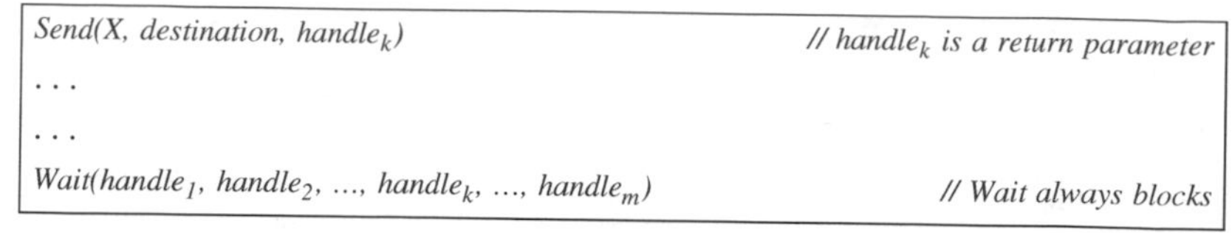

```
Send(X, destination, handle_k)                       // handle_k is a return parameter
...
...
Wait(handle_1, handle_2, ..., handle_k, ..., handle_m)          // Wait always blocks
```

Figure 1.7 A non-blocking *send* primitive. When the *Wait* call returns, at least one of its parameters is posted.

If at the time that *Wait()* is issued, the processing for the primitive (whether synchronous or asynchronous) has completed, the *Wait* returns immediately. The completion of the processing of the primitive is detectable by checking the value of $handle_k$. If the processing of the primitive has not completed, the *Wait* blocks and waits for a signal to wake it up. When the processing for the primitive completes, the communication subsystem software sets the value of $handle_k$ and wakes up (signals) any process with a *Wait* call blocked on this $handle_k$. This is called *posting* the completion of the operation.

There are therefore four versions of the *Send* primitive – synchronous blocking, synchronous non-blocking, asynchronous blocking, and asynchronous non-blocking. For the *Receive* primitive, there are the blocking synchronous and non-blocking synchronous versions. These versions of the primitives are illustrated in Figure 1.8 using a timing diagram. Here, three time lines are

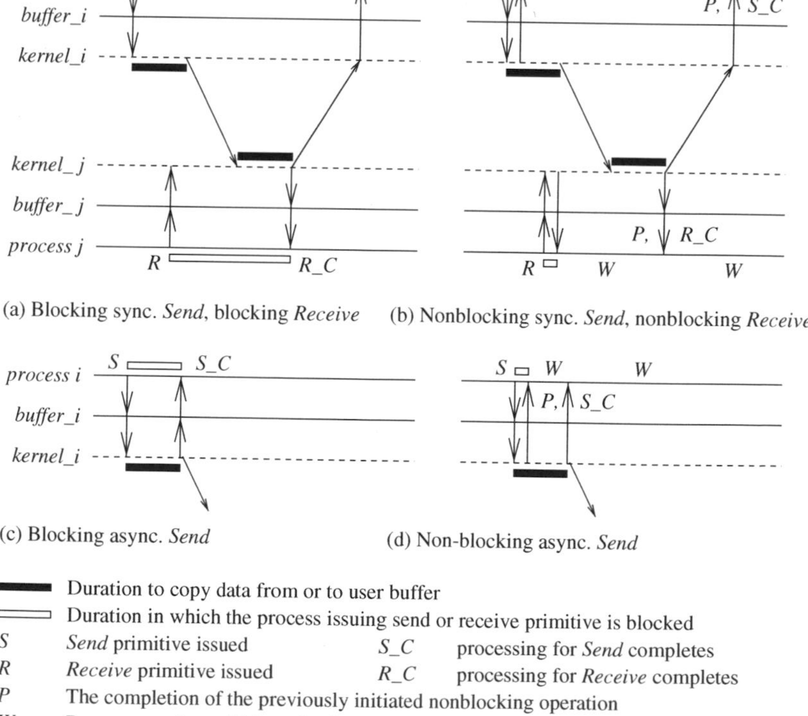

Figure 1.8 Blocking/non-blocking and synchronous/asynchronous primitives [12]. Process P_i is sending and process P_j is receiving. (a) Blocking synchronous *Send* and blocking (synchronous) *Receive*. (b) Non-blocking synchronous *Send* and nonblocking (synchronous) *Receive*. (c) Blocking asynchronous *Send*. (d) Non-blocking asynchronous *Send*.

shown for each process: (1) for the process execution, (2) for the user buffer from/to which data is sent/received, and (3) for the kernel/communication subsystem.

- **Blocking synchronous *Send*** (See Figure 1.8(a)) The data gets copied from the user buffer to the kernel buffer and is then sent over the network. After the data is copied to the receiver's system buffer and a *Receive* call has been issued, an acknowledgement back to the sender causes control to return to the process that invoked the *Send* operation and completes the *Send*.
- **non-blocking synchronous *Send*** (See Figure 1.8(b)) Control returns back to the invoking process as soon as the copy of data from the user buffer to the kernel buffer is initiated. A parameter in the non-blocking call also gets set with the handle of a location that the user process can later check for the completion of the synchronous send operation. The location gets posted after an acknowledgement returns from the receiver, as per the semantics described for (a). The user process can keep checking for the completion of the non-blocking synchronous *Send* by testing the returned handle, or it can invoke the blocking *Wait* operation on the returned handle (Figure 1.8(b)).
- **Blocking asynchronous *Send*** (See Figure 1.8(c)) The user process that invokes the *Send* is blocked until the data is copied from the user's buffer to the kernel buffer. (For the unbuffered option, the user process that invokes the *Send* is blocked until the data is copied from the user's buffer to the network.)
- **non-blocking asynchronous *Send*** (See Figure 1.8(d)) The user process that invokes the *Send* is blocked until the transfer of the data from the user's buffer to the kernel buffer is initiated. (For the unbuffered option, the user process that invokes the *Send* is blocked until the transfer of the data from the user's buffer to the network is initiated.) Control returns to the user process as soon as this transfer is initiated, and a parameter in the non-blocking call also gets set with the handle of a location that the user process can check later using the *Wait* operation for the completion of the asynchronous *Send* operation. The asynchronous *Send* completes when the data has been copied out of the user's buffer. The checking for the completion may be necessary if the user wants to reuse the buffer from which the data was sent.
- **Blocking *Receive*** (See Figure 1.8(a)) The *Receive* call blocks until the data expected arrives and is written in the specified user buffer. Then control is returned to the user process.
- **non-blocking *Receive*** (See Figure 1.8(b)) The *Receive* call will cause the kernel to register the call and return the handle of a location that the user process can later check for the completion of the non-blocking *Receive* operation. This location gets posted by the kernel after the expected data arrives and is copied to the user-specified buffer. The user process can

check for the completion of the non-blocking *Receive* by invoking the *Wait* operation on the returned handle. (If the data has already arrived when the call is made, it would be pending in some kernel buffer, and still needs to be copied to the user buffer.)

A synchronous *Send* is easier to use from a programmer's perspective because the handshake between the *Send* and the *Receive* makes the communication appear instantaneous, thereby simplifying the program logic. The "instantaneity" is, of course, only an illusion, as can be seen from Figure 1.8(a) and (b). In fact, the *Receive* may not get issued until much after the data arrives at P_j, in which case the data arrived would have to be buffered in the system buffer at P_j and not in the user buffer. At the same time, the sender would remain blocked. Thus, a synchronous *Send* lowers the efficiency within process P_i.

The non-blocking asynchronous *Send* (see Figure 1.8(d)) is useful when a large data item is being sent because it allows the process to perform other instructions in parallel with the completion of the *Send*. The non-blocking synchronous *Send* (see Figure 1.8(b)) also avoids the potentially large delays for handshaking, particularly when the receiver has not yet issued the *Receive* call. The non-blocking *Receive* (see Figure 1.8(b)) is useful when a large data item is being received and/or when the sender has not yet issued the *Send* call, because it allows the process to perform other instructions in parallel with the completion of the *Receive*. Note that if the data has already arrived, it is stored in the kernel buffer, and it may take a while to copy it to the user buffer specified in the *Receive* call. For non-blocking calls, however, the burden on the programmer increases because he or she has to keep track of the completion of such operations in order to meaningfully reuse (write to or read from) the user buffers. Thus, conceptually, blocking primitives are easier to use.

1.6.2 Processor synchrony

As opposed to the classification of synchronous and asynchronous communication primitives, there is also the classification of synchronous versus asynchronous processors. *Processor synchrony* indicates that all the processors execute in lock-step with their clocks synchronized. As this synchrony is not attainable in a distributed system, what is more generally indicated is that for a large granularity of code, usually termed as a *step*, the processors are synchronized. This abstraction is implemented using some form of barrier synchronization to ensure that no processor begins executing the next step of code until all the processors have completed executing the previous steps of code assigned to each of the processors.

1.6.3 Libraries and standards

The previous subsections identified the main principles underlying all communication primitives. In this subsection, we briefly mention some publicly available interfaces that embody some of the above concepts.

There exists a wide range of primitives for message-passing. Many commercial software products (banking, payroll, etc., applications) use proprietary primitive libraries supplied with the software marketed by the vendors (e.g., the IBM CICS software which has a very widely installed customer base worldwide uses its own primitives). The message-passing interface (MPI) library [20,30] and the PVM (parallel virtual machine) library [31] are used largely by the scientific community, but other alternative libraries exist. Commercial software is often written using the remote procedure calls (RPC) mechanism [1,6] in which procedures that potentially reside across the network are invoked transparently to the user, in the same manner that a local procedure is invoked [1,6]. Under the covers, socket primitives or socket-like transport layer primitives are invoked to call the procedure remotely. There exist many implementations of RPC [1,7,11] – for example, Sun RPC, and distributed computing environment (DCE) RPC. "Messaging" and "streaming" are two other mechanisms for communication. With the growth of object based software, libraries for remote method invocation (RMI) and remote object invocation (ROI) with their own set of primitives are being proposed and standardized by different agencies [7]. CORBA (common object request broker architecture) [36] and DCOM (distributed component object model) [7] are two other standardized architectures with their own set of primitives. Additionally, several projects in the research stage are designing their own flavour of communication primitives.

1.7 Synchronous versus asynchronous executions

In addition to the two classifications of processor synchrony/asynchrony and of synchronous/asynchronous communication primitives, there is another classification, namely that of *synchronous/asynchronous executions*.

- An *asynchronous execution* is an execution in which (i) there is no processor synchrony and there is no bound on the drift rate of processor clocks, (ii) message delays (transmission + propagation times) are finite but unbounded, and (iii) there is no upper bound on the time taken by a process to execute a step. An example asynchronous execution with four processes P_0 to P_3 is shown in Figure 1.9. The arrows denote the messages; the tail and head of an arrow mark the *send* and *receive* event for that message, denoted by a circle and vertical line, respectively. Non-communication events, also termed as *internal* events, are shown by shaded circles.
- A *synchronous execution* is an execution in which (i) processors are synchronized and the clock drift rate between any two processors is bounded,

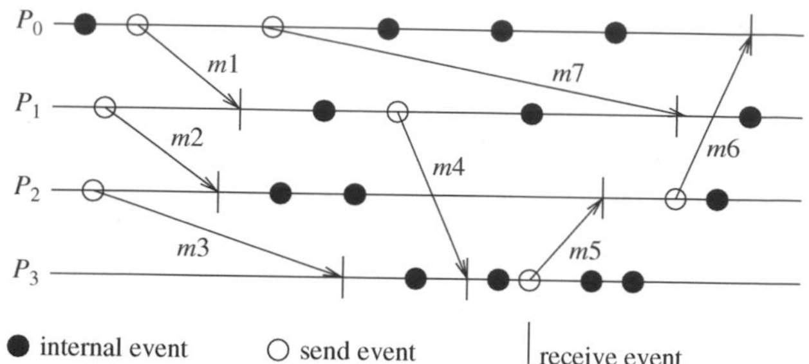

Figure 1.9 An example of an asynchronous execution in a message-passing system. A timing diagram is used to illustrate the execution.

(ii) message delivery (transmission + delivery) times are such that they occur in one logical step or round, and (iii) there is a known upper bound on the time taken by a process to execute a step. An example of a synchronous execution with four processes P_0 to P_3 is shown in Figure 1.10. The arrows denote the messages.

It is easier to design and verify algorithms assuming synchronous executions because of the coordinated nature of the executions at all the processes. However, there is a hurdle to having a truly synchronous execution. It is practically difficult to build a completely synchronous system, and have the messages delivered within a bounded time. Therefore, this synchrony has to be simulated under the covers, and will inevitably involve delaying or blocking some processes for some time durations. Thus, synchronous execution is an abstraction that needs to be provided to the programs. When implementing this abstraction, observe that the fewer the steps or "synchronizations" of the processors, the lower the delays and costs. If processors are allowed to have an asynchronous execution for a period of time and then they synchronize, then the granularity of the synchrony is coarse. This is really a *virtually synchronous execution*, and the abstraction is sometimes termed as *virtual synchrony*. Ideally, many programs want the processes to execute a series of instructions in rounds (also termed as steps or phases) asynchronously, with the requirement that after each round/step/phase, all the processes should be synchronized and all messages sent should be delivered. This is the commonly understood notion of a synchronous execution. Within each round/phase/step, there may be a finite and bounded number of sequential sub-rounds (or sub-phases or sub-steps) that processes execute. Each sub-round is assumed to

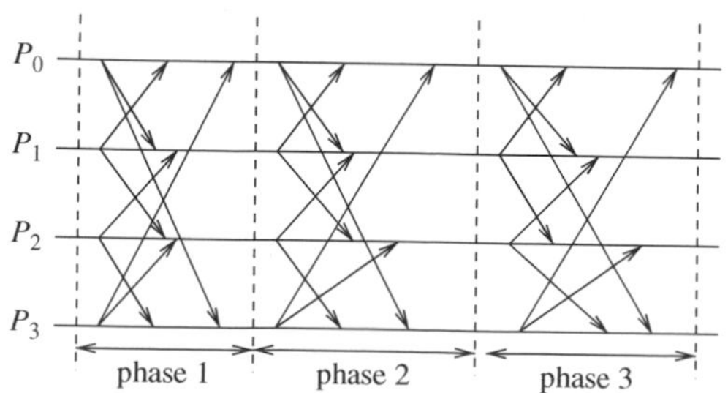

Figure 1.10 An example of a synchronous execution in a message-passing system. All the messages sent in a round are received within that same round.

send at most one message per process; hence the message(s) sent will reach in a single message hop.

The timing diagram of an example synchronous execution is shown in Figure 1.10. In this system, there are four nodes P_0 to P_3. In each round, process P_i sends a message to $P_{(i+1)\, mod\, 4}$ and $P_{(i-1)\, mod\, 4}$ and calculates some application-specific function on the received values.

1.7.1 Emulating an asynchronous system by a synchronous system ($A \rightarrow S$)

An asynchronous program (written for an asynchronous system) can be emulated on a synchronous system fairly trivially as the synchronous system is a special case of an asynchronous system – all communication finishes within the same round in which it is initiated.

1.7.2 Emulating a synchronous system by an asynchronous system ($S \rightarrow A$)

A synchronous program (written for a synchronous system) can be emulated on an asynchronous system using a tool called *synchronizer*, to be studied in Chapter 5.

1.7.3 Emulations

Section 1.5 showed how a shared memory system could be emulated by a message-passing system, and vice-versa. We now have four broad classes of programs, as shown in Figure 1.11. Using the emulations shown, any class can be emulated by any other. If system A can be emulated by system B, denoted A/B, and if a problem is not solvable in B, then it is also not solvable in A. Likewise, if a problem is solvable in A, it is also solvable in B. Hence, in a sense, all four classes are equivalent in terms of "computability" – what can and cannot be computed – in failure-free systems.

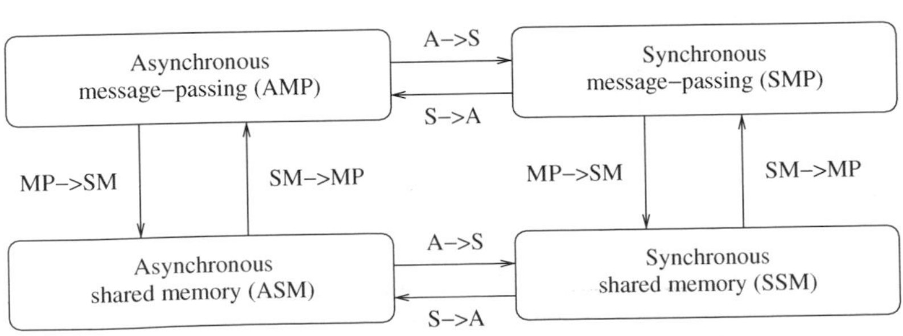

Figure 1.11 Emulations among the principal system classes in a failure-free system.

However, in fault-prone systems, as we will see in Chapter 14, this is not the case; a synchronous system offers more computability than an asynchronous system.

1.8 Design issues and challenges

Distributed computing systems have been in widespread existence since the 1970s when the Internet and ARPANET came into being. At the time, the primary issues in the design of the distributed systems included providing access to remote data in the face of failures, file system design, and directory structure design. While these continue to be important issues, many newer issues have surfaced as the widespread proliferation of the high-speed high-bandwidth internet and distributed applications continues rapidly.

Below we describe the important design issues and challenges after categorizing them as (i) having a greater component related to systems design and operating systems design, or (ii) having a greater component related to algorithm design, or (iii) emerging from recent technology advances and/or driven by new applications. There is some overlap between these categories. However, it is useful to identify these categories because of the chasm among the (i) the systems community, (ii) the theoretical algorithms community within distributed computing, and (iii) the forces driving the emerging applications and technology. For example, the current practice of distributed computing follows the client–server architecture to a large degree, whereas that receives scant attention in the theoretical distributed algorithms community. Two reasons for this chasm are as follows. First, an overwhelming number of applications outside the scientific computing community of users of distributed systems are business applications for which simple models are adequate. For example, the client–server model has been firmly entrenched with the legacy applications first developed by the Blue Chip companies (e.g., HP, IBM, Wang, DEC [now Compaq], Microsoft) since the 1970s and 1980s. This model is largely adequate for traditional business applications. Second, the state of the practice is largely controlled by industry standards, which do not necessarily choose the "technically best" solution.

1.8.1 Distributed systems challenges from a system perspective

The following functions must be addressed when designing and building a distributed system:

- **Communication** This task involves designing appropriate mechanisms for communication among the processes in the network. Some example mechanisms are: remote procedure call (RPC), remote object invocation (ROI), message-oriented communication versus stream-oriented communication.

- **Processes** Some of the issues involved are: management of processes and threads at clients/servers; code migration; and the design of software and mobile agents.
- **Naming** Devising easy to use and robust schemes for names, identifiers, and addresses is essential for locating resources and processes in a transparent and scalable manner. Naming in mobile systems provides additional challenges because naming cannot easily be tied to any static geographical topology.
- **Synchronization** Mechanisms for synchronization or coordination among the processes are essential. Mutual exclusion is the classical example of synchronization, but many other forms of synchronization, such as leader election are also needed. In addition, synchronizing physical clocks, and devising logical clocks that capture the essence of the passage of time, as well as global state recording algorithms, all require different forms of synchronization.
- **Data storage and access** Schemes for data storage, and implicitly for accessing the data in a fast and scalable manner across the network are important for efficiency. Traditional issues such as file system design have to be reconsidered in the setting of a distributed system.
- **Consistency and replication** To avoid bottlenecks, to provide fast access to data, and to provide scalability, replication of data objects is highly desirable. This leads to issues of managing the replicas, and dealing with consistency among the replicas/caches in a distributed setting. A simple example issue is deciding the level of granularity (i.e., size) of data access.
- **Fault tolerance** Fault tolerance requires maintaining correct and efficient operation in spite of any failures of links, nodes, and processes. Process resilience, reliable communication, distributed commit, checkpointing and recovery, agreement and consensus, failure detection, and self-stabilization are some of the mechanisms to provide fault-tolerance.
- **Security** Distributed systems security involves various aspects of cryptography, secure channels, access control, key management – generation and distribution, authorization, and secure group management.
- **Applications Programming Interface (API) and transparency** The API for communication and other specialized services is important for the ease of use and wider adoption of the distributed systems services by non-technical users. Transparency deals with hiding the implementation policies from the user, and can be classified as follows [33]. *Access transparency* hides differences in data representation on different systems and provides uniform operations to access system resources. *Location transparency* makes the locations of resources transparent to the users. *Migration transparency* allows relocating resources without changing names. The ability to relocate the resources as they are being accessed is *relocation*

transparency. *Replication transparency* does not let the user become aware of any replication. *Concurrency transparency* deals with masking the concurrent use of shared resources for the user. *Failure transparency* refers to the system being reliable and fault-tolerant.

- **Scalability and modularity** The algorithms, data (objects), and services must be as distributed as possible. Various techniques such as replication, caching and cache management, and asynchronous processing help to achieve scalability.

Some of the recent experiments in designing large-scale distributed systems include the Globe project at Vrije University [35], and the Globus project [15]. The Grid infrastructure for large-scale distributed computing is a very ambitious project that has gained significant attention to date [16,17]. All these projects attempt to provide the above listed functions as efficiently as possible.

1.8.2 Algorithmic challenges in distributed computing

The previous section addresses the challenges in designing distributed systems from a system building perspective. In this section, we briefly summarize the key algorithmic challenges in distributed computing.

Designing useful execution models and frameworks

The *interleaving* model and *partial order* model are two widely adopted models of distributed system executions. They have proved to be particularly useful for operational reasoning and the design of distributed algorithms. The *input/output automata* model [25] and the *TLA (temporal logic of actions)* are two other examples of models that provide different degrees of infrastructure for reasoning more formally with and proving the correctness of distributed programs.

Dynamic distributed graph algorithms and distributed routing algorithms

The distributed system is modeled as a distributed graph, and the graph algorithms form the building blocks for a large number of higher level communication, data dissemination, object location, and object search functions. The algorithms need to deal with dynamically changing graph characteristics, such as to model varying link loads in a routing algorithm. The efficiency of these algorithms impacts not only the user-perceived latency but also the traffic and hence the load or congestion in the network. Hence, the design of efficient distributed graph algorithms is of paramount importance.

Time and global state in a distributed system

The processes in the system are spread across three-dimensional physical space. Another dimension, time, has to be superimposed uniformly across space. The challenges pertain to providing accurate *physical time*, and to providing a variant of time, called *logical time*. Logical time is relative time, and eliminates the overheads of providing physical time for applications where physical time is not required. More importantly, logical time can (i) capture the logic and inter-process dependencies within the distributed program, and also (ii) track the relative progress at each process.

Observing the *global state* of the system (across space) also involves the time dimension for consistent observation. Due to the inherent distributed nature of the system, it is not possible for any one process to directly observe a meaningful global state across all the processes, without using extra state-gathering effort which needs to be done in a coordinated manner.

Deriving appropriate measures of concurrency also involves the time dimension, as judging the independence of different threads of execution depends not only on the program logic but also on execution speeds within the logical threads, and communication speeds among threads.

Synchronization/coordination mechanisms

The processes must be allowed to execute concurrently, except when they need to synchronize to exchange information, i.e., communicate about shared data. Synchronization is essential for the distributed processes to overcome the limited observation of the system state from the viewpoint of any one process. Overcoming this limited observation is necessary for taking any actions that would impact other processes. The synchronization mechanisms can also be viewed as resource management and concurrency management mechanisms to streamline the behavior of the processes that would otherwise act independently. Here are some examples of problems requiring synchronization:

- **Physical clock synchronization** Physical clocks ususally diverge in their values due to hardware limitations. Keeping them synchronized is a fundamental challenge to maintain common time.
- **Leader election** All the processes need to agree on which process will play the role of a distinguished process – called a leader process. A leader is necessary even for many distributed algorithms because there is often some asymmetry – as in initiating some action like a broadcast or collecting the state of the system, or in "regenerating" a token that gets "lost" in the system.
- **Mutual exclusion** This is clearly a synchronization problem because access to the critical resource(s) has to be coordinated.
- **Deadlock detection and resolution** Deadlock detection should be coordinated to avoid duplicate work, and deadlock resolution should be coordinated to avoid unnecessary aborts of processes.

- **Termination detection** This requires cooperation among the processes to detect the specific global state of quiescence.
- **Garbage collection** Garbage refers to objects that are no longer in use and that are not pointed to by any other process. Detecting garbage requires coordination among the processes.

Group communication, multicast, and ordered message delivery

A group is a collection of processes that share a common context and collaborate on a common task within an application domain. Specific algorithms need to be designed to enable efficient group communication and group management wherein processes can join and leave groups dynamically, or even fail. When multiple processes send messages concurrently, different recipients may receive the messages in different orders, possibly violating the semantics of the distributed program. Hence, formal specifications of the semantics of ordered delivery need to be formulated, and then implemented.

Monitoring distributed events and predicates

Predicates defined on program variables that are local to different processes are used for specifying conditions on the global system state, and are useful for applications such as debugging, sensing the environment, and in industrial process control. On-line algorithms for monitoring such predicates are hence important. An important paradigm for monitoring distributed events is that of *event streaming*, wherein streams of relevant events reported from different processes are examined collectively to detect predicates. Typically, the specification of such predicates uses physical or logical time relationships.

Distributed program design and verification tools

Methodically designed and verifiably correct programs can greatly reduce the overhead of software design, debugging, and engineering. Designing mechanisms to achieve these design and verification goals is a challenge.

Debugging distributed programs

Debugging sequential programs is hard; debugging distributed programs is that much harder because of the concurrency in actions and the ensuing uncertainty due to the large number of possible executions defined by the interleaved concurrent actions. Adequate debugging mechanisms and tools need to be designed to meet this challenge.

Data replication, consistency models, and caching

Fast access to data and other resources requires them to be replicated in the distributed system. Managing such replicas in the face of updates introduces the problems of ensuring consistency among the replicas and cached copies. Additionally, placement of the replicas in the systems is also a challenge because resources usually cannot be freely replicated.

World Wide Web design – caching, searching, scheduling

The Web is an example of a widespread distributed system with a direct interface to the end user, wherein the operations are predominantly read-intensive on most objects. The issues of object replication and caching discussed above have to be tailored to the web. Further, prefetching of objects when access patterns and other characteristics of the objects are known, can also be performed. An example of where prefetching can be used is the case of subscribing to Content Distribution Servers. Minimizing response time to minimize user-perceived latencies is an important challenge. Object search and navigation on the web are important functions in the operation of the web, and are very resource-intensive. Designing mechanisms to do this efficiently and accurately is a great challenge.

Distributed shared memory abstraction

A shared memory abstraction simplifies the task of the programmer because he or she has to deal only with read and write operations, and no message communication primitives. However, under the covers in the middleware layer, the abstraction of a shared address space has to be implemented by using message-passing. Hence, in terms of overheads, the shared memory abstraction is not less expensive.

- **Wait-free algorithms** Wait-freedom, which can be informally defined as the ability of a process to complete its execution irrespective of the actions of other processes, gained prominence in the design of algorithms to control acccess to shared resources in the shared memory abstraction. It corresponds to $n-1$-fault resilience in a n process system and is an important principle in fault-tolerant system design. While wait-free algorithms are highly desirable, they are also expensive, and designing low overhead wait-free algorithms is a challenge.
- **Mutual exclusion** A first course in operating systems covers the basic algorithms (such as the Bakery algorithm and using semaphores) for mutual exclusion in a multiprocessing (uniprocessor or multiprocessor) shared memory setting. More sophisticated algorithms – such as those based on hardware primitives, fast mutual exclusion, and wait-free algorithms – will be covered in this book.
- **Register constructions** In light of promising and emerging technologies of tomorrow – such as biocomputing and quantum computing – that can alter the present foundations of computer "hardware" design, we need to revisit the assumptions of memory access of current systems that are exclusively based on semiconductor technology and the von Neumann architecture. Specifically, the assumption of single/multiport memory with serial access via the bus in tight synchronization with the system hardware clock may not be a valid assumption in the possibility of "unrestricted" and "overlapping" concurrent access to the same memory location. The

study of register constructions deals with the design of registers from scratch, with very weak assumptions on the accesses allowed to a register. This field forms a foundation for future architectures that allow concurrent access even to primitive units of memory (independent of technology) without any restrictions on the concurrency permitted.

- **Consistency models** For multiple copies of a variable/object, varying degrees of consistency among the replicas can be allowed. These represent a trade-off of coherence versus cost of implementation. Clearly, a strict definition of consistency (such as in a uniprocessor system) would be expensive to implement in terms of high latency, high message overhead, and low concurrency. Hence, relaxed but still meaningful models of consistency are desirable.

Reliable and fault-tolerant distributed systems

A reliable and fault-tolerant environment has multiple requirements and aspects, and these can be addressed using various strategies:

- **Consensus algorithms** All algorithms ultimately rely on message-passing, and the recipients take actions based on the contents of the received messages. Consensus algorithms allow correctly functioning processes to reach agreement among themselves in spite of the existence of some malicious (adversarial) processes whose identities are not known to the correctly functioning processes. The goal of the malicious processes is to prevent the correctly functioning processes from reaching agreement. The malicious processes operate by sending messages with misleading information, to confuse the correctly functioning processes.
- **Replication and replica management** Replication (as in having backup servers) is a classical method of providing fault-tolerance. The triple modular redundancy (TMR) technique has long been used in software as well as hardware installations. More sophisticated and efficient mechanisms for replication are the subject of study here.
- **Voting and quorum systems** Providing redundancy in the active (e.g., processes) or passive (e.g., hardware resources) components in the system and then performing voting based on some quorum criterion is a classical way of dealing with fault-tolerance. Designing efficient algorithms for this purpose is the challenge.
- **Distributed databases and distributed commit** For distributed databases, the traditional properties of the transaction (A.C.I.D. – atomicity, consistency, isolation, durability) need to be preserved in the distributed setting. The field of traditional "transaction commit" protocols is a fairly mature area. Transactional properties can also be viewed as having a counterpart for guarantees on message delivery in group communication in the presence of failures. Results developed in one field can be adapted to the other.

- **Self-stabilizing systems** All system executions have associated good (or legal) states and bad (or illegal) states; during correct functioning, the system makes transitions among the good states. Faults, internal or external to the program and system, may cause a bad state to arise in the execution. A *self-stabilizing* algorithm is any algorithm that is guaranteed to eventually take the system to a good state even if a bad state were to arise due to some error. Self-stabilizing algorithms require some in-built redundancy to track additional variables of the state and do extra work. Designing efficient self-stabilizing algorithms is a challenge.
- **Checkpointing and recovery algorithms** Checkpointing involves periodically recording the current state on secondary storage so that, in case of a failure, the entire computation is not lost but can be recovered from one of the recently taken checkpoints. Checkpointing in a distributed environment is difficult because if the checkpoints at the different processes are not coordinated, the local checkpoints may become useless because they are inconsistent with the checkpoints at other processes.
- **Failure detectors** A fundamental limitation of asynchronous distributed systems is that there is no theoretical bound on the message transmission times. Hence, it is impossible to distinguish a sent-but-not-yet-arrived message from a message that was never sent. This implies that it is impossible using message transmission to determine whether some other process across the network is alive or has failed. Failure detectors represent a class of algorithms that probabilistically suspect another process as having failed (such as after timing out after non-receipt of a message for some time), and then converge on a determination of the up/down status of the suspected process.

Load balancing

The goal of load balancing is to gain higher throughput, and reduce the user-perceived latency. Load balancing may be necessary because of a variety of factors such as high network traffic or high request rate causing the network connection to be a bottleneck, or high computational load. A common situation where load balancing is used is in server farms, where the objective is to service incoming client requests with the least turnaround time. Several results from traditional operating systems can be used here, although they need to be adapted to the specifics of the distributed environment. The following are some forms of load balancing:

- **Data migration** The ability to move data (which may be replicated) around in the system, based on the access pattern of the users.
- **Computation migration** The ability to relocate processes in order to perform a redistribution of the workload.
- **Distributed scheduling** This achieves a better turnaround time for the users by using idle processing power in the system more efficiently.

Real-time scheduling

Real-time scheduling is important for mission-critical applications, to accomplish the task execution on schedule. The problem becomes more challenging in a distributed system where a global view of the system state is absent. On-line or dynamic changes to the schedule are also harder to make without a global view of the state.

Furthermore, message propagation delays which are network-dependent are hard to control or predict, which makes meeting real-time guarantees that are inherently dependent on communication among the processes harder. Although networks offering quality-of-service guarantees can be used, they alleviate the uncertainty in propagation delays only to a limited extent. Further, such networks may not always be available.

Performance

Although high throughput is not the primary goal of using a distributed system, achieving good performance is important. In large distributed systems, network latency (propagation and transmission times) and access to shared resources can lead to large delays which must be minimized. The user-perceived turn-around time is very important.

The following are some example issues arise in determining the performance:

- **Metrics** Appropriate metrics must be defined or identified for measuring the performance of theoretical distributed algorithms, as well as for implementations of such algorithms. The former would involve various complexity measures on the metrics, whereas the latter would involve various system and statistical metrics.
- **Measurement methods/tools** As a real distributed system is a complex entity and has to deal with all the difficulties that arise in measuring performance over a WAN/the Internet, appropriate methodologies and tools must be developed for measuring the performance metrics.

1.8.3 Applications of distributed computing and newer challenges

Mobile systems

Mobile systems typically use wireless communication which is based on electromagnetic waves and utilizes a shared broadcast medium. Hence, the characteristics of communication are different; many issues such as range of transmission and power of transmission come into play, besides various engineering issues such as battery power conservation, interfacing with the wired Internet, signal processing and interference. From a computer science perspective, there is a rich set of problems such as routing, location management, channel allocation, localization and position estimation, and the overall management of mobility.

There are two popular architectures for a mobile network. The first is the *base-station* approach, also known as the *cellular approach*, wherein a *cell* which is the geographical region within range of a static but powerful base transmission station is associated with that base station. All mobile processes in that cell communicate with the rest of the system via the base station. The second approach is the *ad-hoc network* approach where there is no base station (which essentially acted as a centralized node for its cell). All responsibility for communication is distributed among the mobile nodes, wherein mobile nodes have to participate in routing by forwarding packets of other pairs of communicating nodes. Clearly, this is a complex model. It poses many graph-theoretical challenges from a computer science perspective, in addition to various engineering challenges.

Sensor networks

A sensor is a processor with an electro-mechanical interface that is capable of sensing physical parameters, such as temperature, velocity, pressure, humidity, and chemicals. Recent developments in cost-effective hardware technology have made it possible to deploy very large (of the order of 10^6 or higher) low-cost sensors. An important paradigm for monitoring distributed events is that of *event streaming*, which was defined earlier. The streaming data reported from a sensor network differs from the streaming data reported by "computer processes" in that the events reported by a sensor network are in the environment, external to the computer network and processes. This limits the nature of information about the reported event in a sensor network.

Sensor networks have a wide range of applications. Sensors may be mobile or static; sensors may communicate wirelessly, although they may also communicate across a wire when they are statically installed. Sensors may have to self-configure to form an ad-hoc network, which introduces a whole new set of challenges, such as position estimation and time estimation.

Ubiquitous or pervasive computing

Ubiquitous systems represent a class of computing where the processors embedded in and seamlessly pervading through the environment perform application functions in the background, much like in sci-fi movies. The intelligent home, and the smart workplace are some example of ubiquitous environments currently under intense research and development. Ubiquitous systems are essentially distributed systems; recent advances in technology allow them to leverage wireless communication and sensor and actuator mechanisms. They can be self-organizing and network-centric, while also being resource constrained. Such systems are typically characterized as having many small processors operating collectively in a dynamic ambient network. The processors may be connected to more powerful networks and processing resources in the background for processing and collating data.

Peer-to-peer computing

Peer-to-peer (P2P) computing represents computing over an application layer network wherein all interactions among the processors are at a "peer" level, without any hierarchy among the processors. Thus, all processors are equal and play a symmetric role in the computation. P2P computing arose as a paradigm shift from client–server computing where the roles among the processors are essentially asymmetrical. P2P networks are typically self-organizing, and may or may not have a regular structure to the network. No central directories (such as those used in domain name servers) for name resolution and object lookup are allowed. Some of the key challenges include: object storage mechanisms, efficient object lookup, and retrieval in a scalable manner; dynamic reconfiguration with nodes as well as objects joining and leaving the network randomly; replication strategies to expedite object search; tradeoffs between object size latency and table sizes; anonymity, privacy, and security.

Publish-subscribe, content distribution, and multimedia

With the explosion in the amount of information, there is a greater need to receive and access only information of interest. Such information can be specified using filters. In a dynamic environment where the information constantly fluctuates (varying stock prices is a typical example), there needs to be: (i) an efficient mechanism for distributing this information (*publish*), (ii) an efficient mechanism to allow end users to indicate interest in receiving specific kinds of information (*subscribe*), and (iii) an efficient mechanism for aggregating large volumes of published information and filtering it as per the user's subscription filter.

Content distribution refers to a class of mechanisms, primarily in the web and P2P computing context, whereby specific information which can be broadly characterized by a set of parameters is to be distributed to interested processes. Clearly, there is overlap between content distribution mechanisms and publish–subscribe mechanisms. When the content involves multimedia data, special requirement such as the following arise: multimedia data is usually very large and information-intensive, requires compression, and often requires special synchronization during storage and playback.

Distributed agents

Agents are software processes or robots that can move around the system to do specific tasks for which they are specially programmed. The name "agent" derives from the fact that the agents do work on behalf of some broader objective. Agents collect and process information, and can exchange such information with other agents. Often, the agents cooperate as in an ant colony, but they can also have friendly competition, as in a free market economy. Challenges in distributed agent systems include coordination mechanisms among the agents, controlling the mobility of the agents, and their software design and interfaces. Research in agents is inter-disciplinary:

spanning artificial intelligence, mobile computing, economic market models, software engineering, and distributed computing.

Distributed data mining

Data mining algorithms examine large amounts of data to detect patterns and trends in the data, to *mine* or extract useful information. A traditional example is: examining the purchasing patterns of customers in order to profile the customers and enhance the efficacy of directed marketing schemes. The mining can be done by applying database and artificial intelligence techniques to a data repository. In many situations, the data is necessarily distributed and cannot be collected in a single repository, as in banking applications where the data is private and sensitive, or in atmospheric weather prediction where the data sets are far too massive to collect and process at a single repository in real-time. In such cases, efficient distributed data mining algorithms are required.

Grid computing

Analogous to the electrical power distribution grid, it is envisaged that the information and computing grid will become a reality some day. Very simply stated, idle CPU cycles of machines connected to the network will be available to others. Many challenges in making grid computing a reality include: scheduling jobs in such a distributed environment, a framework for implementing quality of service and real-time guarantees, and, of course, security of individual machines as well as of jobs being executed in this setting.

Security in distributed systems

The traditional challenges of security in a distributed setting include: confidentiality (ensuring that only authorized processes can access certain information), authentication (ensuring the source of received information and the identity of the sending process), and availability (maintaining allowed access to services despite malicious actions). The goal is to meet these challenges with efficient and scalable solutions. These basic challenges have been addressed in traditional distributed settings. For the newer distributed architectures, such as wireless, peer-to-peer, grid, and pervasive computing discussed in this subsection), these challenges become more interesting due to factors such as a resource-constrained environment, a broadcast medium, the lack of structure, and the lack of trust in the network.

1.9 Selection and coverage of topics

This is a long list of topics and difficult to cover in a single textbook. This book covers a broad selection of topics from the above list, in order to present the fundamental principles underlying the various topics. The goal has been to select topics that will give a good understanding of the field, and of the techniques used to design solutions.

Some topics that have been omitted are interdisciplinary, across fields within computer science. An example is load balancing, which is traditionally covered in detail in a course on parallel processing. As the focus of distributed systems has shifted away from gaining higher efficiency to providing better services and fault-tolerance, the importance of load balancing in distributed computing has diminished. Another example is mobile systems. A mobile system is a distributed system having certain unique characteristics, and there are courses devoted specifically to mobile systems.

1.10 Chapter summary

This chapter first characterized distributed systems by looking at various informal definitions based on functional aspects. It then looked at various architectures of multiple processor systems, and the requirements that have traditionally driven distributed systems. The relationship of a distributed system to "middleware", the operating system, and the network protocol stack provided a different perspective on a distributed system.

The relationship between parallel systems and distributed systems, covering aspects such as degrees of software and hardware coupling, and the relative placement of the processors, memory units, and interconnection networks, was examined in detail. There is some overlap between the fields of parallel computing and distributed computing, and hence it is important to understand their relationhip clearly. For example, various interconnection networks such as the Omega network, the Butterfly network, and the hypercube network, were designed for parallel computing but they are recently finding surprising applications in the design of application-level overlay networks for distributed computing. The traditional taxonomy of multiple processor systems by Flynn [14] was also studied. Important concepts such as the degree of parallelism and of concurrency, and the degree of coupling were also introduced informally.

The chapter then introduced three fundamental concepts in distributed computing. The first concept is the paradigm of shared memory communication versus message-passing communication. The second concept is the paradigm of synchronous executions and asynchronous executions. For both these concepts, emulation of one paradigm by another was studied for error-free systems. The third concept was that of synchronous and asynchronous send communication primitives, of synchronous receive communicaiton primitives, and of blocking and non-blocking send and receive communication primitives.

The chapter then presented design issues and challenges in the field of distributed computing. The challenges were classified as (i) being important from a systems design perspective, or (ii) being important from an algorithmic perspective, or (iii) those that are driven by new applications and emerging technologies. This classification is not orthogonal and is somewhat subjective. The various topics that will be covered in the rest of the book are portrayed on a miniature canvas in the section on the design issues and challenges.

1.11 Exercises

Exercise 1.1 What are the main differences between a parallel system and a distributed system?

Exercise 1.2 Identify some distributed applications in the scientific and commercial application areas. For each application, determine which of the motivating factors listed in Section 1.3 are important for building the application over a distributed system.

Exercise 1.3 Draw the Omega and Butterfly networks for $n = 16$ inputs and outputs.

Exercise 1.4 For the Omega and Butterfly networks shown in Figure 1.4, trace the paths from P_5 to M_2, and from P_6 to M_1.

Exercise 1.5 Formulate the interconnection function for the Omega network having n inputs and outputs, only in terms of the $M = n/2$ switch numbers in each stage. (Hint: Follow an approach similar to the Butterfly network formulation.)

Exercise 1.6 In Figure 1.4, observe that the paths from input 000 to output 111 and from input 101 to output 110 have a common edge. Therefore, simultaneous transmission over these paths is not possible; one path *blocks* another. Hence, the Omega and Butterfly networks are classified as *blocking interconnection networks*.

Let $\Pi(n)$ be any permutation on $\{0 \ldots n-1\}$, mapping the input domain to the output range. A *non-blocking interconnection network* allows simultaneous transmission from the inputs to the outputs for any permutation.

Consider the network built as follows. Take the image of a butterfly in a vertical mirror, and append this mirror image to the output of a butterfly. Hence, for n inputs and outputs, there will be $2log_2n$ stages. Prove that this network is non-blocking.

Exercise 1.7 The Baseline Clos network has a interconnection generation function as follows. Let there be $M = n/2$ switches per stage, and let a switch be denoted by the tuple $\langle x, s \rangle$, where $x \in [0, M-1]$ and stage $s \in [0, log_2n - 1]$.

There is an edge from switch $\langle x, s \rangle$ to switch $\langle y, s+1 \rangle$ if (i) y is the cyclic right-shift of the $(log_2n - s)$ least significant bits of x, (ii) y is the cyclic right-shift of the $(log_2n - s)$ least significant bits of x', where x' is obtained by complementing the LSB of x.

Draw the interconnection diagram for the Clos network having $n = 16$ inputs and outputs, i.e., having 8 switches in each of the 4 stages.

Exercise 1.8 Two interconnection networks are isomorphic if there is a 1:1 mapping f between the switches such that for any switches x and y that are connected to each other in adjacent stages in one network, $f(x)$ and $f(y)$ are also connected in the other network.

Show that the Omega, Butterfly, and Clos (Baseline) networks are isomorphic to each other.

Exercise 1.9 Explain why a *Receive* call cannot be asynchronous.

Exercise 1.10 What are the three aspects of reliability? Is it possible to order them in different ways in terms of importance, based on different applications' requirements? Justify your answer by giving examples of different applications.

Exercise 1.11 Figure 1.11 shows the emulations among the principal system classes in a failure-free system.

1. Which of these emulations are possible in a failure-prone system? Explain.
2. Which of these emulations are not possible in a failure-prone system? Explain.

Exercise 1.12 Examine the impact of unreliable links and node failures on each of the challenges listed in Section 1.8.2.

1.12 Notes on references

The selection of topics and material for this book has been shaped by the authors' perception of the importance of various subjects, as well as the coverage by the existing textbooks.

There are many books on distributed computing and distributed systems. Attiya and Welch [2] and Lynch [25] provide a formal theoretical treatment of the field. The books by Barbosa [3] and Tel [34] focus on algorithms. The books by Chow and Johnson [8], Coulouris *et al.* [11], Garg [18], Goscinski [19], Mullender [26], Raynal [27], Singhal and Shivaratri [29], and Tanenbaum and van Steen [33] provide a blend of theoretical and systems issues.

Much of the material in this introductory chapter is based on well understood concepts and paradigms in the distributed systems community, and is difficult to attribute to any particular source.

A recent overview of the challenges in middleware design from systems' perspective is given in the special issue by Lea *et al.* [24]. An overview of the common object request broker model (CORBA) of the Object Management Group (OMG) is given by Vinoski [36]. The distributed component object model (DCOM) from Microsoft, Sun's Java remote method invocation (RMI), and CORBA are analyzed in perspective by Campbell *et al.* [7]. A detailed treatment of CORBA, RMI, and RPC is given by Coulouris *et al.* [11]. The Open Foundations's distributed computing environment (DCE) is described in [28,33]; DCE is not likely to be enjoy a continuing support base. Descriptions of the Message Passing Interface can be found in Snir *et al.* [30] and Gropp *et al.* [20]. The Parallel Virtual Machine (PVM) framework for parallel distributed programming is described by Sunderam [31].

The discussion of parallel processing, and of the UMA and NUMA parallel architectures, is based on Kumar *et al.* [22]. The properties of the hypercube architecture are surveyed by Feng [13] and Harary *et al.* [21]. The multi-stage interconnection architectures – the Omega (Benes) [4], the Butterfly [10], and Clos [9] were proposed in the papers indicated. A good overview of multistage interconnection networks is given by Wu and Feng [37]. Flynn's taxomomy of multiprocessors is based on [14]. The discussion on blocking/non-blocking primitives as well as synchronous and asynchropnous primitives is extended from Cypher and Leu [12]. The section on design issues and challenges is based on the vast research literature in the area.

The Globe architecture is described by van Steen *et al.* [35]. The Globus architecture is described by Foster and Kesselman [15]. The grid infrastructure and the distributed computng vision for the twenty-first century is described by Foster and Kesselman [16] and by Foster [17]. The World Wide Web is an excellent example of a distributed system that has largely evolved of its own; Tim Berners-Lee is credited with seeding the WWW project; its early description is given by Berners-Lee *et al.* [5].

References

[1] A. Ananda, B. Tay, and E. Koh, A survey of asynchronous remore procedure calls, *ACM SIGOPS Operating Systems Review*, **26**(2), 1992, 92–109.

[2] H. Attiya and J. Welch, *Distributed Computing Fundamentals, Simulations, and Advanced Topics*, 2nd edn, Hoboken, NJ, Wiley Inter-Science, 2004.

[3] V. Barbosa, *An Introduction to Distributed Algorithms*, Cambridge, MA, MIT Press, 1996.

[4] V.E. Benes, *Mathematical Theory of Connecting Networks and Telephone Traffic*, New York, Academic Press, 1965.

[5] T. Berners-Lee, R. Cailliau, A. Luotonen, H. Nielsen, and A. Secret, The World-Wide Web, *Communications of the ACM*, **37**(8), 1994, 76–82.

[6] A. Birrell and B. Nelson, Implementing remote procedure calls, *ACM Transactions on Computer Systems*, **2**(1), 1984, 39–59.

[7] A. Campbell, G. Coulson, and M. Counavis, Managing complexity: middleware explained, *IT Professional Magazine*, October 1999, 22–28.

[8] R. Chow and D. Johnson, *Distributed Operating Systems and Algorithms*, Reading, MA, Harlow, UK, Addison-Wesley, 1997.

[9] C. Clos, A study of non-blocking switching networks, *Bell Systems Technical Journal*, **32**, 1953, 406–424.

[10] J.M. Cooley and J.W. Tukey, An algorithm for the machine calculation of complete Fourier series, *Mathematical Computations*, **19**, 1965, 297–301.

[11] G. Coulouris, J. Dollimore, and T. Kindberg, *Distributed Systems Concepts and Design*, Harlow, UK, 3rd edn, Addison-Wesley, 2001.

[12] R. Cypher and E. Leu, The semantics of blocking and non-blocking send and receive primitives, *Proceedings of the 8th International Symposium on Parallel Processing*, 1994, 729–735.

[13] T.Y. Feng, A survey of interconnection networks, *IEEE Computer*, **14**, 1981, 12–27.

[14] M. Flynn, Some computer organizations and their effectiveness, *IEEE Transactions on Computers*, **C-21**, 1972, 94.

[15] I. Foster and C. Kesselman, Globus: a metacomputing infrastructure toolkit, *International Journal of Supercomputer Applications*, **11**(2), 1997, 115–128.

[16] I. Foster and C. Kesselman, *The Grid: Blueprint for a New Computing Infrastructure*, San Francisco, CA, Morgan Kaufmann, 1998.

[17] I. Foster, The Grid: a new infrastructure for 21st century science, *Physics Today*, **55**(2), 2002, 42–47.

[18] V. Garg, *Elements of Distributed Computing*, New York, John Wiley, 2002.

[19] A. Goscinski, *Distributed Operating Systems: The Logical Design*, Reading, MA, Addison-Wesley, 1991.

[20] W. Gropp, E. Lusk, and A. Skjellum, *Using MPI: Portable Parallel Programming with the Message-passing Interface*, Cambridge, MA, MIT Press, 1994.

[21] F. Harary, J.P. Hayes, and H. Wu, A survey of the theory of hypercube graphs, *Computational Mathematical Applications*, **15**(4), 1988, 277–289.

[22] V. Kumar, A. Grama, A. Gupta, and G. Karypis, *Introduction to Parallel Computing*, 2nd edn, Harlow, UK, Pearson Education 2003.

[23] L. Lamport, Distribution email, May 28, 1987, available at: http://research.microsoft.com/users/lamport/pubs/distributed_systems.txt.

[24] D. Lea, S. Vinoski, and W. Vogels, Guest editors' introduction: asynchronous middleware and services, *IEEE Internet Computing*, **10**(1), 2006, 14–17.

[25] N. Lynch, *Distributed Algorithms*, San Francisco, CA, Morgan Kaufmann, 1996.

[26] S. Mullender, *Distributed Systems*, 2nd edn, New York, ACM Press, 1993.

[27] M. Raynal, *Distributed Algorithms and Protocols*, New York, John Wiley, 1988.

[28] J. Shirley, W. Hu, and D. Magid, *Guide to Writing DCE Applications*, O'Reilly and Associates, Inc., 1992.

[29] M. Singhal and N. Shivaratri, *Advanced Concepts in Operating Systems*, New York, McGraw Hill, 1994.

[30] M. Snir, S. Otto, S. Huss-Lederman, D. Walker, and J. Dongarra, *MPI: The Complete Reference*, Cambridge, MA, MIT Press, 1996.

[31] V. Sunderam, PVM: A framework for parallel distributed computing, *Concurrency – Practice and Experience*, 2(4): 315–339, 1990.

[32] A. Tanenbaum, *Computer Networks*, 3rd edn, New Jersey, Prentice-Hall PTR, 1996.

[33] A. Tanenbaum and M. Van Steen, *Distributed Systems: Principles and Paradigms*, Upper Saddle River, NJ, Prentice-Hall, 2003.

[34] G. Tel, *Introduction to Distributed Algorithms*, Cambridge, Cambridge University Press, 1994.

[35] M. van Steen, P. Homburg, and A. Tanenbaum, Globe: a wide-area distributed system, *IEEE Concurrency*, 1999, 70–78.

[36] S. Vinoski, CORBA: integrating diverse applications within heterogeneous distributed environments, *IEEE Communications Magazine*, **35**(2), 1997, 46–55.

[37] C.L. Wu and T.-Y. Feng, On a class of multistage interconnection networks, *IEEE Transactions on Computers*, **C-29** 1980, 694–702.

CHAPTER

2 A model of distributed computations

A distributed system consists of a set of processors that are connected by a communication network. The communication network provides the facility of information exchange among processors. The communication delay is finite but unpredictable. The processors do not share a common global memory and communicate solely by passing messages over the communication network. There is no physical global clock in the system to which processes have instantaneous access. The communication medium may deliver messages out of order, messages may be lost, garbled, or duplicated due to timeout and retransmission, processors may fail, and communication links may go down. The system can be modeled as a directed graph in which vertices represent the processes and edges represent unidirectional communication channels.

A distributed application runs as a collection of processes on a distributed system. This chapter presents a model of a distributed computation and introduces several terms, concepts, and notations that will be used in the subsequent chapters.

2.1 A distributed program

A distributed program is composed of a set of n asynchronous processes $p_1, p_2, \ldots, p_i, \ldots, p_n$ that communicate by message passing over the communication network. Without loss of generality, we assume that each process is running on a different processor. The processes do not share a global memory and communicate solely by passing messages. Let C_{ij} denote the channel from process p_i to process p_j and let m_{ij} denote a message sent by p_i to p_j. The communication delay is finite and unpredictable. Also, these processes do not share a global clock that is instantaneously accessible to these processes. Process execution and message transfer are asynchronous – a process may execute an action spontaneously and a process sending a message does not wait for the delivery of the message to be complete.

The global state of a distributed computation is composed of the states of the processes and the communication channels [2]. The state of a process is characterized by the state of its local memory and depends upon the context. The state of a channel is characterized by the set of messages in transit in the channel.

2.2 A model of distributed executions

The execution of a process consists of a sequential execution of its actions. The actions are atomic and the actions of a process are modeled as three types of events, namely, internal events, message send events, and message receive events. Let e_i^x denote the xth event at process p_i. Subscripts and/or superscripts will be dropped when they are irrelevant or are clear from the context. For a message m, let $send(m)$ and $rec(m)$ denote its send and receive events, respectively.

The occurrence of events changes the states of respective processes and channels, thus causing transitions in the global system state. An internal event changes the state of the process at which it occurs. A send event (or a receive event) changes the state of the process that sends (or receives) the message and the state of the channel on which the message is sent (or received). An internal event only affects the process at which it occurs.

The events at a process are linearly ordered by their order of occurrence. The execution of process p_i produces a sequence of events $e_i^1, e_i^2, \ldots, e_i^x, e_i^{x+1}, \ldots$ and is denoted by $\mathcal{H}_i$:

$$\mathcal{H}_i = (h_i, \rightarrow_i),$$

where h_i is the set of events produced by p_i and binary relation $\rightarrow_i$ defines a linear order on these events. Relation $\rightarrow_i$ expresses causal dependencies among the events of p_i.

The send and the receive events signify the flow of information between processes and establish causal dependency from the sender process to the receiver process. A relation $\rightarrow_{msg}$ that captures the causal dependency due to message exchange, is defined as follows. For every message m that is exchanged between two processes, we have

$$send(m) \rightarrow_{msg} rec(m).$$

Relation $\rightarrow_{msg}$ defines causal dependencies between the pairs of corresponding send and receive events.

The evolution of a distributed execution is depicted by a space–time diagram. Figure 2.1 shows the space–time diagram of a distributed execution involving three processes. A horizontal line represents the progress of the

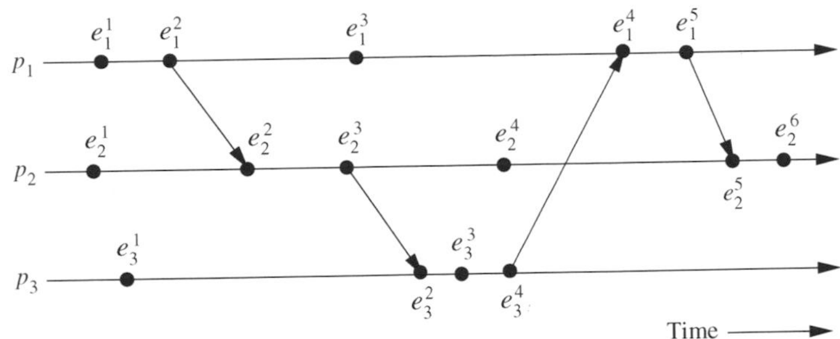

Figure 2.1 The space–time diagram of a distributed execution.

process; a dot indicates an event; a slant arrow indicates a message transfer. Generally, the execution of an event takes a finite amount of time; however, since we assume that an event execution is atomic (hence, indivisible and instantaneous), it is justified to denote it as a dot on a process line. In this figure, for process p_1, the second event is a message send event, the third event is an internal event, and the fourth event is a message receive event.

Causal precedence relation

The execution of a distributed application results in a set of distributed events produced by the processes. Let $H = \cup_i h_i$ denote the set of events executed in a distributed computation. Next, we define a binary relation on the set H, denoted as $\rightarrow$, that expresses causal dependencies between events in the distributed execution.

$$\forall e_i^x,\ \forall e_j^y \in H,\ \ e_i^x \rightarrow e_j^y \quad \Leftrightarrow \quad \begin{cases} e_i^x \rightarrow_i e_j^y \text{ i.e., } (i = j) \wedge (x < y) \\ \text{or} \\ e_i^x \rightarrow_{msg} e_j^y \\ \text{or} \\ \exists e_k^z \in H : e_i^x \rightarrow e_k^z \ \wedge\ e_k^z \rightarrow e_j^y \end{cases}$$

The causal precedence relation induces an irreflexive partial order on the events of a distributed computation [6] that is denoted as $\mathcal{H}=(H, \rightarrow)$.

Note that the relation $\rightarrow$ is Lamport's "happens before" relation [4].[1] For any two events e_i and e_j, if $e_i \rightarrow e_j$, then event e_j is directly or transitively dependent on event e_i; graphically, it means that there exists a path consisting of message arrows and process-line segments (along increasing time) in the space–time diagram that starts at e_i and ends at e_j. For example, in Figure 2.1, $e_1^1 \rightarrow e_3^3$ and $e_3^3 \rightarrow e_2^6$. Note that relation $\rightarrow$ denotes flow of information in a distributed computation and $e_i \rightarrow e_j$ dictates that all the information available

[1] In Lamport's "happens before" relation, an event e_1 *happens before* an event e_2, denoted by $e_i \rightarrow e_j$, if (a) e_1 occurs before e_2 on the same process, or (b) e_1 is the send event of a message and e_2 is the receive event of that message, or (c) $\exists e' |\ e_1$ happens before e' and e' happens before e_2.

at e_i is potentially accessible at e_j. For example, in Figure 2.1, event e_2^6 has the knowledge of all other events shown in the figure.

For any two events e_i and e_j, $e_i \not\rightarrow e_j$ denotes the fact that event e_j does not directly or transitively dependent on event e_i. That is, event e_i does not causally affect event e_j. Event e_j is not aware of the execution of e_i or any event executed after e_i on the same process. For example, in Figure 2.1, $e_1^3 \not\rightarrow e_3^3$ and $e_2^4 \not\rightarrow e_3^1$. Note the following two rules:

- for any two events e_i and e_j, $e_i \not\rightarrow e_j \not\Rightarrow e_j \not\rightarrow e_i$
- for any two events e_i and e_j, $e_i \rightarrow e_j \Rightarrow e_j \not\rightarrow e_i$.

For any two events e_i and e_j, if $e_i \not\rightarrow e_j$ and $e_j \not\rightarrow e_i$, then events e_i and e_j are said to be concurrent and the relation is denoted as $e_i \parallel e_j$. In the execution of Figure 2.1, $e_1^3 \parallel e_3^3$ and $e_2^4 \parallel e_3^1$. Note that relation $\parallel$ is not transitive; that is, $(e_i \parallel e_j) \wedge (e_j \parallel e_k) \not\Rightarrow e_i \parallel e_k$. For example, in Figure 2.1, $e_3^3 \parallel e_2^4$ and $e_2^4 \parallel e_1^5$, however, $e_3^3 \not\parallel e_1^5$.

Note that for any two events e_i and e_j in a distributed execution, $e_i \rightarrow e_j$ or $e_j \rightarrow e_i$, or $e_i \parallel e_j$.

Logical vs. physical concurrency

In a distributed computation, two events are logically concurrent if and only if they do not causally affect each other. Physical concurrency, on the other hand, has a connotation that the events occur at the same instant in physical time. Note that two or more events may be logically concurrent even though they do not occur at the same instant in physical time. For example, in Figure 2.1, events in the set $\{e_1^3, e_2^4, e_3^3\}$ are logically concurrent, but they occurred at different instants in physical time. However, note that if processor speed and message delays had been different, the execution of these events could have very well coincided in physical time. Whether a set of logically concurrent events coincide in the physical time or in what order in the physical time they occur does not change the outcome of the computation.

Therefore, even though a set of logically concurrent events may not have occurred at the same instant in physical time, for all practical and theoretical purposes, we can assume that these events occured at the same instant in physical time.

2.3 Models of communication networks

There are several models of the service provided by communication networks, namely, FIFO (first-in, first-out), non-FIFO, and causal ordering. In the FIFO model, each channel acts as a first-in first-out message queue and thus, message ordering is preserved by a channel. In the non-FIFO model, a channel acts like a set in which the sender process adds messages and the receiver process removes messages from it in a random order. The “causal ordering” model [1] is based on Lamport's "happens before" relation. A system that supports the causal ordering model satisfies the following property:

CO: For any two messages m_{ij} and m_{kj}, if $send(m_{ij}) \longrightarrow send(m_{kj})$, then $rec(m_{ij}) \longrightarrow rec(m_{kj})$.

That is, this property ensures that causally related messages destined to the same destination are delivered in an order that is consistent with their causality relation. Causally ordered delivery of messages implies FIFO message delivery. Furthermore, note that CO $\subset$ FIFO $\subset$ Non-FIFO.

Causal ordering model is useful in developing distributed algorithms. Generally, it considerably simplifies the design of distributed algorithms because it provides a built-in synchronization. For example, in replicated database systems, it is important that every process responsible for updating a replica receives the updates in the same order to maintain database consistency. Without causal ordering, each update must be checked to ensure that database consistency is not being violated. Causal ordering eliminates the need for such checks.

2.4 Global state of a distributed system

The global state of a distributed system is a collection of the local states of its components, namely, the processes and the communication channels [2,3]. The state of a process at any time is defined by the contents of processor registers, stacks, local memory, etc. and depends on the local context of the distributed application. The state of a channel is given by the set of messages in transit in the channel.

The occurrence of events changes the states of respective processes and channels, thus causing transitions in global system state. For example, an internal event changes the state of the process at which it occurs. A send event (or a receive event) changes the state of the process that sends (or receives) the message and the state of the channel on which the message is sent (or received).

Let LS_i^x denote the state of process p_i after the occurrence of event e_i^x and before the event e_i^{x+1}. LS_i^0 denotes the initial state of process p_i. LS_i^x is a result of the execution of all the events executed by process p_i till e_i^x. Let $send(m) \leq LS_i^x$ denote the fact that $\exists y{:}1 \leq y \leq x :: e_i^y = send(m)$. Likewise, let $rec(m) \not\leq LS_i^x$ denote the fact that $\forall y{:}1 \leq y \leq x :: e_i^y \neq rec(m)$.

The state of a channel is difficult to state formally because a channel is a distributed entity and its state depends upon the states of the processes it connects. Let $SC_{ij}^{x,y}$ denote the state of a channel C_{ij} defined as follows:

$$SC_{ij}^{x,y} = \{m_{ij} \mid send(m_{ij}) \leq LS_i^x \bigwedge rec(m_{ij}) \not\leq LS_j^y\}.$$

Thus, channel state $SC_{ij}^{x,y}$ denotes all messages that p_i sent up to event e_i^x and which process p_j had not received until event e_j^y.

2.4.1 Global state

The global state of a distributed system is a collection of the local states of the processes and the channels. Notationally, the global state GS is defined as

$$GS = \{\bigcup_i LS_i^{x_i}, \bigcup_{j,k} SC_{jk}^{y_j,z_k}\}.$$

For a global snapshot to be meaningful, the states of all the components of the distributed system must be recorded at the same instant. This will be possible if the local clocks at processes were perfectly synchronized or there was a global system clock that could be instantaneously read by the processes. However, both are impossible.

However, it turns out that even if the state of all the components in a distributed system has not been recorded at the same instant, such a state will be meaningful provided every message that is recorded as received is also recorded as sent. Basic idea is that an effect should not be present without its cause. A message cannot be received if it was not sent; that is, the state should not violate causality. Such states are called *consistent global states* and are meaningful global states. Inconsistent global states are not meaningful in the sense that a distributed system can never be in an inconsistent state.

A global state $GS = \{\bigcup_i LS_i^{x_i}, \bigcup_{j,k} SC_{jk}^{y_j,z_k}\}$ is a *consistent global state* iff it satisfies the following condition:

$$\forall m_{ij} : send(m_{ij}) \not\leq LS_i^{x_i} \Rightarrow m_{ij} \notin SC_{ij}^{x_i,y_j} \wedge rec(m_{ij}) \not\leq LS_j^{y_j}$$

That is, channel state $SC_{ij}^{x_i,y_j}$ and process state $LS_j^{y_j}$ must not include any message that process p_i sent after executing event $e_i^{x_i}$. A more rigorous definition of the consistency of a global state is given in Chapter 4.

In the distributed execution of Figure 2.2, a global state GS_1 consisting of local states $\{LS_1^1, LS_2^3, LS_3^3, LS_4^2\}$ is inconsistent because the state of p_2 has recorded the receipt of message m_{12}, however, the state of p_1 has not recorded its send. On the contrary, a global state GS_2 consisting of local

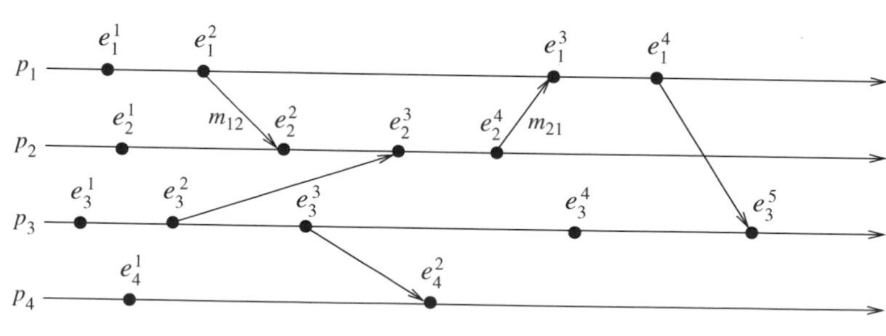

Figure 2.2 The space–time diagram of a distributed execution.

states $\{LS_1^2, LS_2^4, LS_3^4, LS_4^2\}$ is consistent; all the channels are empty except C_{21} that contains message m_{21}.

A global state $GS = \{\bigcup_i LS_i^{x_i}, \bigcup_{j,k} SC_{jk}^{y_j,z_k}\}$ is *transitless* iff

$$\forall i, \forall j \ : 1 \leq i, j \ \leq n \ :: SC_{ij}^{y_i,z_j} = \phi.$$

Thus, all channels are recorded as empty in a transitless global state. A global state is *strongly consistent* iff it is transitless as well as consistent. Note that in Figure 2.2, the global state consisting of local states $\{LS_1^2, LS_2^3, LS_3^4, LS_4^2\}$ is strongly consistent.

Recording the global state of a distributed system is an important paradigm when one is interested in analyzing, monitoring, testing, or verifying properties of distributed applications, systems, and algorithms. Design of efficient methods for recording the global state of a distributed system is an important problem.

2.5 Cuts of a distributed computation

In the space–time diagram of a distributed computation, a zigzag line joining one arbitrary point on each process line is termed a *cut* in the computation. Such a line slices the space–time diagram, and thus the set of events in the distributed computation, into a PAST and a FUTURE. The PAST contains all the events to the left of the cut and the FUTURE contains all the events to the right of the cut. For a cut C, let PAST(C) and FUTURE(C) denote the set of events in the PAST and FUTURE of C, respectively. Every cut corresponds to a global state and every global state can be graphically represented as a cut in the computation's space–time diagram [6].

definition 2.1 *If* $e_i^{Max\_PAST_i(C)}$ *denotes the latest event at process* p_i *that is in the* **PAST** *of a cut C, then the global state represented by the cut is* $\{\bigcup_i LS_i^{Max\_PAST_i(C)}, \bigcup_{j,k} SC_{jk}^{y_j,z_k}\}$ where $SC_{jk}^{y_j,z_k} = \{m \mid send(m) \in \mathbf{PAST(C)} \wedge rec(m) \in \mathbf{FUTURE}(C)\}$.

A consistent global state corresponds to a cut in which every message received in the PAST of the cut was sent in the PAST of that cut. Such a cut is known as a *consistent cut*. All messages that cross the cut from the PAST to the FUTURE are in transit in the corresponding consistent global state. A cut is *inconsistent* if a message crosses the cut from the FUTURE to the PAST. For example, the space–time diagram of Figure 2.3 shows two cuts, C_1 and C_2. C_1 is an inconsistent cut, whereas C_2 is a consistent cut. Note that these two cuts respectively correspond to the two global states GS_1 and GS_2, identified in the previous subsection.

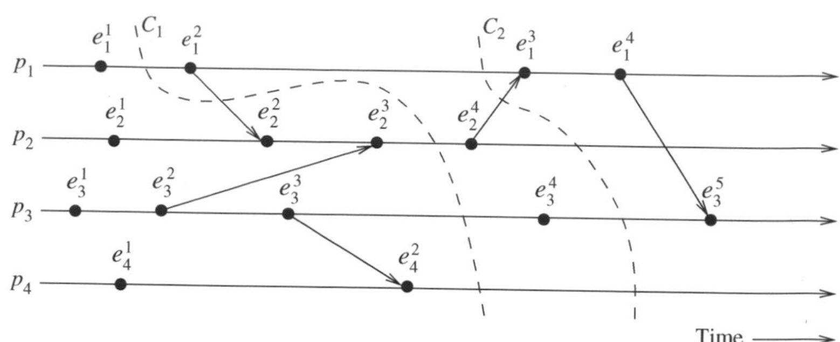

Figure 2.3 Illustration of cuts in a distributed execution.

Cuts in a space–time diagram provide a powerful graphical aid in representing and reasoning about global states of a computation.

2.6 Past and future cones of an event

In a distributed computation, an event e_j could have been affected only by all events e_i such that $e_i \rightarrow e_j$ and all the information available at e_i could be made accessible at e_j. All such events e_i belong to the past of e_j [6]. Let $Past(e_j)$ denote all events in the past of e_j in a computation $(H, \rightarrow)$. Then,

$$Past(e_j) = \{e_i | \forall e_i \in H, e_i \rightarrow e_j\}.$$

Figure 2.4 shows the past of an event e_j. Let $Past_i(e_j)$ be the set of all those events of $Past(e_j)$ that are on process p_i. Clearly, $Past_i(e_j)$ is a totally ordered set, ordered by the relation $\rightarrow_i$, whose maximal element is denoted by $max(Past_i(e_j))$. Obviously, $max(Past_i(e_j))$ is the latest event at process p_i that affected event e_j (see Figure 2.4). Note that $max(Past_i(e_j))$ is always a message send event.

Let $Max\_Past(e_j) = \bigcup_{(\forall i)} \{max(Past_i(e_j))\}$. $Max\_Past(e_j)$ consists of the latest event at every process that affected event e_j and is referred to as the

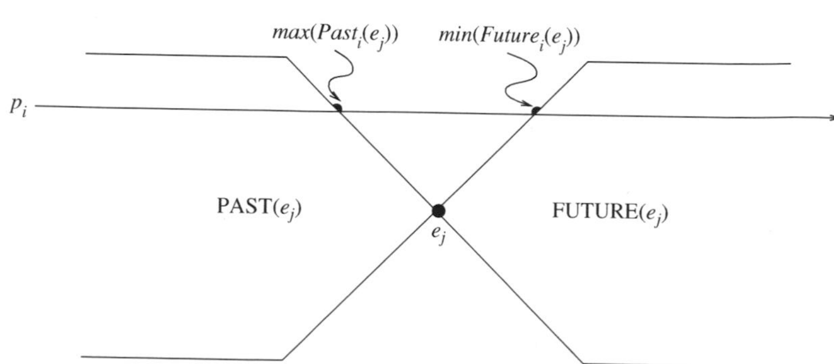

Figure 2.4 Illustration of past and future cones in a distributed computation.

surface of the past cone of e_j [6]. Note that $Max\_Past(e_j)$ is a consistent cut [7]. $Past(e_j)$ represents all events on the past light cone that affect e_j.

Similar to the past is defined the future of an event. The future of an event e_j, denoted by $Future(e_j)$, contains all events e_i that are causally affected by e_j (see Figure 2.4). In a computation $(H, \rightarrow)$, $Future(e_j)$ is defined as:

$$Future(e_j) = \{e_i | \forall e_i \in H, \ e_j \rightarrow e_i\}.$$

Likewise, we can define $Future_i(e_j)$ as the set of those events of $Future(e_j)$ that are on process p_i and $min(Future_i(e_j))$ as the first event on process p_i that is affected by e_j. Note that $min(Future_i(e_j))$ is always a message receive event. Likewise, $Min\_Future(e_j)$, defined as $\bigcup_{(\forall i)}\{min(Future_i(e_j))\}$, consists of the first event at every process that is causally affected by event e_j and is referred to as the *surface of the future cone* of e_j [6]. It denotes a consistent cut in the computation [7]. $Future(e_j)$ represents all events on the future light cone that are affected by e_j.

It is obvious that all events at a process p_i that occurred after $max(Past_i(e_j))$ but before $min(Future_i(e_j))$ are concurrent with e_j. Therefore, all and only those events of computation H that belong to the set "$H - Past(e_j) - Future(e_j)$" are concurrent with event e_j.

2.7 Models of process communications

There are two basic models of process communications [8] – synchronous and asynchronous. The *synchronous* communication model is a blocking type where on a message send, the sender process blocks until the message has been received by the receiver process. The sender process resumes execution only after it learns that the receiver process has accepted the message. Thus, the sender and the receiver processes must synchronize to exchange a message. On the other hand, *asynchronous* communication model is a non-blocking type where the sender and the receiver do not synchronize to exchange a message. After having sent a message, the sender process does not wait for the message to be delivered to the receiver process. The message is bufferred by the system and is delivered to the receiver process when it is ready to accept the message. A buffer overflow may occur if a process sends a large number of messages in a burst to another process.

Neither of the communication models is superior to the other. Asynchronous communication provides higher parallelism because the sender process can execute while the message is in transit to the receiver. However, an implementation of asynchronous communication requires more complex buffer management. In addition, due to higher degree of parallelism and non-determinism, it is much more difficult to design, verify, and implement distributed algorithms for asynchronous communications. The state space of such algorithms are likely to be much larger. Synchronous communication is simpler to handle and implement. However, due to frequent blocking, it is likely to have poor performance and is likely to be more prone to deadlocks.

2.8 Chapter summary

In a distributed system, a set of processes communicate by exchanging messages over a communication network. A distributed computation is spread over geographically distributed processes. The processes do not share a common global memory or a physical global clock, to which processes have instantaneous access.

The execution of a process consists of a sequential execution of its actions (e.g., internal events, message *send* events, and message *receive* events.) The events at a process are linearly ordered by their order of occurrence. Message exchanges between processes signify the flow of information between processes and establish causal dependencies between processes. The causal precedence relation between processes is captured by Lamport's "happens before" relation.

The global state of a distributed system is a collection of the states of its processes and the state of communication channels connecting the processes. A cut in a distributed computation is a zigzag line joining one arbitrary point on each process line. A cut represents a global state in the distributed computation. The past of an event consists of all events that causally affect it and the future of an event consists of all events that are causally affected by it.

2.9 Exercises

Exercise 2.1 Prove that in a distributed computation, for an event, the surface of the past cone (i.e., all the events on the surface) form a consistent cut. Does it mean that all events on the surface of the past cone are always concurrent? Give an example to make your case.

Exercise 2.2 Show that all events on the surface of the past cone of an event are message *send* events. Likewise, show that all events on the surface of the future cone of an event are message *receive* events.

2.10 Notes on references

Lamport in his landmark paper [4] defined the "happens before" relation between events in a distributed systems to capture causality. Other papers on the topic include those by Mattern [6] and by Panengaden and Taylor [7].

References

[1] K. Birman and T. Joseph, Reliable communication in presence of failures, *ACM Transactions on Computer Systems*, **3**, 1987, 47–76.

[2] K. M. Chandy and L. Lamport, Distributed snapshots: determining global states of distributed systems, *ACM Transactions on Computer Systems*, **3**(1), 1985, 63–75.

[3] A. Kshemkalyani, M. Raynal and M. Singhal, Global snapshots of a distributed system, *Distributed Systems Engineering Journal*, **2**(4), 1995, 224–233.

[4] L. Lamport, Time, clocks and the ordering of events in a distributed system, *Communications of the ACM*, **21**, 1978, 558–564.

[5] A. Lynch, Distributed processing solves main-frame problems, *Data Communications*, 1976, 17–22.

[6] F. Mattern, Virtual time and global states of distributed systems, *Proceedings of the Parallel and Distributed Algorithms Conference*, 1988, 215–226.

[7] P. Panengaden and K. Taylor, Concurrent common knowledge: a new definition of agreement for asynchronous events, *Proceedings of the 5th Symposium on Principles of Distributed Computing*, 1988, 197–209.

[8] S. M. Shatz, Communication mechanisms for programming distributed systems, *IEEE Computer*, 1984, 21–28.

CHAPTER

3 Logical time

3.1 Introduction

The concept of causality between events is fundamental to the design and analysis of parallel and distributed computing and operating systems. Usually causality is tracked using physical time. However, in distributed systems, it is not possible to have global physical time; it is possible to realize only an approximation of it. As asynchronous distributed computations make progress in spurts, it turns out that the logical time, which advances in jumps, is sufficient to capture the fundamental monotonicity property associated with causality in distributed systems. This chapter discusses three ways to implement logical time (e.g., scalar time, vector time, and matrix time) that have been proposed to capture causality between events of a distributed computation.

Causality (or the causal precedence relation) among events in a distributed system is a powerful concept in reasoning, analyzing, and drawing inferences about a computation. The knowledge of the causal precedence relation among the events of processes helps solve a variety of problems in distributed systems. Examples of some of these problems is as follows:

- **Distributed algorithms design** The knowledge of the causal precedence relation among events helps ensure liveness and fairness in mutual exclusion algorithms, helps maintain consistency in replicated databases, and helps design correct deadlock detection algorithms to avoid phantom and undetected deadlocks.
- **Tracking of dependent events** In distributed debugging, the knowledge of the causal dependency among events helps construct a consistent state for resuming reexecution; in failure recovery, it helps build a checkpoint; in replicated databases, it aids in the detection of file inconsistencies in case of a network partitioning.

- **Knowledge about the progress** The knowledge of the causal dependency among events helps measure the progress of processes in the distributed computation. This is useful in discarding obsolete information, garbage collection, and termination detection.
- **Concurrency measure** The knowledge of how many events are causally dependent is useful in measuring the amount of concurrency in a computation. All events that are not causally related can be executed concurrently. Thus, an analysis of the causality in a computation gives an idea of the concurrency in the program.

The concept of causality is widely used by human beings, often unconsciously, in the planning, scheduling, and execution of a chore or an enterprise, or in determining the infeasibility of a plan or the innocence of an accused. In day-to-day life, the global time to deduce causality relation is obtained from loosely synchronized clocks (i.e., wrist watches, wall clocks). However, in distributed computing systems, the rate of occurrence of events is several magnitudes higher and the event execution time is several magnitudes smaller. Consequently, if the physical clocks are not precisely synchronized, the causality relation between events may not be accurately captured. Network Time Protocols [15], which can maintain time accurate to a few tens of milliseconds on the Internet, are not adequate to capture the causality relation in distributed systems. However, in a distributed computation, generally the progress is made in spurts and the interaction between processes occurs in spurts. Consequently, it turns out that in a distributed computation, the causality relation between events produced by a program execution and its fundamental monotonicity property can be accurately captured by logical clocks.

In a system of logical clocks, every process has a logical clock that is advanced using a set of rules. Every event is assigned a timestamp and the causality relation between events can be generally inferred from their timestamps. The timestamps assigned to events obey the fundamental monotonicity property; that is, if an event a causally affects an event b, then the timestamp of a is smaller than the timestamp of b.

This chapter first presents a general framework of a system of logical clocks in distributed systems and then discusses three ways to implement logical time in a distributed system. In the first method, Lamport's scalar clocks, the time is represented by non-negative integers; in the second method, the time is represented by a vector of non-negative integers; in the third method, the time is represented as a matrix of non-negative integers. We also discuss methods for efficient implementation of the systems of vector clocks.

The chapter ends with a discussion of virtual time, its implementation using the time-warp mechanism and a brief discussion of physical clock synchronization and the Network Time Protocol.

3.2 A framework for a system of logical clocks

3.2.1 Definition

A system of logical clocks consists of a time domain T and a logical clock C [19]. Elements of T form a partially ordered set over a relation $<$. This relation is usually called the *happened before* or *causal precedence*. Intuitively, this relation is analogous to the *earlier than* relation provided by the physical time. The logical clock C is a function that maps an event e in a distributed system to an element in the time domain T, denoted as $\mathrm{C}(e)$ and called the timestamp of e, and is defined as follows:

$$C : H \mapsto T,$$

such that the following property is satisfied:

$$\text{for two events } e_i \text{ and } e_j,\ e_i \to e_j \Longrightarrow \mathrm{C}(e_i) < \mathrm{C}(e_j).$$

This monotonicity property is called the *clock consistency condition*. When T and C satisfy the following condition,

$$\text{for two events } e_i \text{ and } e_j,\ e_i \to e_j \Leftrightarrow \mathrm{C}(e_i) < \mathrm{C}(e_j),$$

the system of clocks is said to be *strongly consistent*.

3.2.2 Implementing logical clocks

Implementation of logical clocks requires addressing two issues [19]: data structures local to every process to represent logical time and a protocol (set of rules) to update the data structures to ensure the consistency condition.

Each process p_i maintains data structures that allow it the following two capabilities:

- A *local logical clock*, denoted by lc_i, that helps process p_i measure its own progress.
- A *logical global clock*, denoted by gc_i, that is a representation of process p_i's local view of the logical global time. It allows this process to assign consistent timestamps to its local events. Typically, lc_i is a part of gc_i.

The protocol ensures that a process's logical clock, and thus its view of the global time, is managed consistently. The protocol consists of the following two rules:

- **R1** This rule governs how the local logical clock is updated by a process when it executes an event (send, receive, or internal).
- **R2** This rule governs how a process updates its global logical clock to update its view of the global time and global progress. It dictates what information about the logical time is piggybacked in a message and how this information is used by the receiving process to update its view of the global time.

Systems of logical clocks differ in their representation of logical time and also in the protocol to update the logical clocks. However, all logical clock systems implement rules **R1** and **R2** and consequently ensure the fundamental monotonicity property associated with causality. Moreover, each particular logical clock system provides its users with some additional properties.

3.3 Scalar time

3.3.1 Definition

The scalar time representation was proposed by Lamport in 1978 [9] as an attempt to totally order events in a distributed system. Time domain in this representation is the set of non-negative integers. The logical local clock of a process p_i and its local view of the global time are squashed into one integer variable C_i.

Rules **R1** and **R2** to update the clocks are as follows:

- **R1** Before executing a send or internal event, process p_i executes the following:

$$C_i := C_i + d \qquad (d > 0).$$

  In general, every time **R1** is executed, d can have a different value, and this value may be application-dependent. However, typically d is kept at 1 because this is able to identify the time of each event uniquely at a process, while keeping the rate of increase of d to its lowest level.
- **R2** Each message piggybacks the clock value of its sender at sending time. When a process p_i receives a message with timestamp C_{msg}, it executes the following actions:
  1. $C_i := max(C_i, C_{msg})$;
  2. execute **R1**;
  3. deliver the message.

Figure 3.1 shows the evolution of scalar time with $d=1$.

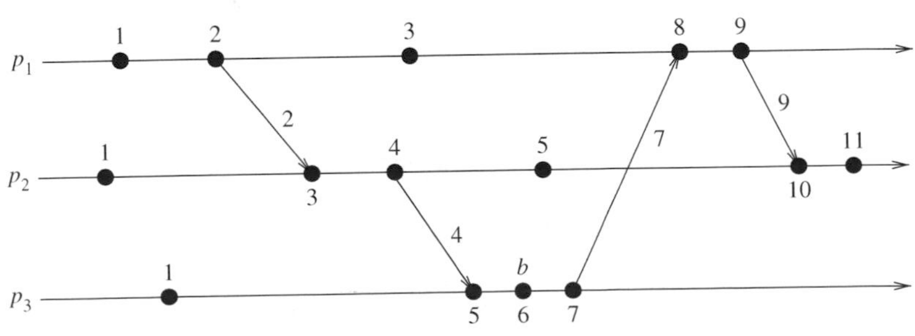

Figure 3.1 Evolution of scalar time [19].

3.3.2 Basic properties

Consistency property

Clearly, scalar clocks satisfy the monotonicity and hence the consistency property:

$$\text{for two events } e_i \text{ and } e_j,\ e_i \rightarrow e_j \Longrightarrow \mathrm{C}(e_i) < \mathrm{C}(e_j).$$

Total Ordering

Scalar clocks can be used to totally order events in a distributed system [9]. The main problem in totally ordering events is that two or more events at different processes may have an identical timestamp. (Note that for two events e_1 and e_2, $\mathrm{C}(e_1) = \mathrm{C}(e_2) \Longrightarrow e_1 \parallel e_2$.) For example, in Figure 3.1, the third event of process P_1 and the second event of process P_2 have identical scalar timestamp. Thus, a tie-breaking mechanism is needed to order such events. Typically, a tie is broken as follows: process identifiers are linearly ordered and a tie among events with identical scalar timestamp is broken on the basis of their process identifiers. The lower the process identifier in the ranking, the higher the priority. The timestamp of an event is denoted by a tuple (t, i) where t is its time of occurrence and i is the identity of the process where it occurred. The total order relation $\prec$ on two events x and y with timestamps (h,i) and (k,j), respectively, is defined as follows:

$$x \prec y \Leftrightarrow (h < k \textbf{ or } (h = k \textbf{ and } i < j)).$$

Since events that occur at the same logical scalar time are independent (i.e., they are not causally related), they can be ordered using any arbitrary criterion without violating the causality relation $\rightarrow$. Therefore, a total order is consistent with the causality relation "$\rightarrow$". Note that $x \prec y \Longrightarrow x \rightarrow y \vee x \parallel y$. A total order is generally used to ensure liveness properties in distributed algorithms. Requests are timestamped and served according to the total order based on these timestamps [9].

Event counting

If the increment value d is always 1, the scalar time has the following interesting property: if event e has a timestamp h, then $h-1$ represents the minimum logical duration, counted in units of events, required before producing the event e [4]; we call it the height of the event e. In other words, $h-1$ events have been produced sequentially before the event e regardless of the processes that produced these events. For example, in Figure 3.1, five events precede event b on the longest causal path ending at b.

No strong consistency

The system of scalar clocks is not strongly consistent; that is, for two events e_i and e_j, $\mathrm{C}(e_i) < \mathrm{C}(e_j) \not\Longrightarrow e_i \rightarrow e_j$. For example, in Figure 3.1, the third event of process P_1 has smaller scalar timestamp than the third event of process P_2. However, the former did not happen before the latter. The reason that scalar clocks are not strongly consistent is that the logical local clock and logical global clock of a process are squashed into one, resulting in the loss of causal dependency information among events at different processes. For example, in Figure 3.1, when process P_2 receives the first message from process P_1, it updates its clock to 3, forgetting that the timestamp of the latest event at P_1 on which it depends is 2.

3.4 Vector time

3.4.1 definition

The system of vector clocks was developed independently by Fidge [4], Mattern [12], and Schmuck [23]. In the system of vector clocks, the time domain is represented by a set of n-dimensional non-negative integer vectors. Each process p_i maintains a vector $vt_i[1..n]$, where $vt_i[i]$ is the local logical clock of p_i and describes the logical time progress at process p_i. $vt_i[j]$ represents process p_i's latest knowledge of process p_j's local time. If $vt_i[j] = x$, then process p_i knows that local time at process p_j has progressed till x. The entire vector vt_i constitutes p_i's view of the global logical time and is used to timestamp events.

Process p_i uses the following two rules **R1** and **R2** to update its clock:

- **R1** Before executing a send or internal event, process p_i updates its local logical time as follows:

$$vt_i[i] := vt_i[i] + d \qquad (d > 0).$$

- **R2** Each message m is piggybacked with the vector clock vt of the sender process at sending time. On the receipt of such a message (m,vt), process p_i executes the following sequence of actions:
  1. update its global logical time as follows:

$$1 \le k \le n \; : \; vt_i[k] := max(vt_i[k], vt[k]);$$

  2. execute **R1**;
  3. deliver the message m.

The timestamp associated with an event is the value of the vector clock of its process when the event is executed. Figure 3.2 shows an example of vector clocks progress with the increment value $d = 1$. Initially, a vector clock is $[0, 0, 0, \ldots, 0]$.

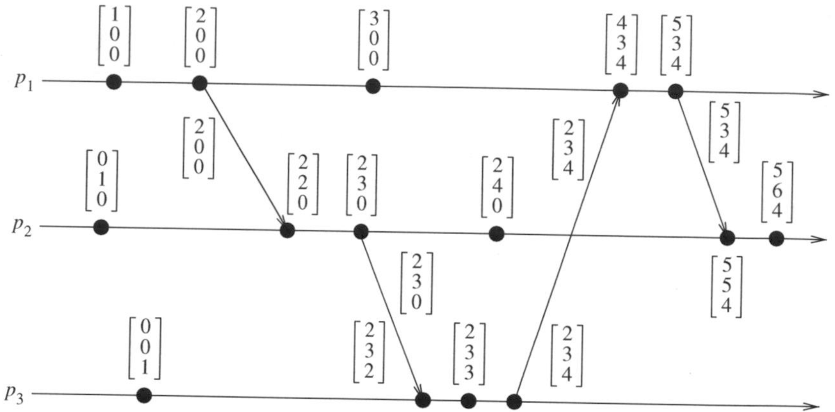

Figure 3.2 Evolution of vector time [19].

The following relations are defined to compare two vector timestamps, vh and vk:

$$vh = vk \Leftrightarrow \forall x : vh[x] = vk[x]$$

$$vh \leq vk \Leftrightarrow \forall x : vh[x] \leq vk[x]$$

$$vh < vk \Leftrightarrow vh \leq vk \text{ and } \exists x : vh[x] < vk[x]$$

$$vh \parallel vk \Leftrightarrow \neg(vh < vk) \wedge \neg(vk < vh).$$

3.4.2 Basic properties

Isomorphism

Recall that relation "$\rightarrow$" induces a partial order on the set of events that are produced by a distributed execution. If events in a distributed system are timestamped using a system of vector clocks, we have the following property.

If two events x and y have timestamps vh and vk, respectively, then

$$x \rightarrow y \Leftrightarrow vh < vk$$

$$x \parallel y \Leftrightarrow vh \parallel vk.$$

Thus, there is an isomorphism between the set of partially ordered events produced by a distributed computation and their vector timestamps. This is a very powerful, useful, and interesting property of vector clocks.

If the process at which an event occurred is known, the test to compare two timestamps can be simplified as follows: if events x and y respectively occurred at processes p_i and p_j and are assigned timestamps vh and vk, respectively, then

$$x \rightarrow y \Leftrightarrow vh[i] \leq vk[i]$$

$$x \parallel y \Leftrightarrow vh[i] > vk[i] \wedge vh[j] < vk[j].$$

Strong consistency

The system of vector clocks is strongly consistent; thus, by examining the vector timestamp of two events, we can determine if the events are causally related. However, Charron–Bost showed that the dimension of vector clocks cannot be less than n, the total number of processes in the distributed computation, for this property to hold [2].

Event counting

If d is always 1 in rule **R1**, then the ith component of vector clock at process p_i, $vt_i[i]$, denotes the number of events that have occurred at p_i until that instant. So, if an event e has timestamp vh, $vh[j]$ denotes the number of events executed by process p_j that causally precede e. Clearly, $\sum vh[j] - 1$ represents the total number of events that causally precede e in the distributed computation.

Applications

Since vector time tracks causal dependencies exactly, it finds a wide variety of applications. For example, it is used in distributed debugging, implementations of causal ordering communication and causal distributed shared memory, establishment of global breakpoints, and in determining the consistency of checkpoints in optimistic recovery.

A brief historical perspective of vector clocks

Although the theory associated with vector clocks was first developed in 1988 independently by Fidge and Mattern, vector clocks were informally introduced and used by several researchers before this. Parker *et al.* [17] used a rudimentary vector clocks system to detect inconsistencies of replicated files due to network partitioning. Liskov and Ladin [11] proposed a vector clock system to define highly available distributed services. Similar system of clocks was used by Strom and Yemini [26] to keep track of the causal dependencies between events in their optimistic recovery algorithm and by Raynal to prevent drift between logical clocks [18]. Singhal [24] used vector clocks coupled with a boolean vector to determine the currency of a critical section execution request by detecting the causality relation between a critical section request and its execution.

3.4.3 On the size of vector clocks

An important question to ask is whether vector clocks of size n are necessary in a computation consisting of n processes. To answer this, we examine the usage of vector clocks.

- A vector clock provides the latest known local time at each other process. If this information in the clock is to be used to explicitly track the progress at every other process, then a vector clock of size n is necessary.

- A popular use of vector clocks is to determine the causality between a pair of events. Given any events e and f, the test for $e \rightarrow f$ if and only if $C(e) < C(f)$, requires a comparison of the vector clocks of e and f. Although it appears that the clock of size n is necessary, that is not quite accurate. It can be shown that a size equal to the dimension of the partial order $(E, \rightarrow)$ is necessary, where the upper bound on this dimension is n. This is explained below.

To understand this result on the size of clocks for determining causality between a pair of events, we first introduce some definitions. A *linear extension* of a partial order $(E, \rightarrow)$ is a linear ordering of E that is consistent with the partial order, i.e., if two events are ordered in the partial order, they are also ordered in the linear order. A linear extension can be viewed as projecting all the events from the different processes on a single time axis. However, the linear order will necessarily introduce ordering between each pair of events, and some of these orderings are not in the partial order. Also observe that different linear extensions are possible in general. Let $\mathcal{P}$ denote the set of tuples in the partial order defined by the causality relation; so there is a tuple (e, f) in $\mathcal{P}$ for each pair of events e and f such that $e \rightarrow f$. Let $\mathcal{L}_1$, $\mathcal{L}_2 \ldots$ denote the sets of tuples in different linear extensions of this partial order. The set $\mathcal{P}$ is contained in the set obtained by taking the intersection of any such collection $\mathcal{L}_1, \mathcal{L}_2 \ldots$. This is because each $\mathcal{L}_i$ must contain all the tuples, i.e., causality dependencies, that are in $\mathcal{P}$. The *dimension* of a partial order is the minimum number of linear extensions whose intersection gives exactly the partial order.

Consider a client–server interaction between a pair of processes. Queries to the server and responses to the client occur in strict alternating sequences. Although $n = 2$, all the events are strictly ordered, and there is only one linear order of all the events that is consistent with the "partial" order. Hence the dimension of this "partial order" is 1. A scalar clock such as one implemented by Lamport's scalar clock rules is adequate to determine $e \rightarrow f$ for any events e and f in this execution.

Now consider an execution on processes p_1 and p_2 such that each sends a message to the other before receiving the other's message. The two send events are concurrent, as are the two receive events. To determine the causality between the send events or between the receive events, it is not sufficient to use a single integer; a vector clock of size $n = 2$ is necessary. This execution exhibits the graphical property called a *crown*, wherein there are some messages $m_0, \ldots m_{n-1}$ such that $Send(m_i) \rightarrow Receive(m_{i+1\, mod\, (n-1)})$ for all i from 0 to $n - 1$. A crown of n messages has dimension n. We introduce the notion of crown and study its properties in Chapter 6.

For a complex execution, it is not straightforward to determine the dimension of the partial order. Figure 3.3 shows an execution involving four processes. However, the dimension of this partial order is two. To see this

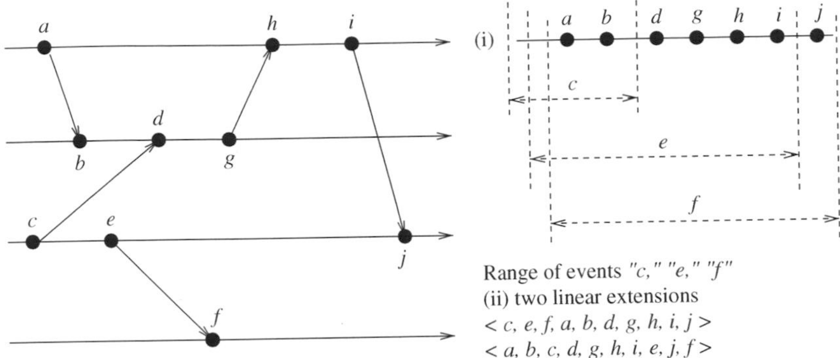

Figure 3.3 Example illustrating dimension of a execution $(E, \rightarrow)$. For $n = 4$ processes, the dimension is 2.

informally, consider the longest chain $\langle a, b, d, g, h, i, j\rangle$. There are events outside this chain that can yield multiple linear extensions. Hence, the dimension is more than 1. The right side of Figure 3.3 shows the earliest possible and the latest possible occurrences of the events not in this chain, with respect to the events in this chain. Let $\mathcal{L}_1$ be the set of tuples in the linear extension $\langle c, e, f, a, b, d, g, h, i, j\rangle$, which contains the following tuples that are not in $\mathcal{P}$:

$$(c, a), (c, b),$$
$$(e, a), (e, b), (e, d), (e, g), (e, h), (e, i),$$
$$(f, a), (f, b), (f, d), (f, g), (f, h), (f, i), (f, j).$$

Let $\mathcal{L}_2$ be the set of tuples in the linear extension $\langle a, b, c, d, g, h, i, e, j, f\rangle$, which contains the following tuples not in $\mathcal{P}$:

$$(a, c), (b, c),$$
$$(a, e), (b, e), (d, e), (g, e), (h, e), (i, e),$$
$$(a, f), (b, f), (d, f), (g, f), (h, f), (i, f), (j, f).$$

Further, observe that $(\mathcal{L}_1 \setminus \mathcal{P})\bigcap\mathcal{L}_2 = \emptyset$ and $(\mathcal{L}_2 \setminus \mathcal{P})\bigcap\mathcal{L}_1 = \emptyset$. Hence, $\mathcal{L}_1\bigcap\mathcal{L}_2 = \mathcal{P}$ and the dimension of the execution is 2 as these two linear extensions are enough to generate $\mathcal{P}$.

Unfortunately, it is not computationally easy to determine the dimension of a partial order. To exacerbate the problem, the above form of analysis has to be completed a posteriori (i.e., off-line), once the entire partial order has been determined after the completion of the execution.

3.5 Efficient implementations of vector clocks

If the number of processes in a distributed computation is large, then vector clocks will require piggybacking of huge amount of information in messages for the purpose of disseminating time progress and updating clocks. The message overhead grows linearly with the number of processors in the system and when there are thousands of processors in the system, the message size becomes huge even if there are only a few events occurring in few processors. In this section, we discuss efficient ways to maintain vector clocks; similar techniques can be used to efficiently implement matrix clocks.

Charron-Bost showed [2] that if vector clocks have to satisfy the strong consistency property, then in general vector timestamps must be at least of size n, the total number of processes. Therefore, in general the size of a vector timestamp is the number of processes involved in a distributed computation; however, several optimizations are possible and next, we discuss techniques to implement vector clocks efficiently [19].

3.5.1 Singhal–Kshemkalyani's differential technique

Singhal–Kshemkalyani's differential technique [25] is based on the observation that between successive message sends to the same process, only a few entries of the vector clock at the sender process are likely to change. This is more likely when the number of processes is large because only a few of them will interact frequently by passing messages. In this technique, when a process p_i sends a message to a process p_j, it piggybacks only those entries of its vector clock that differ since the last message sent to p_j.

The technique works as follows: if entries $i_1, i_2, \ldots, i_{n_1}$ of the vector clock at p_i have changed to $v_1, v_2, \ldots, v_{n_1}$, respectively, since the last message sent to p_j, then process p_i piggybacks a compressed timestamp of the form

$$\{(i_1, v_1), (i_2, v_2), \ldots, (i_{n_1}, v_{n_1})\}$$

to the next message to p_j. When p_j receives this message, it updates its vector clock as follows:

$$vt_j[i_k] = max(vt_j[i_k], v_k) \text{ for } k = 1, 2, \ldots, n_1.$$

Thus this technique cuts down the message size, communication bandwidth and buffer (to store messages) requirements. In the worst case, every element of the vector clock has been updated at p_i since the last message to process p_j, and the next message from p_i to p_j will need to carry the entire vector timestamp of size n. However, on the average the size of the timestamp on a message will be less than n. Note that implementation of this technique requires each process to remember the vector timestamp in the message last sent to every other process. Direct implementation of this will result in $O(n^2)$ storage overhead at each process. This technique also requires that the communication channels follow FIFO discipline for message delivery.

Singhal and Kshemkalyani developed a clever technique that cuts down this storage overhead at each process to $O(n)$. The technique works in the following manner: process p_i maintains the following two additional vectors:

- $LS_i[1 \ldots n]$ ('Last Sent'):
  $LS_i[j]$ indicates the value of $vt_i[i]$ when process p_i last sent a message to process p_j.
- $LU_i[1 \ldots n]$ ('Last Update'):
  $LU_i[j]$ indicates the value of $vt_i[i]$ when process p_i last updated the entry $vt_i[j]$.

Clearly, $LU_i[i] = vt_i[i]$ at all times and $LU_i[j]$ needs to be updated only when the receipt of a message causes p_i to update entry $vt_i[j]$. Also, $LS_i[j]$ needs to be updated only when p_i sends a message to p_j. Since the last communication from p_i to p_j, only those elements k of vector clock $vt_i[k]$ have changed for which $LS_i[j] < LU_i[k]$ holds. Hence, only these elements need to be sent in a message from p_i to p_j. When p_i sends a message to p_j, it sends only a set of tuples,

$$\{(x, vt_i[x]) | LS_i[j] < LU_i[x]\},$$

as the vector timestamp to p_j, instead of sending a vector of n entries in a message.

Thus the entire vector of size n is not sent along with a message. Instead, only the elements in the vector clock that have changed since the last message send to that process are sent in the format $\{(p_1, latest\_value), (p_2, latest\_value), \ldots\}$, where p_i indicates that the p_ith component of the vector clock has changed.

This method is illustrated in Figure 3.4. For instance, the second message from p_3 to p_2 (which contains a timestamp $\{(3, 2)\}$) informs p_2 that the third component of the vector clock has been modified and the new value is 2. This is because the process p_3 (indicated by the third component of

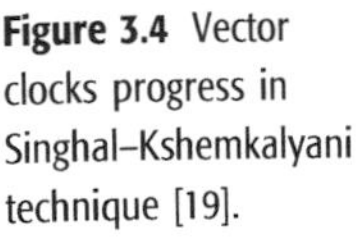
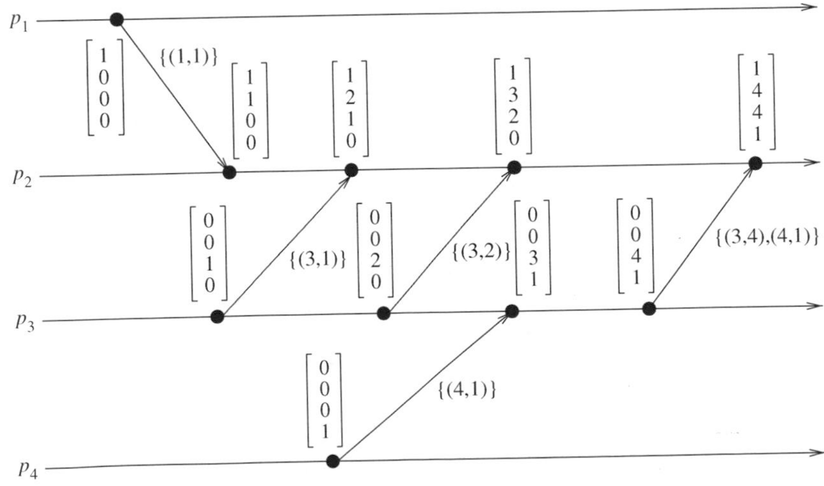

Figure 3.4 Vector clocks progress in Singhal–Kshemkalyani technique [19].

the vector) has advanced its clock value from 1 to 2 since the last message sent to p_2.

The cost of maintaining vector clocks in large systems can be substantially reduced by this technique, especially if the process interactions exhibit temporal or spatial localities. This technique would turn advantageous in a variety of applications including causal distributed shared memories, distributed deadlock detection, enforcement of mutual exclusion and localized communications typically observed in distributed systems.

3.5.2 Fowler–Zwaenepoel's direct-dependency technique

Fowler–Zwaenepoel direct dependency technique [6] reduces the size of messages by transmitting only a scalar value in the messages. No vector clocks are maintained on-the-fly. Instead, a process only maintains information regarding direct dependencies on other processes. A vector time for an event, which represents transitive dependencies on other processes, is constructed off-line from a recursive search of the direct dependency information at processes.

Each process p_i maintains a dependency vector D_i. Initially,

$$D_i[j] = 0 \text{ for } j = 1, \ldots, n.$$

D_i is updated as follows:

1. Whenever an event occurs at p_i, $D_i[i] := D_i[i] + 1$. That is, the vector component corresponding to its own local time is incremented by one.
2. When a process p_i sends a message to process p_j, it piggybacks the updated value of $D_i[i]$ in the message.
3. When p_i receives a message from p_j with piggybacked value d, p_i updates its dependency vector as follows: $D_i[j] := max\{D_i[j], d\}$.

Thus the dependency vector D_i reflects only direct dependencies. At any instant, $D_i[j]$ denotes the sequence number of the latest event on process p_j that *directly* affects the current state. Note that this event may precede the latest event at p_j that *causally* affects the current state.

Figure 3.5 illustrates the Fowler–Zwaenepoel technique. For instance, when process p_4 sends a message to process p_3, it piggybacks a scalar that indicates the direct dependency of p_3 on p_4 because of this message. Subsequently, process p_3 sends a message to process p_2 piggybacking a scalar to indicate the direct dependency of p_2 on p_3 because of this message. Now, process p_2 is in fact indirectly dependent on process p_4 since process p_3 is dependent on process p_4. However, process p_2 is never informed about its indirect dependency on p_4.

Thus although the direct dependencies are duly informed to the receiving processes, the transitive (indirect) dependencies are not maintained by

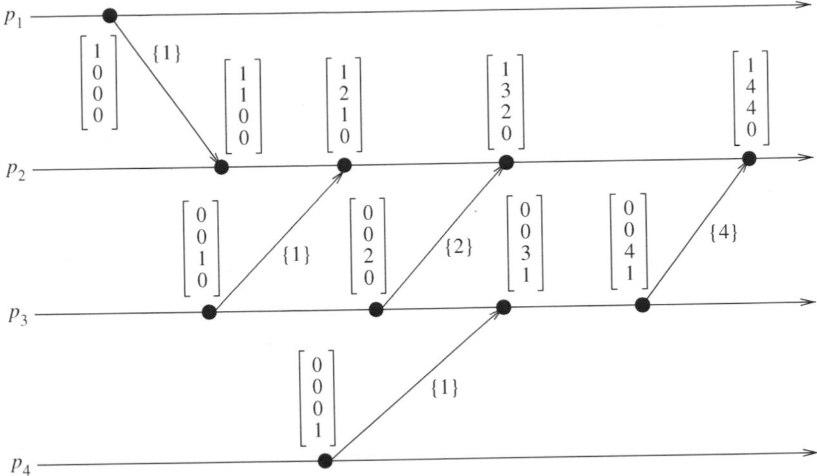

Figure 3.5 Vector clock progress in Fowler–Zwaenepoel technique [19].

this method. They can be obtained only by recursively tracing the direct-dependency vectors of the events off-line. This involves computational overhead and latencies. Thus this method is ideal only for those applications that do not require computation of transitive dependencies on the fly. The computational overheads characteristic of this method makes it best suitable for applications like causal breakpoints and asynchronous checkpoint recovery where computation of causal dependencies is performed offline.

This technique results in considerable saving in the cost; only one scalar is piggybacked on every message. However, the dependency vector does not represent transitive dependencies (i.e., a vector timestamp). The transitive dependency (or the vector timestamp) of an event is obtained by recursively tracing the direct-dependency vectors of processes. Clearly, this will have overhead and will involve latencies. Therefore, this technique is not suitable for applications that require on-the-fly computation of vector timestamps. Nonetheless, this technique is ideal for applications where computation of causal dependencies is performed off-line (e.g., causal breakpoint, asynchronous checkpointing recovery).

The transitive dependencies could be determined by combining an event's direct dependency with that of its directly dependent event. In Figure 3.5, the fourth event of process p_3 is dependent on the first event of process p_4 and the fourth event of process p_2 is dependent on the fourth event of process p_3. By combining these two direct dependencies, it is possible to deduce that the fourth event of process p_2 depends on the first event of process p_4. It is important to note that if event e_j at process p_j occurs before event e_i at process p_i, then all the events from e_0 to e_{j-1} in process p_j also happen before e_i. Hence, it is sufficient to record for e_i the latest event of process p_j that happened before e_i. This way, each event would record its dependencies on the latest event on every other process it depends on and those events maintain their own dependencies. Combining all these dependencies, the entire set of events that a particular event depends on could be determined off-line.

The off-line computation of transitive dependencies can be performed using a recursive algorithm proposed in [6] and is illustrated in a modified form in Algorithm 3.1. *DTV* is the dependency-tracking vector of size n (where n is the number of process) which is supposed to track all the causal dependencies of a particular event e_i in process p_i. The algorithm then needs to be invoked as *DependencyTrack*$(i, D_i^e[i])$. The algorithm initializes *DTV* to the least possible timestamp value which is 0 for all entries except i for which the value is set to $D_i^e[i]$:

$$\text{for all } k = 1, \ldots, n \text{ and } k \neq i,\ DTV[k]{=}0 \text{ and } DTV[i]{=}D_i^e[i].$$

The algorithm then calls the *VisitEvent* algorithm on process p_i and event e_i. *VisitEvent* checks all the entries $(1, \ldots, n)$ of *DTV* and D_i^e and if the value in D_i^e is greater than the value in *DTV* for that entry, then *DTV* assumes the value of D_i^e for that entry. This ensures that the latest event in process j that e_i depends on is recorded in *DTV*. *VisitEvent* is recursively called on all entries that are newly included in *DTV* so that the latest dependency information can be accurately tracked.

Let us illustrate the recursive dependency trace algorithm by tracking the dependencies of the fourth event at process p_2. The algorithm is invoked as

---

DependencyTrack$(i : process,\ \sigma : event\ index)$
* Casual distributed breakpoint for σ_i *\
* *DTV* holds the result *\
for all $k \neq i$ **do**
    $DTV[k]{=}0$
end for
$DTV[i]{=}\sigma$
end DependencyTrack

VisitEvent$(j : process,\ e : event\ index)$
* Place dependencies of τ into *DTV* *\
for all $k \neq j$ **do**
    $\alpha = D_j^e[k]$
    if $\alpha > DTV[k]$ **then**
        $DTV[k]{=}\alpha$
        VisitEvent(k, α)
    end if
end for
end VisitEvent

Algorithm 3.1 Recursive dependency trace algorithm

DependencyTrack(2, 4). *DTV* is initially set to $\langle 0\,4\,0\,0 \rangle$ by *DependencyTrack*. It then calls *VisitEvent*(2, 4). The values held by D_2^4 are $\langle 1\ 4\ 4\ 0 \rangle$. So, *DTV* is now updated to $\langle 1\ 4\ 0\ 0 \rangle$ and *VisitEvent*(1, 1) is called. The values held by D_1^1 are $\langle 1\ 0\ 0\ 0 \rangle$. Since none of the entries are greater than those in *DTV*, the algorithm returns. Again the values held by D_2^4 are checked and this time entry 3 is found to be greater in D_2^4 than *DTV*. So, *DTV* is updated as $\langle 1\ 4\ 4\ 0 \rangle$ and *VisiEvent*(3, 4) is called. The values held by D_3^4 are $\langle 0\ 0\ 4\ 1 \rangle$. Since entry 4 of D_3^4 is greater than that of *DTV*, it is updated as $\langle 1\ 4\ 4\ 1 \rangle$ and *VisitEvent*(4, 1) is called. Since none of the entries in D_4^1: $\langle 1\ 0\ 0\ 0 \rangle$ are greater than those of *DTV*, the algorithm returns to *VisitEvent*(2, 4). Since all the entries have been checked, *VisitEvent*(2, 4) is exited and so is *DependencyTrack*. At this point, *DTV* holds $\langle 1\ 4\ 4\ 1 \rangle$, meaning event 4 of process p_2 is dependent upon event 1 of process p_1, event 4 of process p_3 and event 1 in process p_4. Also, it is dependent on events that precede event 4 of process p_3 and these dependencies could be obtained by invoking the *DependencyTrack* algorithm on the fourth event of process p_3. Thus, all the causal dependencies could be tracked off-line.

This technique can result in a considerable saving of cost since only one scalar is piggybacked on every message. One of the important requirements is that a process updates and records its dependency vectors after receiving a message and before sending out any message. Also, if events occur frequently, this technique will require recording the history of a large number of events.

3.6 Jard–Jourdan's adaptive technique

The Fowler–Zwaenepoel direct-dependency technique does not allow the transitive dependencies to be captured in real time during the execution of processes. In addition, a process must observe an event (i.e., update and record its dependency vector) after receiving a message but before sending out any message. Otherwise, during the reconstruction of a vector timestamp from the direct-dependency vectors, all the causal dependencies will not be captured. If events occur very frequently, this technique will require recording the history of a large number of events.

In the Jard–Jourdan's technique [8], events can be adaptively observed while maintaining the capability of retrieving all the causal dependencies of an observed event. (Observing an event means recording of the information about its dependencies.) This method uses the idea that when an observed event *e* records its dependencies, then events that follow can determine their transitive dependencies, that is, the set of events that they indirectly depend on, by making use of the information recorded about *e*. The reason is that when an event *e* is observed, the information about the send and receive of messages maintained by a process is recorded in that event and the information maintained by the process is then reset and updated. So, when the process propagates information after e, it propagates only history of activities that took place after e. The next observed event either in the same process or in a different one, would then have to look at the information recorded for e to know about the activities that happened before e. This method still does not allow determining all the causal dependencies in real time, but avoids the problem of recording a large amount of history which is realized when using the direct dependency technique.

To implement the technique of recording the information in an observed event and resetting the information managed by a process, Jard–Jourdan defined a *pseudo-direct* relation $\ll$ on the events of a distributed computation as follows:

If events e_i and e_j happen at process p_i and p_j, respectively, then $e_j \ll e_i$ iff there exists a path of message transfers that starts after e_j on the process p_j and ends before e_i on the process e_i such that there is no observed event on the path. The relation is termed pseudo-direct because event e_i may depend upon many unobserved events on the path, say $ue_1, ue_2, \ldots, ue_n$, etc., which are in turn dependent on each other. If e_i happens after ue_n, then e_i is still considered to be directly dependent upon $ue_1, ue_2, \ldots, ue_n$, since these events are unobserved, which is falsely assumed to have direct dependency. If another event e_k happens after e_i, then the transitive dependencies of e_k on $ue_1, ue_2, \ldots, ue_n$ can be determined by using the information recorded at e_i and e_i can do the same with e_j.

The technique is implemented using the following mechanism: the partial vector clock p_vt_i at process p_i is a list of tuples of the form (j, v) indicating that the current state of p_i is pseudo-dependent on the event on process p_j whose sequence number is v. Initially, at a process p_i: p_vt_i={$(i, 0)$}.

Let $p\_vt_i = \{(i_1, v_1), \ldots, (i, v), \ldots (i_n, v_n)\}$ denote the current partial vector clock at process p_i. Let e_vt_i be a variable that holds the timestamp of the observed event.

(i) Whenever an event is observed at process p_i, the contents of the partial vector clock p_vt_i are transferred to e_vt_i and p_vt_i is reset and updated as follows:

$$e\_vt_i = \{(i_1, v_1), \ldots, (i, v), \ldots, (i_n, v_n)\}$$
$$p\_vt_i = \{(i, v+1)\}.$$

(ii) When process p_j sends a message to p_i, it piggybacks the current value of p_vt_j in the message.

(iii) When p_i receives a message piggybacked with timestamp p_vt, p_i updates p_vt_i such that it is the union of the following (let p_vt=$\{(i_{m1}, v_{m1}), \ldots, (i_{mk}, v_{mk})\}$ and $p\_vt_i = \{(i_1, v_1), \ldots, (i_l, v_l)\}$):
- all (i_{mx}, v_{mx}) such that $(i_{mx}, .)$ does not appear in v_pt_i;
- all (i_x, v_x) such that $(i_x, .)$ does not appear in v_pt;
- all $(i_x, max(v_x, v_{mx}))$ for all $(v_x, .)$ that appear in v_pt and v_pt_i.

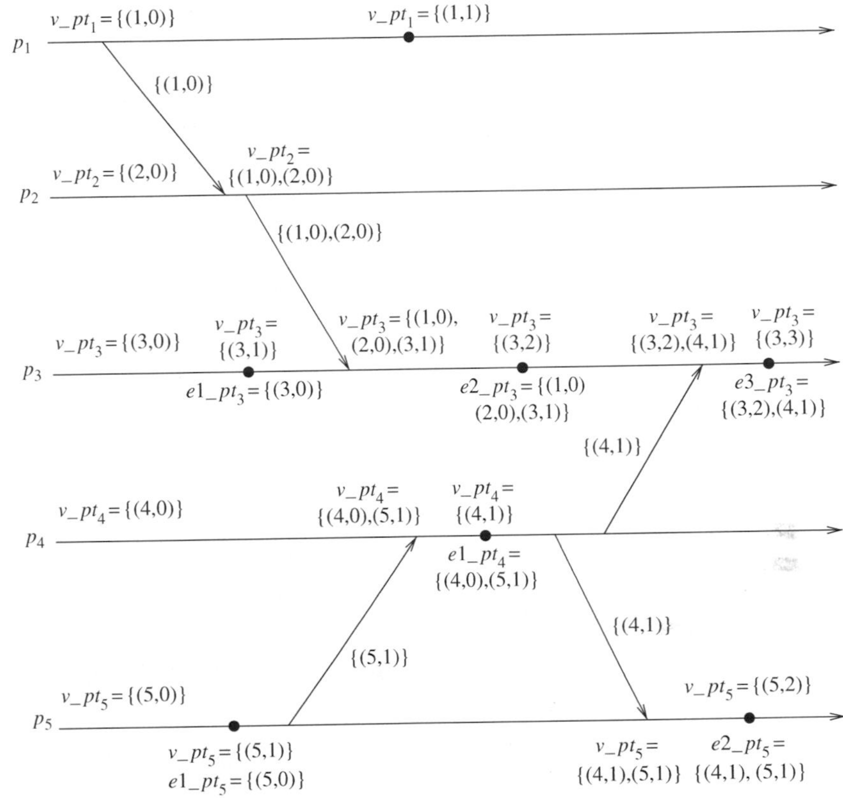

Figure 3.6 Vector clocks progress in the Jard–Jourdan technique [19].

In Figure 3.6, eX_pt_n denotes the timestamp of the Xth observed event at process p_n. For instance, the event 1 observed at p_4 is timestamped $e1\_pt_4 = \{(4, 0), (5, 1)\}$; this timestamp means that the pseudo-direct predecessors of this event are located at process p_4 and p_5, and are respectively the event 0 observed at p_4 and event 1 observed at p_5. v_pt_n denotes a list of timestamps collected by a process p_n for the unobserved events and is reset and updated after an event is observed at p_n. For instance, let us consider v_pt_3. Process p_3 first collects the timestamp of event zero $(3, 0)$ into v_pt_3 and when the observed event 1 occurs, it transfers its content to $e1\_pt_3$, resets its list and updates its value to $(3, 1)$ which is the timestamp of the observed event. When it receives a message from process p_2, it includes those elements that are not already present in its list, namely, $(1, 0)$ and $(2, 0)$ to v_pt_3. Again, when event 2 is observed, it resets its list to $\{(3, 2)\}$ and transfers its content to $e2\_pt_3$ which holds $\{(1, 0), (2, 0), (3, 1)\}$. It can be seen that event 2 at process p_3 is directly dependent upon event 0 on process p_2 and event 1 on process p_3. But, it is pseudo-directly dependent upon event 0 at process p_1. It also depends on event 0 at process p_3 but this dependency information is obtained by examining $e1\_pt_3$ recorded by the observed event. Thus, transitive dependencies of event 2 at process p_3 can be computed by examining the observed events in $e2\_pt_3$. If this is done recursively, then all the causal dependencies of an observed event can be retrieved. It is also pertinent to observe here that these transitive dependencies cannot be determined online but from a log of the events.

This method can help ensure that the list piggybacked on a message is of optimal size. It is also possible to limit the size of the list by introducing a dummy observed event. If the size of the list is to be limited to k, then when timestamps of k events have been collected in the list, a dummy observed event can be introduced to receive the contents of the list. This allows a lot of flexibility in managing the size of messages.

3.7 Matrix time

3.7.1 Definition

In a system of matrix clocks, the time is represented by a set of $n \times n$ matrices of non-negative integers. A process p_i maintains a matrix $mt_i[1..n, 1..n]$ where,

- $mt_i[i, i]$ denotes the local logical clock of p_i and tracks the progress of the computation at process p_i;
- $mt_i[i, j]$ denotes the latest knowledge that process p_i has about the local logical clock, $mt_j[j, j]$, of process p_j (note that row, $mt_i[i, .]$ is nothing but the vector clock $vt_i[.]$ and exhibits all the properties of vector clocks);
- $mt_i[j, k]$ represents the knowledge that process p_i has about the latest knowledge that p_j has about the local logical clock, $mt_k[k, k]$, of p_k.

The entire matrix mt_i denotes p_i's local view of the global logical time. The matrix timestamp of an event is the value of the matrix clock of the process when the event is executed.

Process p_i uses the following rules **R1** and **R2** to update its clock:

- **R1**: Before executing a send or internal event, process p_i updates its local logical time as follows:

$$mt_i[i, i] := mt_i[i, i] + d \qquad (d > 0).$$

- **R2**: Each message m is piggybacked with matrix time mt. When p_i receives such a message *(m,mt)* from a process p_j, p_i executes the following sequence of actions:
  (i) update its global logical time as follows:
    (a) $1 \le k \le n : mt_i[i, k] := max(mt_i[i, k], mt[j, k])$, (that is, update its row $mt_i[i, *]$ with p_j's row in the received timestamp, mt);
    (b) $1 \le k, l \le n : mt_i[k, l] := max(mt_i[k, l], mt[k, l])$;
  (ii) execute **R1**;
  (iii) deliver message m.

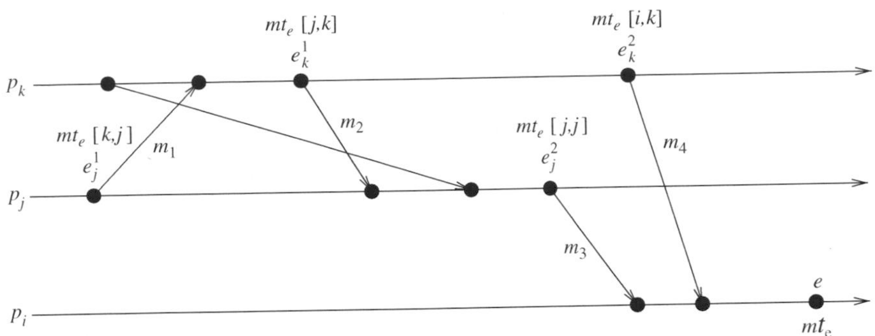

Figure 3.7 Evolution of matrix time [19].

Figure 3.7 gives an example to illustrate how matrix clocks progress in a distributed computation. We assume $d = 1$. Let us consider the following events: e which is the x_ith event at process p_i, e_k^1 and e_k^2 which are the x_k^1th and x_k^2th events at process p_k, and e_j^1 and e_j^2 which are the x_j^1th and x_j^2th events at p_j. Let mt_e denote the matrix timestamp associated with event e. Due to message m_4, e_k^2 is the last event of p_k that causally precedes e, therefore, we have $mt_e[i, k] = mt_e[k, k] = x_k^2$. Likewise, $mt_e[i, j] = mt_e[j, j] = x_j^2$. The last event of p_k known by p_j, to the knowledge of p_i when it executed event e, is e_k^1; therefore, $mt_e[j, k] = x_k^1$. Likewise, we have $mt_e[k, j] = x_j^1$.

A system of matrix clocks was first informally proposed by Michael and Fischer [5] and has been used by Wuu and Bernstein [28] and by Sarin and Lynch [22] to discard obsolete information in replicated databases.

3.7.2 Basic properties

Clearly, vector $mt_i[i, .]$ contains all the properties of vector clocks. In addition, matrix clocks have the following property:

$$\min_k(mt_i[k, l]) \geq t \Rightarrow \text{process } p_i \text{ knows that every other process } p_k \text{ knows that } p_l\text{'s local time has progressed till } t.$$

If this is true, it is clear that process p_i knows that all other processes know that p_l will never send information with a local time $\leq t$. In many applications, this implies that processes will no longer require from p_l certain information and can use this fact to discard obsolete information.

If d is always 1 in the rule **R1**, then $mt_i[k, l]$ denotes the number of events that occurred at p_l and known by p_k as far as p_i's knowledge is concerned.

3.8 Virtual time

The virtual time system is a paradigm for organizing and synchronizing distributed systems using virtual time [7]. This section provides a description of virtual time and its implementation using the time warp mechanism (a lookahead-rollback synchronization mechanism using rollback via antimessages).

The implementation of virtual time using the time warp mechanism works on the basis of an optimistic assumption. Time warp relies on the general lookahead-rollback mechanism where each process executes without regard to other processes having synchronization conflicts. If a conflict is discovered, the offending processes are rolled back to the time just before the conflict and executed forward along the revised path. Detection of conflicts and rollbacks are transparent to users. The implementation of virtual time using the time warp mechanism makes the following optimistic assumption: synchronization conflicts and thus rollback generally occurs rarely.

In the following sections, we discuss in detail virtual time and how the time warp mechanism is used to implement it.

3.8.1 Virtual time definition

Virtual time is a global, one-dimensional, temporal coordinate system on a distributed computation to measure the computational progress and to define synchronization. A virtual time system is a distributed system executing in coordination with an imaginary virtual clock that uses virtual time [7]. Virtual times are real values that are totally ordered by the less than relation, "<". Virtual time is implemented as a collection of several loosely synchronized local virtual clocks. As a rule, these local virtual clocks move forward to higher virtual times; however, occasionally they move backwards.

In a distributed system, processes run concurrently and communicate with each other by exchanging messages. Every message is characterized by four values:

(i) *name of the sender;*
(ii) *virtual send time;*
(iii) *name of the receiver;*
(iv) *virtual receive time.*

Virtual send time is the virtual time at the sender when the message is sent, whereas virtual receive time specifies the virtual time when the message must be received (and processed) by the receiver. Clearly, a big problem arises when a message arrives at a process late, that is, the virtual receive time of the message is less than the local virtual time at the receiver process when the message arrives.

Virtual time systems are subject to two semantic rules similar to Lamport's clock conditions:

Rule 1 Virtual send time of each message < virtual receive time of that message.

Rule 2 Virtual time of each event in a process < virtual time of next event in that process.

The above two rules imply that a process sends all messages in increasing order of virtual send time and a process receives (and processes) all messages in the increasing order of virtual receive time. Causality of events is an important concept in distributed systems and is also a major constraint in the implementation of virtual time. It is important to know which event caused another one and the one that causes another should be completely executed before the caused event can be processed.

The constraint in the implementation of virtual time can be stated as follows:

If an event A causes event B, then the execution of A and B must be scheduled in real time so that A is completed before B starts.

If event A has an earlier virtual time than event B, we need execute A before B provided there is no causal chain from A to B. Better performance can be achieved by scheduling A concurrently with B or scheduling A after B. If A and B have exactly the same virtual time coordinate, then there is no restriction on the order of their scheduling. If A and B are distinct events, they will have different virtual space coordinates (since they occur at different processes) and neither will be a cause for the other. Hence to sum it up, events with virtual time < "t" complete before the starting of events at time "t" and events with virtual time > "t" will start only after events at time "t" are complete.

Characteristics of virtual time

1. Virtual time systems are not all isomorphic; they may be either discrete or continuous.
2. Virtual time may be only partially ordered (in this implementation, total order is assumed.)
3. Virtual time may be related to real time or may be independent of it.
4. Virtual time systems may be visible to programmers and manipulated explicitly as values, or hidden and manipulated implicitly according to some system-defined discipline
5. Virtual times associated with events may be explicitly calculated by user programs or they may be assigned by fixed rules.

3.8.2 Comparison with Lamport's logical clocks

Lamport showed that in real-time temporal relationships "*happens before*" and "*happens after*," operationally definable within a distributed system, form only a partial order, not a total order, and concurrent events are incomparable under that partial order. He also showed that it is always possible to extend partial order to total order by defining artificial clocks. An artificial clock is created for each process with unique labels from a totally ordered set in a manner consistent with partial order. He also provided an algorithm on how to accomplish this task of yielding an assignment of totally ordered clock values. In virtual time, the reverse of the above is done by assuming that every event is labeled with a clock value from a totally ordered virtual time scale satisfying Lamport's clock conditions. Thus the time warp mechanism is an inverse of Lamport's scheme.

In Lamport's scheme, all clocks are conservatively maintained so that they never violate causality. A process advances its clock as soon as it learns of new causal dependency. In virtual time, clocks are optimisticaly advanced and corrective actions are taken whenever a violation is detected.

Lamport's initial idea brought about the concept of virtual time but the model failed to preserve causal independence. It was possible to make an analysis in the real world using timestamps but the same principle could not be implemented completely in the case of asynchronous distributed systems for the lack of a common time base.

The implementation of the virtual time concept using the time warp mechanism is easier to understand and reason about than real time.

3.8.3 Time warp mechanism

In the implementation of virtual time using the time warp mechanism, the virtual receive time of a message is considered as its timestamp. The necessary and sufficient conditions for the correct implementation of virtual time are that each process must handle incoming messages in *timestamp* order. This is highly undesirable and restrictive because process speeds and message delays are likely to be highly variable. So it is natural for some processes to get ahead in virtual time of other processes.

Since we assume that virtual times are real numbers, it is impossible for a process on the basis of local information alone to block and wait for the message with the next timestamp. It is always possible that a message with an earlier timestamp arrives later. So, when a process executes a message, it is very difficult for it determine whether a message with an earlier timestamp will arrive later. This is the central problem in virtual time that is solved by the time warp mechanism.

The advantage of the time warp mechanism is that it doesn't depend on the underlying computer architecture and so portability to different systems is easily achieved. However, message communication is assumed to be reliable, but messages may not be delivered in FIFO order.

The time warp mechanism consists of two major parts: local control mechanism and global control mechanism. The local control mechanism ensures that events are executed and messages are processed in the correct order. The global control mechanism takes care of global issues such as global progress, termination detection, I/O error handling, flow control, etc.

3.8.4 The local control mechanism

There is no global virtual clock variable in this implementation; each process has a *local virtual clock* variable. The local virtual clock of a process doesn't change during an event at that process but it changes only between events. On the processing of the next message from the input queue, the process increases its local clock to the timestamp of the message. At any instant, the value of virtual time may differ for each process but the value is transparent to other processes in the system.

When a message is sent, the virtual send time is copied from the sender's virtual clock while the name of the receiver and virtual receive time are assigned based on the application-specific context.

All arriving messages at a process are stored in an input queue in increasing order of timestamp (receive times). Ideally, no messages from the past (called late messages) should arrive at a process. However, processes will receive late messages due to factors such as different computation rates of processes and network delays. The semantics of virtual time demands that incoming messages be received by each process strictly in timestamp order. The only way to accomplish this is as follows: on the reception of a late message, the receiver rolls back to an earlier virtual time, cancelling all intermediate side effects and then executes forward again by executing the late message in the proper sequence. If all the messages in the input queue of a process are processed, the state of the process is said to *terminate* and its clock is set to +inf. However, the process is not destroyed as a late message may arrive resulting it to rollback and execute again. The situation can be described by saying that each process is doing a constant "lookahead," processing future messages from its input queue.

Over a length computation, each process may roll back several times while generally progressing forward with rollback completely transparent to other processes in the system. Programmers can thus write correct software without paying much attention to late-arriving messages.

Rollback in a distributed system is complicated by the fact that the process that wants to rollback might have sent many messages to other processes, which in turn might have sent many messages to other processes, and so on, leading to deep side effects. For rollback, messages must be effectively "*unsent*" and their side effects should be undone. This is achieved efficiently by using antimessages.

Antimessages and the rollback mechanism

Runtime representation of a process is composed of the following:

1. **Process name** Virtual spaces coordinate which is unique in the system.
2. **Local virtual clock** Virtual time coordinate
3. **State** Data space of the process including execution stack, program counter, and its own variables

4. **State queue** Contains saved copies of process's recent states as rollback with the time warp mechanism requires the state of the process being saved. It is not necessary to retain states all the way from the beginning of the virtual time, however, the reason for which will be explained later in the global control mechanism.
5. **Input queue** Contains all recently arrived messages in order of virtual receive time. Processed messages from the input queue are not deleted as they are saved in the output queue with a negative sign (antimessage) to facilitate future rollbacks.
6. **Output queue** Contains negative copies of messages that the process has recently sent in virtual send time order. They are needed in case of a rollback.

For every message, there exists an antimessage that is the same in content but opposite in sign. Whenever a process sends a message, a copy of the message is transmitted to the receiver's input queue and a negative copy (antimessage) is retained in the sender's output queue for use in sender rollback.

Whenever a message and its antimessage appear in the same queue, regardless of the order in which they arrived, they immediately annihilate each other resulting in shortening of the queue by one message.

Generally when a message arrives at the input queue of a process with timestamp greater than the virtual clock time of its destination process, it is simply enqueued by the interrupt routine and the running process continues. But when the destination process' virtual time is greater than the virtual time of the message received, the process must do a rollback.

The first step in the rollback mechanism is to search the "state queue" for the last saved state with a timestamp that is less than the timestamp of the message received and restore it. We make the timestamp of the received message as the value of the local virtual clock and discard from the state queue all states saved after this time. Then the execution resumes forward from this point. Now all the messages that are sent between the current state and earlier state must be "unsent." This is taken care of by executing a simple rule:

To unsend a message, simply transmit its antimessage.

This results in antimessages following the positive ones to the destination. A negative message causes a rollback at its destination if its virtual receive time is less than the receiver's virtual time (just as a positive message does).

Depending on the timing, there are several possibilities at the receiver's end:

1. If the original (positive) message has arrived but not yet been processed, its virtual receive time must be greater than the value in the receiver's virtual clock. The negative message, having the same virtual receive time, will be enqueued and will not cause a rollback. It will, however, cause annihilation with the positive message leaving the receiver with no record of that message.

2. The second possibility is that the original positive message has a virtual receive time that is now in the present or past with respect to the receiver's virtual clock and it may have already been partially or completely processed, causing side effects on the receiver's state. In this case, the negative message will also arrive in the receiver's past and cause the receiver to rollback to a virtual time when the positive message was received. It will also annihilate the positive message, leaving the receiver with no record that the message existed. When the receiver executes again, the execution will assume that these message never existed. Note that, as a result of the rollback, the process may send antimessages to other processes.
3. A negative message can also arrive at the destination before the positive one. In this case, it is enqueued and will be annihilated when the positive message arrives. If it is the negative message's turn to be executed at a processs' input queue, the receiver may take any action like a no-op. Any action taken will eventually be rolled back when the corresponding positive message arrives. An optimization would be to skip the antimessage from the input queue and treat it as a no-op, and when the corresponding positive message arrives, it will annihilate the negative message, and inhibit any rollback.

The antimessage protocol has several advantages: it is extremely robust and works under all possible circumstances; it is free from deadlocks as there is no blocking; it is also free from domino effects. In the worst case, all processes in the system rollback to the same virtual time as the original and then proceed forward again.

3.8.5 Global control mechanism

The global control mechanism resolves the following issues:

- System global progress amidst rollback activity?
- Detection of global termination?
- Errors, I/O handling on rollbacks?
- Running out of memory while saving copies of messages?

How these issues are resolved by the global control mechanism will be discussed later; first we discuss the important concept of global virtual time.

Global virtual time

The concept of global virtual time (GVT) is central to the global control mechanism. Global virtual time [14] is a property of an instantaneous global snapshot of system at real time "r" and is defined as follows:

Global virtual time (GVT) at real time r is the minimum of:

1. all virtual times in all virtual clocks at time r; and
2. the virtual send times of all messages that have been sent but have not yet been processed at time "r".

GVT is defined in terms of the *virtual send time* of unprocessed messages, instead of the virtual receive time, because of the flow control (discussed below). If every event completes normally, if messages are delivered reliably, if the scheduler does not indefinitely postpone execution of the farthest behind process, and if there is sufficient memory, then GVT will eventually increase.

It is easily shown by induction that the message (sends, arrivals, and receipts) never decreases GVT even though local virtual time clocks roll back frequently. These properties make it appropriate to consider GVT as a virtual clock for the system as a whole and to use it as the measure of system progress. GVT can thus be viewed as a moving commitment horizon: any event with virtual time less than GVT cannot be rolled back and may be committed safely.

It is generally impossible for one time warp mechanism to know at any real time "r," exactly what GVT is. But GVT can be characterized more operationally by its two properties discussed above. This characterization leads to a fast distributed GVT estimation algorithm that takes $O(d)$ time, where "d" is the delay required for one broadcast to all processors in the system. The algorithm runs concurrently with the main computation and returns a value that is between the true GVT at the moment the algorithm starts and the true GVT at the moment of completion. Thus it gives a slightly out-of-date value for GVT which is the best one can get.

During execution of a virtual time system, time warp must periodically estimate GVT. A higher frequency of GVT estimation produces a faster response time and better space utilization at the expense of processor time and network bandwidth.

Applications of GVT

GVT finds several applications in a virtual time system using the time warp mechanism.

Memory management and flow control

An attractive feature of the time warp mechanism is that it is possible to give simple algorithms for managing memory. The time warp mechanism uses the concept of fossil detection where information older than GVT is destroyed to avoid memory overheads due to old states in state queues, messages stored in output queues, "past" messages in input queues that have already been processed, and "future" messages in input queues that have not yet been received.

There is another kind of memory overhead due to future messages in the input queues that have not yet been received. So, if a receiver's memory is full of input messages, the time warp mechanism may be able to recover space by returning an unreceived message to the process that sent it and then rolling back to cancel out the sending event.

Normal termination detection

The time warp mechanism handles the termination detection problem through GVT. A process terminates whenever it runs out of messages and its local virtual clock is set to +inf. Whenever GVT reaches +inf, all local virtual clock variables must read +inf and no message can be in transit. No process can ever again unterminate by rolling back to a finite virtual time. The time warp mechanism signals termination whenever the GVT calculation returns "+inf" value in the system.

Error handling

Not all errors cause termination. Most of the errors can be avoided by rolling back the local virtual clock to some finite value. The error is only "committed" if it is impossible for the process to roll back to a virtual time on or before the error. The committed error is reported to some policy software or to the user.

Input and output

When a process sends a command to an output device, it is important that the physical output activity not be committed immediately because the sending process may rollback and cancel the output request. An output activity can only be performed when GVT exceeds the virtual receive time of the message containing the command.

Snapshots and crash recovery

An entire snapshot of the system at virtual time "*t*" can be constructed by a procedure in which each process "*snapshots*" itself as it passes virtual time *t* in the forward direction and "*unsnapshots*" itself whenever it rolls back over virtual time "*t*". Whenever GVT exceeds "*t*," the snapshot is complete and valid.

Example: distributed discrete event simulations Distributed discrete event simulation [1,16,21] is the most studied example of virtual time systems; every process represents an object in the simulation and virtual time is identified with simulation time. The fundamental operation in discrete event simulation is for one process to schedule an event for execution by another process at a later simulation time. This is emulated by having the first process send a message to the second process with the virtual receive time of the message equal to the event's scheduled time in the simulation. When an event message is received by a process, there are three possibilities: its timestamp is either before, after, or equal to the local value of simulation time.

If its timestamp is after the local time, an input event combination is formed and the appropriate action is taken. However, if the timestamp of the received event message is less than or equal to the local clock value, the process has already processed an event combination with time greater than or equal to the incoming event. The process must then rollback to the time of the incoming message which is done by an elaborate checkpointing mechanism that allows earlier states to be restored. Essentially an earlier state is restored, input event combinations are rescheduled, and output events are cancelled by sending antimessages. The process has buffers that save past inputs, past states, and antimessages.

Distributed discrete event simulation is one of the most general applications of the virtual time paradigm because the virtual times of events are completely under the control of the user, and because it makes use of almost all the degrees of freedom allowed in the definition of a virtual time system.

3.9 Physical clock synchronization: NTP

3.9.1 Motivation

In centralized systems, there is no need for clock synchronization because, generally, there is only a single clock. A process gets the time by simply issuing a system call to the kernel. When another process after that tries to get the time, it will get a higher time value. Thus, in such systems, there is a clear ordering of events and there is no ambiguity about the times at which these events occur.

In distributed systems, there is no global clock or common memory. Each processor has its own internal clock and its own notion of time. In practice, these clocks can easily drift apart by several seconds per day, accumulating significant errors over time. Also, because different clocks tick at different rates, they may not remain always synchronized although they might be synchronized when they start. This clearly poses serious problems to applications that depend on a synchronized notion of time. For most applications and algorithms that run in a distributed system, we need to know time in one or more of the following contexts:

- The time of the day at which an event happened on a specific machine in the network.
- The time interval between two events that happened on different machines in the network.
- The relative ordering of events that happened on different machines in the network.

Unless the clocks in each machine have a common notion of time, time-based queries cannot be answered. Some practical examples that stress the need for synchronization are listed below:

- In database systems, the order in which processes perform updates on a database is important to ensure a consistent, correct view of the database.

To ensure the right ordering of events, a common notion of time between co-operating processes becomes imperative.

- Liskov [10] states that clock synchronization improves the performance of distributed algorithms by replacing communication with local computation. When a node p needs to query node q regarding a property, it can deduce the property with some previous information it has about node q and its knowledge of the local time in node q.
- In distributed sensor networks, sensors sense physical world properties at different locations and a global predicate is evaluated to determine if a global condition was true at some time. To correlate the sensed values reported from different sensors, clock synchronization is useful [27].

Clock synchronization is the process of ensuring that physically distributed processors have a common notion of time. It has a significant effect on many problems like secure systems, fault diagnosis and recovery, scheduled operations, database systems, and real-world clock values. It is quite common that distributed applications and network protocols use timeouts, and their performance depends on how well physically dispersed processors are time-synchronized. Design of such applications is simplified when clocks are synchronized.

Due to different clocks rates, the clocks at various sites may diverge with time, and periodically a clock synchronization must be performed to correct this clock skew in distributed systems. Clocks are synchronized to an accurate real-time standard like UTC (Universal Coordinated Time). Clocks that must not only be synchronized with each other but also have to adhere to physical time are termed *physical clocks*.

3.9.2 Definitions and terminology

We provide the following definitions [13,14]. C_a and C_b are any two clocks.

1. **Time** The time of a clock in a machine p is given by the function $C_p(t)$, where $C_p(t) = t$ for a perfect clock.
2. **Frequency** Frequency is the rate at which a clock progresses. The frequency at time t of clock C_a is $C_a'(t)$.
3. **Offset** Clock offset is the difference between the time reported by a clock and the *real time*. The offset of the clock C_a is given by $C_a(t) - t$. The offset of clock C_a relative to C_b at time $t \geq 0$ is given by $C_a(t) - C_b(t)$.
4. **Skew** The skew of a clock is the difference in the frequencies of the clock and the perfect clock. The skew of a clock C_a relative to clock C_b at time t is $C_a'(t) - C_b'(t)$.

   If the skew is bounded by ρ, then as per Eq.(3.1), clock values are allowed to diverge at a rate in the range of $1-\rho$ to $1+\rho$.

5. **Drift (rate)** The drift of clock C_a is the second derivative of the clock value with respect to time, namely, $C_a''(t)$. The drift of clock C_a relative to clock C_b at time t is $C_a''(t) - C_b''(t)$.

3.9.3 Clock inaccuracies

Physical clocks are synchronized to an accurate real-time standard like UTC (Universal Coordinated Time).

However, due to the clock inaccuracy discussed above, a timer (clock) is said to be working within its specification if

$$1 - \rho \leq \frac{dC}{dt} \leq 1 + \rho, \quad (3.1)$$

where constant ρ is the maximum skew rate specified by the manufacturer. Figure 3.8 illustrates the behavior of fast, slow, and perfect clocks with respect to UTC.

Offset delay estimation method

The *Network Time Protocol (NTP)* [15], which is widely used for clock synchronization on the Internet, uses the the *offset delay estimation* method. The design of NTP involves a hierarchical tree of time servers. The primary server at the root synchronizes with the UTC. The next level contains secondary servers, which act as a backup to the primary server. At the lowest level is the synchronization subnet which has the clients.

Clock offset and delay estimation

In practice, a source node cannot accurately estimate the local time on the target node due to varying message or network delays between the nodes. This protocol employs a very common practice of performing several trials and chooses the trial with the minimum delay. Cristian's remote

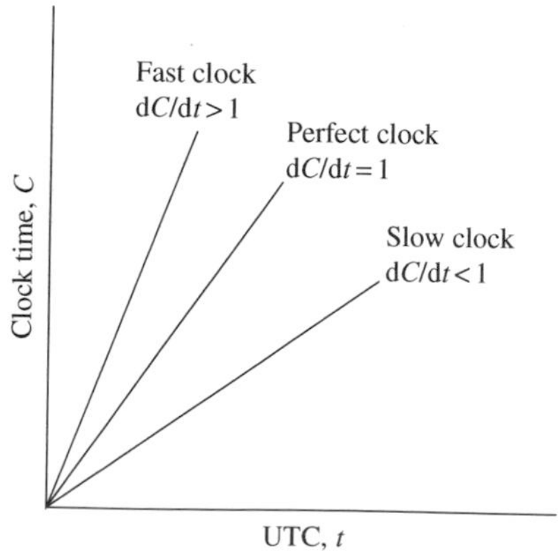

Figure 3.8 The behavior of fast, slow, and perfect clocks with respect to UTC.

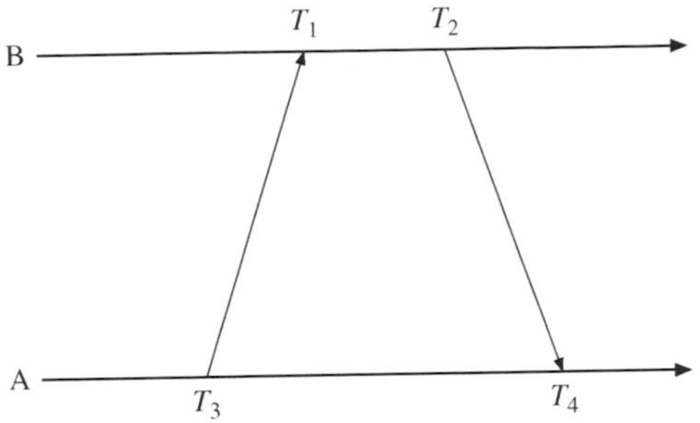

Figure 3.9 Offset and delay estimation [15].

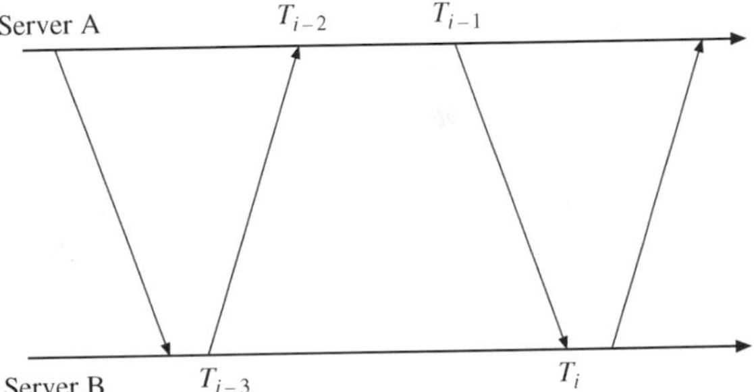

Figure 3.10 Timing diagram for the two servers [15].

clock reading method [3] also relied on the same strategy to estimate message delay.

Figure 3.9 shows how NTP timestamps are numbered and exchanged between peers A and B. Let T_1, T_2, T_3, T_4 be the values of the four most recent timestamps as shown. Assume that clocks A and B are stable and running at the same speed. Let $a = T_1 - T_3$ and $b = T_2 - T_4$. If the network delay difference from A to B and from B to A, called *differential delay*, is small, the clock offset θ and roundtrip delay δ of B relative to A at time T_4 are approximately given by the following:

$$\theta = \frac{a+b}{2}, \qquad \delta = a - b. \tag{3.2}$$

Each NTP message includes the latest three timestamps T_1, T_2, and T_3, while T_4 is determined upon arrival. Thus, both peers A and B can independently calculate delay and offset using a single bidirectional message stream as shown in Figure 3.10. The NTP protocol is shown in Figure 3.11.

3.10 Chapter summary

The concept of causality between events is fundamental to the design and analysis of distributed programs. The notion of time is basic to capture causality between events; however, there is no built-in physical time in distributed

Figure 3.11 The network time protocol (NTP) synchronization protocol [15].

- A pair of servers in symmetric mode exchange pairs of timing messages.
- A store of data is then built up about the relationship between the two servers (pairs of offset and delay).

  Specifically, assume that each peer maintains pairs (O_i, D_i), where:

  O_i – measure of offset (θ)
  D_i – transmission delay of two messages (δ).
- The offset corresponding to the minimum delay is chosen. Specifically, the delay and offset are calculated as follows. Assume that message m takes time t to transfer and m' takes t' to transfer.
- The offset between A's clock and B's clock is O. If A's local clock time is $A(t)$ and B's local clock time is $B(t)$, we have

$$A(t) = B(t) + O. \tag{3.3}$$

Then,

$$T_{i-2} = T_{i-3} + t + O, \tag{3.4}$$

$$T_i = T_{i-1} - O + t'. \tag{3.5}$$

Assuming $t = t'$, the offset O_i can be estimated as

$$O_i = (T_{i-2} - T_{i-3} + T_{i-1} - T_i)/2. \tag{3.6}$$

The round-trip delay is estimated as

$$D_i = (T_i - T_{i-3}) - (T_{i-1} - T_{i-2}). \tag{3.7}$$

- The eight most recent pairs of (O_i, D_i) are retained.
- The value of O_i that corresponds to minimum D_i is chosen to estimate O.

systems and it is possible only to realize an approximation of it. Typically, a distributed computation makes progress in spurts and consequently logical time, which advances in jumps, is sufficient to capture the monotonicity property induced by causality in distributed systems. Causality among events in a distributed system is a powerful concept in reasoning, analyzing, and drawing inferences about a computation.

We presented a general framework of logical clocks in distributed systems and discussed three systems of logical clocks, namely, scalar, vector, and matrix clocks, that have been proposed to capture causality between events of a distributed computation. These systems of clocks have been used to solve a variety of problems in distributed systems such as distributed algorithms design, debugging distributed programs, checkpointing and failure recovery, data consistency in replicated databases, discarding obsolete information, garbage collection, and termination detection.

In scalar clocks, the clock at a process is represented by an integer. The message and the compuatation overheads are small, but the power of scalar clocks is limited – they are not strongly consistent. In vector clocks, the clock at a process is represented by a vector of integers. Thus, the message and the compuatation overheads are likely to be high; however, vector clocks possess a powerful property – there is an isomorphism between the set of partially ordered events in a distributed computation and their vector timestamps. This is a very useful and interesting property of vector clocks that finds applications in several problem domains. In matrix clocks, the clock at a process is represented by a matrix of integers. Thus, the message and the compuatation overheads are high; however, matrix clocks are very powerful – besides containing information about the direct dependencies, a matrix clock contains information about the latest direct dependencies of those dependencies. This information can be very useful in applications such as distributed garbage collection. Thus, the power of systems of clocks increases in the order of scalar, vector, and matrix, but so do the complexity and the overheads.

We discussed three efficient implementations of vector clocks; similar techniques can be used to efficiently implement matrix clocks. Singhal–Kshemkalyani's differential technique exploits the fact that, between successive events at a process, only few entries of its vector clock are likely to change. Thus, when a process p_i sends a message to a process p_j, it piggybacks only those entries of its vector clock that have changed since the last message send to p_j, reducing the communication and buffer (to store messages) overheads. Fowler–Zwaenepoel's direct-dependency technique does not maintain vector clocks on-the-fly. Instead, a process only maintains information regarding direct dependencies on other processes. A vector timestamp for an event, that represents transitive dependencies on other processes, is constructed off-line from a recursive search of the direct dependency information at processes. Thus, the technique has low run-time overhead. In the Fowler–Zwaenepoel technique, however, a process must update and record its dependency vector after receiving a message but before sending out any message. If events occur very frequently, this technique will require recording the history of a large number of events. In the Jard–Jourdan technique, events can be adaptively observed while maintaining the capability of retrieving all the causal dependencies of an observed event.

Virtual time system is a paradigm for organizing and synchronizing distributed systems using virtual time. We discussed virtual time and its implementation using the time warp mechanism.

3.11 Exercises

Exercise 3.1 Why is it difficult to keep a synchronized system of physical clocks in distributed systems?

Exercise 3.2 If events corresponding to vector timestamps Vt_1, Vt_2, ..., Vt_n are mutually concurrent, then prove that

$$(Vt_1[1], Vt_2[2], \ldots . Vt_n[n]) = max(Vt_1, Vt_2, \ldots ., Vt_n).$$

Exercise 3.3 If events e_i and e_j respectively occurred at processes p_i and p_j and are assigned vector timestamps VT_{e_i} and VT_{e_j}, respectively, then show that

$$e_i \rightarrow e_j \Leftrightarrow VT_{e_i}[i] < VT_{e_j}[i].$$

Exercise 3.4 The size of matrix clocks is quadratic with respect to the system size. Hence the message overhead is likely to be substantial. Propose a technique for matrix clocks similar to that of Singhal–Kshemkalyani to decrease the volume of information transmitted in messages and stored at processes.

3.12 Notes on references

The idea of logical time was proposed by Lamport in 1978 [9] in an attempt to order events in distributed systems. He also suggested an implementation of logical time as a scalar time. Vector clocks were developed independently by Fidge [4], Mattern [12], and Schmuck [23]. Charron-Bost formally showed [2] that if vector clocks have to satisfy the strong consistency property, then the length of vector timestamps must be at least n. Efficient implementations of vector clocks can be found in [8,25]. Matrix clocks was informally proposed by Michael and Fischer [7] and used by Wuu and Bernstein [28] and by Lynch and Sarin [22] to discard obsolete information. Raynal and Singhal present a survey of scalar, vector, and matrix clocks in [19]. More details on virtual time can be found in a classical paper by Jefferson [7]. A survey of physical clock synchronization in wireless sensor networks can be found in [27].

References

[1] B. R. Preiss, The Yaddes distributed discrete event simulation specification language and execution environments, *Proceedings of the SCS Multiconference on Distributed Simulation*, 1989, 139–144.

[2] B. Charron-Bost, Concerning the size of logical clocks in distributed systems, *Information Processing Letters*, **39**, 1991, 11–16.

[3] F. Cristian, Probabilistic clock synchronization, *Distributed Computing*, **3**, 1989, 146–158.

[4] C. Fidge, Logical time in distributed computing systems, *IEEE Computer*, August, 1991, 28–33.

[5] M. J. Fischer and A. Michael, Sacrifying serializability to attain hight availability of data in an unreliable network, *Proceedings of the ACM Symposium on Principles of Database Systems*, 1982, 70–75.

[6] J. Fowler and W. Zwaenepoel, Causal distributed breakpoints, *Proceedings of the 10th International Conference on Distributed Computing Systems*, 1990, 134–141.

[7] D. Jefferson, Virtual time, *ACM Toplas*, **7**(3), 1985, 404–425.

[8] C. Jard and G.-C. Jourdan, Dependency tracking and filtering in distributed computations, *Brief Announcements of the ACM Symposium on PODC*, 1994. (A full presentation appeared as IRISA Technical Report No. 851, 1994.)

[9] L. Lamport, Time, clocks and the ordering of events in a distributed system, *Communications of the ACM*, **21**, 1978, 558–564.

[10] B. Liskov, Practical uses of synchronized clocks in distributed systems, *Proceedings of Tenth Annual ACM Symposium on Principles of Distributed Computing*, August 1991, pp. 1–9.

[11] B. Liskov and R. Ladin, Highly available distributed services and fault-tolerant distributed garbage collection, *Proceedings of the 5th ACM Symposium on PODC*, 1986, 29–39.

[12] F. Mattern, Virtual time and global states of distributed systems, in Cosnard, Q and Raynal, R. (eds) *Proceedings of the Parallel and Distributed Algorithms Conference*, North-Holland, 1988, 215–226.

[13] D. L. Mills, *Network Time Protocol (version 3): Specification, Implementation, and Analysis*, Technical Report, Network Information Center, SRI International, Menlo Park, CA, March, 1992.

[14] D. L. Mills, *Modelling and Analysis of Computer Network Clocks*, Technical Report, 92-5-2, Electrical Engineering Department, University of Delaware, May, 1992.

[15] D. L. Mills, Internet time synchronization: the network time protocol, *IEEE Transactions on Communications*, **39**(10), 1991, 1482–1493.

[16] J. Misra, Distributed discrete event simulation, *ACM Computing Surveys*, **18**(1), 1986, 39–65.

[17] D. S. Parker *et al.*, Detection of mutual inconsistency in distributed systems, *IEEE Transactions on Software Engineeing*, **9**(3), 1983, 240–246.

[18] M. Raynal, A distributed algorithm to prevent mutual drift between n logical clocks, *Information Processing Letters*, **24**, 1987, 199–202.

[19] M. Raynal and M. Singhal, Logical time: capturing causality in distributed systems, *IEEE Computer*, **30**(2), 1996, 49–56.

[20] G. Ricart, and A. K. Agrawala, An optimal algorithm for mutual exclusion in computer networks, *Communications of the ACM*, **24**(1), 1981, 9–17

[21] R. Righter and J. C. Walrand, Distributed simulation of discrete event systems, *Proceedings of the IEEE*, 1988, and 99–113.

[22] S. K. Sarin and L. Lynch, Discarding obsolete information in a replicated data base system, *IEEE Transactions on Software Engineering*, **13**(1), 1987, 39–46.

[23] F. Schmuck, The Use of Efficient Broadcast in Asynchronous Distributed Systems, Ph. D. Thesis, Cornell University, TR88-928, 1988.

[24] M. Singhal, A heuristically-aided mutual exclusion algorithm for distributed systems, *IEEE Transactions on Computers*, **38**(5), 1989, 651–662.

[25] M. Singhal and A. Kshemkalyani, An efficient implementation of vector clocks, *Information Processing Letters*, **43**, August, 1992, 47–52.

[26] R. E. Strom and S. Yemini, Optimistic recovery in distributed systems, *ACM Transactions on Computer Systems*, **3**(3), 1985, 204–226.

[27] B. Sundararaman, U. Buy, and A. D. Kshemkalyani, Clock synchronization in wireless sensor networks: a survey, *Ad-Hoc Networks*, **3**(3), 2005, 281–323.

[28] G. T. J. Wuu and A. J. Bernstein, Efficient solutions to the replicated log and dictionary problems, *Proceedings of 3rd ACM Symposium on PODC*, 1984, 233–242.

CHAPTER

4 Global state and snapshot recording algorithms

Recording the global state of a distributed system on-the-fly is an important paradigm when one is interested in analyzing, testing, or verifying properties associated with distributed executions. Unfortunately, the lack of both a globally shared memory and a global clock in a distributed system, added to the fact that message transfer delays in these systems are finite but unpredictable, makes this problem non-trivial.

This chapter first defines consistent global states (also called consistent snapshots) and discusses issues which have to be addressed to compute consistent distributed snapshots. Then several algorithms to determine on-the-fly such snapshots are presented for several types of networks (according to the properties of their communication channels, namely, FIFO, non-FIFO, and causal delivery).

4.1 Introduction

A distributed computing system consists of spatially separated processes that do not share a common memory and communicate asynchronously with each other by message passing over communication channels. Each component of a distributed system has a local state. The state of a process is characterized by the state of its local memory and a history of its activity. The state of a channel is characterized by the set of messages sent along the channel less the messages received along the channel. The global state of a distributed system is a collection of the local states of its components.

Recording the global state of a distributed system is an important paradigm and it finds applications in several aspects of distributed system design. For examples, in detection of stable properties such as deadlocks [17] and termination [22], global state of the system is examined for certain properties;

for failure recovery, a global state of the distributed system (called a checkpoint) is periodically saved and recovery from a processor failure is done by restoring the system to the last saved global state [15]; for debugging distributed software, the system is restored to a consistent global state [8,9] and the execution resumes from there in a controlled manner. A snapshot recording method has been used in the distributed debugging facility of Estelle [11,13], a distributed programming environment. Other applications include monitoring distributed events [30], such as in industrial process control, setting distributed breakpoints [24], protocol specification and verification [4,10,14], and discarding obsolete information [11].

Therefore, it is important that we have efficient ways of recording the global state of a distributed system [6,16]. Unfortunately, there is no shared memory and no global clock in a distributed system and the distributed nature of the local clocks and local memory makes it difficult to record the global state of the system efficiently.

If shared memory were available, an up-to-date state of the entire system would be available to the processes sharing the memory. The absence of shared memory necessitates ways of getting a coherent and complete view of the system based on the local states of individual processes. A meaningful global snapshot can be obtained if the components of the distributed system record their local states at the same time. This would be possible if the local clocks at processes were perfectly synchronized or if there were a global system clock that could be instantaneously read by the processes. However, it is technologically infeasible to have perfectly synchronized clocks at various sites – clocks are bound to drift. If processes read time from a single common clock (maintained at one process), various indeterminate transmission delays during the read operation will cause the processes to identify various physical instants as the same time. In both cases, the collection of local state observations will be made at different times and may not be meaningful, as illustrated by the following example.

Example Let S1 and S2 be two distinct sites of a distributed system which maintain bank accounts A and B, respectively. A site refers to a process in this example. Let the communication channels from site S1 to site S2 and from site S2 to site S1 be denoted by C_{12} and C_{21}, respectively. Consider the following sequence of actions, which are also illustrated in the timing diagram of Figure 4.1:

Time t_0: Initially, Account A = \$600, Account B = \$200, C_{12} = \$0, C_{21} = \$0.

Time t_1: Site S1 initiates a transfer of \$50 from Account A to Account B. Account A is decremented by \$50 to \$550 and a request for \$50 credit to Account B is sent on Channel C_{12} to site S2. Account A = \$550, Account B = \$200, C_{12} = \$50, C_{21} = \$0.

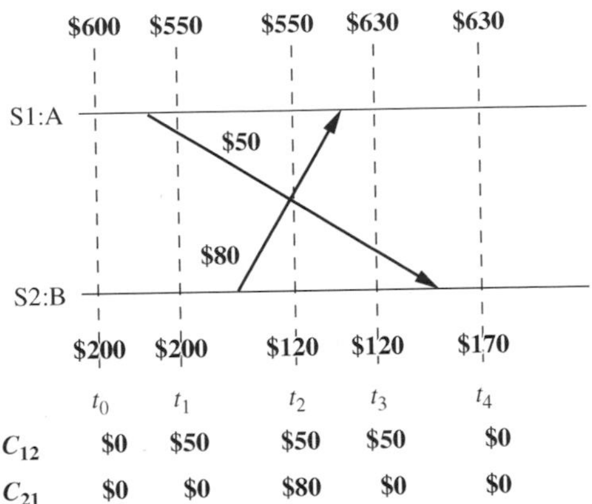

Figure 4.1 A banking example to illustrate recording of consistent states.

Time t_2: Site S2 initiates a transfer of \$80 from Account B to Account A. Account B is decremented by \$80 to \$120 and a request for \$80 credit to Account A is sent on Channel C_{21} to site S1. Account A = \$550, Account B = \$120, C_{12} = \$50, C_{21} = \$80.

Time t_3: Site S1 receives the message for a \$80 credit to Account A and updates Account A. Account A = \$630, Account B = \$120, C_{12} = \$50, C_{21} = \$0.

Time t_4: Site S2 receives the message for a \$50 credit to Account B and updates Account B. Account A = \$630, Account B = \$170, C_{12} = \$0, C_{21} = \$0.

Suppose the local state of Account A is recorded at time t_0 to show \$600 and the local state of Account B and channels C_{12} and C_{21} are recorded at time t_2 to show \$120, \$50, and \$80, respectively. Then the recorded global state shows \$850 in the system. An extra \$50 appears in the system. The reason for the inconsistency is that Account A's state was recorded before the \$50 transfer to Account B using channel C_{12} was initiated, whereas channel C_{12}'s state was recorded after the \$50 transfer was initiated.

This simple example shows that recording a consistent global state of a distributed system is not a trivial task. Recording activities of individual components must be coordinated appropriately. This chapter addresses the fundamental issue of recording a consistent global state in distributed computing systems.

Next section presents the system model and a formal definition of the notion of consistent global state. The subsequent sections present algorithms to record such global states under various communication models such as FIFO communication channels, non-FIFO communication channels, and causal delivery of messages. These algorithms are called snapshot recording algorithms.

4.2 System model and definitions

4.2.1 System model

The system consists of a collection of n processes, $p_1, p_2, \ldots, p_n$, that are connected by channels. There is no globally shared memory and processes communicate solely by passing messages. There is no physical global clock in the system. Message send and receive is asynchronous. Messages are delivered reliably with finite but arbitrary time delay. The system can be described as a directed graph in which vertices represent the processes and edges represent unidirectional communication channels. Let C_{ij} denote the channel from process p_i to process p_j.

Processes and channels have states associated with them. The state of a process at any time is defined by the contents of processor registers, stacks, local memory, etc., and may be highly dependent on the local context of the distributed application. The state of channel C_{ij}, denoted by SC_{ij}, is given by the set of messages in transit in the channel.

The actions performed by a process are modeled as three types of events, namely, internal events, message send events, and message receive events. For a message m_{ij} that is sent by process p_i to process p_j, let $send(m_{ij})$ and $rec(m_{ij})$ denote its send and receive events, respectively. Occurrence of events changes the states of respective processes and channels, thus causing transitions in the global system state. For example, an internal event changes the state of the process at which it occurs. A send event (or a receive event) changes the state of the process that sends (or receives) the message and the state of the channel on which the message is sent (or received). The events at a process are linearly ordered by their order of occurrence.

At any instant, the state of process p_i, denoted by LS_i, is a result of the sequence of all the events executed by p_i up to that instant. For an event e and a process state LS_i, $e \in LS_i$ iff e belongs to the sequence of events that have taken process p_i to state LS_i. For an event e and a process state LS_i, $e \notin LS_i$ iff e does not belong to the sequence of events that have taken process p_i to state LS_i.

A channel is a distributed entity and its state depends on the local states of the processes on which it is incident. For a channel C_{ij}, the following set of messages can be defined based on the local states of the processes p_i and p_j [12]:

Transit: $transit(LS_i, LS_j) = \{m_{ij} \mid send(m_{ij}) \in LS_i \bigwedge rec(m_{ij}) \notin LS_j\}$.

Thus, if a snapshot recording algorithm records the state of processes p_i and p_j as LS_i and LS_j, respectively, then it must record the state of channel C_{ij} as $transit(LS_i, LS_j)$.

There are several models of communication among processes and different snapshot algorithms have assumed different models of communication. In the FIFO model, each channel acts as a first-in first-out message queue and, thus, message ordering is preserved by a channel. In the non-FIFO model, a channel acts like a set in which the sender process adds messages and the receiver process removes messages from it in a random order. A system that supports causal delivery of messages satisfies the following property: "for any two messages m_{ij} and m_{kj}, if $send(m_{ij}) \longrightarrow send(m_{kj})$, then $rec(m_{ij}) \longrightarrow rec(m_{kj})$."

Causally ordered delivery of messages implies FIFO message delivery. The causal ordering model is useful in developing distributed algorithms and may simplify the design of algorithms.

4.2.2 A consistent global state

The global state of a distributed system is a collection of the local states of the processes and the channels. Notationally, global state GS is defined as

$$GS = \{\bigcup_i LS_i, \bigcup_{i,j} SC_{ij}\}.$$

A global state GS is a *consistent global state* iff it satisfies the following two conditions [16]:

C1: $send(m_{ij}) \in LS_i \Rightarrow m_{ij} \in SC_{ij} \oplus rec(m_{ij}) \in LS_j$ ($\oplus$ is the Ex-OR operator).
C2: $send(m_{ij}) \notin LS_i \Rightarrow m_{ij} \notin SC_{ij} \wedge rec(m_{ij}) \notin LS_j$.

Condition **C1** states the law of conservation of messages. Every message m_{ij} that is recorded as sent in the local state of a process p_i must be captured in the state of the channel C_{ij} or in the collected local state of the receiver process p_j. Condition **C2** states that in the collected global state, for every effect, its cause must be present. If a message m_{ij} is not recorded as sent in the local state of process p_i, then it must neither be present in the state of the channel C_{ij} nor in the collected local state of the receiver process p_j.

In a consistent global state, every message that is recorded as received is also recorded as sent. Such a global state captures the notion of causality that a message cannot be received if it was not sent. Consistent global states are meaningful global states and inconsistent global states are not meaningful in the sense that a distributed system can never be in an inconsistent state.

4.2.3 Interpretation in terms of cuts

Cuts in a space–time diagram provide a powerful graphical aid in representing and reasoning about the global states of a computation. A cut is a line joining an arbitrary point on each process line that slices the space–time diagram into a PAST and a FUTURE. Recall that every cut corresponds to a global

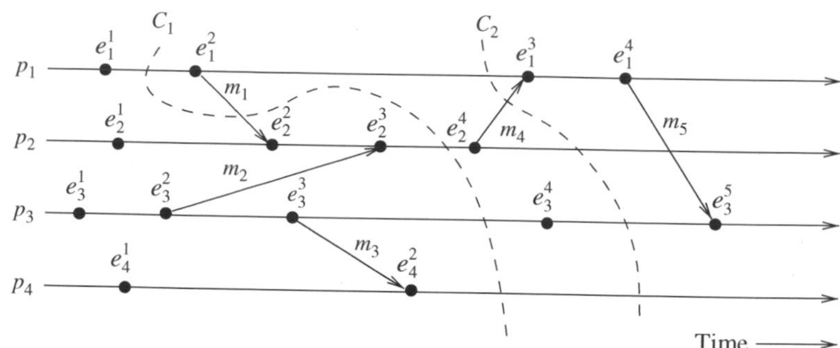

Figure 4.2 An interpretation in terms of a cut.

state and every global state can be graphically represented by a cut in the computation's space–time diagram [3].

A consistent global state corresponds to a cut in which every message received in the PAST of the cut has been sent in the PAST of that cut. Such a cut is known as a *consistent cut*. All the messages that cross the cut from the PAST to the FUTURE are captured in the corresponding channel state. For example, consider the space–time diagram for the computation illustrated in Figure 4.2. Cut C_1 is inconsistent because message m_1 is flowing from the FUTURE to the PAST. Cut C_2 is consistent and message m_4 must be captured in the state of channel C_{21}.

Note that in a consistent snapshot, all the recorded local states of processes are concurrent; that is, the recorded local state of no process casually affects the recorded local state of any other process. (Note that the notion of causality can be extended from the set of events to the set of recorded local states.)

4.2.4 Issues in recording a global state

If a global physical clock were available, the following simple procedure could be used to record a consistent global snapshot of a distributed system. In this, the initiator of the snapshot collection decides a future time at which the snapshot is to be taken and broadcasts this time to every process. All processes take their local snapshots at that instant in the global time. The snapshot of channel C_{ij} includes all the messages that process p_j receives after taking the snapshot and whose timestamp is smaller than the time of the snapshot. (All messages are timestamped with the sender's clock.) Clearly, if channels are not FIFO, a termination detection scheme will be needed to determine when to stop waiting for messages on channels.

However, a global physical clock is not available in a distributed system and the following two issues need to be addressed in recording of a consistent global snapshot of a distributed system [16]:

I1: How to distinguish between the messages to be recorded in the snapshot (either in a channel state or a process state) from those not to be recorded. The answer to this comes from conditions **C1** and **C2** as follows:

Any message that is sent by a process before recording its snapshot, must be recorded in the global snapshot (from **C1**).

Any message that is sent by a process after recording its snapshot, must not be recorded in the global snapshot (from **C2**).

I2: How to determine the instant when a process takes its snapshot. The answer to this comes from condition **C2** as follows:

A process p_j must record its snapshot before processing a message m_{ij} that was sent by process p_i after recording its snapshot.

We next discuss a set of representative snapshot algorithms for distributed systems. These algorithms assume different interprocess communication capabilities about the underlying system and illustrate how interprocess communication affects the design complexity of these algorithms. There are two types of messages: computation messages and control messages. The former are exchanged by the underlying application and the latter are exchanged by the snapshot algorithm. Execution of a snapshot algorithm is transparent to the underlying application, except for occasional delaying some of the actions of the application.

4.3 Snapshot algorithms for FIFO channels

This section presents the Chandy and Lamport algorithm [6], which was the first algorithm to record the global snapshot. We also present three variations of the Chandy and Lamport algorithm.

4.3.1 Chandy–Lamport algorithm

The Chandy-Lamport algorithm uses a control message, called a *marker*. After a site has recorded its snapshot, it sends a *marker* along all of its outgoing channels before sending out any more messages. Since channels are FIFO, a marker separates the messages in the channel into those to be included in the snapshot (i.e., channel state or process state) from those not to be recorded in the snapshot. This addresses issue **I1**. The role of markers in a FIFO system is to act as delimiters for the messages in the channels so that the channel state recorded by the process at the receiving end of the channel satisfies the condition **C2**.

Since all messages that follow a marker on channel C_{ij} have been sent by process p_i after p_i has taken its snapshot, process p_j must record its snapshot no later than when it receives a marker on channel C_{ij}. In general, a process must record its snapshot no later than when it receives a marker on any of its incoming channels. This addresses issue **I2**.

The algorithm

The Chandy–Lamport snapshot recording algorithm is given in Algorithm 4.1. A process initiates snapshot collection by executing the *marker sending rule* by which it records its local state and sends a marker on each outgoing channel. A process executes the *marker receiving rule* on receiving a marker. If the process has not yet recorded its local state, it records the state of the channel on which the marker is received as empty and executes the *marker sending rule* to record its local state. Otherwise, the state of the incoming channel on which the marker is received is recorded as the set of computation messages received on that channel after recording the local state but before receiving the marker on that channel. The algorithm can be initiated by any process by executing the *marker sending rule*. The algorithm terminates after each process has received a marker on all of its incoming channels.

The recorded local snapshots can be put together to create the global snapshot in several ways. One policy is to have each process send its local snapshot to the initiator of the algorithm. Another policy is to have each process send the information it records along all outgoing channels, and to have each process receiving such information for the first time propagate it along its outgoing channels. All the local snapshots get disseminated to all other processes and all the processes can determine the global state.

Multiple processes can initiate the algorithm concurrently. If multiple processes initiate the algorithm concurrently, each initiation needs to be

---

Marker sending rule for process p_i

(1) Process p_i records its state.
(2) For each outgoing channel C on which a marker
  has not been sent, p_i sends a marker along C
  before p_i sends further messages along C.

Marker receiving rule for process p_j
On receiving a marker along channel C:
  if p_j has not recorded its state **then**
    Record the state of C as the empty set
    Execute the "marker sending rule"
  else
    Record the state of C as the set of messages
    received along C after $p_{j's}$ state was recorded
    and before p_j received the marker along C

Algorithm 4.1 The Chandy–Lamport algorithm.

distinguished by using unique markers. Different initiations by a process are identified by a sequence number.

Correctness

To prove the correctness of the algorithm, we show that a recorded snapshot satisfies conditions **C1** and **C2**. Since a process records its snapshot when it receives the first marker on any incoming channel, no messages that follow markers on the channels incoming to it are recorded in the process's snapshot. Moreover, a process stops recording the state of an incoming channel when a marker is received on that channel. Due to FIFO property of channels, it follows that no message sent after the marker on that channel is recorded in the channel state. Thus, condition **C2** is satisfied. When a process p_j receives message m_{ij} that precedes the marker on channel C_{ij}, it acts as follows: if process p_j has not taken its snapshot yet, then it includes m_{ij} in its recorded snapshot. Otherwise, it records m_{ij} in the state of the channel C_{ij}. Thus, condition **C1** is satisfied.

Complexity

The recording part of a single instance of the algorithm requires $O(e)$ messages and $O(d)$ time, where e is the number of edges in the network and d is the diameter of the network.

4.3.2 Properties of the recorded global state

The recorded global state may not correspond to any of the global states that occurred during the computation. Consider two possible executions of the snapshot algorithm (shown in Figure 4.3) for the money transfer example of Figure 4.2:

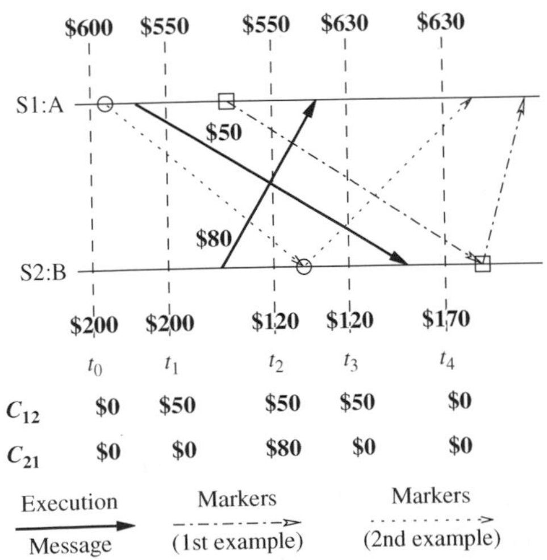

Figure 4.3 Timing diagram of two possible executions of the banking example.

1. (Markers shown using dashed-and-dotted arrows.) Let site S1 initiate the algorithm just after t_1. Site S1 records its local state (account A = \$550) and sends a marker to site S2. The marker is received by site S2 after t_4. When site S2 receives the marker, it records its local state (account B = \$170), the state of channel C_{12} as \$0, and sends a marker along channel C_{21}. When site S1 receives this marker, it records the state of channel C_{21} as \$80. The \$800 amount in the system is conserved in the recorded global state,

$$A = \$550, B = \$170, C_{12} = \$0, C_{21} = \$80.$$

2. (Markers shown using dotted arrows.) Let site S1 initiate the algorithm just after t_0 and before sending the \$50 for S2. Site S1 records its local state (account A = \$600) and sends a marker to site S2. The marker is received by site S2 between t_2 and t_3. When site S2 receives the marker, it records its local state (account B = \$120), the state of channel C_{12} as \$0, and sends a marker along channel C_{21}. When site S1 receives this marker, it records the state of channel C_{21} as \$80. The \$800 amount in the system is conserved in the recorded global state,

$$A = \$600, B = \$120, C_{12} = \$0, C_{21} = \$80.$$

In both these possible runs of the algorithm, the recorded global states never occurred in the execution. This happens because a process can change its state asynchronously before the markers it sent are received by other sites and the other sites record their states.

Nevertheless, as we discuss next, the system could have passed through the recorded global states in some equivalent executions. Suppose the algorithm is initiated in global state S_i and it terminates in global state S_t. Let *seq* be the sequence of events that takes the system from S_i to S_t. Let S^* be the global state recorded by the algorithm. Chandy and Lamport [6] showed that there exists a sequence *seq′* which is a permutation of *seq* such that S^* is reachable from S_i by executing a prefix of *seq′* and S_t is reachable from S^* by executing the rest of the events of *seq′*.

A brief sketch of the proof is as follows: an event *e* is defined as a pre-recording/post-recording event if *e* occurs on a process *p* and *p* records its state after/before *e* in *seq*. A post-recording event may occur after a pre-recording event only if the two events occur on different processes. It is shown that a post-recording event can be swapped with an immediately following pre-recording event in a sequence without affecting the local states of either of the two processes on which the two events occur. By iteratively applying this operation to *seq*, the above-described permutation *seq′* is obtained. It is then shown that S^*, the global state recorded by the algorithm for the processes and channels, is the state after all the pre-recording events have been executed, but before any post-recording event.

Thus, the recorded global state is a valid state in an equivalent execution and if a stable property (i.e., a property that persists such as termination or deadlock) holds in the system before the snapshot algorithm begins, it holds in the recorded global snapshot. Therefore, a recorded global state is useful in detecting stable properties.

A physical interpretation of the collected global state is as follows: consider the two instants of recording of the local states in the banking example. If the cut formed by these instants is viewed as being an elastic band and if the elastic band is stretched so that it is vertical, then recorded states of all processes occur simultaneously at one physical instant, and the recorded global state occurs in the execution that is depicted in this modified space–time diagram. This is called the *rubber-band* criterion. For example, consider the two different executions of the snapshot algorithm, depicted in Figure 4.3. For the execution for which the markers are shown using dashed-and-dotted arrows, the instants of the local state recordings are marked by squares. Applying the rubber-band criterion, these can be stretched to be vertical or instantaneous. Similarly, for the other execution for which the markers are shown using dotted arrows, the instants of local state recordings are marked by circles. Note that the system execution would have been like this, had the processors' speeds and message delays been different. Yet another physical interpretation of the collected global state is as follows: all the recorded process states are mutually concurrent – no recorded process state causally depends upon another. Therefore, logically we can view that all these process states occurred simultaneously even though they might have occurred at different instants in physical time.

4.4 Variations of the Chandy–Lamport algorithm

Several variants of the Chandy–Lamport snapshot algorithm followed. These variants refined and optimized the basic algorithm. For example, the Spezialetti and Kearns algorithm [29] optimizes concurrent initiation of snapshot collection and efficiently distributes the recorded snapshot. Venkatesan's algorithm [32] optimizes the basic snapshot algorithm to efficiently record repeated snapshots of a distributed system that are required in recovery algorithms with synchronous checkpointing.

4.4.1 Spezialetti–Kearns algorithm

There are two phases in obtaining a global snapshot: locally recording the snapshot at every process and distributing the resultant global snapshot to all the initiators. Spezialetti and Kearns [29] provided two optimizations to the Chandy–Lamport algorithm. The first optimization combines snapshots concurrently initiated by multiple processes into a single snapshot. This optimization is linked with the second optimization, which deals with the efficient distribution of the global snapshot. A process needs to take only one snapshot, irrespective of the number of concurrent initiators and all processes are not sent the global snapshot. This algorithm assumes bi-directional channels in the system.

Efficient snapshot recording

In the Spezialetti–Kearns algorithm, a marker carries the identifier of the initiator of the algorithm. Each process has a variable *master* to keep track of the initiator of the algorithm. When a process executes the "marker sending rule" on the receipt of its first marker, it records the initiator's identifier carried in the received marker in the *master* variable. A process that initiates the algorithm records its own identifier in the *master* variable.

A key notion used by the optimizations is that of a *region* in the system. A region encompasses all the processes whose *master* field contains the identifier of the same initiator. A region is identified by the initiator's identifier. When there are multiple concurrent initiators, the system gets partitioned into multiple regions.

When the initiator's identifier in a marker received along a channel is different from the value in the *master* variable, a concurrent initiation of the algorithm is detected and the sender of the marker lies in a different region. The identifier of the concurrent initiator is recorded in a local variable *id-border-set*. The process receiving the marker does not take a snapshot for this marker and does not propagate this marker. Thus, the algorithm efficiently handles concurrent snapshot initiations by suppressing redundant snapshot collections – a process does not take a snapshot or propagate a snapshot request initiated by a process if it has already taken a snapshot in response to some other snapshot initiation.

The state of the channel is recorded just as in the Chandy–Lamport algorithm (including those that cross a border between regions). This enables the snapshot recorded in one region to be merged with the snapshot recorded in the adjacent region. Thus, even though markers arriving at a node contain identifiers of different initiators, they are considered part of the same instance of the algorithm for the purpose of channel state recording.

Snapshot recording at a process is complete after it has received a marker along each of its channels. After every process has recorded its snapshot, the system is partitioned into as many regions as the number of concurrent initiations of the algorithm. The variable *id-border-set* at a process contains the identifiers of the neighboring regions.

Efficient dissemination of the recorded snapshot

The Spezialetti–Kearns algorithm efficiently assembles the snapshot as follows: in the snapshot recording phase, a forest of spanning trees is implicitly created in the system. The initiator of the algorithm is the root of a spanning tree and all processes in its region belong to its spanning tree. If process p_i executed the "marker sending rule" because it received its first marker from process p_j, then process p_j is the parent of process p_i in the spanning tree. When a leaf process in the spanning tree has recorded the states of all incoming channels, the process sends the locally recorded state (local snapshot, *id-border-set*) to its parent in the spanning tree. After an intermediate process in a spanning tree has received the recorded states from all its child processes and has recorded the states of all incoming channels, it forwards its locally recorded state and the locally recorded states of all its descendent processes to its parent.

When the initiator receives the locally recorded states of all its descendents from its children processes, it assembles the snapshot for all the processes in its region and the channels incident on these processes. The initiator knows the identifiers of initiators in adjacent regions using *id-border-set* information it receives from processes in its region. The initiator exchanges the snapshot of its region with the initiators in adjacent regions in rounds. In each round, an initiator sends to initiators in adjacent regions, any new information obtained from the initiator in the adjacent region during the previous round of message exchange. A round is complete when an initiator receives information, or the *blank* message (signifying no new information will be forthcoming) from all initiators of adjacent regions from which it has not already received a *blank* message.

The message complexity of snapshot recording is $O(e)$ irrespective of the number of concurrent initiations of the algorithm. The message complexity of assembling and disseminating the snapshot is $O(rn^2)$ where r is the number of concurrent initiations.

4.4.2 Venkatesan's incremental snapshot algorithm

Many applications require repeated collection of global snapshots of the system. For example, recovery algorithms with synchronous checkpointing need to advance their checkpoints periodically. This can be achieved by repeated invocations of the Chandy–Lamport algorithm. Venkatesan [32] proposed the following efficient approach: execute an algorithm to record an incremental snapshot since the most recent snapshot was taken and combine it with the most recent snapshot to obtain the latest snapshot of the system. The incremental snapshot algorithm of Venkatesan [32] modifies the global snapshot algorithm of Chandy–Lamport to save on messages when computation messages are sent only on a few of the network channels, between the recording of two successive snapshots.

The incremental snapshot algorithm assumes bidirectional FIFO channels, the presence of a single initiator, a fixed spanning tree in the network, and four types of control messages: *init_snap*, *snap_completed*, *regular*, and *ack*. *init_snap* and *snap_completed* messages traverse the spanning tree edges.

regular and *ack* messages, which serve to record the state of non-spanning edges, are not sent on those edges on which no computation message has been sent since the previous snapshot.

Venkatesan [32] showed that the lower bound on the message complexity of an incremental snapshot algorithm is $\Omega(u+n)$, where u is the number of edges on which a computation message has been sent since the previous snapshot. Venkatesan's algorithm achieves this lower bound in message complexity.

The algorithm works as follows: snapshots are assigned version numbers and all algorithm messages carry this version number. The initiator notifies all the processes the version number of the new snapshot by sending *init_snap* messages along the spanning tree edges. A process follows the "marker sending rule" when it receives this notification or when it receives a *regular* message with a new version number. The "marker sending rule" is modified so that the process sends *regular* messages along only those channels on which it has sent computation messages since the previous snapshot, and the process waits for *ack* messages in response to these *regular* messages. When a leaf process in the spanning tree receives all the *ack* messages it expects, it sends a *snap_completed* message to its parent process. When a non-leaf process in the spanning tree receives all the *ack* messages it expects, as well as a *snap_completed* message from each of its child processes, it sends a *snap_completed* message to its parent process.

The algorithm terminates when the initiator has received all the *ack* messages it expects, as well as a *snap_completed* message from each of its child processes. The selective manner in which *regular* messages are sent has the effect that a process does not know whether to expect a *regular* message on an incoming channel. A process can be sure that no such message will be received and that the snapshot is complete only when it executes the "marker sending rule" for the next initiation of the algorithm.

4.4.3 Helary's wave synchronization method

Helary's snapshot algorithm [12] incorporates the concept of message waves in the Chandy–Lamport algorithm. A wave is a flow of control messages such that every process in the system is visited exactly once by a wave control message, and at least one process in the system can determine when this flow of control messages terminates. A wave is initiated after the previous wave terminates. Wave sequences may be implemented by various traversal structures such as a ring. A process begins recording the local snapshot when it is visited by the wave control message.

In Helary's algorithm, the "marker sending rule" is executed when a control message belonging to the wave flow visits the process. The process then forwards a control message to other processes, depending on the wave traversal structure, to continue the wave's progression. The "marker receiving rule" is modified so that if the process has not recorded its state when a marker is received on some channel, the "marker receiving rule" is not executed and no messages received after the marker on this channel are processed until the control message belonging to the wave flow visits the process. Thus, each process follows the "marker receiving rule" only after it is visited by a control message belonging to the wave.

Note that in this algorithm, the primary function of wave synchronization is to evaluate functions over the recorded global snapshot. This algorithm has a message complexity of $O(e)$ to record a snapshot (because all channels need to be traversed to implement the wave).

An example of this function is the number of messages in transit to each process in a global snapshot, and whether the global snapshot is strongly consistent. For this function, each process maintains two vectors, $SENT$ and $RECD$. The ith elements of these vectors indicate the number of messages sent to/received from process i, respectively, since the previous visit of a wave control message. The wave control messages carry a global abstract counter vector whose ith entry indicates the number of messages in transit to process i. These entries in the vector are updated using the $SENT$ and $RECD$ vectors at each node visited. When the control wave terminates, the number of messages in transit to each process as recorded in the snapshot is known.

4.5 Snapshot algorithms for non-FIFO channels

A FIFO system ensures that all messages sent after a marker on a channel will be delivered after the marker. This ensures that condition **C2** is satisfied in the recorded snapshot if LS_i, LS_j, and SC_{ij} are recorded as described in the Chandy–Lamport algorithm. In a non-FIFO system, the problem of global snapshot recording is complicated because a marker cannot be used to delineate messages into those to be recorded in the global state from those not to be recorded in the global state. In such systems, different techniques have to be used to ensure that a recorded global state satisfies condition **C2**.

In a non-FIFO system, either some degree of inhibition (i.e., temporarily delaying the execution of an application process or delaying the send of a computation message) or piggybacking of control information on computation messages to capture out-of-sequence messages is necessary to record a consistent global snapshot [31]. The non-FIFO algorithm by Helary uses message inhibition [12]. The non-FIFO algorithms by Lai and Yang [18], Li *et al.* [20], and Mattern [23] use message piggybacking to distinguish computation messages sent after the marker from those sent before the marker.

The non-FIFO algorithm of Helary [12] uses message inhibition to avoid an inconsistency in a global snapshot in the following way: when a process receives a marker, it immediately returns an acknowledgement. After a process p_i has sent a marker on the outgoing channel to process p_j, it does not send any messages on this channel until it is sure that p_j has recorded its local state. Process p_i can conclude this if it has received an acknowledgement for the marker sent to p_j, or it has received a marker for this snapshot from p_j.

We next discuss snapshot recording algorithms for systems with non-FIFO channels that use piggybacking of computation messages.

4.5.1 Lai–Yang algorithm

Lai and Yang's global snapshot algorithm for non-FIFO systems [18] is based on two observations on the role of a marker in a FIFO system. The first observation is that a marker ensures that condition **C2** is satisfied for LS_i and LS_j when the snapshots are recorded at processes p_i and p_j, respectively. The Lai–Yang algorithm fulfills this role of a marker in a non-FIFO system by using a coloring scheme on computation messages that works as follows:

1. Every process is initially white and turns red while taking a snapshot. The equivalent of the "marker sending rule" is executed when a process turns red.
2. Every message sent by a white (red) process is colored white (red). Thus, a white (red) message is a message that was sent before (after) the sender of that message recorded its local snapshot.
3. Every white process takes its snapshot at its convenience, but no later than the instant it receives a red message.

Thus, when a white process receives a red message, it records its local snapshot before processing the message. This ensures that no message sent by a process after recording its local snapshot is processed by the destination process before the destination records its local snapshot. Thus, an explicit marker message is not required in this algorithm and the "marker" is piggybacked on computation messages using a coloring scheme.

The second observation is that the marker informs process p_j of the value of $\{send(m_{ij})|\ send(m_{ij}) \in LS_i\ \}$ so that the state of the channel C_{ij} can be computed as $transit(LS_i, LS_j)$. The Lai–Yang algorithm fulfills this role of the marker in the following way:

4. Every white process records a history of all white messages sent or received by it along each channel.
5. When a process turns red, it sends these histories along with its snapshot to the initiator process that collects the global snapshot.
6. The initiator process evaluates $transit(LS_i, LS_j)$ to compute the state of a channel C_{ij} as given below:

SC_{ij} = {white messages sent by p_i on C_{ij}} − {white messages received by p_j on C_{ij}}

$$= \{m_{ij} \mid send(m_{ij}) \in LS_i\} - \{m_{ij} \mid rec(m_{ij}) \in LS_j\}.$$

Condition **C2** holds because a red message is not included in the snapshot of the recipient process and a channel state is the difference of two sets of white messages. Condition **C1** holds because a white message m_{ij} is included in the snapshot of process p_j if p_j receives m_{ij} before taking its snapshot. Otherwise, m_{ij} is included in the state of channel C_{ij}.

Though marker messages are not required in the algorithm, each process has to record the entire message history on each channel as part of the local snapshot. Thus, the space requirements of the algorithm may be large. However, in applications (such as termination detection) where the number of messages in transit in a channel is sufficient, message histories can be replaced by integer counters reducing the space requirement. Lai and Yang describe how the size of the local storage and snapshot recording can be reduced by storing only the messages sent and received since the previous snapshot recording, assuming that the previous snapshot is still available. This approach can be very useful in applications that require repeated snapshots of a distributed system.

4.5.2 Li *et al.*'s algorithm

Li *et al.*'s algorithm [20] for recording a global snapshot in a non-FIFO system is similar to the Lai–Yang algorithm. Markers are tagged so as to generalize the red/white colors of the Lai–Yang algorithm to accommodate repeated invocations of the algorithm and multiple initiators. In addition, the algorithm is not concerned with the contents of computation messages and the state of a channel is computed as the number of messages in transit in the channel. A process maintains two counters for each incident channel to record the number of messages sent and received on the channel and reports these counter values with its snapshot to the initiator. This simplification is combined with the incremental technique to compute channel states, which reduces the size of message histories to be stored and transmitted. The initiator computes the state of C_{ij} as: (the number of messages in C_{ij} in the previous snapshot) + (the number of messages sent on C_{ij} since the last snapshot at process p_i) − (the number of messages received on C_{ij} since the last snapshot at process p_j).

Snapshots initiated by an initiator are assigned a sequence number. All messages sent after a local snapshot recording are tagged by a tuple $\langle init\_id, MKNO\rangle$, where $init\_id$ is the initiator's identifier and $MKNO$ is the sequence number of the algorithm's most recent invocation by initiator $init\_id$; to insure liveness, markers with tags similar to the above tags are explicitly sent only on all outgoing channels on which no messages might be sent. The tuple $\langle init\_id, MKNO \rangle$ is a generalization of the red/white colors used in Lai–Yang to accommodate repeated invocations of the algorithm and multiple initiators.

For simplicity, we explain this algorithm using the framework of the Lai–Yang algorithm. The local state recording is done as described by rules 1–3 of the Lai–Yang algorithm.

A process maintains input/output counters for the number of messages sent and received on each incident channel after the last snapshot (by that initiator). The algorithm is not concerned with the contents of computation messages and so the computation of the state of a channel is simplified to computing the number of messages in transit in the channel. This simplification is combined with an incremental technique for computing in-transit messages, also suggested independently by Lai and Yang [18], for reducing the size of the entire message history to be locally stored and to be recorded in a local snapshot to compute channel states. The initiator of the algorithm maintains a variable $TRANSIT_{ij}$ for the number of messages in transit in the channel from process p_i to process p_j, as recorded in the previous snapshot. The channel states are recorded as described in rules 4–6 of the Lai–Yang algorithm:

4. Every white process records a history, as input and output counters, of all white messages sent or received by it along each channel after the previous snapshot (by the same initiator).
5. When a process turns red, it sends these histories (i.e., input and output counters) along with its snapshot to the initiator process that collects the global snapshot.
6. The initiator process computes the state of channel C_{ij} as follows:

$$\begin{aligned} SC_{ij} = transit(LS_i, LS_j) &= TRANSIT_{ij} \\ &+ (\#\text{messages sent on that channel since the last snapshot}) \\ &- (\#\text{messages received on that channel since the last snapshot}). \end{aligned}$$

If the initiator initiates a snapshot before the completion of the previous snapshot, it is possible that some process may get a message with a lower sequence number after participating in a snapshot initiated later. In this case, the algorithm uses the snapshot with the higher sequence number to also create the snapshot for the lower sequence number.

The algorithm works for multiple initiators if separate input/output counters are associated with each initiator, and marker messages and the tag fields carry a vector of tuples, with one tuple for each initiator.

Though this algorithm does not require any additional message to record a global snapshot provided computation messages are eventually sent on each channel, the local storage and size of tags on computation messages are of size $O(n)$, where n is the number of initiators. The Spezialetti and Kearns technique [29] of combining concurrently initiated snapshots can be used with this algorithm.

4.5.3 Mattern's algorithm

Mattern's algorithm [23] is based on vector clocks. Recall that, in vector clocks, the clock at a process in an integer vector of length n, with one component for each process.

Mattern's algorithm assumes a single initiator process and works as follows:

1. The initiator "ticks" its local clock and selects a future vector time s at which it would like a global snapshot to be recorded. It then broadcasts this time s and freezes all activity until it receives all acknowledgements of the receipt of this broadcast.
2. When a process receives the broadcast, it remembers the value s and returns an acknowledgement to the initiator.
3. After having received an acknowledgement from every process, the initiator increases its vector clock to s and broadcasts a dummy message to all processes. (Observe that before broadcasting this dummy message, the local clocks of other processes have a value $\not\geq s$.)
4. The receipt of this dummy message forces each recipient to increase its clock to a value $\geq s$ if not already $\geq s$.
5. Each process takes a local snapshot and sends it to the initiator when (just before) its clock increases from a value less than s to a value $\geq s$. Observe that this may happen before the dummy message arrives at the process.
6. The state of C_{ij} is all messages sent along C_{ij}, whose timestamp is smaller than s and which are received by p_j after recording LS_j.

Processes record their local snapshot as per rule 5. Any message m_{ij} sent by process p_i after it records its local snapshot LS_i has a timestamp $> s$. Assume that this m_{ij} is received by process p_j before it records LS_j. After receiving this m_{ij} and before p_j records LS_j, p_j's local clock reads a value $> s$, as per rules for updating vector clocks. This implies p_j must have already recorded LS_j as per rule 5, which contradicts the assumption. Therefore, m_{ij} cannot be received by p_j before it records LS_j. By rule 6, m_{ij} is not recorded in SC_{ij} and therefore, condition **C2** is satisfied. Condition **C1** holds because each message m_{ij} with a timestamp less than s is included in the snapshot of process p_j if p_j receives m_{ij} before taking its snapshot. Otherwise, m_{ij} is included in the state of channel C_{ij}.

The following observations about the above algorithm lead to various optimizations: (i) The initiator can be made a "virtual" process–so no process has to freeze. (ii) As long as a new higher value of s is selected, the phase of broadcasting s and returning the acks can be eliminated. (iii) Only the initiator's component of s is used to determine when to record a snapshot.

Also, one needs to know only if the initiator's component of the vector timestamp in a message has increased beyond the value of the corresponding component in s. Therefore, it suffices to have just two values of s, say, white and red, which can be represented using one bit.

With these optimizations, the algorithm becomes similar to the Lai–Yang algorithm except for the manner in which $transit(LS_i, LS_j)$ is evaluated for channel C_{ij}. In Mattern's algorithm, a process is not required to store message histories to evaluate the channel states. The state of any channel is the set of all the white messages that are received by a red process on which that channel is incident. A termination detection scheme for non-FIFO channels is required to detect that no white messages are in transit to ensure that the recording of all the channel states is complete. One of the following schemes can be used for termination detection:

1. Each process i keeps a counter $cntr_i$ that indicates the difference between the number of white messages it has sent and received before recording its snapshot. It reports this value to the initiator process along with its snapshot and forwards all white messages, it receives henceforth, to the initiator. Snapshot collection terminates when the initiator has received $\sum_i cntr_i$ number of forwarded white messages.
2. Each red message sent by a process carries a piggybacked value of the number of white messages sent on that channel before the local state recording. Each process keeps a counter for the number of white messages received on each channel. A process can detect termination of recording the states of incoming channels when it receives as many white messages on each channel as the value piggybacked on red messages received on that channel.

The savings of not storing and transmitting entire message histories, over the Lai–Yang algorithm, comes at the expense of delay in the termination of the snapshot recording algorithm and need for a termination detection scheme (e.g., a message counter per channel).

4.6 Snapshots in a causal delivery system

Two global snapshot recording algorithms, namely, Acharya–Badrinath [1] and Alagar–Venkatesan [2] assume that the underlying system supports causal message delivery. The causal message delivery property **CO** provides a built-in message synchronization to control and computation messages. Consequently, snapshot algorithms for such systems are considerably simplified. For example, these algorithms do not send control messages (i.e., markers) on every channel and are simpler than the snapshot algorithms for a FIFO system.

Several protocols exist for implementing causal ordering [5,6,26,28].

4.6.1 Process state recording

Both these algorithms use an identical principle to record the state of processes. An initiator process broadcasts a token, denoted as *token*, to every process including itself. Let the copy of the token received by process p_i be denoted $token_i$. A process p_i records its local snapshot LS_i when it receives $token_i$ and sends the recorded snapshot to the initiator. The algorithm terminates when the initiator receives the snapshot recorded by each process.

These algorithms do not require each process to send markers on each channel, and the processes do not coordinate their local snapshot recordings with other processes. Nonetheless, for any two processes p_i and p_j, the following property (called property **P1**) is satisfied:

$$send(m_{ij}) \notin LS_i \Rightarrow rec(m_{ij}) \notin LS_j.$$

This is due to the causal ordering property of the underlying system as explained next. Let a message m_{ij} be such that $rec(token_i) \longrightarrow send(m_{ij})$. Then $send(token_j) \longrightarrow send(m_{ij})$ and the underlying causal ordering property ensures that $rec(token_j)$, at which instant process p_j records LS_j, happens before $rec(m_{ij})$. Thus, m_{ij}, whose send is not recorded in LS_i, is not recorded as received in LS_j.

Methods of channel state recording are different in these two algorithms and are discussed next.

4.6.2 Channel state recording in Acharya–Badrinath algorithm

Each process p_i maintains arrays $SENT_i[1, \ldots, n]$ and $RECD_i[1, \ldots, n]$. $SENT_i[j]$ is the number of messages sent by process p_i to process p_j and $RECD_i[j]$ is the number of messages received by process p_i from process p_j. The arrays may not contribute to the storage complexity of the algorithm because the underlying causal ordering protocol may require these arrays to enforce causal ordering.

Channel states are recorded as follows: when a process p_i records its local snapshot LS_i on the receipt of $token_i$, it includes arrays $RECD_i$ and $SENT_i$ in its local state before sending the snapshot to the initiator. When the algorithm terminates, the initiator determines the state of channels in the global snapshot being assembled as follows:

1. The state of each channel from the initiator to each process is empty.
2. The state of channel from process p_i to process p_j is the set of messages whose sequence numbers are given by $\{RECD_j[i]+1, \ldots, SENT_i[j]\}$.

We will now show that the algorithm satisfies conditions **C1** and **C2**.

Let a message m_{ij} be such that $rec(token_i) \longrightarrow send(m_{ij})$. Clearly, $send(token_j) \longrightarrow send(m_{ij})$ and the sequence number of m_{ij} is greater than $SENT_i[j]$. Therefore, m_{ij} is not recorded in SC_{ij}. Thus,

$send(m_{ij}) \notin LS_i \Rightarrow m_{ij} \notin SC_{ij}$. This in conjunction with property **P1** implies that the algorithm satisfies condition **C2**.

Consider a message m_{ij} which is the kth message from process p_i to process p_j before p_i takes its snapshot. The two possibilities below imply that condition **C1** is satisfied:

- Process p_j receives m_{ij} before taking its snapshot. In this case, m_{ij} is recorded in p_j's snapshot.
- Otherwise, $RECD_j[i] \leq k \leq SENT_i[j]$ and the message m_{ij} will be included in the state of channel C_{ij}.

This algorithm requires $2n$ messages and 2 time units for recording and assembling the snapshot, where one time unit is required for the delivery of a message. If the contents of messages in channels state are required, the algorithm requires $2n$ messages and 2 time units additionally.

4.6.3 Channel state recording in Alagar–Venkatesan algorithm

A message is referred to as *old* if the send of the message causally precedes the send of the token. Otherwise, the message is referred to as *new*. Whether a message is new or old can be determined by examining the vector timestamp in the message, which is needed to enforce causal ordering among messages.

In the Alagar–Venkatesan algorithm [2], channel states are recorded as follows:

1. When a process receives the *token*, it takes its snapshot, initializes the state of all channels to empty, and returns *Done* message to the initiator. Now onwards, a process includes a message received on a channel in the channel state only if it is an old message.
2. After the initiator has received *Done* message from all processes, it broadcasts a *Terminate* message.
3. A process stops the snapshot algorithm after receiving a *Terminate* message.

An interesting observation is that a process receives all the old messages in its incoming channels before it receives the *Terminate* message. This is ensured by the underlying causal message delivery property.

The causal ordering property ensures that no new message is delivered to a process prior to the *token* and only old messages are recorded in the channel states. Thus, $send(m_{ij}) \notin LS_i \Rightarrow m_{ij} \notin SC_{ij}$. This together with property **P1** implies that condition **C2** is satisfied. Condition **C1** is satisfied because each old message m_{ij} is delivered either before the token is delivered or before the *Terminate* is delivered to a process and thus gets recorded in LS_i or SC_{ij}, respectively.

A comparison of the salient features of the various snapshot recording algorithms discussed is given in Table 4.1.

Table 4.1 A comparison of snapshot algorithms.

Algorithms	Features
Chandy–Lamport [7]	Baseline algorithm. Requires FIFO channels. $O(e)$ messages to record snapshot and $O(d)$ time.
Spezialetti–Kearns [29]	Improvements over [7]: supports concurrent initiators, efficient assembly and distribution of a snapshot. Assumes bidirectional channels. $O(e)$ messages to record, $O(rn^2)$ messages to assemble and distribute snapshot.
Venkatesan [32]	Based on [7]. Selective sending of markers. Provides message-optimal incremental snapshots. $\Omega(n+u)$ messages to record snapshot.
Helary [12]	Based on [7]. Uses wave synchronization. Evaluates function over recorded global state. Adaptable to non-FIFO systems but requires inhibition.
Lai–Yang [18]	Works for non-FIFO channels. Markers piggybacked on computation messages. Message history required to compute channel states.
Li *et al.* [20]	Similar to [18]. Small message history needed as channel states are computed incrementally.
Mattern [23]	Similar to [18]. No message history required. Termination detection (e.g., a message counter per channel) required to compute channel states.
Acharya–Badrinath [1]	Requires causal delivery support. Centralized computation of channel states. Channel message contents need not be known. Requires $2n$ messages, 2 time units.
Alagar-Venkatesan [2]	Requires causal delivery support. Distributed computation of channel states. Requires $3n$ messages, 3 time units, small messages.

n = # processes, u = # edges on which messages were sent after previous snapshot, e = # channels, d = diameter of the network, r = # concurrent initiators.

4.7 Monitoring global state

Several applications such as debugging a distributed program need to detect a system state which is determined by the values of variables on a subset of processes. This state can be expressed as a predicate on variables distributed across the involved processes. Rather than recording and evaluating snapshots at regular intervals, it is more efficient to monitor changes to the variables that affect the predicate and evaluate the predicate only when some component variable changes.

Spezialetti and Kearns [30] proposed a technique, called *simultaneous regions*, for the consistent monitoring of distributed systems to detect global predicates. A process whose local variable is a component of the global predicate informs a monitor whenever the value of the variable changes. This process also coerces other processes to inform the monitor of the values of their variables that are components of the global predicate. The monitor evaluates the global predicate when it receives the next message from each of the involved processes, informing it of the value(s) of their local variable(s). The periods of local computation on each process between the ith and the $i+1$st events at which the values of the local component(s) of the global predicate are reported to the monitor are defined to be the $i+1$st simultaneous regions. The above scheme is extended to arrange multiple monitors hierarchically to evaluate complex global predicates.

4.8 Necessary and sufficient conditions for consistent global snapshots

Many applications (such as transparent failure recovery, distributed debugging, monitoring distributed events, setting distributed breakpoints, protocol specification and verification, etc.) require that local process states are periodically recorded and analyzed during execution or off-line. A saved intermediate state of a process during its execution is called a *local checkpoint* of the process. A global snapshot of a distributed system is a set of local checkpoints one from each process and it represents a snapshot of the distributed computation execution at some instant. A global snapshot is consistent if there is no causal path between any two distinct checkpoints in the global snapshot. Therefore, a consistent snapshot consists of a set of local states that occurred concurrently or had a potential to occur simultaneously. This condition for the consistency of a global snapshot (that no causal path between any two checkpoints) is only the necessary condition but it is not the sufficient condition. In this section, we present the necessary and sufficient conditions under which a local checkpoint or a set of arbitrary collection of local checkpoints can be grouped with checkpoints at other processes to form a consistent global snapshot.

Processes take checkpoints asynchronously. Each checkpoint taken by a process is assigned a unique sequence number. The ith ($i \geq 0$) checkpoint of process p_p is assigned the sequence number i and is denoted by $C_{p,i}$. We assume that each process takes an initial checkpoint before execution begins and takes a *virtual* checkpoint after execution ends. The ith *checkpoint interval* of process p_p consists of all the computation performed between its $(i-1)$th and ith checkpoints (and includes the $(i-1)$th checkpoint but not the ith).

We first show with the help of an example that even if two local checkpoints do not have a causal path between them (i.e., neither happened before the other using Lamport's happen before relation), they may not belong to the same consistent global snapshot. Consider the execution shown in Figure 4.4. Although neither of the checkpoints $C_{1,1}$ and $C_{3,2}$ happened before the other, they cannot be grouped together with a checkpoint on process p_2 to form a

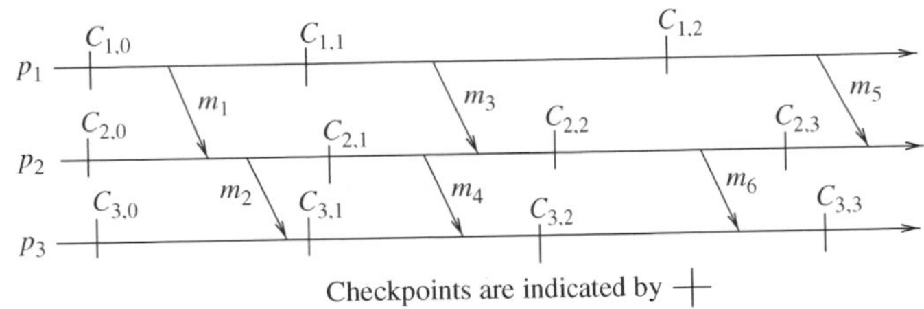

Figure 4.4 An illustration of zigzag paths.

consistent global snapshot. No checkpoint on p_2 can be grouped with both $C_{1,1}$ and $C_{3,2}$ while maintaining the consistency. Because of message m_4, $C_{3,2}$ cannot be consistent with $C_{2,1}$ or any earlier checkpoint in p_2, and because of message m_3, $C_{1,1}$ cannot be consistent with $C_{2,2}$ or any later checkpoint in p_2. Thus, no checkpoint on p_2 is available to form a consistent global snapshot with $C_{1,1}$ and $C_{3,2,}$.

To describe the necessary and sufficient conditions for a consistent snapshot, Netzer and Xu [25] defined a generalization of the Lamport's happens before relation, called a zigzag path. A checkpoint $C_{i,x}$ happens before a checkpoint $C_{j,y}$ (or a causal path exists between two checkpoints) if a sequence of messages exists from $C_{i,x}$ to $C_{j,y}$ such that each message is sent after the previous one in the sequence is received. A zigzag path between two checkpoints is a causal path, however, and allows a message to be sent before the previous one in the path is received. For example, in Figure 4.4, although a causal path does not exist from $C_{1,1}$ to $C_{3,2}$, a zigzag path does exist from $C_{1,1}$ to $C_{3,2}$. This zigzag path is formed by messages m_3 and m_4. This zigzag path means that no consistent snapshot exists in this execution that contains both $C_{1,1}$ and $C_{3,2}$.

Several applications require saving or analyzing consistent snapshots and zigzag paths have implications on such applications. For example, the state from which a distributed computation must restart after a crash must be consistent. Consistency ensures that no process is restarted from a state that has recorded the receipt of a message (called an orphan message) that no other process claims to have sent in the rolled back state. Processes take local checkpoints independently and a consistent global snapshot/checkpoint is found from the local checkpoints for a crash recovery. Clearly, due to zigzag paths, not all checkpoints taken by the processes will belong to a consistent snapshot. By reducing the number of zigzag paths in the local checkpoints taken by processes, one can increase the number of local checkpoints that belong to a consistent snapshot, thus minimizing the roll back necessary to find a consistent snapshot.[1] This can be achieved by tracking zigzag paths online and allowing each process to adaptively take checkpoints at certain

[1] In the worst case, the system would have to restart its execution right from the beginning after repeated rollbacks.

points in the execution so that the number of checkpoints that cannot belong to a consistent snapshot is minimized.

4.8.1 Zigzag paths and consistent global snapshots

In this section, we provide a formal definition of zigzag paths and use zigzag paths to characterize condition under which a set of local checkpoints together can belong to the same consistent snapshot. We then present two special cases: first, the conditions for an arbitrary checkpoint to be useful (i.e., a consistent snapshot exists that contains this checkpoint), and second, the conditions for two arbitrary checkpoints to belong to the same consistent snapshot.

A zigzag path

Recall that if a global snapshot is consistent, then none of its checkpoints happened before the other (i.e., there is no causal path between any two checkpoints in the snapshot). However, as explained earlier using Figure 4.4, if we have two checkpoints such that none of them happened before the other, it is still not sufficient to ensure that they can belong together to the same consistent snapshot. This happens when a zigzag path exists between such checkpoints. A zigzag path is defined as a generalization of Lamport's happens before relation.

definition 4.1 A *zigzag path* exists from a checkpoint $C_{x,i}$ to a checkpoint $C_{y,j}$ iff there exists messages $m_1, m_2, \ldots m_n$ $(n \geq 1)$ such that

1. *m_1 is sent by process p_x after $C_{x,i}$;*
2. *if m_k $(1 \leq k \leq n-1)$ is received by process p_z, then m_{k+1} is sent by p_z in the same or a later checkpoint interval (although m_{k+1} may be sent before or after m_k is received);*
3. *m_n is received by process p_y before $C_{y,j}$.*

For example, in Figure 4.4, a zigzag path exists from $C_{1,1}$ to $C_{3,2}$ due to messages m_3 and m_4. Even though process p_2 sends m_4 before receiving m_3, it does these in the same checkpoint interval. However, a zigzag path does not exist from $C_{1,2}$ to $C_{3,3}$ (due to messages m_5 and m_6) because process p_2 sends m_6 in an earlier checkpoint interval than the one in which it receives m_5.

definition 4.2 *A checkpoint C is involved in a* zigzag cycle *iff there is a zigzag path from C to itself.*

For example, in Figure 4.5, $C_{2,1}$ is on a zigzag cycle formed by messages m_1 and m_2. Note that messages m_1 and m_2 are respectively sent and received in the same checkpoint interval at p_1.

Difference between a zigzag path and a causal path

It is important to understand the difference between a causal path and a zigzag path. A causal path exists from a checkpoint A to another checkpoint B iff there is chain of messages starting after A and ending before B such that each message is sent after the previous one in the chain is received. A zigzag path consists of such a message chain, however, a message in the chain can be sent before the previous one in the chain is received, as long as the send and receive are in the same checkpoint interval. Thus a causal path is always a zigzag path, but a zigzag path need not be a causal path.

Figure 4.4 illustrates the difference between causal and zigzag paths. A causal path exists from $C_{1,0}$ to $C_{3,1}$ formed by chain of messages m_1 and m_2; this causal path is also a zigzag path. Similarly, a zigzag path exists from $C_{1,1}$ to $C_{3,2}$ formed by the chain of messages m_3 and m_4. Since the receive of m_3 happened after the send of m_4, this zigzag path is not a causal path and $C_{1,1}$ does not happen before $C_{3,2}$.

Another difference between a zigzag path and a causal path is that a zigzag path can form a cycle but a causal path never forms a cycle. That is, it is possible for a zigzag path to exist from a checkpoint back to itself, called a zigzag cycle. In contrast, causal paths can never form cycles. A zigzag path may form a cycle because a zigzag path need not represent causality – in a zigzag path, we allow a message to be sent before the previous message in the path is received as long as the send and receive are in the same interval. Figure 4.5 shows a zigzag cycle involving $C_{2,1}$, formed by messages m_1 and m_2.

Consistent global snapshots

Netzer and Xu [25] proved that if no zigzag path (or cycle) exists between any two checkpoints from a set S of checkpoints, then a consistent snapshot can be formed that includes the set S of checkpoints, and vice versa.

For a formal proof, the readers should consult the original paper. Here we give an intuitive explanation. Intuitively, if a zigzag path exists between two checkpoints, and that zigzag path is also a causal path, then the checkpoints are ordered and hence cannot belong to the same consistent snapshot. If the zigzag path between two checkpoints is not a causal path, a consistent snapshot cannot be formed that contains both the checkpoints. The zigzag nature of the path causes any snapshot that includes the two checkpoints to be inconsistent. To visualize the effect of a zigzag path, consider a snapshot

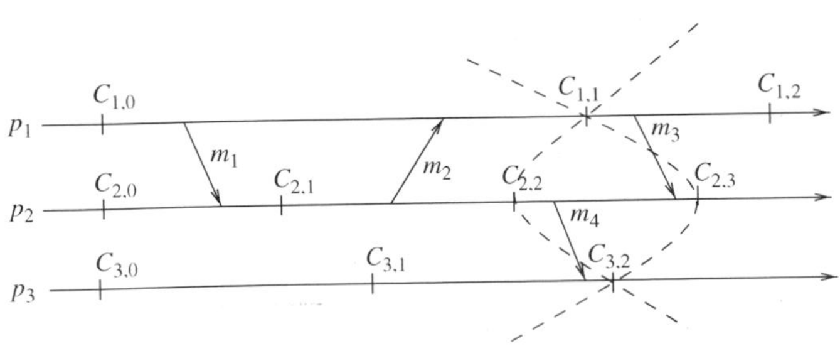

Figure 4.5 A zigzag cycle, inconsistent snapshot, and consistent snapshot.

line[2] through the two checkpoints. Because of the existance of a zigzag path between the two checkpoints, the snapshot line will always cross a message that causes one of the checkpoints to happen before the other, making the snapshot inconsistent. Figure 4.5 illustrates this. Two snapshot lines are drawn from $C_{1,1}$ to $C_{3,2}$. The zigzag path from $C_{1,1}$ to $C_{3,2}$ renders both the snapshot lines inconsistent. This is because messages m_3 and m_4 cross either snapshot line in a way that orders the two of its checkpoints.

Conversely, if no zigzag path (including zigzag cycles) exists between two checkpoints, then it is always possible to construct a consistent snapshot that includes these two checkpoints. We can form a consistent snapshot by including the first checkpoint at every process that has no zigzag path to either checkpoint. Note that messages can cross a consistent snapshot line as long as they do not cause any of the line's checkpoints to happen before each other. For example, in Figure 4.5, $C_{1,2}$ and $C_{2,3}$ can be grouped with $C_{3,1}$ to form a consistent snapshot even though message m_4 crosses the snapshot line.

To summarize:

- the absence of a causal path between checkpoints in a snapshot corresponds to the necessary condition for a consistent snapshot, and the absence of a zigzag path between checkpoints in a snapshot corresponds to the necessary and sufficient conditions for a consistent snapshot;
- a set of checkpoints S can be extended to a consistent snapshot if and only if no checkpoint in S has a zigzag path to any other checkpoint in S;
- a checkpoint can be a part of a consistent snapshot if and only if it is not invloved in a Z-cycle.

4.9 Finding consistent global snapshots in a distributed computation

We now address the problem to determine how individual local checkpoints can be combined with those from other processes to form global snapshots that are consistent. A solution to this problem forms the basis for many algorithms and protocols that must record consistent snapshots on-the-fly or determine post-mortem which global snapshots are consistent.

Netzer and Xu [25] proved the necessary and sufficient conditions to construct a consistent snapshot from a set of checkpoints S. However, they did not define the set of possible consistent snapshots and did not present an algorithm to construct them. Manivannan–Netzer–Singhal [21] analyzed the set of *all* consistent snapshots that can be built from a set of checkpoints S. They proved exactly which sets of local checkpoints from other processes

[2] A snapshot line is a line drawn through a set of checkpoints.

can be combined with those in S to form a consistent snapshot. They also developed an algorithm that enumerates all such consistent snapshots.

We define the following notations due to Wang [33,34].

definition 4.3 *Let A, B be individual checkpoints and R, S be sets of checkpoints. Let* $\rightsquigarrow$ *be a relation defined over checkpoints and sets of checkpoints such that*

1. $A \rightsquigarrow B$ *iff a Z-path exists from A to B;*
2. $A \rightsquigarrow S$ *iff a Z-path exists from A to* some *member of S;*
3. $S \rightsquigarrow A$ *iff a Z-path exists from* some *member of S to A;*
4. $R \rightsquigarrow S$ *iff a Z-path exists from* some *member of R to* some *member of S.*

$S \not\rightsquigarrow S$ defines that no Z-path (including a Z-cycle) exists from any member of S to any other member of S and implies that checkpoints in S are all from different processes.

Using the above notations, the results of Netzer and Xu can be expressed as follows:

Theorem 4.1 *A set of checkpoints S can be extended to a consistent global snapshot if and only if* $S \not\rightsquigarrow S$.

Corollary 4.1 *A checkpoint C can be part of a consistent global snapshot if and only if it is not involved in a Z-cycle.*

Corollary 4.2 *A set of checkpoints S is a consistent global snapshot if and only if* $S \not\rightsquigarrow S$ *and* $|S| = n$*, where n is the number of processes.*

4.9.1 Finding consistent global snapshots

We now discuss exactly which consistent snapshots can be built from a set of checkpoints S. We also present an algorithm to enumerate these consistent snapshots.

Extending S to a consistent snapshot

Given a set S of checkpoints such that $S \not\rightsquigarrow S$, we first discuss what checkpoints from other processes can be combined with S to build a consistent global snapshot. The result is based on the following three observations.

First observation

None of the checkpoints that have a Z-path to or from any of the checkpoints in S can be used. This is because from Theorem 4.1, no checkpoints between which a Z-path exists can ever be part of a consistent snapshot. Thus, only those checkpoints that have no Z-paths to or from any of the checkpoints in S are candidates for inclusion in the consistent snapshot. We call the set of all such candidates the *Z-cone* of S. Similarly, we call the set of all

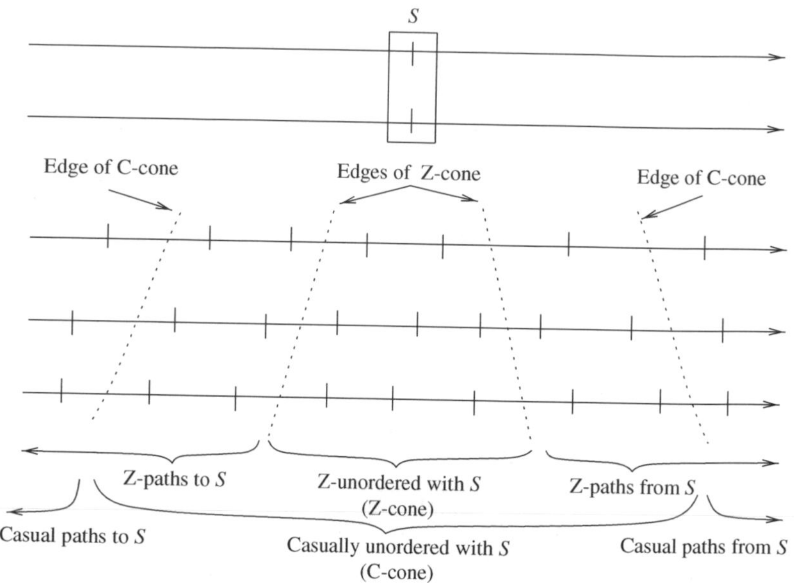

Figure 4.6 The Z-cone and the C-cone associated with a set of checkpoints S [21].

checkpoints that have no causal path to or from any checkpoint in S the *C-cone* of S.[3]

The Z-cone and C-cone help us reason about orderings and consistency. Since a causal path is always a Z-path, the Z-cone of S is a subset of the C-cone of S for an arbitrary S, as shown in Figure 4.6. Note that if a Z-path exists from checkpoint $C_{p,i}$ in process p_p to a checkpoint in S, then a Z-path also exists from every checkpoint in p_p preceding $C_{p,i}$ to the same checkpoint in S (because Z-paths are transitive). Likewise, if a Z-path exists from a checkpoint in S to a checkpoint $C_{q,j}$ in process p_q, then a Z-path also exists from the same checkpoint in S to every checkpoint in p_q following $C_{q,j}$. Causal paths are also transitive and similar results hold for them.

Second observation

Although candidates for building a consistent snapshot from S must lie in the Z-cone of S, not all checkpoints in the Z-cone can form a consistent snapshot with S. From Corollary 4.1, if a checkpoint in the Z-cone is involved in a Z-cycle, then it cannot be part of a consistent snapshot. Lemma 4.1 below states that if we remove from consideration all checkpoints in the Z-cone that are involved in Z-cycles, then each of the remaining checkpoints can be combined with S to build a consistent snapshot.

First we define the set of useful checkpoints with respect to set S.

[3] These terms are inspired by the so-called *light cone* of an event e, which is the set of all events with causal paths from e (i.e., events in e's future). Although the light cone of e contains events ordered *after* e, we define the Z-cone and C-cone of S to be those events with *no* zigzag or causal ordering, respectively, to or from any member of S.

definition 4.4 *Let S be a set of checkpoints such that $S \not\rightsquigarrow S$. Then, for each process p_q, the set S^q_{useful} is defined as*

$$S^q_{useful} = \{C_{q,i} \mid (S \not\rightsquigarrow C_{q,i}) \wedge (C_{q,i} \not\rightsquigarrow S) \wedge (C_{q,i} \not\rightsquigarrow C_{q,i})\}.$$

In addition, we define

$$S_{useful} = \bigcup_q S^q_{useful}.$$

Thus, with respect to set S, a checkpoint C is useful if C does not have a zigzag path to any checkpoint in S, no checkpoint in S has a zigzag path to C, and C is not on a Z-cycle.

Lemma 4.1 *Let S be a set of checkpoints such that $S \not\rightsquigarrow S$. Let $C_{q,i}$ be any checkpoint of process p_q such that $C_{q,i} \notin S$. Then $S \cup \{C_{q,i}\}$ can be extended to a consistent snapshot if and only if $C_{q,i} \in S_{useful}$.*

We omit the proof of the lemma and interested readers can refer to the original paper [21] for a proof.

Lemma 4.1 states that if we are given a set S such that $S \not\rightsquigarrow S$, we are guaranteed that any *single* checkpoint from S_{useful} can belong to a consistent global snapshot that also contains S.

Third observation

However, if we attempt to build a consistent snapshot from S by choosing a *subset* T of checkpoints from S_{useful} to combine with S, there is no guarantee that the checkpoints in T have no Z-paths between them. In other words, although none of the checkpoints in S_{useful} has a Z-path to or from any checkpoint in S, Z-paths may exist between members of S_{useful}. Therefore, we place one final constraint on the set T we choose from S_{useful} to build a consistent snapshot from S: checkpoints in T must have no Z-paths between them. Furthermore, since $S \not\rightsquigarrow S$, from Theorem 4.1, at least one such T must exist.

Theorem 4.2 *Let S be a set of checkpoints such that $S \not\rightsquigarrow S$ and let T be any set of checkpoints such that $S \cap T = \emptyset$. Then, $S \cup T$ is a consistent global snapshot if and only if*

1. $T \subseteq S_{useful}$;
2. $T \not\rightsquigarrow T$;
3. $|S \cup T| = n$.

We omit the proof of the theorem and interested readers can refer to the original paper [21] for a proof.

4.9.2 Manivannan–Netzer–Singhal algorithm for enumerating consistent snapshots

In the previous section, we showed which checkpoints can be used to extend a set of checkpoints S to a consistent snapshot. We now present an algorithm due to Manivannan–Netzer–Singhal [21] that explicitly computes all consistent snapshots that include a given set of checkpoints S. The algorithm restricts its selection of checkpoints to those within the Z-cone of S and it checks for the presence of Z-cycles within the Z-cone. In the next section, we discuss how to detect Z-cones and Z-paths using a graph by Wang [33,34],

```
(1)  ComputeAllCgs(S) {
(2)       let G = ∅
(3)       if S ↛ S then
(4)            let AllProcs be the set of all processes not represented in S
(5)            ComputeAllCgsFrom(S, AllProcs)
(6)       return G
(7)  }
(8)  ComputeAllCgsFrom(T, ProcSet) {
(9)       if (ProcSet = ∅) then
(10)           G = G ∪ T
(11)      else
(12)           let p_q be any process in ProcSet
(13)           for each checkpoint C ∈ T^q_useful do
(14)                ComputeAllCgsFrom(T ∪ {C}, ProcSet \ {p_q})
(15) }
```

Algorithm 4.2 Algorithm for computing all consistent snapshots containing S [21].

The algorithm is shown in Algorithm 4.2 and it computes all consistent snapshots that include a given set S. The function *ComputeAllCgs(S)* returns the set of all consistent checkpoints that contain S. The heart of the algorithm is the function *ComputeAllCgsFrom(T, ProcSet)* which extends a set of checkpoints T in all possible consistent ways, but uses checkpoints only from processes in the set *ProcSet*. After verifying that $S \not\rightsquigarrow S$, *ComputeAllCgs* calls *ComputeAllCgsFrom*, passing a *ProcSet* consisting of the processes not represented in S (lines 2–5). The resulting consistent snapshots are collected in the global variable G that is returned (line 6). It is worth noting that if $S = \emptyset$, the algorithm computes *all* consistent snapshots that exist in the execution.

The recursive function *ComputeAllCgsFrom(T, ProcSet)* works by choosing any process from *ProcSet*, say p_q, and iterating through all checkpoints C in T^q_{useful}. From Lemma 4.1, each such checkpoint extends T toward a consistent snapshot. This means $T \cup \{C\}$ can itself be further extended, eventually arriving at a consistent snapshot. Since this further extension is simply another instance of constructing all consistent snapshots that contain checkpoints from a given set, we make a recursive call (line 14), passing $T \cup \{C\}$ and a *ProcSet* from which process p_q is removed. The recursion eventually terminates when the passed set contains checkpoints from all processes (i.e., *ProcSet* is empty). In this case T is a global snapshot, as it contains one checkpoint from every process, and is added to G (line 10). When the algorithm terminates, all candidates in S_{useful} have been used in extending S, so G contains all consistent snapshots that contain S.

The following theorem argues the correctness of the algorithm.

Theorem 4.3 *Let S be a set of checkpoints and G be the set returned by ComputeAllCgs(S). If $S \not\rightsquigarrow S$, then $T \in G$ if and only if T is a consistent snapshot containing S. That is, G contains exactly the consistent snapshots that contain S.*

We omit the proof of the theorem and interested readers can refer to the original paper [21] for a proof.

4.9.3 Finding Z-paths in a distributed computation

Tracking Z-paths on-the-fly is difficult and remains an open problem. We describe a method for determining the existence of Z-paths between checkpoints in a distributed computation that has terminated or has stopped execution, using the rollback-dependency graph (R-graph) introduced by Wang [33,34]. First, we present the definition of an R-graph.

definition 4.5 *The rollback-dependency graph of a distributed computation is a directed graph $G = (V, E)$, where the vertices V are the checkpoints of the distributed computation, and an edge $(C_{p,i}, C_{q,j})$ from checkpoint $C_{p,i}$ to checkpoint $C_{q,j}$ belongs to E if*

1. *$p = q$ and $j = i + 1$, or*
2. *$p \neq q$ and a message m sent from the ith checkpoint interval of p_p is received by p_q in its jth checkpoint interval $(i, j > 0)$.*

Construction of an R-graph

When a process p_p sends a message m in its ith checkpoint interval, it piggybacks the pair (p, i) with the message. When the receiver p_q receives m in its jth checkpoint interval, it records the existence of an edge from $C_{p,i}$ to $C_{q,j}$. When a process wants to construct the R-graph for finding Z-paths between checkpoints, it broadcasts a request message to collect the existing direct dependencies from all other processes and constructs the complete R-graph. We assume that each process stops execution after it sends a reply to the request so that additional dependencies between checkpoints are not formed while the R-graph is being constructed. For each process, a volatile

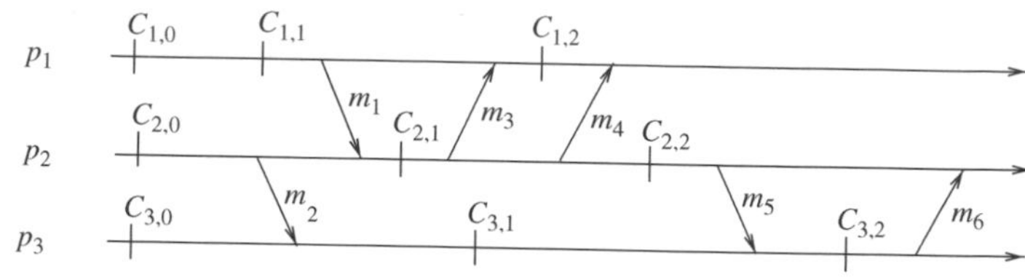

Figure 4.7 A distributed computation.

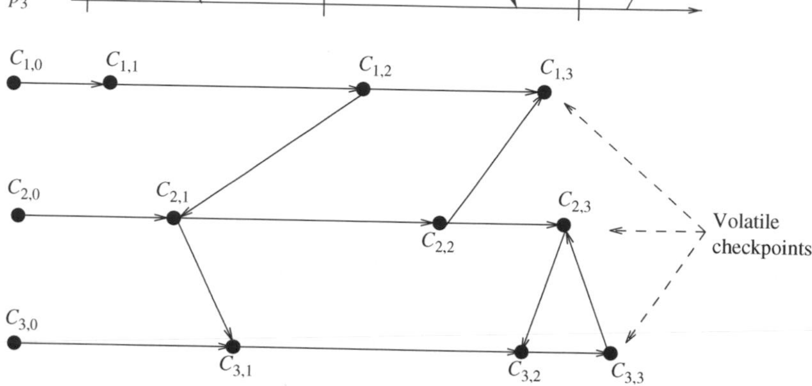

Figure 4.8 The R-graph of the computation in Figure 4.7.

checkpoint is added; the volatile checkpoint represents the volatile state of the process [33,34].

Example 4.1 An R-graph Figure 4.8 shows the R-graph of the computation shown in Figure 4.7. In Figure 4.8, $C_{1,3}$, $C_{2,3}$, and $C_{3,3}$ represent the volatile checkpoints, the checkpoints representing the last state the process attained before terminating.

We denote the fact that there is a path from C to D in the R-graph by $C \overset{\text{rd}}{\rightsquigarrow} D$. It only denotes the existence of a path; it does not specify any particular path. For example, in Figure 4.8, $C_{1,0} \overset{\text{rd}}{\rightsquigarrow} C_{3,2}$. When we need to specify a particular path, we give the sequence of checkpoints that constitute the path. For example, $(C_{1,0}, C_{1,1}, C_{1,2}, C_{2,1}, C_{3,1}, C_{3,2})$ is a path from $C_{1,0}$ to $C_{3,2}$ and $(C_{1,0}, C_{1,1}, C_{1,2}, C_{2,1}, C_{2,2}, C_{2,3}, C_{3,2})$ is also a path from $C_{1,0}$ to $C_{3,2}$.

The following theorem establishes the correspondence between the paths in the R-graph and the Z-paths between checkpoints. This correspondence is very useful in determining whether or not a Z-path exists between two given checkpoints.

Theorem 4.4 *Let $G = (V, E)$ be the R-graph of a distributed computation. Then, for any two checkpoints $C_{p,i}$ and $C_{q,j}$, $C_{p,i} \rightsquigarrow C_{q,j}$ if and only if*

1. $p = q$ *and* $i < j$, *or*
2. $C_{p,i+1} \overset{rd}{\rightsquigarrow} C_{q,j}$ *in* G *(note that in this case* p *could still be equal to* q*).*

For example, in the distributed computation shown in Figure 4.7, a zigzag path exists from $C_{1,1}$ to $C_{3,1}$ because in the corresponding R-graph, shown in Figure 4.8, $C_{1,2} \overset{\text{rd}}{\rightsquigarrow} C_{3,1}$. Likewise, $C_{2,1}$ is on a Z-cycle because in the corresponding R-graph, shown in Figure 4.8, $C_{2,2} \overset{\text{rd}}{\rightsquigarrow} C_{2,1}$.

4.10 Chapter summary

Recording global state of a distributed system is an important paradigm in the design of the distributed systems and the design of efficient methods of recording the global state is an important issue. Recording of global state of a distributed system is complicated due to the lack of both a globally shared memory and a global clock in a distributed system. This chapter first presented a formal definition of the global state of a distributed system and exposed issues related to its capture; it then described several algorithms to record a snapshot of a distributed system under various communication models.

Table 4.1 gives a comparison of the salient features of the various snapshot recording algorithms. Clearly, the higher the level of abstraction provided by a communication model, the simpler the snapshot algorithm. However, there is no best performing snapshot algorithm and an appropriate algorithm can be chosen based on the application's requirement. For examples, for termination detection, a snapshot algorithm that computes a channel state as the number of messages is adequate; for checkpointing for recovery from failures, an incremental snapshot algorithm is likely to be the most efficient; for global state monitoring, rather than recording and evaluating complete snapshots at regular intervals, it is more efficient to monitor changes to the variables that affect the predicate and evaluate the predicate only when some component variable changes.

As indicated in the introduction, the paradigm of global snapshots finds a large number of applications (such as detection of stable properties, checkpointing, monitoring, debugging, analyses of distributed computation, discarding of obsolete information). Moreover, in addition to the problems they solve, the algorithms presented in this chapter are of great importance to people interested in distributed computing as these algorithms illustrate the incidence of properties of communication channels (FIFO, non-FIFO, causal ordering) on the design of a class of distributed algorithms.

We also discussed the necessary and sufficient conditions for consistent snapshots. The non-causal path between checkpoints in a snapshot corresponds to the necessary condition for consistent snapshot, and the non-zigzag path corresponds to the necessary and sufficient conditions for consistent snapshot. Tracking of zigzag path is helpful in forming a global consistent snapshot. The avoidance of zigzag path between any pair of checkpoints from a collection of checkpoints (snapshot) is the necessary and sufficient conditions for a consistent global snapshot. Avoidance of causal paths alone will not be sufficient for consistency.

We also presented an algorithm for finding all consistent snapshots containing a given set S of local checkpoints; if we take $S = \emptyset$, then the algorithm gives the set of all consistent snapshots of a distributed computation. We established the correspondence between the Z-paths and the paths in the R-graph which helps in finding the existence of Z-paths between checkpoints.

4.11 Exercises

Exercise 4.1 Consider the following simple method to collect a global snapshot (it may not always collect a consistent global snapshot): an initiator process takes its snapshot and broadcasts a request to take snapshot. When some other process receives this request, it takes a snapshot. Channels are not FIFO.

Prove that such a collected distributed snapshot will be consistent iff the following holds (assume there are n processes in the system and Vt_i denotes the vector timestamp of the snapshot taken process p_i):

$$(Vt_1[1], Vt_2[2], \ldots., Vt_n[n]) = \max(Vt_1, Vt_2, \ldots., Vt_n).$$

Don't worry about channel states.

Exercise 4.2 What good is a distributed snapshot when the system was never in the state represented by the distributed snapshot? Give an application of distributed snapshots.

Exercise 4.3 Consider a distributed system where every node has its physical clock and all physical clocks are perfectly synchronized. Give an algorithm to record global state assuming the communication network is reliable. (Note that your algorithm should be simpler than the Chandy–Lamport algorithm.)

Exercise 4.4 What modifications should be done to the Chandy–Lamport snapshot algorithm so that it records a strongly consistent snapshot (i.e., all channel states are recorded empty).

Exercise 4.5 Consider two consistent cuts whose events are denoted by $C_1 = C_1(1), C_1(2), \ldots, C_1(n)$ and $C_2 = C_2(1), C_2(2), \ldots, C_2(n)$, respectively.

Define a third cut, $C_3 = C_3(1), C_3(2), \ldots, C_3(n)$, which is the maximum of C_1 and C_2; that is, for every k, $C_3(k) =$ later of C_1(k) and $C_2(k)$.

Define a fourth cut, $C_4 = C_4(1), C_4(2), \ldots, C_4(n)$, which is the minimum of C_1 and C_2; that is, for every k, $C_4(k) =$ earlierof C_1(k) and $C_2(k)$.

Prove that C_3 and C_4 are also consistent cuts.

4.12 Notes on references

The notion of a global state in a distributed system was formalized by Chandy and Lamport [7] who also proposed the first algorithm (CL) for recording the global state, and first studied the various properties of the recorded global state. The space–time diagram, which is a very useful graphical tool to visualize distributed executions, was introduced by Lamport [19]. A detailed survey of snapshot recording algorithms is given by Kshemkalyani *et al.* [16].

Spezialetti and Kearns proposed a variant of the CL algorithm to optimize concurrent initiations by different processes, and to efficiently distribute the recorded snapshot [29]. Venkatesan proposed a variant that handles repeated snapshots efficiently [32]. Helary proposed a variant of the CL algorithm to incorporate message waves in the algorithm [12]. Helary's algorithm is adaptable to a system with non-FIFO channels but requires inhibition [31]. Besides Helary's algorithm [12], the algorithms proposed by Lai and Yang [18], Li *et al.* [20], and by Mattern [23] can all record snapshots in systems with non-FIFO channels. If the underlying network can provide causal order of message delivery [5], then the algorithms by Acharya and Badrinath [1] and by Alagar and Venkatesan [2] can record the global state using $O(n)$ number of messages.

The notion of simultaneous regions for monitoring global state was proposed by Spezialetti and Kearns [30]. The necessary and sufficient conditions for consistent global snapshots were formulated by Netzer and Xu [25] based on the zigzag paths. These have particular application in checkpointing and recovery. Manivannan *et al.* analyzed the set of all consistent snaspshots that can be built from a given set of checkpoints [21]. They also proposed an algorithm to enumerate all such consistent snapshots. The definition of the *R-graph* and other notations and framework used by [21] were proposed by Wang [33,34].

Recording the global state of a distributed system finds applications at several places in distributed systems. For applications in detection of stable properties such as deadlocks, see [17] and for termination, see [22]. For failure recovery, a global state of the distributed system is periodically saved and recovery from a processor failure is done by restoring the system to the last saved global state [15]. For debugging distributed software, the system is restored to a consistent global state [8,9] and the execution resumes from there in a controlled manner. A snapshot recording method has been used in the distributed debugging facility of Estelle [11,13], a distributed programming environment. Other applications include monitoring distributed events [30], setting distributed breakpoints [24], protocol specification and verification [4,10,14], and discarding obsolete information [11].

We will study snapshot algorithms for shared memory in Chapter 12.

References

[1] A. Acharya and B. R. Badrinath, Recording distributed snapshots based on causal order of message delivery, *Information Processing Letters*, **44**, 1992, 317–321.

[2] S. Alagar, and S. Venkatesan, An optimal algorithm for distributed snapshots with causal message ordering, *Information Processing Letters*, **50**, 1994, 311–316.

[3] O. Babaoglu and K. Marzullo, Consistent global states of distributed systems: fundamental concepts and mechanisms, in Mullender, S.J. (ed.) *Distributed Systems*, ACM Press 1993.

[4] O. Babaoglu and M. Raynal, Specification and verification of dynamic properties in distributed computations, *Journal of Parallel and Distributed Systems*, **28**(2), 1995, 173–185.

[5] K. Birman and T. Joseph, Reliable communication in presence of failures, *ACM Transactions on Computer Systems*, **3**, 1987, 47–76.

[6] K. Birman, A. Schiper, and P. Stephenson, Lightweight causal and atomic group multicast, *ACM Transactions on Computer Systems*, **9**(3), 1991, 272–314.

[7] K. M. Chandy and L. Lamport, Distributed snapshots: determining global states of distributed systems, *ACM Transactions on Computer Systems*, **3**(1), 1985, 63–75.

[8] R. Cooper and K. Marzullo, Consistent detection of global predicates, *Proceedings of the ACM/ONR Workshop on Parallel and Distributed Debugging*, May 1991, 163–173.

[9] E. Fromentin, N. Plouzeau, and M. Raynal, An introduction to the analysis and debug of distributed computations, *Proceedings of the 1st IEEE International Conference on Algorithms and Architectures for Parallel Processing*, Brisbane, Australia, April 1995, 545–554.

[10] K. Geihs and M. Seifert, Automated validation of a cooperation protocol for distributed systems, *Proceedings of the 6th International Conference on Distributed Computing Systems*, 1986, 436–443.

[11] O. Gerstel, M. Hurfin, N. Plouzeau, M. Raynal, and S. Zaks, On-the-fly replay: a practical paradigm and its implementation for distributed debugging, *Proceedings of the 6th IEEE International Symposium on Parallel and Distributed Debugging*, Dallas, TX, October 1995, 266–272.

[12] J.-M. Helary, Observing global states of asynchronous distributed applications, *Proceedings of the 3rd International Workshop on Distributed Algorithms*, LNCS 392 1989, 124–134.

[13] M. Hurfin, N. Plouzeau and M. Raynal, A debugging tool for distributed Estelle programs, *Journal of Computer Communications*, **16**(5), 1993, 328–333.

[14] J. Kamal and M. Singhal, *Specification and Verification of Distributed Mutual Exclusion Algorithms*, Technical Report, Department of Computer and Information Science, The Ohio State University, Columbus, OH, 1992.

[15] R. Koo and S. Toueg, Checkpointing and rollback-recovery in distributed systems, *IEEE Transactions on Software Engineering*, January, 1987, 23–31.

[16] A. Kshemkalyani, M. Raynal, and M. Singhal, An introduction to global snapshots of a distributed system, *Distributed Systems Engineering Journal*, **2**(4), 1995, 224–233.

[17] A. Kshemkalyani and M. Singhal, Efficient detection and resolution of generalized distributed deadlocks, *IEEE Transactions on Software Engineering*, **20**(1), 1994, 43–54.

[18] T. H. Lai and T. H. Yang, On distributed snapshots, *Information Processing Letters*, **25**, 1987, 153–158.

[19] L. Lamport, Time, clocks, and the ordering of events in a distributed system, *Communications of the ACM*, **21**(7), 1978, 558–565.

[20] H. F. Li, T. Radhakrishnan, and K. Venkatesh, Global state detection in non-FIFO networks, *Proceedings of the 7th International Conference on Distributed Computing Systems*, 1987, 364–370.

[21] D. Manivannan, R. H. B. Netzer, and M. Singhal, Finding consistent global checkpoints in a distributed computation, *IEEE Transactions of Parallel and Distributed Systems*, June, 1997, 623–627.

[22] F. Mattern, Algorithms for distributed termination detection, *Distributed Computing*, **2**(3), 1987, 161–175.

[23] F. Mattern, Efficient algorithms for distributed snapshots and global virtual time approximation, *Journal of Parallel and Distributed Computing*, **18**, 1993, 423–434.

[24] B. Miller and J. Choi, Breakpoints and halting in distributed programs, *Proceedings of the 8th International Conference on Distributed Computing Systems*, 1988, 316–323.

[25] H. B. R. Netzer and J. Xu, Necessary and sufficient conditions for consistent global snapshots, *IEEE Transactions on Parallel and Distributed Systems*, **6**(2), 1995, 165–169.

[26] M. Raynal, A. Schiper, and S. Toueg, Causal ordering abstraction and a simple way to implement it, *Information Processing Letters*, **39**(6), 1991, 343–350.

[27] S. Sarin and N. Lynch, Discarding obsolete information in a replicated database system, *IEEE Transactions on Software Engineering*, **13**(1), 1987, 39–47.

[28] A. Schiper, J. Eggli, and A. Sandoz, A new algorithm to implement causal ordering, *Proceedings of the 3rd International Workshop on Distributed Algorithms*, LNCS 392, Springer Verlag, 1989, pp. 219–232.

[29] M. Spezialetti and P. Kearns, Efficient distributed snapshots, *Proceedings of the 6th International Conference on Distributed Computing Systems*, 1986, 382–388.

[30] M. Spezialetti and P. Kearns, Simultaneous regions: a framework for the consistent monitoring of distributed systems, *Proceedings of the 9th International Conference on Distributed Computing Systems*, 1989, 61–68.

[31] K. Taylor, The role of inhibition in consistent cut protocols, *Proceedings of the 3rd International Workshop on Distributed Algorithms*, LNCS 392, 1989, 124–134.

[32] S. Venkatesan, Message-optimal incremental snapshots, *Journal of Computer and Software Engineering*, **1**(3), 1993, 211–231.

[33] Y. -M. Wang, Maximum and minimum consistent global checkpoints and their applications, *Proceedings of the* 14*th IEEE Symposium on Reliable Distributed Systems*, Bad Neuenahr, Germany, September 1995, 86–95.

[34] Y. -M. Wang, Consistent global checkpoints that contain a given set of local checkpoints, *IEEE Transactions on Computers*, **46**(4), 1997, 456–468.

CHAPTER

5 Terminology and basic algorithms

In this chapter, we first study a methodical framework in which distributed algorithms can be classified and analyzed. We then consider some basic distributed graph algorithms. We then study *synchronizers*, which provide the abstraction of a synchronous system over an asynchronous system. Finally, we look at some practical graph problems, to appreciate the necessity of designing efficient distributed algorithms.

5.1 Topology abstraction and overlays

The topology of a distributed system can be typically viewed as an undirected graph in which the nodes represent the processors and the edges represent the links connecting the processors. Weights on the edges can represent some cost function we need to model in the application. There are usually three (not necessarily distinct) levels of topology abstraction that are useful in analyzing the distributed system or a distributed application. These are now described using Figure 5.1. To keep the figure simple, only the relevant end hosts participating in the application are shown. The WANs are indicated by ovals drawn using dashed lines. The switching elements inside the WANs, and other end hosts that are not participating in the application, are not shown even though they belong to the physical topological view. Similarly, all the edges connecting all end hosts and all edges connecting to all the switching elements inside the WANs also belong to the physical topology view even though only some edges are shown.

- **Physical topology** The nodes of this topology represent all the network nodes, including switching elements (also called routers), in the WAN and all the end hosts – irrespective of whether the hosts are participating in the application. The edges in this topology represent all the communication links in the WAN in addition to all the direct links between the end hosts.

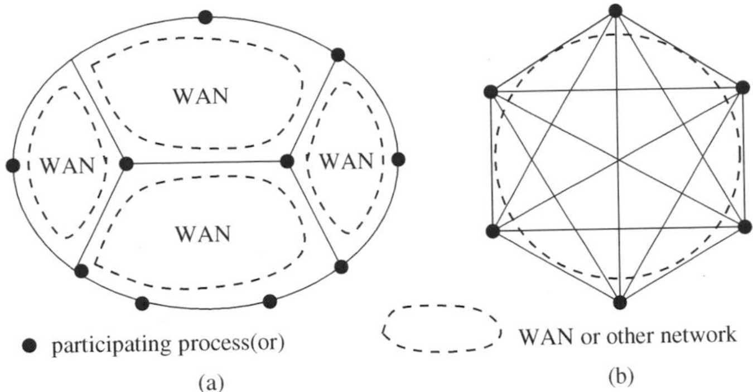

Figure 5.1 Two examples of topological views at different levels of abstraction.

In Figure 5.1(a), the physical topology is not shown explicitly to keep the figure simple.

- **Logical topology** This is usually defined in the context of a particular application. The nodes represent all the end hosts where the application executes. The edges in this topology are logical channels (also termed as logical links) among these nodes. This view is at a higher level of abstraction than that of the physical topology, and the nodes and edges of the physical topology need not be included in this view.

  Often, logical links are modeled between particular pairs of end hosts participating in an application to give a logical topology with useful properties. Figure 5.1(b) shows each pair of nodes in the logical topology is connected to give a fully connected network. Each pair of nodes can communicate directly with each other participant in the application using an incident logical link at this level of abstraction of the topology. However, the logical links may also define some arbitrary connectivity (neighborhood-relation) on the nodes in this abstract view. In Figure 5.1(a), the logical view provides each node with a partial view of the topology, and the connectivity provided is some neighborhood connectivity. To communicate with another application node that is not a logical neighbor, a node may have to use a multi-hop path composed of logical links at this level of abstraction of the topology.

  While the fully connected logical topology in Figure 5.1(b) provides a complete view of the system, updating such a view in a dynamic system incurs an overhead. Neighborhood-based logical topologies as in Figure 5.1(a) are easier to manage.

  We will consider distributed algorithms on logical topologies in this book. Peer-to-peer (P2P) networks (see Chapter 18) are also defined by a logical topology at the application layer. However, the emphasis of P2P networks is on self-organizing networks with built-in functions, e.g., the implementation of application layer functions such as object lookup and location in a distributed manner.

- **Superimposed topology** This is a higher-level topology that is superimposed on the logical topology. It is usually a regular structure such as a tree, ring, mesh, or hypercube. The main reason behind defining such a topology is that it provides a specialized path for efficient information dissemination and/or gathering as part of a distributed algorithm.

  Consider the problem of collecting the sum of variables, one from each node. This can be efficiently solved using n messages by circulating a cumulative counter on a logical ring, or using $n-1$ messages on a logical tree. The ring and tree are examples of superimposed topologies on the underlying logical topology – which may be arbitrary as in Figure 5.1(a) or fully connected as in Figure 5.1(b).

We will encounter various examples of these topologies, A *superimposed topology* is also termed as a *topology overlay*. This latter term is becoming increasingly popular with the spread of the peer-to-peer computing paradigm.

Notation

Whatever the level of topological view we are dealing with, we assume that an undirected graph (N, L) is used to represent the topology. The notation $n = |N|$ and $l = |L|$ will also be used.

5.2 Classifications and basic concepts

5.2.1 Application executions and control algorithm executions

The distributed *application execution* is comprised of the execution of instructions, including the communication instructions, within the distributed application program. The application execution represents the logic of the application. In many cases, a *control algorithm* also needs to be executed in order to monitor the application execution or to perform various auxiliary functions. The control algorithm performs functions such as: creating a spanning tree, creating a connected dominating set, achieving consensus among the nodes, distributed transaction commit, distributed deadlock detection, global predicate detection, termination detection, global state recording, checkpointing, and also memory consistency enforcement in distributed shared memory systems.

The code of the control algorithm is allocated its own memory space. The *control algorithm execution* is *superimposed* on the underlying application execution, but does not interfere with the application execution. In other words, the control algorithm execution including all its send, receive, and internal events are transparent to (or not visible to) the application execution.

The distributed *control algorithm* is also sometimes termed as a *protocol*; although the term *protocol* is also loosely used for any distributed algorithm.

In the literature on formal modeling of network algorithms, the term *protocol* is more commonly used.

5.2.2 Centralized and distributed algorithms

In a distributed system, a *centralized algorithm* is one in which a predominant amount of work is performed by one (or possibly a few) processors, whereas other processors play a relatively smaller role in accomplishing the joint task. The roles of the other processors are usually confined to requesting information or supplying information, either periodically or when queried.

A typical system configuration suited for centralized algorithms is the *client–server* configuration. Presently, much commercial software is written using this configuration, and is adequate. From a theoretical perspective, the single server is a potential bottleneck for both processing and bandwidth access on the links. The single server is also a single point of failure. Of course, these problems are alleviated in practice by using replicated servers distributed across the system, and then the overall configuration is not as centralized any more.

A *distributed algorithm* is one in which each processor plays an equal role in sharing the message overhead, time overhead, and space overhead. It is difficult to design a purely distributed algorithm (that is also efficient) for some applications. Consider the problem of recording a global state of all the nodes. The well-known Chandy–Lamport algorithm which we studied in Chapter 4 is distributed – yet one node, which is typically the initiator, is responsible for assembling the local states of the other nodes, and hence plays a slightly different role. Algorithms that are designed to run on a logical-ring superimposed topology tend to be fully distributed to exploit the symmetry in the connectivity. Algorithms that are designed to run on the logical tree and other asymmetric topologies with a predesignated root node tend to have some asymmetry that mirrors the asymmetric topology. Although fully distributed algorithms are ideal, partly distributed algorithms are sometimes more practical to implement in real systems. At any rate, the advances in peer-to-peer networks, ubiquitous and ad-hoc networks, and mobile systems will require distributed solutions.

5.2.3 Symmetric and asymmetric algorithms

A *symmetric algorithm* is an algorithm in which all the processors execute the same logical functions. An *asymmetric algorithm* is an algorithm in which different processors execute logically different (but perhaps partly overlapping) functions.

A centralized algorithm is always asymmetric. An algorithm that is not fully distributed is also asymmetric. In the client–server configuration, the clients and the server execute asymmetric algorithms. Similarly, in a tree configuration, the root and the leaves usually perform some functions that are different from each other, and that are different from the functions of the internal nodes of the tree. Applications where there is inherent asymmetry in the roles of the cooperating processors will necessarily have asymmetric algorithms. A typical example is where one processor initiates the computation of some global function (e.g., min, sum).

5.2.4 Anonymous algorithms

An *anonymous system* is a system in which neither processes nor processors use their process identifiers and processor identifiers to make any execution decisions in the distributed algorithm. An *anonymous algorithm* is an algorithm which runs on an anonymous system and therefore does not use process identifiers or processor identifiers in the code.

An anonymous algorithm possesses structural elegance. However, it is equally hard, and sometimes provably impossible, to design – as in the case of designing an anonymous leader election algorithm on a ring [1]. If we examine familiar examples of multiprocess algorithms, such as the famous Bakery algorithm for mutual exclusion in a shared memory system, or the "wait-wound" or "wound-die" algorithms used for transaction serializability in databases, we observe that the process identifier is used in resolving ties or contentions that are otherwise unresolved despite the symmetric and noncentralized nature of the algorithms.

5.2.5 Uniform algorithms

A *uniform algorithm* is an algorithm that does not use n, the number of processes in the system, as a parameter in its code. A uniform algorithm is desirable because it allows scalability transparency, and processes can join or leave the distributed execution without intruding on the other processes, except its immediate neighbors that need to be aware of any changes in their immediate topology. Algorithms that run on a logical ring and have nodes communicate only with their neighbors are uniform. In Section 5.10, we will study a uniform algorithm for leader election.

5.2.6 Adaptive algorithms

Consider the context of a problem X. In a system with n nodes, let k, $k \leq n$ be the number of nodes "participating" in the context of X when the algorithm to solve X is executed. If the complexity of the algorithm can be expressed in terms of k rather than in terms of n, the algorithm is *adaptive*. For example, if the complexity of a mutual exclusion algorithm can be expressed in terms of the actual number of nodes contending for the critical section when the algorithm is executed, then the algorithm would be adaptive.

5.2.7 Deterministic versus non-deterministic executions

A *deterministic receive* primitive specifies the source from which it wants to receive a message. A *non-deterministic receive* primitive can receive a message from any source – the message delivered to the process is the first message that is queued in the local incoming buffer, or the first message that comes in subsequently if no message is queued in the local incoming buffer. A distributed program that contains no non-deterministic receives has a *deterministic execution*; otherwise, if it contains at least one non-deterministic receive primitive, it is said to have a *non-deterministic execution*.

Each execution defines a partial order on the events in the execution. Even in an asynchronous system (defined formally in Section 5.2.9), for any deterministic (asynchronous) execution, repeated re-execution will reproduce the same partial order on the events. This is a very useful property for applications such as debugging, detection of unstable predicates, and for reasoning about global states.

Given any non-deterministic execution, any re-execution of that program may result in a very different outcome, and any assertion about a non-deterministic execution can be made only for that particular execution. Different re-executions may result in different partial orders because of variable factors such as (i) lack of an upper bound on message delivery times and unpredictable congestion; and (ii) local scheduling delays on the CPUs due to timesharing. As such, non-deterministic executions are difficult to reason with.

5.2.8 Execution inhibition

Blocking communication primitives freeze the local execution[1] until some actions connected with the completion of that communication primitive have occurred. But from a logical perspective, is the process really prevented from executing further? The non-blocking flavors of those primitives can be used to eliminate the freezing of the execution, and the process invoking that primitive may be able to execute further (from the perspective of the program logic) until it reaches a stage in the program logic where it cannot execute further until the communication operation has completed. Only now is the process really frozen.

Distributed applications can be analyzed for freezing. Often, it is more interesting to examine the control algorithm for its freezing/inhibitory effect on the application execution. Here, inhibition refers to protocols delaying actions of the underlying system execution for an interval of time. In the literature on inhibition, the term "*protocol*" is used synonymously with the term "*control algorithm*." Protocols that require processors to suspend their

[1] The OS dispatchable entity – the process or the thread – is frozen.

normal execution until some series of actions stipulated by the protocol have been performed are termed as *inhibitory* or *freezing protocols* [10].

Different executions of a distributed algorithm can result in different interleavings of the events. Thus, there are multiple executions associated with each algorithm (or protocol). Protocols can be classified as follows, in terms of inhibition:

- A protocol is *non-inhibitory* if no system event is disabled in any execution of the protocol. Otherwise, the protocol is *inhibitory*.
- A disabled event e in an execution is said to be *locally delayed* if there is *some* extension of the execution (beyond the current state) such that: (i) the event becomes enabled after the extension; and (ii) there is no intervening receive event in the extension, Thus, the interval of inhibition is under local control. A protocol is *locally inhibitory* if any event disabled in any execution of the protocol is locally delayed.
- An inhibitory protocol for which there is some execution in which some delayed event is not locally delayed is said to be *globally inhibitory*. Thus, in some (or all) execution of a globally inhibitory protocol, at least one event is delayed waiting to receive communication from another processor.

An orthogonal classification is that of *send inhibition*, *receive inhibition*, and *internal event inhibition*:

- A protocol is *send inhibitory* if some delayed events are send events.
- A protocol is *receive inhibitory* if some delayed events are receive events.
- A protocol is *internal event inhibitory* if some delayed events are internal events.

These classifications help to characterize the degree of inhibition necessary to design protocols to solve various problems. Problems can be theoretically analyzed in terms of the possibility or impossibility of designing protocols to solve them under the various classes of inhibition. These classifications also serve as a yardstick to evaluate protocols. The more stringent the class of inhibition, the less desirable is the protocol. In the study of algorithms for recording global states and algorithms for checkpointing, we have the opportunity to analyze the protocols in terms of inhibition.

5.2.9 Synchronous and asynchronous systems

A *synchronous system* is a system that satisfies the following properties:

- There is a known upper bound on the message communication delay.
- There is a known bounded drift rate for the local clock of each processor with respect to real-time. The drift rate between two clocks is defined as the rate at which their values diverge.
- There is a known upper bound on the time taken by a process to execute a logical step in the execution.

An *asynchronous system* is a system in which none of the above three properties of synchronous systems are satisfied. Clearly, systems can be designed that satisfy some combination but not all of the criteria that define a synchronous system. The algorithms to solve any particular problem can vary drastically, based on the model assumptions; hence it is important to clearly identify the system model beforehand. Distributed systems are inherently asynchronous; later in this chapter, we will study synchronizers that provide the abstraction of a synchronous execution.

5.2.10 Online versus offline algorithms

An *on-line* algorithm is an algorithm that executes as the data is being generated. An *off-line* algorithm is an algorithm that requires all the data to be available before algorithm execution begins. Clearly, on-line algorithms are more desirable. Debugging and scheduling are two example areas where on-line algorithms offer clear advantages. On-line scheduling allows for dynamic changes to the schedule to account for newly arrived requests with closer deadlines. On-line debugging can detect errors when they occur, as opposed to collecting the entire trace of the execution and then examining it for errors.

5.2.11 Failure models

A failure model specifies the manner in which the component(s) of the system may fail. There exists a rich class of well-studied failure models. It is important to specify the failure model clearly because the algorithm used to solve any particular problem can vary dramatically, depending on the failure model assumed. A system is *t-fault tolerant* if it continues to satisfy its specified behavior as long as no more than t of its components (whether processes or links or a combination of them) fail. The *mean time between failures (MTBF)* is usually used to specify the expected time until failure, based on statistical analysis of the component/system.

Process failure models [26]

- **Fail-stop** [31] In this model, a properly functioning process may fail by stopping execution from some instant thenceforth. Additionally, other processes can learn that the process has failed. This model provides an abstraction – the exact mechanism by which other processes learn of the failure can vary.
- **Crash** [21] In this model, a properly functioning process may fail by stopping to function from any instance thenceforth. Unlike the fail-stop model, other processes do not learn of this crash.
- **Receive omission** [27] A properly functioning process may fail by intermittently receiving only some of the messages sent to it, or by crashing.

- **Send omission** [16] A properly functioning process may fail by intermittently sending only some of the messages it is supposed to send, or by crashing.
- **General omission** [27] A properly functioning process may fail by exhibiting either or both of send omission and receive omission failures.
- **Byzantine or malicious failure, with authentication** [22] In this model, a process may exhibit any arbitrary behavior. However, if a faulty process claims to have received a specific message from a correct process, then that claim can be verified using authentication, based on unforgeable signatures.
- **Byzantine or malicious failure** [22] In this model, a process may exhibit any arbitrary behavior and no authentication techniques are applicable to verify any claims made.

The above process failure models, listed in order of increasing severity (except for send omissions and receive omissions, which are incomparable with each other), apply to both synchronous and asynchronous systems.

Timing failures can occur in synchronous systems, and manifest themselves as some or all of the following at each process: (i) general omission failures; (ii) process clocks violating their prespecified drift rate; (iii) the process violating the bounds on the time taken for a step of execution. In term of severity, timing failures are more severe than general omission failures but less severe than Byzantine failures with message authentication.

The failure models less severe than Byzantine failures, and timing failures, are considered "benign" because they do not allow processes to arbitrarily change state or send messages that are not to be sent as per the algorithm. Benign failures are easier to handle than Byzantine failures.

Communication failure models

- **Crash failure** A properly functioning link may stop carrying messages from some instant thenceforth.
- **Omission failures** A link carries some messages but not the others sent on it.
- **Byzantine failures** A link can exhibit any arbitrary behavior, including creating spurious messages and modifying the messages sent on it.

The above link failure models apply to both synchronous and asynchronous systems. *Timing failures* can occur in synchronous systems, and manifest themselves as links transporting messages faster or slower than their specified behavior.

5.2.12 Wait-free algorithms

A *wait-free algorithm* is an algorithm that can execute (synchronization operations) in an $(n-1)$-process fault tolerant manner, i.e., it is resilient to

$n-1$ process failures [18,20]. Thus, if an algorithm is wait-free, then the (synchronization) operations of any process must complete in a bounded number of steps irrespective of the failures of all the other processes.

Although the concept of a k-fault-tolerant system is very old, wait-free algorithm design in distributed computing received attention in the context of mutual exclusion synchronization for the distributed shared memory abstraction. The objective was to enable a process to access its critical section, even if the process in the critical section fails or misbehaves by not exiting from the critical section. Wait-free algorithms offer a very high degree of robustness. Designing a wait-free algorithm is usually very expensive and may not even be possible for some synchronization problems, e.g., the simple producer–consumer problem. Wait-free algorithms will be studied in Chapters 12 and 14. Wait-free algorithms can be viewed as a special class of fault-tolerant algorithms.

5.2.13 Communication channels

Communication channels are normally first-in first-out queues (FIFO). At the network layer, this property may not be satisfied, giving non-FIFO channels. These and other properties such as causal order of messages will be studied in Chapter 6.

5.3 Complexity measures and metrics

The performance of sequential algorithms is measured using the time and space complexity in terms of the lower bounds (Ω, ω) representing the best case, the upper bounds (O, o) representing the worst case, and the exact bound (θ). For distributed algorithms, the definitions of space and time complexity need to be refined, and additionally, message complexity also needs to be considered for message-passing systems. At the appropriate level of abstraction at which the algorithm is run, the system topology is usually assumed to be an undirected unweighted graph $G = (N, L)$. We denote $|N|$ as n, $|L|$ as l, and the diameter of the graph as d. The *diameter* of a graph is the minimum number of edges that need to be traversed to go from any node to any other node. More formally, the diameter is $max_{i,j \in N}\{$length of the shortest path between i and $j\}$. For a tree embedded in the graph, its depth is denoted as h. Other graph parameters, such as eccentricity and degree of edge incidence, can be used when they are required. It is also assumed that identical code runs at each processor; if this assumption is not valid, then different complexities need to be stated for the different codes. The complexity measures are as follows:

- **Space complexity per node** This is the memory requirement at a node. The best case, average case, and worst case memory requirement at a node can be specified.

- **Systemwide space complexity** The system space complexity (best case, average case, or worst case) is not necessarily n times the corresponding space complexity (best case, average case, or worst case) per node. For example, the algorithm may not permit all nodes to achieve the best case at the same time. We will later study a distributed predicate detection algorithm (Algorithm 11.6 in Chapter 11) for which both the worst case space complexity per node as well as the worst case systemwide space complexity are proportional to $O(n^2)$. If during execution, the worst case occurs at one node, then the worst case will not occur at all the other nodes in that execution.
- **Time complexity per node** This measures the processing time per node, and does not explicitly account for the message propagation/transmission times, which are measured as a separate metric.
- **Systemwide time complexity** If the processing in the distributed system occurs at all the processors concurrently, then the system time complexity is not n times the time complexity per node. However, if the executions by the different processes are done serially, as in the case of an algorithm in which only the unique token-holder is allowed to execute, then the overall time complexity is additive.
- **Message complexity** This has two components – a space component and a time component.
  - **Number of messages** The number of messages contributes directly to the space complexity of the message overhead.
  - **Size of messages** This size, in conjunction with the number of messages, measures the space component on messages. Further, for very large messages, this also contributes to the time component via the increased transmission time.
  - **Message time complexity** The number of messages contributes to the time component indirectly, besides affecting the count of the send events and message space overhead. Depending on the degree of concurrency in the sending of the messages – i.e., whether all messages are sequentially sent (with reference to the execution partial order), or all processes can send concurrently, or something in between – the time complexity is affected. For asynchronous executions, the time complexity component is measured in terms of *sequential message hops*, i.e., the length of the longest chain in the partial order $(E, \prec)$ on the events. For synchronous executions, the time complexity component is measured in terms of rounds (also termed as steps or phases).

It is usually difficult to determine all of the above complexities for most algorithms. Nevertheless, it is important to be aware of the different factors that contribute towards the overhead. When stating the complexities, it should also be specified whether the algorithm has a synchronous or asynchronous execution. Depending on the algorithm, further metrics such as the number of send events, or the number of receive events, may be of interest. If message multicast is allowed, it should be stated whether a multicast send event is counted as a single event. Also, whether the message multicast is counted as a single message or as multiple messages needs to be clarified. This would depend on whether or not hardware multicasting is used by the lower layers of the network protocol stack.

For shared memory systems, the message complexity is not an issue if the shared memory is not being provided by the distributed shared memory abstraction over a message-passing system. The following additional changes in the emphasis on the usual complexity measures would need to be considered:

- The size of shared memory, as opposed to the size of local memory, is important. The justification is that shared memory is expensive, local memory is not.
- The number of synchronization operations using synchronization variables is a useful metric because it affects the time complexity.

5.4 Program structure

Hoare, who pioneered programming language support for concurrent processes, designed concurrent sequential processes (CSP), which allows communicating processes to synchronize efficiently. The typical program structure for any process in a distributed application is based on CSP's repetitive command over the alternative command on multiple guarded commands, and is as follows:

$$*[\, G_1 \longrightarrow CL_1 \,||\, G_2 \longrightarrow CL_2 \,||\, \cdots \,||\, G_k \longrightarrow CL_k \,].$$

The *repetitive* command (denoted by "*") denotes an infinite loop. Inside the *repetitive* command is the *alternative* command over *guarded* commands. The *alternative* command, denoted by a sequence of "||" separating guarded commands, specifies execution of exactly one of its constituent guarded commands. The *guarded* command has the syntax "$G \longrightarrow CL$" where the guard G is a boolean expression and CL is a list of commands that are only executed if G is true. The guard expression may contain a term to check if a message from a/any other process has arrived. The alternative command over the guarded commands fails if all the guards fail; if more than one guard is true, one of those successful guarded commands is nondeterministically chosen for execution. When a *guarded* command $G_m \longrightarrow CL_m$ does get executed, the execution of CL_m is atomic with the execution of G_m.

The structure of distributed programs has similar semantics to that of CSP although the syntax has evolved to something very different. The format for the pseudo-code used in this book is as indicated below. Algorithm 5.2 serves to illustrate this format.

1. The process-local variables whose scope is global to the process, and message types, are declared first.
2. Shared variables, if any, (for distributed shared memory systems) are explicitly labeled as such.
3. This is followed by any initialization code.
4. The *repetitive* and the *alternative* commands are not explicitly shown.
5. The *guarded* commands are shown as explicit modules or procedures (e.g., lines 1–4 in Algorithm 5.2). The guard usually checks for the arrival of a message of a certain type, perhaps with additional conditions on some parameter values and other local variables.
6. The body of the procedure gives the list of commands to be executed if the guard evaluates to *true*.
7. Process termination may be explicitly stated in the body of any procedure(s).
8. The symbol $\perp$ is used to denote an undefined value. When used in a comparison, its value is $-\infty$.

5.5 Elementary graph algorithms

This section examines elementary distributed algorithms on graphs. The reader is assumed to be familiar with the centralized algorithms to solve these basic graph problems. The distributed algorithms here introduce the reader to the difficulty of designing distributed algorithms wherein each node has only a partial view of the graph (system), which is confined to its immediate neighbors. Further, a node can communicate with only its immediate neighbors along the incident edges. Unless otherwise specified, we assume unweighted undirected edges, and asynchronous execution by the processors. Communication is by message-passing on the edges.

The first algorithm is a synchronous spanning tree algorithm. The next three are asynchronous algorithms to construct spanning trees. These elementary algorithms are theoretically important from a practical perspective because spanning trees are a very efficient form of information distribution and collection in distributed systems.

5.5.1 Synchronous single-initiator spanning tree algorithm using flooding

The code for all processes is not only symmetrical, but also proceeds in rounds. This algorithm assumes a designated root node, *root*, which initiates the algorithm. The pseudo-code for each process P_i is shown in Algorithm 5.1. The root initiates a flooding of QUERY messages in the graph to identify tree edges. The parent of a node is that node from which a QUERY is first received; if multiple QUERYs are received in the same round, one of the senders is randomly chosen as the parent. Exercise 5.1 asks you to modify

```
(local variables)
int visited, depth ⟵ 0
int parent ⟵⊥
set of int Neighbors ⟵ set of neighbors
(message types)
QUERY

(1)  if i = root then
(2)        visited ⟵ 1;
(3)        depth ⟵ 0;
(4)        send QUERY to Neighbors;
(5)  for round = 1 to diameter do
(6)        if visited = 0 then
(7)              if any QUERY messages arrive then
(8)                    parent ⟵ randomly select a node from which
                          QUERY was received;
(9)                    visited ⟵ 1;
(10)                   depth ⟵ round;
(11)                   send QUERY to Neighbors \ {senders of
                          QUERYs received in this round};
(12)       delete any QUERY messages that arrived in this round.
```

Algorithm 5.1 Spanning tree algorithm: the synchronous breadth-first search (BFS) spanning tree algorithm. The code shown is for processor P_i, $1 \leq i \leq n$.

the algorithm so that each node identifies not only its parent node but also all its children nodes.

Example Figure 5.2 shows an example execution of the algorithm with node A as initiator. The resulting tree is shown in boldface, and the round numbers in which the QUERY messages are sent are indicated next to the messages. The reader should trace through this example for clarity. For example, at the end of round 2, E receives a QUERY from B and F and randomly chooses F as the parent. A total of nine QUERY messages are sent in the network which has eight links.

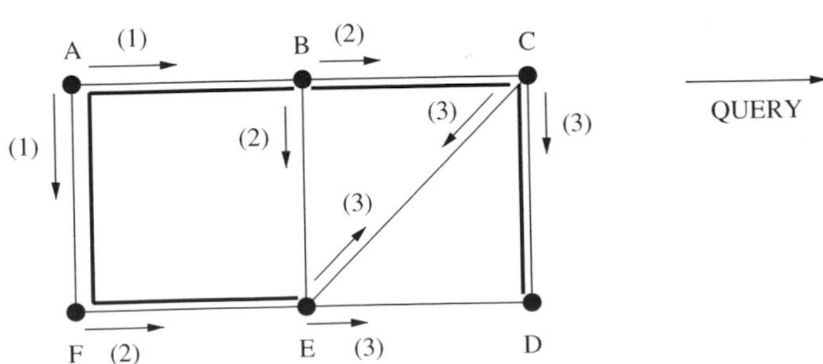

Figure 5.2 Example execution of the synchronous BFS spanning tree algorithm (Algorithm 5.1).

Termination

The algorithm terminates after all the rounds are executed. It is straightforward to modify the algorithm so that a process exits after the round in which it sets its *parent* variable (see Exercise 5.1).

Complexity

- The local space complexity at a node is of the order of the degree of edge incidence.
- The local time complexity at a node is of the order of (diameter + degree of edge incidence).
- The global space complexity is the sum of the local space complexities.
- This algorithm sends at least one message per edge, and at most two messages per edge. Thus the number of messages is between l and $2l$.
- The message time complexity is d rounds or message hops.

The spanning tree obtained is a breadth-first tree (BFS). Although the code is the same for all processes, the predesignated root executes a different logic to being with. Hence, in the strictest sense, the algorithm is asymmetric.

5.5.2 Asynchronous single-initiator spanning tree algorithm using flooding

This algorithm assumes a designated root node which initiates the algorithm. The pseudo-code for each process P_i is shown in Algorithm 5.2. The root initiates a flooding of QUERY messages in the graph to identify tree edges. The parent of a node is that node from which a QUERY is first received; an ACCEPT message is sent in response to such a QUERY. Other QUERY messages received are replied to by a REJECT message. Each node terminates its algorithm when it has received from all its non-parent neighbors a response to the QUERY sent to them. Procedures 1, 2, 3, and 4 are each executed atomically.

In this asynchronous system, there is no bound on the time it takes to propagate a message, and hence no notion of a message round. Unlike in the synchronous algorithm, each node here needs to track its neighbors to determine which nodes are its children and which nodes are not. This tracking is necessary in order to know when to terminate. After sending QUERY messages on the outgoing links, the sender needs to know how long to keep waiting. This is accomplished by requiring each node to return an "acknowledgement" for each QUERY it receives. The acknowledgement message has to be of a different type than the QUERY type. The algorithm in the figure uses two messages types – called as ACCEPT (+ ack) and REJECT (- ack) – besides the QUERY to distinguish between the child nodes and non-child nodes.

(local variables)
int $parent \longleftarrow \perp$
set of int $Children, Unrelated \longleftarrow \emptyset$
set of int $Neighbors \longleftarrow$ set of neighbors
(message types)
QUERY, ACCEPT, REJECT

(1) When the predesignated root node wants to initiate the algorithm:
(1a) **if** $(i = root$ **and** $parent = \perp)$ **then**
(1b)   **send** QUERY to all neighbors;
(1c) $parent \longleftarrow i$.

(2) When QUERY arrives from j:
(2a) **if** $parent = \perp$ **then**
(2b) $parent \longleftarrow j$;
(2c)   **send** ACCEPT to j;
(2d)   **send** QUERY to all neighbors except j;
(2e)   **if** $(Children \cup Unrelated) = (Neighbors \setminus \{parent\})$ **then**
(2f)   **terminate**.
(2g) **else send** REJECT to j.

(3) When ACCEPT arrives from j:
(3a) $Children \longleftarrow Children \cup \{j\}$;
(3b) **if** $(Children \cup Unrelated) = (Neighbors \setminus \{parent\})$ **then**
(3c)   **terminate**.

(4) When REJECT arrives from j:
(4a) $Unrelated \longleftarrow Unrelated \cup \{j\}$;
(4b) **if** $(Children \cup Unrelated) = (Neighbors \setminus \{parent\})$ **then**
(4c)   **terminate**.

Algorithm 5.2 Spanning tree algorithm: the asynchronous algorithm assuming a designated root that initiates a flooding. The code shown is for processor P_i, $1 \le i \le n$.

Termination

The termination condition is given above. Some notes on distributed algorithms are in place. In some algorithms such as this algorithm, it is possible to locally determine the termination condition; however, for some algorithms, the termination condition is not locally determinable and an explicit termination detection algorithm needs to be executed.

Complexity

- The local space complexity at a node is of the order of the degree of edge incidence.

- The local time complexity at a node is also of the order of the degree of edge incidence.
- The global space complexity is the sum of the local space complexities.
- This algorithm sends at least two messages (QUERY and its response) per edge, and at most four messages per edge (when two QUERIES are sent concurrently, each will have a REJECT response). Thus the number of messages is between $2l$ and $4l$.
- The message time complexity is $(d+1)$ message hops, assuming synchronous communication. In an asynchronous system, we cannot make any claim about the tree obtained, and its depth may be equal to the length of the longest path from the root to any other node, which is bounded only by $n-1$ corresponding to a depth-first tree.

Example Figure 5.3 shows an example execution of the asynchronous algorithm (i.e., in an asynchronous system). The resulting spanning tree rooted at A is shown in boldface. The numbers next to the QUERY messages indicate the approximate chronological order in which messages get sent. Recall that each procedure is executed atomically; hence the sending of a message sent at a particular time is triggered by the receipt of a corresponding message at the same time. The same numbering used for messages sent by different nodes implies that those actions occur concurrently and independently. ACCEPT and REJECT messages are not shown to keep the figure simple. It does not matter when the ACCEPT and REJECT messages are delivered.

1. A sends a QUERY to B and F.
2. F receives QUERY from A and determines that AF is a *tree edge*. F forwards the QUERY to E and C.
3. E receives a QUERY from F and determines that FE is a *tree edge*. E forwards the QUERY to B and D. C receives a QUERY from F and determines that FC is a *tree edge*. C forwards the QUERY to B and D.
4. B receives a QUERY from E and determines that EB is a *tree edge*. B forwards the QUERY to A, C, and D.
5. D receives a QUERY from E and determines that ED is a *tree edge*. D forwards the QUERY to B and C.

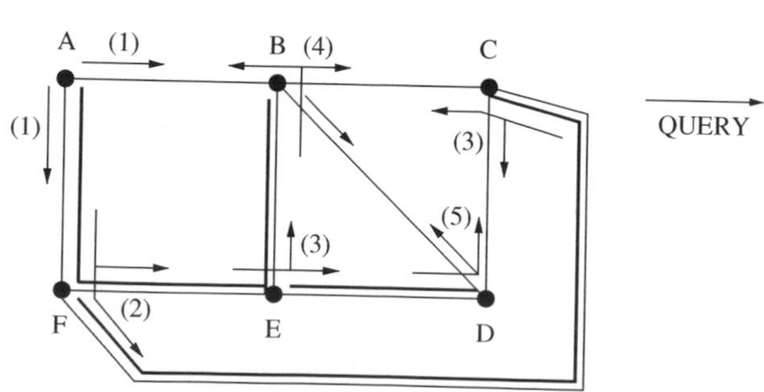

Figure 5.3 Example execution of the asynchronous flooding-based single initiator spanning tree algorithm (Algorithm 5.2).

Each node sends an ACCEPT message (not shown in Figure 5.3 for simplicity) back to the parent node from which it received its first QUERY. This is to enable the parent, i.e., the sender of the QUERY, to recognize that the edge is a tree edge, and to identify its child. All other QUERY messages are negatively acknowledged by a REJECT (also not shown for simplicity). Thus, a REJECT gets sent on each back edge (such as BA) and each cross edge (such as BD, BC, and CD) to enable the sender of the QUERY on that edge to recognize that that edge does not lead to a child node. We can also observe that on each tree edge, two messages (a QUERY and an ACCEPT) get sent. On each cross-edge and each back-edge, four messages (two QUERY and two REJECT) get sent.

Note that this algorithm does not guarantee a breadth-first tree. Exercise 5.3 asks you to modify this algorithm to obtain a BFS tree.

5.5.3 Asynchronous concurrent-initiator spanning tree algorithm using flooding

We modify Algorithm 5.2 by assuming that any node may spontaneously initiate the spanning tree algorithm provided it has not already been invoked locally due to the receipt of a QUERY message. The resulting algorithm is shown in Algorithm 5.3. The crucial problem to handle is that of dealing with concurrent initiations, where two or more processes that are not yet participating in the algorithm initiate the algorithm concurrently. As the objective is to construct a single spanning tree, two options seem available when concurrent initiations are detected. Note that even though there can be multiple concurrent initiations, along any single edge, only two concurrent initiations will be detected.

Design 1

When two concurrent initiations are detected by two adjacent nodes that have sent a QUERY from different initiations to each other, the two partially computed spanning trees can be merged. However, this merging cannot be done based only on local knowledge or there might be cycles.

Example In Figure 5.4, consider that the algorithm is initiated concurrently by A, G, and J. The dotted lines show the portions of the graphs covered by the three algorithms. At this time, the initiations by A and G are detected along edge BD, the initiations by A and J are detected along edge CF, the initiations by G and J are detected along edge HI. If the three partially computed spanning trees are merged along BD, CF, and HI, there is no longer a spanning tree.

```
(local variables)
int parent, myroot ⟵ ⊥
set of int Children, Unrelated ⟵ ∅
set of int Neighbors ⟵ set of neighbors
(message types)
QUERY, ACCEPT, REJECT
(1)   When the node wants to initiate the algorithm as a root:
(1a)  if (parent = ⊥) then
(1b)      send QUERY(i) to all neighbors;
(1c)      parent, myroot ⟵ i.

(2)   When QUERY(newroot) arrives from j:
(2a)  if myroot < newroot then     // discard earlier partial execution due
                                   // to its lower priority
(2b)      parent ⟵ j; myroot ⟵ newroot; Children, Unrelated ⟵ ∅;
(2c)      send QUERY(newroot) to all neighbors except j;
(2d)      if Neighbors = {j} then
(2e)          send ACCEPT(myroot) to j; terminate.    // leaf node
(2f)  else send REJECT(newroot) to j.
      // if newroot = myroot then parent is already identified.
      // if newroot < myroot ignore the QUERY. j will update its root
      // when it receives QUERY(myroot).

(3)   When ACCEPT(newroot) arrives from j:
(3a)  if newroot = myroot then
(3b)      Children ⟵ Children ∪ {j};
(3c)      if (Children ∪ Unrelated) = (Neighbors\{parent}) then
(3d)          if i = myroot then
(3e)              terminate.
(3f)          else send ACCEPT(myroot) to parent.
      // if newroot < myroot then ignore the message. newroot > myroot
      // will never occur.

(4)   When REJECT(newroot) arrives from j:
(4a)  if newroot = myroot then
(4b)      Unrelated ⟵ Unrelated ∪ {j};
(4c)      if (Children ∪ Unrelated) = (Neighbors\{parent}) then
(4d)          if i = myroot then
(4e)              terminate.
(4f)          else send ACCEPT(myroot) to parent.
      // if newroot < myroot then ignore the message. newroot > myroot
      // will never occur.
```

Algorithm 5.3 Spanning tree algorithm (asynchronous) without assuming a designated root. Initiators use flooding to start the algorithm. The code shown is for processor P_i, $1 \leq i \leq n$.

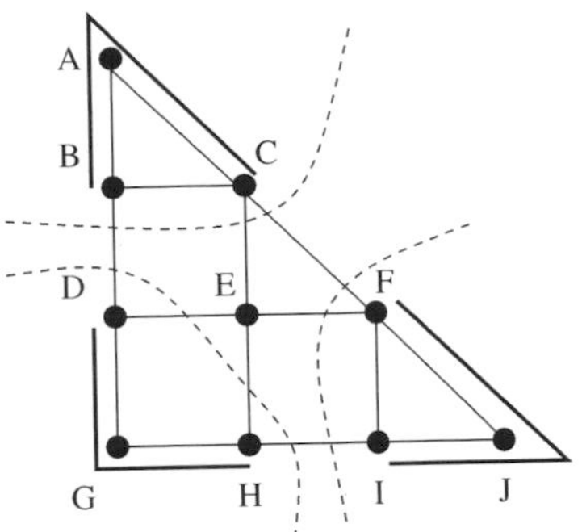

Figure 5.4 Example execution of the asynchronous flooding-based concurrent initiator spanning tree algorithm (Algorithm 5.3).

Interestingly, even if there are just two initiations, the two partially computed trees may "meet" along multiple edges in the graph, and care must be taken not to introduce cycles during the merger of the trees.

Design 2

Suppress the instance initiated by one root and continue the instance initiated by the other root, based on some rule such as tie-breaking using the processor identifier. Again, it must be ensured that the rule is correct.

Example In Figure 5.4, if A's initiation is suppressed due to the conflict detected along BD, G's initiation is suppressed due to the conflict detected along HI, and J's initiation is suppressed due to the conflict detected along CF, the algorithm hangs.

Algorithm 5.3 uses the second design option, allowing only the algorithm initiated by the root with the higher processor identifier to continue. To implement this, the messages need to be enhanced with a parameter that indicates the root node which initiated that instance of the algorithm. It is relatively more difficult to use the first option to merge partially computed spanning trees.

When a QUERY(*newroot*) from j arrives at i, there are three possibilities:

$newroot > myroot$: Process i should suppress its current execution due to its lower priority. It reinitializes the data structures and joins j's subtree with *newroot* as the root.

$newroot = myroot$: j's execution is initiated by the same root as i's initiation, and i has already identified its parent. Hence a REJECT is sent to j.

$newroot < myroot$: j's root has a lower priority and hence i does not join j's subtree. i sends a REJECT. j will eventually receive a QUERY(*myroot*) from i; and abandon its current execution in favour of i's *myroot* (or a larger value).

When an ACCEPT(*newroot*) from j arrives at i, there are three possibilities:

newroot = *myroot*: The ACCEPT is in response to a QUERY sent by i. The ACCEPT is processed normally.

newroot < *myroot*: The ACCEPT is in response to a QUERY i had sent to j earlier, but i has updated its *myroot* to a higher value since then. Ignore the ACCEPT message.

newroot > *myroot*: The ACCEPT is in response to a QUERY i had sent earlier. But i never updates its *myroot* to a lower value. So this case cannot arise.

The three possibilities when a REJECT(*newroot*) from j arrives at i are the same as for the ACCEPT message.

Termination

A serious drawback of the algorithm is that only the root knows when its algorithm has terminated. To inform the other nodes, the root can send a special message along the newly constructed spanning tree edges.

Complexity

The time complexity of the algorithm is $O(l)$ messages, and the number of messages is $O(nl)$.

5.5.4 Asynchronous concurrent-initiator depth first search spanning tree algorithm

As in Algorithm 5.3, this algorithm assumes that any node may spontaneously initiate the spanning tree algorithm provided it has not already been invoked locally due to the receipt of a QUERY message. It differs from Algorithm 5.3 in that it is based on a depth-first search (DFS) of the graph to identify the spanning tree. The algorithm should handle concurrent initiations (when two or more processes that are not yet participating in the algorithm initiate the algorithm concurrently). The pseudo-code for each process P_i is shown in Algorithm 5.4. The parent of each node is that node from which a QUERY is first received; an ACCEPT message is sent in response to such a QUERY. Other QUERY messages received are replied to by a REJECT message. The actions to execute when a QUERY, ACCEPT, or REJECT arrives are nontrivial and the analysis for the various cases (*newroot* <, =, > *myroot*) are similar to the analysis of these cases for Algorithm 5.3.

Termination

The analysis is the same as for Algorithm 5.3.

Complexity

The time complexity of the algorithm is $O(l)$ messages, and the number of messages is $O(nl)$.

(local variables)
int *parent*, *myroot* $\longleftarrow \perp$
set of int *Children* $\longleftarrow \emptyset$
set of int *Neighbors*, *Unknown* $\longleftarrow$ set of neighbors
(message types)
QUERY, ACCEPT, REJECT

(1) When the node wants to initiate the algorithm as a root:
(1a) **if** $(parent = \perp)$ **then**
(1b) **send** QUERY(i) to i (itself).

(2) When QUERY(*newroot*) arrives from j:
(2a) **if** $myroot < newroot$ **then**
(2b) $parent \longleftarrow j$; $myroot \longleftarrow newroot$; $Unknown \longleftarrow$ set of neighbors;
(2c) $Unknown \leftarrow Unknown \setminus \{j\}$;
(2d) **if** $Unknown \neq \emptyset$ **then**
(2e) **delete** some x from *Unknown*;
(2f) **send** QUERY(*myroot*) to x;
(2g) **else send** ACCEPT(*myroot*) to j;
(2h) **else if** $myroot = newroot$ **then**
(2i) **send** REJECT to j. // if $newroot < myroot$ ignore the query.
// j will update its root to a higher root identifier when it receives its
// QUERY.

(3) When ACCEPT(*newroot*) or REJECT(*newroot*) arrives from j:
(3a) **if** $newroot = myroot$ **then**
(3b) **if** ACCEPT message arrived **then**
(3c) $Children \longleftarrow Children \cup \{j\}$;
(3d) **if** $Unknown = \emptyset$ **then**
(3e) **if** $parent \neq i$ **then**
(3f) **send** ACCEPT(*myroot*) to *parent*;
(3g) **else** set i as the root; **terminate**.
(3h) **else**
(3i) **delete** some x from *Unknown*;
(3j) **send** QUERY(*myroot*) to x.
// if $newroot < myroot$ ignore the query. Since sending QUERY to j, i
// has updated its *myroot*.
// j will update its *myroot* to a higher root identifier when it receives a
// QUERY initiated by it.
// $newroot > myroot$ will never occur.

Algorithm 5.4 Spanning tree algorithm (DFS, asynchronous). The code shown is for processor P_i, $1 \leq i \leq n$.

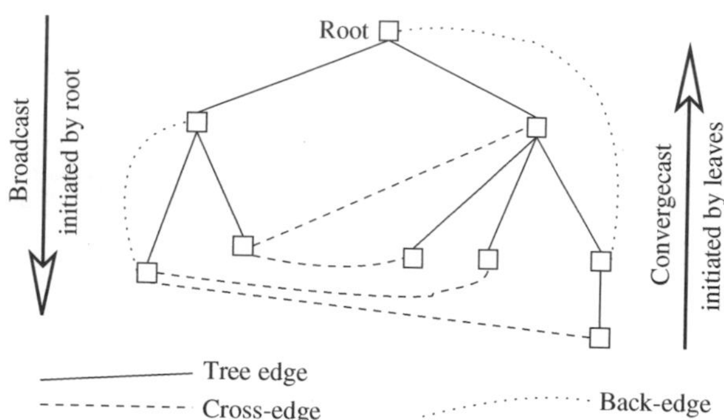

Figure 5.5 A generic spanning tree on a graph. The broadcast and convergecast operations are indicated.

5.5.5 Broadcast and convergecast on a tree

A spanning tree is useful for distributing (via a broadcast) and collecting (via a convergecast) information to/from all the nodes. A generic graph with a spanning tree, and the convergecast and broadcast operations are illustrated in Figure 5.5.

A *broadcast algorithm* on a spanning tree can be specified by two rules:

BC1: The root sends the information to be broadcast to all its children. Terminate.

BC2: When a (nonroot) node receives information from its parent, it copies it and forwards it to its children. Terminate.

A *convergecast algorithm* collects information from all the nodes at the root node in order to compute some *global function*. It is initiated by the leaf nodes of the tree, usually in response to receiving a request sent by the root using a broadcast. The algorithm is specified as follows:

CVC1: Leaf node sends its report to its parent. Terminate.

CVC2: At a nonleaf node that is not the root: When a report is received from all the child nodes, the collective report is sent to the parent. Terminate.

CVC3: At the root: When a report is received from all the child nodes, the global function is evaluated using the reports. Terminate.

Termination

The termination condition for each node in a broadcast as well as in a convergecast is self-evident.

Complexity

Each broadcast and each convergecast requires $n-1$ messages and time equal to the maximum height h of the tree, which is $O(n)$.

An example of the use of convergecast is as follows. Suppose each node has an integer variable associated with the application, and the objective is to compute the minimum of these variables. Each leaf node can report its local value to its parent. When a non-leaf node receives a report from all its children, it computes the minimum of those values, and sends this minimum value to its parent.

Another example of the use of convergecast is in solving the *leader election* problem in Section 5.10. Leader election requires that all the processes agree on a common distinguished process, also termed as the *leader*. A leader is required in many distributed systems and algorithms because algorithms are typically not completely symmetrical, and some process has to take the lead in initiating the algorithm; another reason is that we would not want all the processes to replicate the algorithm initiation, to save on resources.

5.5.6 Single source shortest path algorithm: synchronous Bellman–Ford

Given a weighted graph, with potentially unidirectional links, representing the network topology, the Bellman–Ford sequential shortest path algorithm [4,12] finds the shortest path from a given node, say i_0, to all other nodes. The algorithm is correct when there are no cyclic paths having negative weight.

A synchronous distributed algorithm to compute the shortest path is given in Algorithm 5.5. It is assumed that the topology (N, L) is not known to any process; rather, each process can communicate only with its neighbors and is aware of only the incident links and their weights. It is also assumed that the processes know the number of nodes $|N| = n$, i.e., the algorithm is not uniform. This assumption on n is required for termination.

```
(local variables)
int length ⟵ ∞
int parent ⟵ ⊥
set of int Neighbors ⟵ set of neighbors
set of int {weight_{i,j}, weight_{j,i} | j ∈ Neighbors} ⟵ the known values of
    the weights of incident links
(message types)
UPDATE
if i = i_0 then length ⟵ 0;
for round = 1 to n − 1 do
        send UPDATE(i, length) to all neighbors;
        await UPDATE(j, length_j) from each j ∈ Neighbors;
        for each j ∈ Neighbors do
                if (length > (length_j + weight_{j,i}) then
                        length ⟵ length_j + weight_{j,i}; parent ⟵ j.
```

Algorithm 5.5 The single source synchronous distributed Bellman–Ford shortest path algorithm. The source is i_0. The code shown is for processor P_i, $1 \le i \le n$.

The following features can be observed from the algorithm:

- After k rounds, each node has its *length* variable set to the length of the shortest path consisting of at most k hops. The *parent* variable points to the parent node along such a path. This *parent* field is used in the routing table to route to i_0.
- After the first round, the *length* variable of all nodes one hop away from the root in the final minimum spanning tree (MST) would have stablized; after k rounds, the *length* variable of all the nodes up to k hops away in the final MST would have stabilized.

Termination

As the longest path can be of length $n-1$, the values of all variables stabilize after $n-1$ rounds.

Complexity

The time complexity of this synchronous algorithm is: $n-1$ rounds. The message complexity of this synchronous algorithm is: $(n-1)l$ messages.

5.5.7 Distance vector routing

When the network graph is dynamically changing, as in a real communication network wherein the link weights model the delays or loads on the links, the shortest paths are required for routing. The classic distance vector routing algorithm (DVR) [33] used in the ARPANET up to 1980, is based on the above synchronous algorithm (Algorithm 5.5) and requires the following changes.

- The outer **for** loop runs indefinitely, and the *length* and *parent* variables never stabilize, because of the dynamic nature of the system.
- The variable *length* is replaced by array $LENGTH[1..n]$, where $LENGTH[k]$ denotes the length measured with node k as source/root. The $LENGTH$ vector is also included on each UPDATE message. Now, the kth component of the $LENGTH$ received from node m indicates the length of the shortest path from m to the root k. For each destination k, the triangle inequality of the Bellman–Ford algorithm is applied over all the $LENGTH$ vectors received in a round.
- The variable *parent* is replaced by array $PARENT[1..n]$, where $PARENT[k]$ denotes the next hop to which to route a packet destined for k. The array $PARENT$ serves as the routing table.
- The processes exchange their distance vectors periodically over a network that is essentially asynchronous. If a message does not arrive within the period, the algorithm assumes a default value, and moves to the next round. This makes it virtually synchronous. Besides, if the period between exchanges is assumed to be much larger than the propagation time from a neighbor and the processing time for the received message, the algorithm is effectively synchronous.

5.5.8 Single source shortest path algorithm: asynchronous Bellman–Ford

The asynchronous version of the Bellman–Ford algorithm [4,5,12] is shown in Algorithm 5.6. It is assumed that there are no negative weight cycles in (N, L).

The algorithm does not give the termination condition for the nodes. Exercise 5.14 asks you to modify the algorithm so that each node knows when the length of the shortest path to itself has been computed.

This algorithm, unfortunately, has been shown to have an exponential $\Omega(c^n)$ number of messages and exponential $\Omega(c^n \cdot d)$ time complexity in the worst case, where c is some constant (see Exercise 5.16).

(local variables)
int $length \longleftarrow \infty$
set of int $Neighbors \longleftarrow$ set of neighbors
set of int $\{weight_{i,j}, weight_{j,i} \mid j \in Neighbors\} \longleftarrow$ the known values of the weights of incident links

(message types)
UPDATE

(1) **if** $i = i_0$ **then**
(1a) $length \longleftarrow 0$;
(1b) **send** UPDATE(i_0, 0) to all neighbors; **terminate**.

(2) When UPDATE(i_0, $length_j$) arrives from j:
(2a) **if** $(length > (length_j + weight_{j,i}))$ **then**
(2b) $length \longleftarrow length_j + weight_{j,i}$; $parent \longleftarrow j$;
(2c) **send** UPDATE(i_0, $length$) to all neighbors;

Algorithm 5.6 The asynchronous distributed Bellman–Ford shortest path algorithm for a given source i_0. The code shown is for processor P_i, $1 \le i \le n$.

If all links are assumed to have equal weight, the algorithm that computes the shortest path effectively computes the minimum-hop path; the minimum-hop routing tables to all destinations are computed using $O(n^2 \cdot l)$ messages (see Exercise 5.17).

5.5.9 All sources shortest paths: asynchronous distributed Floyd–Warshall

The Floyd–Warshall algorithm [9] computes all-pairs shortest paths in a graph in which there are no negative weight cycles. It is briefly summarized first, before a distributed version is studied. The centralized algorithm shown in Algorithm 5.7 uses $n \times n$ matrices $LENGTH$ and VIA:

$LENGTH[i, j]$ is the length of the shortest path from i to j. $LENGTH[i, j]$ is initialized to the initial known conditions: (i) $weight_{i,j}$ if i and j are neighbors, (ii) 0 if $i = j$, and (iii) ∞ otherwise.

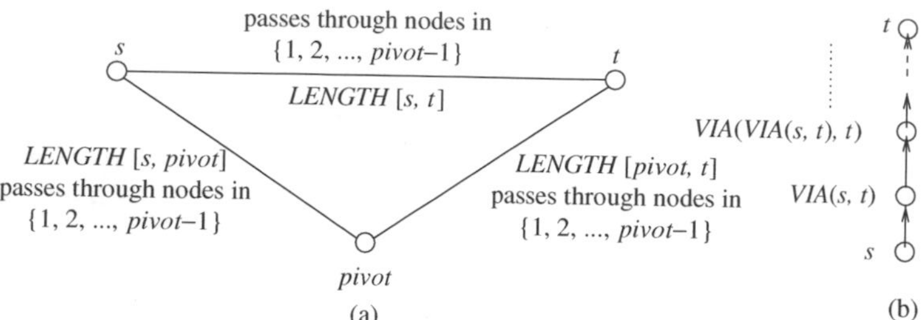

Figure 5.6 The all-pairs shortest paths algorithm by Floyd–Warshall. (a) Triangle inequality used in iteration *pivot* uses paths via $\{1, \ldots, pivot-1\}$. (b) The *VIA* relationships along a branch of the sink tree for a given (s, t) pair.

$VIA[i, j]$ is the first hop on the shortest path from i to j. $VIA[i, j]$ is initialized to the initial known conditions: (i) j if i and j are neighbors, (ii) 0 if $i = j$, and (iii) ∞ otherwise.

After *pivot* iterations of the outer loop, the following **invariant** holds:

$LENGTH[i, j]$ is the shortest path going through intermediate nodes from the set $\{1, \ldots, pivot\}$. $VIA[i, j]$ is the corresponding first hop.

Convince yourself of this invariant using Algorithm 5.7 and Figure 5.6. In this figure, the *LENGTH* is for the paths that pass through nodes from $\{1 \ldots pivot-1\}$. The time complexity of the centralized algorithm is $O(n^3)$.

The distributed asynchronous algorithm by Toueg [34] is shown in Algorithm 5.8. Row i of the *LENGTH* and *VIA* data structures is stored at node i which is responsible for updating this row. To avoid ambiguity, we rename these data structures as *LEN* and *PARENT*, respectively. When the algorithm terminates, the final values of row i of *LENGTH* is available at node i as *LEN*.

There are two challenges in making the Floyd–Warshall algorithm distributed:

1. How to access the remote datum $LENGTH[pivot, t]$ for each execution of line (4) in the centralized algorithm of Algorithm 5.7, now being executed by i?
2. How to synchronize the execution at the different nodes? If the different nodes are not executing the same iteration of the outermost loop of Algorithm 5.7, the distributed algorithm becomes incorrect.

```
(1)  for pivot = 1 to n do
(2)      for s = 1 to n do
(3)          for t = 1 to n do
(4)              if LENGTH[s, pivot] + LENGTH[pivot, t]
                   < LENGTH[s, t] then
(5)                  LENGTH[s, t] ⟵ LENGTH[s, pivot]
                       +LENGTH[pivot, t];
(6)                  VIA[s, t] ⟵ VIA[s, pivot].
```

Algorithm 5.7 The centralized Floyd–Warshall all-pairs shortest paths routing algorithm.

```
(local variables)
int LEN[1..n]          // LEN[j] is the length of the shortest known
                       // path from i to node j.
                       // LEN[j] = weight_ij for neighbor j, 0 for
                       // j = i, ∞ otherwise
int PARENT[1..n]       // PARENT[j] is the parent of node i (myself)
                       // on the sink tree rooted at j.
                       // PARENT[j] = j for neighbor j, ⊥ otherwise

set of int Neighbors ⟵ set of neighbors
int pivot, nbh ⟵ 0

(message types)
IN_TREE(pivot), NOT_IN_TREE(pivot),
PIV_LEN(pivot, PIVOT_ROW[1..n])
    // PIVOT_ROW[k] is LEN[k] of node pivot, which is LEN[pivot, k] in
                                          // the central algorithm.
                 // the PIV_LEN message is used to convey PIVOT_ROW.

for pivot = 1 to n do
    for each neighbor nbh ∈ Neighbors do
        if PARENT[pivot] = nbh then
            send IN_TREE(pivot) to nbh;
        else send NOT_IN_TREE(pivot) to nbh;
    await IN_TREE or NOT_IN_TREE message from each neighbor;
    if LEN[pivot] ≠ ∞ then
        if pivot ≠ i then
            receive PIV_LEN(pivot, PIVOT_ROW[1..n]) from
              PARENT[pivot];
        for each neighbor nbh ∈ Neighbors do
            if IN_TREE message was received from nbh then
                if pivot = i then
                    send PIV_LEN(pivot, LEN[1..n]) to nbh;
                else send PIV_LEN(pivot, PIVOT_ROW[1..n])
                        to nbh;
        for t = 1 to n do
            if LEN[pivot] + PIVOT_ROW[t] < LEN[t] then
                LEN[t] ⟵ LEN[pivot] + PIVOT_ROW[t];
                PARENT[t] ⟵ PARENT[pivot].
```

Algorithm 5.8 Toueg's asynchronous distributed Floyd–Warshall all-pairs shortest paths routing algorithm. The code shown is for processor P_i, $1 \leq i \leq n$.

The problem of accessing the remote datum *LENGTH*[*pivot*, *t*] is solved by using the idea of the distributed *sink tree*. In the centralized algorithm, after each iteration *pivot* of the outermost loop, if $LENGTH[s, t] \neq \infty$, then

$VIA[s, t]$ points to the parent node on the path to t and this is the shortest path going through nodes $\{1 \ldots pivot\}$. Observe that $VIA[VIA[s, t], t]$ will also point to $VIA[s, t]$'s parent node on the shortest path to t, and so on. Effectively, tracing through the VIA nodes gives the shortest path to t; this path is acyclic because of the "shortest path" property (see **invariant, p. 152**). Thus, all nodes s for which $LENGTH[s, t] \neq \infty$ are part of a tree to t, and this tree is termed as a *sink tree*, with t as the root or the *sink node*. In the distributed algorithm, the parent of any node on the sink tree for t is stored in $PARENT[t]$.

Applying the sink tree idea to node *pivot* in iteration *pivot* of the distributed algorithm, we have the following observations for any node i in any iteration *pivot*.

- If $LEN[pivot] = \infty$, then i will not update its LEN and $PARENT$ arrays in this iteration. Hence there is no need for i to receive the remote data $PIV\_ROW[1, \ldots, n]$. In fact, there is no known path from i to *pivot* at this stage.
- If $LEN[pivot] \neq \infty$, then the remote data $PIVOT\_ROW[1, \ldots, n]$ is distributed to all the nodes lying on the sink tree of *pivot*. Observe that i necessarily lies on the sink tree of *pivot*. The parent of i, and its parent's parent, and so on, all lie on that sink tree.

The asynchronous distributed algorithm proceeds as follows. In iteration *pivot*, node *pivot* broadcasts its LEN vector along its sink tree. To implement this broadcast, the parent-child edges of the sink tree need to be identified. Note that any node on the sink tree of *pivot* does not know which of its neighbors are its children. Hence, each node awaits a IN_TREE or NOT_IN_TREE message from each of its neighbors (lines 2–6) to identify it children. These flows seen at node i are illustrated in Figure 5.7. The broadcast of the pivot's LEN vector is initiated by node *pivot* in lines 10–13. For example, consider the first iteration, where $pivot = 1$:

Node 1 The node executes lines 1, 2–5 by sending NOT_IN_TREE, line 6 in which it gets IN_TREE messages from its neighbors, and lines 10–13, wherein the node sends its LEN vector to its neighbors.

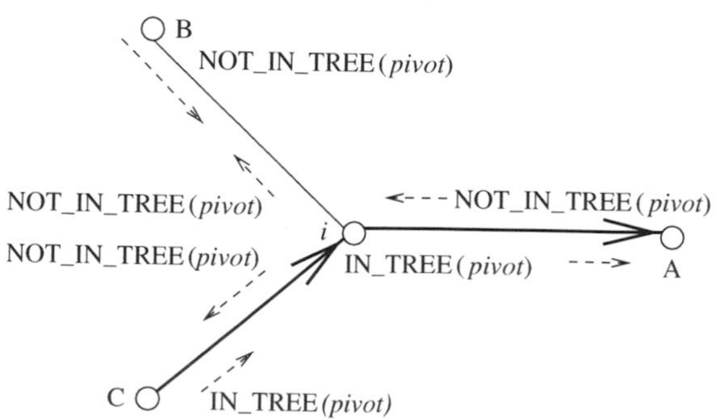

Figure 5.7 Message flows to determine how to selectively distribute *PIV_ROW* in iteration *pivot* in Toueg's distributed Floyd–Warshall algorithm.

Node > 1 In lines 1–4, the neighbors of node 1 send IN_TREE to node 1. In line 9, the neighbors receive *PIVOT_LEN* from the pivot, i.e., node 1. The reader can step through the remainder of the protocol.

When i receives *PIV_LEN* message containing the pivot's *PIVOT_ROW* $[1..n]$ from its parent (line 9), it forwards it to its children (lines 10–11 and 14). The two inner loops of the centralized algorithm are then executed in lines 15–18 of the distributed algorithm.

The inherent distribution of *PIVOT_ROW* via the **receive** from the parent (line 9) and **send** to the children (line 14), as well as the synchronization of the **send** (lines 4–5) and **receive** (line 6) of IN_TREE and NOT_IN_TREE messages among neighbor nodes ensures that the asynchronous execution of the nodes gets synchronized and all nodes are forced to execute the innermost nested iteration concurrently with each other. Notice the dependence between the **send** of lines 4–5 and **receive** of line 6, and between the **receive** of line 9 and the **send** of lines 13 or 14.

The techniques for synchronization used here will be formalized in Section 5.6 under the subject of synchronizers.

Complexity

In each of the n iterations of the outermost loop, two IN_TREE or NOT_IN_TREE messages are sent per edge, and at most $n-1$ PIV_LEN messages are sent. The overall number of messages is $n \cdot (2l+n)$. The PIV_LEN is of size n while the IN_TREE and NOT_IN_TREE messages are of size $O(1)$. The execution time complexity per node is $O(n^2)$, plus the time for n convergecast–broadcast phases.

5.5.10 Asynchronous and synchronous constrained flooding (w/o a spanning tree)

Asynchronous algorithm (Algorithm 5.9)

This algorithm allows any process to initiate a broadcast via (constrained) flooding along the edges of the graph [33]. It is assumed that all channels are FIFO. Duplicates are detected by using sequence numbers. Each process uses the $SEQNO[1..n]$ vector, where $SEQNO[k]$ tracks the latest sequence number of the update initiated by process k. If the sequence number on a newly arrived message is not greater than the sequence numbers already seen for that initiator, the message is simply discarded; otherwise, it is flooded on all other outgoing links. This mechanism is used by the link state routing protocol in the Internet to distribute any updates about the link loads and the network topology.

Complexity

The message complexity is: $2l$ messages in the worst case, where each message M has overhead $O(1)$. The time complexity is: diameter d number of sequential hops.

```
(local variables)
int SEQNO[1..n] ⟵ 0̄
set of int Neighbors ⟵ set of neighbors
(message types)
UPDATE

    To send a message M:
if i = root then
        SEQNO[i] ⟵ SEQNO[i] + 1;
        send UPDATE(M, i, SEQNO[i]) to each j ∈ Neighbors.

    When UPDATE(M, j, seqno_j) arrives from k:
if SEQNO[j] < seqno_j then
        Process the message M;
        SEQNO[j] ⟵ seqno_j;
        send UPDATE(M, j, seqno_j) to Neighbors/{k};
else discard the message.
```

Algorithm 5.9 The asynchronous flooding algorithm. The code shown is for processor P_i, $1 \leq i \leq n$. Any and all nodes can initiate the algorithm spontaneously.

Synchronous algorithm (Algorithm 5.10)

This algorithm [33] allows all processes to flood a local value throughout the network. The local array $STATEVEC[1..n]$ is such that $STATEVEC[k]$ is the estimate of the local value of process k. After d number of rounds, it is guaranteed that the local value of each process has propagated throughout the network.

Complexity

The time complexity is: diameter d rounds, and the message complexity is: $2l \cdot d$ messages, each of size n.

```
(local variables)
int STATEVEC[1..n] ⟵ 0̄
set of int Neighbors ⟵ set of neighbors
(message types)
UPDATE

STATEVEC[i] ⟵ local value;
for round = 1 to diameter d do
        send UPDATE(STATEVEC[1..n]) to each j ∈ Neighbors;
        for count = 1 to |Neighbors| do
                await UPDATE(SV[1..n]) from some j ∈ Neighbors;
                STATEVEC[1..n] ⟵ max(STATEVEC[1..n], SV[1..n]).
```

Algorithm 5.10 The synchronous flooding algorithm for learning all node's identifiers. The code shown is for processor P_i, $1 \leq i \leq n$.

5.5.11 Minimum-weight spanning tree (MST) algorithm in a synchronous system

A minimum-weight spanning tree (MST) minimizes the cost of transmission from any node to any other node in the graph. The classical centralized MST algorithms such as those by Prim, Dijkstra, and Kruskal [9] assume that the entire weighted graph is available for examination.

- Kruskal's algorithm begins with a forest of graph components. In each iteration, it identifies the minimum-weight edge that connects two different components, and uses this edge to merge two components. This continues until all the components are merged into a single component.
- In Prim's algorithm and Dijkstra's algorithm, a single-node component is selected. In each iteration, a minimum-weight edge incident on the component is identified, and the component expands to include that edge and the node at the other end of that edge. After $n-1$ iterations, all the nodes are included. The MST is defined by the edges that are identified in each iteration to expand the initial component.

In a distributed algorithm, each process can communicate only with its neighbors and is aware of only the incident links and their weights. It is also assumed that the processes know the value of $|N| = n$. The weight of each edge is unique in the network, which is necessary to guarantee a unique MST. (If weights are not unique, the IDs of the nodes on which they are incident can be used as tie-breakers by defining a well-formed order.)

A distributed algorithm by Gallagher, Humblet, and Spira [14] that generalizes the strategy of Kruskal's centralized algorithm is given after reviewing some definitions. A *forest* (i.e., a disjoint union of trees) is a graph in which any pair of nodes is connected by at most one path. A *spanning forest* of an undirected graph (N, L) is a maximal forest of (N, L), i.e., an acyclic and not necessarily connected graph whose set of vertices is N. When a spanning forest is connected, it becomes a *spanning tree*.

A spanning forest of G is a subgraph G' of G having the same node set as G; the spanning forest can be viewed as a set of spanning trees, one spanning tree per "connected component" of G'. All MST algorithms begin with a spanning forest having n nodes (or connected components) and without any edges. They then add a "*minimum-weight outgoing edge*" (MWOE) between two components.[2] The spanning trees of the combining connected components combine with the MWOE to form a single spanning tree for the combined connected component. The addition of the MWOE is repeated until a spanning

[2] Note that this is an undirected graph. The direction of the "outgoing" edge is logical in the sense that it identifies the direction of expansion of the connected component under consideration.

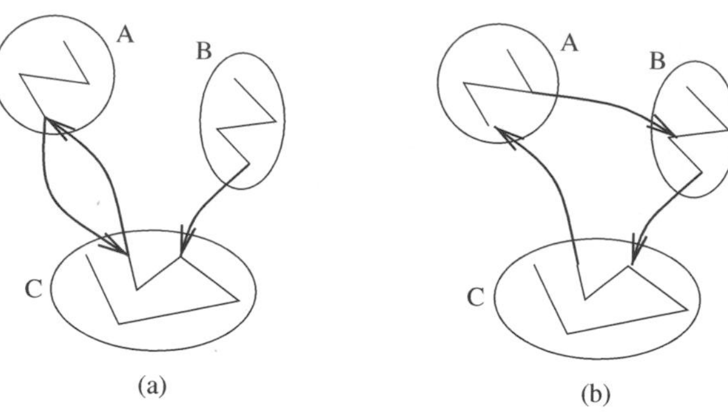

Figure 5.8 Merging of MWOE components. (a) A cycle of length 2 is possible. (b) A cycle of length greater than 2 is not possible.

tree is produced for the entire graph (N, L). Such algorithms are correct because of the following observation.

Observation 5.1 *For any spanning forest* $\{(N_i, L_i) \mid i = 1 \ldots k\}$ *of a weighted undirected graph G, consider any component* (N_j, L_j). *Denote by* λ_j, *the edge having the smallest weight among those that are incident on only one node in* N_j. *Then an MST for the graph G that includes all the edges in each* L_i *in the spanning forest, must also include edge* λ_i.

This observation says that for any "minimum-weight" component created so far, when it grows by joining another component, the growth must be via the MWOE for that component under consideration. Intuitively, the logic is as follows. For any component containing node set N_j, if edge x is used instead of the MWOE λ_j to connect with nodes in $N \setminus N_j$, then the resulting tree cannot be a MST because edge x can always be replaced with the MWOE that was not chosen to yield a lower cost tree.

Consider Figure 5.8(a) where three components have been identified and are encircled. The MWOE for each component is marked by an outgoing edge (other outgoing edges are not shown). Each of the three components shown must grow only by merging with the component at the other end of the MWOE.

In a distributed algorithm, the addition of the edges should be done concurrently by having all the components identify their respective minimum-weight outgoing edge. The synchronous algorithm of Gallagher–Humblet–Spira [14] uses this above observation, and is given in Algorithm 5.11. Initially, each node is the leader of its component which contains only that node. The algorithm uses $log(n)$ iterations. In each iteration, each component merges with at least one other component. Hence, $log(n)$ iterations guarantee termination with a single component.

```
(message types)
SEARCH_MWOE(leader)   // broadcast by current leader on tree edges
EXAMINE(leader)       // sent on non-tree edges after receiving
                      // SEARCH_MWOE
REPLY_MWOE(local_ID, remote_ID)   // details of potential MWOEs
                                  // are convergecast to leader
ADD_MWOE(local_ID, remote_ID)   // sent by leader to add MWOE
                                // and identify new leader
NEW_LEADER(leader)   // broadcast by new leader after merging
                     // components

leader = i;
for round = 1 to log(n) do   // each merger in each iteration involves at
                             // least two components

1. if leader = i then
        broadcast SEARCH_MWOE(leader) along marked edges of tree
        (Section 5.5.5).
2. On receiving a SEARCH_MWOE(leader) message that was broadcast on
   marked edges:
   (a) Each process i (including leader) sends an EXAMINE message along
       unmarked (i.e., non-tree) edges to determine if the other end of the
       edge is in the same component (i.e., whether its leader is the same).
   (b) From among all incident edges at i, for which the other end
       belongs to a different component, process i picks its incident
       MWOE(localID,remoteID).
3. The leaf nodes in the MST within the component initiate the convergecast
   (Section 5.5.5) using REPLY_MWOEs, informing their parent of their
   MWOE(localID,remoteID). All the nodes participate in this convergecast.
4. if leader = i then
        await convergecast replies along marked edges.
        Select the minimum MWOE(localID,remoteID) from all the replies.
        broadcast ADD_MWOE(localID,remoteID) along marked
          edges of tree (Section 5.5.5).
        // To ask process localID to mark the (localID, remoteID)
        // edge, i.e., include it in MST of component.
5. if an MWOE edge gets marked by both the components on which it is
   incident then
   (a) Define new_leader as the process with the larger ID on which that
       MWOE is incident (i.e., process whose ID is max(localID, remoteID)).
   (b) new_leader identifies itself as the leader for the next round.
   (c) new_leader broadcasts NEW_LEADER in the newly formed compo-
       nent along the marked edges (Section 5.5.5) announcing itself as the
       leader for the next round.
```

Algorithm 5.11 The synchronous MST algorithm by Gallagher–Humblet–Spira (GHS algorithm). The code shown is for processor P_i, $1 \leq i \leq n$.

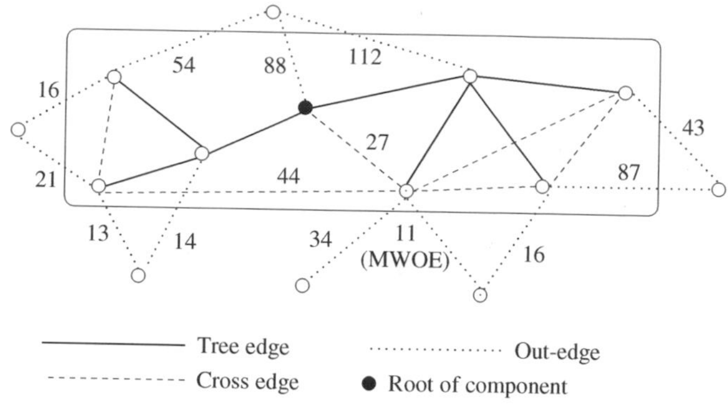

Figure 5.9 The phases within an iteration in a component.

Each iteration goes through a broadcast–convergecast–broadcast sequence to identify the MWOE of the component, and to select the *leader* for the next iteration. The MWOE is identified after the broadcast (steps 1 and 2) and convergecast (step 3) by the current leader, which then does a second broadcast (step 4). The leader is selected at the end of this second broadcast (step 4); among all the components that merge in an iteration, a single leader is selected, and it identifies itself among all the nodes in the newly forming component by doing a third broadcast (step 5). This sequence of steps can be visualized using the connected component enclosed within a rectangle in Figure 5.9, using the following narrative: (a) root broadcasts SEARCH_MWOE; (b) convergecast REPLY_MWOE occurs; (c) root broadcasts ADD_MWOE; (d) if the MWOE is also chosen as the MWOE by the component at the other end of the MWOE, the incident process with the higher ID is the leader for the next iteration and broadcasts NEW_LEADER.

The correctness of the above algorithm hinges on the fact that in any iteration, when each component of the spanning forest joins with one or more other components of the spanning forest, the result is still a spanning forest! Observe that each component picks exactly *one* MWOE with which it connects to another component. However, more than two components can join together in one iteration. If multiple components join, we need to observe that the resulting component is still a spanning forest. To do so, model a directed graph (P, M) where P is the set of components at the start of an iteration and M is the set of $|P|$ MWOE edges chosen by the components in P. In this graph, there is exactly one outgoing edge from each node in P. Recall that the direction of the MWOE is logical; the underlying graph remains undirected. If component A chooses to include a MWOE leading to component B, then directed edge (A, B) exists in (P, M). By tracing any path in this graph, observe that MWOE weights must be monotonically decreasing. To see that (i) the merging of components retains the spanning forest property, and (ii) there is a unique leader in each component after the merger in the previous round, consider the following two cases:

1. If two components join, then each must have picked the other to join with, and we have a cycle of length two. As each component was a spanning forest, joining via the common MWOE still retains the spanning forest property, and there is a unique leader in the merged component.
2. If three or more components join, then two sub-cases are possible:
   - There is some cycle of length three or more (see Figure 5.8(b)). But as any path in (P, M) follows MWOEs of monotonically decreasing weights, this implies a contradiction because at least one node must have chosen an incorrect MWOE.
   - There is no cycle of length 3 or more, and at least one node in (P, M) will have two or more incoming edges (component C in Figure 5.8(a)). Further, there must exist a cycle of length two. Exercise 5.22 asks you to prove this formally. As the graph has a cycle of length at most two (case 1), the resulting component after the merger of all the involved components is still a spanning component, and there is a unique leader in the merged component. That leader is the node with the larger PID incident on the MWOE that gets marked by both components on which it is incident.

Complexity

- In each of the $log(n)$ iterations, each component merges with at least one other component. So after the first iteration, there are at most $n/2$ components, after the second, at most $n/4$ components, and so on. Hence, at most $log(n)$ iterations are needed and the number of nodes in each component after iteration k is at least 2^k. In each iteration, the time complexity is $O(n)$ because the time complexity for broadcast and convergecast is bounded by $O(n)$. Hence the time complexity is $O(n \cdot log(n))$.
- In each of the $log(n)$ iterations, $O(n)$ messages are sent along the marked tree edges (steps 1, 3, 4, and 5). There may be up to $l = |L|$ EXAMINE messages to determine the MWOEs in step 2 of each iteration. Hence, the total message complexity is $O((n+l) \cdot log(n))$.

The correctness of the GHS algorithm hinges on the fact that the execution occurs in synchronous rounds. This is necessary in step 2, where a process sends EXAMINE messages to its unmarked neighbors to determine whether those neighbors belong to the same or a different component than itself. If the neighbor is not synchronized, problems can occur. For example, consider edge (j, k), where j and k become a part of the same component in "iteration" x. From j's perspective, the neighbor k may not yet have received its leader's ID that was broadcast in step 5 of the previous iteration; hence k replies to the EXAMINE message sent by j based on an older ID for its leader. The testing process j may (incorrectly) include k in the same component as itself, thereby creating cycles in the graph. As the distance from the leader to any node in its component is not known, this needs to be dealt with even in a synchronous system. One way to enforce the synchronicity is to wait for $O(n)$ number of communication steps; this way, all communication within the round would have completed in the synchronous model.

5.5.12 Minimum-weight spanning tree (MST) in an asynchronous system

There are two approaches to designing the asynchronous MST algorithm.

In the first approach, the synchronous GHS algorithm is *simulated* in an asynchronous setting. In such a simulation, the same synchronous algorithm is run, but is augmented by additional protocol steps and control messages to provide the synchronicity. Observe from the synchronous GHS that the difficulty in making it asynchronous lies in step 2. If the two nodes at the ends of an unmarked edge are in different levels, the algorithm can go wrong. Two possible ways to deal with this problem are as follows:

- After each round, an additional broadcast and convergecast on the marked edges are serially done. The newly identified leader broadcasts its ID and round number on the tree edges; the convergecast is then initiated by the leaves to acknowledge this broadcast. When the convergecast completes at the leader, it then begins the next round. Now in step 2, if the recipient of an EXAMINE message is in an earlier round, it simply delays the response to the EXAMINE, thus forcing synchrony.

  This costs $n \cdot log(n)$ extra messages.
- When a node gets involved in a new round, it simply informs each neighbor (reachable along unmarked or non-tree edges) of its new level. Only when the neighbors along unmarked edges are all in the same round does the node send the EXAMINE message in step 2.

  This costs $|L| \cdot log(n)$ extra messages.

The second approach to designing the asynchronous MST is to directly address all the difficulties that arise due to lack of synchrony. The original asynchronous GHS algorithm uses this approach even though it is patterned along the synchronous GHS algorithm. By carefully engineering the asynchronous algorithm, it achieves the same message complexity $O(n \cdot log(n) + l)$ as the synchronous algorithm and a time complexity $O(n \cdot log(n) \cdot (l + d))$. We do not present the algorithm here because it is a well-engineered algorithm with intricate details; rather, we only point out some of the difficulties in designing this algorithm:

- In step 2, if the two nodes are in different components or in different levels, there needs to be a mechanism to determine this.
- If the combining of components at different levels is permitted, then some component may keep combining with only single-node components in the worst case, thereby increasing the complexity by changing the $log(n)$ factor to the factor n.

- The search for MWOEs by adjacent components at different levels needs to be coordinated carefully. Specifically, the rules for merging such components, as well as the rules for the concurrent search for the MWOE by these two components, need to be specified.

5.6 Synchronizers

General observations on synchronous and asynchronous algorithms

From the spanning tree algorithms, shortest path routing algorithms, constrained flooding algorithms, and the MST algorithms, it can be observed that it is much more difficult to design the algorithm for an asynchronous system, than for a synchronous system. This can be generalized to all algorithms, with few exceptions. The example algorithms also suggest that simulating synchronous behavior (of an algorithm designed for a synchronous system) on an asynchronous system is often a direct way to realize the algorithms on asynchronous systems.

Given that typical distributed systems are asynchronous, the logical question to address is whether there is a general technique to convert an algorithm designed for a synchronous system, to run on an asynchronous system. The generic class of transformation algorithms to run synchronous algorithms on asynchronous systems are called *synchronizers*. We make the following observations. (i) We consider only failure-free systems, whether synchronous or asynchronous. We will see later (in Chapter 14) that such transformations may not be possible in asynchronous systems in which either processes fail or channels are unreliable. (ii) Using a synchronizer provides a sure way to obtain an asynchronous algorithm. However, such an algorithm may have high complexity. Although more difficult, it may be possible to design more efficient asynchronous algorithms from scratch, rather than transforming the synchronous algorithms to run on asynchronous systems. (This was seen in the case of the GHS algorithm.) Thus, the field of systematic algorithm design for asynchronous systems is an open and challenging field.

Practically speaking, in an asynchronous system, a synchronizer is a mechanism that indicates to each process when it is safe to proceed to the next round of execution of the "synchronous" algorithm. Conceptually, the synchronizer signals to each process when it is sure that all messages to be received in the current round have arrived.

The mesage complexity M_a and time complexity T_a of the asynchronous algorithm are as follows:

$$M_a = M_s + (M_{init} + rounds \cdot M_{round}), \tag{5.1}$$

$$T_a = T_s + (T_{init} + rounds \cdot T_{round}), \tag{5.2}$$

Table 5.1 The message and time complexities for the *simple*, α, β, and γ synchronizers. h_c is the greatest height of a tree among all the clusters. L_c is the number of tree edges and designated edges in the clustering scheme for the γ synchronizer. d is the graph diameter.

	Simple synchronizer	α synchronizer	β synchronizer	γ synchronizer
M_{init}	0	0	$O(n \cdot log(n) + \|L\|)$	$O(kn^2)$
T_{init}	d	0	$O(n)$	$n \cdot log(n)/log(k)$
M_{round}	$2\|L\|$	$O(\|L\|)$	$O(n)$	$O(L_c)$ $(\leq O(kn))$
T_{round}	1	$O(1)$	$O(n)$	$O(h_c)$ $(\leq O(log(n)/log(k)))$

where:

- M_s is the number of messages in the synchronous algorithm;
- *rounds* is the number of rounds in the synchronous algorithm;
- T_s is the time for the synchronous algorithm. Assuming one unit (message hop) per round, this equals *rounds*;
- M_{round} is the number of messages needed to simulate a round;
- T_{round} is the number of sequential message hops needed to simulate a round;
- M_{init} and T_{init} are the number of messages and the number of sequential message hops, respectively, in the initialization phase in the asynchronous system.

We now look at four standard synchronizers: the simple, the α, the β, and the γ synchronizers, proposed by Awerbuch [3]. The message and time complexities of these are summarized in Table 5.1.

The α, β, and γ synchronizers use the notion of process safety, defined as follows. A process i is said to be *safe* in round r if all messages sent by i in round r have been received. The α and β synchronizers are extreme cases of the γ synchronizer and form its building blocks.

A simple synchronizer

This synchronizer requires each process to send every neighbor one and only one message in each round. If no message is to be sent in the synchronous algorithm, an empty dummy message is sent in the asynchronous algorithm; if more than one message are sent in the synchronous algorithm, they are combined into one message in the asynchronous algorithm. In any round, when a process receives a message from each neighbor, it moves to the next round.

We make the following observations about this synchronizer.

- In physical time, any two processes may be only one round apart. Thus, if process i is in round $round_i$, any other adjacent process j must be in rounds $round_i - 1$, $round_i$, or $round_i + 1$ only.
- When process i is in round $round_i$, it can receive messages only from rounds $round_i$ or $round_i + 1$ from its neighbors.

Initialization

Any process may start round i. Within d time units, all processes will participate in that round. Hence, $T_{init} = d$. $M_{init} = 0$ because no explicit messages are required solely for initialization.

Complexity

Each round requires a message to be sent on each incident link in each direction. Hence, $M_{round} = 2|L|$ and $T_{round} = 1$.

The α synchronizer

At any process i, the α synchronizer in round r moves the process to the next round $r + 1$ if all the neighboring processes are *safe* for round r.

A process can learn about the safety of its neighbor if any message sent by this process is required to be acknowledged. Once a neighbor j has received acknowledgements for all the messages it sent, it sends a message informing i (and all its other neighbors) that it is safe.

Example The operation is illustrated in Figure 5.10. (step 1) Node A sends a message to nodes C and E, and receives messages from B and E in the same round. (step 2) These messages are acknowledged after they are received. (step 3) Once node A receives the acknowledgements from C and E, it sends a message to all its neighbors to notify them that node A is safe. This allows the neighbors to not wait on A before proceeding to the next round. Node A itself can proceed to the next round only after it receives a safety notification from each of its neighbors, whether or not there was any exchange of application execution messages with them in that round.

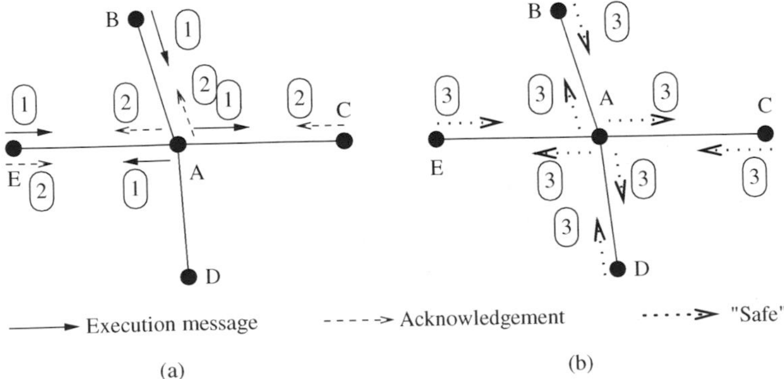

Figure 5.10 An example showing steps of the α synchronizer. (a) Execution messages (step 1) and their acknowledgements (step 2). (b) "I am safe" messages (step 3).

Complexity

For every message sent ($\leq |L|$) in a round, an ack is required. If $l'(< |L|)$ messages are sent in a round, l' acks are needed, giving a message overhead of $2l'$ thus far; but it is assumed that an underlying transport layer (or equivalent) protocol uses acks, and hence these come for free. But additionally, $2|L|$ messages are required so that each process can inform all its neighbors that it is safe. Thus the message complexity $M_{round} = 2|L| + 2l' = O(|L|)$. The time complexity $T_{round} = O(1)$.

Initialization

No explicit initialization is needed. A process that spontaneously wakes up and initializes the algorithm sends messages to (some of) its neighbors, who then acknowledge any message received, and also reply that they are safe.

The β synchronizer

This synchronizer assumes a rooted spanning tree. Safe leaf nodes initiate a convergecast; an intermediate node propagates the convergecast to its parent when all the nodes in its subtree, including itself, are safe. When the root becomes safe and receives the convergecast from all its children, it uses a tree broadcast to inform all the nodes to move to the next phase.

Example Compared to the α synchronizer, steps 1 and 2 as described with respect to Figure 5.10 are the same to determine when to notify others about safety. The actual notification about safety uses the convergecast–broadcast sequence on a pre-established tree, instead of using step 3 of Figure 5.10.

Complexity

Just as for the α synchronizer, an ack is required by the β synchronizer for each message of the l' messages sent in a round; hence l' acks are required, but these can be assumed to come for free, thanks to the transport layer or an equivalent lower layer protocol. Now instead of $2l$ further messages as in the α synchronizer, only $2(n-1)$ further messages are required for the convergecast and broadcast. Hence, $M_{round} = 2(n-1)$. For each round, there is an average case $2 \cdot log(n)$ delay for T_{round} and a worst-case $2n$ delay for T_{round}, incurred by the convergecast and the broadcast.

Initialization

There is an initialization cost, incurred by the set up of the spanning tree (the Algorithms in Section 5.5). As noted in Section 5.5, this cost is: $O(n \cdot log(n) + |L|)$ messages and $O(n)$ time.

The γ synchronizer

The network is organized into a set of clusters, as shown in Figure 5.11. Within a cluster, a spanning tree hierarchy exists with a distinguished root node. The

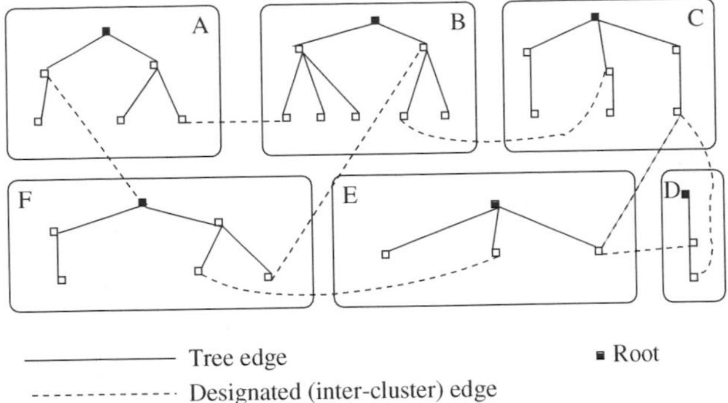

Figure 5.11 Cluster organization for the γ synchronizer, showing six clusters A–F. Only the tree edges within each cluster, and the inter-cluster *designated* edges are shown.

height of a clustering scheme, $h(c)$, is the maximum height of the spanning trees across all of the clusters. Two clusters are neighbors if there is at least one edge between one node in each of the two clusters; one of such multiple edges is the *designated* edge for that pair of clusters. Within a cluster, the β synchronizer is executed; once a cluster is "stabilized," the α synchronizer is executed among the clusters, over the *designated* edges. To convey the results of the stabilization of the inter-cluster α synchronizer, within each cluster, a convergecast and broadcast phase is then executed. Over the *designated* inter-cluster edges, two types of messages are exchanged for the α synchronizer: *My_cluster_safe*, and *Neighboring_cluster_safe*, with semantics that are self evident. The details of the algorithm are given in Algorithm 5.12.

Complexity

- Let L_c be the total number of tree edges plus designated edges in the clustering scheme. In each round, there are four messages – *Subtree_safe*, *This_cluster_safe*, *Neighboring_cluster_safe*, and *Next_round* – per tree edge, and two *My_cluster_safe* messages over each designated edge. Hence, M_{round} is $O(L_c)$.
- Let h_c be the maximum height of any tree among the clusters, then the time complexity component T_{round} is $O(h_c)$. This is due to the four phases – convergecast, broadcast, convergecast, and broadcast – contributing $4h_c$ time, the two units of time needed for all processes to become safe, and one unit of time needed for the inter-cluster messages *My_cluster_safe*.

Exercise 5.25 asks you to work out a formal design of how to partition the nodes into clusters, how to choose a root and a spanning tree of appropriate depth for each cluster, and how to designate the preferred edges. The requirements on the design scheme are to be able to control the complexity by suitably tuning a parameter k. The $\gamma(k)$ synchronizer reduces to the α synchronizer when $k = n - 1$, i.e., each cluster contains a single node. The

$\gamma(k)$ synchronizer reduces to the β synchronizer when $k = 2$, i.e., there is a single cluster. The construction will allow the $\gamma(k)$ synchronizer to be viewed as a parameterized synchronizer based on clustering.

(message types)
Subtree_safe // β synchronizer phase's convergecast within cluster
This_cluster_safe // β synchronizer phase's broadcast within cluster
My_cluster_safe // embedded inter-cluster α synchronizer's messages
// across cluster boundaries
Neighboring_cluster_safe // Convergecast following inter-cluster α
// synchronizer phase
Next_round // Broadcast following inter-cluster α synchronizer phase

for each *round* **do**

1. (β **synchronizer phase**) This phase aims to detect when all the nodes within a cluster are safe, and inform all the nodes in that cluster.
   (a) Using the spanning tree, leaves initiate the **convergecast** of the "*Subtree_safe*" message towards the root of the cluster.
   (b) After the convergecast completes, the root initiates a **broadcast** of "*This_cluster_safe*" on the spanning tree within the cluster.
   (c) **(Embedded α synchronizer)**
      (i) During this broadcast in the tree, as the nodes get engaged, the nodes also send "*My_cluster_safe*" messages on any incident *designated* inter-cluster edges.
      (ii) Each node also awaits "*My_cluster_safe*" messages along any such incident *designated* edges.
2. **(Convergecast and broadcast phase)** This phase aims to detect when all neighboring clusters are safe, and to inform every node within this cluster.
   (a) **(Convergecast)**
      (i) After the broadcast of the earlier phase (1(b)) completes, the leaves initiate a convergecast using "*Neighboring_cluster_safe*" messages once they receive any expected "*My_cluster_safe*" messages (step 1(c)) on all the *designated* incident edges.
      (ii) An intermediate node propagates the convergecast once it receives the "*Neighboring_cluster_safe*" message from all its children, and also any expected "*My_cluster_safe*" message (as per step 1(c)) along *designated* edges incident on it.
   (b) **(Broadcast)** Once the convergecast completes at the root of the cluster, a "*Next_round*" message is broadcast in the cluster's tree to inform all the tree nodes to move to the next round.

Algorithm 5.12 The γ synchronizer.

5.7 Maximal independent set (MIS)

For a graph (N, L), an *independent set* of nodes N', where $N' \subset N$, is such that for each i and j in N', $(i, j) \notin L$. An independent set N' is a *maximal independent set* if no strict superset of N' is an independent set. A graph may have multiple maximal independent sets; all of which may not be of the same size.[3]

The maximal independent set problem requires that adjacent nodes must not be chosen. This has application in wireless broadcast where it is required that transmitters must not broadcast on the same frequency within range of each other. More generally, for any shared resources (the radio frequency bandwidth in the above example) to allow a maximum concurrent use while avoiding interference or conflicting use, a maximal independent set is required.

Computing a maximal independent set in a distributed manner is challenging. The problem becomes further interesting when a maximal independent set must be maintained when processes join and leave, and links can go down, or new links between existing nodes can be established.

A simple and elegant distributed algorithm for the MIS problem in a static system, proposed by Luby [24], is presented in Algorithm 5.13 for an asynchronous system. The idea is as follows. In each iteration, each node P_i selects a random number $random_i$ and exchanges this value with its neighbors using the RANDOM message. If $random_i$ is less than the random numbers chosen by all its neighbors, the node includes itself in the MIS and exits. However, whether or not a node gets included in the MIS, it informs its neighbors via the indicator parameter on the SELECTED message. On receiving SELECTED messages from all the neighbors, if a node finds that at least one of its neighbors has been selected for inclusion in the MIS, the node eliminates itself from the candidate set for inclusion. However, whether or not an unselected node eliminates itself from the candidate set, it informs its neighbors via the indicator parameter on the ELIMINATED message. If a node learns that a neighbor j is eliminated from candidature, the node deletes j from *Neighbors*, and proceeds to the next iteration.

The algorithm constructs an IS because once a node is selected to be in the IS, all its neighbors are deleted from the set of remaining candidate nodes for inclusion in the IS. The algorithm constructs an MIS because only the neighbors of the selected nodes are eliminated from being candidates.

Example Figure 5.12(a) and (b) show the first two rounds in the execution of the MIS algorithm. The winners have a check mark and the losers have a

[3] The problem of finding the largest sized independent set is the *maximum independent set* problem. This is NP-hard.

cross next to them. In the third round, the node labeled I includes itself as a winner. The MIS is $\{C, E, G, I, K\}$.

```
(variables)
set of integer Neighbors   // set of neighbors
real random_i   // random number from a sufficiently large range
boolean selected_i   // becomes true when P_i is included in the MIS
boolean eliminated_i   // becomes true when P_i is eliminated from the
                       // candidate set
(message types)
RANDOM(real random)   // a random number is sent
SELECTED(integer pid, boolean indicator)   // whether sender was
                                           // selected in MIS
ELIMINATED(integer pid, boolean indicator)   // whether sender was
                                             // removed from candidates
(1a) repeat
(1b)   if Neighbors = ∅ then
(1c)        selected_i ⟵ true; exit();
(1d)   random_i ⟵ a random number;
(1e)   send RANDOM(random_i) to each neighbor;
(1f)   await RANDOM(random_j) from each neighbor j ∈ Neighbors;
(1g)   if random_i < random_j (∀j ∈ Neighbors) then
(1h)        send SELECTED(i, true) to each j ∈ Neighbors;
(1i)        selected_i ⟵ true; exit();                    // in MIS
(1j)   else
(1k)        send SELECTED(i, false) to each j ∈ Neighbors;
(1l)        await SELECTED(j, ⋆) from each j ∈ Neighbors;
(1m)        if SELECTED(j, true) arrived from some j ∈ Neighbors
              then
(1n)             for each j ∈ Neighbors from which SELECTED
                   (⋆, false) arrived do
(1o)                  send ELIMINATED(i, true) to j;
(1p)             eliminated_i ⟵ true; exit();            // not in MIS
(1q)        else
(1r)             send ELIMINATED(i, false) to each j ∈ Neighbors;
(1s)             await ELIMINATED(j, ⋆) from each j ∈ Neighbors;
(1t)             for all j ∈ Neighbors do
(1u)                  if ELIMINATED(j, true) arrived then
(1v)                       Neighbors ⟵ Neighbors \ {j};
(1w) forever.
```

Algorithm 5.13 Luby's algorithm for the maximal independent set in an asynchronous system. Code shown is for process P_i, $1 \le i \le n$.

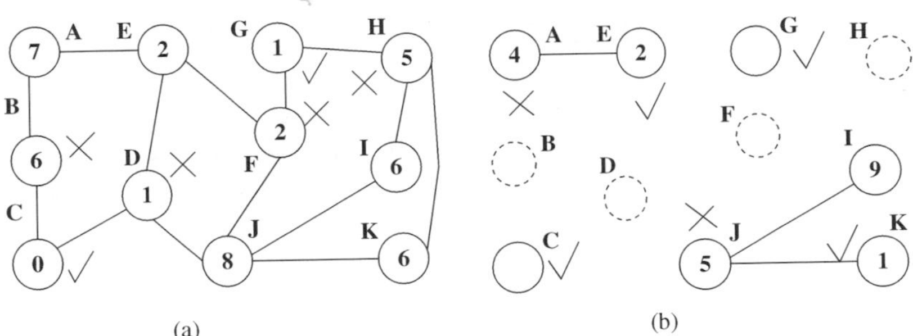

Figure 5.12 An example showing the execution of the MIS algorithm. (a) Winners and losers in round 1. (b) Winners up to round 2, and the losers in round 2.

Complexity

It is evident that in each iteration, at least one node will be included in the MIS, and at least one node will be eliminated from the candidate set. So at most $n/2$ iterations of the **repeat** loop are required. In fact, the expected number of iterations is $O(\log n)$. The reader is referred to the paper by Luby [24] for the proof of this bound.

5.8 Connected dominating set

A *dominating set* of graph (N, L) is a set $N' \subseteq N$ such that each node in $N \setminus N'$ has edge to some node in N'. Determining whether there exists a dominating set of size $k < |N|$ is NP-complete. A *connected dominating set* (CDS) of (N, L) is a dominating set N' such that the subgraph induced by the nodes in N' is connected.

Finding the miminum connected dominating set (MCDS) is NP-complete, and hence polynomial time heuristics are used to design approximation algorithms. In addition to the time and message complexities, the *approximation factor* becomes an important metric. The approximation factor is the worst case ratio of the size of the CDS obtained by the algorithm to the size of the MCDS. Another useful metric is the *stretch factor*. This is the worst-case ratio of the length of the shortest route between the dominators of two nodes in the CDS overlay, to the length of the shortest routes between the two nodes in the underlying graph.

The connected dominating set can form a backbone along which a broadcast can be performed. All nodes are guaranteed to be within range of the backbone and can hence receive the broadcast. The set is thus useful for routing, particularly in the wide-area network and also in wireless networks.

A simple heuristic is to create a spanning tree and delete the edges to the leaf nodes to get a CDS. Another heuristic is to create an MIS and add edges to create a CDS. However, designing an algorithm with a low approximation factor is non-trivial. Section 5.15 points to a couple of sources for efficient distributed CDS algorithms.

5.9 Compact routing tables

Routing tables are traditionally as large as the number of destinations n. This can have high storage requirements as well as table lookup and processing overheads when routing each packet. If the table can be reorganized such that it is indexed by the incident incoming link, and the table entry gives the outgoing link, then the table size becomes the degree of the node, which can be much smaller than n. Further efficiency would depend on how the destinations reachable per channel are represented and accessed. Some of the approaches to designing compact routing tables include the following:

- **Hierarchical routing schemes [33]** The network graph is organized into clusters in a hierarchical manner, with each cluster having one clusterhead designated node that represents the cluster at the next higher level in the hierarchy. There is detailed information about routing within a cluster, at all the routers within that cluster. If the destination does not lie in the same cluster as the source, the packet is sent to the clusterhead and up the hierarchy as appropriate. Once the clusterhead of the destination is found in the routing tables, then the packet is sent across the network at that level of the hierarchy, and then down the hierarchy in the destination cluster. This form of routing is widely used in the Internet.
- **Tree-labeling schemes [15]** This family of schemes uses a logical tree topology for routing. The routing scheme requires labeling the nodes of the graph in such a way that all the destinations reachable via any link can be represented as a range of contiguous addresses $[x, y]$. A node with degree *deg* need only maintain *deg* entries in its routing table, where each entry is a range of contiguous addresses. For all the address intervals $[x, y]$ except at most one, the scheme must satisfy $x < y$.

  Example Figure 5.13 shows tree labeling on a tree with seven nodes. The tree edge labels are enclosed in rectangles. Non-tree edges are in dashed lines.

  Tree-labeling can provide great savings, compared to a table of size n at each node. Unfortunately, all traffic is confined to the logical tree edges.

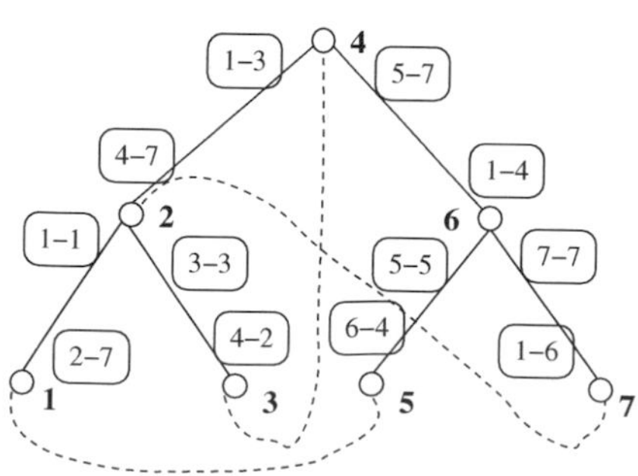

Figure 5.13 Tree labeling on a graph with seven nodes.

Exercise 5.26 asks you to show that it is always possible to generate a tree-labeling scheme.

- **Interval routing schemes** [15,35] The tree-labeling schemes suffer from the fact that data can be sent only over tree edges, wasting the remaining bandwidth in the system. Interval routing extends the tree labeling so that the data packets need not be sent only on the edges of a tree.

  Formally, given a graph (N, L), an interval routing scheme is a tuple $(\mathcal{B}, \mathcal{I})$, where:

  1. node labeling: $\mathcal{B}$ is a 1:1 mapping on N, which assigns labels to nodes;
  2. edge labeling: the mapping $\mathcal{I}$ labels each edge in L by some subset of node labels $\mathcal{B}(N)$ such that for any node x, all destinations are covered $(\cup_{y \in Neighbors} \mathcal{I}(x, y) \cup \mathcal{B}(x) = N)$ and there is no duplication of coverage $(\mathcal{I}(x, w) \cap \mathcal{I}(x, y) = \emptyset$ for $w, y \in Neighbors)$;
  3. for any source s and destination t nodes, there must exist a sequence of nodes $\langle s = x_0, x_1 \ldots x_{k-1}, x_k = t \rangle$ where $\mathcal{B}(t) \in \mathcal{I}(x_{i-1}, x_i)$ for each i between 1 and k. Therefore, for each source and destination pair, there must exist a path under the new mapping.

  To show that an interval labeling scheme is possible for every graph, a tree with the following property is constructed: "there are no cross-edges in the corresponding graph." The tree generated by a depth-first traversal always satisfies this property. Nodes are labeled by a preorder traversal whereas the edges are labeled by a more detailed scheme, see [35].

  Two drawbacks of interval routing schemes are that: (i) they do not give any guarantees on the efficiency (lengths) of the routing paths that get chosen, and (ii) they are not robust to small changes in the topology.

- **Prefix routing schemes** [15] Prefix routing schemes overcome the drawbacks of interval routing. (This prefix routing is not to be confused with the CIDR routing used in the internet. CIDR also uses the prefixes of the destination IP address.) In prefix routing, the node labels as well as the channel labels are drawn from the same domain and are viewed as strings. The routing decision at a router is as follows: identify the channels whose label is the longest prefix of the address of the destination. This is the channel on which to route the packet for that particular destination.

The *stretch factor* of a routing scheme r is defined as $max_{i,j \in N}\{\frac{distance_r(i,j)}{distance_{opt}(i,j)}\}$. This is an important metric in evaluating a compact routing scheme.

All the above approaches for compact routing are rich in distributed graph algorithmic problems and challenges, including identifying and proving bounds on the efficiency of computed routes. Different graph topologies yield interesting results for these routing schemes.

5.10 Leader election

We have seen the role of a leader process in several algorithms such as the minimum spanning tree and broadcast/convergecast to compute a function over all the participating processes.

Leader election requires that all the processes agree on a common distinguished process, also termed as the *leader*. A leader is required in many distributed systems because algorithms are typically not completely symmetrical, and some process has to take the lead in initiating the algorithm; another reason is that we would not want all the processes to replicate the algorithm initiation, to save on resources.

Typical algorithms for leader election assume a ring topology is available. Each process has a left neighbor and a right neighbor. The Lelang, Chang, and Roberts (LCR) algorithm [6,23] assumes an asynchronous unidirectional ring. It also assumes that all processes have unique identifiers. Each process in the ring sends its identifier to its left neighbor. When a process P_i receives the identifier k from its right neighbor P_j, it acts as follows:

- $i < k$: forward the identifier k to its left neighbor;
- $i > k$: ignore the message received from neighbor j;
- $i = k$: due to the assumption on nonanonymity, P_i's identifier must have circluated across the entire ring. Hence P_i can declare itself the leader.

P_i can then send another message around the ring announcing that it has been chosen as the leader. The algorithm is given in Algorithm 5.14.

Complexity

The LCR algorithm (Algorithm 5.14) is in its simplest form. Several optimizations are possible. For example, if i has forwarded a probe with value z and a probe with value x, where $i < x < z$ arrives, no forwarding action on the probe needs to be taken. Despite this, it is straightforward to see that the message complexity of this algorithm is $n \cdot (n-1)/2$ and the time complexity is $O(n)$.

The $O(n^2)$ message cost can be reduced to $O(n\, log\, n)$ by using a binary search in both directions as proposed by Hirschberg and Sinclair [19]. In round k, the token is circulated to 2^k neighbors on both the left and right sides. To cover the entire ring, a logarithmic number of steps are needed. Consider that in each round, a process tries to become a leader, and only the winners in round k can proceed to round $k+1$. In effect, a process i is a leader in round k if and only if i is the highest identifier among 2^k neighbors in both directions. Hence, any pair of leaders after round k are at least 2^k apart. Hence the number of leaders diminishes logarithmically as $n/2^k$. Observe that in each round, there are at most n messages sent, using the supression technique of the LCR algorithm. Thus the overall complexity is $O(n \cdot log\, n)$.

(variables)
boolean *participate* $\leftarrow$ *false* // becomes true when P_i is participates in
// leader election

(message types)
PROBE **integer** // contains a node identifier
SELECTED **integer** // announcing the result

(1) When a process wakes up to participate in leader election:
(1a) **send** PROBE(i) to right neighbor;
(1b) *participate* $\longleftarrow$ *true*.

(2) When a PROBE(k) message arrives from the left neighbor P_j:
(2a) **if** $participate = false$ **then** execute step (1) first.
(2b) **if** $i > k$ **then**
(2c) discard the probe;
(2d) **else if** $i < k$ **then**
(2e) **forward** PROBE(k) to right neighbor;
(2f) **else if** $i = k$ **then**
(2g) declare i is the leader;
(2h) circulate SELECTED(i) to right neighbor;

(3) When a SELECTED(x) message arrives from left neighbor:
(3a) **if** $x \neq i$ **then**
(3b) note x as the leader and forward message to right neighbor;
(3c) **else** do not forward the SELECTED message.

Algorithm 5.14 The LCR leader election algorithm in a synchronous system. Code shown is for process P_i, $1 \leq i \leq n$.

It has been shown that there cannot exist a deterministic leader election algorithm for anonymous rings. Hence, the assumption about node identifiers is necessary in this model. However, the algorithm can be uniform, i.e., the total number of processes need not be known.

5.11 Challenges in designing distributed graph algorithms

We have thus far considered some elementary but important graph problems, and seen how to solve them in distributed algorithms. The algorithms either fail or require a more complicated redesign if we assume that the graph topology changes dynamically, which happens in mobile systems.

- The graph (N, L) changes dynamically in the normal course of execution of a distributed execution. An example is the load on a network link, which is really determined as the aggregate of many different flows. It is unrealistic to expect that this will ever be static. All of a sudden, the MST algorithms (and others) need a complete overhaul.

- The graph can change if either there are link or node failures, or worse still, partitions in the network. The graph can also change when new links and new nodes are added to the network. Again, the algorithms seen thus far need to be redesigned to accommodate such changes.

The challenge posed by mobile systems additionally needs to deal with the new communication model. Here, each node is capable of transmitting data wirelessly, and all nodes within a certain radius can receive it. This is the unit-disk radius model.

5.12 Object replication problems

We now describe a real-life graph problem based on web/data replication, which also requires dynamic distributed solutions.

1. Consider a weighted graph (N, L), wherein k users are situated at some $N_k \subseteq N$ nodes, and r replicas of a data item can be placed at some $N_r \subseteq N$. What is the optimal placement of the replicas if $k > r$ and the users access the data item in read-only mode?
   A solution requires evaluating all placements of N_r among the nodes in N to identify $\min(\sum_{i \in N_k, r_i \in N_r} dist_{i,r_i})$, where $dist_{i,r_i}$ is the cost from node i to r_i, the replica nearest to i.
2. If we assume that the read accesses from each of the users in N_k have a certain frequency (or weight), the minimization function would change.
3. If each edge has a certain bandwidth or capacity, that too has to be taken into account in identifying a feasible solution.
4. Now assume that a user access to the shared data is a read operation with probability x, and an update operation with probability $1-x$. An update operation also requires all replicas to be updated. What is the optimal placement of the replicas if $k > r$?

Many such graph problems do not always have polynomial solutions even in the static case. With dynamically changing input parameters, the case appears even more hopeless for an optimal solution. Fortunately, heuristics can often be used to provide good solutions.

5.12.1 Problem definition

In a large distributed system, data replication is useful for rapid access to data and for fault-tolerance. Here we look at Wolfson *et al.*'s optimal data replication strategy that is dynamic in that it adapts to the read and write patterns from the different nodes [37]. Let the network be modeled by the graph (V, E), and let us focus on a single object for simplicity. Define a *replication scheme* as a subset R of V such that each node in R has a replica of the object. Let r_i and w_i denote the rates of reads and writes issued by node i. Let $c_r(i)$ and $c_w(i)$ denote the cost of a read and write issued by node i. Let $\mathcal{R}$ denote the set of all possible replication schemes. The goal is to minimize the cost of the replication scheme:

$$\min_{R\in\mathcal{R}} \left[\sum_{i\in V} r_i \cdot c_r(i) + \sum_{i\in V} w_i \cdot c_w(i) \right]. \tag{5.3}$$

The algorithm assumes one copy serializability, which can be implemented by the read-one-write-all (ROWA) policy. ROWA can be strictly implemented in conjunction with a concurrency control mechanism such as two-phase locking; however, lazy propagation can also be used for weaker semantics.

5.12.2 Algorithm outline

For arbitrary graph topologies, minimizing the cost as in Eq. (5.3) is NP-complete. So we assume a tree topology T, as shown in Figure 5.14. The nodes in the replication scheme R are shown in the ellipse. If T is allowed to be a *tree overlay* T on the network topology, then all algorithm communication is confined to the overlay. Conceptually, the set of nodes R containing the replicas is an amoeba-like connected subgraph that moves around the overlay tree T towards the "center of gravity" of the read and write activity. The amoeba-like subgraph expands when the relative cost of the reads is more than that of writes, and shrinks as the relative cost of writes is more than that of reads, reaching an equilibrium under steady state activity. This equilibrium-state subgraph for the replication scheme is optimal. The algorithm executes in steps that are separated by predetermined time periods or "epochs." Irrespective of the initial replication scheme, the algorithm converges to the optimal replication scheme in $(diameter + 1)$ number of steps once the read-and-write pattern stabilizes.

5.12.3 Reads and writes

Read

A read operation is performed from the closest replica on the tree T. If the node issuing the read query or receiving a forwarded read query is not in

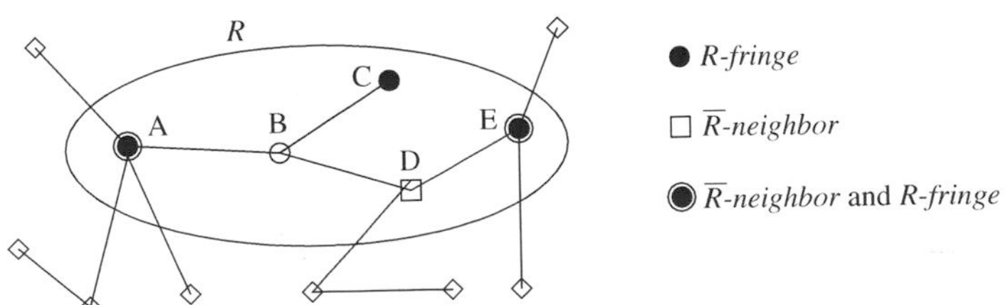

Figure 5.14 The tree topology and the replication scheme R. Nodes inside the ellipse belong to the replication scheme.

R, it forwards the query towards the nodes in R along the tree edges – for this, it suffices that a *parent* pointer point in the direction of the subgraph R. Once the query reaches a node in R, the value read is returned along the same path.

Write

A write is performed to every replica in the current replication scheme R. If a write operation is issued by a node not in R, the operation request is propagated to the closest node in R, like for the read operation request. Once a write operation reaches a node i in R, the local replica is updated, and the operation is propagated to all neighbors of i that belong to R. To implement this, a node needs to track the set of its neighbors that belong to R. This is done using a variable, *R-neighbor*.

Implementation

To execute a read or write operation, a node needs to know (i) whether it is in R (so it can read/write from the local replica), (ii) which of its neighbors are in R (to propagate write requests), and (iii) if the node is not in R, then which of its neighbors is the unique node that leads on the tree to R (so it can propagate read and write requests). After appropriate initialization, this information is always locally available by tracking the status of the neighbor nodes.

5.12.4 Converging to an replication scheme

Within the replication scheme R, three types of nodes are defined:

- $\overline{R}$*-neighbor*: Such a node i belongs to R but has at least one neighbor j that does not belong to R.
- *R-fringe*: Such a node i belongs to R and has only one neighbor j that belongs to R. Thus, i is a leaf node in the subgraph of T induced by R and j is the parent of i.
- *singleton*: $|R| = 1$ and $i \in R$.

Example In Figure 5.14, node C is an *R-fringe* node, nodes A and E are both *R-fringe* and $\overline{R}$*-neighbor* nodes, and node D is an $\overline{R}$*-neighbor* node.

The algorithm uses the following three tests to adjust the replication scheme to converge to the optimal scheme:

- **Expansion test** An $\overline{R}$*-neighbor* node i examines each such neighbor j to determine whether j can be included in the replication scheme, using an *expansion test*. Node j is included in the replication scheme if the volume of reads coming from and via j is more than the volume of writes that would have to be propagated to j from i if j were included in the replication scheme.

(variables)

integer $Neighbors[1 \ldots b_i]$; // b_i neighbors in tree T topology

integer $Read\_Received[1 \ldots b_i]$; // jth element gives # reads // from $Neighbors[j]$

integer $Write\_Received[1 \ldots b_i]$; // jth element gives # writes // from $Neighbors[j]$

integer $write_i$, $read_i$; // # writes and # reads issued locally

boolean *success*;

(1) P_i determines which tests to execute at the end of each epoch:

(1a) **if** i is $\overline{R}$-*neighbor* and *R-fringe* **then**

(1b) **if** *expansion test* fails **then**

(1c) *reduction test*

(1d) **else if** i is $\overline{R}$-*neighbor* and *singleton* **then**

(1e) **if** *expansion test* fails **then**

(1f) *switch test*

(1g) **else if** i is $\overline{R}$-*neighbor* and not *R-fringe* and not *singleton* **then**

(1h) *expansion test*

(1i) **else if** i is $\overline{\overline{R}}-neighbor$ and *R-fringe* **then**

(1j) *contraction test*.

(2) P_i executes *expansion test:*

(2a) **for** j **from** 1 **to** b_i **do**

(2b) **if** $Neighbors[j]$ not in R **then**

(2c) **if** $Read\_Received[j] > (write_i + \sum_{k=1 \ldots b_i, k \neq j} Write\_Received[k])$ **then**

(2d) send a copy of the object to $Neighbors[j]$; $success \longleftarrow 1$;

(2e) **return**(*success*).

(3) P_i executes *contraction test:*

(3a) let $Neighbors[j]$ be the only neighbor in R;

(3b) **if** $Write\_Received[j] > (read_i + \sum_{k=1 \ldots b_i, k \neq j} Read\_Received[k])$ **then**

(3c) seek permission from $Neighbors[j]$ to exit from R;

(3d) **if** permission received **then**

(3e) $success \longleftarrow 1$; inform all neighbors;

(3f) **return**(*success*).

(4) P_i executes *switch test:*

(4a) **for** j **from** 1 **to** b_i **do**

(4b) **if** $(Read\_Received[j] + Write\_Received[j]) > [\sum_{k=1 \ldots b_i, k \neq j}(Read\_Received[k] + Write\_Received[k]) + read_i + write_i]$ **then**

(4c) transfer object copy to $Neighbors[j]$; $success \longleftarrow 1$; inform all neighbors;

(4d) **return**(*success*).

Algorithm 5.15 Adaptive data replication algorithm executed by a node P_i in replication scheme R. All variables except *Neighbors* are reset at the end of each epoch. R stabilizes in $diameter + 1$ epochs after the read–write rates stabilize.

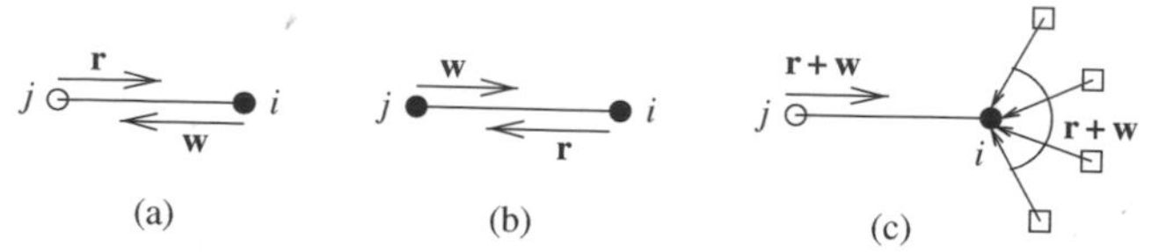

Figure 5.15 Adaptive data replication tests executed by node *i*. (a) Expansion test. (b) Contraction test. (c) Switch test.

Example In Figure 5.15(a), node i includes j in the replication scheme if $r > w$.

- **Contraction test** An *R-fringe* node i examines whether it can exclude itself from the replication scheme, using a *contraction test*. Node i excludes itself from the replication scheme if the volume of writes being propagated to it from j is more than the volume of reads that i would have to forward to j if i were to exit the replication scheme. Before exiting, node i must seek permission from j to prevent a situation where $R = \{i, j\}$ and both i and j simultaneously have a successful *contraction test* and exit, leaving no copies of the object.

  Example In Figure 5.15(b), node i excludes itself from the replication scheme if $w > r$.

- **Switch test** A *singleton* node i executes the *switch test* to determine if it can transfer its replica to some neighbor to optimize the objective function. A singleton node transfers its replica to a neighbor j if the volume of requests being forwarded by that neighbor is greater than the volume of requests the node would have to forward to that neighbor if the replica were shifted from itself to that neighbor. If such a node j exists, observe that it is uniquely identified among the neighbors of node i.

  Example In Figure 5.15(c), node i transfers its replica to j if $r+w$ being forwarded by j is greater than $r+w$ that node i receives from all other nodes.

The various tests are executed at the end of each "epoch." An $\overline{R}$-*neighbor* node may also be an *R-fringe* node or a *singleton* node; in either case, the *expansion test* is executed first and if it fails, then the *contraction test* or the *switch test* is executed. Note that a singleton node cannot be an *R-fringe* node. The code is given in Algorithm 5.15.

Implementation

Each node needs to be able to determine whether it is in R, whether it is an $\overline{R}$-*neighbor* node, an *R-fringe* node, or a *singleton* node. This can be determined if a node knows whether it is in R, the set of neighbor nodes, and for each such neighbor, whether it is in R. This is a subset of the information required for implementing read and write operations, and can be tracked easily using local exchanges. Hence, these operations are not shown in the code in Algorithm 5.15. The actions to service read and write requests described earlier are also straightforward and are not shown code.

Correctness
Given an initial connected replication scheme, the replication scheme after each epoch remains connected, and the replication schemes in two consecutive epochs either intersect or are adjacent singletons. This property follows from the fact that for each node $i \in R$, in each epoch, at most one of the three tests – expansion, contraction, and switch – succeeds, and the corresponding transformation satisfies the above property. Given two disconnected components of a replication scheme, it is easy to see that adding nodes to combine the components can never increase the cost (Eq. (5.3)) of the replication scheme.

Once the read–write pattern stabilizes, the replication scheme stabilizes within $diameter + 1$ number of epochs, and the resulting replication scheme is optimal. The proof is fairly complex; below are the main steps to show termination, and these can be validated intuitively. For the optimality argument, note that each change in an epoch reduces the cost. The proof that the replication scheme on termination is globally optimal and not just locally optimal is given in the full paper [37].

Termination

- After a *switch test* succeeds, no other *expansion test* can succeed.
- If a node exits the replication scheme in a *contraction test*, it cannot re-enter the replication scheme via an *expansion test*.
- If a node exits the replication scheme in a *switch test*, it cannot re-enter the replication scheme again.

Thus, if a node exits the replication scheme, it can re-enter only by a *switch test*, and that too if the exit was via a *contraction test*. But then, no further *expansion test* can succeed. Hence, a node can exit the replication scheme at most once more – via a *switch test*. Each node can exit the replication scheme at most twice, and after the first *switch test*, no expansion can occur. Hence the replication scheme stabilizes.

It can be seen that the replication scheme first expands wherever possible, and then contracts. If it becomes a *singleton*, then the only changes possible are switches.

Arbitrary graphs

The algorithm so far assumes the graph was a tree, on which the replication scheme "amoeba" moves into optimal position. For arbitrary graphs, a tree overlay can be used. However, the tree structure also has to change dynamically because the shortest path in the spanning tree between two arbitrary nodes is not always the shortest path between the nodes in the graph. Modified versions of the three tests can now be used, but the structure of the graph does not guarantee the global optimum solution, but only that a local optimum is reached.

5.13 Chapter summary

This chapter first examined various views of the distributed system at different levels of abstraction of the topology of the system graph. It then introduced basic terminology for classifying distributed algorithms and distributed executions. This covered failure models of nodes and links. It then examined several performance metrics for distributed algorithms.

The chapter then examined several traditional distributed algorithms on graphs. The most basic of such algorithms are the spanning tree, minimum-weight spanning tree, and the shortest path algorithms – both single source and multi-source. The importance of these algorithms lies in the fact that spanning trees are used for information distribution and collection via *broadcast* and *convergecast*, respectively, and these functions need to be performed by a wide range of distributed applications. The convergecast and broadcast performed on the spanning trees also allow the repeated computation of a global function such as *min*, *max*, and $\sum$. Some of the shortest path routing algorithms studied are seen to be used in the Internet at the network layer. In all cases, the synchronous version and then the asynchronous version of the algorithms were examined.

The various examples of algorithm design showed that it is often easier to construct an algorithm for a synchronous system than it is for an asynchronous system. The chapter then studied synchronizers, which are transformations that allow any algorithm designed for a synchronous system to run in an asynchronous system. Specifically, four synchronizers, in the order of increasing complexity, were studied – the simple synchronizer, the α synchronizer, the β synchronizer, and the γ synchronizer.

A distributed randomized algorithm for the maximal independent set problem was studied, and then the problem of determining a connected dominating set was examined. The chapter then examined several compact routing schemes. These aim to trade-off routing table size for slightly longer routes. The leader election problem was then considered. The chapter concluded by taking a look at the problem of dynamic replication of read/write objects to minimize traffic.

5.14 Exercises

Exercise 5.1 Adapt the synchronous BFS spanning tree algorithm (Algorithm 5.1) to satisfy the following properties:

1. The root node can detect once the entire algorithm has terminated. The root should then terminate.
2. Each node is able to identify its child nodes without using any additional messages.
3. A process exits after the round in which it sets its *parent* variable.

What is the resulting space, time, and message complexity in each case?

Exercise 5.2 What is the exact number of messages sent in the spanning tree algorithm (Algorithm 5.2)? You may want to use additional parameters to characterize the graph. Is it possible to reduce the number of messages to exactly $2l$?

Exercise 5.3 Modify Algorithm 5.2 to obtain a BFS tree with the asynchronous system, while retaining the framework of the flooding mechanism.

Exercise 5.4 Modify the asynchronous spanning tree algorithm (Algorithm 5.2) to eliminate the use of REJECT messages. What is the message overhead of the modified algorithm?

Exercise 5.5 What is the maximum distance between any two nodes in the tree obtained by running Algorithm 5.3?

Exercise 5.6 For Algorithm 5.3, show each of the performance complexities introduced in Section 5.3.

Exercise 5.7 For Algorithm 5.4, show each of the performance complexities introduced in Section 5.3.

Exercise 5.8 (Based on Cheung [7]) Simplify Algorithm 5.4 to deal with only a single initiator. What is the message complexity and the time complexity of the resulting algorithm?

Exercise 5.9 (Based on [2]) Modify the algorithm derived in Exercise 5.8 to obtain a depth-first search tree but with time complexity $O(n)$. (Assuming a single intiator for simplicity does not reduce the time complexity. A different strategy needs to be used.)

Exercise 5.10 Formally write the convergecast algorithm of Section 5.5.5 using the style for the other algorithms in this chapter.

Modify your algorithm to satisfy the following property. Each node has a sensed temperature reading. The maximum temperature reading is to be collected by the root.

Exercise 5.11 Modify the synchronous flooding algorithm (Algorithm 5.10) so as to reduce the complexity, assuming that all the processes only need to know the highest process identifier among all the processes in the network. For this adapted algorithm, what are the lowered complexity measures?

Exercise 5.12 Adapt Algorithms 5.5 and 5.10 to design a synchronous algorithm that achieves the following property: "in each round, each node may or may not generate a new update that it wants to distribute throughout the network. If such an update is locally generated within a round, it should be synchronously propagated in the network."

Exercise 5.13 In the synchronous distributed Bellman–Ford algorithm (Algorithm 5.5), the termination condition for the algorithm assumed that each process knew the number of nodes in the graph. If this number is not known, what can be done to find it?

Exercise 5.14 In the asynchronous Bellman–Ford algorithm (Algorithm 5.6), what can be said about the termination conditions when (i) n is not known, and when (ii) n is known?

For each of these two cases, modify the asynchronous Bellman–Ford algorithm to allow each process to determine when to terminate.

Exercise 5.15 Modify the asynchronous Bellman–Ford algorithm (Algorithm 5.6) to devise the distance vector routing algorithm outlined in Section 5.5.7.

Exercise 5.16 For the asynchronous Bellman–Ford algorithm (Algorithm 5.6), show that it has an exponential $\Omega(c^n)$ number of messages and exponential $\Omega(c^n \cdot d)$ time complexity in the worst case, where c is some constant [25].

Exercise 5.17 For the asynchronous Bellman–Ford algorithm (Algorithm 5.6), if all links are assumed to have equal weight, the algorithm effectively computes the minimum-hop path. Show that under this assumption, the minimum-hop routing tables to all destinations are computed using $O(n^2 \cdot l)$ messages.

Exercise 5.18 For the asynchronous Bellman–Ford algorithm (Algorithm 5.6):

1. If some of the links may have negative weights, what would be the impact on the shortest paths? Explain your answer.
2. If the link weights can keep changing (as in the Internet), can cycles be formed during routing based on the computed next hop?

Exercise 5.19 In the distributed Floyd–Warshall algorithm (Algorithm 5.8), consider iteration k at node i and iteration $k+1$ at node j. Examine the dependencies in the code of i and j in these two iterations.

Exercise 5.20 In the distributed Floyd–Warshall algorithm (Algorithm 5.8):

1. Show that the parameter *pivot* is redundant on all the message types when the communication channels are FIFO.
2. Show that the parameter *pivot* is required on all the message types when the communication channels are non-FIFO.

Exercise 5.21 In the synchronous distributed GHS algorithm (Algorithm 5.11), it was assumed that all the edge weights were unique. Explain why this assumption was necessary, and give a way to make the weights unique if they are not so.

Exercise 5.22 In the synchronous GHS MST algorithm, prove that when several components join to form a single component, there must exist a cycle of length two in the component graph of MWOE edges.

Exercise 5.23 Identify how the complexity of the synchronous GHS algorithm can be reduced from $O((n+|L|)log\ n)$ to $O((n\ log\ n)+|L|)$. Explain and prove your answer.

Exercise 5.24 Consider the simple, α, and β synchronizers. Identify some algorithms or application areas where you can identify one synchronizer as being more efficient than the others.

Exercise 5.25 For the γ synchronizer, significant flexibility can be achieved by varying a parameter k that is used to give a bound on L_c (sum of the number of tree edges and clustering edges) and h_c (maximum height of any tree in any cluster). Visually, this parameter determines the flatness of the cluster hierarchy.

Show that for every k, $2 \leq k < n$, a clustering scheme can be designed so as to satisfy the following bounds: (1) $L_c < k \cdot n$, and (2) $h_c \leq (log\ n)/(log\ k)$.

Exercise 5.26 1. For the tree-labeling scheme for compact routing, show that a pre-order traversal of the tree generates a numbering that always permits tree-labeled routing.
2. Will post-order traversal always generate a valid tree-labeling scheme?
3. Will in-order traversal always generate a valid tree-labeling scheme?

Exercise 5.27 1. For the tree-labeling schemes, show that there is no *uniform* bound on the *dialation*, which is defined as the ratio of the length of the tree path to the optimal path, between any pair of nodes and an arbitrary tree.
2. Is it possible to bound the dialation by choosing a tree for any given graph? Explain your answer.

Exercise 5.28 Examine all the algorithms in this chapter, and classify them using the classifications introduced in Sections (5.2.1–5.2.10).

Exercise 5.29 Examine the impact of both fail-stop process failures and of crash process failures on all the algorithms described in this chapter. Explain your answers in each case.

Exercise 5.30 (Adaptive data replication) In the adaptive data replication scheme (Section 5.12), consider a node that is both an $\overline{R}$*-neighbor* and an *R-fringe* node.

1. Can the *expansion test* and the *reduction test* both be successful? Prove your answer.
2. The algorithm first performs the *expansion test*, and if it fails, then it performs the *reduction test*. Is it possible to restructure the algorithm to perform the *reduction test* first, and then the *expansion test*? Prove your answer.

Exercise 5.31 Modify the rules of the *expansion*, *contraction*, and *switch* tests in the adaptive dynamic replication algorithm of Section 5.12 to adapt to tree overlays on arbitrary graphs, rather than to tree graphs. Justify the correctness of the modified tests.

5.15 Notes on references

The discussion on the classification of distributed algorithms is based on the vast literature, and many of the definitions are difficult to attribute to a particular source. The discussion on execution inhibition is based on Critchlow and Taylor [10]. The discussion on failure models is based on Hadzilacos and Toueg [17]. Crash failures were proposed by Lamport and Fischer [21]. Failstop failures were introduced by Schlichting and Schneider [30]. Send omission failures were introduced by Hadzilacos [16]. General omission failures and timing failures were introduced by Perry and

Toueg [27] and Christian *et al.* [8], respectively. The notion of wait-freedom was introduced by Lamport [20] and later developed by Herlihy [18]. The notions of the space, message, and time complexities have been around for a long time. The time and message complexity measures were formalized by Peterson and Fischer [28] and later by Awerbuch [3].

The various spanning tree algorithms are common knowledge and have been used informally in many contexts. Broadcast, convergecast, and distributed spanning trees are listed as part of a suite of elementary algorithms [13]. Segall [32] formally presented the broadcast and convergecast algorithms, and the breadth-first search spanning tree algorithm, on which Algorithm 5.1 is based. Algorithms 5.3 and 5.4, which compute flooding-based and depth-first search based spanning trees, respectively, in the face of concurrent initiators, use the technique of supressing lower priority initiations. This technique has been used in many other contexts in computer science (e.g., database transaction serialization, deadlock detection). An asynchronous DFS algorithm with a specified root was given by Cheung [7]. Algorithm 5.4 adapts this to handle concurrent initiators. The solution to Exercise 5.9, which asks for a linear-time DFS tree, was given by Awerbuch [2].

The synchronous Bellman–Ford algorithm is derived from the Bellman–Ford shortest path algorithm [4,12]. The asynchronous Bellman–Ford was formalized by Chandy and Misra [5]. The distance vector routing algorithm and synchronous flooding algorithm of Algorithm 5.10 are based on the Arpanet protocols [33]. The Floyd–Warshall algorithm is from [9] and its distributed version was given by Toueg [34]. The asynchronous flooding algorithm outlined in Algorithm 5.9 is based on the link state routing protocol used in the Internet [33].

The synchronous distributed minimum spanning tree algorithm was given by Gallagher *et al.* [14]. Its asynchronous version was also proposed by the same authors. The notion of synchronizers, and the α, β, and γ synchronizers were introduced by Awerbuch [3]. The randomized algorithm for the maximal independent set (MIS) was proposed by Luby [24]. Several distributed algorithms to create connected dominating sets with a low approximation factor are surveyed by Wan *et al.* [36]. The randomized algorithm for connected dominating set by Dubhashi *et al.* [11] has an approximation factor of $O(log\Delta)$, where Δ is the maximum degree of the network. This algorithm also has a stretch factor of $O(log\,n)$. Compact routing based on the tree topology was introduced by Santoro and Khatib [29]. Its generalization to interval routing was introduced by van Leeuwen and Tan [35]. A survey of interval routing mechanisms is given by Gavoille [15]. The LCR algorithm for leader election was proposed by LeLann [23] and Chang and Roberts who provided several optimizations [6]. The $O(n\,log\,n)$ alogrithm for leader election was given by Hirschberg and Sinclair [19]. The result on the impossibility of election on anonymous rings was shown by Angluin [1]. The adaptive replication algorithm was proposed by Wolfson *et al.* [37].

References

[1] D. Angluin, Local and global properties in networks of processors, *Proceedings of the 12th ACM Symposium on Theory of Computing*, 1980, 82–93.

[2] B. Awerbuch, Optimal distributed algorithms for minimum weight spanning tree, counting, leader election, and related problems, *Proceedings of 19th ACM Symposium on Principles of Theory of Computing (STOC)*, 1987, 230–240.

[3] B. Awerbuch, Complexity of network synchronization, *Journal of the ACM*, **32**(4), 1985, 804–823.

[4] R. Bellman, *Dynamic Programming*, Princeton, NJ, Princeton University Press, 1957.

[5] K. M. Chandy and J. Misra, Distributed computations on graphs: shortest path algorithms, *Communications of the ACM*, **25**(11), 1982, 833–838.

[6] E. Chang and R. Roberts, An improved algorithm for decentralized extrema-finding in circular configurations of processes, *Communications of the ACM*, **22**(5), 1979, 281–283.

[7] T.-Y. Cheung, Graph traversal techniques and the maximum flow problem in distributed computation, *IEEE Transactions on Software Engineering*, **9**(4), 1983, 504–512.

[8] F. Christian, H. Aghili, H. Strong, and D. Dolev, Atomic broadcast: from simple message diffusion to Byzantine agreement, *Proceedings of the 15th International Symposium on Fault-Tolerant Computing*, 1985, 200–206.

[9] T. Cormen, C. Leiserson, R. Rivest, and C. Stein, *An Introduction to Algorithms*, 2nd edn, Cambridge, MA, MIT Press, 2001.

[10] C. Critchlow and K. Taylor, The inhibition spectrum and the achievement of causal consistency, *Distributed Computing*, **10**(1), 1996, 11–27.

[11] D. Dubhashi, A. Mei, A. Panconesi, J. Radhakrishnan, and A. Srinivasan, Fast distributed algorithms for (weakly) connected dominating sets and linear-size skeletons, *Proceedings of the 14th Annual Symposium on Discrete Algorithms*, 2003, 717–724.

[12] L. Ford and D. Fulkerson, *Flows in Networks*, Princeton, NJ, Princeton University Press, 1962.

[13] E. Gafni, Perspectives on distributed network protocols: a case for building blocks, *Proceedings of the IEEE MILCOM*, Monterey, CA, 1986.

[14] R. Gallagher, P. Humblet, and P. Spira, A distributed algorithm for minimum-weight spanning trees, *ACM Transactions on Programming Languages and Systems*, **5**(1), 1983, 66–77.

[15] C. Gavoille, A survey on interval routing, *Theoretical Computer Science*, **245**(2), 2000, 217–253.

[16] V. Hadzilacos, *Issues of Fault Tolerance in Concurrent Computations*, Ph.D. dissertation, Harvard University, Computer Science Technical Report, 11-84, 1984.

[17] V. Hadzilacos and S. Toueg, Fault-tolerant broadcasts and related problems, in Mullender, S. (ed.) *Distributed Systems*, Addison-Wesley, 1993, 97–146.

[18] M. Herlihy, Wait-free synchronization, *ACM Transactions on Programming Languages and Systems*, **15**(5), 1991, 745–770.

[19] D. Hirschberg and J. Sinclair, Decentralized extrema-finding in circular configurations of processors, *Communications of the ACM*, **23**(11), 1980, 627–628.

[20] L. Lamport, Concurrent reading and writing, *Communications of the ACM*, **20**(11), 1977, 806–811.

[21] L. Lamport and M. Fischer, *Byzantine Generals and Transaction Commit Protocols*, SRI International, Technical Report 62, 1982.

[22] L. Lamport, R. Shostak, and M. Pease, The Byzantine generals problem, *ACM Transactions on Programming Languages and Systems*, **4**(3), 1982, 382–401.

[23] G. LeLann, Distributed systems, towards a formal approach, *IFIP Congress Proceedings*, 1977, 155–160.

[24] M. Luby, A simple parallel algorithm for the maximal independent set problem, *SIAM Journal of Computing*, **15**(4), 1986, 1036–1053.

[25] N. Lynch, *Distributed Algorithms*, San Francisco, CA, Morgan Kaufmann, 1996.

[26] S. Mullender, *Distributed Systems*, 2nd edn, Addison–Wesley, 1993.

[27] K. Perry and S. Toueg, Distributed agreement in the presence of processor and communication faults, *IEEE Transactions on Software Engineering*, **12**(3), 1986, 477–482.

[28] G. Peterson and M. Fischer, Economical solutions for the critical section problem in a distributed system, *Proceedings of the 9th ACM Symposium on Theory of Computing*, Boulder, CO, May, 1977, 91–97.

[29] N. Santoro and R. Khatib, Labelling and implicit routing in networks, *The Computer Journal*, **28**, 1985, 5–8.

[30] R. Schlichting and F. Schneider, Fail-stop processors: an approach to designing fault-tolerant computing systems, *ACM Transactions on Computer Systems*, **1**(3), 1983, 222–238.

[31] F. B. Schneider, Byzantine generals in action: implementing fail-stop processors, *ACM Transactions on Computer Systems*, **2**(2), 1984, 145–154.

[32] A. Segall, Distributed network protocols, *IEEE Transactions on Information Theory*, **29**(1), 1983, 23–35.

[33] A. Tanenbaum, *Computer Networks*, 3rd edn, NJ, Prentice-Hall PTR, 1996.

[34] S. Toueg, *An All-pairs Shortest Path Distributed Algorithm*, IBM Technical Report RC 8327, 1980.

[35] J. van Leeuwen and R. Tan, Interval routing, *The Computer Journal*, **30**, 1987, 298–307.

[36] P. Wan, K. Alzoubi, and O. Frieder, Distributed construction of connected dominating set in wireless ad-hoc networks, *Proceedings of the IEEE Infocom*, New York, June 2002, 1597–1604.

[37] O. Wolfson, S. Jajodia, and Y. Huang, An adaptive data replication algorithm, *ACM Transactions on Database Systems*, **22**(2), 1997, 255–314.

CHAPTER

6 Message ordering and group communication

Inter-process communication via message-passing is at the core of any distributed system. In this chapter, we will study non-FIFO, FIFO, causal order, and synchronous order communication paradigms for ordering messages. We will then examine protocols that provide these message orders. We will also examine several semantics for group communication with multicast – in particular, causal ordering and total ordering. We will then look at how exact semantics can be specified for the expected behavior in the face of processor or link failures. Multicasts are required at the application layer when superimposed topologies or overlays are used, as well as at the lower layers of the protocol stack. We will examine some popular multicast algorithms at the network layer. An example of such an algorithm is the Steiner tree algorithm, which is useful for setting up multi-party teleconferencing and videoconferencing multicast sessions.

Notation

As before, we model the distributed system as a graph (N, L). The following notation is used to refer to messages and events:

- When referring to a message without regard for the identity of the sender and receiver processes, we use m^i. For message m^i, its send and receive events are denoted as s^i and r^i, respectively.
- More generally, send and receive events are denoted simply as s and r. When the relationship between the message and its send and receive events is to be stressed, we also use M, *send*(M), and *receive*(M), respectively.

For any two events a and b, where each can be either a send event or a receive event, the notation $a \sim b$ denotes that a and b occur at the same process, i.e., $a \in E_i$ and $b \in E_i$ for some process i. The send and receive event pair for a message is said to be a pair of *corresponding* events. The send event corresponds to the receive event, and vice-versa. For a given execution E, let the set of all send–receive event pairs be denoted as $\mathcal{T} = \{(s, r) \in E_i \times E_j \mid s$ corresponds to $r\}$. When dealing with message ordering definitions, we will consider only send and receive events, but not internal events, because only communication events are relevant.

6.1 Message ordering paradigms

The order of delivery of messages in a distributed system is an important aspect of system executions because it determines the messaging behavior that can be expected by the distributed program. Distributed program logic greatly depends on this order of delivery. To simplify the task of the programmer, programming languages in conjunction with the middleware provide certain well-defined message delivery behavior. The programmer can then code the program logic with respect to this behavior.

Several orderings on messages have been defined: (i) non-FIFO, (ii) FIFO, (iii) causal order, and (iv) synchronous order. There is a natural hierarchy among these orderings. This hierarchy represents a trade-off between concurrency and ease of use and implementation. After studying the definitions of and the hierarchy among the ordering models, we will study some implementations of these orderings in the middleware layer. This section is based on Charron-Bost *et al.* [7].

6.1.1 Asynchronous executions

Definition 6.1 (*A*-**execution**) An asynchronous execution (or *A*-execution) is an execution $(E, \prec)$ for which the causality relation is a partial order.

There cannot exist any causality cycles in any real asynchronous execution because cycles lead to the absurdity that an event causes itself. On any logical link between two nodes in the system, messages may be delivered in any order, *not necessarily* first-in first-out. Such executions are also known as *non-FIFO executions*. Although each physical link typically delivers the messages sent on it in FIFO order due to the physical properties of the medium, a logical link may be formed as a composite of physical links and multiple paths may exist between the two end points of the logical link. As an example, the mode of ordering at the Network Layer in connectionless networks such as IPv4 is non-FIFO. Figure 6.1(a) illustrates an *A*-execution under non-FIFO ordering.

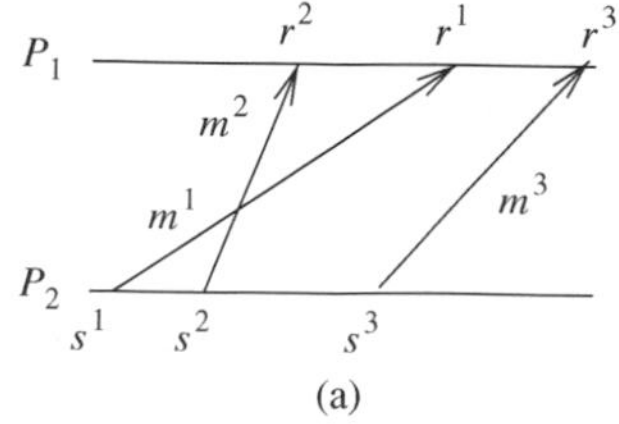

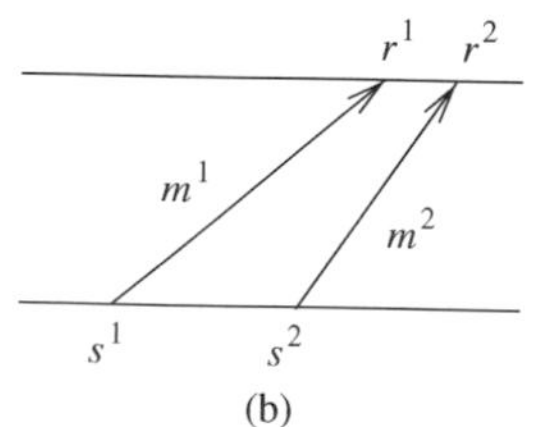

Figure 6.1 Illustrating FIFO and non-FIFO executions. (a) An *A*-execution that is not a FIFO execution. (b) An *A*-execution that is also a FIFO execution.

6.1.2 FIFO executions

Definition 6.2 (FIFO executions) A FIFO execution is an A-execution in which,
for all (s, r) and $(s', r') \in \mathcal{T}$, $(s \sim s'$ and $r \sim r'$ and $s \prec s') \Longrightarrow r \prec r'$.

On any logical link in the system, messages are necessarily delivered in the order in which they are sent. Although the logical link is inherently non-FIFO, most network protocols provide a connection-oriented service at the transport layer. Therefore, FIFO logical channels can be realistically assumed when designing distributed algorithms. A simple algorithm to implement a FIFO logical channel over a non-FIFO channel would use a separate numbering scheme to sequence the messages on each logical channel. The sender assigns and appends a $\langle sequence\_num,\ connection\_id \rangle$ tuple to each message. The receiver uses a buffer to order the incoming messages as per the sender's sequence numbers, and accepts only the "next" message in sequence. Figure 6.1(b) illustrates an A-execution under FIFO ordering.

6.1.3 Causally ordered (CO) executions

Definition 6.3 (Causal order (CO)) A CO execution is an A-execution in which,
for all (s, r) and $(s', r') \in \mathcal{T}$, $(r \sim r'$ and $s \prec s') \Longrightarrow r \prec r'$.

If two send events s and s' are related by causality ordering (not physical time ordering), then a causally ordered execution requires that their corresponding receive events r and r' occur in the same order at all common destinations. Note that if s and s' are not related by causality, then CO is vacuously satisfied because the antecedent of the implication is false.

Examples

- **Figure 6.2(a)** shows an execution that violates CO because $s^1 \prec s^3$ and at the common destination P_1, we have $r^3 \prec r^1$.
- **Figure 6.2(b)** shows an execution that satisfies CO. Only s^1 and s^2 are related by causality but the destinations of the corresponding messages are different.

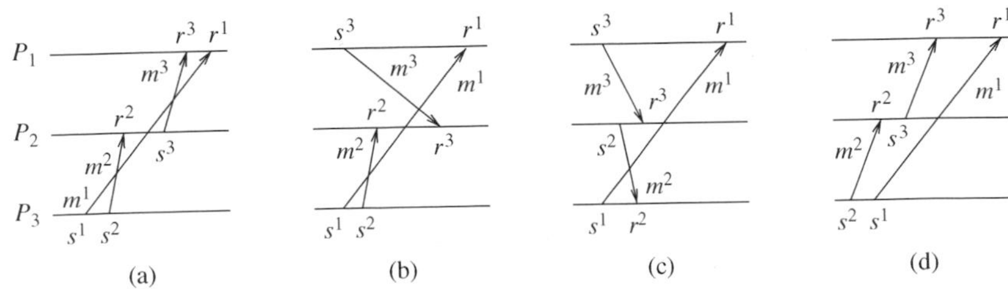

Figure 6.2 Illustration of causally ordered executions. (a) Not a CO execution. (b), (c), and (d) CO executions.

- **Figure 6.2(c)** shows an execution that satisfies CO. No send events are related by causality.
- **Figure 6.2(d)** shows an execution that satisfies CO. s^2 and s^1 are related by causality but the destinations of the corresponding messages are different. Similarly for s^2 and s^3.

Causal order is useful for applications requiring updates to shared data, implementing distributed shared memory, and fair resource allocation such as granting of requests for distributed mutual exclusion. Some of these uses will be discussed in detail in Section 6.5 on ordering message broadcasts and multicasts.

To implement CO, we distinguish between the arrival of a message and its delivery. A message m that arrives in the local OS buffer at P_i may have to be delayed until the messages that were sent to P_i causally before m was sent (the "overtaken" messages) have arrived and are processed by the application. The delayed message m is then given to the application for processing. The event of an application processing an arrived message is referred to as a *delivery* event (instead of as a *receive* event) for emphasis.

Example Figure 6.2(a) shows an execution that violates CO. To enforce CO, message m^3 should be kept pending in the local buffer after it arrives at P_1, until m^1 arrives and m^1 is delivered.

Definition 6.4 (Definition of causal order (CO) for implementations) If $send(m^1) \prec send(m^2)$ then for each common destination d of messages m^1 and m^2, $deliver_d(m^1) \prec deliver_d(m^2)$ must be satisfied.

Observe that if the definition of causal order is restricted so that m^1 and m^2 are sent by the same process, then the property degenerates into the FIFO property. In a FIFO execution, no message can be overtaken by another message between the same (sender, receiver) pair of processes. The FIFO property which applies on a per-logical channel basis can be extended globally to give the CO property. In a CO execution, no message can be overtaken by a chain of messages between the same (sender, receiver) pair of processes.

Example Figure 6.2(a) shows an execution that violates CO. Message m^1 is overtaken by the messages in the chain $\langle m^2, m^3 \rangle$.

CO executions can also be alternatively characterized by Definition 6.5 by simultaneously dropping the requirement from the implicand of Definition 6.3 that the receive events be on the same process, and relaxing the consequence from $(r \prec r')$ to $\neg(r' \prec r)$, i.e., the message m' sent causally later than m is not received causally earlier at the common destination. This ordering is known as message ordering (MO).

Definition 6.5 (Message order (MO)) A MO execution is an A-execution in which,
for all (s, r) and $(s', r') \in \mathcal{T}$, $s \prec s' \Longrightarrow \neg(r' \prec r)$.

Example Consider any message pair, say m^1 and m^3 in Figure 6.2(a). $s^1 \prec s^3$ but $\neg(r^3 \prec r^1)$ is false. Hence, the execution does not satisfy MO.

You are asked to prove the equivalence of MO executions and CO executions in Exercise 6.1. This will show that in a CO execution, a message cannot be overtaken by a chain of messages.

Another characterization of a CO execution in terms of the partial order $(E, \prec)$ is known as the empty-interval (EI) property.

Definition 6.6 (Empty-interval execution) An execution $(E, \prec)$ is an empty-interval (EI) execution if for each pair of events $(s, r) \in \mathcal{T}$, the open interval set $\{x \in E \mid s \prec x \prec r\}$ in the partial order is empty.

Example Consider any message, say m^2, in Figure 6.2(b). There does not exist any event x such that $s^2 \prec x \prec r^2$. This holds for all messages in the execution. Hence, the execution is EI.

You are asked to prove the equivalence of EI executions and CO executions in Exercise 6.1. A consequence of the EI property is that for an empty interval $\langle s, r\rangle$, there exists some *linear* extension[1] $<$ such that the corresponding interval $\{x \in E \mid s < x < r\}$ is also empty. An empty $\langle s, r\rangle$ interval in a linear extension indicates that the two events may be arbitrarily close and can be represented by a vertical arrow in a timing diagram, which is a characteristic of a synchronous message exchange. Thus, an execution E is CO if and only if for each message, there exists *some* space–time diagram in which that message can be drawn as a vertical message arrow. This, however, does not imply that *all* messages can be drawn as vertical arrows in the *same* space–time diagram. If all messages could be drawn vertically in an execution, all the $\langle s, r\rangle$ intervals would be empty in the *same* linear extension and the execution would be synchronous.

Another characterization of CO executions is in terms of the causal past/future of a send event and its corresponding receive event. The following corollary can be derived from the EI characterization above (Definition 6.6).

Corollary 6.1 *An execution $(E, \prec)$ is CO if and only if for each pair of events $(s, r) \in \mathcal{T}$ and each event $e \in E$,*

- *weak common past*: $e \prec r \Longrightarrow \neg(s \prec e)$;
- *weak common future*: $s \prec e \Longrightarrow \neg(e \prec r)$.

Example Corollary 6.1 can be observed for the executions in Figures 6.2(b)–(d).

[1] A linear extension of a partial order $(E, \prec)$ is any total order $(E, <)$ such that each ordering relation of the partial order is preserved.

If we require that the past of both the s and r events are identical (and analogously for the future), viz., $e \prec r \Longrightarrow e \prec s$ and $s \prec e \Longrightarrow r \prec e$, we get a subclass of CO executions, called *synchronous executions*.

6.1.4 Synchronous execution (SYNC)

When all the communication between pairs of processes uses synchronous send and receive primitives, the resulting order is the synchronous order. As each synchronous communication involves a handshake between the receiver and the sender, the corresponding send and receive events can be viewed as occuring instantaneously and atomically. In a timing diagram, the "instantaneous" message communication can be shown by bidirectional vertical message lines. Figure 6.3(a) shows a synchronous execution on an asynchronous system. Figure 6.3(b) shows the equivalent timing diagram with the corresponding instantaneous message communication.

The "instantaneous communication" property of synchronous executions requires a modified definition of the causality relation because for each $(s, r) \in \mathcal{T}$, the send event is not causally ordered before the receive event. The two events are viewed as being atomic and simultaneous, and neither event precedes the other.

Definition 6.7 (Causality in a synchronous execution) The synchronous causality relation $\ll$ on E is the smallest transitive relation that satisfies the following:

S1: If x occurs before y at the same process, then $x \ll y$.
S2: If $(s, r) \in \mathcal{T}$, then for all $x \in E$, $[(x \ll s \Longleftrightarrow x \ll r)$ and $(s \ll x \Longleftrightarrow r \ll x)]$.
S3: If $x \ll y$ and $y \ll z$, then $x \ll z$.

We can now formally define a synchronous execution.

Definition 6.8 (Synchronous execution) A synchronous execution (or S-execution) is an execution $(E, \ll)$ for which the causality relation $\ll$ is a partial order.

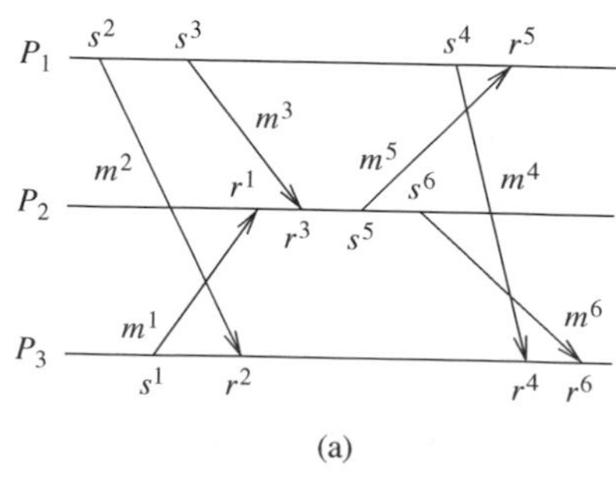

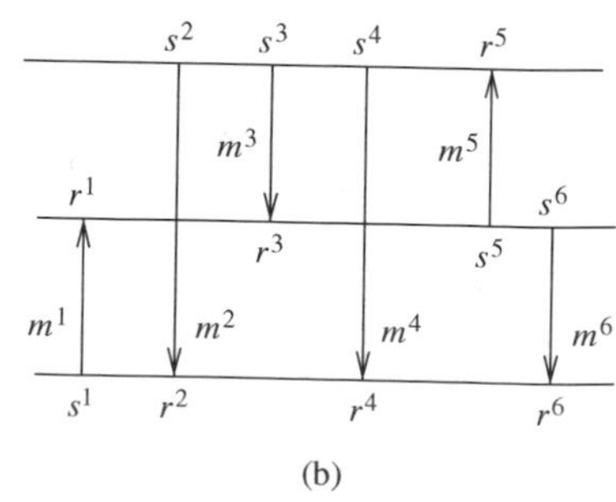

Figure 6.3 Illustration of a synchronous communication. (a) Execution in an asynchronous system. (b) Equivalent instantaneous communication.

We now show how to timestamp events in synchronous executions.

Definition 6.9 (Timestamping a synchronous execution) An execution $(E, \prec)$ is synchronous if and only if there exists a mapping from E to T (scalar timestamps) such that

- for any message M, $T(s(M)) = T(r(M))$;
- for each process P_i, if $e_i \prec e_i'$ then $T(e_i) < T(e_i')$.

By assuming that a send event and its corresponding receive event are viewed atomically, i.e., $s(M) \prec r(M)$ and $r(M) \prec s(M)$, it follows that for any events e_i and e_j that are not the send event and the receive event of the same message, $e_i \prec e_j \Longrightarrow T(e_i) < T(e_j)$.

6.2 Asynchronous execution with synchronous communication

When all the communication between pairs of processes is by using synchronous send and receive primitives, the resulting order is synchronous order. The send and receive events of a message appear instantaneous, see the example in Figure 6.3. We now address the following question:

- If a program is written for an asynchronous system, say a FIFO system, will it still execute correctly if the communication is done by synchronous primitives instead? There is a possibility that the program may *deadlock*, as shown by the code in Figure 6.4.

Charron-Bost *et al.* [7] observed that a distributed algorithm designed to run correctly on asynchronous systems (called *A-executions*) may not run correctly on synchronous systems. An algorithm that runs on an asynchronous system may *deadlock* on a synchronous system.

Examples The asynchronous execution of Figure 6.4, illustrated in Figure 6.5(a) using a timing diagram, will deadlock if run with synchronous primitives. The executions in Figure 6.5(b)–(c) will also deadlock when run on a synchronous system.

Figure 6.4 A communication program for an asynchronous system deadlocks when using synchronous primitives.

Process i	**Process** j
...	...
$Send(j)$	$Send(i)$
$Receive(j)$	$Receive(i)$
...	...

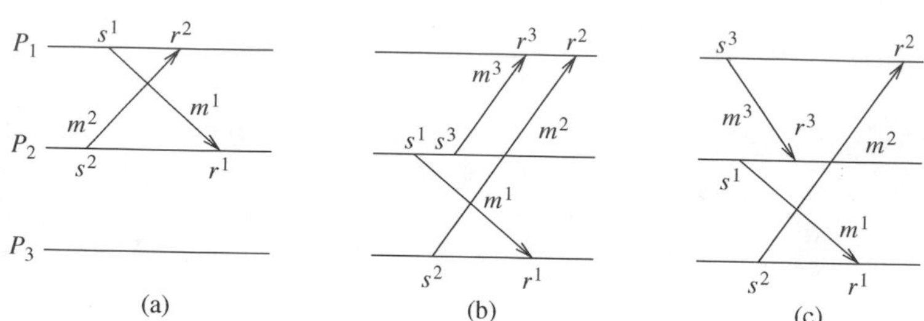

Figure 6.5 Illustrations of asynchronous executions and of crowns. (a) Crown of size 2. (b) Another crown of size 2. (c) Crown of size 3.

6.2.1 Executions realizable with synchronous communication (RSC)

An execution can be modeled (using the interleaving model) as a feasible schedule of the events to give a total order that extends the partial order $(E, \prec)$. In an A-execution, the messages can be made to appear instantaneous if there exists a linear extension of the execution, such that each send event is immediately followed by its corresponding receive event in this linear extension. Such an A-execution can be realized under synchronous communication and is called a *realizable with synchronous communication* (RSC) execution.

Definition 6.10 (Non-separated linear extension) A non-separated linear extension of $(E, \prec)$ is a linear extension of $(E, \prec)$ such that for each pair $(s, r) \in \mathcal{T}$, the interval $\{ x \in E \mid s < x < r \}$ is empty.

Examples

- Figure 6.2(d): $\langle s^2, r^2, s^3, r^3, s^1, r^1 \rangle$ is a linear extension that is non-separated. $\langle s^2, s^1, r^2, s^3, r^3, r^1 \rangle$ is a linear extension that is separated.
- Figure 6.3(b): $\langle s^1, r^1, s^2, r^2, s^3, r^3, s^4, r^4, s^5, r^5, s^6, r^6 \rangle$ is a linear extension that is non-separated. $\langle s^1, s^2, r^1, r^2, s^3, s^4, r^4, r^3, s^5, s^6, r^6, r^5 \rangle$ is a linear extension that is separated.

Definition 6.11 (RSC execution) [7] An A-execution $(E, \prec)$ is an RSC execution if and only if there exists a non-separated linear extension of the partial order $(E, \prec)$.

In the non-separated linear extension, if the adjacent send event and its corresponding receive event are viewed atomically, then that pair of events shares a common past and a common future with each other. The various other characterizations of S-executions seen in Section 6.1.4 are also seen to hold.

To use Definition 6.11 requires checking for all the linear extensions, incurs exponential overhead. You can verify this by trying to create and examine all the linear extensions of the execution in Figure 6.5(b) or (c). Thus, Definition 6.11 does not provide a practical test to determine whether a program written for a non-synchronous system, say a FIFO system, will still execute correctly if the communication is done by synchronous primitives.

We now study a characterization of the execution in terms of a graph structure called a *crown*; the crown leads to a feasible test for a RSC execution.

Definition 6.12 (Crown) Let E be an execution. A crown of size k in E is a sequence $\langle (s^i, r^i), i \in \{0, \ldots, k-1\} \rangle$ of pairs of corresponding send and receive events such that: $s^0 \prec r^1$, $s^1 \prec r^2$, $\ldots$, $s^{k-2} \prec r^{k-1}$, $s^{k-1} \prec r^0$.

Examples

- Figure 6.5(a): The crown is $\langle (s^1, r^1), (s^2, r^2) \rangle$ as we have $s^1 \prec r^2$ and $s^2 \prec r^1$. This execution represents the program execution in Figure 6.4.
- Figure 6.5(b): The crown is $\langle (s^1, r^1), (s^2, r^2) \rangle$ as we have $s^1 \prec r^2$ and $s^2 \prec r^1$.
- Figure 6.5(c): The crown is $\langle (s^1, r^1), (s^3, r^3), (s^2, r^2) \rangle$ as we have $s^1 \prec r^3$ and $s^3 \prec r^2$ and $s^2 \prec r^1$.
- Figure 6.2(a): The crown is $\langle (s^1, r^1), (s^2, r^2), (s^3, r^3) \rangle$ as we have $s^1 \prec r^2$ and $s^2 \prec r^3$ and $s^3 \prec r^1$.

In a crown, the send event s^i and receive event r^{i+1} may lie on the same process (e.g., Figure 6.5(c)) or may lie on different processes (e.g., Figure 6.5(a)). We can also make the following observations:

- In an execution that is not CO (see the example in Figure 6.2(a)), there must exist pairs (s, r) and (s', r') such that $s \prec r'$ and $s' \prec r$. It is possible to generalize this to state that a non-CO execution must have a crown of size at least 2. (Exercise 6.4 asks you to prove that in a non-CO execution, there must exist a crown of size exactly 2.)
- CO executions that are not synchronous, also have crowns, e.g., the execution in Figure 6.2(b) has a crown of size 3.

Intuitively, the cyclic dependencies in a crown indicate that it is not possible to find a linear extension in which all the (s, r) event pairs are adjacent. In other words, it is not possible to schedule entire messages in a serial manner, and hence the execution is not RSC.

To determine whether the RSC property holds in $(E, \prec)$, we need to determine whether there exist any cyclic dependencies among messages. Rather than incurring the exponential overhead of checking all linear extensions of E, we can check for crowns by using the test in Figure 6.6. On the set of messages $\mathcal{T}$, we define an ordering $\hookrightarrow$ such that $m \hookrightarrow m'$ if and only if $s \prec r'$.

Example By drawing the directed graph $(\mathcal{T}, \hookrightarrow)$ for each of the executions in Figures 6.2, 6.3, and 6.5, it can be seen that the graphs for Figures 6.2(d) and Figure 6.3 are acyclic. The other graphs have a cycle.

Figure 6.6 The crown test to determine the existence of cyclic dependencies among messages.

1. Define the $\hookrightarrow: \mathcal{T} \times \mathcal{T}$ relation on messages in the execution $(E, \prec)$ as follows. Let $\hookrightarrow ([s, r], [s', r'])$ if and only if $s \prec r'$. Observe that the condition $s \prec r'$ (which has the form used in the definition of a crown) is implied by all the four conditions: (i) $s \prec s'$, or (ii) $s \prec r'$, or (iii) $r \prec s'$, and (iv) $r \prec r'$.
2. Now define a *directed* graph $G_{\hookrightarrow} = (\mathcal{T}, \hookrightarrow)$, where the vertex set is the set of messages $\mathcal{T}$ and the edge set is defined by $\hookrightarrow$.
   Observe that the relation $\hookrightarrow: \mathcal{T} \times \mathcal{T}$ is a partial order if and only if $G_{\hookrightarrow}$ has no cycle, i.e., there must not be a cycle with respect to $\hookrightarrow$ on the set of corresponding (s, r) events.
3. It can be seen from the definition of a crown (Definition 6.12) that $G_{\hookrightarrow}$ has a directed cycle if and only if $(E, \prec)$ has a crown.

This test leads to the following theorem [7].

Theorem 6.1 (Crown criterion) *The* crown criterion *states that an A-computation is RSC, i.e., it can be realized on a system with synchronous communication, if and only if it contains no crown.*

Example Using the directed graph $(\mathcal{T}, \hookrightarrow)$ for each of the executions in Figures 6.2, 6.3(a), and 6.5, it can be seen that the executions in Figures 6.2(d) and Figure 6.3(a) are RSC. The others are not RSC.

Although checking for a non-separated linear extension of $(E, \prec)$ has exponential cost, checking for the presence of a crown based on the message scheduling test of Figure 6.6 can be performed in time that is linear in the number of communication events (see Exercise 6.3). An execution is not RSC and its graph $G_{\hookrightarrow}$ contains a cycle if and only if in the corresponding space–time diagram, it is possible to form a cycle by (i) moving along message arrows in either direction, but (ii) always going left to right along the time line of any process.

As an RSC execution has a non-separated linear extension, it is possible to assign scalar timestamps to events, as it was assigned for a synchronous execution (Definition 6.9), as follows.

Definition 6.13 (Timestamps for a RSC execution) An execution $(E, \prec)$ is RSC if and only if there exists a mapping from E to T (scalar timestamps) such that

- for any message M, $T(s(M)) = T(r(M))$;
- for each (a, b) in $(E \times E) \setminus \mathcal{T}$, $a \prec b \Longrightarrow T(a) < T(b)$.

From the acyclic message scheduling criterion (Theorem 6.1) and the timestamping property above, it can be observed that an A-execution is RSC if and only if its timing diagram can be drawn such that all the message arrows are vertical.

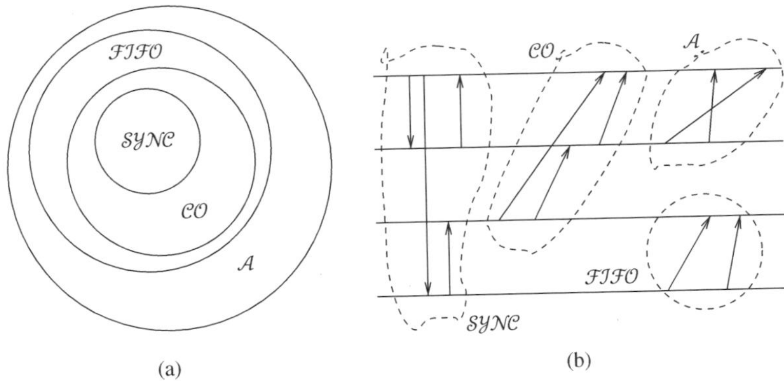

Figure 6.7 Hierarchy of execution classes. (a) Venn diagram. (b) Example executions.

6.2.2 Hierarchy of ordering paradigms

Let $\mathcal{SYNC}$ (or $\mathcal{RSC}$), $\mathcal{CO}$, $\mathcal{FIFO}$, and $\mathcal{A}$ denote the set of all possible executions ordered by synchronous order, causal order, FIFO order, and non-FIFO order, respectively. We have the following results:

- For an A-execution, A is RSC if and only if A is an S-execution.
- $\mathcal{RSC} \subset \mathcal{CO} \subset \mathcal{FIFO} \subset \mathcal{A}$. This hierarchy is illustrated in Figure 6.7(a), and example executions of each class are shown side-by-side in Figure 6.7(b). Figure 6.1(a) shows an execution that belongs to $\mathcal{A}$ but not to $\mathcal{FIFO}$. Figure 6.2(a) shows an execution that belongs to $\mathcal{FIFO}$ but not to $\mathcal{CO}$. Figures 6.2(b) and (c) show executions that belong to $\mathcal{CO}$ but not to $\mathcal{RSC}$.
- The above hierarchy implies that some executions belonging to a class X will not belong to any of the classes included in X. Thus, there are more restrictions on the possible message orderings in the smaller classes. Hence, we informally say that the included classes have less concurrency. The degree of concurrency is most in $\mathcal{A}$ and least in $\mathcal{SYNC}$.
- A program using synchronous communication is easiest to develop and verify. A program using non-FIFO communication, resulting in an $\mathcal{A}$-execution, is hardest to design and verify. This is because synchronous order offers the most simplicity due to the restricted number of possibilities, whereas non-FIFO order offers the greatest difficulties because it admits a much larger set of possibilities that the developer and verifier need to account for.

Thus, there is an inherent trade-off between the amount of concurrency provided, and the ease of designing and verifying distributed programs.

6.2.3 Simulations

Asynchronous programs on synchronous systems

Theorem 6.1 indicates that an A-execution can be run using synchronous communication primitives if and only if it is an RSC execution. The events in

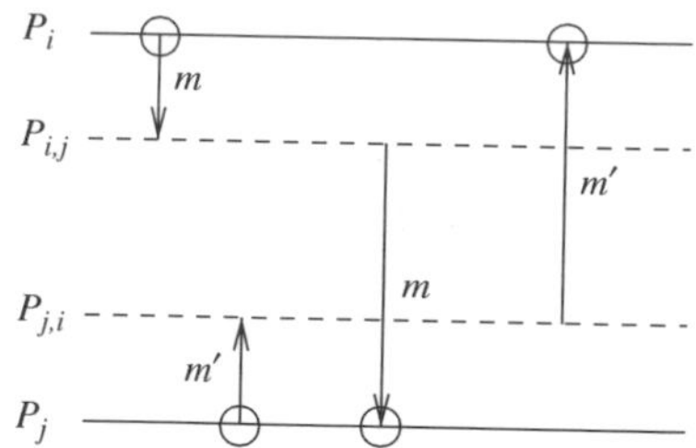

Figure 6.8 Modeling channels as processes to simulate an execution using asynchronous primitives on an synchronous system.

the RSC execution are scheduled as per some nonseparated linear extension, and adjacent (s, r) events in this linear extension are executed sequentially in the synchronous system. The partial order of the asynchronous execution remains unchanged.

If an A-execution is not RSC, then there is no way to schedule the events to make them RSC, without actually altering the partial order of the given A-execution. However, the following indirect strategy that does not alter the partial order can be used. Each channel $C_{i,j}$ is modeled by a control process $P_{i,j}$ that simulates the channel buffer. An asynchronous communication from i to j becomes a synchronous communication from i to $P_{i,j}$ followed by a synchronous communication from $P_{i,j}$ to j. This enables the decoupling of the sender from the receiver, a feature that is essential in asynchronous systems. This approach is illustrated in Figure 6.8. The communication events at the application processes P_i and P_j are encircled. Observe that it is expensive to implement the channel processes.

Synchronous programs on asynchronous systems

A (valid) S-execution can be trivially realized on an asynchronous system by scheduling the messages in the order in which they appear in the S-execution. The partial order of the S-execution remains unchanged but the communication occurs on an asynchronous system that uses asynchronous communication primitives. Once a message send event is scheduled, the middleware layer waits for an acknowledgement; after the ack is received, the synchronous send primitive completes.

6.3 Synchronous program order on an asynchronous system

There do not exist real systems with instantaneous communication that allows for synchronous communication to be naturally realized. We need to address the basic question of how a system with synchronous communication can be implemented. We first examine non-determinism in program execution, and CSP as a representative synchronous programming language, before examining an implementation of synchronous communication.

Non-determinism

The discussions on the message orderings and their characterizations so far assumed a given partial order. This suggests that the distributed programs are *deterministic*, i.e., repeated runs of the same program will produce the same partial order. In many cases, programs are *non-deterministic* in the following senses (we are not considering here the unpredictable message delays that cause different runs to non-deterministically have different global orderings of the events in physical time:)

1. A receive call can receive a message from any sender who has sent a message, if the expected sender is not specified. The receive calls in most of the algorithms in Chapter 5 are non-deterministic in this sense – the receiver is willing to perform a rendezvous with any willing and ready sender.
2. Multiple send and receive calls which are enabled at a process can be executed in an interchangeable order.
   If i sends to j, and j sends to i concurrently using blocking synchronous calls, there results a deadlock, similar to the one in Figure 6.4. However, there is no semantic dependency between the send and the immediately following receive at each of the processes. If the receive call at one of the processes can be scheduled before the send call, then there is no deadlock.
   In this section, we consider scheduling synchronous communication events (over an asynchronous system).

6.3.1 Rendezvous

One form of group communication is called *multiway rendezvous*, which is a synchronous communication among an arbitrary number of asynchronous processes. All the processes involved "meet with each other," i.e., communicate "synchronously" with each other at one time. The solutions to this problem are fairly complex, and we will not consider them further as this model of synchronous communication is not popular. Here, we study rendezvous between a pair of processes at a time, which is called *binary rendezvous* as opposed to the *multiway rendezvous*.

Support for *binary rendezvous* communication was first provided by programming languages such as CSP and Ada. We consider here a subset of CSP. In these languages, the repetitive command (the $*$ operator) over the alternative command (the $||$ operator) on multiple guarded commands (each having the form $G_i \longrightarrow CL_i$) is used, as follows:

$$*[G_1 \longrightarrow CL_1 \ || \ G_2 \longrightarrow CL_2 \ || \ \cdots \ || \ G_k \longrightarrow CL_k].$$

Each communication command may be a part of a guard G_i, and may also appear within the statement block CL_i. A guard G_i is a boolean expression. If a guard G_i evaluates to true then CL_i is said to be *enabled*, otherwise CL_i is said to be *disabled*. A send command of local variable x to process P_k is denoted as "$x\,!\,P_k$." A receive from process P_k into local variable x is denoted as "$P_k\,?\,x$." Some typical observations about synchronous communication under *binary rendezvous* are as follows:

- For the receive command, the sender must be specified. However, multiple recieve commands can exist. A type check on the data is implicitly performed.
- Send and received commands may be individually disabled or enabled. A command is disabled if it is guarded and the guard evaluates to *false*. The guard would likely contain an expression on some local variables.
- Synchronous communication is implemented by *scheduling* messages under the covers using asynchronous communication. Scheduling involves pairing of matching send and receive commands that are both enabled. The communication events for the control messages under the covers do not alter the partial order of the execution.

The concept underlying *binary rendezvous*, which provides synchronous communication, differs from the concept underlying the classification of synchronous send and receive primitives as blocking or non-blocking (studied in Chapter 1). *Binary rendezvous* explicitly assumes that multiple send and receives are enabled. Any send or receive event that can be "matched" with the corresponding receive or send event can be scheduled. This is dynamically scheduling the ordering of events and the partial order of the execution.

6.3.2 Algorithm for binary rendezvous

Various algorithms were proposed to implement *binary rendezvous* in the 1980s [1,16]. These algorithms typically share the following features. At each process, there is a set of tokens representing the current interactions that are enabled locally. If multiple interactions are enabled, a process chooses one of them and tries to "synchronize" with the partner process. The problem reduces to one of scheduling messages satisfying the following constraints:

- Schedule on-line, atomically, and in a distributed manner, i.e., the scheduling code at any process does not know the application code of other processes.
- Schedule in a deadlock-free manner (i.e., crown-free), such that both the sender and receiver are enabled for a message when it is scheduled.
- Schedule to satisfy the progress property (i.e., find a schedule within a bounded number of steps) in addition to the safety (i.e., correctness) property.

Additional features of a good algorithm are: (i) symmetry or some form of fairness, i.e., not favoring particular processes over others during scheduling, and (ii) efficiency, i.e., using as few messages as possible, and involving as low a time overhead as possible.

We now outline a simple algorithm by Bagrodia [1] that makes the following assumptions:

1. Receive commands are forever enabled from all processes.
2. A send command, once enabled, remains enabled until it completes, i.e., it is not possible that a send command gets disabled (by its guard getting falsified) before the send is executed.
3. To prevent deadlock, process identifiers are used to introduce asymmetry to break potential crowns that arise.
4. Each process attempts to schedule only one *send* event at any time.

The algorithm illustrates how crown-free message scheduling is achieved on-line.

The message types used are: (i) M, (ii) $ack(M)$, (iii) $request(M)$, and (iv) $permission(M)$. A process blocks when it knows that it can successfully synchronize the current message with the partner process. Each process maintains a queue that is processed in FIFO order only when the process is unblocked. When a process is blocked waiting for a particular message that it is currently synchronizing, any other message that arrives is queued up.

Execution events in the synchronous execution are only the *send* of the message M and *receive* of the message M. The send and receive events for the other message types – $ack(M)$, $request(M)$, and *permission*(M) which are control messages – are under the covers, and are not included in the synchronous execution. The messages $request(M)$, $ack(M)$, and $permission(M)$ use M's unique tag; the message M is not included in these messages. We use capital SEND(M) and RECEIVE(M) to denote the primitives in the application execution, the lower case send and receive are used for the control messages.

The algorithm to enforce synchronous order is given in Algorithm 6.1. The key rules to prevent cycles among the messages are summarized as follows and illustrated in Figure 6.9:

- To send to a lower priority process, messages M and $ack(M)$ are involved in that order. The sender issues $send(M)$ and blocks until $ack(M)$ arrives. Thus, when sending to a lower priority process, the sender blocks waiting for the partner process to synchronize and send an acknowledgement.
- To send to a higher priority process, messages $request(M)$, $permission(M)$, and M are involved, in that order. The sender issues $send(request(M))$, does not block, and awaits permission. When $permission(M)$ arrives, the sender issues $send(M)$.

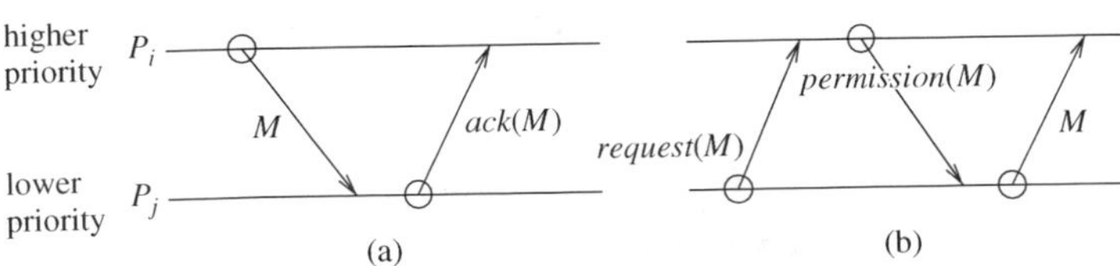

Figure 6.9 Messages used to implement synchronous order. P_i has higher priority than P_j. (a) P_i issues SEND(M). (b) P_j issues SEND(M).

(message types)
M, $ack(M)$, $request(M)$, $permission(M)$

(1) **P_i wants to execute SEND(M) to a lower priority process P_j:**
P_i executes $send(M)$ and blocks until it receives $ack(M)$ from P_j. The send event SEND(M) now completes.

Any M' message (from a higher priority processes) and $request(M')$ request for synchronization (from a lower priority processes) received during the blocking period are queued.

(2) **P_i wants to execute SEND(M) to a higher priority process P_j:**

(2a) P_i seeks permission from P_j by executing $send(request(M))$.
// to avoid deadlock in which cyclically blocked processes queue
// messages.

(2b) While P_i is waiting for permission, it remains unblocked.

(i) If a message M' arrives from a higher priority process P_k, P_i accepts M' by scheduling a RECEIVE(M') event and then executes $send(ack(M'))$ to P_k.

(ii) If a $request(M')$ arrives from a lower priority process P_k, P_i executes $send(permission(M'))$ to P_k and blocks waiting for the message M'. When M' arrives, the RECEIVE(M') event is executed.

(2c) When the $permission(M)$ arrives, P_i knows partner P_j is synchronized and P_i executes $send(M)$. The SEND(M) now completes.

(3) ***request*(M) arrival at P_i from a lower priority process P_j:**
At the time a $request(M)$ is processed by P_i, process P_i executes $send(permission(M))$ to P_j and blocks waiting for the message M. When M arrives, the RECEIVE(M) event is executed and the process unblocks.

(4) **Message M arrival at P_i from a higher priority process P_j:**
At the time a message M is processed by P_i, process P_i executes RECEIVE(M) (which is assumed to be always enabled) and then $send(ack(M))$ to P_j.

(5) **Processing when P_i is unblocked:**
When P_i is unblocked, it dequeues the next (if any) message from the queue and processes it as a message arrival (as per rules 3 or 4).

Algorithm 6.1 A simplified implementation of synchronous order. Code shown is for process P_i, $1 \le i \le n$.

Thus, when sending to a higher priority process, the sender asks the higher priority process via the $request(M)$ to give permission to send. When the higher priority process gives permission to send, the higher priority process, which is the intended receiver, blocks.

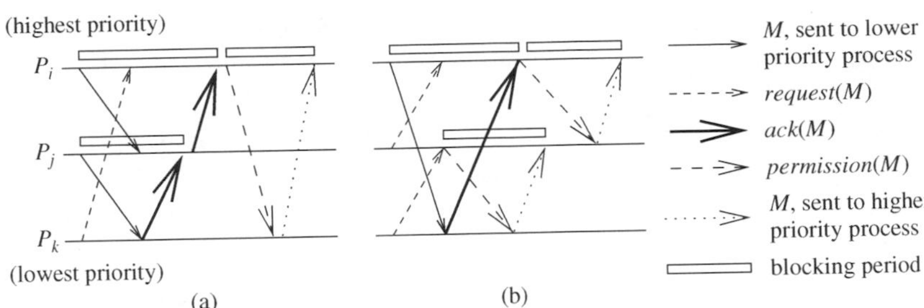

Figure 6.10 Examples showing how to schedule messages sent with synchronous primitives.

In either case, a higher priority process blocks on a lower priority process. So cyclic waits are avoided.

In more detail, a cyclic wait is prevented because before sending a message M to a higher priority process, a lower priority process requests the higher priority process for permission to synchronize on M, in a non-blocking manner. While waiting for this permission, there are two possibilities:

1. If a message M' from a higher priority process arrives, it is processed by a receive (assuming receives are always enabled) and $ack(M')$ is returned. Thus, a cyclic wait is prevented.
2. Also, while waiting for this permission, if a $request(M')$ from a lower priority process arrives, a $permission(M')$ is returned and the process blocks until M' actually arrives.

Note that the $receive(M')$ event effectively gets permuted before the $send(M)$ event (steps 2(bi) and 2(bii)).

Examples: Figure 6.10 shows two examples of how the algorithm breaks cyclic waits to schedule messages. Observe that in all cases in the algorithm, a higher priority process blocks on lower priority processes, irrespective of whether the higher priority process is the intended sender or the receiver of the message being scheduled. In Figure 6.10(a), at process P_k, the receive of the message from P_j effectively gets permuted before P_k's own $send(M)$ event due to step 2(bi). In Figure 6.10(b), at process P_j, the receive of the $request(M')$ message from P_k effectively causes M' to be permuted before P_j's own message that it was attempting to schedule with P_i, due to step 2(bii).

6.4 Group communication

Processes across a distributed system cooperate to solve a joint task. Often, they need to communicate with each other as a group, and therefore there needs to be support for *group communication*. A *message broadcast* is the sending of a message to all members in the distributed system. The notion of a system can be confined only to those sites/processes participating in the joint application. Refining the notion of *broadcasting*, there is *multicasting* wherein a message is sent to a certain subset, identified as a *group*, of the processes in the system. At the other extreme is *unicasting*, which is the familiar point-to-point message communication.

Broadcast and multicast support can be provided by the network protocol stack using variants of the spanning tree. This is an efficient mechanism for distributing information. However, the hardware-assisted or network layer protocol assisted multicast cannot efficiently provide features such as the following:

- Application-specific ordering semantics on the order of delivery of messages.
- Adapting groups to dynamically changing membership.
- Sending multicasts to an arbitrary set of processes at each send event.
- Providing various fault-tolerance semantics.

If a multicast algorithm requires the sender to be a part of the destination group, the multicast algorithm is said to be a *closed group* algorithm. If the sender of the multicast can be outside the destination group, the multicast algorithm is said to be an *open group* algorithm. Open group algorithms are more general, and therefore more difficult to design and more expensive to implement, than closed group algorithms. Closed group algorithms cannot be used in several scenarios such as in a large system (e.g., on-line reservation or Internet banking systems) where client processes are short-lived and in large numbers. It is also worth noting that, for multicast algorithms, the number of groups may be potentially exponential, i.e., $O(2^n)$, and algorithms that have to explicitly track the groups can incur this high overhead.

In the remainder of this chapter we will examine multicast and broadcast mechanisms under varying degrees of strictness of assumptions on the order of delivery of messages. Two popular orders for the delivery of messages were proposed in the context of group communication: *causal order* and *total order*. Much of the seminal work on group communication was initiated by the ISIS project [4,5].

6.5 Causal order (CO)

Causal order has many applications such as updating replicated data, allocating requests in a fair manner, and synchronizing multimedia streams. We explain here the use of causal order in updating replicas of a data item in the system. Consider Figure 6.11(a), which shows two processes P_1 and P_2 that issue updates to the three replicas $R1(d)$, $R2(d)$, and $R3(d)$ of data item d. Message m creates a causality between $send(m1)$ and $send(m2)$. If P_2 issues its update causally after P_1 issued its update, then P_2's update should be seen by the replicas after they see P_1's update, in order to preserve the semantics

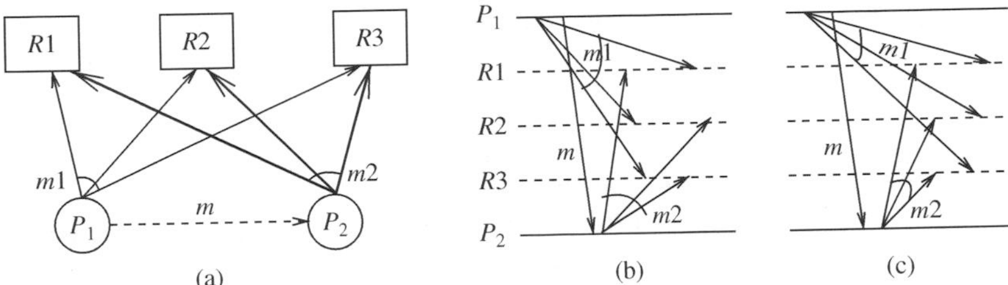

Figure 6.11 Updates to object replicas are issued by two processes.

of the application. (In this case, CO is satisfied.) However, this may happen at some, all, or none of the replicas. Figure 6.11(b) shows that $R1$ sees P_2's update first, while $R2$ and $R3$ see P_1's update first. Here, CO is violated. Figure 6.11(c) shows that all replicas see P_2's update first. However, CO is still violated. If message m did not exist as shown, then the executions shown in Figure 6.11(b) and (c) would satisfy CO.

Given a system with FIFO channels, causal order needs to be explicitly enforced by a protocol. The following two criteria must be met by a causal ordering protocol:

- **Safety** In order to prevent causal order from being violated, a message M that arrives at a process may need to be buffered until all systemwide messages sent in the causal past of the $send(M)$ event to that same destination have already arrived.
  Therefore, we distinguish between the arrival of a message at a process (at which time it is placed in a local system buffer) and the event at which the message is given to the application process (when the protocol deems it safe to do so without violating causal order). The arrival of a message is transparent to the application process. The delivery event corresponds to the *receive* event in the execution model.
- **Liveness** A message that arrives at a process must eventually be delivered to the process.

Both the algorithms we will study in this section allow each send event to unicast, multicast, or broadcast a message in the system.

6.5.1 The Raynal–Schiper–Toueg algorithm [22]

Intuitively, it seems logical that each message M should carry a log of all other messages, or their identifiers, sent causally before M's send event, and sent to the same destination $dest(M)$. This log can then be examined to ensure whether it is safe to deliver a message. All algorithms aim to reduce this log overhead, and the space and time overhead of maintaining the log information at the processes. Algorithm 6.2 gives a canonical algorithm that is representative of several algorithms that try to reduce the size of the local space and message space overhead by various techniques. In order to implement safety, the messages piggyback the control information that helps to determine when it is safe to deliver the message to the destination. FIFO channels are assumed in the system. Each process maintains an $n \times n$ array *SENT* and a size n array *DELIV*. $SENT_i[j, k]$ at process P_i gives the number of messages sent by P_j to P_k, as known to P_i. $DELIV_i[j]$ gives the number of messages from P_j that have been delivered to P_i. Safety is implemented primarily by step (2a). Liveness is implemented under the assumption that there are no failures, and that message propagation/transmission times are finite.

(local variables)
int $SENT[1 \ldots n, 1 \ldots n]$
int $DELIV[1 \ldots n]$ // $DELIV[k]$ = # messages sent by k that are
// delivered locally

(1) **send event**, where P_i wants to send message M to P_j:
(1a) **send** $(M, SENT)$ to P_j;
(1b) $SENT[i, j] \longleftarrow SENT[i, j] + 1$.

(2) **message arrival,** when (M, ST) arrives at P_i from P_j:
(2a) **deliver** M to P_i when **for each** process x,
(2b) $DELIV[x] \geq ST[x, i]$;
(2c) $\forall x, y$, $SENT[x, y] \longleftarrow \max(SENT[x, y], ST[x, y])$;
(2d) $DELIV[j] \longleftarrow DELIV[j] + 1$.

Algorithm 6.2 Canonical algorithm by Raynal–Schiper–Toueg (RST) to implement causal ordering of messages. Code for P_i, $1 \leq i \leq n$.

Complexity

This algorithm takes $O(n^2)$ integers space at each process, and the message space overhead is also n^2 integers. The time complexity at each process for each send and deliver event is $O(n^2)$.

6.5.2 The Kshemkalyani–Singhal optimal algorithm [20,21]

The space and time optimal algorithm of Kshemkalyani and Singhal (KS) [20,21] uses the following notation. The ath multicast message sent by process i is denoted $M_{i,a}$, and the set of destinations of this multicast is denoted $M_{i,a}.Dests$. The algorithm uses the following *Delivery Condition* for correctness: a message M^* that carries information "$d \in M.Dests$," where message M was sent to d in the causal past of $Send(M^*)$, is not delivered to d if M has not yet been delivered to d.

A natural question to address to obtain optimality is: for how long should the information "$d \in M_{i,a}.Dests$" be stored in the log at a process, and piggybacked on messages? The following are the necessary and sufficient conditions on how long this information should be stored.

An optimal CO algorithm stores in local message logs and propagates on messages, information of the form "d is a destination of M" about a message M sent in the causal past, *as long as* and *only as long as*:

(*Propagation Constraint I*) it is not known that the message M is delivered to d, and

(*Propagation Constraint II*) it is not known that a message has been sent to d in the causal future of $Send(M)$, and hence it is not guaranteed using a reasoning based on transitivity that the message M will be delivered to d in CO.

The Propagation Constraints also imply that if either (I) or (II) is false, the information "$d \in M.Dests$" must *not* be stored or propagated, even to remember that (I) or (II) has been falsified. Stated differently, the information "$d \in M_{i,a}.Dests$" must be available in the causal future of event $e_{i,a}$, but:

- not in the causal future of $Deliver_d(M_{i,a})$, and
- not in the causal future of $e_{k,c}$, where $d \in M_{k,c}.Dests$ and there is no other message sent causally between $M_{i,a}$ and $M_{k,c}$ to the same destination d.

In the causal future of $Deliver_d(M_{i,a})$, and $Send(M_{k,c})$, the information is redundant; elsewhere, it is necessary. Additionally, to maintain optimality, no other information should be stored, including information about what messages have been delivered. As information about what messages have been delivered (or are guaranteed to be delivered without violating causal order) is necessary for the Delivery Condition, this information is inferred using a set-operation based logic.

The Propagation Constraints are illustrated with the help of Figure 6.12. The message M is sent by process i at event e to process d. The information "$d \in M.Dests$":

- must exist at $e1$ and $e2$ because (I) and (II) are true;
- must not exist at $e3$ because (I) is false;
- must not exist at $e4$, $e5$, $e6$ because (II) is false;
- must not exist at $e7$, $e8$ because (I) and (II) are false.

Information about messages (i) not known to be delivered and (ii) not guaranteed to be delivered in CO, is *explicitly* tracked by the algorithm using (*source, timestamp, destination*) information. The information must be deleted as soon as either (i) or (ii) becomes false. The key problem in designing an optimal CO algorithm is to identify the events at which (i) or (ii) becomes false. Information about messages already delivered and messages guaranteed to be delivered in CO is *implicitly* tracked without storing or propagating it, and is derived from the explicit information. Such implicit information is used for determining when (i) or (ii) becomes false for the explicit information being stored or carried in messages.

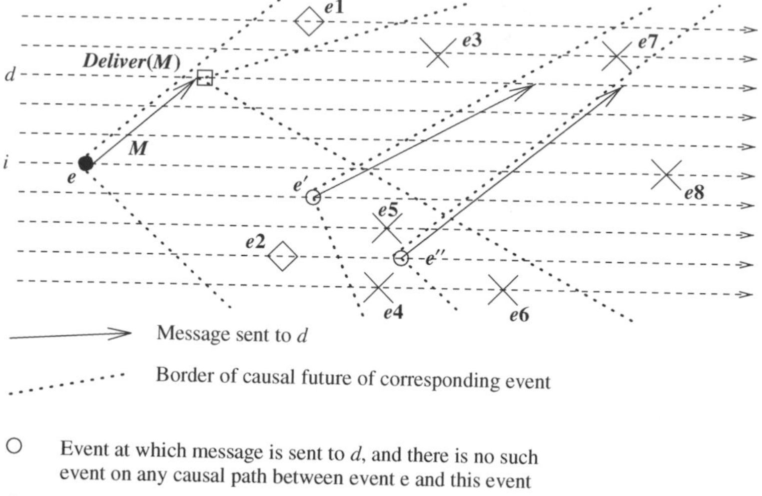

Figure 6.12 Illustrating the necessary and sufficient conditions for causal ordering [21].

The algorithm is given in Algorithm 6.3. Procedure SND is executed atomically. Procedure RCV is executed atomically except for a possible interruption in line 2a where a non-blocking wait is required to meet the Delivery Condition. Note that the pseudo-code can be restructured to complete the processing of each invocation of SND and RCV procedures in a single pass of the data structures, by always maintaining the data structures sorted row–major and then column–major.

1. **Explicit tracking** Tracking of (source, timestamp, destination) information for messages (i) not known to be delivered and (ii) not guaranteed to be delivered in CO, is done explicitly using the $l.Dests$ field of entries in local logs at nodes and $o.Dests$ field of entries in messages. Sets $l_{i,a}.Dests$ and $o_{i,a}.Dests$ contain explicit information of destinations to which $M_{i,a}$ is not guaranteed to be delivered in CO and is not known to be delivered. The information about "$d \in M_{i,a}.Dests$" is propagated up to the earliest events on all causal paths from (i, a) at which it is known that $M_{i,a}$ is delivered to d or is guaranteed to be delivered to d in CO.
2. **Implicit tracking** Tracking of messages that are either (i) already delivered, or (ii) guaranteed to be delivered in CO, is performed implicitly. The information about messages (i) already delivered or (ii) guaranteed to be delivered in CO is deleted and not propagated because it is redundant as far as enforcing CO is concerned. However, it is useful in determining what information that is being carried in other messages and is being stored in logs at other nodes has become redundant and thus can be purged. The semantics are implicitly stored and propagated. This information about messages that are (i) already delivered or (ii) guaranteed to be delivered in

(local variables)

$clock_j \longleftarrow 0$; // local counter clock at node j

$SR_j[1 \ldots n] \longleftarrow \overline{0}$; // $SR_j[i]$ is the timestamp of last msg. from i delivered to j

$LOG_j = \{(i, clock_i, Dests)\} \longleftarrow \{\forall i, (i, 0, \emptyset)\}$;

// Each entry denotes a message sent in the causal past, by i at $clock_i$. $Dests$ is the set of
// remaining destinations for which it is not known that
// $M_{i,clock_i}$ (i) has been delivered, or (ii) is guaranteed to be delivered in CO.

(1) **SND:** j **sends a message** M **to** $Dests$:

(1a) $clock_j \longleftarrow clock_j + 1$;

(1b) **for all** $d \in M.Dests$ **do**:

$O_M \longleftarrow LOG_j$; // O_M denotes $O_{M_{j.clock_j}}$

for all $o \in O_M$, modify $o.Dests$ as follows:

if $d \notin o.Dests$ **then** $o.Dests \longleftarrow (o.Dests \setminus M.Dests)$;

if $d \in o.Dests$ **then** $o.Dests \longleftarrow (o.Dests \setminus M.Dests) \bigcup \{d\}$;

// Do not propagate information about indirect dependencies that are
// guaranteed to be transitively satisfied when dependencies of M are satisfied.

for all $o_{s,t} \in O_M$ **do**

if $o_{s,t}.Dests = \emptyset \bigwedge (\exists o'_{s,t'} \in O_M \mid t < t')$ **then** $O_M \longleftarrow O_M \setminus \{o_{s,t}\}$;

// do not propagate older entries for which $Dests$ field is $\emptyset$

send $(j, clock_j, M, Dests, O_M)$ to d;

(1c) **for all** $l \in LOG_j$ **do** $l.Dests \longleftarrow l.Dests \setminus Dests$;

// Do not store information about indirect dependencies that are guaranteed
// to be transitively satisfied when dependencies of M are satisfied.

Execute $PURGE\_NULL\_ENTRIES(LOG_j)$; // purge $l \in LOG_j$ if $l.Dests = \emptyset$

(1d) $LOG_j \longleftarrow LOG_j \bigcup \{(j, clock_j, Dests)\}$.

(2) **RCV:** j **receives a message** $(k, t_k, M, Dests, O_M)$ **from** k:

(2a) // Delivery Condition: ensure that messages sent causally before M are delivered.

for all $o_{m,t_m} \in O_M$ **do**

if $j \in o_{m,t_m}.Dests$ **wait until** $t_m \leq SR_j[m]$;

(2b) Deliver M; $SR_j[k] \longleftarrow t_k$;

(2c) $O_M \longleftarrow \{(k, t_k, Dests)\} \bigcup O_M$;

for all $o_{m,t_m} \in O_M$ **do** $o_{m,t_m}.Dests \longleftarrow o_{m,t_m}.Dests \setminus \{j\}$;

// delete the now redundant dependency of message represented by o_{m,t_m} sent to j

(2d) // Merge O_M and LOG_j by eliminating all redundant entries.
// Implicitly track "already delivered" & "guaranteed to be delivered in CO"
// messages.

for all $o_{m,t} \in O_M$ **and** $l_{s,t'} \in LOG_j$ **such that** $s = m$ **do**

if $t < t' \bigwedge l_{s,t} \notin LOG_j$ **then** mark $o_{m,t}$;

// $l_{s,t}$ had been deleted or never inserted, as $l_{s,t}.Dests = \emptyset$ in the causal past

if $t' < t \bigwedge o_{m,t'} \notin O_M$ **then** mark $l_{s,t'}$;

// $o_{m,t'} \notin O_M$ because $l_{s,t'}$ had become $\emptyset$ at another process in the causal past

Delete all marked elements in O_M and LOG_j ;

// delete entries about redundant information

for all $l_{s,t'} \in LOG_j$ **and** $o_{m,t} \in O_M$, **such that** $s = m \bigwedge t' = t$ **do**

$l_{s,t'}.Dests \longleftarrow l_{s,t'}.Dests \bigcap o_{m,t}.Dests$;

// delete destinations for which Delivery
// Condition is satisfied or guaranteed to be satisfied as per $o_{m,t}$

Delete $o_{m,t}$ from O_M; // information has been incorporated in $l_{s,t'}$

$LOG_j \longleftarrow LOG_j \bigcup O_M$; // merge non-redundant information of O_M into LOG_j

(2e) $PURGE\_NULL\_ENTRIES(LOG_j)$. // Purge older entries l for which $l.Dests = \emptyset$

PURGE_NULL_ENTRIES(Log_j): // Purge older entries l for which $l.Dests = \emptyset$ is
// implicitly inferred

for all $l_{s,t} \in Log_j$ **do**

if $l_{s,t}.Dests = \emptyset \bigwedge (\exists l'_{s,t'} \in Log_j \mid t < t')$ **then** $Log_j \longleftarrow Log_j \setminus \{l_{s,t}\}$.

Algorithm 6.3 The algorithm by Kshemkalyani–Singhal to optimally implement causal ordering of messages. Code for P_j, $1 \leq j \leq n$.

CO is tracked without explicitly storing it. Rather, the algorithm derives it from the existing explicit information about messages (i) not known to be delivered and (ii) not guaranteed to be delivered in CO, by examining only $o_{i,a}.Dests$ or $l_{i,a}.Dests$, which is a part of the explicit information. There are two types of implicit tracking:

- The absence of a node i.d. from destination information – i.e., $\exists d \in M_{i,a}.Dests \mid d \notin l_{i,a}.Dests \bigvee d \notin o_{i,a}.Dests$ – implicitly contains information that the message has been already delivered or is guaranteed to be delivered in CO to d. Clearly, $l_{i,a}.Dests = \emptyset$ or $o_{i,a}.Dests = \emptyset$ implies that message $M_{i,a}$ has been delivered or is guaranteed to be delivered in CO to *all* destinations in $M_{i,a}.Dests$. An entry whose $.Dests = \emptyset$ is maintained because of the implicit information in it, viz., that of known delivery or guaranteed CO delivery to all destinations of the multicast, is useful to purge redundant information as per the Propagation Constraints.
- As the distributed computation evolves, several entries $l_{i,a_1}, l_{i,a_2}, \ldots$ such that $\forall p$, $l_{i,a_p}.Dests = \emptyset$ may exist in a node's log and a message may be carrying several entries $o_{i,a_1}, o_{i,a_2}, \ldots$ such that $\forall p$, $o_{i,a_p}.Dests = \emptyset$. The second implicit tracking uses a mechanism to prevent the proliferation of such entries. The mechanism is based on the following observation: "*For any two multicasts M_{i,a_1}, M_{i,a_2} such that $a_1 < a_2$, if $l_{i,a_2} \in LOG_j$, then $l_{i,a_1} \in LOG_j$. (Likewise for any message.)*" Therefore, if $l_{i,a_1}.Dests$ becomes $\emptyset$ at a node j, then it can be deleted from LOG_j provided $\exists\ l_{i,a_2} \in LOG_j$ such that $a_1 < a_2$. The presence of such l_{i,a_1}s in LOG_j is automatically implied by the presence of entry l_{i,a_2} in LOG_j. Thus, for a multicast $M_{i,z}$, if $l_{i,z}$ does not exist in LOG_j, then $l_{i,z}.Dests = \emptyset$ implicitly exists in LOG_j iff $\exists\ l_{i,a} \in LOG_j \mid a > z$. As a result of the second implicit tracking mechanism, a node does not keep (and a message does not carry) entries of type $l_{i,a}.Dests = \emptyset$ in its log. However, note that a node must always keep at least one entry of type $l_{i,a}$ (the one with the highest timestamp) in its log for each sender node i. The same holds for messages.

The information tracked implicitly is useful in purging information explicitly carried in other $O_{M''}$s and stored in LOG entries about "yet to be delivered to" destinations for the same message $M_{i,a}$ as well as for messages $M_{i,a'}$, where $a' < a$. Thus, whenever $o_{i,a}$ in some $O_{M'}$ propagates to node j, in line (2d), (i) the implicit information in $o_{i,a}.Dests$ is used to eliminate redundant information in $l_{i,a}.Dests \in LOG_j$; (ii) the implicit information in $l_{i,a}.Dests \in LOG_j$ is used to eliminate redundant information in $o_{i,a}.Dests$; (iii) the implicit information in $o_{i,a}$ is used to eliminate redundant information $l_{i,a'} \in LOG_j$ if $\nexists\ o_{i,a'} \in O_{M'}$ and $a' < a$; (iv) the implicit information in $l_{i,a}$ is used to eliminate redundant information $o_{i,a'} \in O_{M'}$ if $\nexists\ l_{i,a'} \in LOG_j$ and $a' < a$; and (v) only non-redundant information remains in $O_{M'}$ and LOG_j; this is merged together into an updated LOG_j.

Example [6] In the example in Figure 6.13, the timing diagram illustrates (i) the propagation of explicit information "$P_6 \in M_{5,1}.Dests$" and (ii) the inference of implicit information that "$M_{5,1}$ has been delivered to P_6, or is guaranteed to be delivered in causal order to P_6 with respect to any future messages." A thick arrow indicates that the corresponding message contains the explicit information piggybacked on it. A thick line during some interval of the time line of a process indicates the duration in which this information resides in the log local to that process. The number "a" next to an event indicates that it is the ath event at that process.

Multicasts $M_{5,1}$ and $M_{4,2}$

Message $M_{5,1}$ sent to processes P_4 and P_6 contains the piggybacked information "$M_{5,1}.Dests = \{P_4, P_6\}$." Additionally, at the send event (5, 1), the information "$M_{5,1}.Dests = \{P_4, P_6\}$" is also inserted in the local log Log_5. When $M_{5,1}$ is delivered to P_6, the (new) piggybacked information "$P_4 \in M_{5,1}.Dests$" is stored in Log_6 as "$M_{5,1}.Dests = \{P_4\}$"; information about "$P_6 \in M_{5,1}.Dests$," which was needed for routing, must *not* be stored in Log_6 because of constraint I. Symmetrically, when $M_{5,1}$ is delivered to process P_4 at event (4, 1), *only* the new piggybacked information "$P_6 \in M_{5,1}.Dests$" is inserted in Log_4 as "$M_{5,1}.Dests = \{P_6\}$," which is later propagated during multicast $M_{4,2}$.

Multicast $M_{4,3}$

At event (4, 3), the information "$P_6 \in M_{5,1}.Dests$" in Log_4 is propagated on multicast $M_{4,3}$ only to process P_6 to ensure causal delivery using the Delivery Condition. The piggybacked information on message $M_{4,3}$ sent to process P_3 must not contain this information because of constraint II. (The piggybacked information contains "$M_{4,3}.Dests = \{P_6\}$." As long as any future message

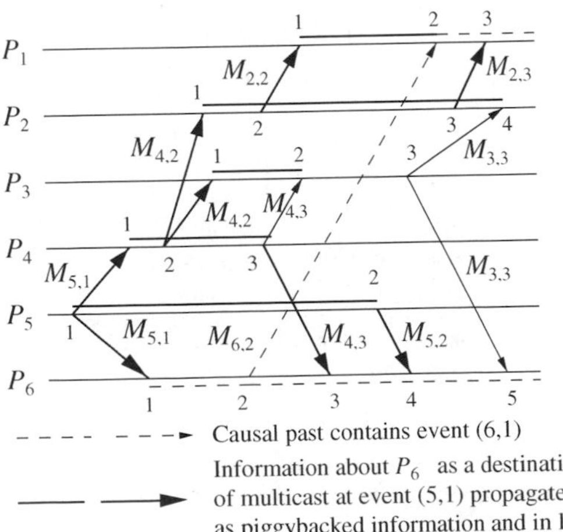

Figure 6.13 An example to illustrate the propagation constraints [6].

Message to dest.	Piggybacked $M_{5,1}$.Dests
$M_{5,1}$to P_4,P_6	$\{P_4,P_6\}$
$M_{4,2}$to P_3,P_2	$\{P_6\}$
$M_{2,2}$to P_1	$\{P_6\}$
$M_{6,2}$to P_1	$\{P_4\}$
$M_{4,3}$to P_6	$\{P_6\}$
$M_{4,3}$to P_3	$\{\}$
$M_{5,2}$to P_6	$\{P_4,P_6\}$
$M_{2,3}$to P_1	$\{P_6\}$
$M_{3,3}$to P_2,P_6	$\{\}$

sent to P_6 is delivered in causal order w.r.t. $M_{4,3}$ sent to P_6, it will also be delivered in causal order w.r.t. $M_{5,1}$ sent to P_6.) And as $M_{5,1}$ is already delivered to P_4, the information "$M_{5,1}.Dests = \emptyset$" is piggybacked on $M_{4,3}$ sent to P_3. Similarly, the information "$P_6 \in M_{5,1}.Dests$" must be deleted from Log_4 as it will no longer be needed, because of constraint II. "$M_{5,1}.Dests = \emptyset$" is stored in Log_4 to remember that $M_{5,1}$ has been delivered or is guaranteed to be delivered in causal order to all its destinations.

Learning implicit information at P_2 and P_3

When message $M_{4,2}$ is received by processes P_2 and P_3, they insert the (new) piggybacked information in their local logs, as information "$M_{5,1}.Dests = \{P_6\}$." They both continue to store this in Log_2 and Log_3 and propagate this information on multicasts until they "learn" at events (2, 4) and (3, 2) on receipt of messages $M_{3,3}$ and $M_{4,3}$, respectively, that any future message is guaranteed to be delivered in causal order to process P_6, w.r.t. $M_{5,1}$ sent to P_6. Hence by constraint II, this information must be deleted from Log_2 and Log_3. The logic by which this "learning" occurs is as follows:

- When $M_{4,3}$ with piggybacked information "$M_{5,1}.Dests = \emptyset$" is received by P_3 at (3, 2), this is inferred to be valid current *implicit* information about multicast $M_{5,1}$ because the log Log_3 already contains explicit information "$P_6 \in M_{5,1}.Dests$" about that multicast. Therefore, the explicit information in Log_3 is inferred to be old and must be deleted to achieve optimality. $M_{5,1}.Dests$ is set to $\emptyset$ in Log_3.
- The logic by which P_2 learns this implicit knowledge on the arrival of $M_{3,3}$ is identical.

Processing at P_6

Recall that when message $M_{5,1}$ is delivered to P_6, only "$M_{5,1}.Dests = \{P_4\}$" is added to Log_6. Further, P_6 propagates only "$M_{5,1}.Dests = \{P_4\}$" (from Log_6) on message $M_{6,2}$, and this conveys the current *implicit* information "$M_{5,1}$ has been delivered to P_6," by its very absence in the explicit information.

- When the information "$P_6 \in M_{5,1}.Dests$" arrives on $M_{4,3}$, piggybacked as "$M_{5,1}.Dests = \{P_6\}$," it is used only to ensure causal delivery of $M_{4,3}$ using the Delivery Condition, and is not inserted in Log_6 (constraint I) – further, the presence of "$M_{5,1}.Dests = \{P_4\}$" in Log_6 implies the *implicit* information that $M_{5,1}$ has already been delivered to P_6. Also, the absence of P_4 in $M_{5,1}.Dests$ in the explicit piggybacked information implies the *implicit* information that $M_{5,1}$ has been delivered or is guaranteed to be delivered in causal order to P_4, and, therefore, $M_{5,1}.Dests$ is set to $\emptyset$ in Log_6.
- When the information "$P_6 \in M_{5,1}.Dests$" arrives on $M_{5,2}$, piggybacked as "$M_{5,1}.Dests = \{P_4, P_6\}$," it is used only to ensure causal delivery of

$M_{4,3}$ using the Delivery Condition, and is not inserted in Log_6 because Log_6 contains "$M_{5,1}.Dests = \emptyset$," which gives the *implicit* information that $M_{5,1}$ has been delivered or is guaranteed to be delivered in causal order to both P_4 and P_6. (Note that at event (5, 2), P_5 changes $M_{5,1}.Dests$ in Log_5 from $\{P_4, P_6\}$ to $\{P_4\}$, as per constraint II, and inserts "$M_{5,2}.Dests = \{P_6\}$" in Log_5.)

Processing at P_1

We have the following processing:

- When $M_{2,2}$ arrives carrying piggybacked information "$M_{5,1}.Dests = \{P_6\}$," this (new) information is inserted in Log_1.
- When $M_{6,2}$ arrives with piggybacked information "$M_{5,1}.Dests = \{P_4\}$," P_1 "learns" *implicit* information "$M_{5,1}$ has been delivered to P_6" by the very absence of explicit information "$P_6 \in M_{5,1}.Dests$" in the piggybacked information, and hence marks information "$P_6 \in M_{5,1}.Dests$" for deletion from Log_1. Simultaneously, "$M_{5,1}.Dests = \{P_6\}$" in Log_1 implies the *implicit* information that $M_{5,1}$ has been delivered or is guaranteed to be delivered in causal order to P_4. Thus, P_1 also "learns" that the explicit piggybacked information "$M_{5,1}.Dests = \{P_4\}$" is outdated. $M_{5,1}.Dests$ in Log_1 is set to $\emptyset$.
- Analogously, the information "$P_6 \in M_{5,1}.Dests$" piggybacked on $M_{2,3}$, which arrives at P_1, is inferred to be outdated (and hence ignored) using the *implicit* knowledge derived from "$M_{5,1}.Dests = \emptyset$" in Log_1.

6.6 Total order

While causal order has many uses, there are other orderings that are also useful. *Total order* is such an ordering [4,5]. Consider the example of updates to replicated data, as shown in Figure 6.11. As the replicas are of just one data item d, it would be logical to expect that all replicas see the updates in the same order, whether or not the issuing of the updates are causally related. This way, the issue of coherence and consistency of the replica values goes away. Such a replicated system would still be useful for fault-tolerance, as well as for easy availability for "read" operations. Total order, which requires that all messages be received in the same order by the recipients of the messages, is formally defined as follows:

Definition 6.14 (Total order) For each pair of processes P_i and P_j and for each pair of messages M_x and M_y that are delivered to both the processes, P_i is delivered M_x before M_y if and only if P_j is delivered M_x before M_y.

Example The execution in Figure 6.11(b) does not satisfy total order. Even if the message m did not exist, total order would not be satisfied. The execution in Figure 6.11(c) satisfies total order.

6.6.1 Centralized algorithm for total order

Assuming all processes broadcast messages, the centralized solution shown in Algorithm 6.4 enforces total order in a system with FIFO channels. Each process sends the message it wants to broadcast to a centralized process, which simply relays all the messages it receives to every other process over FIFO channels. It is straightforward to see that total order is satisfied. Furthermore, this algorithm also satisfies causal message order.

---

(1) When process P_i wants to multicast a message M to group G:
(1a) **send** $M(i, G)$ to central coordinator.

(2) When $M(i, G)$ arrives from P_i at the central coordinator:
(2a) **send** $M(i, G)$ to all members of the group G.

(3) When $M(i, G)$ arrives at P_j from the central coordinator:
(3a) **deliver** $M(i, G)$ to the application.

Algorithm 6.4 A centralized algorithm to implement total order and causal order of messages.

Complexity
Each message transmission takes two message hops and exactly n messages in a system of n processes.

Drawbacks
A centralized algorithm has a single point of failure and congestion, and is therefore not an elegant solution.

6.6.2 Three-phase distributed algorithm

A distributed algorithm that enforces total and causal order for closed groups is given in Algorithm 6.5. The three phases of the algorithm are first described from the viewpoint of the sender, and then from the viewpoint of the receiver.

Sender

Phase 1 In the first phase, a process multicasts (line 1b) the message M with a locally unique tag and the local timestamp to the group members.

Phase 2 In the second phase, the sender process awaits a reply from all the group members who respond with a tentative proposal for a revised timestamp for that message M. The **await** call in line 1d is non-blocking,

```
record Q_entry
    M: int;                         // the application message
    tag: int;                       // unique message identifier
    sender_id: int;                 // sender of the message
    timestamp: int;                 // tentative timestamp assigned to message
    deliverable: boolean;           // whether message is ready for delivery
(local variables)
queue of Q_entry: temp_Q, delivery_Q
int: clock                // Used as a variant of Lamport's scalar clock
int: priority             // Used to track the highest proposed timestamp
(message types)
REVISE_TS(M, i, tag, ts)
            // Phase 1 message sent by P_i, with initial timestamp ts
PROPOSED_TS(j, i, tag, ts)
        // Phase 2 message sent by P_j, with revised timestamp, to P_i
FINAL_TS(i, tag, ts)    // Phase 3 message sent by P_i, with final timestamp

(1)   When process P_i wants to multicast a message M with a tag tag:
(1a)  clock ← clock + 1;
(1b)  send REVISE_TS(M, i, tag, clock) to all processes;
(1c)  temp_ts ← 0;
(1d)  await PROPOSED_TS(j, i, tag, ts_j) from each process P_j;
(1e)  ∀j ∈ N, do temp_ts ← max(temp_ts, ts_j);
(1f)  send FINAL_TS(i, tag, temp_ts) to all processes;
(1g)  clock ← max(clock, temp_ts).

(2)   When REVISE_TS(M, j, tag, clk) arrives from P_j:
(2a)  priority ← max(priority + 1, clk);
(2b)  insert (M, tag, j, priority, undeliverable) in temp_Q;
                                                   // at end of queue
(2c)  send PROPOSED_TS(i, j, tag, priority) to P_j.

(3)   When FINAL_TS(j, x, clk) arrives from P_j:
(3a)  Identifyentry Q_e in temp_Q, where Q_e.tag = x and Q_e.sender_id = j;
(3b)  mark Q_e.deliverable as true;
(3c)  Update Q_e.timestamp to clk and re-sort temp_Q based on the
        timestamp field;
(3d)  if (head(temp_Q)).tag = Q_e.tag then
(3e)     move Q_e from temp_Q to delivery_Q;
(3f)     while (head(temp_Q)).deliverable is true do
(3g)            dequeue head(temp_Q) and insert in delivery_Q.

(4)   When P_i removes a message (M, tag, j, ts, deliverable) from
        head(delivery_Q_i):
(4a)  clock ← max(clock, ts) + 1.
```

Algorithm 6.5 A distributed algorithm to implement total order and causal order of messages. Code at P_i, $1 \leq i \leq n$.

i.e., any other messages received in the meanwhile are processed. Once all expected replies are received, the process computes the maximum of the proposed timestamps for M, and uses the maximum as the final timestamp.

Phase 3 In the third phase, the process multicasts the final timestamp to the group in line (1f).

Receivers

Phase 1 In the first phase, the receiver receives the message with a tentative/proposed timestamp. It updates the variable *priority* that tracks the highest proposed timestamp (line 2a), then revises the proposed timestamp to the *priority*, and places the message with its tag and the revised timestamp at the tail of the queue *temp_Q* (line 2b). In the queue, the entry is marked as undeliverable.

Phase 2 In the second phase, the receiver sends the revised timestamp (and the tag) back to the sender (line 2c). The receiver then waits in a non-blocking manner for the final timestamp (correlated by the message tag).

Phase 3 In the third phase, the final timestamp is received from the multicaster (line 3). The corresponding message entry in *temp_Q* is identified using the tag (line 3a), and is marked as deliverable (line 3b) after the revised timestamp is overwritten by the final timestamp (line 3c). The queue is then resorted using the timestamp field of the entries as the key (line 3c). As the queue is already sorted except for the modified entry for the message under consideration, that message entry has to be placed in its sorted position in the queue. If the message entry is at the head of the *temp_Q*, that entry, and all consecutive subsequent entries that are also marked as deliverable, are dequeued from *temp_Q*, and enqueued in *deliver_Q* in that order (the loop in lines 3d–3g).

Complexity

This algorithm uses three phases, and, to send a message to $n-1$ processes, it uses $3(n-1)$ messages and incurs a delay of three message hops.

Example An example execution to illustrate the algorithm is given in Figure 6.14. Here, A and B multicast to a set of destinations and C and D are the common destinations for both multicasts.

- **Figure 6.14(a)** The main sequence of steps is as follows:
  1. A sends a *REVISE_TS*(7) message, having timestamp 7. B sends a *REVISE_TS*(9) message, having timestamp 9.
  2. C receives A's *REVISE_TS*(7), enters the corresponding message in *temp_Q*, and marks it as undeliverable; *priority* = 7. C then sends *PROPOSED_TS*(7) message to A.

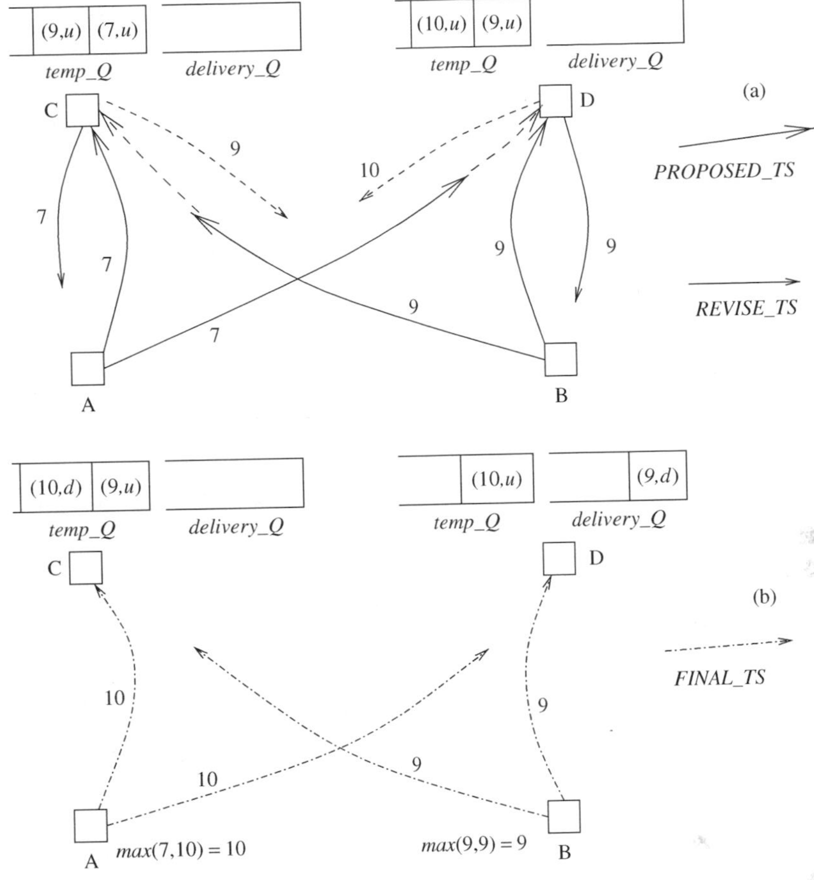

Figure 6.14 An example to illustrate the three-phase total ordering algorithm. (a) A snapshot for *PROPOSED_TS* and *REVISE_TS* messages. The dashed lines show the further execution after the snapshot. (b) The *FINAL_TS* messages in the example.

3. D receives B's *REVISE_TS*(9), enters the corresponding message in *temp_Q*, and marks it as undeliverable; *priority* = 9. D then sends *PROPOSED_TS*(9) message to B.
4. C receives B's *REVISE_TS*(9), enters the corresponding message in *temp_Q*, and marks it as undeliverable; *priority* = 9. C then sends *PROPOSED_TS*(9) message to B.
5. D receives A's *REVISE_TS*(7), enters the corresponding message in *temp_Q*, and marks it as undeliverable; *priority* = 10. D assigns a tentative timestamp value of 10, which is greater than all of the timestamps on *REVISE_TS*s seen so far, and then sends *PROPOSED_TS*(10) message to A.

The state of the system is as shown in the figure.

- **Figure 6.14(b)** The continuing sequence of main steps is as follows:

6. When A receives *PROPOSED_TS*(7) from C and *PROPOSED_TS*(10) from D, it computes the final timestamp as $max(7, 10) = 10$, and sends *FINAL_TS*(10) to C and D.

7. When B receives *PROPOSED_TS*(9) from C and *PROPOSED_TS*(9) from D, it computes the final timestamp as $max(9, 9) = 9$, and sends *FINAL_TS*(9) to C and D.
8. C receives *FINAL_TS*(10) from A, updates the corresponding entry in *temp_Q* with the timestamp, resorts the queue, and marks the message as deliverable. As the message is not at the head of the queue, and some entry ahead of it is still undeliverable, the message is not moved to *delivery_Q*.
9. D receives *FINAL_TS*(9) from B, updates the corresponding entry in *temp_Q* by marking the corresponding message as deliverable, and resorts the queue. As the message is at the head of the queue, it is moved to *delivery_Q*.

This is the system snapshot shown in Figure 6.14(b). The following further steps will occur:

10. When C receives *FINAL_TS*(9) from B, it will update the corresponding entry in *temp_Q* by marking the corresponding message as deliverable. As the message is at the head of the queue, it is moved to the *delivery_Q*, and the next message (of A), which is also deliverable, is also moved to the *delivery_Q*.
11. When D receives *FINAL_TS*(10) from A, it will update the corresponding entry in *temp_Q* by marking the corresponding message as deliverable. As the message is at the head of the queue, it is moved to the *delivery_Q*.

Algorithm 6.5 is closely structured along the lines of Lamport's algorithm for mutual exclusion. We will later see that Lamport's mutual exclusion algorithm has the property that when a process is at the head of its own queue and has received a REPLY from all other processes, the REQUEST of that process is at the head of all the queues. This can be exploited to deliver the message by all the processes in the same total order (instead of entering the critical section).

6.7 A nomenclature for multicast

In this section, we systematically classify the various kinds of multicast algorithms possible [9]. Observe that there are four classes of source–destination relationships, as illustrated in Figure 6.15, for open groups:

- **SSSG** Single source and single destination group.
- **MSSG** Multiple sources and single destination group.
- **SSMG** Single source and multiple, possibly overlapping, groups.
- **MSMG** Multiple sources and multiple, possibly overlapping, groups.

The SSSG and SSMG classes are straightforward to implement, assuming the presence of FIFO channels between each pair of processes. Both total

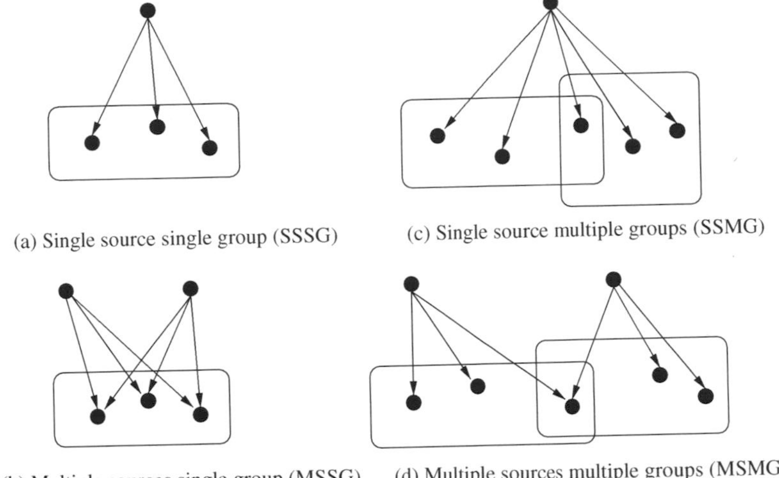

Figure 6.15 Four classes of source–destination relationships for open-group multicasts. For closed-group multicasts, the sender needs to be part of the recipient group.

order and causal order are guaranteed. The MSSG class is also straightforward to handle; the centralized implementation in Algorithm 6.4 provides both total and causal order. The central coordinator effectively converts this class to the SSSG class.

We now consider a design approach for the MSMG class. This approach, commonly termed as the *propagation tree* approach, uses a semi-centralized structure that adapts the centralized algorithm of Algorithm 6.4 and was proposed by Chiu and Hsaio [9] and Jia [16].

6.8 Propagation trees for multicast

To manage the complications of delivery order across multiple overlapping groups $\mathcal{G} = \{G_1 \ldots G_g\}$, the algorithm first identifies a set of *metagroups* $\mathcal{MG} = \{MG_1, \ldots MG_h\}$ with the following properties: (i) each process belongs to a single metagroup, and has the exact same group membership as every other process in that metagroup; (ii) no other process outside that metagroup has that exact group membership.

Example Figure 6.16(a) shows some groups and their metagroups. $\langle ABC \rangle$, $\langle AB \rangle$, $\langle AC \rangle$, and $\langle A \rangle$ are the metagroups of user group $\langle A \rangle$.

The definition of metagroups transforms the problem of MSMG multicast to groups, to the problem of MSSG multicast to metagroups, which is easier to solve.

A distinguished node in each metagroup acts as the manager for that metagroup. For each user group G_i, one of its metagroups is chosen to be its *primary metagroup (PM)* and denoted as $PM(G_i)$. All the metagroups are

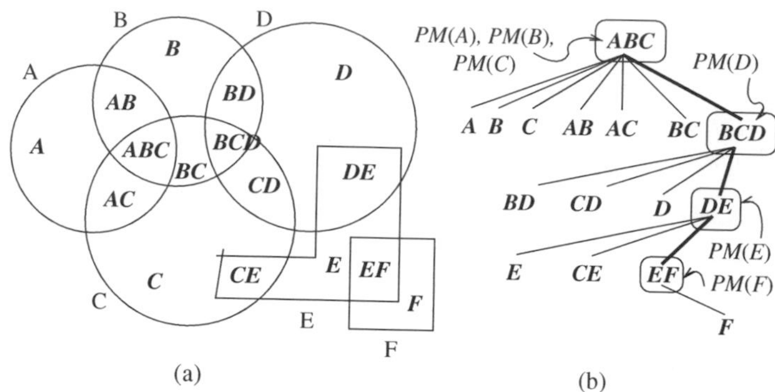

Figure 6.16 Example illustrating a propagation tree [9]. Metagroups are shown in boldface. (a) Groups A, B, C, D, E, and F, and their metagroups. (b) A *propagation tree*, with the primary meta-groups labeled.

organized in a *propagation forest* or *tree* structure satisfying the following property: for user group G_i, its primary metagroup $PM(G_i)$ is at the lowest possible level (i.e., farthest from the root) of the tree such that all the metagroups whose destinations contain any nodes of G_i belong to the subtree rooted at $PM(G_i)$.

Example In Figure 6.16, $\langle ABC \rangle$ is the primary metagroup of A, B, and C. $\langle B, C, D \rangle$ is the primary metagroup of D. $\langle D, E \rangle$ is the primary metagroup of E. $\langle E, F \rangle$ is the primary metagroup of F.

The following properties can be seen to be satisfied by the *propagation tree*:

1. The primary metagroup $PM(G)$, is the ancestor of all the other metagroups of G in the propagation tree.
2. $PM(G)$ is uniquely defined.
3. For any metagroup MG, there is a unique path to it from the PM of any of the user groups of which the metagroup MG is a subset.
4. In addition, for any two primary metagroups $PM(G_1)$ and $PM(G_2)$, they should either lie on the same branch of a tree, or be in disjoint trees. In the latter case, their groups membership sets are necessarily disjoint.

Key idea

The metagroup $PM(G_i)$ of user group G_i, is useful for multicasts, as follows: *multicasts to G_i are sent first to the metagroup $PM(G_i)$ as only the subtree rooted at $PM(G_i)$ can contain the nodes in G_i*. The message is then propagated down the subtree rooted at $PM(G_i)$.

The following definitions are useful to understand and explain the algorithm:

- MG_1 *subsumes* MG_2 (where $MG_1 \neq MG_2$) if for each group G such that a member of MG_2 is a member of G, we have that some member of MG_1 is also a member of G. In other words, MG_1 is a subset of each user group G of which MG_2 is a subset.

Example In Figure 6.16, $\langle AB \rangle$ subsumes $\langle A \rangle$. Any member of $MG_2 = \langle A \rangle$ is a member of A and each member of $\langle AB \rangle$ is also a member of A. Similarly, $\langle AB \rangle$ subsumes $\langle B \rangle$.

- MG_1 *is joint with* MG_2 if neither metagroup subsumes the other and there is some group G such that $MG_1, MG_2 \subset G$.

  Example In Figure 6.16, $\langle ABC \rangle$ is joint with $\langle CD \rangle$. Neither subsumes the other and both are a subset of C.

Example Figure 6.16 shows some groups, their metagroups, and their *propagation tree*. Metagroup $\langle ABC \rangle$ is the primary metagroup $PM(A)$, $PM(B)$, $PM(C)$. Meta-group $\langle BCD \rangle$ is the primary metagroup $PM(D)$. Thus, a multicast to group D will be sent to $\langle BCD \rangle$.

We note that the propagation tree is not unique because it depends on the order in which metagroups are processed. Various optimizations on the propagation tree can also be performed, but we require that features (1)–(4) above should be satisfied by the tree. Exercise 6.10 asks you to design an algorithm to construct a propagation tree. A metagroup that has members from multiple user groups is desirable as the root in order to have a tree with low height.

Correctness

The rules for forwarding messages during a multicast are given in Algorithm 6.6. Each process needs to know the propagation tree, computed at a central location. Each metagroup has a distinguished process which acts as the *manager* or representative of that metagroup.

The array $SV[1 \ldots h]$ kept by each process P_i tracks in $SV[k]$, the number of messages multicast by P_i that will traverse through primary metagroup $PM(G_k)$. This array is piggybacked on each message multicast by process P_i.

The manager of each primary metagroup keeps an array $RV[1 \ldots n]$ that tracks in $RV[k]$, the number of messages sent by process P_k that have been received by this primary metagroup.

As in the CO algorithms, a message from P_i can be processed by a primary metagroup j if $RV_j[i] = SV_i[j]$; otherwise it buffers the message until this condition is satisfied (lines 2a–2c). At a non-primary metagroup, this check need not be performed because it never receives a message directly from the sender of the multicast. The multicast sender always sends the message to the primary metagroup first. At the non-primary metagroup, the relative order

(local variables)
integer: $SV[1 \ldots h]$; //kept by each process. h is #(primary //metagroups), $h \leq |\mathcal{G}|$
integer: $RV[1 \ldots n]$; //kept by each primary metagroup manager. //n is #(processes)
set of integers: PM_set; //set of primary metagroups through which //message must traverse

(1) When process P_i wants to multicast message M to group G:
(1a) **send** $M(i, G, SV_i)$ to manager of $PM(G)$, primary metagroup of G;
(1b) $PM\_set \longleftarrow$ { primary metagroups through which M must traverse };
(1c) **for all** $PM_x \in PM\_set$ **do**
(1d) $SV_i[x] \longleftarrow SV_i[x] + 1$.

(2) When P_i, the manager of a metagroup MG receives $M(k, G, SV_k)$ from P_j:
// Note: P_i may not be a manager of any metagroup
(2a) **if** MG is a primary metagroup **then**
(2b) **buffer** the message **until** $(SV_k[i] = RV_i[k])$;
(2c) $RV_i[k] \longleftarrow RV_i[k] + 1$;
(2d) **for each** child metagroup that is subsumed by MG **do**
(2e) **send** $M(k, G, SV_k)$ to the manager of that child metagroup;
(2f) **if** there are no child metagroups **then**
(2g) **send** $M(k, G, SV_k)$ to each process in this metagroup.

Algorithm 6.6 Protocol to enforce total and causal order using propagation trees.

of messages has already been determined by some ancestor metagroup; so it simply forwards the message as per lines 2d–2g.

- The logic behind why total order is maintained is straightforward. For any metagroups MG_1 and MG_2, and any groups G_x and G_y of which the metagroups are a subset, the primary metagroups $PM(G_x)$ and $PM(G_y)$ both subsume MG_1 and MG_2, and both lie on the same branch of the *propagation tree* to either MG_1 or MG_2. The primary metagroup that is lower in the tree will necessarily receive the two multicasts in some order. The assumption of FIFO channels guarantees that all processes in metagroups subsumed by this lower primary metagroup will receive the messages sent to the two groups in a common order.
- Causal order is guaranteed because of the check made by managers of the primary metagroups in lines 2a–2c. Assume that messages M and M' are multicast to G and G', respectively. For nodes in $G \cap G'$, there are two cases, as shown in Figure 6.17. In each case, the sequence numbers next to messages indicate the order in which the messages are sent.

  Case Figure 6.17(a) and (b): Here, the senders of M and M' are different. P_k sends M to G. After $P_i \in G$ receives M, P_i sends M' to G'.

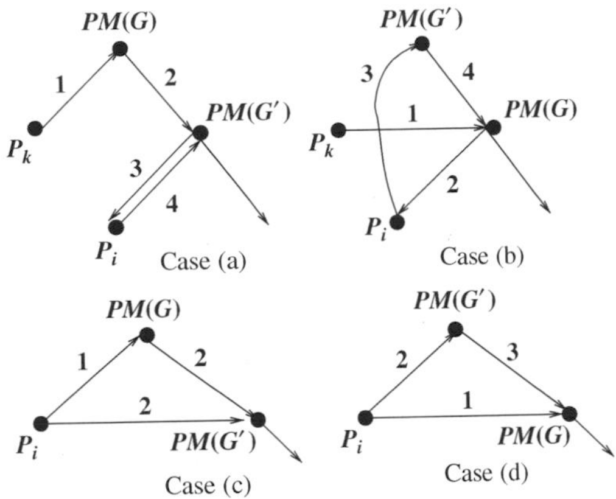

Figure 6.17 The four cases for the correctness of causal ordering using *propagation trees*. The sequence numbers indicate the order in which the messages are sent.

Thus, we have the causal chain $Send_k(k, M, G)$, $Deliver_i(k, M, G)$, $Send_i(i, M', G')$. For any destination MG_q such that $MG_q \subset G \cap G'$, the primary metagroup of G and G' must both be ancestors of the metagroup of P_i because of the assumption of *closed groups*.

Case (a): $PM(G')$ will have already received and processed M (flow 2) before it receives M' (flow 4).

Case (b): $PM(G)$ will have already received and processed M (flow 1) before it receives M' (flow 4). Assuming FIFO channels, CO is guaranteed for all processes in $G \cap G'$.

Case Figure 6.17(c) and (d): P_i sends M to G and then P_i sends M' to G'. Thus, we have the causal chain $Send_i(i, M, G)$, $Send_i(i, M', G')$.

Case (c): The check in lines 2a–2c by $PM(G')$ ensures that $PM(G')$ will not process M' before it processes M.

Case (d): The check in lines 2a–2c by $PM(G)$ ensures that $PM(G)$ will not process M' before it processes M. Assuming FIFO channels, CO is guaranteed for all processes in $G \cap G'$.

6.9 Classification of application-level multicast algorithms

We have seen some algorithmically challenging techniques in the design of multicast algorithms. The most general scenario allows each process to multicast to an arbitrary and dynamically changing group of processes at each step. As this generality incurs more overhead, algorithms implemented on real systems tend to be more "centralized" in one sense or another: Defago *et al.* give an exhaustive survey and this section is based on this survey [11]. For details of the various protocols, please refer to the survey. Many multicast protocols have been developed and deployed, but they can all be classified as belonging to one of the following five classes.

Communication history-based algorithms

Algorithms in this class use a part of the communication history to guarantee ordering requirements.

The RST [22] and KS [20,21] algorithms belong to this class, and provide only causal ordering. They do not need to track separate groups, and hence work for open-group multicasts.

Lamport's algorithm, wherein messages are assigned scalar timestamps and a process can deliver a message only when it knows that no other message with a lower timestamp can be multicast, also belongs to this class. The NewTop protocol [12], which extends Lamport's algorithm to overlapping groups, also guarantees both total and causal ordering. Both these algorithms use closed-group configurations.

Privilege-based algorithms

The operation of such algorithms is illustrated in Figure 6.18(a). A token circulates among the sender processes. The token carries the sequence number for the next message to be multicast, and only the token-holder can multicast. After a multicast send event, the sequence number is updated. Destination processes deliver messages in the order of increasing sequence numbers. Senders need to know the other senders, hence closed groups are assumed. Such algorithms can provide total ordering, as well as causal ordering using a closed group configuration (see Exercise 6.12).

Examples of specific algorithms are On-Demand, and Totem. They differ in implementation details such as whether a token ring topology is assumed

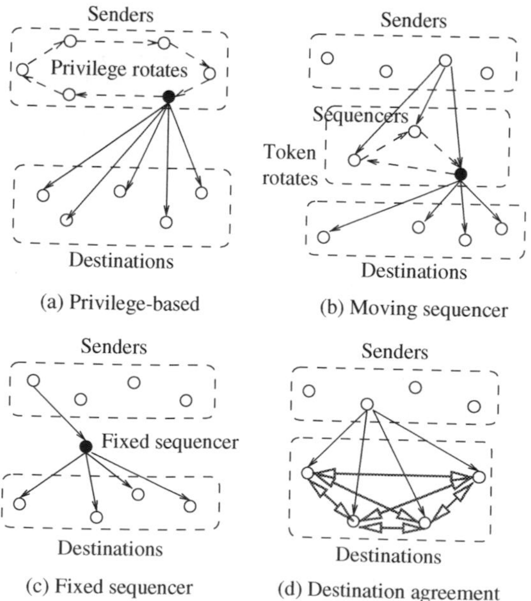

Figure 6.18 Models for sequencing messages. (a) Privilege-based algorithms. (b) Moving sequencer algorithms. (c) Fixed sequencer algorithms. (d) Destination agreement algorithms.

(Totem) or not (On-Demand). Such algorithms are not scalable because they do not permit concurrent send events. Hence they are of limited use in large systems.

Moving sequencer algorithms

The operation of such algorithms is illustrated in Figure 6.18(b). The original algorithm was proposed by Chang and Maxemchuck [8]; various variants of it were given by the Pinwheel and RMP algorithms. These algorithms work as follows. (1) To multicast a message, the sender sends the message to all the sequencers. (2) Sequencers circulate a token among themselves. The token carries a sequence number and a list of all the messages for which a sequence number has already been assigned – such messages have been sent already. (3) When a sequencer receives the token, it assigns a sequence number to all received but unsequenced messages. It then sends the newly sequenced messages to the destinations, inserts these messages in to the token list, and passes the token to the next sequencer. (4) Destination processes deliver the messages received in the order of increasing sequence number.

Moving sequencer algorithms guarantee total ordering.

Fixed sequencer algorithms

The operation of such algorithms is illustrated in Figure 6.18(c). This class is a simplified version of the previous class. There is a single sequencer (unless a failure occurs), which makes this class of algorithms essentially centralized.

The propagation tree approach studied earlier, belongs to this class. Other algorithms are the ISIS sequencer, Amoeba, Phoenix, and Newtop's asymmetric algorithm. Let us look briefly at Newtop's asymmetric algorithm. All processes maintain logical clocks, and each group has an independent sequencer. The unicast from the sender to the sequencer, as well as the multicast from the sequencer are timestamped. A process that belongs to multiple groups must delay the sending of the next message (to the relevant sequencer) until it has received and processed all messages, from the various sequencers, corresponding to the previous messages it sent. Assuming FIFO channels, it can be shown that total order is maintained.

Destination agreement algorithms

The operation of such algorithms is illustrated in Figure 6.18(d). In this class of algorithms, the destinations receive the messages with some limited ordering information. They then exchange information among themselves to define an order. There are two sub-classes here: (i) the first sub-class uses timestamps (Lamport's three-phase algorithm (Algorithm 6.5) belongs to this sub-class); (ii) the second sub-class uses an agreement or "consensus" protocol among the processes. We will study agreement protocols in Chapter 14.

6.10 Semantics of fault-tolerant group communication

A failure-free system can be assumed only in an ideal world. When a system component fails in the midst of the multicast operation, which is a non-atomic operation that spans across time and across multiple links and nodes, the behavior of a multicast protocol must adhere to a well-defined specification, and, correspondingly, the protocol must ensure that the specification under the failure mode is also implemented. This enables well-defined actions during recovery after the failure. This section is based on the results of Hadzilacos and Toueg [15]. Questions such as the following need to be addressed:

- For a multicast, if one correct process delivers the message M, what can be said about the other correct processes and faulty processes that also deliver M?
- For a multicast, if one faulty process delivers the message M, what can be said about the other correct processes and faulty processes that also deliver M?
- For causal or total order multicast, if one correct or faulty process delivers M, what can be said about other correct processes and faulty processes that also deliver M?

There are two broad flavors of the specifications. In the regular flavor, there are no conditions on the messages delivered to faulty processors (because they are faulty). However, assuming the benign failure model, under some conditions, it may be useful to specify and control the behavior of such faulty processes also. Therefore, the second flavor of specifications, termed as the *uniform* specifications, also states the expected behavior of faulty processes. In the following description of the specifications [15], the regular flavor and the uniform flavor are stated. To parse for the regular flavor, the parenthesized words should be omitted. To parse for the uniform flavor, the *italicized* and parenthesized modifiers to the definitions of the regular flavor are included.

(***Uniform***) Reliable multicast of M.

Validity If a correct process multicasts M, then all correct processes will eventually deliver M.

(***Uniform***) **agreement** If a correct (*or faulty*) process delivers M, then all correct processes will eventually deliver M.

(***Uniform***) **integrity** Every correct (*or faulty*) process delivers M at most once, and only if M was previously multicast by $sender(M)$.

The validity property states that once the multicast is initiated by a correct process, it will go to completion. The agreement property states that all correct processes get the same view of a message, irrespective of whether a correct process or a faulty process broadcasts it. The integrity property states that correct processes have non-duplicate delivery of messages, and that they are not delivered spurious messages. While the regular agreement property permits a faulty process to deliver a message that is never delivered to any correct process, this undesirable behavior can be problematic in applications such as atomic commit in database protocols, and is explicitly ruled out by uniform agreement. While the regular Integrity property permits a faulty process to deliver a message multiple times, and to deliver a message that was never sent, this behavior is explicitly ruled out by uniform integrity.

The orderings FIFO order, causal order, and total order are now defined for multicasts, in both the regular and uniform flavors. The uniform flavor requires that even faulty processes do not violate the ordering properties. These definitions of the regular and uniform flavors are superimposed on the basic definition of a (uniform) reliable multicast, given above. The regular flavor and the uniform flavor of each definition is read using the semantics above for parsing the corresponding flavors of multicast. In these definitions which deal with the relative order of messages, it is important that the multicast groups are identical, in which case the messages get broadcast within the common group.

(***Uniform***) **FIFO order** If a process broadcasts M before it broadcasts M', then no correct (*or faulty*) process delivers M' unless it previously delivered M.

(***Uniform***) **causal order** If M is broadcast causally before M' is broadcast, then no correct (*or faulty*) process delivers M' unless it previously delivered M.

(***Uniform***) **total order** If correct (*or faulty*) processes a and b both deliver M and M', then a delivers M before M' if and only if b delivers M before M'.

It is time to remember the folklore result that any protocol or implementation that deals with fault-tolerance incurs a greater cost than what it would in a failure-free environment. In some case, this extra cost can be substantial. Nevertheless, it is important to formally specify the behavior in the face of faults, and to provide the implementations that can realize such behavior. We will not deal with implementations of the above fault-tolerant specifications of multicasts.

Excessive delay in delivering a multicast message can also be viewed as a fault. Applications with real-time constraints require that if a message is delivered, it should be within a bounded period Δ, termed the latency, after it was multicast. This specification can be based on either a global observer's notion of time, or the local time at each process, leading to real-time Δ-timeliness and local-time Δ-timeliness, respectively:

(***Uniform***) **real-time Δ-timeliness** For some known constant Δ, if M is multicast at real-time t, then no correct (*or faulty*) process delivers M after real-time $t+\Delta$.

(*Uniform*) local Δ-timeliness For some known constant Δ, if M is multicast at local time t_m, then no correct (*or faulty*) process i delivers M after its local time $t_m + \Delta$ on i's clock.

Specifying local-time Δ-timeliness requires care because the local clocks at processes can vary. It is assumed that the sender timestamps the message multicast with its local time t_m, and any receiver should receive the message within $t_m + \Delta$ on its local clock. The efficacy of this specification depends on how closely the local clocks are synchronized. A protocol to synchronize physical clocks was studied in Chapter 3.

6.11 Distributed multicast algorithms at the network layer

Several applications can interface directly with the network layer and the lower hardware-related layers to exploit the physical connectivity and the physical topology for group communication. The network is viewed as a graph (N, L), and various graph algorithms – centralized or distributed – are run to establish and maintain efficient routing structures. For example,

- LANs connected by bridges maintain spanning trees for distributing information and for forward/backward learning of destinations;
- the network layer of the Internet has a rich suite of multicast algorithms.

In this section, we will study the principles underlying several such algorithms. Some of the algorithms in this section may not be distributed. Nevertheless, they are intended for a distributed setting, namely the LAN or the WAN.

6.11.1 Reverse path forwarding (RPF) for constrained flooding

As studied in Chapter 5, broadcasting data using flooding in a network (N, L) requires up to $2|L|$ messages. Reverse path forwarding (RPF) is a simple but elegant technique that brings down the overhead significantly at very little cost. Network nodes are assumed to run the distance vector routing (DVR) algorithm (Chapter 5), which was used in the Internet until 1983. (Since 1983, the LSR-based algorithms described in Chapter 5 have been used. These are more sophisticated and provide more information than that required by DVR.)

The simple DVR algorithm assumes that each node knows the next hop on the path to each destination x. This path is assumed to be the approximation to the "best" path. Let $Next\_hop(x)$ denote the function that gives the next hop on the "best" path to x. The RPF algorithm leverages the DVR algorithm for point-to-point routing, to achieve constrained flooding. The RPF algorithm for constrained flooding is shown in Algorithm 6.7.

(1) When process P_i wants to multicast message M to group $Dests$:
(1a) **send** $M(i, Dests)$ on all outgoing links.

(2) When a node i receives message $M(x, Dests)$ from node j:
(2a) **if** $Next\_hop(x) = j$ **then** // this will necessarily be a new message
(2b) **forward** $M(x, Dests)$ on all other incident links besides (i, j);
(2c) **else** ignore the message.

Algorithm 6.7 Reverse path forwarding (RPF).

This simple RPF algorithm has been experimentally shown to be effective in bringing the number of messages for a multicast closer to $|N|$ than to $|L|$. Actually, the algorithm does a broadcast to all the nodes, and this broadcast is smartly curtailed to approximate a spanning tree. The curtailed broadcast is effective because, implicitly, an approximation to a tree rooted at the source is identified, without it being computed or stored at any node.

Pruning of the implicit broadcast tree can be used to deal with unwanted multicast packets. If a node receives the packets but the application running on it does not need the packets, and all "downstream" (in the implicit tree) nodes also do not need the packets, the node can send a *prune* message to the parent in the tree indicating that packets should not be forwarded on that edge. Implementing this in a dynamic network where the tree periodically changes and the application's node membership also changes dynamically is somewhat tricky (see Exercise 6.14).

6.11.2 Steiner trees

The problem of finding an optimal "spanning" tree that spans only all nodes participating in a multicast group, known as the *Steiner tree problem*, is formalized as follows.

Steiner tree problem

Given a weighted graph (N, L) and a subset $N' \subseteq N$, identify a subset $L' \subseteq L$ such that (N', L') is a subgraph of (N, L) that connects all the nodes of N'.

A *minimal Steiner tree* is a minimal-weight subgraph (N', L'). The minimal Steiner tree problem has been well-studied and is known to be NP-complete. When the link weights change, the tree has to be recomputed to obtain the new minimal Steiner tree, making it even more difficult to use in dynamic networks.

Several heuristics have been proposed to construct an approximation to the minimal Steiner tree. A simple heuristic constructs a MST, and deletes edges that are not necessary. This algorithm is given by the first three steps of Algorithm 6.8. The worst case cost of this heuristic is twice the cost of the optimal solution. Algorithm 6.8 can show better performance when using the heuristic by Kou *et al.* [19], given by steps 4 and 5 in the algorithm.

The resulting Steiner tree cost is also at most twice the cost of the minimal Steiner tree, but behaves better on average.

Input: weighted graph $G = (N, L)$, and $N' \subseteq N$, where N' is the set of Steiner points

(1) Construct the complete undirected distance graph $G' = (N', L')$ as follows:
$L' = \{(v_i, v_j) \mid v_i, v_j \text{ in } N'\}$, and $wt(v_i, v_j)$ is the length of the shortest path from v_i to v_j in (N, L).
(2) Let T' be the minimal spanning tree of G'. If there are multiple minimum spanning trees, select one randomly.
(3) Construct a subgraph G_s of G by replacing each edge of the MST T' of G', by its corresponding shortest path in G. If there are multiple shortest paths, select one randomly.
(4) Find the minimum spanning tree T_s of G_s. If there are multiple minimum spanning trees, select one randomly.
(5) Using T_s, delete edges as necessary so that all the leaves are the Steiner points N'. The resulting tree, $T_{Steiner}$, is the heuristic's solution.

Algorithm 6.8 The Kou–Markowsky–Berman heuristic for a minimum Steiner tree.

Cost The time complexity of the heuristic algorithm for each of the five steps is as follows: step 1: $O(|N'| \cdot |N|^2)$; step 2: $O(|N'|^2)$; step 3: $O(|N|)$; step 4: $O(|N|^2)$; step 5: $O(|N|)$. Step 1 dominates, hence the time complexity is $O(|N'| \cdot |N|^2)$.

6.11.3 Multicast cost functions

Consider a source node s that has to do a multicast to Steiner nodes. As before, we are given the weighted graph (N, L) and the Steiner node set N'. We can define several cost functions [3]. For example, let $cost(i)$ be the cost of the path from s to i in the routing scheme R.

The *destination cost of R* is defined as $\frac{1}{|N'|} \sum_{i \in N'} cost(i)$. This represents the average cost of the routing. If the cost is measured in time delay, this routing function metric gives the shortest average time for the multicast to reach nodes in N'.

As a variant, a link is counted only once even if it is used on the minimum cost path to multiple destinations. This variant reduces to the Steiner tree problem of Section 6.11.2. The sum of the costs of the edges in the Steiner tree routing scheme R is defined as the *network cost*.

6.11.4 Delay-bounded Steiner trees

Multimedia networks and interactive applications have given rise to the need for a minimum Steiner tree that also satisfies delay constraints on the transmission. Thus now, the goal is not only to minimize the cost of the tree (measured in terms of a parameter such as the link weight, which models the available bandwidth or a similar cost measure) but also to minimize the delay (propagation delay). The problem is formalized as follows.

Delay-bounded minimal Steiner tree problem

Given a weighted graph (N, L), there are two weight functions $\mathcal{C}(l)$ and $\mathcal{D}(l)$ for each edge in L. $\mathcal{C}(l)$ is a positive real cost function on $l \in L$ and $\mathcal{D}(l)$ is a positive integer delay function on $l \in L$. For a given delay tolerance Δ, a given source s and a destination set $Dest$, where $\{s\} \cup Dest = N' \subseteq N$, identify a spanning tree T covering all the nodes in N', subject to the constraints below. Here, we let $path(s, v)$ denote the path from s to v in T.

- $\sum_{l \in T} \mathcal{C}(l)$ is minimized, subject to
- $\forall v \in N'$, $\sum_{l \in path(s,v)} \mathcal{D}(l) < \Delta$.

Finding such a minimal Steiner tree, subject to another parameter, is at least as difficult as finding a Steiner tree. It can be shown that this problem reduces to the Steiner tree problem. A detailed study of two heuristics to solve this problem is presented by Kompella *et al.* [18]. A *constrained cheapest path* between x and y is the cheapest path between x and y that has delay less than Δ. The cost and delay on such a path are denoted by $\mathcal{C}(x, y)$ and $\mathcal{D}(x, y)$, respectively. If two or more paths have the lowest cost, the lowest delay path is chosen. The steps to compute the constrained Steiner tree are shown in Algorithm 6.9. Step 1 computes the complete closure graph G' on nodes in N'. The two heuristics given below are used in Step 2 to greedily build a constrained Steiner tree on G'. Step 3 expands the tree edges in G' to their original paths in G. An example of a constrained Steiner tree for the input graph in Figure 6.19(a) is given in Figure 6.19(b).

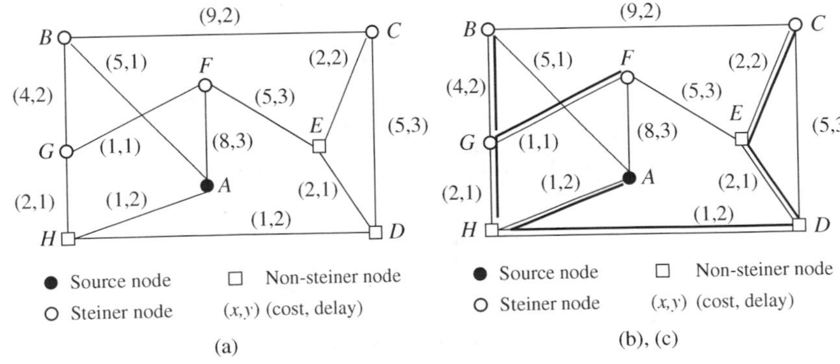

Figure 6.19 Constrained Steiner tree example [18]. (a) Network graph. (b) and (c) MST and Steiner tree (optimal) are the same and shown in thick lines.

$\mathcal{C}(l)$ // cost of edge l
$\mathcal{D}(l)$ // delay of edge l
T; // constrained spanning tree to be constructed
$\mathcal{P}_C(x, y)$; // cost of constrained cheapest path from x to y
$\mathcal{P}_D(x, y)$; // delay on constrained cheapest path from x to y
$\mathcal{C}_d(x, y)$; // cost of the cheapest path with delay exactly d

Input: weighted graph $G = (N, L)$, and $N' \subseteq N$, where N' is the set of Steiner points, source is s, and Δ is the constraint on the delay.

1. Compute the closure graph G' on (N', L), to be the complete graph on N'. The closure graph is computed using the all-pairs constrained cheapest paths using a dynamic programming approach analogous to Floyd's algorithm. For any pair of nodes $x, y \in N'$:

   - $\mathcal{P}_C(x, y) = min_{d<\Delta}\mathcal{C}_d(x, y)$. This selects the cheapest constrained path, satisfying the condition of Δ, among the various paths possible between x and y. The various $\mathcal{C}_d(x, y)$ can be calculated using DP as follows:
   - $\mathcal{C}_d(x, y) = min_{z\in N}\{\mathcal{C}_{d-\mathcal{D}(z,y)}(x, z) + \mathcal{C}(z, y)\}$. For a candidate path from x to y passing through z, the path with weight exactly d must have a delay of $d - \mathcal{D}(z, y)$ for x to z when the edge (z, y) has delay $\mathcal{D}(z, y)$.

   In this manner, the complete closure graph G' is computed. $\mathcal{P}_D(x, y)$ is the delay on the constrained cheapest path that corresponds to a cost of $\mathcal{P}_C(x, y)$.
2. Construct a constrained spanning tree of G' using a greedy approach that sequentially adds edges to the subtree of the constrained spanning tree T (thus far) until all the Steiner points are included. The initial value of T is the singleton s. Consider that node u is in the tree and we are considering whether to add edge (u, v).
   The following two edge selection criteria (heuristics) can be used to decide whether to include edge (u, v) in the tree:

   - CST_{CD}: $f_{CD}(u, v) = \begin{cases} \dfrac{\mathcal{C}(u, v)}{\Delta - (\mathcal{P}_D(s, u) + \mathcal{D}(u, v))}, & \text{if } \mathcal{P}_D(s, u) + \mathcal{D}(u, v) < \Delta \\ \infty, & \text{otherwise.} \end{cases}$

     The numerator is the "incremental cost" of adding (u, v) and the denominator is the "residual delay" that could be afforded. The goal is to minimize the incremental cost, while also maximizing the residual delay by choosing an edge that has low delay. Thus, the heuristic picks the neighbor v that minimizes f_{CD}, for all u in T and all v adjacent to T.

   - CST_C: $f_c = \begin{cases} \mathcal{C}(u, v), & \text{if } \mathcal{P}_D(s, u) + \mathcal{D}(u, v) < \Delta \\ \infty, & \text{otherwise.} \end{cases}$

     This heuristic picks the lowest cost edge between the already included tree edges and their nearest neighbor, as long as the total delay is less than Δ.

   The chosen node v is included in T. This step 2 is repeated until T includes all $|N'|$ nodes in G'.
3. Expand the edges of the constrained spanning tree T on G' into the constrained cheapest paths they represent in the original graph G. Delete/break any loops introduced by this expansion.

Algorithm 6.9 The constrained minimum Steiner tree algorithm using the CST_{CD} and CST_C heuristics.

- **Heuristic** CST_{CD} This heuristic tries to choose low-cost edges, while also trying to pick edges that maximize the remaining allowable delay. The motivation is to try to reduce the tree cost by path sharing, by extending the path beyond the selected edge. This heuristic has the tendency to optimize on delay also, while adding to the cost.
- **Heuristic** CST_C This heuristic simply minimizes the cost while ensuring that the delay bound is met.

Complexity Assuming integer-valued Δ, step 1, which finds the constrained cheapest shortest paths over all the nodes, has $O(n^3\Delta)$ time complexity. This is because all pairs of end and intermediate nodes have to be examined, for all integer delay values from 1 to Δ. Step 2, which constructs the constrained MST on the closure graph having k nodes, has $O(k^3)$ time complexity. Step 3, which expands the constrained spanning tree, involves expanding the k edges to up to $n-1$ edges each and then eliminating loops. This has $O(kn)$ time overhead. The dominating step is step 1.

6.11.5 Core-based trees

In the core-based tree approach, each group has a center node, or *core* node. A multicast tree is constructed dynamically, and grows on-demand, as follows. (i) A node wishing to join the tree as a receiver sends a unicast "join" message to the core node. (ii) The join message marks the edges as it travels; it either reaches the core node, or some node which is already a part of the multicast tree. The path followed by the "join" message from its source till the core/multicast tree is grafted to the multicast tree, and defines the path to the "core." (iii) A node on the tree multicasts a message by using a flooding on the core tree. (iv) A node not on the tree sends a message towards the core node; as soon as the message reaches any node on the tree, the message is flooded on the tree. In a network with a dynamically changing topology, care needs to be taken to maintain the tree structure and prevent messages from looping. This problem also exists for normal routing algorithms, such as the LSR and DVR algorithms (Chapter 5), in dynamic networks.

Current systems do not widely implement the Steiner tree for group multicast, even though it is more efficient after the initial cost to construct the Steiner tree. They prefer the simpler core-based tree (CBT) approach.

Core-based trees have various variants. A multi-core-based tree has more than one core node. For all CBT algorithms, high-bandwidth links can be specially chosen over others for forming the tree. Core-based trees have a natural analog in wireless networks, wherein it is reasonable to constitute the core tree of high-bandwidth wired links or high-power wireless links.

6.12 Chapter summary

At the core of distributed computing is the communication by message-passing among the processes participating in the application. This chapter studied several message ordering paradigms for communication, such as synchronous, FIFO, causally ordered, and non-FIFO orderings. These orders form a hierarchy. The chapter then examined several algorithms to implement these orderings. Group communication is an important aspect of communication in distributed systems. Causal order and total order are the popular forms of ordering when doing group multicasts and broadcasts. Algorithms to implement these orderings in group communication were also studied.

Maintaining communication in the presence of faults is necessary in real-world systems. Faults and their impacts are unpredictable. However, the behavior in the presence of faults needs to be clearly specified so that the application knows what to expect in terms of message delivery and message ordering in the presence of potential faults. The chapter studied some formal specifications of the expected behavior of group communication when faults might occur.

This chapter also studied some distributed multicast algorithms at the network layer. These algorithms include reverse path forwarding, multicast along Steiner trees and delay-bounded Steiner trees, and multicast based on core-based trees over the network graph. The solutions to some of these problems are NP-complete. Hence, only heuristics for polynomial time solutions are examined assuming a centralized setting to perform the computation.

6.13 Exercises

Exercise 6.1 (Characterizing causal ordering)

1. Prove that the CO property (Definition 6.3) and the message order property (Definition 6.5) characterize an identical class of executions.
2. Prove that the CO property (Definition 6.3) and the empty interval property (Definition 6.6) characterize an identical class of executions.

Exercise 6.2 Draw the directed graph $(\mathcal{T}, \hookrightarrow)$ for each of the executions in Figures 6.2, 6.3, and 6.5.

Exercise 6.3 Give a linear time algorithm to determine whether an A-execution $(E, \prec)$ is RSC.

Hint: Use the definition of a crown and perform a topological sort on the messages using the $\hookrightarrow$ relation.

Exercise 6.4 Show that a non-CO execution must have a crown of size 2.

Exercise 6.5 Synchronous systems were defined in Chapter 5. Synchronous send and receive primitives were also introduced in Chapter 1. Synchronous executions were defined formally in Definition 6.8.

These concepts are closely related. Explain carefully the differences and relationships between: (i) a synchronous execution, (ii) an (asynchronous) execution that uses synchronous communication, and (iii) a synchronous system.

Exercise 6.6 Rewrite the spanning tree algorithm of Figure 5.3 using CSP-like notation. You can assume a wildcard operator in a receive call to specify that any sender can be matched.

Exercise 6.7 The algorithm to implement synchronous order by scheduling messages, as given in Algorithm 6.1, uses process identifiers to break cyclic waits.

1. Analyze the fairness of this algorithm.
2. If the algorithm is not fair, suggest some ways to make it fair.
3. Will the use of rotating logical identifiers increase the fairness of the algorithm?

Exercise 6.8 Show the following containment relationships between causally ordered and totally ordered multicasts (hint: you may use Figure 6.11):

1. Show that a causally ordered multicast need not be a total order multicast.
2. Show that a total order multicast need not be a causal order multicast.

Exercise 6.9 Assume that all messages are being broadcast. Justify your answers to each of the following:

1. Modify the causal message ordering algorithm (Algorithm 6.2) so that processes use only two vectors of size n, rather than the $n \times n$ array.
2. Is it possible to implement total order using a vector of size n?
3. Is it possible to implement total order using a vector of size $O(1)$?
4. Is it possible to implement causal order using a vector of size $O(1)$?

Exercise 6.10 Design a (centralized) algorithm to create a propagation tree satisfying the properties given in Section 6.8.

Exercise 6.11 For the multicast algorithm based on propagation trees, answer the following:

1. What is a tight upper bound on the number of multicast groups?
2. What is a tight upper bound on the number of metagroups of the multicast groups?
3. Examine and justify in detail, the impact (to the propagation tree) of (i) an existing process departing from one of the multiple groups of which it is a member; (ii) an existing process joining another group; (iii) the formation of a new group containing new processes; (iv) the formation of a new group containing processes that are already part of various other groups.

Exercise 6.12 For multicast algorithms, show the following.

1. Privilege-based multicast algorithms provide (i) causal ordering if closed groups are assumed, and (ii) total ordering.

2. Moving sequencer algorithms, which work with open groups, provide total ordering.
3. Fixed sequencer algorithms provide total ordering.

Exercise 6.13 In the example of Figure 6.16, draw the propagation tree that would result if ⟨CE⟩ were considered before ⟨BCD⟩ as a child of ⟨ABC⟩.

Exercise 6.14 Consider the reverse path forwarding algorithm (Algorithm 6.7) for doing a multicast.

1. Modify the code to perform *pruning* of the multicast tree.
2. Now modify the code of (1) to also deal with dynamic changes to the network topology (use the algorithms in Chapter 5).
3. Now modify the code to deal with dynamic changes in the membership of the application at the various nodes.

Exercise 6.15 Give a (centralized) algorithm for creating a propagation tree, for any set of groups.

Exercise 6.16 Prove that the propagation tree for a given set of groups is not unique.

Exercise 6.17 For the graph in Figure 6.19, compute the following spanning trees:

1. Steiner tree (based on the KMB heuristic).
2. Delay-bounded Steiner (heuristic CST_{CD}), with a delay bound of 8 units.
3. Delay-bounded Steiner (heuristic CST_C), with a delay bound of 8 units.

Exercise 6.18 Design a graph for which the CST_{CD} and CST_C heuristics yield different delay-bounded Steiner trees.

Exercise 6.19 The algorithms for creating the propagation tree, the Steiner tree, and the delay-bounded Steiner tree are centralized. Identify the exact challenges in making these algorithms distributed.

6.14 Notes on references

The discussion on synchronous, asynchronous, and RSC-executions is based on Charron-Bost *et al.* [7]. The CSP language for synchronous communication was first proposed and formalized by Hoare [16]. The discussion on implementing synchronous order is based on Bagrodia [1]. The discussion on the group communication paradigm, as well as on total order and causal order is based on Birman and Joseph [4,5]. The algorithm for causal order (Algorithm 6.2) is given by Raynal *et al.* [22]. The space and time optimal algorithm for causal order is given by Kshemkalyani and Singhal [20,21]. The example to illustrate this algorithm is taken from [6]. The algorithm for total order (Algorithm 6.5) is taken from the ISIS project by Birman and Joseph [4,5]. The algorithm for total order using propagation trees is based on Garcia-Molina and Spauster [13], Jia [17], and Chiu and Hsiao [9]. The classification of application-level multicast algorithms was given by Defago *et al.* [11]. The moving sequencer algorithms were proposed by Chang and Maxemchuk [8]. An efficient fault-tolerant group communication protcol is given in [12]. A comprehensive survey of group communication specifications given by Chockler *et al.* [10] as well as the survey in [11] discuss the systems Totem, Pinwheel, RMP, On-Demand, Isis, Amoeba, Phoenix, and Newtop. The Steiner tree problem was named after

Steiner and developed in [14]. The Steiner tree heuristic discussed was proposed by Kou *et al.* [19]. The network cost and destination cost metrics were introduced by [3]. They further showed a detailed analysis of the bounds on the metrics. The discussion on the delay-bounded minimum Steiner tree is based on Kompella *et al.* [18]. The discussion on the semantics of fault-tolerant group communication is given by Hadzilacos and Toueg [15]. Core-based trees were proposed by Ballardie *et al.* [2].

References

[1] R. Bagrodia, Synchronization of asynchronous processes in CSP, *ACM Transactions in Programming Languages and Systems*, **11**(4), 1989, 585–597.

[2] T. Ballardie, P. Francis, and J. Crowcroft, Core based trees (CBT), *ACM SIGCOMM Computer Communication Review*, **23**(4), 1993, 85–95.

[3] K. Bharath-Kumar and J. Jaffe, Routing to multiple destinations in computer networks, *IEEE Transactions on Communications*, **31**(3) 1983, 343–351.

[4] K. Birman and T. Joseph, Reliable communication in the presence of failures, *ACM Transactions on Computer Systems*, **5**(1), 1987, 47–76.

[5] K. Birman, A. Schiper, and P. Stephenson, Lightweight causal and atomic group multicast, *ACM Transactions on Computer Systems*, **9**(3), 1991, 272–314.

[6] P. Chandra, P. Gambhire, and A. D. Kshemkalyani, Performance of the optimal causal multicast algorithm: a statistical analysis, *IEEE Transactions on Parallel and Distributed Systems*, **15**(1), 2004, 40–52.

[7] B. Charron-Bost, G. Tel, and F. Mattern, Synchronous, asynchronous, and causally ordered communication, *Distributed Computing*, **9**(4), 1996, 173–191.

[8] J.-M. Chang and N. Maxemchuk, Reliable broadcast protocols, *ACM Transactions on Computer Systems*, **2**(3), 1984, 251–273.

[9] G.-M. Chiu and C.-M. Hsiao, A note on total ordering multicast using propagation trees, *IEEE Transactions on Parallel and Distributed Systems*, **9**(2), 1998, 217–223.

[10] G. Chockler, I. Keidar, and R. Vitenberg, Group communication specifications: a comprehensive study, *ACM Computing Surveys*, **33**(4), 2001, 1–43.

[11] X. Defago, A. Schiper, and P. Urban, Total order broadcast and multicast algorithms: taxonomy and survey, *ACM Computing Surveys*, **36**(4), 2004, 372–421.

[12] P. Ezhilchelvan, R. Macdo, and S. Shrivastava, Newtop: a fault-tolerant group communication protocol, *Proceedings of the 15th IEEE International Conference on Distributed Computing Systems*, Vancouver, Canada, May, 1995, 296–306.

[13] H. Garcia-Molina and A. Spauster, Ordered and reliable multicast communication, *ACM Transactions on Computer Systems*, **9**(3), 1991, 242–271.

[14] E. Gilbert and H. Pollack, Steiner minimal trees, *SIAM Journal of Applied Mathematics*, **16**(1), 1968, 1–29.

[15] V. Hadzilacos and S. Toueg, Fault-tolerant broadcasts and related problems in Mullender, S. (ed.), *Distributed Systems*, New York, Addison-Wesley, 1993, 97–146.

[16] C. A. R. Hoare, Communicating sequential processes, *Communications of the ACM*, **21**(8), 1978, 666–677.

[17] X. Jia, A total ordering multicast protocol using propagation trees, *IEEE Transactions on Parallel and Distributed Systems*, **6**(6), 1995, 617–627.

[18] V. Kompella, J. Pasquale, and G. Polyzos, Multcast routing for multimedia communication, *IEEE/ACM Transactions on Networking*, **1**(3), 1993, 86–92.
[19] L. Kou, G. Markowsky, and L. Berman, A fast algorithm for Steiner trees, *Acta Informatica*, **15**, 1981, 141–145.
[20] A. D. Kshemkalyani and M. Singhal, An optimal algorithm for generalized causal message ordering, *Proceedings of the 15th ACM Symposium on Principles of Distributed Computing*, May 1996, 87.
[21] A. D. Kshemkalyani and M. Singhal, Necessary and sufficient conditions on information for causal message ordering and their optimal implementation, *Distributed Computing*, **11**(2), 1998, 91–111.
[22] M. Raynal, A. Schiper, and S. Toueg, The causal ordering abstraction and a simple way to implement it, *Information Processing Letters*, **39**, 1991, 343–350.

CHAPTER

7 Termination detection

7.1 Introduction

In distributed processing systems, a problem is typically solved in a distributed manner with the cooperation of a number of processes. In such an environment, inferring if a distributed computation has ended is essential so that the results produced by the computation can be used. Also, in some applications, the problem to be solved is divided into many subproblems, and the execution of a subproblem cannot begin until the execution of the previous subproblem is complete. Hence, it is necessary to determine when the execution of a particular subproblem has ended so that the execution of the next subproblem may begin. Therefore, a fundamental problem in distributed systems is to determine if a distributed computation has terminated.

The detection of the termination of a distributed computation is non-trivial since no process has complete knowledge of the global state, and global time does not exist. A distributed computation is considered to be globally terminated if every process is locally terminated and there is no message in transit between any processes. A "locally terminated" state is a state in which a process has finished its computation and will not restart any action unless it receives a message. In the termination detection problem, a particular process (or all of the processes) must infer when the underlying computation has terminated.

When we are interested in inferring when the underlying computation has ended, a termination detection algorithm is used for this purpose. In such situations, there are two distributed computations taking place in the distributed system, namely, the *underlying computation* and the *termination detection algorithm*. Messages used in the underlying computation are called *basic* messages, and messages used for the purpose of termination detection (by a termination detection algorithm) are called *control* messages.

A termination detection (TD) algorithm must ensure the following:

1. Execution of a TD algorithm cannot indefinitely delay the underlying computation; that is, execution of the termination detection algorithm must not freeze the underlying computation.
2. The termination detection algorithm must not require addition of new communication channels between processes.

7.2 System model of a distributed computation

A distributed computation consists of a fixed set of processes that communicate solely by message passing. All messages are received correctly after an arbitrary but finite delay. Communication is *asynchronous*, i.e., a process never waits for the receiver to be ready before sending a message. Messages sent over the same communication channel may not obey the FIFO ordering.

A distributed computation has the following characteristics:

1. At any given time during execution of the distributed computation, a process can be in only one of the two states: *active*, where it is doing local computation and *idle*, where the process has (temporarily) finished the execution of its local computation and will be reactivated only on the receipt of a message from another process. The active and idle states are also called the *busy* and *passive* states, respectively.
2. An active process can become idle at any time. This corresponds to the situation where the process has completed its local computation and has processed all received messages.
3. An idle process can become active only on the receipt of a message from another process. Thus, an idle process cannot spontaneously become active (except when the distributed computation begins execution).
4. Only active processes can send messages. (Since we are not concerned with the initialization problem, we assume that all processes are initially idle and a message arrives from outside the system to start the computation.)
5. A message can be received by a process when the process is in either of the two states, i.e., *active* or *idle*. On the receipt of a message, an *idle* process becomes *active*.
6. The sending of a message and the receipt of a message occur as atomic actions.

We restrict our discussion to executions in which every process eventually becomes idle, although this property is in general undecidable. If a termination detection algorithm is applied to a distributed computation in which some processes remain in their active states forever, the TD algorithm itself will not terminate.

Definition of termination detection

Let $p_i(t)$ denote the state (active or idle) of process p_i at instant t and $c_{i,j}(t)$ denote the number of messages in transit in the channel at instant t from process p_i to process p_j. A distributed computation is said to be terminated at time instant t_0 iff:

$$(\forall i :: p_i(t_0) = idle) \wedge (\forall i, j :: c_{i,j}(t_0) = 0).$$

7.3 Termination detection using distributed snapshots

The algorithm uses the fact that a consistent snapshot of a distributed system captures stable properties. Termination of a distributed computation is a stable property. Thus, if a consistent snapshot of a distributed computation is taken after the distributed computation has terminated, the snapshot will capture the termination of the computation.

The algorithm assumes that there is a logical bidirectional communication channel between every pair of processes. Communication channels are reliable but non-FIFO. Message delay is arbitrary but finite.

7.3.1 Informal description

The main idea behind the algorithm is as follows: when a computation terminates, there must exist a unique process which became idle last. When a process goes from active to idle, it issues a request to all other processes to take a local snapshot, and also requests itself to take a local snapshot. When a process receives the request, if it agrees that the requester became idle after itself, it grants the request by taking a local snapshot for the request. A request is said to be *successful* if all processes have taken a local snapshot for it. The requester or any external agent may collect all the local snapshots of a request. If a request is successful, a global snapshot of the request can thus be obtained and the recorded state will indicate termination of the computation, viz., in the recorded snapshot, all the processes are idle and there is no message in transit to any of the processes.

7.3.2 Formal description

The algorithm needs logical time to order the requests. Each process i maintains an *logical clock* denoted by x, which is initialized to zero at the start of the computation. A process increments its x by one each time it becomes idle. A basic message sent by a process at its logical time x is of the form $B(x)$. A control message that requests processes to take local snapshot issued by process i at its logical time x is of the form $R(x, i)$. Each process synchronizes its logical clock x loosely with the logical clocks x's on other processes in such a way that it is the maximum of clock values ever received or sent in messages. Besides logical clock x, a process maintains a variable k such that when the process is idle, (x,k) is the maximum of the values (x, k) on all messages $R(x, k)$ ever received or sent by the process. Logical time is compared as follows: $(x, k) > (x', k')$ iff $(x > x')$ or $((x = x')$ and $(k > k'))$, i.e., a tie between x and x' is broken by the process identification numbers k and k'.

The algorithm is defined by the following four rules [8]. We use guarded statements to express the conditions and actions. Each process i applies one of the rules whenever it is applicable.

R1: When process i is active, it may send a basic message to process j at any time by doing

send a $B(x)$ *to* j.

R2: Upon receiving a $B(x')$, process i does

let $x := x' + 1$;
if(i is *idle*) $\rightarrow$ go *active*.

R3: When process i goes *idle*, it does

let $x := x + 1$;
let $k := i$;
send message $R(x, k)$ to all other processes;
take a local snapshot for the request by $R(x, k)$.

R4: Upon receiving message $R(x', k')$, process i does

$[((x', k') > (x, k)) \wedge$ (i is *idle*) $\rightarrow$ let$(x, k) := (x', k')$;
take a local snapshot for the request by$R(x', k')$;
□
$((x', k') \leq (x, k)) \wedge$ (i is *idle*) $\rightarrow$ do nothing;
□
(i is *active*) $\rightarrow$ let $x := max(x', x)]$.

7.3.3 Discussion

As per rule R1, when a process sends a basic message to any other process, it sends its logical clock value in the message. From rule R2, when a process receives a basic message, it updates its logical clock based on the clock value contained in the message. Rule R3 states that when a process becomes idle, it updates its local clock, sends a request for snapshot $R(x, k)$ to every other process, and takes a local snapshot for this request.

Rule R4 is the most interesting. On the receipt of a message $R(x', k')$, the process takes a local snapshot if it is idle and $(x', k') > (x, k)$, i.e., timing in the message is later than the local time at the process, implying that the sender of $R(x', k')$ terminated after this process. In this case, it is likely that the sender is the last process to terminate and thus, the receiving process takes a snapshot for it. Because of this action, every process will eventually take a local snapshot for the last request when the computation has terminated, that is, the request by the latest process to terminate will become successful.

In the second case, $(x', k') \leq (x, k)$, implying that the sender of $R(x', k')$ terminated before this process. Hence, the sender of $R(x', k')$ cannot be the last process to terminate. Thus, the receiving process does not take a snapshot for it. In the third case, the receiving process has not even terminated. Hence, the sender of $R(x', k')$ cannot be the last process to terminate and no snapshot is taken.

The last process to terminate will have the largest clock value. Therefore, every process will take a snapshot for it; however, it will not take a snapshot for any other process.

7.4 Termination detection by weight throwing

In termination detection by weight throwing, a process called *controlling agent*[1] monitors the computation. A communication channel exists between each of the processes and the controlling agent and also between every pair of processes.

Basic idea

Initially, all processes are in the idle state. The weight at each process is zero and the weight at the controlling agent is 1. The computation starts when the controlling agent sends a basic message to one of the processes. The process becomes active and the computation starts. A non-zero weight W $(0 < W \leq 1)$ is assigned to each process in the active state and to each message in transit in the following manner: When a process sends a message, it sends a part of its weight in the message. When a process receives a message, it add the weight received in the message to its weight. Thus, the sum of weights on all the processes and on all the messages in transit

[1] The controlling agent can be one of the processes in the computation.

is always 1. When a process becomes passive, it sends its weight to the controlling agent in a control message, which the controlling agent adds to its weight. The controlling agent concludes termination if its weight becomes 1.

Notation

- The weight on the controlling agent and a process is in general represented by W.
- $B(DW)$: A basic message B is sent as a part of the computation, where DW is the weight assigned to it.
- $C(DW)$: A control message C is sent from a process to the controlling agent where DW is the weight assigned to it.

7.4.1 Formal description

The algorithm is defined by the following four rules [9]:

Rule 1: The controlling agent or an active process may send a basic message to one of the processes, say P, by splitting its weight W into $W1$ and $W2$ such that $W1 + W2 = W$, $W1 > 0$ and $W2 > 0$. It then assigns its weight $W := W1$ and sends a basic message $B(DW := W2)$ to P.

Rule 2: On the receipt of the message $B(DW)$, process P adds DW to its weight W ($W := W + DW$). If the receiving process is in the idle state, it becomes active.

Rule 3: A process switches from the active state to the idle state at any time by sending a control message $C(DW := W)$ to the controlling agent and making its weight $W := 0$.

Rule 4: On the receipt of a message $C(DW)$, the controlling agent adds DW to its weight ($W := W + DW$). If W = 1, then it concludes that the computation has terminated.

7.4.2 Correctness of the algorithm

To prove the correctness of the algorithm, the following sets are defined:

A: set of weights on all active processes;
B: set of weights on all basic messages in transit;
C: set of weights on all control messages in transit;
W_c: weight on the controlling agent.

Two invariants I_1 and I_2 are defined for the algorithm:

$$I_1: W_c + \sum_{W \in (A \cup B \cup C)} W = 1.$$
$$I_2: \forall W \in (A \cup B \cup C),\ W > 0.$$

Invariant I_1 states that the sum of weights at the controlling process, at all active processes, on all basic messages in transit, and on all control messages in transit is always equal to 1. Invariant I_2 states that weight at each active process, on each basic message in transit, and on each control message in transit is non-zero.

Hence,

$$\begin{aligned} & W_c = 1 \\ & \Longrightarrow \textstyle\sum_{W \in (A \cup B \cup C)} W = 0 \text{ (by } I_1) \\ & \Longrightarrow (A \cup B \cup C) = \phi \text{ (by } I_2) \\ & \Longrightarrow (A \cup B) = \phi. \end{aligned}$$

Note that $(A \cup B) = \phi$ implies that the computation has terminated. Therefore, the algorithm never detects a false termination.

Further,

$$\begin{aligned} & (A \cup B) = \phi \\ & \Longrightarrow W_c + \textstyle\sum_{W \in C} W = 1 \text{ (by } I_1). \end{aligned}$$

Since the message delay is finite, after the computation has terminated, eventually $W_c = 1$. Thus, the algorithm detects a termination in finite time.

7.5 A spanning-tree-based termination detection algorithm

The algorithm assumes there are n processes P_i, $0 \le i \le n-1$, which are modeled as the nodes i, $0 \le i \le n-1$, of a fixed connected undirected graph. The edges of the graph represent the communication channels, through which a process sends messages to neighboring processes in the graph. The algorithm uses a fixed spanning tree of the graph with process P_0 at its root which is responsible for termination detection. Process P_0 communicates with other processes to determine their states and the messages used for this purpose are called signals. All leaf nodes report to their parents, if they have terminated. A parent node will similarly report to its parent when it has completed processing and all of its immediate children have terminated, and so on. The root concludes that termination has occurred, if it has terminated and all of its immediate children have also terminated.

The termination detection algorithm generates two waves of signals moving inward and outward through the spanning tree. Initially, a contracting wave of signals, called *tokens*, moves inward from leaves to the root. If this token wave reaches the root without discovering that termination has occurred, the root initiates a second outward wave of *repeat* signals. As this repeat wave reaches leaves, the token wave gradually forms and starts moving inward again. This sequence of events is repeated until the termination is detected.

7.5.1 Definitions

1. Tokens: a contracting wave of signals that move inward from the leaves to the root.
2. Repeat signal: if a token wave fails to detect termination, node $P0$ initiates another round of termination detection by sending a signal called Repeat, to the leaves.
3. The nodes which have one or more tokens at any instant form a set S.
4. A node j is said to be outside of set S if j does not belong to S and the path (in the tree) from the root to j contains an element of S. Every path from the root to a leaf may not contain a node of S.
5. Note that all nodes outside S are idle. This is because, any node that terminates, transmits a token to its parent. When a node transmits the token, it goes out of the set S.

We first give a simple algorithm for termination detection and discuss a problem associated with it. Then we provide the correct algorithm.

7.5.2 A simple algorithm

Initially, each leaf process is given a token. Each leaf process, after it has terminated, sends its token to its parent. When a parent process terminates and after it has received a token from each of its children, it sends a token to its parent. This way, each process indicates to its parent process that the subtree below it has become idle. In a similar manner, the tokens get propagated to the root. The root of the tree concludes that termination has occurred, after it has become idle and has received a token from each of its children.

A problem with the algorithm

This simple algorithm fails under some circumstances. After a process has sent its token to its parent, it should remain idle. However, this is not the case. The problem arises when a process after it has sent a token to its parent, receives a message from some other process. Note that this message could cause the process (that has already sent a token to its parent) to again become active. Hence the simple algorithm fails since the process that indicated to its parent that it has become idle, is now active because of the message it

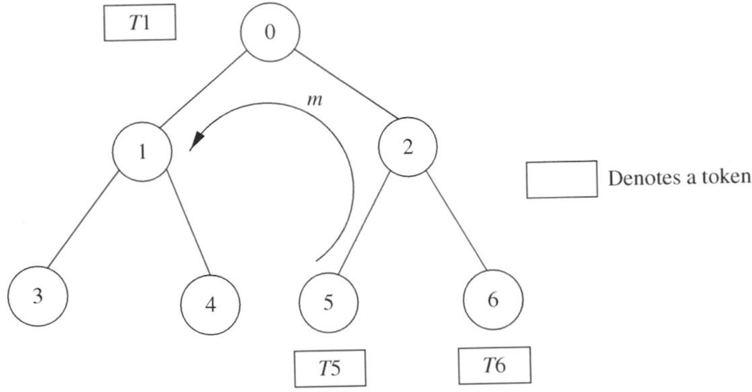

Figure 7.1 An example of the problem.

received from an active process. Hence, the root node just because it received a token from a child, can't conclude that all processes in the child's subtree have terminated. The algorithm has to be reworked to accommodate such message-passing scenarios.

The problem is explained with the example shown in Figure 7.1. Assume that process 1 has sent its token ($T1$) to its parent, namely, process 0. On receiving the token, process 0 concludes that process 1 and its children have terminated. Process 0 if it is idle, can conclude that termination has occurred, whenever it receives a token from process 2. But now assume that just before process 5 terminates, it sends a message m to process 1. On the reception of this message, process 1 becomes active again. Thus, the information that process 0 has about process 1 (that it is idle) becomes void. Therefore, this simple algorithm does not work.

7.5.3 The correct algorithm

We now present the correct algorithm that was developed by Topor [19] and it works even when messages such as the one if Figure 7.1 are present. The main idea is to color the processes and tokens and change the color when such messages are involved.

The basic idea

In order to enable the root node to know that a node in its children's subtree, that was assumed to be terminated, has become active due to a message, a coloring scheme for tokens and nodes is used. The root can determine that an idle process has been activated by a message, based on the color of the token it receives from its children. All tokens are initialized to white. If a process had sent a message to some other process, it sends a black token to its parent on termination; otherwise, it sends a white token on termination. Hence, the parent process on getting the black token knows that its child had sent a message to some other process. The parent, when sending its token (on terminating) to its parent, sends a black token only if it received a black token from one of its children. This way, the parent's parent knows that one of the processes in its child's subtree had sent a message to some other process. This gets propagated and finally the root node knows that message-passing was involved when it receives a black token from one of its children. In this case, the root asks all nodes in the system to restart the termination detection. For this, the root sends a Repeat signal to all other process. After receiving the Repeat signal, all leaves will restart the termination detection algorithm.

The algorithm description

The algorithm works as follows:

1. Initially, each leaf process is provided with a token. The set S is used for book-keeping to know which processes have the token. Hence S will be the set of all leaves in the tree.
2. Initially, all processes and tokens are white. As explained above, coloring helps the root know if a message-passing was involved in one of the subtrees.
3. When a leaf node terminates, it sends the token it holds to its parent process.
4. A parent process will collect the token sent by each of its children. After it has received a token from all of its children and after it has terminated, the parent process sends a token to its parent.
5. A process turns black when it sends a message to some other process. This coloring scheme helps a process remember that it has sent a message. When a process terminates, if it is black, it sends a black token to its parent.
6. A black process turns back to white after it has sent a black token to its parent.
7. A parent process holding a black token (from one of its children), sends only a black token to its parent, to indicate that a message-passing was involved in its subtree.
8. Tokens are propagated to the root in this fashion. The root, upon receiving a black token, will know that a process in the tree had sent a message to some other process. Hence, it restarts the algorithm by sending a Repeat signal to all its children.
9. Each child of the root propagates the Repeat signal to each of its children and so on, until the signal reaches the leaves.
10. The leaf nodes restart the algorithm on receiving the Repeat signal.
11. The root concludes that termination has occurred, if:
    (a) it is white;
    (b) it is idle; and
    (c) it has received a white token from each of its children.

7.5.4 An example

We now present an example to illustrate the working of the algorithm.

1. Initially, all nodes 0 to 6 are white (Figure 7.2). Leaf nodes 3, 4, 5, and 6 are each given a token. Node 3 has token $T3$, node 4 has token T4, node 5 has token $T5$, and node 6 has token $T6$. Hence, S is {3, 4, 5, 6}.
2. When node 3 terminates, it transmits $T3$ to node 1. Now S changes to 1, 4, 5, 6. When node 4 terminates, it transmits $T4$ to node 1 (Figure 7.3). Hence, S changes to {1, 5, 6}.
3. Node 1 has received a token from each of its children and, when it terminates, it transmits a token $T1$ to its parent (Figure 7.4). S changes to {0, 5, 6}.
4. After this, suppose node 5 sends a message to node 1, causing node 1 to again become active (Figure 7.5). Since node 5 had already sent a token to its parent node 0 (thereby making node 0 assume that node 5 had terminated), the new message makes the system inconsistent as far as termination detection is concerned. To deal with this, the algorithm executes the following steps.
5. Node 5 is colored black, since it sent a message to node 1.

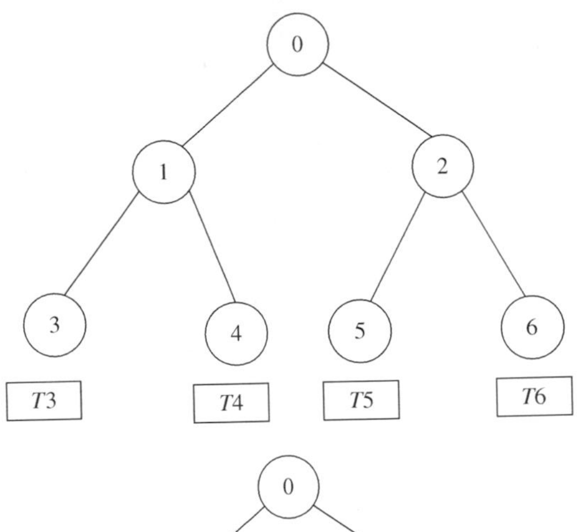

Figure 7.2 All leaf nodes have tokens. S = {3, 4, 5, 6}.

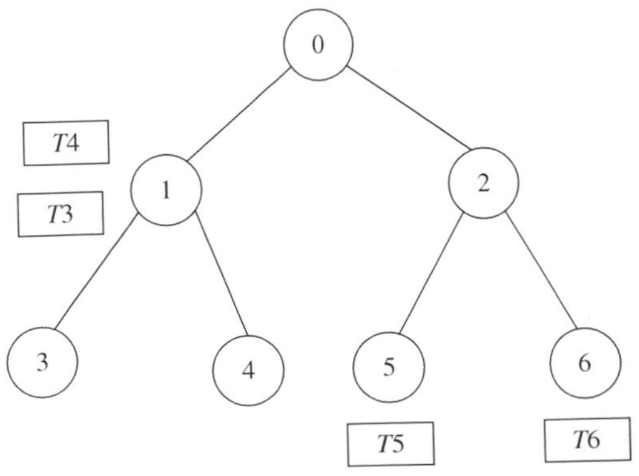

Figure 7.3 Nodes 3 and 4 become idle. S = {1, 5, 6}.

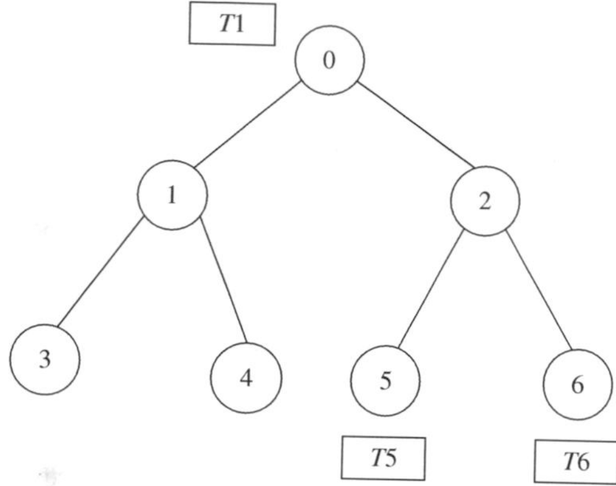

Figure 7.4 Node 1 becomes idle. $S = \{0, 5, 6\}$.

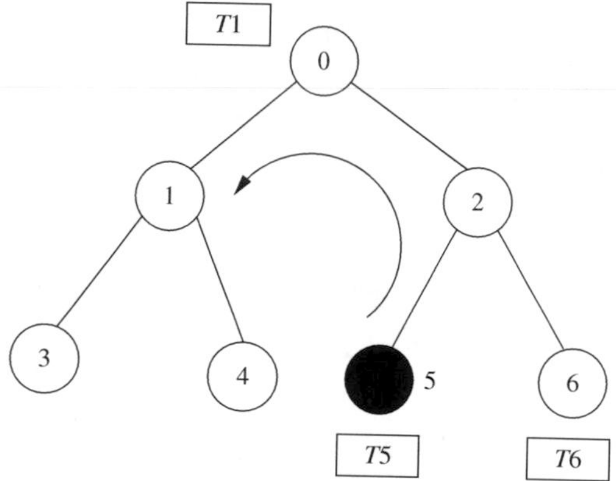

Figure 7.5 Node 5 sends a message to node 1.

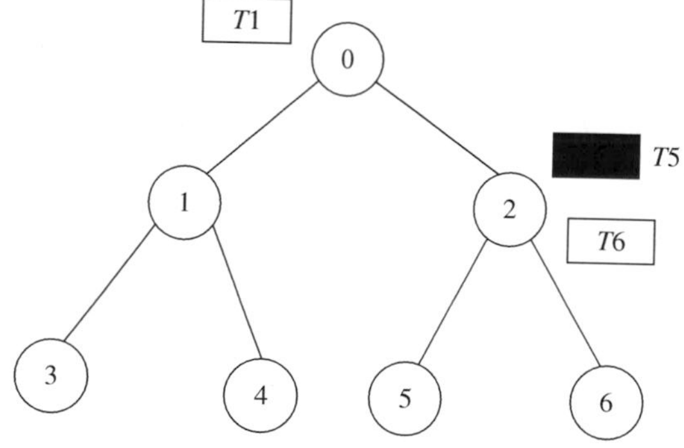

Figure 7.6 Nodes 5 and 6 become idle. $S = \{0, 2\}$.

6. When node 5 terminates, it sends a black token $T5$ to node 2. So, S changes to $\{0, 2, 6\}$. After node 5 sends its token, it turns white (Figure 7.6). When node 6 terminates, it sends the white token $T6$ to node 2. Hence, S changes to $\{0, 2\}$.
7. When node 2 terminates, it sends a black token $T2$ to node 0, since it holds a black token $T5$ from node 5 (Figure 7.7).

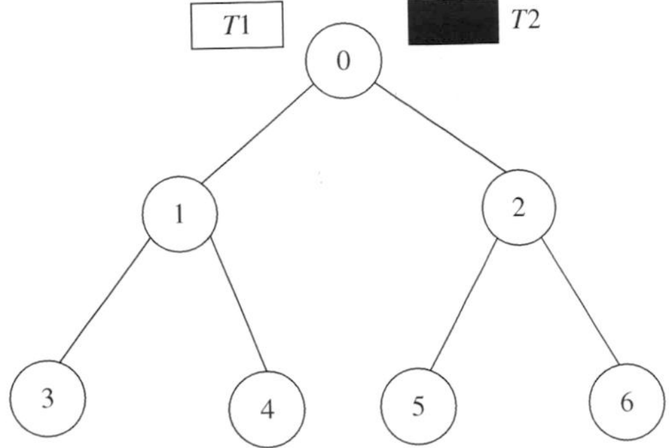

Figure 7.7 Node 2 becomes idle. $S = \{0\}$. Node 0 initiates a repeat signal.

8. Since node 0 has received a black token $T2$ from node 2, it knows that there was a message sent by one or more of its children in the tree and hence sends a Repeat signal to each of its children.
9. The Repeat signal is propagated to the leaf nodes and the algorithm is repeated. Node 0 concludes that termination has occurred if it is white, it is idle, and it has received a white token from each of its children.

7.5.5 Performance

The best case message complexity of the algorithm is $O(n)$, where n is the number of processes in the computation. The best case occurs when all nodes send all computation messages in the first round. Therefore, the algorithm executes only twice and the message complexity depends only on the number of nodes.

However, the worst case complexity of the algorithm is $O(n^* M)$, where M is the number of computation messages exchanged. The worst case occurs when only computation message is exchanged every time the algorithm is executed. This causes the root to restart termination detection as many times as there are computation messages. Hence, the worst case complexity is $O(n^* M)$.

7.6 Message-optimal termination detection

Now we discuss a message optimal termination detection algorithm by Chandrasekaran and Venkatesan [2]. The network is represented by a graph $G = (V, E)$, where V is the set of nodes, and $E \subseteq V \times V$ is the set of edges or communication links. The communication links are bidirectional and exhibit FIFO property. The processors and communication links incur arbitrary but finite delays in executing their functions. The algorithm assumes the existence of a leader and a spanning tree in the network. If a leader is not available, the minimum spanning tree algorithm of Gallager *et al.* [7] can be used to elect a leader and find a spanning tree using $O(|E| + |V| \log |V|)$ messages.

7.6.1 The main idea

Let us reconsider the method for termination detection disussed in the previous section. The root of the tree initiates one phase of termination detection by turning white. An interior node, on receiving a white token from its parent, turns white and transmits a white token to all of its children. Eventually each leaf receives a white token and turns white. When a leaf node becomes idle, it transmits a token to its parent and the token has the same color as that of the leaf node. An interior node waits for a token from each of its children. It also waits until it becomes idle. It then sends a white token to its parent if its color is white and it received a white token from each of its children. Finally, the root node infers the termination of the underlying computation if it receives a white token from each child, its color is white, and it is idle.

This simple algorithm is inefficient in terms of message complexity due to the following reasons. Consider the scenario shown in Figure 7.8, where node p sends a message m to node q. Before node q received the message m, it had sent a white token to its parent (because it was idle and it had received a white token from each of its children). In this situation, node p cannot send a white token to its parent until node q becomes idle. To insure this, in Topor's algorithm, node p changes its color to black and sends a black token to its parent so that termination detection is performed once again. Thus, every message of the underlying computation can potentially cause the execution of one more round of the termination detection algorithm, resulting in significant message traffic.

The main idea behind the message-optimal algorithm is as follows: when a node p sends a message m to node q, p should wait until q becomes idle and only after that, p should send a white token to its parent. This rule ensures that if an idle node q is restarted by a message m from from a node p, then the sender p waits till q terminates before p can send a white token to its parent. To achieve this, when node q terminates, it sends an acknowledgement (a control message) to node p informing node p that the set of actions triggered

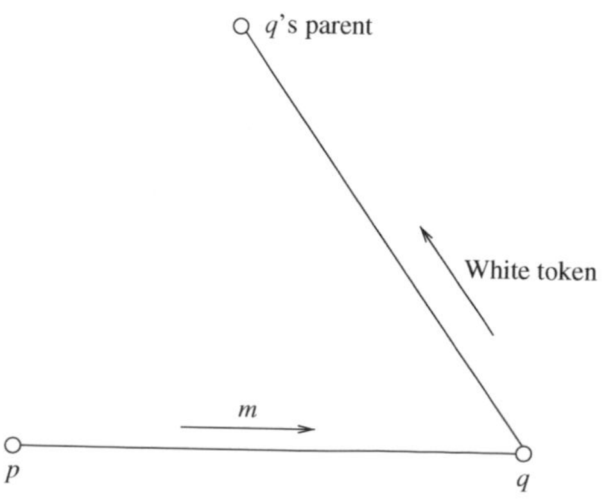

Figure 7.8 Node p sends a message m to node q that has already sent a white token to its parent [2].

by message m has been completed and that node p can send a white token to its parent. However, note that node q, after being woken up by message m from node p, may wake up another idle node r, which in turn may wake up other nodes. Therefore, node q should not send an acknowledgement to p until it receives acknowledgement messages for all of the messages it sent after it received message m from node p. This restriction also applies to node r and other nodes. Clearly, both the sender and the receiver keep track of each message, and a node will send a white token to its parent only after it has received an acknowledgement for every message it has sent and has received a white token from each of its children.

7.6.2 Formal description of the algorithm

Initially, all nodes in the network are in state NDT (not detecting termination) and all links are uncolored. For termination detection, the root node changes its state to DT (detecting termination) and sends a warning message on each of its outgoing edges. When a node p receives a warning message from its neighbor, say q, it colors[2] the incoming link (q, p) and if it is in state NDT, it changes its state to DT, colors each of its outgoing edges, and sends a warning message on each of its outgoing edges.

When a node p in state DT sends a basic message to its neighbor q, it keeps track of this information by pushing the entry $TO(q)$ on its local stack.

When a node x receives a basic message from node y on the link (y, x) that is colored by x, node x knows that the sender node y will need an acknowledgement for this message from it. The receiver node x keeps track of this information by pushing the entry $FROM(y)$ on its local stack. Procedure receive_message is given in Algorithm 7.1.

```
Procedure receive_message(y: neighbor);
(* performed when a node x receives a message from its neighbor y on the link
(y,x) that was colored by x *)
            begin
                  receive message from y on the link (y,x)
                  if (link (y,x) has been colored by x) then
                        push FROM(y) on the stack
            end;
```

Algorithm 7.1 Procedure *receive_message*.

[2] All links are uncolored or colored. The shade of the color does not matter.

Eventually, every node in the network will be in the state DT as the network is connected. Note that both sender and receiver keep track of every message in the system.

When a node p becomes idle, it calls procedure *stack_cleanup*, which is defined in Algorithm 7.2. Procedure *stack_cleanup* examines its stack from the top and, for every entry of the form *FROM*(q), deletes the entry and sends the *remove_entry* message to node q. Node p repeats this until it encounters an entry of the form *TO*(x) on the stack. The idea behind this step is to inform those nodes that sent a message to p that the actions triggered by their messages to p are complete.

Procedure *stack_cleanup*;
begin
    while (top entry on stack is not of the form "*TO*()") **do**
        begin
            pop the entry on the top of the stack;
            let the entry be *FROM*(q);
            send a *remove_entry* message to q
        end
end;

Algorithm 7.2 Procedure *stack_cleanup*.

When a node x receives a *remove_entry* message from its neighbor y, node x infers that the operations triggered by its last message to y have been completed and hence it no longer needs to keep track of this information. Node x on receipt of the control message *remove_entry* from node y, examines its stack from the top and deletes the first entry of the form *TO*(y) from the stack. If node x is idle, it also performs the *stack_cleanup* operation. The procedure *receive_remove_entry* is defined in Algorithm 7.3.

Procedure *receive_remove_entry*(y: neighbor);
(* performed when a node x receives a *remove_entry* message from its neighbor y *)
begin
    scan the stack and delete the first entry of the form *TO*(y);
    if idle **then**
        stack_cleanup
end;

Algorithm 7.3 Procedure *receive_remove_entry*.

A node sends a *terminate* message to its parent when it satisfies all the following conditions:

1. It is idle.
2. Each of its incoming links is colored (it has received a warning message on each of its incoming links).
3. Its stack is empty.
4. It has received a *terminate* message from each of its children (this rule does not apply to leaf nodes).

When the root node satisfies all of the above conditions, it concludes that the underlying computation has terminated.

7.6.3 Performance

We analyze the number of control messages used by the algorithm in the worst case. Each node in the network sends one warning message on each outgoing link. Thus, each link carries two warning messages, one in each direction. Since there are $|E|$ links, the total number of warning messages generated by the algorithm is $2*|E|$. For every message generated by the underlying computation (after the start of the termination detection algorithm), exactly one *remove_entry* message is sent on the network. If M is the number of messages sent by the underlying computation, then at most M *remove_entry* messages are used. Finally, each node sends exactly one *terminate* message to its parent (on the tree edge) and since there are only $|V|$ nodes and $|V|-1$ tree edges, only $|V|-1$ *terminate* messages are sent. Hence, the total number of messages generated by the algorithm is $2*|E|+|V|-1+M$. Thus, the message complexity of the algorithm is $O(|E|+M)$ as $|E|>|V|-1$ for any connected network. The algorithm is asymptotically optimal in the number of messages.

7.7 Termination detection in a very general distributed computing model

So far we assumed that the reception of a single message is enough to activate a passive process. Now we consider a general model of distributed computing where a passive process does not necessarily become active on the receipt of a message [1]. Instead, the condition of activation of a passive process is more general and a passive process requires a set of messages to become active. This requirement is expressed by an *activation condition* defined over the set DS_i of processes from which a passive process P_i is expecting messages. The set DS_i associated with a passive process P_i is called the *dependent set* of P_i. A passive process becomes active only when its activation condition is fulfilled.

7.7.1 Model definition and assumptions

The distributed computation consists of a finite set P of processes P_i, $i = 1, \ldots, n$, interconnected by unidirectional communication channels. Communication channels are reliable, but they do not obey FIFO property. Message transfer delay is finite but unpredictable.

A passive process that has terminated its computation by executing for example an end or stop statement is said to be individually terminated; its dependent set is empty and therefore, it can never be activated.

AND, OR, and AND-OR models

There are several request models, such as AND, OR, AND-OR models. In the AND model, a passive process P_i can be activated only after a message from every process belonging to DS_i has arrived. In the OR model, a passive process P_i can be activated when a message from any process belonging to DS_i has arrived. In the AND-OR model, the requirement of a passive process P_i is defined by a set R_i of sets $DS_i^1, DS_i^2, \ldots, DS_i^{q_i}$, such that for all r, $1 \leq r \leq q_i$, $DS_i^r \subseteq P$. The dependent set of P_i is $DS_i = DS_i^1 \cup DS_i^2 \cup \ldots DS_i^{q_i}$. Process P_i waits for messages from all processes belonging to DS_i^1 or for messages from all processes belonging to $DS_i^2 \ldots$ or for messages from all processes belonging to $DS_i^{q_i}$.

The k out of n model

In the k out of n model, the requirement of a passive process P_i is defined by the set DS_i and an integer k_i, $1 \leq k_i \leq |DS_i| = n_i$. Process P_i becomes active when it receives messages from k_i distinct processes in DS_i. Note that a more general k out of n model can be constructed as disjunctions of several k out of n requests.

Predicate fulfilled

To abstract the activation condition of a passive process P_i, a predicate $fulfilled_i(A)$ is introduced, where A is a subset of P. Predicate $fulfilled_i(A)$ is true if and only if messages arrived (and not yet consumed) from all processes belonging to set A are sufficient to activate process P_i.

7.7.2 Notation

The following notation will be used to define the termination of a distributed computation:

- $passive_i$: true iff P_i is passive.
- $empty(j, i)$: true iff all messages sent by P_j to P_i have arrived at P_i; the messages not yet consumed by P_i are in its local buffer.
- $arr_i(j)$: true iff a message from P_j to P_i has arrived at P_i and has not yet been consumed by P_i.

- ARR_i = {processes P_j such that $arr_i(j)$}.
- NE_i = {processes P_j such that $\neg$ $empty(j, i)$}.

7.7.3 Termination definitions

Two different types of terminations are defined, dynamic termination and static termination:

- **Dynamic termination** The set of processes P is said to be dynamically terminated at some instant if and only if the predicate *Dterm* is true at that moment where:

$$Dterm \equiv \forall P_i \in P : passive_i \wedge \neg fulfilled_i(ARR_i \cup NE_i).$$

  Dynamic termination means that no more activity is possible from processes, though messages of the underlying computation can still be in transit. This definition is useful in "early" detection of termination as it allows us to conclude whether a computation has terminated even if some of its messages have not yet arrived.
  Note that dynamic termination is a stable property because once *Dterm* is true, it remains true.
- **Static termination** The set of processes P is said to be statically terminated at some instant if and only if the predicate *Sterm* is true at that moment where:

$$Sterm \equiv \forall P_i \in P : passive_i \wedge (NE_i = \emptyset) \wedge \neg fulfilled_i(ARR_i).$$

  Static termination means all channels are empty and none of the processes can be activated. Thus, static termination is focused on the state of both channels and processes. When compared to *Dterm*, the predicate *Sterm* corresponds to "late" detection as, additionally, all channels must be empty.

7.7.4 A static termination detection algorithm

Informal description

A control process C_i, called a *controller*, is associated with each application process P_i. Its role is to observe the behavior of process P_i and to cooperate with other controllers C_j to detect occurrence of the predicate *Sterm*. In order to detect static termination, a controller, say C_α, initiates detection by sending a control message *query* to all controllers (including itself). A controller C_i responds with a message $reply(ld_i)$, where ld_i is a Boolean value. C_α combines all the Boolean values received in *reply* messages to compute $td := \bigwedge_{1\leq i\leq n} ld_i$. If td is true, C_α concludes that termination has occurred. Otherwise, it sends new *query* messages. The basic sequence of sending of *query* messages followed by the reception of associated *reply* messages is called a *wave*.

The core of the algorithm is the way a controller C_i computes the value ld_i sent back in a *reply* message. To ensure safety, the values $ld_1, \ldots ld_n$ must be such that:

$$\bigwedge_{1\leq i\leq n} ld_i \Longrightarrow Sterm$$

$$\Longrightarrow \forall P_i \in P : passive_i \wedge (NE_i = \emptyset) \wedge \neg fulfilled_i(ARR_i).$$

A controller C_i delays a response to a *query* as long as the following local predicate is false: $passive_i \wedge (notack_i = 0) \wedge \neg fulfilled_i(ARR_i)$. When this predicate is false, the static termination cannot be guaranteed.

For correctness, the values reported by a wave must not miss the activity of processes "in the back" of the wave. This is achieved in the following manner: each controller C_i maintains a Boolean variable cp_i (initialized to true iff P_i is initially passive) in the following way:

- When P_i becomes active, cp_i is set to false.
- When C_i sends a reply message to C_α, it sends the current value of cp_i with this message, and then sets cp_i to true.

Thus, if a reply message carries value true from C_i to C_α, it means that P_i has been continuously passive since the previous wave, and the messages arrived and not yet consumed are not sufficient to activate P_i, and all output channels of P_i are empty.

Formal description

The algorithm for static termination detection is as follows. By a *message*, we mean any message of the underlying computation; *queries* and *replies* are called *control* messages.

S1: When P_i sends a message to P_j

$$notack_i := notack_i + 1$$

S2: When a message from P_j arrives at P_i

send *ack* to C_j

S3: When C_i receives *ack* from C_j

$$notack_i = notack_i - 1$$

S4: When P_i becomes active

$$cp_i := false.$$

(* A passive process can only become active when its activation condition is true; this activation is under the control of the underlying operating system, and the termination detection algorithm only observes it. *)

S5: When C_i receives query from C_α

> **Wait until**
> $((passive_i \wedge (notack_i = \emptyset) \neg fulfilled_i(ARR_i))$;
> $ld_i := cp_i$;
> $cp_i := true$;
> send $reply(ld_i)$ to C_α

S6: When controller C_α decides to detect static termination

> **repeat** send *query* to all C_i;
> receive $reply(ld_i)$ from all C_i;
> $td := \bigwedge_{1 \le i \le n} ld_i$;
> **until** td;
> *claim static termination*

Performance

The efficiency of this algorithm depends on the implementation of waves. Two waves are in general necessary to detect static termination. A wave needs two types of messages: n queries and n replies, each carrying one bit. Thus, $4n$ control messages of two distinct types carrying at most one bit each are used to detect the termination once it has occurred. If waves are supported by a ring, this complexity reduces to $2n$. The detection delay is equal to duration of two sequential wave executions.

7.7.5 A dynamic termination detection algorithm

Recall that dynamic termination can occur before all messages of the computation have arrived. Thus, termination of the computation can be detected sooner than in static termination.

Informal description

Let C_α denote the controller that launches the waves. In addition to cp_i, each controller C_i has the following two vector variables, denoted as s_i and r_i, that count messages, respectively, sent to and received from every other process:

- $s_i[j]$ denotes the number of messages sent by P_i to P_j;
- $r_i[j]$ denotes the number of messages received by P_i from P_j.

Let S denote an $n \times n$ matrix of counters used by C_α; entry $S[i, j]$ represents C_α's knowledge about the number of messages sent by P_i to P_j.

First, C_α sends to each C_i a query message containing the vector $(S[1,i], \ldots, S[n,i])$, denoted by $S[.,i]$. Upon receiving this query message, C_i computes the set of its non-empty channels. This set is denoted by ANE_i because it is an approximate knowledge but is sufficient to ensure correctness. Then C_i computes ld_i, which is true if and only if P_i has been continuously passive since the previous wave and its requirement cannot be fulfilled by all the messages arrived and not yet consumed (ARR_i) and all messages potentially in its input channels (ANE_i). C_i sends to C_α a reply message carrying the values ld_i and vector s_i. Vector s_i is used by C_α to update row $S[i,.]$ and thus gain more accurate knowledge. If $\bigwedge_{1 \le i \le n} ld_i$ evaluates to true, C_α claims dynamic termination of the underlying computation. Otherwise, C_α launches a new wave by sending *query* messages.

Vector variables s_i and r_i allow C_α to update its (approximate) global knowledge about messages sent by each P_i to each P_j and get an approximate knowledge of the set of non-empty input channels.

Formal description

All controllers C_i execute statements S1 to S4. Only the initiator C_α executes S5. Local variables s_i, r_i, and S are initialized to 0.

S1: When P_i sends a message to P_j

$$s_i[j] := s_i[j] + 1$$

S2: When a message from P_j arrives at P_i

$$r_i[j] := r_i[j] + 1$$

S3: When P_i becomes active

$$cp_i := false$$

S4: When C_i receives $query(VC[1...n])$ from C_α

(*$VC[1...n] = S[1...n, i]$ is the ith column of S*)

$ANE_i := \{P_j : VC[j] > r_i[j]\}$;

$ld_i := cp_i \wedge \neg fulfilled_i(ARR_i \cup ANE_i)$;

$cp_i := (state_i = \text{passive})$;

send $reply(ld_i, s_i)$ to C_α

S5: When controller C_α decides to detect dynamic termination
  repeat for each C_i
    send $query(S[1\dots n, i])$ to C_i;
    (* the ith column of S is sent to C_i *)
    receive $reply(ld_i, s_i)$ from all C_i;
    $\forall i \in [1..n] : S[i, .] := s_i$;
    $td := \bigwedge_{1 \le i \le n} ld_i$
  until td;
  claim dynamic termination

Performance

The dynamic termination detection algorithm needs two waves after dynamic termination has occurred to detect it. Since no acknowledgements are necessary, its message complexity is $4n$, which is lower than the static termination detection algorithm. However, messages are composed of n monotonically increasing counters. As waves are sequential, *query* (and *reply*) messages between C_α and each C_i are received and processed in their sending order; this FIFO property can be used in conjunction with Singhal–Kshemkalyani's differential technique to decrease the size of the control messages. The detection delay is two waves but is shorter than the delay of the static termination algorithm as acknowledgements are not used.

7.8 Termination detection in the atomic computation model

Mattern [12] developed several algorithm for termination detection in the atomic computation model.

Assumptions

1. Processes communicate solely by messages. Messages are received correctly after an arbitrary but finite delay. Messages sent over the same communication channel may not obey the FIFO rule.
2. A *time cut* is a line crossing all process lines. A time line can be a straight vertical line or a zigzag line, crossing all process lines. The time cut of a distributed computation is a set of actions characterized by a fact that whenever an action of a process belongs to that set, all previous actions of the same process also belong to the set.
3. We assume that all atomic actions are totally globally ordered i.e., no two actions occur at the same time instant.

7.8.1 The atomic model of execution

In the *atomic model* of the distributed computation, a process may at any time take any message from one of its incoming communication channels, immediately change its internal state, and at the same instant send out zero or more messages. All local actions at a process are performed in zero time. Thus, consideration of process states is eliminated when performing termination detection.

In the atomic model, a distributed computation has terminated at time instant t if at this instant all communications channels are empty. This is because execution of an internal action at a process is instantaneous.

A dedicated process, P_1, the initiator, determines if the distributed computation has terminated. The initiator P_1 starts termination detection by sending control messages directly or indirectly to all other processes. Let us assume that processes $P_1, \ldots, P_n$ are ordered in sequence of the arrival of the control message.

7.8.2 A naive counting method

To find out if there are any messages in transit, an obvious solution is to let every process *count* the number of basic messages sent and received. We denote the total number of basic messages P_i has sent at (global) time instant t by $s_i(t)$, and the number of messages received by $r_i(t)$. The values of the two local counters are communicated to the initiator upon request. Having directly or indirectly received these values from all processes, the initiator can accumulate the counters. Figure 7.9 shows an example, where the time instants at which the processes receive the control messages and communicate the values of their counters to the initiator are symbolized by striped dots. These are connected by a line representing a "control wave," which induces a time cut.

If the accumulated values at the initiator indicate that the sum of all the messages received by all processes is the same as the sum of all messages

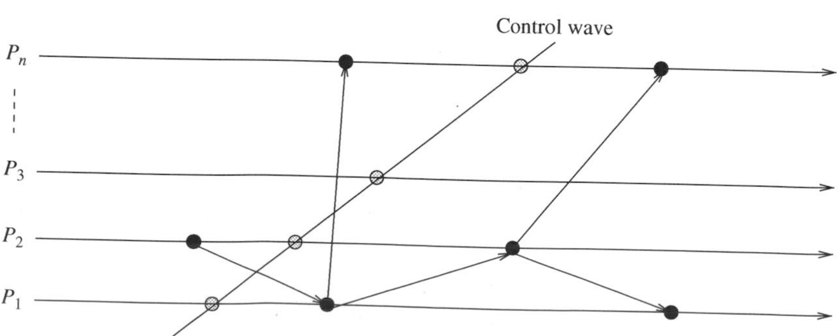

Figure 7.9 An example showing a control wave with a backward communication [12].

sent by all processes, it may give an impression that all the messages sent have been received, i.e., there is no message in transit.

Unfortunately because of the time delay of the control wave, this simple method is not correct. The example in Figure 7.9 shows that the counters can become corrupted by messages "from the future," crossing from the right side of the control wave to its left.

The accumulated result indicates that one message was sent and one received although the computation has not terminated. This misleading result is caused by the fact that the time cut is inconsistent. A time cut is considered to be inconsistent, if when the diagonal line representing it is made vertical by compressing or expanding the local time scales, a message crosses the control wave backwards.

However, this naive method for termination detection works if the time cut representing the control wave is consistent.

Various strategies can be applied to correct the deficiencies of the naive counting method:

- If the time cut is inconsistent, restart the algorithm later.
- Design techniques that will only provide consistent time cuts.
- Do not lump the count of all messages sent and all messages received. Instead, relate the messages sent and received between pairs of processes.
- Use techniques like freezing the underlying computation.

7.8.3 The four counter method

A very simple solution consists of counting twice using the naive counting method and comparing the results. After the initiator has received the response from the last process and accumulated the values of the counters R^* and S^* (where $R^* := \sum_{\forall i} r_i(t_i)$ and $S^* := \sum_{\forall i} s_i(t_i)$), it starts a second control wave (see Figure 7.10), resulting in values R'^* and S'^*. The system is terminated if values of the four counters are equal, i.e., $R^* = S^* = R'^* = S'^*$. In fact, a slightly stronger result exists: if $R^* = S'^*$, then the system terminated at the end of the first wave (t_2 in Figure 7.10).

Let t_2 denote the time instant at which the first wave is finished, and t_3 $(\geq t_2)$ denote the starting time of the second wave (see Figure 7.10).

1. Local message counters are monotonic, that is, $t \leq t'$ implies $s_i(t) \leq s_i(t')$ and $r_i(t) \leq r_i(t')$. This follows from the definition.
2. The total number of messages sent or received is monotonic, that is, $t \leq t'$ implies $S(t) \leq S(t')$ and $R(t) \leq R(t')$.
3. $R^* \leq R(t_2)$. This follows from (1) and the fact that all values r_i are collected before t_2.
4. $S'^* \geq S(t_3)$. This follows from (1) and the fact that all values s_i are collected after t_3.
5. For all t, $R(t) \leq S(t)$. This is because the number of messages in transit $D(t) := S(t) - R(t) \geq 0$.

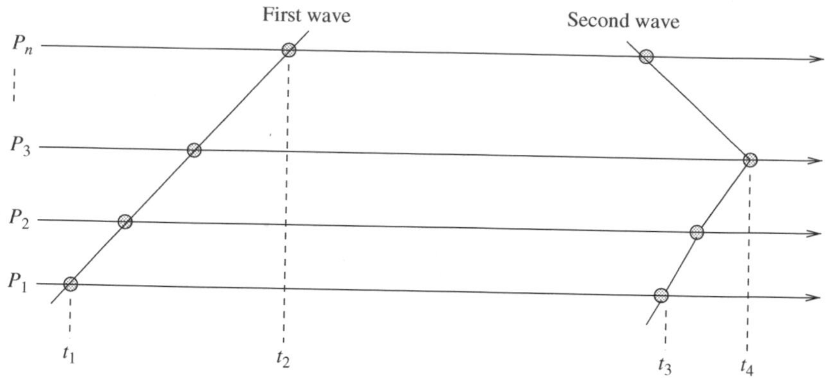

Figure 7.10 An example showing two control waves [12].

Now we show that if $R^* = S'^*$, then the computation had terminated at the end of the first wave:

$$\begin{aligned} R^* = S'^* &\Longrightarrow R(t_2) \geq S(t_3) \\ &\Longrightarrow R(t_2) \geq S(t_2) \\ &\Longrightarrow R(t_2) = S(t_2) \end{aligned}$$

That is, the computation terminated at t_2 (at the end of the first wave).

If the system terminated before the start of the first wave, it is trivial that all messages arrived before the start of the first wave, and hence the values of the accumulated counters will be identical. Therefore, termination is detected by the algorithm in two "rounds" after it had occurred. Note that the second wave of an unsuccessful termination test can be used as the first wave of the next termination test. However, a problem with this method is to decide when to start the next wave after an unsuccessful test – there is a danger of an unbounded control loop.

7.8.4 The sceptic algorithm

Note that the values of the counters obtained by the first wave of the four counter method can become corrupted if there is some activity at the right of the wave. To detect such activity, we use *flags* which are initialized by the first wave, and set by the processes when they receive (or alternatively when they send) messages. The second wave checks if any of the flags have been set, in which case a possible corruption is indicated. A general drawback is that at least two waves are necessary to detect the termination.

It is possible to devise several variants based on the *logical control topology*. If the initiator asks every process individually, it corresponds to a star topology. It is possible to implement the sceptic algorithm on a ring; however, symmetry is not easily achieved since different waves may interfere when a single flag is used at each process. A spanning tree is also an interesting control configuration. Echo algorithms used as a parallel graph traversal method induce two phases. The "down" phase is characterized by the receipt of a first control message which is propagated to all other neighbors, and the "up" phase by the receipt of the last of the echoes from its neighboring nodes. These two phases can be used as two necessary waves of the sceptic method for termination detection.

7.8.5 The time algorithm

The time algorithm is a single wave detection algorithm where termination can be detected in one single wave after its occurrence at the expense of increased amount of control information or augmenting every message with a timestamp. In the time algorithm, each process has a *local clock* represented by a counter initialized to 0.

A control wave started by the initiator at time i, accumulates the values of the counters and "synchronizes" the local clocks by setting them to $i+1$. Thus, the control wave separates "past" from "future." If a process receives a message whose *timestamp* is greater than its own local time, the process has received a message from the future (i.e., the message crossed the wave from right to left) and the message has corrupted the counters. After such a message has been received, the current control wave is nullified on arrival at the process.

Formal description

Every process P_j $(1 \leq j \leq n)$ has a local message counter *COUNT* (initialized to 0) that holds the value $s_j - r_j$, a local discrete *CLOCK* (initialized to 0), and a variable *TMAX* (also initialized to 0) that holds the latest send time of all messages received by P_j.

The psuedo code for process P_j is shown in Algorithm 7.4.

A control message consists of four parameters: the (local) time at which the control round was started, the accumulator for the message counters, a flag which is set when a process has received a basic message from the future ($TMAX \geq TIME$), and the identification of the initiating process. The first component of a basic message is always the timestamp.

For each single control wave, any basic message that crosses the wave from the right side of its induced cut to its left side is detected. Note that different control waves do not interfere; they merely advance the local clocks further. Once the system is terminated, the values of the *TMAX* variables remain fixed and since for every process P_j, $TMAX_j \leq max\ CLOCK_i$ $(1 \leq i \leq n)$, the process with the maximum clock value can detect global termination in one round. Other processes may need more rounds.

```
(a) When sending a basic message to P_i:
(1)        COUNT ← COUNT +1;
(2)        send ⟨CLOCK,...⟩ to P_i;
       /* timestamped basic message */

(b) When receiving a basic message ⟨TSTAMP,...⟩:
(3)        COUNT←COUNT −1;
(4)        TMAX←max(TSTAMP, TMAX);
(5)        /* process the message */

(c) When receiving a control message ⟨TIME, ACCU, INVALID, INIT⟩:
(6)        CLOCK ← max(TIME, CLOCK): /* synchronize the local clock */
(7)        if INIT = j /* complete round? */
(8)              then if ACCU = 0 and not INVALID
(9)                    then "terminated" else "try again";
(10)                 end_if;
(11)             else send ⟨TIME, ACCU + COUNT, INVALID or
                            TMAX ≥ TIME, INIT⟩ to P_((j mod n)+1);
(12)       end_if;

(d) When starting a control round:
(13)       CLOCK ← CLOCK+1;
(14)       send ⟨CLOCK, COUNT, false, j⟩ to P_((j mod n)+1);
```

Algorithm 7.4 The time algorithm at P_j [12].

7.8.6 Vector counters method

Vector counters method of termination detection consists of counting messages in such a way that it is not possible to mislead the accumulated counters.

The configuration used is the ring with n processes where every process P_j $(1 \leq j \leq n)$ has a *COUNT* vector of length n, where $COUNT[i]$ $(1 \leq i \leq n)$ denotes the ith component of the vector. A circulating control message also consists of a vector of length n. For each process P_j, the local variable $COUNT[i]$ $(i \neq j)$ holds the number of basic messages that have been sent to process P_i since the last visit of the control message. Likewise, the negative value of $COUNT[j]$ indicates how many messages have been received from any other process. At any (global) time instant, the sum of the kth components of all n *COUNT* vectors including the circulating control vector equals the number of messages currently on their way to process P_k, $1 \leq k \leq n$. This property is maintained invariant by the implementation given below. For simplicity, we assume that no process communicates with itself, P_{n+1} is identical to P_1, an operation on a vector is defined by the operating on each of its components, and 0* denotes the null vector.

The psuedo code for process P_j is shown in Algorithm 7.5.

```
COUNT is initialized to 0*
(a) When sending a basic message to P_i (i ≠ j):
(1)      COUNT[i] ← COUNT[i] + 1;

(b) The following instructions are executed at the end of all local actions
    triggered by the receipt of a basic message:
(2)      COUNT[j] ← COUNT[j] − 1;
(3)      if COUNT[j] = 0 then
(4)            if COUNT = 0*
                     then "system terminated"
(5)                  else send accumulate ⟨COUNT⟩ to P_{j+1};
(6)                  COUNT ← 0*;
(7)            end_if;
(8)      end_if;

(c) When receiving a control message "accumulate ⟨ACCU⟩":
(9)      COUNT ← COUNT + ACCU;
(10)     if COUNT [j] ≤ 0 then
(11)           if COUNT = 0*
                     then "system terminated"
(12)                 else send accumulate ⟨COUNT⟩ to P_{j+1};
(13)                 COUNT ← 0*;
(14)           end_if;
(15)     end_if;
```

Algorithm 7.5 Vector counters algorithm at P_j [12].

An initiator P_i starts the algorithm by sending the control message "accumulate $\langle 0^* \rangle$" to P_{i+1}. A mechanism is needed to ensure that every process is visited at least once by the control message, i.e., that the control vector makes at least one complete round after the start of the algorithm.

Every process counts the number of outgoing messages individually by incrementing the counter indexed by the receiver's process number (line 1); the counter indexed by its own number is decremented on receipt of a message (line 2). When a process receives the circulating control message, it accumulates the values in the message to its *COUNT* vector (line 9). A check is then made (line 10) to determine whether any basic messages known to the control message have still not arrived at P_j. If this is the case ($COUNT[j]>0$), the control message is removed from the ring and regenerated at a later time (line 5) when all expected messages have been received by P_j. For this purpose, every time a basic message is received by a process P_j, a test is made to check whether $COUNT[j]$ is equal to 0 (line 3). Note that lines 4–15 are only executed when the control vector is at P_j. Note that there is at most one process P_j with $COUNT[j]>0$, and if this is the case at P_j, the control vector "waits" at process P_j (lines 11–13 are not executed and the control vector remains at P_j).

If the control message is not required to wait at nodes for outstanding basic messages, the algorithm can be simplified considerably by removing lines 3–8 as well as lines 10 and 15.

Performance

The number of control messages exchanged by this algorithm is bounded by $n(m+1)$, where m denotes the number of basic messages, because at least one basic message is received in every round of the control message, excluding the first round. Therefore, the worst case communication complexity for this algorithm is $O(mn)$.

7.8.7 A channel counting method

The channel counting method is a refinement of the vector counter method in the following way: a process keeps track of the number of messages sent to each process and keeps track of the number of messages received from each process, using appropriate counters.

Each process P_j has n counters, $C_{j1}^+, \ldots, C_{jn}^+$, for outgoing messages and n counters, $C_{1j}^-, \ldots, C_{nj}^-$, for incoming messages. C_{ij}^- is incremented when P_j receives a message from process P_i, and C_{jk}^+ is incremented when P_j sends a message to P_k. Upon demand, each process informs the values of the counters to the initiator. The initiator reports termination if $C_{ij}^- = C_{ij}^+$ for all i,j.

The method becomes more practical if it is combined with the echo algorithm, where test messages flow down on every edge of the graph and echoes proceed in the opposite direction. The value of C_{ij}^- is transmitted upwards from process P_j to P_i in an echo; whereas, a test message sent by P_i to P_j carries the value of C_{ij}^+ with it. A process receiving a test message from another process (the activator), propagates it in parallel with any other process to which it sent basic messages whose receipts have not yet been confirmed. If it has already done this, or if all basic messages sent out have been confirmed, an echo is immediately sent to the activator. There are no special acknowledgement messages. A process P_i receiving the value of C_{ij}^- in an echo knows that all messages it sent to P_j have arrived if the value of C_{ij}^- equals the value of its own counter C_{ij}^+. An echo is only propagated towards the activator if an echo has been received from each subtree and all channels in the subtrees are empty.

Formal description

Each process P_j has the following arrays of counters:

1. $OUT[i]$: counts the number of basic messages sent to P_i.
2. $IN[i]$: counts the number of basic messages received from P_i.

3. *REC*[*i*]: records the number of its messages that P_j knows have been received by P_i.

OUT[*i*] corresponds to C^+_{ji} and *IN*[*i*] to C^-_{ij}. A variable *ACTIVATOR* is used to hold the index number of the activating process and a counter *DEGREE* indicates how many echoes are still missing.

The psuedo code for process P_j is shown in Algorithm 7.6.

```
/*OUT, IN, and REC are initialized to 0* and DEGREE to 0.*/

(a) When sending a basic message to P_i:
(1)        OUT[i] → OUT[i]+1;

(b) When receiving a basic message from P_i:
(2)        IN[i] ← IN[i]+1;

(c) On the receipt of a control message test ⟨m⟩ from P_i where m ≤ IN[i]:
(3)        if DEGREE > 0 or OUT = REC /* already engaged or
            subtree is quiet */
(4)                then send echo ⟨IN[i]⟩ to P_i;
(5)                else ACTIVATOR ← i; /* trace activating process */
(6)                PROPOGATE /* and test all subtrees */
(7)        end_if;

(d) On the receipt of a control message echo ⟨m⟩ from P_i:
(8)        REC[i] ← m;
(9)        DEGREE ← DEGREE − 1; /* decrease missing echoes counter */
(10)        if DEGREE = 0 then
                        /* last echo checks whether all subtrees are quiet */
(11)                PROPAGATE;
(12)        end_if;
(13)        if DEGREE = 0 then /*all echoes arrived, everything quiet */
(14)                send echo ⟨IN[ACTIVATOR]⟩ to P_ACTIVATOR;
(15)        end_if;

(e) The procedure PROPAGATE called at lines 6 and 11 is defined as follows:
(16) procedure PROPAGATE:
(17)        loop for K = 1 to n do
(18)                if OUT[K] ≠ REC[K] then /* confirmation missing */
(19)                        send test ⟨OUT[K]⟩ to P_k; /* check subtree */
(20)                        DEGREE ← DEGREE + 1;
(21)                end_if;
(22)        end_loop;
(23) end_procedure;
```

Algorithm 7.6 Channel counting algorithm [12].

Variable *DEGREE* is incremented when a process sends a test message (line 20) and it is decremented when a process receives an *ECHO* message (line 9). If $DEGREE > 0$, it means the node is "engaged" and a test message is immediately responded to with an echo message (line 4). An echo is also returned for a test message if $OUT = REC$ (line 3), i.e., if process sent no messages at all or if all messages sent out by it have been acknowledged. Lines 10–15 insure that an echo is only returned if the arrival of all basic messages has been confirmed and all computations in the subtree finished. This is done by sending further test messages (via procedure *PROPAGATE*) after the last echo has arrived (lines 10–12). These test messages visit any of the subtree root processes that have not yet acknowledged all basic messages sent to them. The procedure *PROPAGATE* increases the value of the variable *DEGREE* if any processes are visited, thus preventing the generation of an echo (lines 13–15).

To minimize the number of control messages, test messages should not overtake basic messages. To achieve this, test messages carry with them a count of the number of basic messages sent over the communication channel (line 19). If a test messages overtakes some basic messages (and it is not overtaken by basic messages), its count will be greater than the value of the IN-counter of the receiver process. In this case, the test message is put on hold and delivered later when all basic messages with lower count have been received (guard $m \leq IN[i]$ in point (c) insures this).

The initiator starts the termination test only once, as if it had received a test $\langle 0 \rangle$ message from some imaginary process P_0. On termination detection, instead of eventually sending an echo to P_0, it reports termination. Test messages only travel along those channels that have been used by basic messages; processes that did not participate in the distributed computation are not visited by test messages. For each test message, an echo is eventually sent in the opposite direction.

Performance

At least one basic message must have been sent between the two test messages along the same channel. This results in an upper bound of $2m$ control messages, where m denotes the number of basic messages. Hence, the worst case communication complexity is $O(m)$. However, the worst case should rarely occur, particularly if the termination test is started well after the computation started. In many situations, the number of control messages should be much smaller than m. The exact number of control messages involved in channel counting is difficult to estimate because it is highly dependent on communication patterns of the underlying computation.

7.9 Termination detection in a faulty distributed system

An algorithm is presented that detects termination in distributed systems in which processes fail in a fail-stop manner. The algorithm is based on the weight-throwing method. In such a distributed system, a computation is said to be terminated if and only if each healthy process is idle and there is no basic message in transit whose destination is a healthy process. This is independent of faulty processes and undeliverable messages (i.e., whose destination is a faulty process). Based on the weight-throwing scheme, a scheme called flow detecting scheme is developed by Tseng [20] to derive a fault-tolerant termination detection algorithm.

Assumptions

Let $S = \{P_1, P_2, \ldots, P_n\}$ be the set of processes in the distributed computation. C_{ij} represents the bidirectional channel between P_i and P_j. The communication network is asynchronous. Communications channels are reliable, but they are non-FIFO. At any time, an arbitrary number of processes may fail. However, the network remains connected in the presence of faults. The fail-stop model implies that a failed process stops all activities and cannot rejoin the computation in the current session. Detection of faults takes a finite amount of time.

7.9.1 Flow detecting scheme

Weights may be lost because a process holding a non-zero weight may crash or a message destined to a crashed process is carrying a weight. Therefore, due to faulty processes and undeliverable messages carrying weights, it may not be possible for the leader to accumulate the total weight of 1 to declare termination. In the case of a process crash, the lost weight must be calculated. To solve this problem, the concept of flow invariant is used.

The concept of flow invariant

Define $H \subseteq S$ as the set of all healthy processes. Define *subsystem* H to be part of the system containing all processes in H and communication channels connecting two processes in H. According to the concept of flow invariant, the weight change of the subsystem during time interval I, during which the system is doing computation, is equal to (weights flowing into H during I) – (weights flowing out of H during I). To implement this concept, a variable called net_i is assigned to each process P_i belonging to H. This variable records the total weight flowing into and out of the subsystem H. Initially, $\forall i \; net_i = 0$. The following flow-detecting rules are defined:

Rule 1: Whenever a process P_i which belongs to H receives a message with weight x from another process P_j which does not belong to H, x is added to net_i.

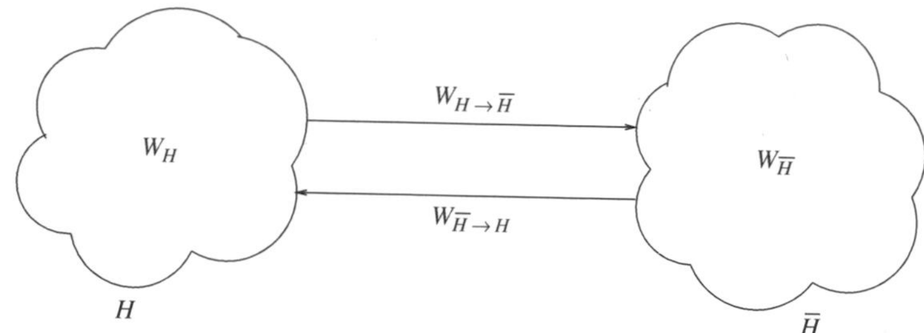

Figure 7.11 Healthy and faulty process sets and message flow between them [20].

Rule 2: Whenever a process P_i which belongs to H sends a message with weight x to a process P_j which does not belong to H, x is subtracted from net_i.

Let W_H be the sum of the weights of all processes in H and all in-transit messages transmitted between processes in H:

$$W_H = \sum_{P_i \in H} (net_i + 1/n),$$

where $1/n$ is the initial weight held by each process P_i.

Let $\overline{H} = S\text{–}H$ be the set of faulty processes. The distribution of weights is divided into four parts:

W_H: weights of processes in H.
$W_{\overline{H}}$: weights of processes in $\overline{H}$.
$W_{H\rightarrow\overline{H}}$: weights held by in-transit messages from H to $\overline{H}$.
$W_{\overline{H}\rightarrow H}$: weights held by in-transit messages from $\overline{H}$ to H.

This is shown in Figure 7.11. $W_{\overline{H}}$ and $W_{H\rightarrow\overline{H}}$ are lost and cannot be used in the termination detection.

7.9.2 Taking snapshots

In distributed systems, due to the lack of a perfectly synchronized global clock, it is not possible to get a global view of the subsystem H and hence it may not possible to determine W_H. We obtain $\overline{W}_H$, which is an estimated value of W_H, by taking snapshots on the subsystem H and by using the above equation for W_H.

However, note that weights in $W_{\overline{H}\rightarrow H}$ carried by in-transit messages may join H and change W_H. To obtain a stable value of W_H, channels from $\overline{H}$ to H are disconnected before taking snapshots of H. Once a channel is disconnected, a healthy process can no longer send or receive messages along it.

A snapshot on H is the collection of net_i's from all processes in H. A snapshot is said to be consistent if all channels from H to $\overline{H}$ are disconnected before taking the snapshot (i.e., recording the values of net_i).

A snapshot is taken upon a snapshot request by the leader process. The leader uses the information in a consistent snapshot and the equation to compute W_H to calculate $\overline{W}_H$. Snapshots are requested when a new faulty process is found or when a new leader is elected. It should be noted that $\overline{W}_H$ is an estimate of the weight remaining in the system. This is because processes can fail and stop any time and there may not exist any point in real time in the computation where H is the healthy set of processes. Suppose H' is the set of healthy processes at some point in time in the computation after taking the snapshot. If $H = H'$, then $\overline{W}_H = W_{H'}$; otherwise, $\overline{W}_H \geq W_{H'}$ must be true, because of the fail-stop model of processes. This eliminates the possibility of declaring termination falsely. Thus, the leader can safely declare termination after it has collected $\overline{W}_H$ of weight.

7.9.3 Description of the algorithm

The algorithm combines the weight-throwing scheme, the flow detecting scheme and a snapshot-recording scheme [20]. Process P_i elects itself the leader if it knows that all P_j, $j < i$, are faulty. The leader process takes snapshots and estimates remaining weight in the system.

Data structures

The following data structures are used at process P_i, $i = 1, ..., n$:

- l_i is the identity of the leader known to P_i. Initially $l_i = 1$.
- w_i is the weight currently held by P_i. Initially $w_i = 1/n$.
- s_i is the system's total weight assumed by P_i. P_i will try to collect this amount of weight. Initially, s_i=1.
- $NET_i[1, \ldots, n]$ is an array of real numbers. $NET_i[j]$ keeps track of the total weight flowing into P_i from P_j. Initially, $NET_i[j] = 0$ for all $j = 1, \ldots, n$.
- F_i is a set of faulty processes. A process P_j belongs to F_i if and only if P_i knows that P_j is faulty and P_i has disconnected its channel to P_j. Initially, F_i is a null set.
- SN_i is a set of processes. When P_i initiates a snapshot, SN_i is a set of processes to which P_i sends snapshot requests. A process P_j belonging to SN_i is removed from SN_i if P_i receives a reply from P_j or if P_i finds P_j is faulty. No new snapshot is started unless SN_i is an empty set. Initially, SN_i is a null set, which implies no snapshot is in progress.
- t_i is used for temporarily calculating the total remaining weight while a snapshot is in progress.
- c_i is a boolean, used for temporarily calculating the consistency of a snapshot.

Types of messages

The following four types of messages are exchanged by the algorithm:

- $B(x)$ is a basic message B with weight x.
- $C(x)$ is a control message that is used to report weight x to the leader process.
- *Request*(F_i) is a snapshot requesting message sent by the leader process P_i. The set F_i is to inform the receiver the set of faulty processes known to P_i.
- *Reply*(F_j, NET_j) is the state reporting message sent by P_j in reply to the leader's *Request*() message.

The algorithm

The algorithm is described for process P_i. The algorithm consists of nine event-driven atomic actions, each having the format "(guard) → (actions)." Actions are triggered by sending/receiving messages, changing local states, or detecting new faulty processes. Actions A1 to A5 in Algorithm 7.7 implement weight-throwing and flow-detecting schemes.

A1: (P_i sending a basic message B to P_j) →
  w_i is partitioned into x and y such that $x > 0, y > 0$ and $x + y = w_i$;
  $B(x)$ is sent to P_j;
  $NET_i[j] := NET_i[j] - x$;
  $w_i := y$;
A2: (P_i receiving a basic message $B(x)$ from P_j) →
  $NET_i[j] := NET_i[j] + x$;
  $w_i := w_i + x$;
  pass the basic message to the underlying system;
A3: (P_i becoming idle) →
  if $l_i \neq i$ **then**
    send $C(w_i)$ to P_{l_i};
    $NET_i[l_i] := NET_i[l_i] - w_i$;
    $w_i := 0$;
  end if;
A4: (P_i receiving a control message $C(x)$ from P_j) →
  $NET_i[j] := NET_i[j] + x$;
  $w_i := w_i + x$;
A5: (P_i is *idle*) $\wedge$ ($s_i = w_i$) →
  announce "termination";

Algorithm 7.7 Termination detection algorithm at P_i [20].

A1 is activated when P_i sends a basic message to another process. A2 is triggered by receiving a basic message. A3 is the weight-reporting action.

When P_i is not the leader, it sends its weight to the leader process in a control message. A4 describes P_i's response on receiving a control message. In all actions A1–A4, $NET_i[1 \ldots n]$ records the weight-flowing information. In A5, leader P_i announces the termination.

Actions F1 to F4 in Algorithm 7.8 deal with faults and take snapshots of the system.

(* Actions for detecting a fault when no snapshot is in progress *)
F1: $(P_i$ detecting P_j faulty$) \wedge (P_j \notin F_i) \wedge (SN_i = \emptyset) \rightarrow$
    disconnect the channel from P_i to P_j;
    $F_i := F_i \cup \{P_j\}$;
    $l_i = min\{k \mid P_k \in S - F_i\}$;
    if $(l_i = i)$, **then** call *snapshot*(); **end if**;

(* Actions on receiving a snapshot request *)
F2: $(P_i$ receiving $Request(F_j)$ from $P_j) \rightarrow$
    $l_i := j$;
    for every P_f belonging to $F_j - F_i$, disconnect the channel $C_{i,f}$;
    $F_i := F_i \cup F_j$;
    Send a $Reply(F_i, NET_i[1 \ldots n])$ to P_j;

(* Actions on receiving a snapshot response *)
F3: $(P_i$ receiving $Reply(F_j, NET_j[1 \ldots n]$ from $P_j) \rightarrow$
    if $(F_i \neq F_j) \vee \neg c_i$ **then**
        for every P_f belonging to $F_j - F_i$, disconnect the channel $C_{i,f}$;
            $F_i = F_i \cup F_j$;
            $c_i = false$;
    else
            $t_i = t_i + 1/n + \sum_{P_f \in F_j} NET_j[f]$;
    end if;
    $SN_i = SN_i - \{P_j\}$;
    if $SN_i = \emptyset$ **then**
            if c_i **then** $s_i := t_i$ **else** call *snapshot*(); **end if**;
    end if;

(* Actions for detecting a fault when a snapshot is in progress *)
F4: $(P_i$ detecting P_j faulty$) \wedge (SN_i \neq \emptyset) \rightarrow$
    Disconnect the channel $C_{i,j}$;
    $F_i := F_i \cup \{P_j\}$;
    $c_i := false$;
    $SN_i := SN_i - \{P_j\}$;
    if $SN_i = \emptyset$ **then** call *snapshot*(); **end if**;

(* Snapshot-taking procedure *)
Procedure *snapshot*() (* assuming the caller is P_i *)
    begin

$SN_i = S - F_i - \{P_i\}$; (* processes that will receive requests *)
$\forall P_k \in SN_i$, send a *Request*(F_i) to P_k;
$t_i := 1/n + \sum_{P_f \in F_i} NET_i[f]$;
$c_i := true$;
end;

Algorithm 7.8 Snapshot algorithm at P_i [20].

F1 is triggered when P_i detects for the first time that a process P_j is faulty and no snapshot is currently in progress. The channel from P_i to P_j is disconnected. Then P_i elects a healthy process with least i.d. as its leader. If process P_i itself is the leader, then it invokes a snapshot procedure to initiate a snapshot.

In the *snapshot*() procedure, first SN_i is set to the set of processes to which the Request()s are to be sent and sends a *Request*() to these processes. This prevents F1 from being executed until the snapshot finishes. Assuming that the current healthy process set is $S - F_i$ and this snapshot is consistent, more weight is added to t_i as P_i receives *Reply*() messages from other processes.

F2 describes P_i's response on receiving a *Request*() message from P_j. P_i disconnects channels to faulty processes and sends a *Reply*() message to P_j, which sent the *Request*() message.

The initiator of the snapshot P_i waits for each P_j belonging to SN_i for either a *Reply*() coming from P_j or P_j being detected as faulty.

If a *Reply*() is received from P_j, F3 is executed. F3 describes P_i's actions on receiving such a snapshot response. The consistency of the snapshot is checked. If the snapshot is still consistent, t_i is updated. Then the barrier SN_i is reduced by one. If the barrier becomes null and the snapshot is consistent, s_i is updated to t_i. If the snapshot is not consistent, another snapshot is initiated.

The snapshot initiator P_i executes F4 when it detects a process $P_j \in SN_i$, is faulty and a snapshot is in progress. Another snapshot is started only when $SN_i = \emptyset$. Such a procedure is repeated until a consistent snapshot is obtained. Because of the fail-stop model of processes, the number of healthy processes is a non-increasing function of time and eventually the procedure will terminate.

7.9.4 Performance analysis

If k processes become faulty, at most $2k$ snapshots will be taken. Each snapshot costs at most $n-1$ *Request*()s and $n-1$ *Reply*()s. Thus, the message overhead due to snapshots is bounded by $4kn$.

If M basic messages are issued, processes will be activated at most M times. So processes will not turn idle more than $M + n$ times. So at most $M+n$ control messages $C(x)$ will be issued.

Thus, the message complexity of the algorithm is $O(M + kn + n)$.

The termination detection delay is bounded by $O(k+1)$. The termination detection delay is defined as the maximum number of message hops needed, after the termination has occurred, by the algorithm to detect the termination.

7.10 Chapter summary

A distributed computation is terminated if every process is locally terminated and there is no message in transit between any processes. Determining if a distributed computation has terminated is a fundamental problem in distributed systems. Detection of the termination of a distributed computation is a non-trivial task since no process has complete knowledge of the global state.

A number of algorithms have been developed to detect the termination of a distributed computation. These algorithms are based on the concepts of snapshot collection, weight throwing, spanning-tree, etc. In this chapter, we described a set of representative termination detection algorithms. Brzezinski *et al.* developed a very general model of the termination a distributed computation where the reception of a single message may not be enough to activate a passive process. Instead, the condition of activation is more general and a passive process requires reception of a set of messages to become active. Mattern developed several algorithms for termination detection in the atomic computation model. Tseng developed a weight-throwing algorithm to detect termination in distributed systems which allows processes to fail in a fail-stop manner.

Termination detection is a fundamental problem and it finds applications at several places in distributed systems.

7.11 Exercises

Exercise 7.1 Huang's termination detection algorithm could be redesigned using a counter to avoid the need of splitting weights. Present an algorithm for termination detection that uses counters instead of weights.

Exercise 7.2 Design a termination detection algorithm that is based on the concept of weight throwing and is tolerant to message losses. Assume that processes do not crash.

Exercise 7.3 Termination detection algorithms assume that an idle process can only be activated on the reception of a message. Consider a system where an idle process can become active spontaneously without receiving a message. Do you think a termination detection algorithm can be designed for such a system? Give reasons for your answer.

Exercise 7.4 Design an efficient termination detection algorithm for a system where the communication delay is zero.

Exercise 7.5 Design an efficient termination detection algorithm for a system where the computation at a process is instantaneous (that is, all proceses are always in the idle state.)

7.12 Notes on references

The termination detection problem was brought to prominence in 1980 by Francez [5] and by Dijkstra and Scholten [4]. Since then, a large number of termination detection algorithms having different features and for a variety of logical system configurations have been developed. A termination detection algorithm that uses distributed snapshot is discussed in [8]. A termination detection algorithm based on weight throwing is discussed in [9]. A termination detection algorithm based on weight throwing was first developed by Mattern [13]. Dijkstra *et al.* [3] present a ring-based termination detection algorithm. Topor [19] adapts this algorithm to a spanning tree configuration. Chandrasekaran and Venkatesan [2] present a message optimal termination detection algorithm. Brzezinski *et al.* [1] define a very general model of the termination problem, introduce the concept of static and dynamic terminations, and develop algorthms to detect static and dynamic terminations. Mattern developed [12] several algorithms for termination detection for the atomic model of computation. An algorithm for termination detection under faulty processes is given by Tseng [20]. Mayo and Kearns [14,15] present efficient termination detection based on roughly synchronized clocks. Other algorithms for termination detction can be found in [6,10,11,16–18,21].

Many termination detection algorithms use a spanning tree configuration. An efficient distributed algorithm to construct a minimum-weight spanning tree is given in [7].

References

[1] J. Brzezinski, J. M. Helary, and M. Raynal, Termination detection in a very general distributed computing model, *Proceedings of the International Conference on Distributed Computing Systems*, Poland, 1993, 374–381.

[2] S. Chandrasekaran and S. Venkatesan, A message-optimal algorithm for distributed termination detection. *Journal of Parallel and Distributed Computing*, 1990, 245–252.

[3] E. W. Dijikstra, W. H. J. Feijen, and A. J. M. van Gasteren, Derivations of a termination detection algorithm for distributed computations, *Information Processing Letters*, **16**(5), 1983, 217–219.

[4] E. W. Dijkstra and C. S. Scholten, Termination detection for distributed computations, *Information Processing Letters*, **11**(1), 1980, 1–4.

[5] N. Francez, Distributed termination, *ACM Transactions on Programming Langauges*, **2**(1), 1980, 42–55.

[6] N. Francez and M. Rodeh, Achieving distributed termination without freezing, *IEEE Transaction on Software Engineering*, May, 1982, 287–292.

[7] R. G. Gallager, P. Humblet, and P. Spira, A distributed algorithm for minimum weight spanning trees, *ACM Transactions on Programming Langauges and Systems*, January, 1983, 66–77.

[8] Shing-Tsaan Huang, Termination detection by using distributed snapshots, *Information Processing Letters*, **32**, 1989, 113–119.

[9] S. T. Huang, Detecting termination of distributed computations by external agents, *Proceedings of the 9th International Conference on Distributed Computing Systems*, 1989, 79–84.

[10] D. Kumar, A class of termination detection algorithms for distributed computations, *Proceedings of the 5th Conference on Foundation of Software Technology and Theoretical Computer Science*, New Delhi, LNCS 206, 1985, 73–100.

[11] T. H. Lai, Termination detection for dynamically distributed systems with non-first-in-first-out communication, *Journal of Parallel and Distributed Computing*, December, 1986, 577–599.

[12] F. Mattern, Algorithms for distributed termination detection, *Distributed Computing*, **2**, 1987, 161–175.

[13] F. Mattern, Global quiescence detection based on credit distribution and recovery, *Information Processing Letters*, **30**(4), 1989, 195–200.

[14] J. Mayo and P. Kearns, Distributed termination detection with roughly synchronized clocks, *Information Processing Letters*, **52**(2), 1994, 105–108.

[15] J. Mayo and P. Kearns, Efficient distributed termination detection with roughly synchronized clocks, *Parallel and Distributed Computing and Systems*, 1995, 305–307.

[16] J. Misra and K. M. Chandy, Termination detection of diffusing computations in communication sequential processes, *ACM Transactions on Programming Languages and Systems*, January, 1982, 37–42.

[17] S. P. Rana, A distributed solution of the distributed termination problem, *Information Processing Letters*, **17**(1), 43–46.

[18] S. Ronn and H. Saikkonen, Distributed termination detection with counters, *Information Processing Letters*, **34**(5), 1990, 223–227.

[19] R. W. Topor, Termination detection for distributed computations, *Information Processing Letters*, **18**(1), 1984, 33–36.

[20] Yu-Chee Tseng, Detecting termination by weight-throwing in a faulty distributed system, *Journal of Parallel Distributed Computing*, **25**(1), 1995, 7–15.

[21] Yu-Chee Tseng and Cheng-Chung Tan, On termination detection protocols in a mobile distributed computing environment, *Proceedings of ICPADS*, 1998, 156–163.

CHAPTER

8 Reasoning with knowledge

In a distributed system, processes make local decisions based on their limited view of the system state. A process learns of new facts when it receives messages from other processes, and can reason only with the additional knowledge available to it. This chapter provides a formal framework in which it is easier to understand the role of knowledge in the system, and how processes can reason with such knowledge. The first three sections are based on the book by Fagin *et al.* [3]. The logic of knowledge, classically termed as *epistemic logic*, is the formal logical analysis of reasoning about knowledge. Epistemic knowledge first received much attention from philosophers in the mid-twentieth century.

8.1 The muddy children puzzle

Consider the classical "muddy children" puzzle of Halpern and Moses [5] and Halpern and Fagin [4]. Imagine there are n children who return from playing outdoors, and k, $k \geq 1$, of the n children have mud on their foreheads. Let Ψ denote the fact "at least one child has a muddy forehead." Assume that each child can see all other children and their foreheads, but not their own forehead. We also assume that the children are intelligent and truthful, and answer any question asked of them, simultaneously. We now consider two scenarios.

In Scenario A, the father who now shows up on the scene, first makes a statement announcing Ψ. We assume that this announcement is heard by everyone, and that everyone is aware that the announcement is being made in their common presence. The father now repeatedly asks the children, "Do you have mud on your forehead?" The first $k-1$ times that the father asks the question, all the children will say "No" and the kth time the father asks the question, the children with mud on their foreheads (henceforth, referred to as the muddy children) will all say "Yes." This can be proved by induction on k.

- If $k = 1$, the single muddy child, seeing no other muddy children and knowing the announcement of Ψ, will conclude on hearing the father's question that he/she is the muddy child.
- If $k = 2$, let the two muddy children be $m1$ and $m2$. The first time the question is asked, neither can answer in the affirmative. But when $m1$ hears the negative answer of $m2$, $m1$ can reason that $m1$ himself must be muddy because otherwise $m2$ would have answered "Yes" in the first round using the logic for the $k = 1$ case. Hence, $m1$ answers "Yes" the second time, and $m2$ who uses analogous reasoning, also answers "Yes."
- We assume the induction hypothesis is true for $k = x$ muddy children.
- For $k = x + 1$ muddy children, the proof is as follows. Each muddy child reasons in the following manner. "If there were x muddy children, then they would all have answered 'Yes' when the question is asked for the xth time. As that did not happen, there must be more than x muddy children, and as I can see only x other muddy children, I myself must also be muddy. So I will answer 'Yes' when the question is asked for the $(x + 1)$th time."

In Scenario B, the father who now shows up on the scene, does *not* make the announcement of Ψ, but repeatedly asks the children, "Do you have mud on your forehead?" All the children repeatedly respond with a "No." This can be shown by induction on q, the number of times the father asks the question, that "no matter how many muddy children there are, all children answer 'No' to the first q questions." For $q = 1$, each child answers "No" because they cannot distinguish between the two situations wherein they do and do not have mud on their forehead. Assume the hypothesis is true for $q = x$. For $q = x + 1$, the situation is unchanged because each child has no further knowledge to distinguish the two situations wherein they do and do not have mud on their forehead.

In Scenario A, the father announced Ψ whereas in Scenario B, the father did not announce Ψ, and the responses of the children were very different. The announcement of Ψ effectively made Ψ *common knowledge* among the children, and this enabled the children to reason differently. The above puzzle introduces the notions of knowledge, levels of knowledge, and common knowledge in a system. We now define these formally and consider how such logic can be adapted to computing systems.

8.2 Logic of knowledge

8.2.1 Knowledge operators

A definition of knowledge requires the identification of an appropriate set of *possible worlds* (also called possible universes or possible configurations), and a family of possible relations between those worlds [3]. In a given global state, the possible worlds at a process denote all the global states that the process believes may be consistent with its local state. These states are expressible as logical formulas.

Fact ϕ can be a primitive proposition or a formula using the usual logical connectives ($\wedge, \vee, \neg$) on primitive propositions, the "knowledge operator" K, and the "everyone knows" operator E. Propositional logic is adequate to cover many interesting scenarios that occur in distributed executions, although first-order and higher-order logics can also be used. The traditional semantics of knowledge, using the K and E operators, were first based on *timed executions*. Intuitively, a process i that knows a fact ϕ is said to have knowledge $K_i(\phi)$, and if "every process in the system knows ϕ," then the system exhibits knowledge $E^1(\phi) = \bigwedge_{i\in N} K_i(\phi)$. A knowledge level of $E^2(\phi)$ indicates that every process knows $E^1(\phi)$, i.e., $E^2(\phi) = E\ (E^1(\phi))$. Inductively, $E^k(\phi) = E^{k-1}\ (E^1(\phi))$ for $k > 1$. Thus, a hierarchy of levels of knowledge $E^j(\phi)(j \in Z^*)$ gets defined, where Z^* is used to denote the set of whole numbers $\{0, 1, 2, 3, \ldots\}$. It can be seen that $E^{k+1}(\phi) \Longrightarrow E^k(\phi)$. Each level in the hierarchy represents a different level of group knowledge among the processes.

In the limiting case, we have the $\bigwedge_{j\in Z*} E^j(\phi)$. Informally, this knowledge of a fact ϕ stands for "everyone knows that everyone knows that everyone knows... (infinitely often) the fact ϕ." This limit is informally called common knowledge of ϕ. Strictly speaking, the epistemic logic is finitary and hence does not allow such infinite conjunctions. On a more formal note, common knowledge of ϕ, denoted as $C(\phi)$, is defined as the knowledge X that is the greatest fixed point of $E(\phi \wedge X)$. Stated differently, common knowledge is a state of knowledge X satisfying the equality, $X = E(\phi \wedge X)$. The theory of fixed points is quite intricate. For our purposes, it suffices if we informally view the fixed point as implying the infinite conjunction $\bigwedge_{j\in Z^*} E^j(\phi)$. Common knowledge of a fact captures the notion of everyone agreeing on the fact, and is therefore an important notion in distributed systems.

8.2.2 The muddy children puzzle again

We now revisit the muddy children puzzle. Assume there are k children $m1, \ldots, mk$ with mud on their forehead. In this system, $E^{k-1}(\Psi)$ is true, but not $E^k(\Psi)$.

In Scenario A, we have the following:

- Consider $k = 1$. Here, the child with the mud on the forehead does not see any muddy child, and hence $E(\Psi)$ is false.
- Consider $k = 2$. Here, $E^1(\Psi)$ is true because every child can see at least one muddy child, and thus $\wedge_{i\in N} K_i(\Psi)$. However, $m1$ can see only one muddy child $m2$ and therefore, in some possible world, $m1$ believes that at $m2$, $K_{m2}(\Psi)$ may be false, i.e., $\neg K_{m1} K_{m2}(\Psi)$, and hence $E^2(\Psi)$ is false.

- Generalizing this reasoning for k muddy children, $E^{k-1}(\Psi)$ is true because $K_{i_1} \dots K_{i_{k-1}}(\Psi)$ is true for all instantations of $i_1, \dots, i_{k-1}$. This is so because everyone can see at least $k-1$ muddy children. However, $E^k(\Psi)$ is false because $K_{i_1} \dots K_{i_k}(\Psi)$ is false when i_1 is instantiated by any of $m1, \dots, mk$. This is so because everyone can see $k-1$ muddy children, and only the $n-k$ clean children can see the k muddy children.

In Scenario B, the only knowledge available in the system is $E^{k-1}(\Psi)$, and this is not enough for the children with mud on their forehead to ever respond affirmatively to the father. In order for the children with mud to be able to respond affirmatively, $E^k(\Psi)$ needs to be true in the system so that the children can use the knowledge progressively step-by-step and answer correctly in the kth round of questioning. How was this $E^k(\Psi)$ achievable in Scenario A? When the father announced Ψ in everyone's common presence, he provided the system $C(\Psi)$ and hence $E^k(\Psi)$. Thus, every child knew Ψ, and every child knew that every child knew Ψ, and every child knew that every child knew that every child knew Ψ, and so on. With this $E^k(\Psi)$ being present in the system, the children could use it progressively round-by-round until, in the kth round, they could answer the father's question correctly.

8.2.3 Kripke structures

A popular approach to defining semantics is in terms of *possible worlds*. This approach is formalized using *Kripke structures* [3].

Definition 8.1 (Kripke structure) *A Kripke structure M for n agents and a set of primitive propositions Φ is a tuple $(S, \pi, \mathcal{K}_1, \dots \mathcal{K}_n)$, where the components of this tuple are as follows:*

1. *S is the set of all consistent states (or possible worlds), with respect to an execution.*
2. *π is an interpretation that associates a truth assignment to each primitive proposition in Φ, for each state $s \in S$. Thus, $\forall s \in S,\ \pi(s) : \Phi \rightarrow \{0, 1\}$.*
3. *$\mathcal{K}_i$ is a binary relation on S giving all the pairs of states that are indistinguishable by P_i.*

A Kripke structure is conveniently viewed as a graph with labeled nodes connected by labeled edges. The set of nodes is the set of states S; the label of node $s \in S$ also gives the primitive propositions that are true and false at s. In our simple example, we assume that Φ contains a single proposition. The logic can be extended to multiple propositions in a straightforward manner. The edge (s, t) is labeled by the identity of every process P_i such that $(s, t) \in \mathcal{K}_i$, i.e., every process P_i that cannot distinguish between states s and t. We assume (for simplicity) that edges are bidirectional and that the $\mathcal{K}$ relations

Figure 8.1 The definitions of regular knowledge [3].

Definition 8.2 (Regular knowledge)

$(M, s) \models \phi$ if and only if ϕ is true in state s in Kripke structure M, i.e., $\pi(s)(\phi) = true$.

Analogously, we can define formulae using conjunctions and negations over primitive propositions.

$(M, s) \models K_i(\phi)$ if and only if $(M, t) \models \phi$, for all states t such that $(s, t) \in \mathcal{K}_i$

$(M, s) \models E^1(\phi)$ if and only if $(M, s) \models \bigwedge_{i\in N} K_i(\phi)$

$(M, s) \models E^{k+1}(\phi)$ for $k \geq 1$ if and only if $(M, s) \models \bigwedge_{i\in N} K_i(E^k(\phi))$, for $k \geq 1$

$(M, s) \models C(\phi)$ if and only if $(M, s) \models \bigwedge_{k\in Z*} E^k(\phi)$

(Distributed Knowledge D**)** $(M, s) \models D(\phi)$ if and only if $(M, t) \models \phi$ for each state t such that $(s, t) \in \cap_i \mathcal{K}_i$

are reflexive, i.e., there is a self-loop at each node. In Section 8.2.4, the muddy children puzzle is used to illustrate the definitions and concepts of this section.

The formal definition of knowledge is now given in Figure 8.1 [3]. Here, "$\models$" denotes the "satisfaction" operator.

This definition of levels of knowledge has a very convenient and useful graph-theoretic representation, as we will illustrate for the muddy children puzzle.

Definition 8.3 (Reachability of states)

1. *A state t is reachable from state s in k steps if there exist states $s_0, s_1, \ldots, s_k$ such that $s_0 = s$, $s_k = t$,* and for all $j \in [0, k-1]$, *there exists some P_i such that $(s_j, s_{j+1}) \in \mathcal{K}_i$.*
2. *A state t is reachable from state s if t is reachable from s in k steps, for some $k > 1$.*

Definition 8.3 defines state reachability in the Kripke structure. The following definitions of knowledge are expressed in terms of reachability of states within the Kripke structure, and can be readily seen to mirror the original definition of knowledge in Figure 8.1.

Theorem 8.1 (Knowledge in terms of reachability of states)

1. $(M, s) \models E^k(\phi)$ *if and only if* $(M, t) \models \phi$ *for each state t that is reachable from state s in k steps.*
2. $(M, s) \models C(\phi)$ *if and only if* $(M, t) \models \phi$ *for each state t that is reachable from state s.*

8.2.4 Muddy children puzzle using Kripke structures

We now illustrate the Kripke structure for the muddy children puzzle. The definitions and concepts of the previous section will be clarified by the example.

Let us assume there are $n = 3$ children; and $k = 2$ children have mud on their forehead. Each of the eight states can be described by a boolean n-vector, where a clean child is denoted by a 0 and a child with mud on their forehead is denoted by a 1. Let us further assume that the actual state is $(1, 1, 0)$. The Kripke structure M is illustrated in Figure 8.2(a).

In the world $(1, 1, 0)$, each child can see that there is at least one other child who has mud on the forehead, and hence $(M, (1, 1, 0)) \models E(\psi)$. From Theorem 8.1, it follows that $(M, (1, 1, 0)) \models \neg E^2(\Psi)$ because the world $(0, 0, 0)$ is 2-reachable from $(1, 1, 0)$ and Ψ is not true in this world. Generalizing this observation in terms of Kripke structures assuming k muddy children, we have that $E^{k-1}(\Psi)$ is true because each world reachable in $k - 1$ hops has at least one "1", implying there is at least one child with a muddy forehead. However, $E^k(\Psi)$ is false because the world $(0, \ldots, 0)$ is reachable in k hops.

Scenario A

Fact Ψ is already known to all children in the state $(1, 1, 0)$. Still, when the father announces Ψ in Scenario A, the state of knowledge changes. Before the father's announcement, child 2 believes the state $(1, 0, 0)$ possible, and in that state $(1, 0, 0)$, child 1 considers the state $(0, 0, 0)$ possible. After the father announces Ψ, it becomes *common knowledge* that one child has a muddy forehead – this change in the group's state of knowledge can be graphically depicted by deleting all the edges connecting state $(0, 0, 0)$. After the father has announced Ψ, even if one child has a muddy forehead, he will not consider the state $(0, 0, 0)$ possible. When the father asks the question the first time and all children respond "No," then all edges connecting to all possible worlds with a single "1" in the state tuple get deleted – this is because if there were only a single child with mud on his/her forehead,

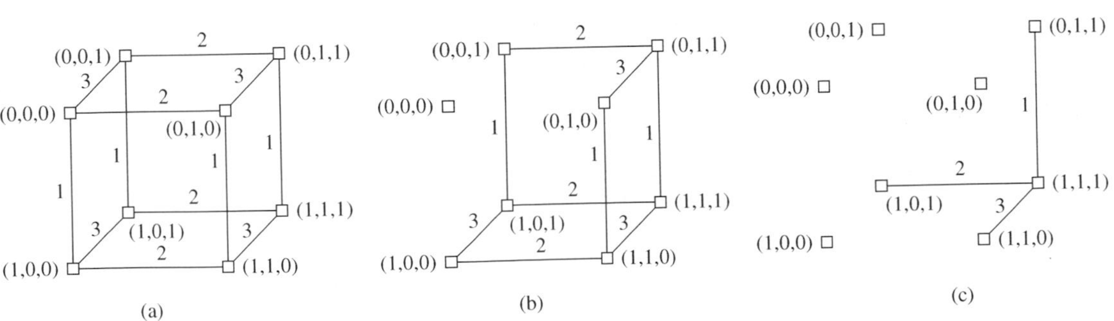

Figure 8.2 Kripke structure for the $n = 3$ muddy children puzzle [3]. Note that the actual state is (1, 1, 0). (a) The entire Kripke structure. (b) After the father announces Ψ. (c) After the first round of questioning and its answers.

he/she would have answered "Yes" in response to the first question. Thus, it is now common knowledge that there are at least two children with muddy foreheads. Generalizing this by induction, when the father asks the question the xth time and all children respond "No," then all edges connecting to all possible worlds with x or fewer than x "1"s in the state tuple get deleted – this is because if there were only x children with mud on their forehead, they would all have answered "Yes" in response to the xth question. It is now common knowledge that there are at least $x+1$ children with mud on their forehead. If there are exactly $x+1$ children with mud on their forehead, those children will all answer "Yes" the $(x+1)$th time the question is asked because they can see exactly x other children with mud on their foreheads. They could not answer "Yes" earlier because they considered a world possible in which they did not have mud on their own forehead.

Graphically, the Kripke structure gets modified in each iteration as follows. If in the iteration, it becomes common knowledge that world t is impossible, then for each node s reachable from the actual state r, any edge (s, t) is deleted. The Kripke structure shown in Figure 8.2(a) gets modified to that shown in part (b) after the father's announcement. That further gets modified as shown in part (c), after the first time the question is asked and all the children reply "No."

Scenario B

In Scenario B, the childrens' state of knowledge never changes and hence the Kripke structure in Figure 8.2(a) never changes, no matter how often the father askes the question. When the father asks the question the first time, all children answer "No" because they each consider both worlds possible – one in which they have mud on their forehead, and one in which they do not. As it is common knowledge even before the father asks the question that the answer is going to be "No," no knowledge is added by the question or the response. Inductively, this argument holds for each round of questioning. Hence, the Kripke structure never changes.

8.2.5 Properties of knowledge

In a formal framework to reason about knowledge, the properties of knowledge must also be specified formally. Although these properties can be specified with different semantics, the most common semantics that are adequate for modeling real distributed systems are given by the axiom system that is historically termed as the S5 system [3]. We first characterize the formulae that are always true. A formula ψ is *valid* in Kripke structure M, denoted $M \models \psi$, if $(M, s) \models \psi$, for all $s \in S$. A formula ψ is *satisfiable* in M if $(M, s) \models \psi$, for some $s \in S$. A formula is *valid* if it is valid in all structures, and it is *satisfiable* if it is satisfiable in some structure. The five axioms of modal logic S5, given in Figure 8.3, are satisfied for all formulas, all structures, and all processes.

Figure 8.3 The axioms of the S5 modal logic [3].

- **Distribution axiom**: $K_i\psi \wedge K_i(\psi \Longrightarrow \phi) \Longrightarrow K_i\phi$.
  Each process knows the logical consequences of its local knowledge. The knowledge operator gets distributed over the implication relation.
- **Knowledge axiom**: $K_i\psi \Longrightarrow \psi$.
  If a process knows a fact, then the fact is necessarily true. If $K_i\psi$ is true in a particular state, then ψ is true in all states that the process considers as possible.
- **Positive introspection axiom**: $K_i\psi \Longrightarrow K_iK_i\psi$.
  A process knows what it knows.
- **Negative introspection axiom**: $\neg K_i\psi \Longrightarrow K_i\neg K_i\psi$.
  A process knows what it does not know.
- **Knowledge generalization rule**: For a valid formula or fact ψ, $K_i\psi$.
  If ψ is true in all possible worlds, then ψ must be true in all the possible worlds with respect to any process and any given world. Hence, $K_i\psi$ must be true in all possible worlds. Here, it is assumed that a process knows all *valid* formulas, which are necessarily true. Note that this rule is different from the rule "$\psi \Longrightarrow K_i\psi$."

8.3 Knowledge in synchronous systems

Classical problems such as the "muddy children" problem and the "cheating husbands" problem, which are widely used to illustrate the theory of knowledge, have been explained in the synchronous system model. The definitions and the treatment of knowledge we have seen thus far was again for synchronous systems.

Common knowledge captures the notion of agreement in distributed systems. What are the various ways by which common knowledge can be attained in a synchronous system?

- By initializing all the processes with common knowledge of the fact.
- By broadcasting the fact to every process in a round of communication, and having all the processes know that the fact is being broadcast. Each process can begin supporting common knowledge from the next round. This is the mechanism that was used by the father when he announced Ψ in the muddy children puzzle in Scenario A.

8.4 Knowledge in asynchronous systems

Here, we adapt the definitions of knowledge given in Figure 8.1 to asynchronous systems [9].

8.4.1 Logic and definitions

In the system model, the possible worlds are the consistent cuts of the set of possible executions in an asynchronous system. Let (a, c) denote a cut c in asynchronous execution a. As each cut also identifies a global state, (a, c) is also used to denote the state of the system after (a, c). $(a, c)_i$ denotes the projection of c on process i, and is also used to denote the state of process i after (a, c). Two cuts c and c' are indistinguishable by process i, denoted $(a, c) \sim_i (a', c')$, if and only if $(a, c)_i = (a', c')_i$. The semantics of knowledge are based on *asynchronous executions*, instead of timed executions.

The modal operator $K_i(\phi)$ means "ϕ is true in all possible consistent global states (cuts) that include process i's local state." Observe that $K_i(\phi)$ is implicitly quantified over all consistent states over all runs, that include i's local state. Similar meanings hold for $E(\phi)$ and $E^k(\phi)$, for $k > 1$. We also define $E^0(\phi)$ to be ϕ for simplicity. Formal definitions of knowledge for asynchronous distributed systems are given in Definition 8.4.

Definition 8.4 (Knowledge in asynchronous systems defined using consistent cuts)

$(a, c) \models \phi$ *if and only if* ϕ *is true in cut* c *of asynchronous execution* a.
$(a, c) \models K_i(\phi)$ *if and only if* $\forall (a', c'), ((a', c') \sim_i (a, c) \Longrightarrow (a', c') \models \phi)$.
$(a, c) \models E^0(\phi)$ *if and only if* $(a, c) \models \phi$.
$(a, c) \models E^1(\phi)$ *if and only if* $(a, c) \models \bigwedge_{i\in N} K_i(\phi)$.
$(a, c) \models E^{k+1}(\phi)$ for $k \geq 1$ *if and only if* $(a, c) \models \bigwedge_{i\in N} K_i(E^k(\phi))$, for $k \geq 1$.
$(a, c) \models C(\phi)$ *if and only if* $(a, c) \models$ *the greatest fixed point knowledge* X *satisfying* $X = E(X \wedge \phi)$. $C(\phi)$ *implies* $\bigwedge_{k\in Z*} E^k(\phi)$.

As knowledge can be known by a process only in a specific local state, we also say that "i knows ϕ in state s_i^x," denoted $s_i^x \models \phi$, as a shorthand for $(\forall (a, c))\ ((a, c)_i = s_i^x \Longrightarrow (a, c) \models \phi)$. Analogously, we define $s_i^x \models K_i(\phi)$ to be $(\forall (a, c))\ ((a, c)_i = s_i^x \Longrightarrow (a, c) \models K_i(\phi))$. Recall that ϕ can be of the form $E^k(\psi)$, for any fact ψ.

Definition 8.5 (Learning) [1,9] Process i learns ϕ *in state* s_i^x *of execution* a *if* i *knows* ϕ *in* s_i^x *and, for all states* s_i^y *in execution* a *such that* $y < x$, i *does not know* ϕ.

We also say that a *process attains* ϕ (in some state) if the process learns ϕ in the present or an earlier state. A fact ϕ is *attained in an execution* a if $\exists c, (a, c) \models \phi$. Observe that a process cannot attain a fact before the fact is attained in an execution. This corresponds to the intuition that even though a fact becomes true in an execution, information may need to be propagated for a process to learn the fact.

Definition 8.6 (Local fact) *A fact* ϕ *is local to process* i *in system* A *if* $A \models (\phi \Longrightarrow K_i\phi)$.

A fact that is not local is a *global fact*. The state of a process, the local clock value of a process, and the local component of the vector timestamp of an event at a process are examples of local facts. The global state of a system and the timestamp of a cut are examples of global facts.

8.4.2 Agreement in asynchronous systems

We consider the following problem: "Two processes that communicate by asynchronous message-passing need to agree on a binary value. Does there exist a protocol that they can follow to reach consensus?" Reaching consensus among a group of processes implies the attainment of common knowledge among that group of processes. We first consider a system where communication is not reliable, implying that messages may be lost in transit.

Theorem 8.2 *There does not exist any protocol for two processes to reach common knowledge about a binary value in an asynchronous message-passing system with unreliable communication.*

An informal argument is as follows. Without loss of generality, we assume that the fact is true at P_i, and the processes P_i and P_j follow a protocol in which they send messages to each other serially. Thus, P_i first sends a message M to P_j informing it of the fact, and because communication is not reliable, P_i needs an acknowledgement ACK1 back to know that P_j has received the message. But then, P_j does not know whether P_i has received the acknowledgement ACK1. Hence, it does not know whether or when P_i will begin supporting the common knowledge, and hence it itself cannot begin supporting the common knowledge. Therefore, P_i needs to send back an acknowledgement ACK2 to P_j to acknowledge the receipt of ACK1. However, P_i now needs an acknowledgement of the delivery of ACK2, similar to its need for an acknowledgement for M. This is a non-terminating argument, and hence this protocol will not work to achieve common knowledge.

More generally, let there exist a protocol with k messages being sent between P_i and P_j, and let this be the *minimal* protocol in the sense of using the minimum number of messages. Then, the sender of the last message asserts common knowledge of the fact even if it does not know whether the message was delivered. Hence, the kth message is redundant, which implies there is a protocol with $k-1$ messages to attain common knowledge. This contradicts the assumption that the minimal protocol requires k messages. Hence, such a protocol does not exist. Using similar reasoning, we also have the following similar impossibility result for reliable asynchronous systems.

Theorem 8.3 *There does not exist any protocol for two processes to reach common knowledge about a binary value in a reliable asynchronous message-passing system without an upper bound on message transmission times.*

Furthermore, even though the upper bound on message transmission time can be guaranteed, a process does not know when to support the common knowledge and hence requires an acknowledgement, and the sender of that acknowledgement will itself require an acknowledgement, and so on.

8.4.3 Variants of common knowledge

Common knowledge captures the notion of agreement among the processes, and hence attaining common knowledge is a fundamental problem in distributed systems. Common knowledge requires the notion of simultaneity of action across the processes. The instantaneous simultaneity attained by tightly synchronized clocks has some margin of error. Given the impossibility of achieving simultaneity, and hence of attaining common knowledge in reliable asynchronous systems, what hope is there? Fortunately, there are weaker versions of common knowledge that can be substituted for regular common knowledge, and these are described below [9].

Epsilon common knowledge

This form of common knowledge corresponds to the processes reaching agreement within ϵ time units. This definition implicitly assumes timed runs as it is not possible to exactly define time units in an asynchronous system. This common knowledge is defined using E^ϵ which denotes "everyone knows within a time duration of ϵ units." Epsilon common knowledge $C^\epsilon(\phi)$ is the greatest fixed point of $X = E^\epsilon(\phi \wedge X)$, where X is the free variable in the greatest fixed-point operator.

Eventual common knowledge

This form of common knowledge corresponds to the processes reaching agreement at some (not necessarily consistent) global state in the execution.

$E^\diamond$ denotes "everyone will eventually know (at some point in their execution)." Eventual common knowledge $C^\diamond(\phi)$ is the greatest fixed point of $X = E^\diamond(\phi \wedge X)$.

Timestamped common knowledge

This form of common knowledge corresponds to the processes reaching agreement at local states having the same local clock value. It is applicable to asynchronous systems. Let $K_i^T(\phi)$ denote the fact that process i knows ϕ at local clock value T. Then $E^T(\phi) = \bigwedge_i K_i^T(\phi)$ and timestamped common knowledge $C^T(\phi)$ is the greatest fixed point of $X = E^T(\phi \wedge X)$. If it is common knowledge that all clocks are always perfectly synchronized, then timestamped common knowledge is equivalent to regular common knowledge.

Concurrent common knowledge

This form of common knowledge corresponds to the processes reaching agreement at local states that belong to a consistent cut. When a process P_i attains concurrent common knowledge of a fact ϕ, it also knows that each other process P_j has also attained the same concurrent common knowledge in its local state which is consistent with P_i's local state.

This form of knowledge is applicable to asynchronous systems and is the most popular form of common knowledge in real systems. Hence, we define it in detail below, and give four protocols for attaining such common knowledge. Here, we note that this variant of common knowledge is incomparable with C^ϵ, $C^\diamond$, and C^T.

8.4.4 Concurrent common knowledge

Concurrent common knowledge is based on the notion of the various processes attaining the common knowledge on a consistent cut [9]. The *possibly* operator[1] P_i in conjunction with the K_i operator is used to formally define such knowlege. $P_i(\phi)$ means "ϕ is true in *some* consistent state in the same asynchronous run, that includes process i's local state." $E^C(\phi)$ is defined as $\bigwedge_{i \in N} K_i(P_i(\phi))$. $E^C(\phi)$ means that every process at the (given) cut knows only that ϕ is true in *some* cut that is consistent with its own local state. By induction, similar meanings can be assigned for higher levels of knowledge. The formal definition of levels of concurrent knowledge E^C is as shown in Definition 8.7.

[1] The notation P_i for this operator is not to be confused with P_i used to denote process i. Also, the semantics of this operator is different from the *Possibly* modality defined on global predicates.

Definition 8.7 (Concurrent knowledge for asynchronous distributed systems) [7,9]

$(a, c) \models \phi$ *if and only if* ϕ *is true in cut* c *of execution* a.
$(a, c) \models K_i(\phi)$ *if and only if* $\forall(a', c'), ((a', c') \sim_i (a, c) \Longrightarrow (a', c') \models \phi)$.
$(a, c) \models P_i(\phi)$ *if and only if* $\exists(a, c'), ((a, c') \sim_i (a, c) \wedge (a, c') \models \phi)$.
$(a, c) \models E^{C^0}(\phi)$ *if and only if* $(a, c) \models \phi$.
$(a, c) \models E^{C^1}(\phi)$ *if and only if* $(a, c) \models \bigwedge_{i \in N} K_i P_i(\phi)$.
$(a, c) \models E^{C^{k+1}}(\phi)$ *for* $k \geq 1$ *if and only if* $(a, c) \models \bigwedge_{i \in N} K_i P_i(E^{C^k}(\phi))$, *for* $k \geq 1$.
$(a, c) \models C^C(\phi)$ *if and only if* $(a, c) \models$ *the greatest fixed point knowledge* X *satisfying* $X = E^C(X \wedge \phi)$.
$C^C(\phi)$ *implies* $\bigwedge_{k \in Z*} (E^C)^k(\phi)$.

The concurrent knowledge definitions are weaker than the corresponding knowledge definitions in Definition 8.4. But for a *local*, *stable* fact, and assuming other processes learn the fact via message chains, it can be seen that the two definitions become equivalent [1,7,9].

If concurrent common knowledge $C^C(\phi)$ is attained at a consistent cut, then (informally speaking) each process at its local cut state knows that "in some state consistent with its own local cut state, ϕ is true *and that* all other process know all this same knowledge (described within quotes)."

Concurrent common knowledge is a necessary and sufficient condition for performing concurrent actions in asynchronous distributed systems, analogous to simultaneous actions and common knowledge in synchronous systems. The form of knowledge underlying many existing protocols involves processes reaching agreement about some property of a consistent global state, defined using *logical time* and *causality*, and can be easily understood in terms of concurrent common knowledge.

Global snapshot algorithms (Chapter 4) can be run concurrently with the underlying computation and can be used to achieve concurrent common knowledge. Snapshot algorithms typically require $|L|$ messages and d time steps, where d is the diameter of the network. More message-efficient snapshot algorithms that need only $O(|N|)$ messages use certain forms of computation inhibition [2] – local or global inhibition, and send inhibition and/or receive inhibition based on network characteristics such as availability of FIFO channels – to reduce the number of messages needed to take a snapshot. Nevertheless, each snapshot requires at least $O(|N|)$ messages and possibly inhibitory delay as overhead.

Specifically, concurrent common knowledge can be attained in an asynchronous system, as shown by the protocols in Algorithms 8.1–8.4. In these protocols, each process P_i must attain $C^C(\phi)$ on a consistent cut, (i) by learning

ϕ, and (ii) by learning that each other process P_j will also attain that exact state of knowledge in a local state that is consistent with P_i's local state in which P_i attains $C^C(\phi)$.

Snapshot-based algorithm

Protocol 1 (Algorithm 8.1) is a form of a global snapshot algorithm, where each process knows that all processes are participating in the same algorithm. It can also be viewed as a variant of the distributed asynchronous breadth-first search algorithm (seen in Chapter 5). Observe that the set of states denoted as *cut state* at each process, and at which the processes begin supporting $C^C(\phi)$, indeed form a consistent set of states.

Complexity

Protocol 1 uses $2l$ messages and a time complexity equal to the diameter of the network.

Three-phase send-inhibitory algorithm

Protocol 2 (Algorithm 8.2) has three phases and uses send-inhibition. It also assumes that the predicate ϕ that becomes true when the protocol is initiated remains true for the duration of the protocol. Observe that send inhibition is necessary to ensure that the set of *cut states* at which the processes begin supporting $C^C(\phi)$ are consistent. A process P_i does not send any message between receiving the *PREPARE* and sending the *CUT* (when it reaches its *cut state*), and receiving the *RESUME* control message. Any message sent by P_i after receiving the *RESUME* message will necessarily be received by any other process P_j after P_j has reached its *cut state*. Hence the *cut states* are guaranteed to be consistent with each other.

Protocol 1 (Snapshot-based algorithm)

(1) At some time when the initiator I knows ϕ:

- it sends a marker $MARKER(I, \phi, CCK)$ to each neighbor P_j, and atomically reaches its *cut state*.

(2) When a process P_i receives for the first time, a message $MARKER(I, \phi, CCK)$ from a process P_j:

- process P_i forwards the message to all of its neighbors except P_j, and atomically reaches its *cut state*.

Algorithm 8.1 Snapshot-based protocol to attain concurrent common knowledge. A process attains $C^C(\phi)$ when it reaches its *cut state*.

Protocol 2 (Three-phase send-inhibitory algorithm).

(1) At some time when the initiator I knows ϕ:

- it sends a marker $PREPARE(I, \phi, CCK)$ to each process P_j.

(2) When a (non-initiator) process receives a marker $PREPARE(I, \phi, CCK)$:

- it begins send-inhibition for non-protocol events;
- it sends a marker $CUT(I, \phi, CCK)$ to the initiator I;
- it reaches its *cut state* at which it attains $C^C(\phi)$.

(3) When the initiator I receives a marker $CUT(I, \phi, CCK)$ from each other process:

- the initiator reaches its *cut state*;
- it sends a marker $RESUME(I, \phi, CCK)$ to all other processes.

(4) When a (non-initiator) process receives a marker $RESUME(I, \phi, CCK)$:

- it resumes sending its non-protocol messages that had been inhibited in step 2.

Algorithm 8.2 Three-phase send-inhibitory protocol to attain concurrent common knowledge. A process attains $C^C(\phi)$ when it reaches its *cut state*.

Complexity

Protocol 2 uses $3(n-1)$ messages and a time complexity of three message hops. However, it is send-inhibitory and requires FIFO channels.

The three-phase send-inhibitory tree algorithm

Protocol 3 (Algorithm 8.3) is a variant of protocol 2. It uses a (Broadcast – Convergecast – Broadcast) sequence on a spanning tree (ST) on the network topology to record the global state along a consistent cut.

Complexity

This message-send inhibitory algorithm requires a total of $3(n-1)$ messages and works in a system with non-FIFO channels.

Inhibitory ring algorithm

Protocol 4 (Algorithm 8.4) assumes that a logical ring is superimposed on the network topology. Each process records its *cut state* when it receives the CUT message, and begins send-inhibition. Therefore a process can infer $(E^C)^i(\phi)$ (for any i) is attained by processes along a consistent cut including the current local state.

Complexity

This message-send inhibitory algorithm requires $2n$ messages, and works in a system with FIFO channels. The time complexity is $2n$ hops.

Algorithms such as the above variants of the classical snapshot algorithm require at least $O(l)$ messages, or $O(n)$ messages and varying degrees of message inhibition *each* time there is a need to achieve concurrent knowledge of some fact.

Protocol 3 (Three-phase send-inhibitory tree algorithm)

Phase I (broadcast) The root initiates *PREPARE* control messages down the ST (spanning tree); when a process receives such a message, it inhibits computation message sends and propagates the received control message down the ST.

Phase II (convergecast) A leaf node initiates this phase after it receives the *PREPARE* control message broadcast in phase I. The leaf reaches and records its *cut state*, and sends a *CUT* control message up the ST. An intermediate (and the root) node reaches and records its *cut state* when it receives such a *CUT* control message from each of its children, and then propagates the control message up the ST.

Phase III (broadcast) The root initiates a broadcast of a *RESUME* control message down the ST after phase II terminates. On receiving such a *RESUME* message, a process resumes inhibited computation message send activity and propagates the control message down the ST.

Algorithm 8.3 Three-phase send-inhibitory tree protocol to attain concurrent common knowledge. A process attains $C^C(\phi)$ when it reaches its *cut state*.

Protocol 4 (Send-inhibitory ring algorithm)

1. Once a fact ϕ about the system state is known to some process, the process atomically reaches its *cut state* and begins supporting $C^C(\phi)$, begins send inhibition, and sends a control message $CUT(\phi)$ along the ring.
2. This $CUT(\phi)$ message announces ϕ. When a process receives the $CUT(\phi)$ message, it reaches its *cut state* and begins supporting $C^C(\phi)$, begins send inhibition, and forwards the message along the ring.
3. When the initiator gets back $CUT(\phi)$, it stops send inhibition, and forwards a *RESUME* message along the ring.
4. When a process receives the *RESUME* message, it stops send-inhibition, and forwards the *RESUME* message along the ring. The protocol terminates when the initiator gets back the *RESUME* it initiated.

Algorithm 8.4 Send-inhibitory ring protocol to attain concurrent common knowledge. A process attains $C^C(\phi)$ when it reaches its *cut state*.

8.5 Knowledge transfer

Formalizing how processes learn facts is done by relating knowledge gain to message chains in the execution [1].

Definition 8.8 (Message chain) *A* message chain *in an execution is a sequence of messages* $\langle m_{i_k}, m_{i_{k-1}}, m_{i_{k-2}}, \ldots, m_{i_1}\rangle$ *such that for all* $0 < j \leq k$, m_{i_j} *is sent by process* i_j *to process* i_{j-1} *and* $receive(m_{i_j}) \prec send(m_{i_{j-1}})$. *A message chain identifies the corresponding process chain* $\langle i_0, i_1, \ldots, i_{k-2}, i_{k-1}, i_k\rangle$.

Definition 8.8 adopts the convention that a process chain lists the processes in an order which is the reverse of the order in which they send the messages in the corresponding message chain. Furthermore, a process chain includes the recipient of the last message sent in the corresponding message chain, and this is the first process in the process chain. A message chain with k messages thus identifies a process chain with $k+1$ processes. Knowledge can be transferred among processes only if a process chain exists among those processes. If ϕ is false in an execution and later P_1 knows that P_2 knows that... P_k knows ϕ, then there must exist a process chain $\langle i_1, i_2, \ldots, i_k\rangle$.

In the system model used thus far, $(a, c)_i$ denotes the projection of c on process i, and is also used to denote the state of process i after (a, c). Two cuts c and c' are indistinguishable by process i, denoted $(a, c) \sim_i (a', c')$, if and only if $(a, c)_i = (a', c')_i$. In the *interleaving model* of the distributed system execution, wherein all the events at the different processes are interleaved to form a global total order, the indistinguishability of different views can be expressed using *isomorphism of executions*. In the following explanation, we assume x, y, z denote executions or execution prefixes in the interleaving model. We let x_p denote the projection of execution x on process p.

Definition 8.9 (Isomorphism of executions) [1]

1. *For all executions x and y, relation* $x[p]y$ *is defined to be true if and only if* $x_p = y_p$.
2. *For all executions x and y and a process group G, relation* $x[G]y$ *is defined to be true if and only if, for all* $p \in G$, $x_p = y_p$.
3. *Let* G_i *be process group i and let* $k > 1$. *Then,* $x[G_0, G_1, \ldots, G_k]z$ *if and only if* $x[G_0, G_1, \ldots, G_{k-1}]y$ *and* $y[G_k]z$.

Two executions are isomorphic with respect to a group of processes if and only if none of the processes in the group can distinguish between the two executions. For Definition 8.9(1) and (2), drawing an analogy with Kripke structures (Definition 8.1), the edges connecting two state nodes (which would correspond to the states after executions x and y) are labeled by all the processes that cannot distinguish between the two states. Thus, for all i

such that $(x, y) \in \mathcal{K}_i$, the edge connecting (x, y) is labeled with P_i. For Definition 8.9(3), analogously in Kripke structures (Definition 8.1), the set of states reachable from x in k steps, denoted z, can be expressed in terms of the set of states reachable from x in $k-1$ steps, denoted y, and the set of states z reachable from states in y in one step. The definition of isomorphism of executions allows an alternate way of reasoning with local views of processes, tailored more for asynchronous distributed computing systems. When a message is received in an execution, the set of executions that are isomorphic can only decrease because now executions that do not contain the corresponding send event can be ruled out. The knowledge operator in the interleaving model is defined as follows [1].

Definition 8.10 (Knowledge operator in the interleaving model) *p knows ϕ at execution x if and only if, for all executions y such that $x[p]y$, ϕ is true at y.*

Theorem 8.3 formally shows in the interleaving model that knowledge is gained sequentially [1].

Theorem 8.3 (Knowledge transfer) *For process groups $G_1, \ldots, G_k$, and executions x and y, $(K_{G_1}K_{G_2}\ldots K_{G_k}(\phi)$ at x and $x[G_1, \ldots, G_k]y) \implies K_{G_k}(\phi)$ at y.*

The theorem can be shown to be true by induction on k, along the lines of the following argument. For $k = 1$, the result is straightforward. Assume the induction hypothesis for $k-1$. For k, we can infer there exists some z such that $x[G_1, \ldots, G_{k-1}]z$ and $z[G_k]y$. From $K_{G_1}K_{G_2}\ldots K_{G_{k-1}}[K_{G_k}(\phi)]$ at x, and from the induction hypothesis, it can be inferred that $K_{G_{k-1}}[K_{G_k}(\phi)]$ at z. Hence, $K_{G_k}(\phi)$ at z. As $z[G_k]y$, $K_{G_k}(\phi)$ at y.

In terms of Kripke structures, Theorem 8.3 states that there is a path from state node $x = s_0$ to state node $y = s_k$, via state nodes $s_1, s_2, \ldots, s_{k-1}$, such that the k edges (s_i, s_{i+1}), $0 \leq i \leq k-1$, on the path are labeled by G_{i+1}.

Theorem 8.4 formalizes the observation that there must exist a message chain $\langle m_{i_k}, m_{i_{k-1}}, m_{i_{k-2}}, \ldots, m_{i_1}\rangle$ in order that a fact ϕ that becomes known to P_k after execution prefix x of y, leads to the state of knowledge $K_1K_2\ldots K_k(\phi)$ after execution y [1].

Theorem 8.4 (Knowledge gain theorem) *For processes $P_1, \ldots, P_k$, and executions x and y, where x is a prefix of y, let*

- *$\neg K_k(\phi)$ at x and $K_1K_2\ldots K_k(\phi)$ at y.*

Then there is a process chain $\langle i_1, \ldots, i_{k-1}, i_k\rangle$ in (x, y).

8.6 Knowledge and clocks

We assume all facts are timestamped (physically or logically) by the time of their becoming true and by the process at which they became true. A *full-information protocol* (FIP) is a protocol in which a process piggybacks all the knowledge it has on outgoing messages, and in which a process adds to its knowledge all the knowledge that is piggybacked on any message it receives [4]. Thus, knowledge always increases when a message is received. The amount of knowledge would keep increasing as the execution proceeds, which may not make FIP protocols a practical way to distribute knowledge.

Facts can always be appropriately encoded as integers. Monotonic facts are facts about a property that keep increasing monotonically (e.g., the latest time of taking a checkpoint at a process). By using a mapping between logical clocks and monotonic facts, information about the monotonic facts can be communicated between processes using logical clock values piggybacked on messages. Being monotonic, all earlier facts can be inferred from the fixed amount of information that is maintained and piggybacked on messages. As a specific example, the vector clock $Clk_i[j]$ indicates the local time at each process P_j, and implicitly that all lower clock values at P_j have occurred. With appropriate encoding, facts about a monotonic property can be represented using vector clocks.

Matrix clocks [6–8,10–12] are an extension of the idea behind vector clocks and contain information about other processes' views of the system execution. A matrix clock is an array of size $n \times n$. Matrix clocks are used to design distributed database protocols [6], fault-tolerant protocols, and protocols to discard obsolete information in distributed databases [11]. They are also used to solve the distributed dictionary and distributed log problems [12]. The rules that process P_i executes atomically to maintain its matrix clock using the *matrix clock protocol* are given in Algorithm 8.5.

Vector clocks can be thought of as imparting knowledge to a process: when $Clk[i] = x$ at process h, process h knows that process i has executed at least x events. Matrix clocks impart one more level of knowledge: when $Clk[i, j] = x$ at process h, process h knows that process i knows that process j has executed at least x events.

1. The jth row of the matrix clock at process P_i, indicated by $Clk_i[j, \cdot]$, gives the latest vector clock value of P_j's clock, as known to P_i.
2. The jth column of the matrix clock at process P_i, indicated by $Clk_i[\cdot, j]$, gives the latest scalar clock values of process P_j, i.e., $Clk[j, j]$, as known to each process in the system.

(local variables)
int $Clk_i[1 \ldots n, 1 \ldots n]$

MC0: $Clk_i[j, k]$ is initialized to 0 for all j and k

MC1: Before process i executes an internal event, it does the following:
$Clk_i[i, i] = Clk_i[i, i] + 1$

MC2: Before process i executes a send event, it does the following:
$Clk_i[i, i] = Clk_i[i, i] + 1$
Send message timestamped by Clk_i.

MC3: When process i receives a message with timestamp T from process j, it does the following:
$(k \in N)$ $Clk_i[i, k] = max(Clk_i[i, k], T[j, k])$;
$(l \in N \setminus \{i\})$ $(k \in N)$, $Clk_i[l, k] = max(Clk_i[l, k], T[l, k])$;
$Clk_i[i, i] = Clk_i[i, i] + 1$;
deliver the message.

Algorithm 8.5 Matrix clocks.

For a vector clock Clk_i, the jth entry $Clk_i[j]$ represents the knowledge $K_i K_j(\phi_j)$, where ϕ_j is the local component of process P_j's clock. For a matrix clock Clk_i, the $[j, k]$th entry $Clk_i[j, k]$ represents the knowledge $K_i K_j K_k(\phi_k)$, where ϕ_k is the local component $Clk_k[k, k]$ of process P_k's clock [7,8].

Vector and matrix clocks are convenient because they are updated without sending any additional messages; knowledge is imparted via the *inhibition-free ambient message-passing* that (i) *eliminates protocol messages* by using piggybacking, and (ii) *diffuses* the latest *knowledge* using only messages, whenever sent, by the underlying execution.

Observe that the vector clock at a process provides knowledge $E^0(\phi)$, where ϕ is a property of the global state (namely, the local scalar clock value of each process). Analogously, observe that a matrix clock at a process P_j gives the knowledge

$$K_j(E^1(\phi)) = K_j(\bigwedge_{i \in N} K_i(\phi)),$$

where ϕ is a property of the global state, namely, the local scalar clock value of each process.

8.7 Chapter summary

Processes in a distributed system can reason only with the partial view they have of the computation. The knowledge at a process is based on the values of its variables and any messages received by the process. The chapter first discussed the role of knowledge by using the muddy children puzzle of Halpern and Moses [5]. To formalize the role of knowledge, several knowledge operators – E (every process knows), K (the process knows), and C (common knowledge) – were introduced, Kripke structures were introduced to formalize these semantics in terms of *possible worlds*. The muddy children puzzle was recast in terms of the more formal Kripke structures.

The definitions of knowledge in synchronous systems and in asynchronous systems were then studied. The fundamental result that common knowledge cannot be attained in an error-free message-passing asynchronous system was then examined. Four weaker versions of common knowledge – epsilion common knowledge, eventual common knowledge, timestamped common knowledge, and concurrent common knowledge – that are achievable in asynchronous systems were then examined. Concurrent common knowledge underlies most of the protocols in asynchronous systems. Several algorithms to achieve concurrent common knowledge were then studied – the snapshot based algorithm, a three-phase send-inhibitory algorithm, an algorithm that use the tree overlay, and one algorithm that uses a logical ring. A section on how processes learn new information (viz, gain new knowledge) considered knowledge transfer and knowledge gain in terms of process chains and isomorphism of execution views. Finally, the relationship between the level of knowledge in message-passing asynchronous systems and size of matrix logical clocks was studied.

8.8 Exercises

Exercise 8.1 In the muddy children puzzle (Section 8.1), if Ψ = "At most k children have mud on the forehead," will the muddy children be able to identify themselves? If yes, in how many rounds of questioning? If not, why not? Analyze this scenario in detail.

Exercise 8.2 There are two black hats and two white hats. One of these hats is hidden away and the color of this hat is not known to anybody. The remaining three hats are placed on the heads of three persons A, B, and C in such a way that none of the persons knows the color of the hat placed on his/her head. Draw a Kripke structure that describes this situation.

Exercise 8.3 In a failure-free asynchronous message-passing system of n processes, process P_i learns a fact ϕ.

1. Devise a simple non-inhibitory protocol using a logical ring along which to pass control messages to achieve the following, and justify your answers. Use timing diagrams to illustrate your answers.
   (a) A protocol to attain $E^2(\phi)$ in the system.
   (b) A protocol so that each process knows $E^2(\phi)$.
2. What is the earliest global time at which all processes know that everyone knows $E^2(\phi)$? How can all the processes know about this time?

Exercise 8.4 In Theorem 8.3, assume that there exists an upper bound on message transmission times. Which (if any) variant of common knowledge can hold in the system? Please state your assumptions clearly to justify the reasoning used in your answer.

Exercise 8.5 Consider the matrix clocks given in Algorithm 8.5. At any point in time after the execution of atomic steps MC0, MC1, MC2, or MC3, what is the minimum number of entries among the n^2 entries of Clk_i that are guaranteed to be replicas of other entries in Clk_i? Identify the exact set(s) of elements of the array Clk_i that will necessarily be identical.

Exercise 8.6 Prove the following. For the equalities, you need to prove the implication in both directions. For each part, first prove the results using the interleaving model, and then prove the results using the partial order model.

(a) $K_i \neg\phi$ implies that $\neg K_i\phi$.
(b) $K_i\ \phi \vee \neg K_i\ \phi$.
(c) $K_i\ \phi \vee K_i\ \neg\phi$, if ϕ is a constant.
(d) $K_i\phi \wedge K_i\psi = K_i(\phi \wedge \psi)$.
(e) $K_i\phi \vee K_i\psi = K_i(\phi \vee \psi)$.
(f) $K_i(\neg K_i\phi) = \neg K_i\phi$.

8.9 Notes on references

The muddy children example is taken from Halpern and Moses [5] and Halpern and Fagin [4]. The discussions on Kripke structures, S5 modal logic, and the definitions of regular knowledge and (regular) common knowledge in synchronous systems (Figure 8.1) are based on an excellent text by Fagin *et al.* [3]. The discussion on local facts, learning, knowledge transfer, and isomorphisms is based on the work by Chandy and Misra [1]. The results (Theorems 8.1 and 8.2) on agreement in asynchronous message-passing systems appear to be folklore. The notion of inhibition was formalized by Critchlow and Taylor [2]. Concurrent common knowledge and protocols to attain it for asynchronous systems were formalized by Panangaden and Taylor [9]. The definitions of epsilon, eventual, and timestamped common knowledge for asynchronous systems are also based on [9]. The definition of knowledge for asynchronous systems (Definition 8.4) is based on Kshemkalyani [7,8]. Matrix clocks are first used by Krishnakumar and Bernstein [6], Wuu and Bernstein [12], and Sarin and Lynch [11], and also studied by Ruget [10]. The relationship between clocks of various dimensions and knowledge was formalized by Kshemkalyani [7,8].

References

[1] K. M. Chandy and J. Misra, How processes learn, *Distributed Computing*, **1**, 1986, 40–52.
[2] C. Critchlow and K. Taylor, The inhibition spectrum and the achievement of causal consistency, *Distributed Computing*, **10**(1), 1996, 11–27.

[3] R. Fagin, J. Halpern, Y. Moses, and M. Vardi, *Reasoning about Knowledge*, Cambridge, MA, MIT Press, 1995.

[4] J. Halpern and R. Fagin, Modeling knowledge and action in distributed systems, *Distributed Computing*, **3**(4), 1989, 139–179.

[5] J. Halpern and Y. Moses, Knowledge and common knowledge in a distributed environment, *Journal of the ACM*, **37**(3), 1990, 549–587.

[6] A. Krishnakumar and A. Bernstein, Bounded ignorance: a technique for increasing concurrency in a replicated system, *ACM Transactions on Database Systems*, **19**(4), 1994, 586–625.

[7] A. Kshemkalyani, The power of logical clock abstractions, *Distributed Computing*, **17**(2), 2004, 131–151.

[8] A. Kshemkalyani, Concurrent knowledge and logical clock abstractions, *Proceedings of the 20th Conference on Foundations of Software Technology and Theoretical Computer Science*, Lecture Notes in Computer Science, 1974, Springer-Verlag, 2000, 489–502.

[9] P. Panangaden and K. Taylor, Concurrent common knowledge: defining agreement for asynchronous systems, *Distributed Computing*, **6**(2), 1992, 73–94.

[10] F. Ruget, Cheaper matrix clocks, *Proceedings of the 8th Workshop on Distributed Algorithms*, 1994, 355–369.

[11] S. Sarin and N. Lynch, Discarding obsolete information in a distributed database system, *IEEE Transactions on Software Engineering*, **13**(1), 1987, 39–46.

[12] G. Wuu and A. Bernstein, Efficient solutions to the replicated log and dictionary problems, *Proceedings of the 3rd ACM Symposium on Principles of Distributed Computing*, 1984, 232–242.

CHAPTER

9 Distributed mutual exclusion algorithms

9.1 Introduction

Mutual exclusion is a fundamental problem in distributed computing systems. Mutual exclusion ensures that concurrent access of processes to a shared resource or data is serialized, that is, executed in a mutually exclusive manner. Mutual exclusion in a distributed system states that only one process is allowed to execute the critical section (CS) at any given time. In a distributed system, shared variables (semaphores) or a local kernel cannot be used to implement mutual exclusion. Message passing is the sole means for implementing distributed mutual exclusion. The decision as to which process is allowed access to the CS next is arrived at by message passing, in which each process learns about the state of all other processes in some consistent way. The design of distributed mutual exclusion algorithms is complex because these algorithms have to deal with unpredictable message delays and incomplete knowledge of the system state. There are three basic approaches for implementing distributed mutual exclusion:

1. Token-based approach.
2. Non-token-based approach.
3. Quorum-based approach.

In the token-based approach, a unique token (also known as the PRIVILEGE message) is shared among the sites. A site is allowed to enter its CS if it possesses the token and it continues to hold the token until the execution of the CS is over. Mutual exclusion is ensured because the token is unique. The algorithms based on this approach essentially differ in the way a site carries out the search for the token. In the non-token-based approach, two or more successive rounds of messages are exchanged among the sites to determine which site will enter the CS next. A site enters the critical section (CS) when an assertion, defined on its local variables, becomes true. Mutual exclusion is enforced because the assertion becomes true only at one site at any given time. In the quorum-based approach, each site requests permission to execute the CS from a subset of sites (called a quorum). The quorums are formed in such a way that when two sites concurrently request access to the CS, at least one site receives both the requests and this site is responsible to make sure that only one request executes the CS at any time.

In this chapter, we describe several distributed mutual exclusion algorithms and compare their features and performance. We discuss relationship among various mutual exclusion algorithms and examine trade-offs among them.

9.2 Preliminaries

In this section, we describe the underlying system model, discuss the requirements that mutual exclusion algorithms should satisfy, and discuss what metrics we use to measure the performance of mutual exclusion algorithms.

9.2.1 System model

The system consists of n sites, $S_1, S_2, \ldots, S_n$. Without loss of generality, we assume that a single process is running on each site. The process at site S_i is denoted by p_i. All these processes communicate asynchronously over an underlying communication network. A process wishing to enter the CS requests all other or a subset of processes by sending REQUEST messages, and waits for appropriate replies before entering the CS. While waiting the process is not allowed to make further requests to enter the CS. A site can be in one of the following three states: requesting the CS, executing the CS, or neither requesting nor executing the CS (i.e., idle). In the "requesting the CS" state, the site is blocked and cannot make further requests for the CS. In the "idle" state, the site is executing outside the CS. In the token-based algorithms, a site can also be in a state where a site holding the token is executing outside the CS. Such state is referred to as the *idle token* state. At any instant, a site may have received several pending requests for CS. A site queues up these requests and serves them one at a time.

We do not make any assumption regarding communication channels if they are FIFO or not. This is algorithm specific. We assume that channels reliably deliver all messages, sites do not crash, and the network does not get partitioned. Some mutual exclusion algorithms are designed to handle such situations. Many algorithms use Lamport-style logical clocks to assign a timestamp to critical section requests. Timestamps are used to decide the priority of requests in case of a conflict. The general rule followed is that the smaller the timestamp of a request, the higher its priority to execute the CS.

We use the following notation: N denotes the number of processes or sites involved in invoking the critical section, T denotes the average message delay, and E denotes the average critical section execution time.

9.2.2 Requirements of mutual exclusion algorithms

A mutual exclusion algorithm should satisfy the following properties:

1. **Safety property** The safety property states that at any instant, only one process can execute the critical section. This is an essential property of a mutual exclusion algorithm.
2. **Liveness property** This property states the absence of deadlock and starvation. Two or more sites should not endlessly wait for messages that will never arrive. In addition, a site must not wait indefinitely to execute the CS while other sites are repeatedly executing the CS. That is, every requesting site should get an opportunity to execute the CS in finite time.
3. **Fairness** Fairness in the context of mutual exclusion means that each process gets a fair chance to execute the CS. In mutual exclusion algorithms, the fairness property generally means that the CS execution requests are executed in order of their arrival in the system (the time is determined by a logical clock).

The first property is absolutely necessary and the other two properties are considered important in mutual exclusion algorithms.

9.2.3 Performance metrics

The performance of mutual exclusion algorithms is generally measured by the following four metrics:

- **Message complexity** This is the number of messages that are required per CS execution by a site.
- **Synchronization delay** After a site leaves the CS, it is the time required before the next site enters the CS (see Figure 9.1). Note that normally one or more sequential message exchanges may be required after a site exits the CS and before the next site can enter the CS.
- **Response time** This is the time interval a request waits for its CS execution to be over after its request messages have been sent out (see Figure 9.2). Thus, response time does not include the time a request waits at a site before its request messages have been sent out.
- **System throughput** This is the rate at which the system executes requests for the CS. If SD is the synchronization delay and E is the average critical section execution time, then the throughput is given by the following equation:

$$\text{System throughput} = \frac{1}{(SD + E)}.$$

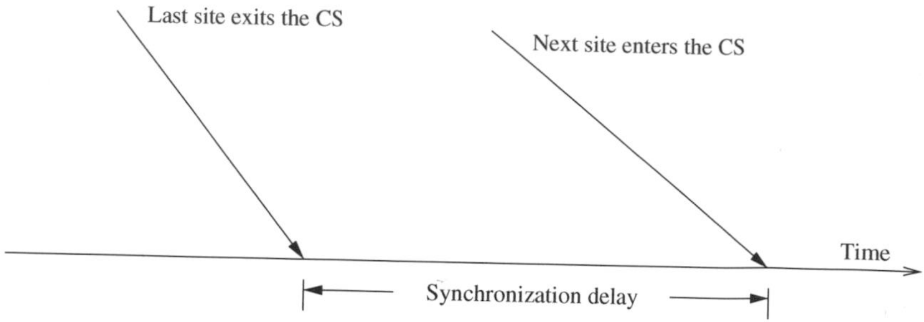

Figure 9.1 Synchronization delay.

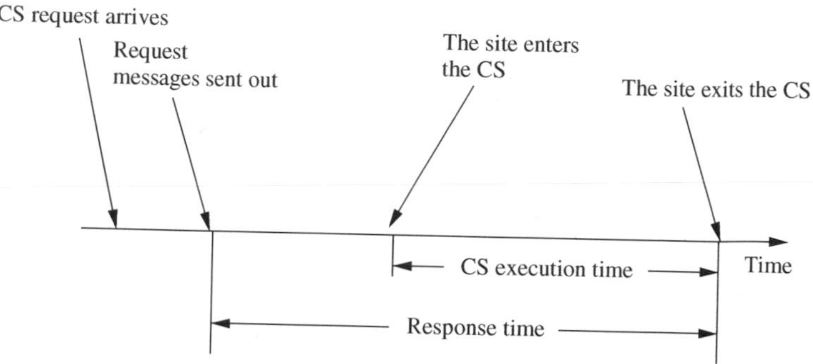

Figure 9.2 Response time.

Generally, the value of a performance metric fluctuates statistically from request to request and we generally consider the average value of such a metric.

Low and high load performance

The load is determined by the arrival rate of CS execution requests. Performance of a mutual exclusion algorithm depends upon the load and we often study the performance of mutual exclusion algorithms under two special loading conditions, viz., "low load" and "high load." Under *low load* conditions, there is seldom more than one request for the critical section present in the system simultaneously. Under *heavy load* conditions, there is always a pending request for critical section at a site. Thus, in heavy load conditions, after having executed a request, a site immediately initiates activities to execute its next CS request. A site is seldom in the idle state in heavy load conditions. For many mutual exclusion algorithms, the performance metrics can be computed easily under low and heavy loads through a simple mathematical reasoning.

Best and worst case performance

Generally, mutual exclusion algorithms have best and worst cases for the performance metrics. In the best case, prevailing conditions are such that a performance metric attains the best possible value. For example, in most mutual exclusion algorithms the best value of the response time is a round-trip message delay plus the CS execution time, $2T + E$. Often for mutual exclusion algorithms, the best and worst cases coincide with low and high loads, respectively. For examples, the best and worst values of the response time are achieved when load is, respectively, low and high; in some mutual exclusion algorithms the best and the worse message traffic is generated at low and heavy load conditions, respectively.

9.3 Lamport's algorithm

Lamport developed a distributed mutual exclusion algorithm (Algorithm 9.1) as an illustration of his clock synchronization scheme [12]. The algorithm is fair in the sense that requests for CS are executed in the order of their timestamps and time is determined by logical clocks. When a site processes a request for the CS, it updates its local clock and assigns the request a timestamp. The algorithm executes CS requests in the increasing order of timestamps. Every site S_i keeps a queue, *request_queue*$_i$, which contains mutual exclusion requests ordered by their timestamps. (Note that this queue is different from the queue that contains local requests for CS execution awaiting their turn.) This algorithm requires communication channels to deliver messages in FIFO order.

Requesting the critical section

- When a site S_i wants to enter the CS, it broadcasts a REQUEST(ts_i, i) message to all other sites and places the request on *request_queue*$_i$. ((ts_i, i) denotes the timestamp of the request.)
- When a site S_j receives the REQUEST(ts_i, i) message from site S_i, it places site S_i's request on *request_queue*$_j$ and returns a timestamped REPLY message to S_i.

Executing the critical section

Site S_i enters the CS when the following two conditions hold:

L1: S_i has received a message with timestamp larger than (ts_i, i) from all other sites.

L2: S_i's request is at the top of *request_queue*$_i$.

Releasing the critical section

- Site S_i, upon exiting the CS, removes its request from the top of its request queue and broadcasts a timestamped RELEASE message to all other sites.
- When a site S_j receives a RELEASE message from site S_i, it removes S_i's request from its request queue.

Algorithm 9.1 Lamport's algorithm.

When a site removes a request from its request queue, its own request may come at the top of the queue, enabling it to enter the CS. Clearly, when a site receives a REQUEST, REPLY, or RELEASE message, it updates its clock using the timestamp in the message.

Correctness

Theorem 9.1 *Lamport's algorithm achieves mutual exclusion.*

Proof Proof is by contradiction. Suppose two sites S_i and S_j are executing the CS concurrently. For this to happen conditions L1 and L2 must hold at both the sites *concurrently*. This implies that at some instant in time, say t, both S_i and S_j have their own requests at the top of their *request_queues* and condition L1 holds at them. Without loss of generality, assume that S_i's request has smaller timestamp than the request of S_j. From condition L1 and FIFO property of the communication channels, it is clear that at instant t the request of S_i must be present in $request\_queue_j$ when S_j was executing its CS. This implies that S_j's own request is at the top of its own *request_queue* when a smaller timestamp request, S_i's request, is present in the $request\_queue_j$ – a contradiction! Hence, Lamport's algorithm achieves mutual exclusion. □

Theorem 9.2 *Lamport's algorithm is fair.*

Proof A distributed mutual exclusion algorithm is fair if the requests for CS are executed in the order of their timestamps. The proof is by contradiction. Suppose a site S_i's request has a smaller timestamp than the request of another site S_j and S_j is able to execute the CS before S_i. For S_j to execute the CS, it has to satisfy the conditions L1 and L2. This implies that at some instant in time S_j has its own request at the top of its queue and it has also received a message with timestamp larger than the timestamp of its request from all other sites. But *request_queue* at a site is ordered by timestamp, and according to our assumption S_i has lower timestamp. So S_i's request must be placed ahead of the S_j's request in the $request\_queue_j$. This is a contradiction. Hence Lamport's algorithm is a fair mutual exclusion algorithm. □

Example In Figures 9.3 to 9.6, we illustrate the operation of Lamport's algorithm. In Figure 9.3, sites S_1 and S_2 are making requests for the CS and send out REQUEST messages to other sites. The timestamps of the requests are (1,1) and (1,2), respectively. In Figure 9.4, both the sites S_1 and S_2 have received REPLY messages from all other sites. S_1 has its request at the top of its *request_queue* but site S_2 does not have its request at the top of its *request_queue*. Consequently, site S_1 enters the CS. In Figure 9.5, S_1 exits and sends RELEASE mesages to all other sites. In Figure 9.6, site S_2 has received REPLY from all other sites and also received a RELEASE message

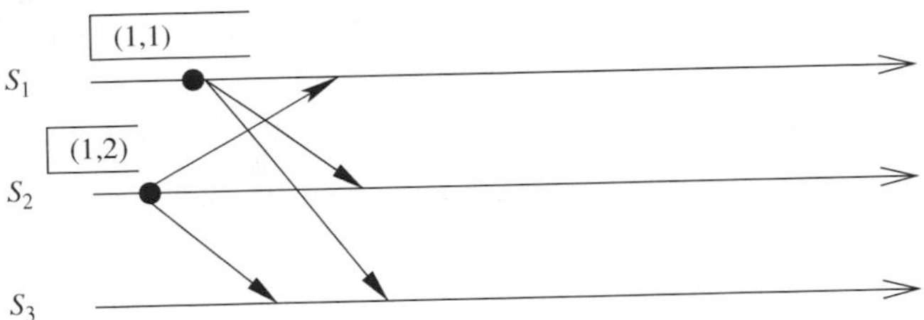

Figure 9.3 Sites S_1 and S_2 are Making Requests for the CS.

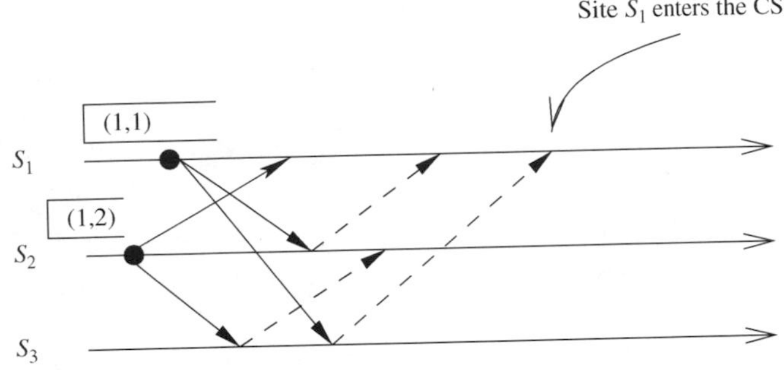

Figure 9.4 Site S_1 enters the CS.

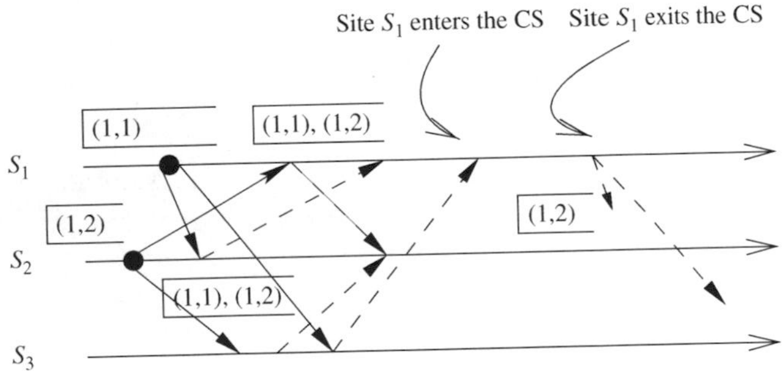

Figure 9.5 Site S_1 exits the CS and sends RELEASE messages.

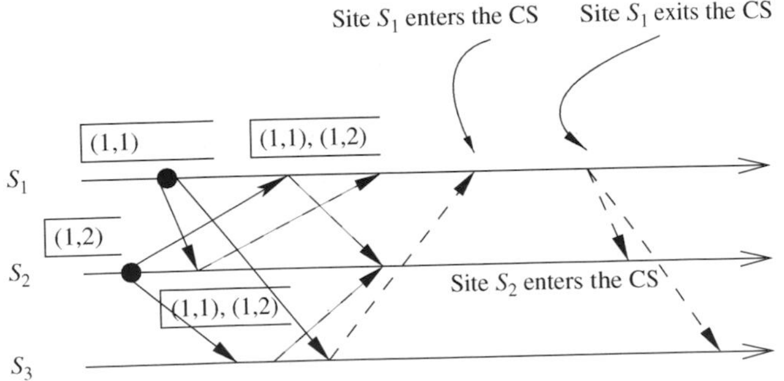

Figure 9.6 Site S_2 enters the CS.

from site S_1. Site S_2 updates its *request_queue* and its request is now at the top of its *request_queue*. Consequently, it enters the CS next.

Performance

For each CS execution, Lamport's algorithm requires $(N-1)$ REQUEST messages, $(N-1)$ REPLY messages, and $(N-1)$ RELEASE messages. Thus, Lamport's algorithm requires $3(N-1)$ messages per CS invocation. The synchronization delay in the algorithm is T.

An optimization

In Lamport's algorithm, REPLY messages can be omitted in certain situations. For example, if site S_j receives a REQUEST message from site S_i after it has sent its own REQUEST message with a timestamp higher than the timestamp of site S_i's request, then site S_j need not send a REPLY message to site S_i. This is because when site S_i receives site S_j's request with a timestamp higher than its own, it can conclude that site S_j does not have any smaller timestamp request which is still pending (because communication channels preserve FIFO ordering).

With this optimization, Lamport's algorithm requires between $3(N-1)$ and $2(N-1)$ messages per CS execution.

9.4 Ricart–Agrawala algorithm

The Ricart–Agrawala [21] algorithm (Algorithm 9.2) does not require communication channels to be FIFO. The algorithm uses two types of messages: REQUEST and REPLY. A process sends a REQUEST message to all other processes to request their permission to enter the critical section. A process sends a REPLY message to a process to give its permission to that process. Processes use Lamport-style logical clocks to assign a timestamp to critical section requests. Timestamps are used to decide the priority of requests in case of conflict – if a process p_i that is waiting to execute the critical section receives a REQUEST message from process p_j, then if the priority of p_j's request is lower, p_i defers the REPLY to p_j and sends a REPLY message to p_j only after executing the CS for its pending request. Otherwise, p_i sends a REPLY message to p_j immediately, provided it is currently not executing the CS. Thus, if several processes are requesting execution of the CS, the highest priority request succeeds in collecting all the needed REPLY messages and gets to execute the CS.

Each process p_i maintains the request-deferred array, RD_i, the size of which is the same as the number of processes in the system. Initially, $\forall i\ \forall j$:

$RD_i[j] = 0$. Whenever p_i defers the request sent by p_j, it sets $RD_i[j] = 1$, and after it has sent a REPLY message to p_j, it sets $RD_i[j] = 0$.

Requesting the critical section

(a) When a site S_i wants to enter the CS, it broadcasts a timestamped REQUEST message to all other sites.

(b) When site S_j receives a REQUEST message from site S_i, it sends a REPLY message to site S_i if site S_j is neither requesting nor executing the CS, or if the site S_j is requesting and S_i's request's timestamp is smaller than site S_j's own request's timestamp. Otherwise, the reply is deferred and S_j sets $RD_j[i] := 1$.

Executing the critical section

(c) Site S_i enters the CS after it has received a REPLY message from every site it sent a REQUEST message to.

Releasing the critical section

(d) When site S_i exits the CS, it sends all the deferred REPLY messages: $\forall j$ if $RD_i[j] = 1$, then S_i sends a REPLY message to S_j and sets $RD_i[j] := 0$.

Algorithm 9.2 The Ricart–Agrawala algorithm.

When a site receives a message, it updates its clock using the timestamp in the message. Also, when a site takes up a request for the CS for processing, it updates its local clock and assigns a timestamp to the request. In this algorithm, a site's REPLY messages are blocked only by sites that are requesting the CS with higher priority (i.e., smaller timestamp). Thus, when a site sends out deferred REPLY messages, the site with the next highest priority request receives the last needed REPLY message and enters the CS. Execution of the CS requests in this algorithm is always in the order of their timestamps.

Correctness

Theorem 9.3 *Ricart–Agrawala algorithm achieves mutual exclusion.*

Proof Proof is by contradiction. Suppose two sites S_i and S_j are executing the CS concurrently and S_i's request has higher priority (i.e., smaller timestamp) than the request of S_j. Clearly, S_i received S_j's request after it has made its own request. (Otherwise, S_i's request will have lower priority.) Thus, S_j can concurrently execute the CS with S_i only if S_i returns a REPLY to S_j (in response to S_j's request) before S_i exits the CS. However, this is impossible because S_j's request has lower priority. Therefore, the Ricart–Agrawala algorithm achieves mutual exclusion. □

In the Ricart–Agrawala algorithm, for every requesting pair of sites, the site with higher priority request will always defer the request of the lower priority site. At any time only the highest priority request succeeds in getting all the needed REPLY messages.

Example Figures 9.7 to 9.10 illustrate the operation of the Ricart–Agrawala algorithm. In Figure 9.7, sites S_1 and S_2 are each making requests for the

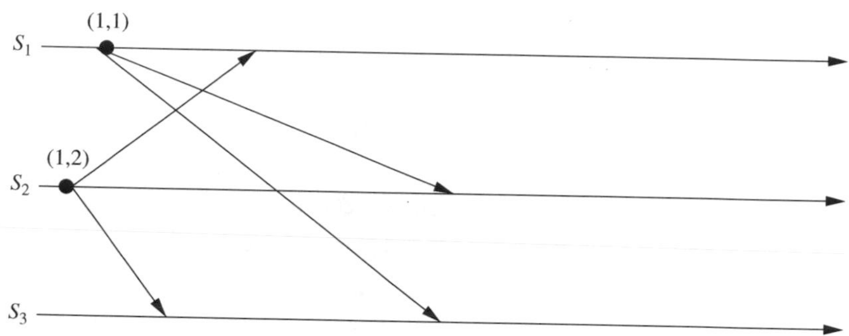

Figure 9.7 Sites S_1 and S_2 each make a request for the CS.

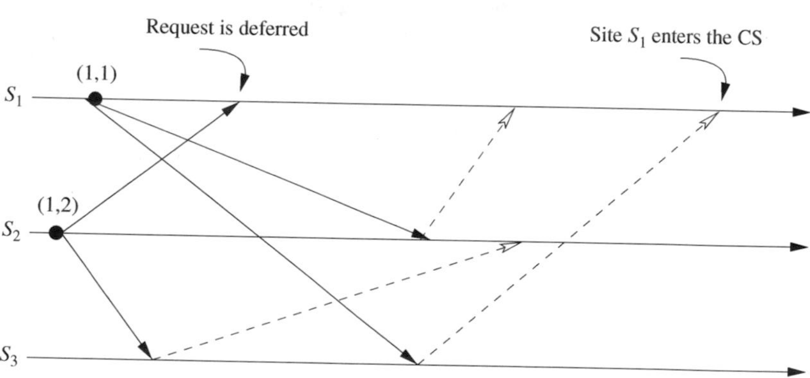

Figure 9.8 Site S_1 enters the CS.

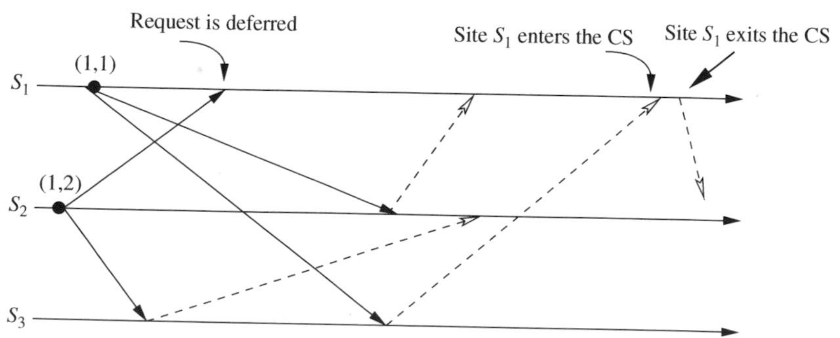

Figure 9.9 Site S_1 exits the CS and sends a REPLY message to S_2's deferred request.

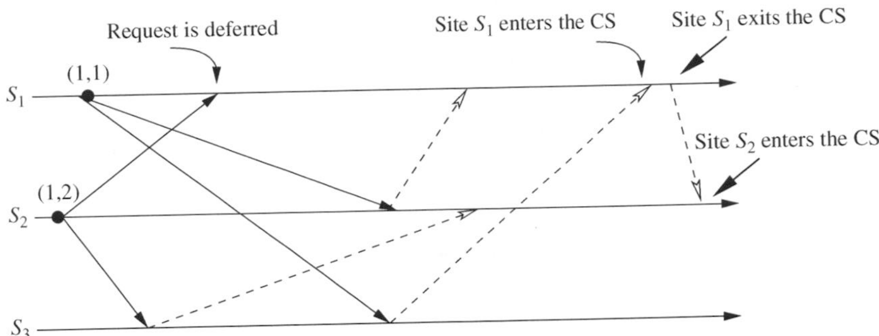

Figure 9.10 Site S_2 enters the CS.

CS and sending out REQUEST messages to other sites. The timestamps of the requests are (1,1) and (1,2), respectively. In Figure 9.8, S_1 has received REPLY messages from all other sites and, consequently, enters the CS. In Figure 9.9, S_1 exits the CS and sends a REPLY mesage to site S_2. In Figure 9.10, site S_2 has received REPLY from all other sites and enters the CS next.

Performance

For each CS execution, the Ricart–Agrawala algorithm requires $(N-1)$ REQUEST messages and $(N-1)$ REPLY messages. Thus, it requires $2(N-1)$ messages per CS execution. The synchronization delay in the algorithm is T.

9.5 Singhal's dynamic information-structure algorithm

Most mutual exclusion algorithms use a static approach to invoke mutual exclusion, i.e., they always take the same course of actions to invoke mutual exclusion no matter what is the state of the system. A problem with these algorithms is the lack of efficiency because these algorithms fail to exploit the changing conditions in the system. Note that an algorithm can exploit dynamic conditions of the system to optimize the performance.

For example, if few sites are invoking mutual exclusion very frequently and other sites invoke mutual exclusion much less frequently, then a frequently invoking site need not ask for the permission of less frequently invoking site every time it requests an access to the CS. It only needs to take permission from all other frequently invoking sites. Singhal [28] developed an adaptive mutual exclusion algorithm based on this observation. The information-structure of the algorithm evolves with time as sites learn about the state of the system through messages. Dynamic information-structure mutual exclusion algorithms are attractive because they can adapt to fluctuating system conditions to optimize the performance.

The design of such adaptive mutual exclusion algorithms is challenging and we list some of the design challenges next:

- How does a site efficiently know what sites are currently actively invoking mutual exclusion?
- When a less frequently invoking site needs to invoke mutual exclusion, how does it do it?
- How does a less frequently invoking site makes a transition to more frequently invoking site and vice-versa?
- How do we ensure that mutual exclusion is guaranteed when a site does not take the permission of every other site?
- How do we ensure that a dynamic mutual exclusion algorithm does not waste resources and time in collecting systems state, offsetting any gain?

System model

We consider a distributed system consisting of n autonomous sites, say S_1, $S_2, \ldots, S_n$, which are connected by a communication network. We assume that the sites communicate completely by message passing. Message propagation delay is finite but unpredictable and, between any pair of sites, messages are delivered in the order they are sent. For the ease of presentation, we assume that the underlying communication network is reliable and sites do not crash. However, methods have been proposed for recovery from message losses and site failures.

Data structures

The information-structure at a site S_i consists of two sets. The first set R_i, called the *request set*, contains the sites from which S_i must acquire permission before executing CS. The second set I_i, called the *inform set*, contains the sites to which S_i must send its permission to execute CS after executing its CS.

Every site S_i maintains a logical clock C_i, which is updated according to Lamport's rules. Every request for CS execution is assigned a timestamp which is used to determine its priority. The smaller the timestamp of a request, the higher its priority. Every site maintains three boolean variables to denote the state of the site: *Requesting*, *Executing*, and *My_priority*. *Requesting* and *Executing* are true if and only if the site is requesting or executing CS, respectively. *My_priority* is true if the pending request of S_i has priority over the current incoming request.

Initialization

The system starts in the following initial state:

For a site S_i ($i = 1$ to n),

$$R_i := \{S_1, S_2, \ldots, S_i - 1, S_i\}$$
$$I_i := \{S_i\}$$
$$C_i := 0$$
$$Requesting = Executing := \text{False}$$

Thus, initially site S_i, $1 \leq i \leq n$, sends request messages only to sites S_i, $S_i - 1, \ldots, S_1$. If we stagger sites S_n to S_1 from left to right, then the initial system state has the following two properties:

1. Each site requests permission from all the sites to its right and from no site to its left. Conversely, for a site, all the sites to its left asks for its permission and no site to its right asks for its permission. Or putting together, for a site, only all the sites to its left will ask for its permission and it will ask for the permission of only all the sites to its right. Therefore, every site S_i divides all the sites into two disjoint groups: all the sites in the first group request permission from S_i, and S_i requests permission from all the sites in the second group. This property is important for enforcing mutual exclusion.
2. The cardinality of R_i decreases in a stepwise manner from left to right. Due to this reason, this configuration has been called "staircase pattern" in topological sense [26].

9.5.1 Description of the algorithm

Site S_i executes the three steps shown in Algorithm 9.3 to invoke mutual exclusion. The REQUEST message handler at a site processes incoming REQUEST messages. It takes actions such as updating the information-structure and sending REQUEST/REPLY messages to other sites. The REQUEST message handler at site S_i is given in Algorithm 9.3. The REPLY message handler at a site processes incoming REPLY messages. It updates the information-structure. The REPLY message handler at site S_i is given in Algorithm 9.3. Note that the REQUEST and REPLY message handlers and the steps of the algorithm access shared data structures, viz., C_i, R_i, and I_i. To guarantee the correctness, it's important that execution of the REQUEST and REPLY message handlers and all three steps of the algorithm (except "wait for $R_i = \emptyset$ to hold" in step 1) mutually exclude each other.

```
Step 1: (Request critical section)
          Requesting := true;
          C_i := C_i + 1;
          Send REQUEST(C_i, i) message to all sites in R_i;
          Wait until R_i = Ø; /* Wait until all sites in R_i have sent
                                 a reply to S_i */
          Requesting := false;
Step 2: (Execute critical section)
          Executing := true;
          Execute CS;
          Executing := false;
Step 3: (Release critical section)
          For every site S_k in I_i (except S_i) do
                Begin
                      I_i := I_i - {S_k};
                      Send REPLY(C_i, i) message to S_k;
                      R_i := R_i + {S_k}
                End
REQUEST message handler:
   /* Site S_i is handling message REQUEST(c, j) */
C_i := max{C_i, c};
Case

Requesting = true:
Begin
if My_priority then I_i := I_i + {S_j}
/*My_Priority is true if the pending request of S_i has priority over the incoming
request */
Else
   Begin
       Send REPLY(C_i, i) message to S_j;
       If not (S_j ∈ R_i) then
           Begin
               R_i := R_i + {S_j};
               Send REQUEST(C_i, i) message to site S_j;
           End;
   End;
End;

Executing = true: I_i := I_i + {S_j};
Executing = false ∧ Requesting = false:
Begin
R_i := R_i + {S_j};
Send REPLY(C_i, i) message to S_j;
End;

REPLY message handler:
  /* Site S_i is handling a message REPLY(c, j) */
Begin
C_i := max{C_i, c};
R_i := R_i - {S_j};
End;
```

Algorithm 9.3 Singhal's dynamic information-structure algorithm [28].

An explanation of the algorithm

At high level, S_i acquires permission to execute the CS from all sites in its request set R_i and it releases the CS by sending a REPLY message to all sites in its inform set I_i.

If site S_i, which itself is requesting the CS, receives a higher priority REQUEST message from a site S_j, then S_i takes the following actions: (i) S_i immediately sends a REPLY message to S_j, (ii) if S_j is not in R_i, then[1] S_i also sends a REQUEST message to S_j, and (iii) S_i places an entry for S_j in R_i. Otherwise (i.e., if the request of S_i has priority over the request of S_j), S_i places an entry for S_j into I_i so that S_j can be sent a REPLY message when S_i finishes with the execution of the CS.

If S_i receives a REQUEST message from S_j when it is executing the CS, then it simply puts S_j in I_i so that S_j can be sent a REPLY message when S_i finishes with the execution of the CS. If S_i receives a REQUEST message from S_j when it is neither requesting nor executing the CS, then it places an entry for S_j in R_i and sends S_j a REPLY message.

Rules for information exchange and updating request and inform sets are such that the staircase pattern is preserved in the system even after the sites have executed the CS any number of times. However, the positions of sites in the staircase pattern change as the system evolves. (For a proof of this, see [28].) The site to execute CS last positions itself at the right end of the staircase pattern.

9.5.2 Correctness

We informally discuss why the algorithm achieves mutual exclusion and why it is free from deadlocks. For a formal proof, the readers are referred to [27].

Achieving mutual exclusion

Note that the initial state of the information-structure satisfies the following condition: for every S_i and S_j, either $S_j \in R_i$ or $S_i \in R_j$. Therefore, if two sites request CS, one of them will always ask for the permission of the another. However, this is not sufficient for mutual exclusion [28]. Whenever there is a conflict between two sites (i.e., they concurrently invoke mutual exclusion), the sites dynamically adjust their request sets such that both request permission of each other satisfying the condition for mutual exclusion. This is a nice feature of the algorithm because if the information-structures of the sites satisfy the condition for mutual exclusion all the

[1] Absence of S_j from R_i implies that S_i has not previously sent a REQUEST message to S_j. This is the reason why S_i also sends a REQUEST message to S_j when it receives a REQUEST message from S_j. This step is also required to preserve the staircase pattern of the information-structure of the system.

time, the sites will exchange more messages. Instead, it is more desirable to dynamically adjust the request set of the sites as and when needed to insure mutual exclusion because it optimizes the number of messages exchanged.

Freedom from deadlocks

In the algorithm, each request is assigned a globally unique timestamp which determines its priority. The algorithm is free from deadlocks because sites use timestamp ordering (which is unique system wide) to decide request priority and a request is blocked only by higher priority requests.

Example Consider a system with five sites $S_1, \ldots, S_5$. Suppose S_2 and S_3 want to enter the CS concurrently, and they both send appropriate request messages. S_3 sends a request message to sites in its Request set – $\{S_1, S_2\}$, and S_2 sends a request message to the only site in its Request set – $\{S_1\}$. There are three possible scenarios:

1. If timestamp of S_3's request is smaller, then on receiving S_3's request, S_2 sends a REPLY message to S_3. S_2 also adds S_3 to its Request set and sends S_3 a REQUEST message. On receiving a REPLY message from S_2, S_3 removes S_2 from its Request set. S_1 sends a REPLY to both S_2 and S_3 because it is neither requesting to enter the CS nor executing the CS. S_1 adds S_2 and S_3 to its Request set because any one of these sites could possibly be in the CS when S_1 requests for an entry into CS in the future. On receiving S_1's REPLY message, S_3 removes S_1 from its Request set and, since it has REPLY messages from all sites in its (initial) Request set, it enters the CS.
2. If timestamp of S_3's request is larger, then on receiving S_3's request, S_2 adds S_3 to its Inform set. When S_2 gets a REPLY from S_1, it enters the CS. When S_2 relinquishes the CS, it informs S_3 (the i.d. of S_3 is present in S_2's Inform set) about its consent to enter the CS. Then, S_2 removes S_3 from its Inform set and add S_3 to its Request set. This is logical because S_3 could be executing in CS when S_2 requests a "CS entry" permission in the future.
3. If S_2 receives a REPLY from S_1 and starts executing CS before S_3's REQUEST reaches S_2, S_2 simply adds S_3 to its Inform set, and sends S_3 a REPLY after exiting the CS.

9.5.3 Performance analysis

The synchronization delay in the algorithm is T. Below, we compute the message complexity in low and heavy loads.

Low load condition

In the case of low traffic of CS requests, most of the time only one or no request for the CS will be present in the system. Consequently, the staircase pattern will re-establish between two sucssive requests for CS and there will seldom be an interference among the CS requests from different sites. In the staircase configuration, the cardinality of the request sets of the sites will be 1, 2, . . . , $(n-1)$, n, respectively, from right to left. Therefore, when the traffic of requests for CS is low, sites will send 0, 1, 2, . . . , $(n-1)$ number of REQUEST messages with equal likelihood (assuming uniform traffic of CS requests at sites). Therefore, the mean number of REQUEST messages sent per CS execution for this case will be $= (0+1+2+ \ldots +(n-1))/n = (n-1)/2$. Since a REPLY message is returned for every REQUEST message, the average number of messages exchanged per CS execution will be $2^*(n-1)/2 = (n-1)$.

Heavy load condition

When the rate of CS requests is high, all the sites will always have a pending request for CS execution. In this case, a site receives on average $(n-1)/2$ REQUEST messages from other sites while waiting for its REPLY messages. Since a site sends REQUEST messages only in response to REQUEST messages of higher priority, on average it will send $(n-1)/4$ REQUEST messages while waiting for REPLY messages. Therefore, the average number of messages exchanged per CS execution in high demand will be $2^*[(n-1)/2+(n-1)/4] = 3^*(n-1)/2$.

9.5.4 Adaptivity in heterogeneous traffic patterns

An interesting feature of the algorithm is that its information-structure adapts itself to the environments of heterogeneous traffic of CS requests and to statistical fluctuations in traffic of CS requests to optimize the performance (the number of messages exchanged per CS execution). In non-uniform traffic environments, sites with higher traffic of CS requests will position themselves towards the right end of the staircase pattern. That is, sites with higher traffic of CS requests will tend to have lower cardinality of their request sets. Also, at a high traffic site S_i, if $S_j \in R_i$, then S_j is also a high traffic site (this comes intuitively because all high traffic sites will cluster towards the right end of the staircase). Consequently, high traffic sites will mostly send REQUEST messages only to other high traffic sites and will seldom send REQUEST messages to sites with low traffic. This adaptivity results in a reduction in the number of messages as well as in the delay in granting CS in environments of heterogeneous traffic.

9.6 Lodha and Kshemkalyani's fair mutual exclusion algorithm

Lodha and Kshemkalyani's algorithm [13] (Algorithm 9.4) decreases the message complexity of the Ricart–Agrawala algorithm by using the following interesting observation: when a site is waiting to execute the CS, it need not receive REPLY messages from every other site. To enter the CS, a site only needs to receive a REPLY message from the site whose request just precedes its request in priority. For example, if sites $S_{i_1}, S_{i_2}, ..S_{i_j}$ have a pending request for CS and the request of S_{i_1} has the highest priority and that of S_{i_j} has the lowest priority and the priority of requests decreases from S_{i_1} to S_{i_j}, then a site S_{i_k} only needs a REPLY message from site $S_{i_{k-1}}$, $1 < k \leq j$ to enter the CS.

9.6.1 System model

Each request is assigned a priority *ReqID* and requests for CS access are granted in the order of decreasing priority. We will defer the details of what *ReqID* is composed of to later sections. The underlying communication network is assumed to be error free.

Definition 9.1 *R_i and R_j are concurrent iff P_i's REQUEST message is received by P_j after P_j has made its request and P_j's REQUEST message is received by P_i after P_i has made its request.*

Definition 9.2 *Given R_i, we define the concurrency set of R_i as follows: $CSet_i = \{R_j \mid R_i \text{ is concurrent with } R_j\} \cup \{R_i\}$.*

9.6.2 Description of the algorithm

Algorithm 9.4 uses three types of messages (REQUEST, REPLY, and FLUSH) and obtains savings on the number of messages exchanged per CS access by assigning multiple purposes to each. For the purpose of blocking a mutual exclusion request, every site S_i has a data structure called *local_request_queue* (denoted as LRQ_i), which contains all concurrent requests made with respect to S_i's request, and these requests are ordered with respect to their priority.

All requests are totally ordered by their priorities and the priority is determined by the timestamp of the request. Hence, when a process receives a REQUEST message from some other process, it can immediately determine if it is allowed to access the CS before the requesting process or after it.

In this algorithm, messages play multiple roles and this will be discussed first.

Multiple uses of a REPLY message

1. A REPLY message acts as a reply from a process that is not requesting.
2. A REPLY message acts as a collective reply from processes that have higher priority requests.

A REPLY(R_j) from a process P_j indicates that R_j is the request made by P_j for which it has executed the CS. It also indicates that all the requests with priority $\geq$ priority of R_j have finished executing CS and are no longer in contention.

Thus, in such situations, a REPLY message is a logical reply and denotes a collective reply from all processes that had made higher priority requests.

Uses of a FLUSH message

Similar to a REPLY message, a FLUSH message is a logical reply and denotes a collective reply from all processes that had made higher priority requests. After a process has exited the CS, it sends a FLUSH message to a process requesting with the next highest priority, which is determined by looking up the process's local request queue. When a process P_i finishes executing the CS, it may find a process P_j in one of the following states:

1. R_j is in the local queue of P_i and located in some position after R_i, which implies that R_j is concurrent with R_i.
2. P_j had replied to R_i and P_j is now requesting with a lower priority. (Note that in this case R_i and R_j are not concurrent.)
3. P_j's request had higher priority than P_i's (implying that it had finished the execution of the CS) and is now requesting with a lower priority. (Note that in this case R_i and R_j are not concurrent.)

A process P_i, after executing the CS, sends a FLUSH message to a process identified in state 1 above, which has the next highest priority, whereas it sends REPLY messages to the processes identified in states 2 and 3 as their requests are not concurrent with R_i (the requests of processes in states 2 and 3 were deferred by P_i till it exits the CS). Now it is up to the process receiving the FLUSH message and the processes recieving REPLY messages in states 2 and 3 to determine who is allowed to enter the CS next.

Consider a scenario where we have a set of requests R_3, R_0, R_2, R_4, R_1 ordered in decreasing priority, where R_0, R_2, R_4 are concurrent with one another. P_0 maintains a local queue of [R_0, R_2, R_4] and, when it exits the CS, it sends a FLUSH (only) to P_2.

Multiple uses of a REQUEST message

Considering two processes P_i and P_j, there can be two cases:

Case 1 P_i and P_j are not concurrently requesting. In this case, the process which requests first will get a REPLY message from the other process.

Case 2 P_i and P_j are concurrently requesting. In this case, there can be two subcases:

1. P_i is requesting with a higher priority than P_j. In this case, P_j's REQUEST message serves as an implicit REPLY message to P_i's request. Also, P_j should wait for REPLY/FLUSH message from some process to enter the CS.
2. P_i is requesting with a lower priority than P_j. In this case, P_i's REQUEST message serves as an implicit REPLY message to P_j's request. Also, P_i should wait for REPLY/FLUSH message from some process to enter the CS.

---

(1) *Initial local state for process* P_i:
- **int** $My\_Sequence\_Number_i = 0$
- **array of boolean** $RV_i[j] = 0, \forall j \in \{1...N\}$
- **queue of ReqID** LRQ_i is NULL
- **int** $Highest\_Sequence\_Number\_Seen_i = 0$

(2) *InvMutEx*: Process P_i executes the following to invoke mutual exclusion:
(2a) $My\_Sequence\_Number_i = Highest\_Sequence\_Number\_Seen_i + 1$.
(2b) $LRQ_i = NULL$.
(2c) Make REQUEST(R_i) message, where $R_i = (My\_Sequence\_Number_i, i)$.
(2d) Insert this REQUEST in LRQ_i in sorted order.
(2e) Send this REQUEST message to all other processes.
(2f) $RV_i[k] = 0 \forall k \in \{1 \ldots N\} - \{i\}$. $RV_i[i]=1$.

(3) *RcvReq*: Process P_i receives REQUEST(R_j), where $R_j = (SN, j)$, from process P_j:
(3a) $Highest\_Sequence\_Number\_Seen_i = max(Highest\_Sequence\_Number\_Seen_i, SN)$.
(3b) If P_i is requesting:
(3bi) If $RV_i[j] = 0$, then insert this request in LRQ_i (in sorted order) and mark $RV_i[j] = 1$. If (*CheckExecuteCS*), then execute CS.
(3bii) If $RV_i[j] = 1$, then defer the processing of this request, which will be processed after P_i executes CS.
(3c) If P_i is not requesting, then send a REPLY(R_i) message to P_j. R_i denotes the ReqID of the last request made by P_i that was satisfied.

(4) *RcvReply*: Process P_i receives REPLY(R_j) message from process P_j. R_j denotes the ReqID of the last request made by P_j that was satisfied:
(4a) $RV_i[j]=1$.
(4b) Remove all requests from LRQ_i that have a priority $\geq$ the priority of R_j.
(4c) If (*CheckExecuteCS*), then execute CS.

(5) *FinCS*: Process P_i finishes executing CS:
(5a) Send FLUSH(R_i) message to the next candidate in LRQ_i. R_i denotes the ReqID that was satisfied.
(5b) Send REPLY(R_i) to the deferred requests. R_i is the ReqID corresponding to which P_i just executed the CS.

(6) *RcvFlush*: Process P_i receives a FLUSH(R_j) message from a process P_j:
(6a) $RV_i[j] = 1$
(6b) Remove all requests in LRQ_i that have the priority $\geq$ the priority of R_j.
(6c) If (*CheckExecuteCS*) then execute CS.

(7) *CheckExecuteCS*: If ($RV_i[k] = 1, \forall k \in \{1 \ldots N\}$) and P_i's request is at the head of LRQ_i, then return *true*, else return *false*.

Algorithm 9.4 Lodha and Kshemkalyani's fair mutual exclusion algorithm [13].

Examples

- **Figure 9.11** Processes P_1 and P_2 make concurrent requests. They send out REQUESTs to all other processes. The REQUEST sent by P_1 to P_3 is delayed and hence is not shown until in Figure 9.13.
- **Figure 9.12** When P_3 receives the REQUEST from P_2, it sends REPLY to P_2.
- **Figure 9.13** The delayed REQUEST of P_1 arrives at P_3 just after P_3 sends out its REQUEST for CS, which makes it concurrent with the request of P_1.
- **Figure 9.14** P_1 exits the CS and sends out a FLUSH message to P_2.
- **Figure 9.15** Since the requests of P_2 and P_3 are not concurrent, P_2 sends a FLUSH message to P_3. P_3 removes (1,1) from its local queue and enters the CS.

The data structures LRQ and RV are updated in each step as discussed previously.

9.6.3 Safety, fairness and liveness

Proofs for safety, fairness and liveness are quite involved and interested readers are referred to the original paper for detailed proofs.

9.6.4 Message complexity

To execute the CS, a process P_i sends $(N-1)$ REQUEST messages. It receives $(N-|CSet_i|)$ REPLY messages. There are two cases to consider:

1. $|CSet_i| \geq 2$. There are two subcases here:

   (a) There is at least one request in $CSet_i$ whose priority is smaller than that of R_i. So P_i will send one FLUSH message. In this case the total number of messages for CS access is $2N-|CSet_i|$. When all the requests are concurrent, this reduces to N messages.

   (b) There is no request in $CSet_i$, whose priority is less than the priority of R_i. P_i will not send a FLUSH message. In this case, the total number of messages for CS access is $2N-1-|CSet_i|$. When all the requests are concurrent, this reduces to $N-1$ messages.

2. $|CSet_i| = 1$. This is the worst case, implying that all requests are satisfied serially. P_i will not send a FLUSH message. In this case, the total number of messages for CS access is $2(N-1)$ messages.

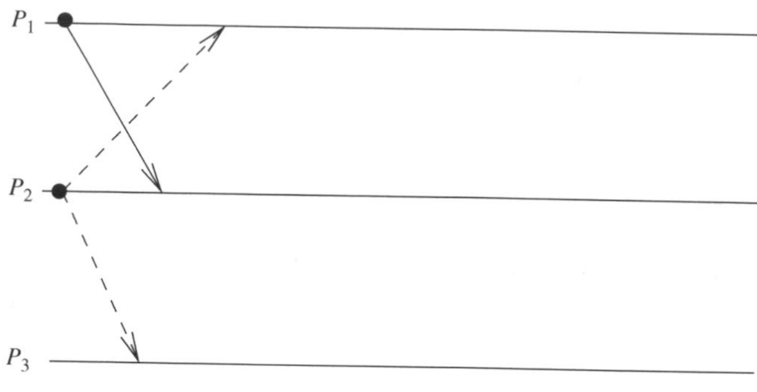

Figure 9.11 Processes P_1 and P_2 send out REQUESTs.

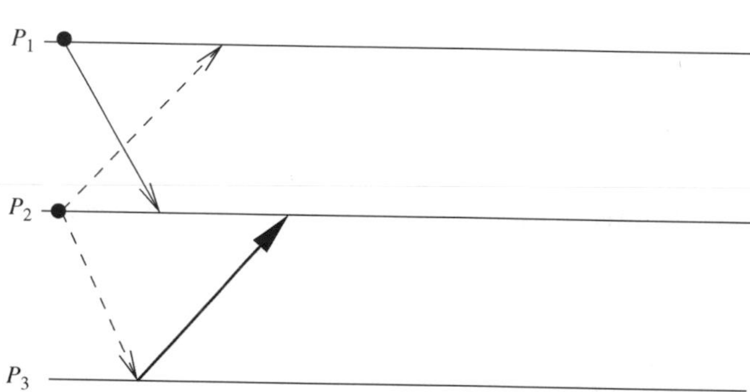

Figure 9.12 P_3 sends a REPLY message to P_2 only.

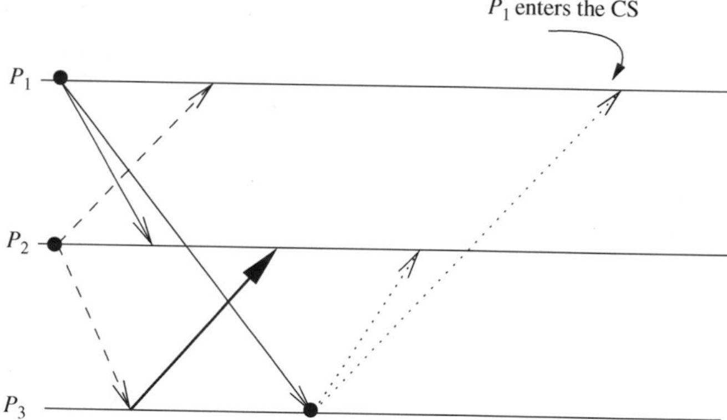

Figure 9.13 P_3 sends out a REQUEST message.

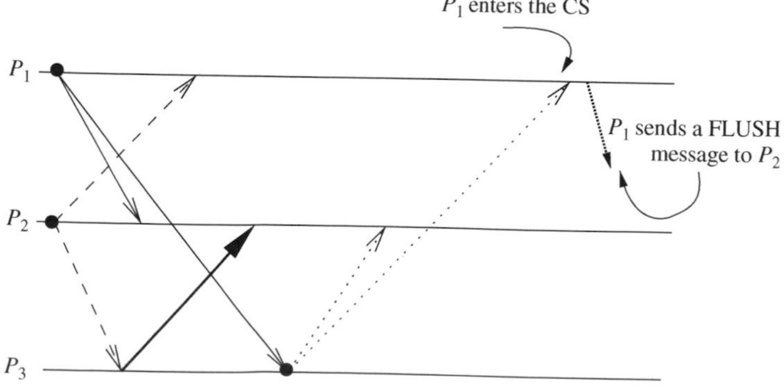

Figure 9.14 P_1 exits the CS and sends a FLUSH message to P_2.

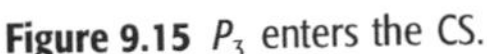

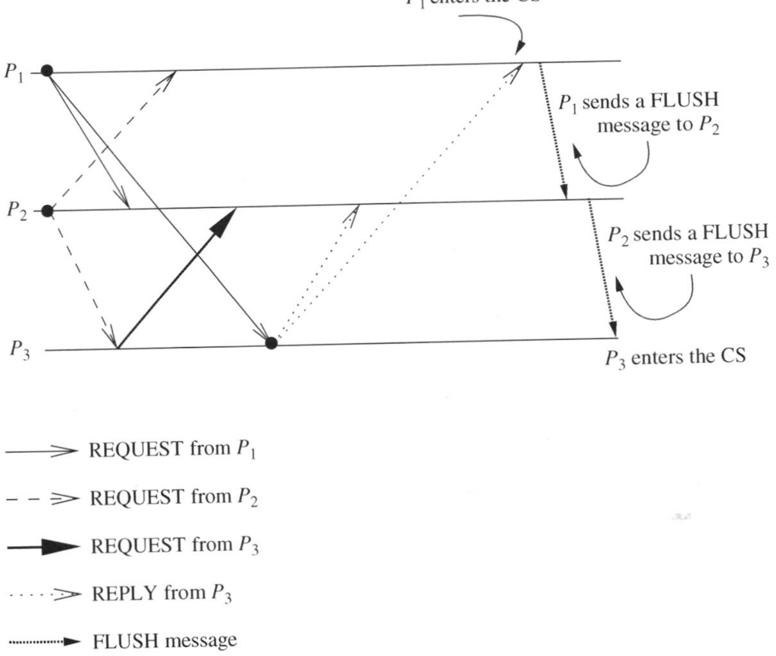

Figure 9.15 P_3 enters the CS.

9.7 Quorum-based mutual exclusion algorithms

Quorum-based mutual exclusion algorithms respresented a departure from the trend in the following two ways:

1. A site does not request permission from all other sites, but only from a subset of the sites. This is a radically different approach as compared to the Lamport and Ricart–Agrawala algorithms, where all sites participate in conflict resolution of all other sites. In quorum-based mutual exclusion algorithms, the request set of sites are chosen such that $\forall i\ \forall j : 1 \leq i, j \leq N :: R_i \cap R_j \neq \emptyset$. Consequently, every pair of sites has a site which mediates conflicts between that pair.
2. In quorum-based mutual exclusion algorithms, a site can send out only one REPLY message at any time. A site can send a REPLY message only after it has received a RELEASE message for the previous REPLY message. Therefore, a site S_i *locks* all the sites in R_i in exclusive mode before executing its CS.

Quorum-based mutual exclusion algorithms significantly reduce the message complexity of invoking mutual exclusion by having sites ask permission from only a subset of sites.

Since these algorithms are based on the notion of "Coteries" and "Quorums," we first describe the idea of coteries and quorums. A coterie C is defined as a set of sets, where each set $g \in C$ is called a quorum. The following properties hold for quorums in a coterie:

- **Intersection property** For every quorum g, $h \in C$, $g \cap h \neq \emptyset$. For example, sets $\{1,2,3\}$, $\{2,5,7\}$, and $\{5,7,9\}$ cannot be quorums in a coterie because the first and third sets do not have a common element.
- **Minimality property** There should be no quorums g, h in coterie C such that $g \supseteq h$. For example, sets $\{1,2,3\}$ and $\{1,3\}$ cannot be quorums in a coterie because the first set is a superset of the second.

Coteries and quorums can be used to develop algorithms to ensure mutual exclusion in a distributed environment. A simple protocol works as follows: let "a" be a site in quorum "A." If "a" wants to invoke mutual exclusion, it requests permission from all sites in its quorum "A." Every site does the same to invoke mutual exclusion. Due to the Intersection property, quorum "A" contains at least one site that is common to the quorum of every other site. These common sites send permission to only one site at any time. Thus, mutual exclusion is guaranteed.

Note that the Minimality property ensures efficiency rather than correctness. In the simplest form, quorums are formed as sets that contain a majority of sites. There exists a variety of quorums and a variety of ways to construct quorums. For example, Maekawa [14] used the theory of projective planes to develop quorums of size $\sqrt{N}$.

9.8 Maekawa's algorithm

Maekawa's algorithm [14] was the first quorum-based mutual exclusion algorithm. The request sets for sites (i.e., quorums) in Maekawa's algorithm are constructed to satisfy the following conditions:

M1 $(\forall i \; \forall j : i \neq j, \; 1 \leq i, j \leq N :: R_i \cap R_j \neq \emptyset)$.
M2 $(\forall i : 1 \leq i \leq N :: S_i \in R_i)$.
M3 $(\forall i : 1 \leq i \leq N :: |R_i| = K)$.
M4 Any site S_j is contained in K number of R_is, $1 \leq i, j \leq N$.

Maekawa used the theory of projective planes and showed that $N = K(K-1)+1$. This relation gives $|R_i| = \sqrt{N}$.

Since there is at least one common site between the request sets of any two sites (condition M1), every pair of sites has a common site which mediates conflicts between the pair. A site can have only one outstanding REPLY message at any time; that is, it grants permission to an incoming request if it has not granted permission to some other site. Therefore, mutual exclusion is guaranteed. This algorithm requires delivery of messages to be in the order they are sent between every pair of sites.

Conditions M1 and M2 are necessary for correctness; whereas conditions M3 and M4 provide other desirable features to the algorithm. Condition M3 states that the size of the requests sets of all sites must be equal, which implies that all sites should have to do an equal amount of work to invoke mutual exclusion. Condition M4 enforces that exactly the same number of sites should request permission from any site, which implies that all sites have "equal responsibility" in granting permission to other sites.

In Maekawa's algorithm, a site S_i executes the steps shown in Algorithm 9.5 to execute the CS.

Requesting the critical section:

(a) A site S_i requests access to the CS by sending REQUEST(i) messages to all sites in its request set R_i.

(b) When a site S_j receives the REQUEST(i) message, it sends a REPLY(j) message to S_i provided it hasn't sent a REPLY message to a site since its receipt of the last RELEASE message. Otherwise, it queues up the REQUEST(i) for later consideration.

Executing the critical section:

(c) Site S_i executes the CS only after it has received a REPLY message from every site in R_i.

Releasing the critical section:

(d) After the execution of the CS is over, site S_i sends a RELEASE(i) message to every site in R_i.

(e) When a site S_j receives a RELEASE(i) message from site S_i, it sends a REPLY message to the next site waiting in the queue and deletes that entry from the queue. If the queue is empty, then the site updates its state to reflect that it has not sent out any REPLY message since the receipt of the last RELEASE message.

Algorithm 9.5 Maekawa's algorithm.

Correctness

Theorem 9.3 *Maekawa's algorithm achieves mutual exclusion.*

Proof Proof is by contradiction. Suppose two sites S_i and S_j are concurrently executing the CS. This means site S_i received a REPLY message from all sites in R_i and concurrently site S_j was able to receive a REPLY message from all sites in R_j. If $R_i \cap R_j = \{S_k\}$, then site S_k must have sent REPLY messages to both S_i and S_j concurrently, which is a contradiction. □

Performance

Note that the size of a request set is $\sqrt{N}$. Therefore, an execution of the CS requires $\sqrt{N}$ REQUEST, $\sqrt{N}$ REPLY, and $\sqrt{N}$ RELEASE messages, resulting in $3\sqrt{N}$ messages per CS execution. Synchronization delay in this algorithm is $2T$. This is because after a site S_i exits the CS, it first releases all the sites in R_i and then one of those sites sends a REPLY message to the next site that executes the CS. Thus, two sequential message transfers are required between two successive CS executions. As discussed next, Maekawa's algorithm is deadlock-prone. Measures to handle deadlocks require additional messages.

9.8.1 Problem of deadlocks

Maekawa's algorithm can deadlock because a site is exclusively locked by other sites and requests are not prioritized by their timestamps [14,22]. Thus, a site may send a REPLY message to a site and later force a higher priority request from another site to wait.

Without loss of generality, assume three sites S_i, S_j, and S_k simultaneously invoke mutual exclusion. Suppose $R_i \cap R_j = \{S_{ij}\}$, $R_j \cap R_k = \{S_{jk}\}$, and $R_k \cap R_i = \{S_{ki}\}$. Since sites do not send REQUEST messages to the sites in their request sets in any particular order and message delays are arbitrary, the following scenario is possible: S_{ij} has been locked by S_i (forcing S_j to wait at S_{ij}), S_{jk} has been locked by S_j (forcing S_k to wait at S_{jk}), and S_{ki} has been locked by S_k (forcing S_i to wait at S_{ki}). This state represents a deadlock involving sites S_i, S_j, and S_k.

Handling deadlocks

Maekawa's algorithm handles deadlocks by requiring a site to yield a lock if the timestamp of its request is larger than the timestamp of some other request waiting for the same lock (unless the former has succeeded in acquiring locks on all the needed sites) [14,22]. A site suspects a deadlock (and initiates message exchanges to resolve it) whenever a higher priority request arrives and waits at a site because the site has sent a REPLY message to a lower priority request.

Deadlock handling requires the following three types of messages:

FAILED A FAILED message from site S_i to site S_j indicates that S_i cannot grant S_j's request because it has currently granted permission to a site with a higher priority request.

INQUIRE An INQUIRE message from S_i to S_j indicates that S_i would like to find out from S_j if it has succeeded in locking all the sites in its request set.

YIELD A YIELD message from site S_i to S_j indicates that S_i is returning the permission to S_j (to yield to a higher priority request at S_j).

Details of how Maekawa's algorithm handles deadlocks are as follows:

- When a REQUEST(ts, i) from site S_i blocks at site S_j because S_j has currently granted permission to site S_k, then S_j sends a FAILED(j) message to S_i if S_i's request has lower priority. Otherwise, S_j sends an INQUIRE(j) message to site S_k.
- In response to an INQUIRE(j) message from site S_j, site S_k sends a YIELD(k) message to S_j provided S_k has received a FAILED message from a site in its request set and if it sent a YIELD to any of these sites, but has not received a new REPLY from it.
- In response to a YIELD(k) message from site S_k, site S_j assumes as if it has been released by S_k, places the request of S_k at an appropriate location in the request queue, and sends a REPLY(j) to the site whose request is at the top of the queue.

Thus, Maekawa-type algorithms require extra messages to handle deadlocks and may exchange these messages even though there is no deadlock. The maximum number of messages required per CS execution in this case is $5\sqrt{N}$.

9.9 Agarwal–El Abbadi quorum-based algorithm

Agarwal and El Abbadi [1] developed a simple and efficient mutual exclusion algorithm by introducing tree quorums. They gave a novel algorithm for constructing tree-structured quorums in the sense that it uses the hierarchical structure of a network. The mutual exclusion algorithm is independent of the underlying topology of the network and there is no need for a multicast facility in the network. However, such facility will improve the performance of the algorithm. The mutual exclusion algorithm assumes that sites in the distributed system can be organized into a structure such as tree, grid, binary tree, etc. and there exists a routing mechanism to exchange messages between different sites in the system.

The Agarwal–El Abbadi quorum-based algorithm, however, constructs quorums from trees. Such quorums are called "tree-structured quorums." The following sections describe an algorithm for constructing tree-structured quorums and present an analysis of the algorithm and a protocol for mutual exclusion in distributed systems using tree-structured quorums.

9.9.1 Constructing a tree-structured quorum

All the sites in the system are logically organized into a complete binary tree. To build such a tree, any site could be chosen as the root, any other two sites may be chosen as its children, and so on. For a complete binary tree with level "k," we have $2^{k+1}-1$ sites with its root at level k and leaves at level 0. The number of sites in a path from the root to a leaf is equal to the level of the tree $k+1$, which is equal to $O(\log n)$. There will be 2^k leaves in the tree. A path in a binary tree is the sequence $a_1, a_2, \ldots, a_i, a_{i+1}, \ldots, a_k$ such that a_i is the parent of a_{i+1}.

The algorithm for constructing structured quorums from the tree is given in Algorithm 9.6. For the purpose of presentation, we assume that the tree is complete, however, the algorithm works for any arbitrary binary tree.

```
FUNCTION GetQuorum(Tree: NetworkHierarchy): QuorumSet;
   VAR left, right: QuorumSet;
     BEGIN
     IF Empty (Tree) THEN
         RETURN ({});
     ELSE IF GrantsPermission(Tree↑.Node) THEN
           RETURN((Tree↑.Node) ∪ GetQuorum (Tree↑.LeftChild));
           OR
           RETURN((Tree↑.Node) ∪ GetQuorum (Tree↑.RightChild));
           ELSE
             left ← GetQuorum(Tree↑.left);
             right ← GetQuorum(Tree↑.right);
             IF (left = ∅ ∨ right = ∅) THEN
                  (* Unsuccessful in establishing a quorum *)
                  EXIT(-1);
             ELSE
                  RETURN(left ∪ right);
             END; (* IF *)
        END; (* IF *)
    END; (* IF *)
END GetQuorum
```

Algorithm 9.6 Algorithm for constructing a tree-structured quorum [1].

The algorithm for constructing tree-structured quorums uses two functions called *GetQuorum*(*Tree*) and *GrantsPermission*(*site*) and assumes that there is a well-defined root for the tree. *GetQuorum* is a recursive function that takes a tree node "x" as the parameter and calls *GetQuorum* for its child node provided that the *GrantsPermission*(x) is true. The *GrantsPermission*(x) is true only when the node "x" agrees to be in the quorum. If the node "x" is down due to a failure, then it may not agree to be in the quorum and the value of *GrantsPermission*(x) will be false. The algorithm tries to construct quorums in a way that each quorum represents any path from the root to a leaf, i.e., in this case the (no failures) quorum is any set $a_1, a_2, \ldots, a_i, a_{i+1}, \ldots, a_k$, where a_1 is the root and a_k is a leaf, and for all $i < k$, a_i is the parent of a_{i+1}. If it fails to find such a path (say, because node "x" has failed), the control goes to the ELSE block which specifies that the failed node "x"

is substituted by two paths both of which start with the left and right children of "x" and end at leaf nodes. Note that each path must terminate in a leaf site. If the leaf site is down or inaccessible due to any reason, then the quorum cannot be formed and the algorithm terminates with an error condition. The sets that are constructed using this algorithm are termed as *tree quorums*.

9.9.2 Analysis of the algorithm for constructing tree-structured quorums

The best case scenario of the algorithm takes $O(\log n)$ sites to form a tree quorum. There are certain cases where even in the event of a failure, $O(\log n)$ sites are sufficient to form a tree quorum. For example, if the site that is parent of a leaf node fails, then the number of sites that are necessary for a quorum will be still $O(\log n)$. Thus, the algorithm requires very few messages in a relatively fault-free environment. It can tolerate the failure up to $n - \mathrm{O}(\log n)$ sites and still form a tree quorum. In the worst case, the algorithm requires the majority of sites to construct a tree quorum and the number of sites is same for all cases (faults or no faults). The worst case tree quorum size is determined as $O((n+1)/2)$ by induction.

9.9.3 Validation

The tree quorums constructed by the above algorithm are valid, i.e., they conform to the coterie properties such as Intersection property and Minimality property. To prove the correctness of the algorithm, consider a binary tree with level $k+1$. Assume that root of the tree is a_1. The tree can be viewed as consisting of a root, a left subtree, and a right subtree. According to Algorithm 9.6, the constructed quorums contain one of the following:

1. $\{a_1\} \cup$ {sites from the left subtree};
2. $\{a_1\} \cup$ {sites from the right subtree};
3. {sites from the quorum set of left subtree} $\cup$ {sites from the quorum set of right subtree}.

Clearly, the quorum of type 1 has non-empty intersection with those quorums formed using types 2 or 3, which shows that the Intersection property holds true. Also, the members in the quorum of type 1 are not contained in quorums of types 2 and 3. Thus, the Minimality property holds true. Similar conditions exist for quorums of types 2 and 3. This forms as the basis for proving correctness of the algorithm based on induction.

9.9.4 Examples of tree-structured quorums

Now we present examples of tree-structured quorums for a better understanding of the algorithm. In the simplest case, when there is no node failure, the number of quorums formed is equal to the number of leaf sites.

Figure 9.16 A tree of 15 sites.

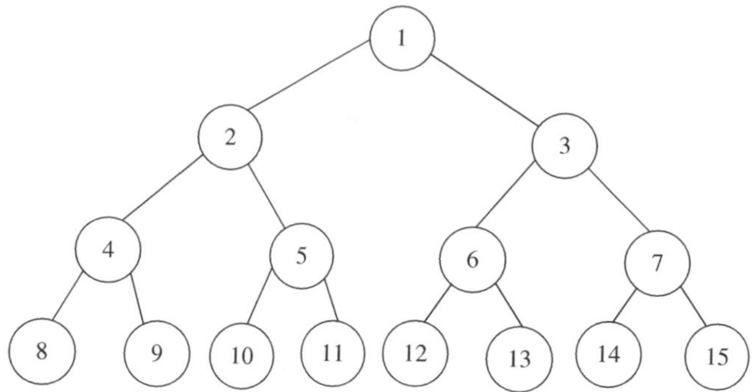

Consider the tree of height 3 shown in Figure 9.16 constructed from 15 ($2^{3+1} - 1$) sites. Now, a quorum has all sites along any path from root to leaf. In this case eight quorums are formed from eight possible root-leaf paths: 1–2–4–8, 1–2–4–9, 1–2–5–10, 1–2–5–11, 1–3–6–12, 1–3–6–13, 1–3–7–14 and 1–3–7–15. If any site fails, the algorithm substitutes for that site two possible paths starting from the site's two children and ending in leaf nodes. For example, when node 3 fails, we consider the possible paths starting from children 6 and 7 and ending at the leaf nodes. The possible paths starting from child 6 are 6–12 and 6–13, while the possible paths starting from child 7 are 7–14 and 7–15. So, when node 3 fails, the following eight quorums can be formed: {1,6,12,7,14}, {1,6,12,7,15}, {1,6,13,7,14}, {1,6,13,7,15}, {1,2,4,8}, {1,2,4,9}, {1,2,5,10}, {1,2,5,11}.

If a failed site is a leaf node, the operation has to be aborted and a tree-structured quorum cannot be formed (see lines 13–15 of the algorithm above). However, quorum formation can continue with other working nodes. Since the number of nodes from root to leaf in an "n" node complete tree is log n, the best case for quorum formation, i.e, the least number of nodes needed for a quorum is log n. In the worst case, a majority of sites are needed for mutual exclusion. For example, if sites 1 and 2 are down in Figure 9.16, the quorums that are formed must include either {4,8} or {4,9} and either {5,10} or {5,11} and one of the four paths {3,6,12}, {3,6,13} {3,7,14} or {3,7,15}. In this case, the following are the candidates for quorums: {4,5,3,6,8,10,12}, {4,5,3,6,8,10,13}, {4,5,3,6,8,11,12}, {4,5,3,6,8,11,13}, {4,5,3,6,9,10,12}, {4,5,3,6,9,10,13}, {4,5,3,6,9,11,12}, {4,5,3,6,9,11,13}, {4,5,3,7,8,10,14}, {4,5,3,7,8,10,15}, {4,5,3,7,8,11,14}, {4,5,3,7,8,11,15}, {4,5,3,7,9,10,14}, {4,5,3,7,9,10,15}, {4,5,3,7,9,11,14}, and {4,5,3,7,9,11,15}.

When the number of node failures is greater than or equal to *log n*, the algorithm may not be able to form tree-structured quorum. For example when sites 1, 2, 4, and 8 are inaccessible, the set of sites {3,5,6,7,8,9,10,11,12,13,14,15} form a majority of sites but not a structured quorum. So, as long as the number of site failures is less than log n, the tree quorum algorithm guarantees the formation of a quorum and it exhibits the property of "graceful degradation," which is useful in distributed fault tolerance. As failures occur and increase, the probability of forming quorums decreases and mutual exclusion is achieved at increasing costs because when a node fails, instead of one path from node, the quorum must include two paths starting from the node's children. For example, in a tree of level k, the size of quorum is $(k+1)$. If a node failure occurs at level $i > 0$, then the quorum size increases to $(k-i)$ $+2i$. The penalty is severe when the failed node is near the root. Thus, the tree quorum algorithm may still allow quorums to be formed even after the failures of $n- \lfloor \log n \rfloor$ sites.

9.9.5 The algorithm for distributed mutual exclusion

We now describe the algorithm for achieving distributed mutual exclusion using tree-structured quorums. Suppose a site s wants to enter the critical section (CS). The following events should occur in the order given:

1. Site s sends a "Request" message to all other sites in the structured quorum it belongs to.
2. Each site in the quorum stores incoming requests in a *request queue*, ordered by their timestamps.
3. A site sends a "Reply" message, indicating its consent to enter CS, only to the request at the head of its *request queue*, having the lowest timestamp.
4. If the site s gets a "Reply" message from all sites in the structured quorum it belongs to, it enters the CS.
5. After exiting the CS, s sends a "Relinquish" message to all sites in the structured quorum. On the receipt of the "Relinquish" message, each site removes s's request from the head of its *request queue*.
6. If a new request arrives with a timestamp smaller than the request at the head of the queue, an "Inquire" message is sent to the process whose request is at the head of the queue and the site waits for a "Yield" or "Relinquish" message.
7. When a site s receives an "Inquire" message, it acts as follows:
    - If s has acquired all of its necessary replies to access the CS, then it simply ignores the "Inquire" message and proceeds normally and sends a "Relinquish" message after exiting the CS.
    - If s has not yet collected enough replies from its quorum, then it sends a "Yield" message to the inquiring site.
8. When a site gets the "Yield" message, it puts the pending request (on behalf of which the "Inquire" message was sent) at the head of the queue and sends a "Reply" message to the requestor.

9.9.6 Correctness proof

Mutual exclusion is guaranteed because the set of quorums satisfy the Intersection property. Proof for freedom from deadlock is similar to that of Maekawa's algorithm. The readers are referred to the original source [14].

Example Consider a coterie C which consists of quorums {1,2,3}, {2,4,5}, and {4,1,6}. Suppose nodes 3, 5, and 6 want to enter CS, and they send requests to sites (1, 2), (2, 4), and (1, 4), respectively. Suppose site 3's request arrives at site 2 before site 5's request. In this case, site 2 will grant permission to site 3's request and reject site 5's request. Similarly, suppose site 3's request arrives at site 1 before site 6's request. So site 1 will grant permission to site 3's request and reject site 6's request. Since sites 5 and 6 did not get consent from all sites in their quorums, they do not enter the CS. Since site 3 alone gets consent from all sites in its quorum, it enters the CS and mutual exclusion is achieved.

9.10 Token-based algorithms

In token-based algorithms, a unique token is shared among the sites. A site is allowed to enter its CS if it possesses the token. A site holding the token can enter its CS repeatedly until it sends the token to some other site. Depending upon the way a site carries out the search for the token, there are numerous token-based algorithms. Next, we discuss two token-based mutual exclusion algorithms.

Before we start with the discussion of token-based algorithms, two comments are in order. First, token-based algorithms use sequence numbers instead of timestamps. Every request for the token contains a sequence number and the sequence numbers of sites advance independently. A site increments its sequence number counter every time it makes a request for the token. (A primary function of the sequence numbers is to distinguish between old and current requests.) Second, the correctness proof of token-based algorithms, that they enforce mutual exclusion, is trivial because an algorithm guarantees mutual exclusion so long as a site holds the token during the execution of the CS. Instead, the issues of freedom from starvation, freedom from deadlock, and detection of the token loss and its regeneration become more prominent.

9.11 Suzuki–Kasami's broadcast algorithm

In Suzuki–Kasami's algorithm [29] (Algorithm 9.7), if a site that wants to enter the CS does not have the token, it broadcasts a REQUEST message for the token to all other sites. A site that possesses the token sends it to the requesting site upon the receipt of its REQUEST message. If a site receives a REQUEST message when it is executing the CS, it sends the token only after it has completed the execution of the CS.

Although the basic idea underlying this algorithm may sound rather simple, there are two design issues that must be efficiently addressed:

1. **How to distinguishing an outdated REQUEST message from a current REQUEST message** Due to variable message delays, a site may receive a token request message after the corresponding request has been satisfied. If a site cannot determine if the request corresponding to a token request has been satisfied, it may dispatch the token to a site that does not need it. This will not violate the correctness, however, but it may seriously degrade the performance by wasting messages and increasing the delay at sites that are genuinely requesting the token. Therefore, appropriate mechanisms should be implemented to determine if a token request message is outdated.
2. **How to determine which site has an outstanding request for the CS** After a site has finished the execution of the CS, it must determine what sites have an outstanding request for the CS so that the token can be dispatched to one of them. The problem is complicated because when a site S_i receives a token request message from a site S_j, site S_j may have an outstanding request for the CS. However, after the corresponding request for the CS has been satisfied at S_j, an issue is how to inform site S_i (and all other sites) efficiently about it.

Outdated REQUEST messages are distinguished from current REQUEST messages in the following manner: a REQUEST message of site S_j has the form REQUEST(j, sn) where sn ($sn = 1, 2, \ldots$) is a sequence number that indicates that site S_j is requesting its snth CS execution. A site S_i keeps an array of integers $RN_i[1, \ldots, n]$ where $RN_i[j]$ denotes the largest sequence number received in a REQUEST message so far from site S_j. When site S_i receives a REQUEST(j, sn) message, it sets $RN_i[j] := max(RN_i[j], sn)$. Thus, when a site S_i receives a REQUEST(j, sn) message, the request is outdated if $RN_i[j] > sn$.

Sites with outstanding requests for the CS are determined in the following manner: the token consists of a queue of requesting sites, Q, and an array of integers $LN[1, \ldots, n]$, where $LN[j]$ is the sequence number of the request which site S_j executed most recently. After executing its CS, a site S_i updates $LN[i] := RN_i[i]$ to indicate that its request corresponding to sequence number $RN_i[i]$ has been executed. Token array $LN[1, \ldots, n]$ permits a site to determine if a site has an outstanding request for the CS. Note that at site S_i if $RN_i[j] = LN[j] + 1$, then site S_j is currently requesting a token. After executing the CS, a site checks this condition for all the j's to determine all the sites that are requesting the token and places their i.d.'s in queue Q if these i.d.'s are not already present in Q. Finally, the site sends the token to the site whose i.d. is at the head of Q.

Requesting the critical section:

(a) If requesting site S_i does not have the token, then it increments its sequence number, $RN_i[i]$, and sends a REQUEST(i, sn) message to all other sites. ("sn" is the updated value of $RN_i[i]$.)

(b) When a site S_j receives this message, it sets $RN_j[i]$ to $max(RN_j[i], sn)$. If S_j has the idle token, then it sends the token to S_i if $RN_j[i] = LN[i] + 1$.

Executing the critical section:

(c) Site S_i executes the CS after it has received the token.

Releasing the critical section: Having finished the execution of the CS, site S_i takes the following actions:

(d) It sets $LN[i]$ element of the token array equal to $RN_i[i]$.

(e) For every site S_j whose i.d. is not in the token queue, it appends its i.d. to the token queue if $RN_i[j] = LN[j] + 1$.

(f) If the token queue is nonempty after the above update, S_i deletes the top site i.d. from the token queue and sends the token to the site indicated by the i.d.

Algorithm 9.7 Suzuki–Kasami's broadcast algorithm.

Thus, as shown in Algorithm 9.7, after executing the CS, a site gives priority to other sites with outstanding requests for the CS (over its pending requests for the CS). Note that Suzuki–Kasami's algorithm is not symmetric because a site retains the token even if it does not have a request for the CS, which is contrary to the spirit of Ricart and Agrawala's definition of symmetric algorithm: "no site possesses the right to access its CS when it has not been requested."

Correctness

Mutual exclusion is guaranteed because there is only one token in the system and a site holds the token during the CS execution.

Theorem 9.3 *A requesting site enters the CS in finite time.*

Proof Token request messages of a site S_i reach other sites in finite time. Since one of these sites will have token in finite time, site S_i's request will be placed in the token queue in finite time. Since there can be at most $N - 1$ requests in front of this request in the token queue, site S_i will get the token and execute the CS in finite time. □

Performance

The beauty of the Suzuki–Kasami algorithm lies in its simplicity and efficiency. No message is needed and the synchronization delay is zero if a site holds the idle token at the time of its request. If a site does not hold the token when it makes a request, the algorithm requires N messages to obtain the token. The synchronization delay in this algorithm is 0 or T.

9.12 Raymond's tree-based algorithm

Raymond's tree-based mutual exclusion algorithm [19] uses a spanning tree of the computer network to reduce the number of messages exchanged per critical section execution. The algorithm exchanges only $O(\log N)$ messages under light load, and approximately four messages under heavy load to execute the CS, where N is the number of nodes in the network.

The algorithm assumes that the underlying network guarantees message delivery. The time or order of message arrival cannot be predicted. All nodes of the network are completely reliable. (Only for the initial part of the discussion, i.e., until node failure is discussed.) If the network is viewed as a graph, where the nodes in the network are the vertices of the graph, and the links between nodes are the edges of the graph, a spanning tree of a network of N nodes will be a tree that contains all N nodes. A minimal spanning tree is one such tree with minimum cost. Typically, this cost function is based on the network link characteristics. The algorithm operates on a minimal spanning tree of the network topology or logical structure imposed on the network.

The algorithm considers the network nodes to be arranged in an unrooted tree structure as shown in Figure 9.17. Messages between nodes traverse along the undirected edges of the tree in the Figure 9.17. The tree is also a spanning tree of the seven nodes A, B, C, D, E, F, and G. It also turns out to be a minimal spanning tree because it is the only spanning tree of these seven nodes. A node needs to hold information about and communicate only to its immediate-neighboring nodes. In Figure 9.17, for example, node C holds information about and communicates only to nodes B, D, and G; it does not need to know about the other nodes A, E, and F for the operation of the algorithm.

Similar to the concept of tokens used in token-based algorithms, this algorithm uses a concept of privilege to signify which node has the privilege to enter the critical section. Only one node can be in possession of the privilege (called the privileged node) at any time, except when the privilege is in transit

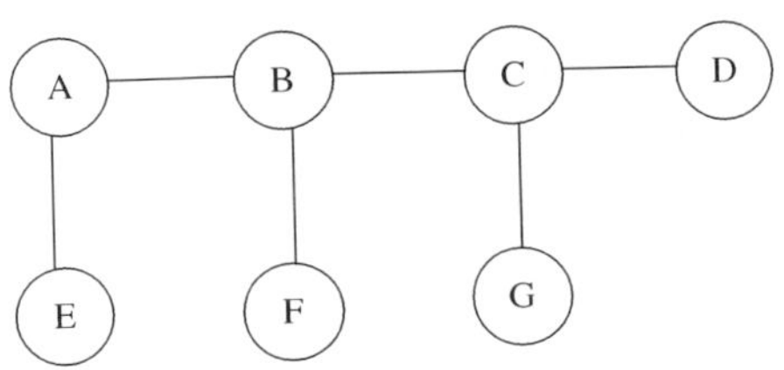

Figure 9.17 Nodes with an unrooted tree structure.

from one node to another in the form of a PRIVILEGE message. When there are no nodes requesting for the privilege, it remains in possession of the node that last used it.

9.12.1 The HOLDER variables

Each node maintains a HOLDER variable that provides information about the placement of the privilege in relation to the node itself. A node stores in its HOLDER variable the identity of a node that it thinks has the privilege or leads to the node having the privilege. The HOLDER variables of all the nodes maintain directed paths from each node to the node in the possession of the privilege.

For two nodes X and Y, if $\text{HOLDER}_X = \text{Y}$, we could redraw the undirected edge between the nodes X and Y as a directed edge from X to Y. Thus, for instance, if node G holds the privilege, Figure 9.17 can be redrawn with logically directed edges as shown in Figure 9.18. The shaded node represents the privileged node. The following will be the values of the HOLDER variables of various nodes:

$\text{HOLDER}_A = \text{B}$(Since the privilege is located in a sub-tree of A denoted by B.

Proceeding with similar reasoning, we have

$$\text{HOLDER}_B = \text{C},$$

$$\text{HOLDER}_C = \text{G},$$

$$\text{HOLDER}_D = \text{C},$$

$$\text{HOLDER}_E = \text{A},$$

$$\text{HOLDER}_F = \text{B},$$

$$\text{HOLDER}_G = \text{self}.$$

Now suppose that node B, which does not hold the privilege, wants to execute the critical section. Then B sends a REQUEST message to HOLDER_B, i.e., C, which in turn forwards the REQUEST message to HOLDER_C, i.e., G. So a series of REQUEST messages flow between the node making the request for the privilege and the node having the privilege.

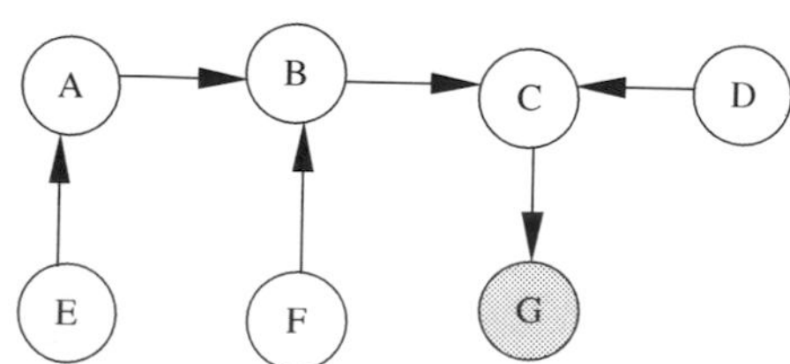

Figure 9.18 Tree with logically directed edges, all pointing in a direction towards node G – the privileged node.

Table 9.1 Variables used in the algorithm.

Variable name	Possible values	Comments
HOLDER	"self" or the identity of one of the immediate neighbors.	Indicates the location of the privileged node in relation to the current node.
USING	True or false.	Indicates if the current node is executing the critical section.
REQUEST_Q	A FIFO queue that could contain "self" or the identities of immediate neighbors as elements.	The REQUEST_Q of a node consists of the identities of those immediate neighbors that have requested for privilege but have not yet been sent the privilege.
ASKED	True or false.	Indicates if node has sent a request for the privilege.

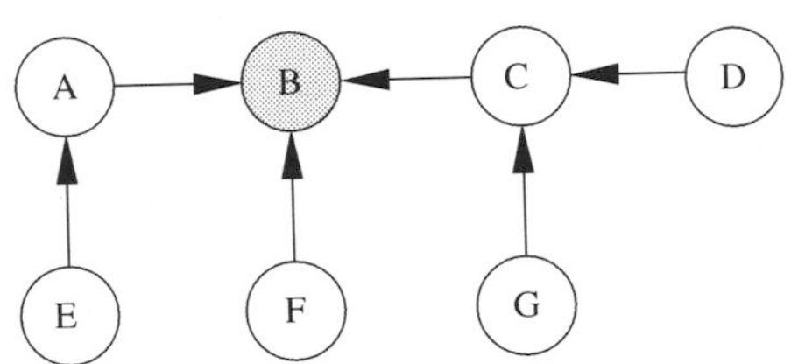

Figure 9.19 Tree with logically directed edges when node B holds the privilege.

The privileged node G, if it no longer needs the privilege, sends the PRIVILEGE message to its neighbor C, which made a request for the privilege, and resets $HOLDER_G$ to C. Node C, in turn, forwards the PRIVILEGE to node B, since it had requested the privilege on behalf of B. Node C also resets $HOLDER_C$ to B. The tree in Figure 9.18 will now look as shown in Figure 9.19.

Thus, at any stage, except when the PRIVILEGE message is in transit, the HOLDER variables collectively make sure that directed paths are maintained from each of the $N - 1$ nodes to the privileged node in the network.

9.12.2 The operation of the algorithm

Data structures

Each node maintains variables that are defined in Table 9.1. The value "self" is placed in REQUEST_Q if the node makes a request for the privilege for its own use. The maximum size of REQUEST_Q of a node is the number of immediate neighbors + 1 (for "self"). ASKED prevents the sending of duplicate requests for privilege, and also makes sure that the REQUEST_Qs of the various nodes do not contain any duplicate elements.

9.12.3 Description of the algorithm

The algorithm consists of the following parts:

- ASSIGN_PRIVILEGE;
- MAKE_REQUEST;
- events;
- message overtaking.

ASSIGN_PRIVILEGE

This is a routine to effect the sending of a PRIVILEGE message. A privileged node will send a PRIVILEGE message if:

- it holds the privilege but is not using it;
- its REQUEST_Q is not empty; and
- the element at the head of its REQUEST_Q is not "self." That is, the oldest request for privilege must have come from another node.

A situation where "self" is at the head of REQUEST_Q may occur immediately after a node receives a PRIVILEGE message. The node will enter into the critical section after removing "self" from the head of REQUEST_Q. If the i.d. of another node is at the head of REQUEST_Q, then it is removed from the queue and a PRIVILEGE message is sent to that node. Also, the variable ASKED is set to false since the currently privileged node will not have sent a request to the node (called HOLDER-to-be) that is about to receive the PRIVILEGE message.

MAKE_REQUEST

This is a routine to effect the sending of a REQUEST message. An unprivileged node will send a REQUEST message if:

- it does not hold the privilege;
- its REQUEST_Q is not empty, i.e., it requires the privilege for itself, or on behalf of one of its immediate neighboring nodes; and
- it has not sent a REQUEST message already.

The variable ASKED is set to true to reflect the sending of the REQUEST message. The MAKE_REQUEST routine makes no change to any other variables. The variable ASKED will be true at a node when it has sent REQUEST message to an immediate neighbor and has not received a response. The variable will be false otherwise. A node does not send any REQUEST messages, if ASKED is true at that node. Thus the variable ASKED makes sure that unnecessary REQUEST messages are not sent from the unprivileged node, and consequently ensures that the REQUEST_Q of an immediate neighbor does not contain duplicate entries of a neighboring node. This makes the REQUEST_Q of any node bounded, even when operating under heavy load.

Table 9.2 Events in the algorithms.

Event	Algorithm functionality
A node wishes to execute critical section.	Enqueue (REQUEST_Q, Self); ASSIGN_PRIVILEGE; MAKE_REQUEST
A node receives a REQUEST message from one of its immediate neighbors X.	Enqueue(REQUEST_Q, X); ASSIGN_PRIVILEGE; MAKE_REQUEST
A node receives a PRIVILEGE message.	HOLDER := self; ASSIGN_PRIVILEGE; MAKE_REQUEST
A node exits the critical section.	USING := false; ASSIGN_PRIVILEGE; MAKE_REQUEST

Events

The four events that constitute the algorithm are shown in Table 9.2.

- **A node wishes critical section entry** If it is the privileged node, the node could enter the critical section using the ASSIGN_PRIVILEGE routine. If not, the node could send a REQUEST message using the MAKE_REQUEST routine in order to get the privilege.
- **A node receives a REQUEST message from one of its immediate neighbors** If this node is the current HOLDER, it may send the PRIVILEGE to a requesting node using the ASSIGN_PRIVILEGE routine. If not, it could forward the request using the MAKE_REQUEST routine.
- **A node receives a PRIVILEGE message** The ASSIGN_PRIVILEGE routine could result in the execution of the critical section at the node, or may forward the privilege to another node. After the privilege is forwarded, the MAKE_REQUEST routine could send a REQUEST message to reacquire the privilege, for a pending request at this node.
- **A node exits the critical section** On exit from the critical section, this node may pass the privilege on to a requesting node using the ASSIGN_PRIVILEGE routine. It may then use the MAKE_REQUEST routine to get back the privilege, for a pending request at this node.

Message overtaking

This algorithm does away with the use of sequence numbers because of its inherent operations and by the acyclic structure it employs. Figure 9.20 shows the logical pattern of message flow between any two neighboring nodes (nodes A and B here).

If any message overtaking occurs between nodes A and B, it can occur when a PRIVILEGE message is sent from node A to node B, which is then very closely followed by a REQUEST message from node A to node B. In other words, node A sends the privilege and immediately wants it back. Such

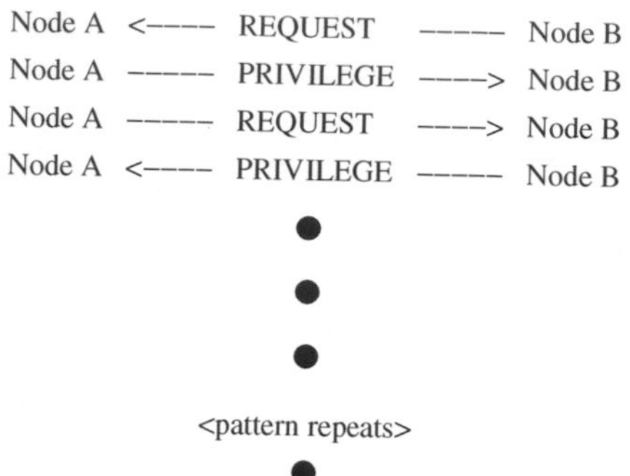

Figure 9.20 Logical pattern of message flow between neighboring nodes A and B.

message overtaking as described above will not affect the operation of the algorithm. If node B receives the REQUEST message from node A before receiving the PRIVILEGE message from node A, A's request will be queued in REQUEST_Q_B. Since B is not a privileged node, it will not be able to send a privilege to node A in reply. When node B receives the PRIVILEGE message from A after receiving the REQUEST message, it could enter the critical section or could send a PRIVILEGE message to an immediate neighbor at the head of REQUEST_Q_B, which need not be node A. So message overtaking does not affect the algorithm.

9.12.4 Correctness

The algorithm provides the following guarantees:

- mutual exclusion is guaranteed;
- deadlock is impossible;
- starvation is impossible.

Mutual exclusion is guaranteed

The algorithm ensures that, at any instant of time, no more than one node holds the privilege, which is a necessity for mutual exclusion. Whenever a node receives a PRIVILEGE message, it becomes privileged. Similarly, whenever a node sends a PRIVILEGE message, it becomes unprivileged. Between the instants one node becomes unprivileged and another node becomes privileged, there is no privileged node. Thus, there is at most one privileged node at any point of time in the network.

Deadlock is impossible

When the critical section is free, and one or more nodes want to enter the critical section but are not able to do so, a deadlock may occur. This could happen due to any of the following scenarios:

1. The privilege cannot be transferred to a node because no node holds the privilege.

2. The node in possession of the privilege is unaware that there are other nodes requiring the privilege.
3. The PRIVILEGE message does not reach the requesting unprivileged node.

None of the above three scenarios can occur in this algorithm, thus guarding against deadlocks. Scenario 1 can never occur in this algorithm because we have assumed that nodes do not fail and messages are not lost. There can never be a situation where REQUEST messages do not arrive at the privileged node. The logical pattern established using HOLDER variables ensures that a node that needs the privilege sends a REQUEST message either to a node holding the privilege or to a node that has a path to a node holding the privilege. Thus scenario 2 can never occur in this algorithm. The series of REQUEST messages are enqueued in the REQUEST_Qs of various nodes such that the REQUEST_Qs of those nodes collectively provide a logical path for the transfer of the PRIVILEGE message from the privileged node to the requesting unprivileged nodes. So scenario 3 can never occur in this algorithm.

Starvation is impossible

When node A holds the privilege, and node B requests the privilege, the identity of B or the i.d.s of the proxy nodes for node B will be present in the REQUEST_Qs of various nodes in the path connecting the requesting node to the currently privileged node. So, depending upon the position of the i.d. of node B in those REQUEST_Qs, node B will sooner or later receive the privilege. Thus once node B's REQUEST message reaches the privileged node A, node B is sure to receive the privilege.

To better illustrate, let us consider Figure 9.19. Node B is the current holder of the privilege. Suppose that node C is already at the head of REQUEST_Q_B. Assume that the REQUEST_Qs of all other nodes are empty. Now if node E wants to enter the critical section, it will send a REQUEST message to its immediate neighbor, node A. We will show that node E does not starve. Assume that B is executing the critical section by the time E's REQUEST is propagated to node B. At this instance, the REQUEST_Qs of E, A, and B will be as follows:

$$\text{REQUEST\_Q}_E = \text{self},$$

$$\text{REQUEST\_Q}_A = \text{E},$$

$$\text{REQUEST\_Q}_B = \text{C, A}.$$

When node B exits the critical section, it removes the node at the head of REQUEST_Q_B, i.e., node C, and send the privilege to node C. Node B will then send a REQUEST to node C on behalf of node A, which requested privilege on behalf of node E. After node C receives the privilege and completes executing the critical section, the REQUEST_Qs of nodes C, B, A, and E will look as follows:

$$\text{REQUEST\_Q}_C = \text{B},$$
$$\text{REQUEST\_Q}_B = \text{A},$$
$$\text{REQUEST\_Q}_A = \text{E},$$
$$\text{REQUEST\_Q}_E = \text{self}.$$

Now, the next node to receive the privilege will be node E, a fact that is represented by the logical path "BAE" that the REQUEST_Qs of nodes C, B, and A form. Since node B had requested privilege on behalf of node A, and node A on behalf of node E, the PRIVILEGE ultimately gets propagated to node E. Thus, a node never starves.

9.12.5 Cost and performance analysis

The algorithm exhibits the following worst-case cost: (2 * longest path length of the tree) messages per critical section entry. This happens when the privilege is to be passed between nodes at either end of the longest path of the minimal spanning tree. Thus the worst possible network topology for this algorithm will be one where all nodes are arranged in a straight line. In a straight line the longest path length will be $N - 1$, and thus the algorithm will exchange $2 * (N - 1)$ messages per CS execution. However, if all nodes generate equal number of REQUEST messages for the privilege, the average number of messages needed per critical section entry will be approximately $2N/3$ because the average distance between a requesting node and a privileged node is $(N + 1)/3$.

The best topology for the algorithm is the radiating star topology. The worst-case cost of this algorithm for this topology is $O(\log_{K-1} N)$. Even among radiating star topologies, trees with higher fan-outs are preferred. The longest path length of such trees is typically $O(\log N)$. Thus, on average, this algorithm involves the exchange of $O(\log N)$ messages per critical section execution.

When under heavy load, the algorithm exhibits an interesting property: "as the number of nodes requesting the privilege increases, the number of messages exchanged per critical section entry decreases." In fact, it requires the exchange of only four messages per CS execution as explained below.

When all nodes are sending privilege requests, PRIVILEGE messages travel along all $N-1$ edges of the minimal spanning tree exactly twice to give the privilege to all N nodes. Each of these PRIVILEGE messages travel in response to a REQUEST message. Thus, a total of $4 * (N-1)$ messages travel across the minimal spanning tree. Hence, the total number of messages exchanged per critical section execution is $4(N-1)/\text{N}$, which is approximately 4.

9.12.6 Algorithm initialization

Algorithm initialization begins with one node being chosen as the privileged node. This node then sends INITIALIZE messages to its immediate neighbors. On receiving the INITIALIZE message, a node sets its HOLDER variable to the node that sent the INITIALIZE message, and send INITIALIZE messages to its own immediate neighbors. Once INITIALIZE message is received, a node can start making privilege requests even if the entire tree is not initialized.

The initialization of the following variables is the same at all nodes:

$$\text{USING} := \text{false},$$

$$\text{ASKED} := \text{false},$$

$$\text{REQUEST\_}Q := \text{empty}.$$

9.12.7 Node failures and recovery

If a node fails, lost information can be reconstructed on restart. Once a node restarts, it enters into a recovery phase and selects a delay period for the recovery phase in order to get back all the lost information. It sends RESTART messages to its immediate neighbors and waits for ADVISE messages. During the recovery phase, the node can still receive REQUEST and PRIVILEGE messages; it acts as any normal node would act in response to those messages except that ASSIGN_PRIVILEGE and MAKE_REQUEST routines are not executed.

The ADVISE message that a recovering node A receives from each immediate neighbor B will contain information on the HOLDER, ASKED, and REQUEST_Q variables of B, from which A can reconstruct its own HOLDER, ASKED, and REQUEST_Q variables.

For example, if $\text{HOLDER}_B = \text{A}$ for all immediate neighbors B of node A, it means node A holds the privilege, and hence $\text{HOLDER}_A = \text{self}$. Similar reasoning can be applied to determine value of ASKED_A and the elements of REQUEST_Q_A. REQUEST_Q_A can be reconstructed but the elements may not be in proper order. To ensure proper order, the ADVISE messages could provide real or logical timestamps for its REQUEST messages. USING_A can be set to false.

The recovering node's REQUEST_Q can have duplicates if it processes REQUEST messages sent currently and the ones it receives in the ADVISE messages. However, this does not affect the working of the algorithm as long as the REQUEST_Q is large enough to accommodate such situations. A node can also possibly fail when recovering from an earlier failure. In such a case, ASSIST messages related to the first recovery phase can be identified by making use of the delay chosen for recovery or unique identifiers, and those messages can be discarded.

9.13 Chapter summary

Mutual exclusion is a fundamental problem in distributed computing systems, where concurrent access to a shared resource or data is serialized. Mutual exclusion in a distributed system requires that only one process be allowed to execute the critical section at any given time. Mutual exclusion algorithms for distributed computing systems have been designed based on three approaches: token-based approach, non-token-based approach, and quorum-based approach. In token-based algorithms, a unique token is shared among the sites and a site is allowed to enter its critical section only if it possesses the token. Depending upon the way the token is managed in the system, there are several token-based algorithms.

In the non-token-based approach, sites exchange two or more rounds of messages to determine which site will enter the critical section next. In the quorum-based approach, each site requests permission from a subset of sites (called a quorum). The quorums are formed in such a way that when two sites concurrently request access to the CS, at least one site receives both the requests and which is responsible to make sure that only one request executes the critical section at any time.

A large number of mutual exclusion algorithms based on these approaches have been developed. In this chapter, we described a set of representative mutual exclusion algorithms. Early mutual exclusion algorithms were static in the sense they always take the same course of actions to invoke mutual exclusion regardless of the state of the system. These algorithms lack efficiency because these algorithms fail to exploit the changing conditions in the system. Lately, dynamic mutual exclusion algorithms have been developed. Such algorithms exploit dynamic conditions of the system to optimize the performance.

9.14 Exercises

Exercise 9.1 Consider the following simple method to enforce mutual exclusion: all sites are arranged in a logical ring fashion and a unique token circulates around the ring hopping from a site to another site. When a site needs to executes its CS, it waits for the token, grabs the token, executes the CS, and then dispatches the token to the next site on the ring. If a site does not need the token on its arrival, it immediately dispatches the token to the next site (in zero time).

1. What is the reponse time when the load is low?
2. What is the reponse time when the load is heavy?

Assume there are N sites, the message/token delay is T, and the CS execution time is E.

Exercise 9.2 In Lamport's algorithm, condition L1 can hold concurrently at several sites. Why do we need this condition for guaranteeing mutual exclusion?

Exercise 9.3 Show that in Lamport's algorithm if a site S_i is executing the critical section, then S_i's request need not be at the top of the *request_queue* at another site S_j. Is this still true when there are no messages in transit?

Exercise 9.4 What is the purpose of a REPLY message in Lamport's algorithm? Note that it is not necessary that a site must always return a REPLY message in response to a REQUEST message. State the condition under which a site does not have to return REPLY message. Also, give the new message complexity per critical section execution in this case.

Exercise 9.5 Show that in the Ricart–Agrawala algorithm the critical section is accessed in increasing order of timestamp. Does the same hold in Maekawa's algorithm?

Exercise 9.6 Mutual exclusion can be achieved using the following simple method in a distributed system (called the "centralized" mutual exclusion algorithm): to access the shared resource, a site sends the request to the site that contains the resource. This site executes the requests using any classical methods for mutual exclusion (like semaphores).

Discuss what prompted Lamport's mutual exclusion algorithm even though it requires many more messages ($3(N-1)$ as compared to only 3).

Exercise 9.7 Show that in Lamport's algorithm the critical section is accessed in increasing order of timestamp.

Exercise 9.8 Show by examples that the staircase configuration among sites is preserved in Singhal's dynamic mutual exclusion algorithm when two or more sites request the CS concurrently and have executed the CSs.

9.15 Notes on references

Singhal gives a taxonomy on distributed mutual exclusion in [24]. Raynal presents a survey of mutual exclusion algorithms in [20]. A large number of token-based mutual exclusion algorithms have appeared in the last several years, e.g., mutual exclusion algorithms by Ahamad and Bernabeu [2], Helary *et al.* [10], Naimi and Trehel [15], Chang *et al.* [6], and Neilsen and Mizuno [16]. In [23], Saxena and Rai present a survey of permission-based distributed mutual exclusion algorithms.

Nishio *et al.* [18] presented a technique for generation of a unique token in case of a token loss. A dynamic heuristic-based token mutual exclusion algorithm is given in [26]. Snepscheut [30] extended tree-based algorithms to handle a connected network of any topology (i.e., graphs). Due to network topology, such algorithms are fault-tolerant to site and link failures. Chang *et al.* [7] present a fault-tolerant mutual exclusion algorithm. Goscinski [8] has presented two mutual exclusion algorithms for real-time distributed systems. Coterie-based mutual exclusion algorithms, which are a generalization of Maekawa's $\sqrt{N}$ algorithm, have lately attracted considerable attention. Barbara and Garcia-Molina [9] and Ibaraki and Kameda [11] have discussed theoretical aspects of coteries. Cao and Singhal developed a delay optimal coterie-based mutual exclusion algorithm [5].

Sanders [22] gave the concept of information structures to develop a generalized mutual exclusion algorithm. Other mutual exclusion algorithms can be found in [3,4,17,25].

References

[1] D. Agrawal and A. E. Abbadi, An efficient and fault-tolerant solution for distributed mutual exclusion, *ACM Transactions on Computer Systems*, **9**(1), 1991, 1–20.

[2] J. M. Bernabeu-Auban and M. Ahamad, Applying a path-compression technique to obtain an effective distributed mutual exclusion algorithm, *Proceedings of the 3rd International Workshop on Distributed Algorithms*, September 1989, 33–44.

[3] G. Buckley and A. Silberschatz, A failure tolerant centralized mutual exclusion algorithm, *Proceedings of the 4th International Conference on Distributed Computing Systems*, May 1984, 347–356.

[4] O. S. F. Carvalho and G. Roucairol, On mutual exclusion in real-time distributed computing systems, Technical Correspondence, *Communications of the ACM*, **26**(2), 1983, 146–147.

[5] G. Cao and M. Singhal, A delay-optimal quorum-based mutual exclusion algorithm for distributed systems, *IEEE Transactions on Parallel and Distributed Systems*, **12**(12), 2001, 1256–1268.

[6] Y. Chang, M. Singhal, and M. Liu, A dynamic token-based distributed mutual exclusion algorithm, *Proceedings of the 10th IEEE International Phoenix Conference on Computer and Communications*, March 1991, 240–246.

[7] Y. Chang, M. Singhal, and M. Liu, A fault-tolerant mutual exclusion algorithm for distributed systems, *Proceedings of the 9th Symposium on Reliable Distributed Software and Systems*, October 1990, 146–154.

[8] A. Goscinski, Two algorithms for mutual exclusion in real-time distributed computing systems, *Journal of Parallel and Distributed Computing*, **9**(1), 1990, 77–82.

[9] H. Garcia-Molina and D. Barbara, How to assign votes in a distributed system, *Journal of the ACM*, 1985.

[10] M. Helary, N. Plouzeau, and M. Raynal A distributed algorithm for mutual exclusion in an arbitrary network, *Computing Journal*, **31**(4), 1988, 289–295.

[11] T. Ibaraki and T. Kameda, *Theory of Coteries*, Technical Report, CSS/LCCR TR90-09, University of Kyoto, Kyoto, Japan, 1990.

[12] L. Lamport Time, clocks and ordering of events in distributed systems, *Communications of the ACM*, **21**(7), 1978, 558–565.

[13] S. Lodha and A. Kshemkalyani, A fair distributed mutual exclusion algorithm, *IEEE Transactions on Parallel and Distributed Systems*, **11**(6), 2000, 537–549.

[14] M. Maekawa, A $\sqrt{N}$ algorithm for mutual exclusion in decentralized systems, *ACM Transactions on Computer Systems*, **3**(2), 1995, 145–159.

[15] M. Naimi and M. Trehel, An improvement of the $\log N$ distributed algorithm for mutual exclusion, *Proceedings of the 7th International Conference on Distributed Computing Systems*, September 23–25, 1987, 371–377.

[16] M. L. Neilsen and M. Mizuno, A DAG-based algorithm for distributed mutual exclusion, *Proceedings of the 11th International Conference on Distributed Computing Systems*, May 21–23, 1991, 354–360.

[17] M. Nesterenko and M. Mizuno, A quorum-based self-stabilizing distributed mutual exclusion algorithm, *Journal of Parallel and Distributed Computing*, **62**(2), 2002, 284–305.

[18] S. Nishio, K. F. Li, and E. G. Manning, A resilient mutual exclusion algorithm for computer networks, *IEEE Transactions on Parallel and Distributed Systems*, **1**(3), 1990, 344–356.

[19] K. Raymond, Tree-based algorithm for distributed mutual exclusion, *ACM Transactions on Computer Systems*, **7**, 1989, 61–77.

[20] M. Raynal, A simple taxonomy of distributed mutual exclusion algorithms, *Operating Systems Review*, **25**(2), 1991, 47–50.

[21] G. Ricart and A. K. Agrawala, An optimal algorithm for mutual exclusion in computer networks, *Communications of the ACM*, **24**(1), 1981, 9–17.

[22] B. Sanders, The information structure of distributed mutual exclusion algorithms, *ACM Transactions on Computer Systems*, **5**(3), 1987, 284–299.

[23] P. C. Saxena and J. Rai, A survey of permission-based distributed mutual exclusion algorithms, *Computer Standards and Interfaces*, **25**(2), 2003, 159–181.

[24] M. Singhal, A taxonomy of distributed mutual exclusion, *Journal of Parallel and Distributed Computing*, **18**(1), 1993, 94–101.

[25] M. Singhal, "A class of deadlock-free Maekawa type mutual exclusion algorithms for distributed systems", *Distributed Computing*, **4**(3), 1991, 131–138.

[26] M. Singhal, A heuristically-aided algorithm for mutual exclusion in distributed systems, *IEEE Transactions on Computers*, **38**(5), 1989, 651–662.

[27] M. Singhal, A dynamic information structure mutual exclusion algorithm for distributed systems, *Proceedings of the 9th International Conference on Distributed Computing Systems*, June 5–9, 1989, Newport Beach, CA, 70–78.

[28] M. Singhal, A dynamic information-structure mutual exclusion algorithm for distributed systems, *IEEE Transactions on Parallel and Distributed Systems*, **3**(1), 1992, 121–125.

[29] I. Suzuki and T. Kasami, A distributed mutual exclusion algorithm, *ACM Transactions on Computer Systems*, **3**(4), 1985, 344–349.

[30] J. L. A. van de Snepscheut, Fair mutual exclusion on a graph of processes, *Distributed Computing*, **2**, 1987, 113–115.

CHAPTER

10 Deadlock detection in distributed systems

10.1 Introduction

Deadlocks are a fundamental problem in distributed systems and deadlock detection in distributed systems has received considerable attention in the past. In distributed systems, a process may request resources in any order, which may not be known a priori, and a process can request a resource while holding others. If the allocation sequence of process resources is not controlled in such environments, deadlocks can occur. A deadlock can be defined as a condition where a set of processes request resources that are held by other processes in the set.

Deadlocks can be dealt with using any one of the following three strategies: deadlock prevention, deadlock avoidance, and deadlock detection. Deadlock prevention is commonly achieved by either having a process acquire all the needed resources simultaneously before it begins execution or by pre-empting a process that holds the needed resource. In the deadlock avoidance approach to distributed systems, a resource is granted to a process if the resulting global system is safe. Deadlock detection requires an examination of the status of the process–resources interaction for the presence of a deadlock condition. To resolve the deadlock, we have to abort a deadlocked process.

In this chapter, we study several distributed deadlock detection techniques based on various strategies.

10.2 System model

A distributed system consists of a set of processors that are connected by a communication network. The communication delay is finite but unpredictable. A distributed program is composed of a set of n asynchronous processes P_1, $P_2, \ldots, P_i, \ldots, P_n$ that communicate by message passing over the communication network. Without loss of generality we assume that each process is running on a different processor. The processors do not share a common global memory and communicate solely by passing messages over the communication network. There is no physical global clock in the system to which processes have instantaneous access. The communication medium may deliver messages out of order, messages may be lost, garbled, or duplicated due to timeout and retransmission, processors may fail, and communication links may go down. The system can be modeled as a directed graph in which vertices represent the processes and edges represent unidirectional communication channels.

We make the following assumptions:

- The systems have only reusable resources.
- Processes are allowed to make only exclusive access to resources.
- There is only one copy of each resource.

A process can be in two states, *running* or *blocked*. In the running state (also called *active* state), a process has all the needed resources and is either executing or is ready for execution. In the blocked state, a process is waiting to acquire some resource.

10.2.1 Wait-for graph (WFG)

In distributed systems, the state of the system can be modeled by a directed graph, called a *wait-for graph* (WFG). In a WFG, nodes are processes and there is a directed edge from node P_1 to node P_2 if P_1 is blocked and is waiting for P_2 to release some resource. A system is deadlocked if and only if there exists a directed cycle or knot in the WFG.

Figure 10.1 shows a WFG, where process P_{11} of site 1 has an edge to process P_{21} of site 1 and an edge to process P_{32} of site 2. Process P_{32} of site 2 is waiting for a resource that is currently held by process P_{33} of site 3. At the same time process P_{21} at site 1 is waiting on process P_{24} at site 4 to release a resource, and so on. If P_{33} starts waiting on process P_{24}, then processes in the WFG are involved in a deadlock depending upon the request model.

10.3 Preliminaries

10.3.1 Deadlock handling strategies

There are three strategies for handling deadlocks, *viz.*, deadlock prevention, deadlock avoidance, and deadlock detection. Handling of deadlocks becomes highly complicated in distributed systems because no site has accurate knowledge of the current state of the system and because every inter-site communication involves a finite and unpredictable delay. Deadlock prevention is commonly achieved either by having a process acquire all the needed

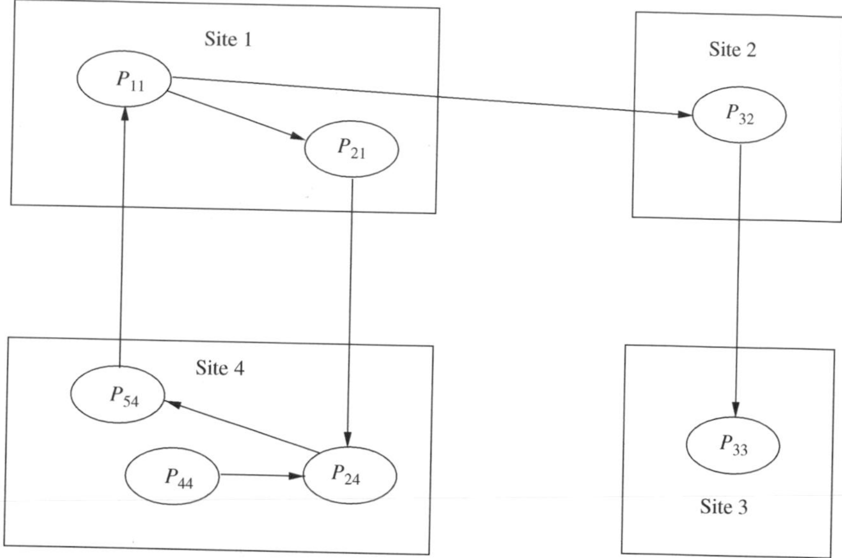

Figure 10.1 Example of a WFG.

resources simultaneously before it begins executing or by pre-empting a process that holds the needed resource. This approach is highly inefficient and impractical in distributed systems.

In the deadlock avoidance approach to distributed systems, a resource is granted to a process if the resulting global system state is safe (note that a global state includes all the processes and resources of the distributed system). Due to several problems, however, deadlock avoidance is impractical in distributed systems.

Deadlock detection requires an examination of the status of process–resource interactions for the presence of cyclic wait. Deadlock detection in distributed systems seems to be the best approach to handle deadlocks in distributed systems. In this chapter, we limit the discussion to deadlock detection techniques in distributed systems.

10.3.2 Issues in deadlock detection

Deadlock handling using the approach of deadlock detection entails addressing two basic issues: first, detection of existing deadlocks and, second, resolution of detected deadlocks.

Detection of deadlocks

Detection of deadlocks involves addressing two issues: maintenance of the WFG and searching of the WFG for the presence of cycles (or knots). Since, in distributed systems, a cycle or knot may involve several sites, the search for cycles greatly depends upon how the WFG of the system is represented across the system. Depending upon the way WFG information is maintained and the search for cycles is carried out, there are centralized, distributed, and hierarchical algorithms for deadlock detection in distributed systems [43].

Correctness criteria

A deadlock detection algorithm must satisfy the following two conditions:

- **Progress (no undetected deadlocks)** The algorithm must detect all existing deadlocks in a finite time. Once a deadlock has occurred, the deadlock detection activity should continuously progress until the deadlock is detected. In other words, after all wait-for dependencies for a deadlock have formed, the algorithm should not wait for any more events to occur to detect the deadlock.
- **Safety (no false deadlocks)** The algorithm should not report deadlocks that do not exist (called *phantom or false* deadlocks). In distributed systems where there is no global memory and there is no global clock, it is difficult to design a correct deadlock detection algorithm because sites may obtain an out-of-date and inconsistent WFG of the system. As a result, sites may detect a cycle that never existed but whose different segments existed in the system at different times. This is the main reason why many deadlock detection algorithms reported in the literature are incorrect.

Resolution of a detected deadlock

Deadlock resolution involves breaking existing wait-for dependencies between the processes to resolve the deadlock. It involves rolling back one or more deadlocked processes and assigning their resources to blocked processes so that they can resume execution. Note that several deadlock detection algorithms propagate information regarding wait-for dependencies along the edges of the wait-for graph. Therefore, when a wait-for dependency is broken, the corresponding information should be immediately cleaned from the system. If this information is not cleaned in a timely manner, it may result in detection of phantom deadlocks. Untimely and inappropriate cleaning of broken wait-for dependencies is the main reason why many deadlock detection algorithms reported in the literature are incorrect.

10.4 Models of deadlocks

Distributed systems allow many kinds of resource requests. A process might require a single resource or a combination of resources for its execution. This section introduces a hierarchy of request models starting with very restricted forms to the ones with no restrictions whatsoever. This hierarchy shall be used to classify deadlock detection algorithms based on the complexity of the resource requests they permit.

10.4.1 The single-resource model

The single-resource model is the simplest resource model in a distributed system, where a process can have at most one outstanding request for only one unit of a resource. Since the maximum out-degree of a node in a WFG for the single resource model can be 1, the presence of a cycle in the WFG shall indicate that there is a deadlock. In a later section, an algorithm to detect deadlock in the single-resource model is presented.

10.4.2 The AND model

In the AND model, a process can request more than one resource simultaneously and the request is satisfied only after all the requested resources are granted to the process. The requested resources may exist at different locations. The out degree of a node in the WFG for AND model can be more than 1. The presence of a cycle in the WFG indicates a deadlock in the AND model. Each node of the WFG in such a model is called an AND node.

Consider the example WFG described in the Figure 10.1. Process P_{11} has two outstanding resource requests. In case of the AND model, P_{11} shall become active from idle state only after both the resources are granted. There is a cycle $P_{11} \rightarrow P_{21} \rightarrow P_{24} \rightarrow P_{54} \rightarrow P_{11}$, which corresponds to a deadlock situation.

In the AND model, if a cycle is detected in the WFG, it implies a deadlock but not vice versa. That is, a process may not be a part of a cycle, it can still be deadlocked. Consider process P_{44} in Figure 10.1. It is not a part of any cycle but is still deadlocked as it is dependent on P_{24}, which is deadlocked. Since in the single-resource model, a process can have at most one outstanding request, the AND model is more general than the single-resource model.

10.4.3 The OR model

In the OR model, a process can make a request for numerous resources simultaneously and the request is satisfied if any one of the requested resources is granted. The requested resources may exist at different locations. If all requests in the WFG are OR requests, then the nodes are called OR nodes. Presence of a cycle in the WFG of an OR model does not imply a deadlock in the OR model. To make it more clear, consider Figure 10.1. If all nodes are OR nodes, then process P_{11} is not deadlocked because once process P_{33} releases its resources, P_{32} shall become active as one of its requests is satisfied. After P_{32} finishes execution and releases its resources, process P_{11} can continue with its processing.

In the OR model, the presence of a knot indicates a deadlock [19]. In a WFG, a vertex v is in a knot if for all $u :: u$ is reachable from $v : v$ is reachable from u. No paths originating from a knot shall have dead ends.

A deadlock in the OR model can be intuitively defined as follows [6]: a process P_i is blocked if it has a pending OR request to be satisfied. With every blocked process, there is an associated set of processes called dependent set. A process shall move from an *idle* to an *active* state on receiving a grant message from any of the processes in its dependent set. A process is permanently blocked if it never receives a grant message from any of the processes in its dependent set. Intuitively, a set of processes S is deadlocked if all the processes in S are permanently blocked. To formally state that a set of processes is deadlocked, the following conditions hold true:

1. Each of the processes in the set S is blocked.
2. The dependent set for each process in S is a subset of S.
3. No grant message is in transit between any two processes in set S.

We now show that a set of processes S shall remain permanently blocked in the OR model if the above conditions are met. A blocked process in the set S becomes *active* only after receiving a grant message from a process in its dependent set, which is a subset of S. Note that no grant message can be expected from any process in S because they are all blocked. Also, the third condition states that no grant messages in transit between any two processes in set S. So, all the processes in set S are permanently blocked.

Hence, deadlock detection in the OR model is equivalent to finding knots in the graph. Note that, there can be a deadlocked process that is not a part of a knot. Consider Figure 10.1, where P_{44} can be deadlocked even though it is not in a knot. So, in the OR model, a blocked process P is deadlocked if it is either in a knot or it can only reach processes on a knot.

10.4.4 The AND-OR model

A generalization of the previous two models (OR model and AND model) is the AND-OR model. In the AND-OR model, a request may specify any combination of *and* and *or* in the resource request. For example, in the AND-OR model, a request for multiple resources can be of the form *x and* (*y or z*). The requested resources may exist at different locations. To detect the presence of deadlocks in such a model, there is no familiar construct of graph theory using WFG. Since a deadlock is a stable property (i.e., once it exists, it does not go away by itself), this property can be exploited and a deadlock in the AND-OR model can be detected by repeated application of the test for OR-model deadlock. However, this is a very inefficient strategy. Efficient algorithms to detect deadlocks in AND-OR model are discussed in [16].

10.4.5 The $\binom{p}{q}$ model

Another form of the AND-OR model is the $\binom{p}{q}$ model (called the *P*-out-of-*Q* model), which allows a request to obtain any *p* available resources from a pool of q resources. Both the models are the same in expressive power. However, $\binom{p}{q}$ model lends itself to a much more compact formation of a request.

Every request in the $\binom{p}{q}$ model can be expressed in the AND-OR model and vice-versa. Note that AND requests for p resources can be stated as $\binom{p}{p}$ and OR requests for p resources can be stated as $\binom{1}{p}$.

10.4.6 Unrestricted model

In the unrestricted model, no assumptions are made regarding the underlying structure of resource requests. In this model, only one assumption that the deadlock is stable is made and hence it is the most general model. This way of looking at the deadlock problem helps in separation of concerns: concerns about properties of the problem (stability and deadlock) are separated from underlying distributed systems computations (e.g., message passing versus synchronous communication). Hence, these algorithms can be used to detect other stable properties as they deal with this general model. But, these algorithms are of more theoretical value for distributed systems since no further assumptions are made about the underlying distributed systems computations which leads to a great deal of overhead (which can be avoided in simpler models like AND or OR models).

10.5 Knapp's classification of distributed deadlock detection algorithms

Distributed deadlock detection algorithms can be divided into four classes [22]: path-pushing, edge-chasing, diffusion computation, and global state detection.

10.5.1 Path-pushing algorithms

In path-pushing algorithms, distributed deadlocks are detected by maintaining an explicit global WFG. The basic idea is to build a global WFG for each site of the distributed system. In this class of algorithm, whenever deadlock computation is performed, each site sends its local WFG to all the neighboring sites. After the local data structure of each site is updated, this updated WFG is then passed along to other sites, and the procedure is repeated until one site has a sufficiently complete picture of the global state to announce deadlock or to establish that no deadlocks are present. This feature of sending around the paths of the global WFG has led to the term path-pushing algorithms.

Examples of such algorithms are Menasce-Muntz [33], Gligor and Shattuck [11], Ho and Ramamoorthy [18], and Obermarck [38].

10.5.2 Edge-chasing algorithms

In an edge-chasing algorithm, the presence of a cycle in a distributed graph structure is verified by propagating special messages called probes along the edges of the graph. These probe messages are different from the request and reply messages. The formation of a cycle can be detected by a site if it receives the matching probe sent by it previously.

Whenever a process that is executing receives a probe message, it simply discards this message and continues. Only blocked processes propagate probe messages along their outgoing edges. An interesting variation of this method can be found in Mitchell [35], where probes are sent upon request and in the opposite direction of the edges.

The main advantage of edge-chasing algorithms is that probes are fixed size messages that are normally very short. Examples of such algorithms include the Chandy *et al.* [6], Choudhary *et al.* [7], Kshemkalyani–Singhal [27], and Sinha–Natarajan [42] algorithms.

10.5.3 Diffusing computation-based algorithms

In *diffusion computation*-based distributed deadlock detection algorithms, deadlock detection computation is diffused through the WFG of the system. These algorithms make use of echo algorithms to detect deadlocks [5]. This computation is superimposed on the underlying distributed computation. If this computation terminates, the initiator declares a deadlock. The main feature of the superimposed computation is that the global WFG is implicitly reflected in the structure of the computation. The actual WFG is never built explicitly.

To detect a deadlock, a process sends out query messages along all the outgoing edges in the WFG. These queries are successively propagated (i.e., diffused) through the edges of the WFG. Queries are discarded by a running process and are echoed back by blocked processes in the following way: when a blocked process first receives a query message for a particular deadlock detection initiation, it does not send a reply message until it has received a reply message for every query it sent (to its successors in the WFG). For all subsequent queries for this deadlock detection initiation, it immediately sends back a reply message. The initiator of a deadlock detection detects a deadlock when it has received a reply for every query it has sent out. Examples of these types of deadlock detection algorithms include the Chandy–Misra–Haas algorithm for one OR model [6] and the Chandy–Herman algorithm [16].

10.5.4 Global state detection-based algorithms

Global state detection-based deadlock detection algorithms exploit the following facts: (i) a consistent snapshot of a distributed system can be obtained without freezing the underlying computation, and (ii) a consistent snapshot may not represent the system state at any moment in time, but if a stable property holds in the system before the snapshot collection is initiated, this property will still hold in the snapshot.

Therefore, distributed deadlocks can be detected by taking a snapshot of the system and examining it for the condition of a deadlock. Examples of these types of algorithms include the Bracha–Toueg [2], Wang *et al.* [45], and Kshemkalyani–Singhal [26] algorithms.

10.6 Mitchell and Merritt's algorithm for the single-resource model

Mitchell and Merritt's algorithm [35] belongs to the class of edge-chasing algorithms where probes are sent in the opposite direction to the edges of the WFG. When a probe initiated by a process comes back to it, the process declares deadlock. The algorithm has many good features, such as:

1. Only one process in a cycle detects the deadlock. This simplifies the deadlock resolution – this process can abort itself to resolve the deadlock. This algorithm can be improvised by including priorities, and the lowest priority process in a cycle detects deadlock and aborts.
2. In this algorithm, a process that is detected in deadlock is aborted spontaneously, even though under this assumption phantom deadlocks cannot be excluded. It can be shown, however, that only genuine deadlocks will be detected in the absence of spontaneous aborts.

Each node of the WFG has two local variables, called labels: a private label, which is unique to the node at all times, though it is not constant, and a public label, which can be read by other processes and which may not be unique. Each process is represented as u/v, where u and v are the public and private labels, respectively. Initially, private and public labels are equal for each process.

A global WFG is maintained and it defines the entire state of the system. The algorithm is defined by the four state transitions shown in Figure 10.2, where $z = inc(u, v)$, and $inc(u, v)$ yields a unique label greater than both u and v. Labels that are not shown do not change. Block creates an edge in the WFG. Two messages are needed: one resource request and one message back to the blocked process to inform it of the public label of the process it is waiting for. Activate denotes that a process has acquired the resource from the process it was waiting for. Transmit propagates larger labels in the opposite direction to the edges by sending a probe message. Whenever a process receives a probe that is less than its public label, it simply ignores that probe. Detect means that the probe with the private label of some process has returned to it, indicating a deadlock.

Mitchell and Merritt showed that every deadlock is detected. Next, we show that, in the absence of spontaneous aborts, only genuine deadlocks are detected. As there are no spontaneous aborts, we have the following invariant:

$$\text{for all processes } u/v\text{: } v \leqslant u.$$

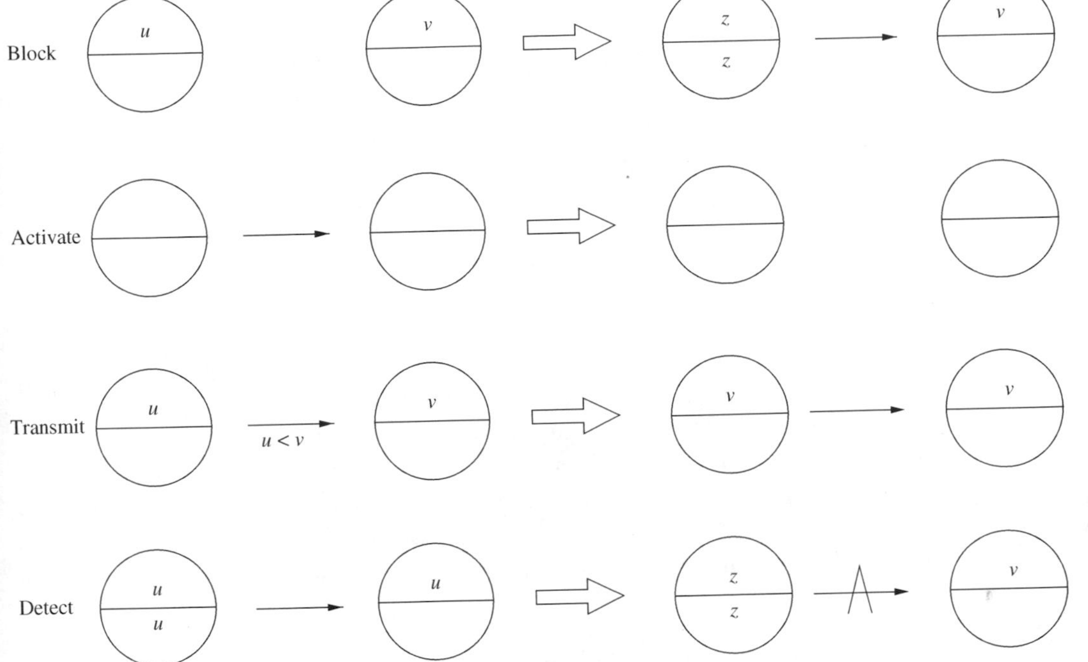

Figure 10.2 The four possible state transitions [22].

Proof Initially $u = v$ for all processes. The only requests that change u or v are:

1. Block: u and v are set such that $u = v$.
2. Transmit: u is increased.

Hence, the invariant follows. □

From the previous invariant, we have the following lemmas.

Lemma 10.1 *For any process u/v, if $u > v$, then u was set by a Transmit step.*

Theorem 10.1 *If a deadlock is detected, a cycle of blocked nodes exists.*

Proof A deadlock is detected if the following edge $p \rightarrow p'$ exists:

We have the following claims:

1. u has been propagated from p to p' via a sequence of transmits.
2. p has been continuously blocked since it transmitted u.
3. All intermediate nodes in the Transmit path of (1), including p', have been continuously blocked since they transmitted u.

From the above claims, the proof for the theorem follows as discussed below.

From the invariant and the uniqueness of private label v of $p' : v < u$. By Lemma 10.1, u was set by a Transmit step. From the semantics of Transmit, there is some p'' with private label u and public label w. If $w = u$, then $p'' = p$, and it is a success. Otherwise, if $w < u$, we repeat the argument. Since there is only one process with $u = v$, it is p. If p is active then it indicates that it has transmitted u else it is blocked if it detects deadlock. Hence upon blocking it incremented its private label. But private and public labels cannot be equal. Consider a process that has been active since it transmitted u. Clearly, its predecessor is also active, as Transmits migrate in opposite direction. By repeating this argument, we can show that p has been active since it transmitted u. □

The above algorithm can easily be extended to include priorities, so that whenever a deadlock occurs, the lowest priority process gets aborted. This algorithm has two phases. The first phase is almost identical to the algorithm. In the second phase the smallest priority is propagated around the circle. The propagation stops when one process recognizes the propagated priority as its own.

Message complexity

Now we calculate the complexity of the algorithm. If we assume that a deadlock persists long enough to be detected, the worst-case complexity of the algorithm is $s(s-1)/2$ Transmit steps, where s is the number of processes in the cycle.

10.7 Chandy–Misra–Haas algorithm for the AND model

We now discuss Chandy–Misra–Haas's distributed deadlock detection algorithm for the AND model [6], which is based on edge-chasing.

The algorithm uses a special message called *probe*, which is a triplet (i, j, k), denoting that it belongs to a deadlock detection initiated for process P_i and it is being sent by the home site of process P_j to the home site of process P_k. A probe message travels along the edges of the global WFG graph, and a deadlock is detected when a probe message returns to the process that initiated it.

A process P_j is said to be *dependent* on another process P_k if there exists a sequence of processes P_j, P_{i1}, P_{i2}, ..., P_{im}, P_k such that each process except

P_k in the sequence is blocked and each process, except P_j, holds a resource for which the previous process in the sequence is waiting. Process P_j is said to be *locally dependent* upon process P_k if P_j is dependent upon P_k and both the processes are on the same site.

Data structures

Each process P_i maintains a boolean array, $dependent_i$, where $dependent_i(j)$ is true only if P_i knows that P_j is dependent on it. Initially, $dependent_i(j)$ is false for all i and j.

The algorithm

Algorithm 10.1 is executed to determine if a blocked process is deadlocked. Therefore, a probe message is continuously circulated along the edges of the global WFG graph and a deadlock is detected when a probe message returns to its initiating process.

---

if P_i is locally dependent on itself
  then declare a deadlock
  else for all P_j and P_k such that
    (a) P_i is locally dependent upon P_j, and
    (b) P_j is waiting on P_k, and
    (c) P_j and P_k are on different sites,
  send a probe (i, j, k) to the home site of P_k

On the receipt of a probe (i, j, k), the site takes
  the following actions:

if
    (d) P_k is blocked, and
    (e) $dependent_k(i)$ is false, and
    (f) P_k has not replied to all requests by P_j,
  then
    begin
      $dependent_k(i) = true$;
      if $k = i$
        then declare that P_i is deadlocked
      else for all P_m and P_n such that
        (a′) P_k is locally dependent upon P_m, and
        (b′) P_m is waiting on P_n, and
        (c′) P_m and P_n are on different sites,
      send a probe (i, m, n) to the home site of P_n
    end.

Algorithm 10.1 Chandy–Misra–Haas algorithm for the AND model [6].

Performance analysis

In the algorithm, one probe message (per deadlock detection initiation) is sent on every edge of the WFG which connects processes on two sites. Thus, the algorithm exchanges at most $m(n-1)/2$ messages to detect a deadlock that involves m processes and spans over n sites. The size of messages is fixed and is very small (only three integer words). The delay in detecting a deadlock is $O(n)$.

10.8 Chandy–Misra–Haas algorithm for the OR model

We now discuss Chandy–Misra–Haas's distributed deadlock detection algorithm for the OR model [6], which is based on the approach of diffusion-computation (see Algorithm 10.2).

A blocked process determines if it is deadlocked by initiating a diffusion computation. Two types of messages are used in a diffusion computation: *query*(i, j, k) and *reply*(i, j, k), denoting that they belong to a diffusion computation initiated by a process P_i and are being sent from process P_j to process P_k.

Basic idea

A blocked process initiates deadlock detection by sending query messages to all processes in its dependent set (i.e., processes from which it is waiting to receive a message). If an active process receives a *query* or *reply* message, it discards it. When a blocked process P_k receives a *query*(i, j, k) message, it takes the following actions:

1. If this is the first *query* message received by P_k for the deadlock detection initiated by P_i (called the *engaging query*), then it propagates the *query* to all the processes in its dependent set and sets a local variable $num_k(i)$ to the number of *query* messages sent.
2. If this is not the engaging *query*, then P_k returns a *reply* message to it immediately provided P_k has been continuously blocked since it received the corresponding engaging *query*. Otherwise, it discards the *query*.

Process P_k maintains a boolean variable $wait_k(i)$ that denotes the fact that it has been continuously blocked since it received the last engaging *query* from process P_i. When a blocked process P_k receives a *reply*(i, j, k) message, it decrements $num_k(i)$ only if $wait_k(i)$ holds. A process sends a reply message in response to an engaging *query* only after it has received a *reply* to every *query* message it has sent out for this engaging *query*.

The initiator process detects a deadlock when it has received *reply* messages to all the *query* messages it has sent out.

The algorithm

The algorithm works as shown in Algorithm 10.2. For ease of presentation, we have assumed that only one diffusion computation is initiated for a process. In practice, several diffusion computations may be initiated for a process (a diffusion computation is initiated every time the process gets blocked), but at any time only one diffusion computation is current for any process. However, messages for outdated diffusion computations may still be in transit. The current diffusion computation can be distinguished from outdated ones by using sequence numbers.

Initiate a diffusion computation for a blocked process P_i:
send *query*(i, i, j) to all processes P_j in the dependent set DS_i of P_i;
$num_i(i) := |DS_i|$; $wait_i(i) := true$;

When a blocked process P_k receives a *query*(i, j, k):
if this is the engaging *query* for process P_i then
send *query*(i, k, m) to all P_m in its dependent set DS_k;
$num_k(i) := |DS_k|$; $wait_k(i) := true$
else if $wait_k(i)$ then send a *reply*(i, k, j) to P_j.

When a process P_k receives a *reply*(i, j, k):
if $wait_k(i)$ then
$num_k(i) := num_k(i) - 1$;
if $num_k(i) = 0$ then
if $i = k$ then **declare a deadlock**
else send *reply*(i, k, m) to the process P_m
which sent the engaging query.

Algorithm 10.2 Chandy–Misra–Haas algorithm for the OR model [6].

Performance analysis

For every deadlock detection, the algorithm exchanges e *query* messages and e *reply* messages, where $e = n(n-1)$ is the number of edges.

10.9 Kshemkalyani–Singhal algorithm for the *P*-out-of-*Q* model

The Kshemkalyani–Singhal algorithm [26] (Algorithm 10.3) to detect deadlocks in the *P*-out-of-*Q* model (also called the generalized distributed deadlocks) is based on the global state detection approach. The Kshemkalyani–Singhal algorithm [26] is a single-phase algorithm, which consists of a fan-out sweep of messages outwards from an initiator process and a fan-in sweep of messages inwards to the initiator process. A *sweep*

Data structures: a node i has the following local variables:

$wait_i$: boolean (:= $false$); /*records the current status.*/
t_i : integer (:= 0); /*denotes the current time.*/
t_block_i : real; /*denotes the local time when i blocked last.*/
$in(i)$: set of nodes whose requests are outstanding at node i.
$out(i)$: set of nodes on which node i is waiting.
p_i : integer (:= 0); /*the number of replies required for unblocking.*/
w_i : real (:= 1.0); /*keeps weight to detect the termination of the algorithm.*/

Computation events:

REQUEST_SEND(i):
/*Executed by node i when it blocks on a p_i-out-of-q_i request.*/
For every node j on which i is blocked do
    $out(i) \leftarrow out(i) \cup \{j\}$;
    send *REQUEST(i)* to j;
set p_i to the number of replies needed;
$t\_block_i \leftarrow t_i$;
$wait_i \leftarrow true$;

REQUEST_RECEIVE(j):
/*Executed by node i when it receives a request made by j. */
$in(i) \leftarrow in(i) \cup \{j\}$.

REPLY_SEND(j):
/*Executed by node i when it replies to a request by j.*/
$in(i) \leftarrow in(i) - \{j\}$;
send *REPLY(i)* to j.

REPLY_RECEIVE(j):
/*Executed by node i when it receives a reply from j to its request.*/
if valid reply for the current request
then
  begin
    $out(i) \leftarrow out(i) - \{j\}$;
    $p_i \leftarrow p_i - 1$;
    $p_i = 0 \rightarrow$
        $\{wait_i \leftarrow false$;
        $\forall k \in out(i)$, **send** *CANCEL(i)* to k;
        $out(i) \leftarrow \emptyset.\}$
  end

CANCEL_RECEIVE(j):
/*Executed by node i when it receives a cancel from j.*/
if $j \in in(i)$ then $in(i) \leftarrow in(i) - \{j\}$.

Algorithm 10.3 Kshemkalyani–Singhal algorithm for the *P*-out-of-*Q* model.

of a WFG is a traversal of the WFG in which all messages are sent in the direction of the WFG edges (outward sweep) or all messages are sent against the direction of the WFG edges (inward sweep). In the outward sweep, the algorithm records a snapshot of a distributed WFG. In the inward sweep, the recorded distributed WFG is reduced to determine if the initiator is deadlocked. Both the outward and the inward sweeps are executed concurrently in the algorithm. Complications are introduced because the two sweeps can overlap in time at a process, i.e., the reduction of the WFG at a process can begin before the WFG at that process has been completely recorded. The algorithm deals with these complications.

System model

The system has n nodes, and every pair of nodes is connected by a logical channel. An event in a computation can be an internal event, a message send event, or a message receive event. Events are assigned timestamps using Lamport's clocks [29].

The computation messages can be either *REQUEST*, *REPLY*, or *CANCEL* messages. To execute a p_i-out-of-q_i request, an active node i sends q_i *REQUEST*s to q_i other nodes and remains blocked until it receives sufficient number of *REPLY* messages. When node i blocks on node j, node j becomes a successor of node i and node i becomes a predecessor of node j in the WFG. A *REPLY* message denotes the granting of a request. A node i unblocks when p_i out of its q_i requests have been granted. When a node unblocks, it sends *CANCEL* messages to withdraw the remaining q_i–p_i requests it had sent.

Sending and receiving of *REQUEST*, *REPLY*, and *CANCEL* messages are *computation events*. The sending and receiving of deadlock detection algorithm messages are *algorithmic* or *control events*.

10.9.1 Informal description of the algorithm

When a node *init* blocks on a *P*-out-of-*Q* request, it initiates the deadlock detection algorithm. The algorithm records the part of the WFG that is reachable from *init* (henceforth, called the *init*'s WFG) in a distributed snapshot [4]; the distributed snapshot includes only those dependency edges and nodes that form *init*'s WFG.

The distributed WFG is recorded using *FLOOD* messages in the outward sweep and the recorded WFG is examined for deadlocks using *ECHO* messages in the inward sweep. To detect a deadlock, the initiator *init* records its local state and sends *FLOOD* messages along all of its outward dependencies. When node i receives the first *FLOOD* message along an existing inward dependency, it records its local state. If node i is blocked at this time, it sends out *FLOOD* messages along all of its outward dependencies to continue the recording of the WFG in the outward sweep. If node i is active at this time (i.e., it does not have any outward dependencies and is a leaf node in the WFG), then it initiates reduction of the WFG by returning an *ECHO* message along the incoming dependency even before the states of all incoming dependencies have been recorded in the WFG snapshot at the leaf node.

ECHO messages perform reduction of the recorded WFG by simulating the granting of requests in the inward sweep. A node i in the WFG is reduced if it receives *ECHO*s along p_i out of its q_i outgoing edges indicating that p_i of its requests can be granted. An edge is reduced if an *ECHO* is received on the edge indicating that the request it represents can be granted. After a local snapshot has been recorded at node i, any transition made by i from idle to active state is captured in the process of reduction. The nodes that can be reduced do not form a deadlock whereas the nodes that cannot be reduced are deadlocked. The order in which reduction of the nodes and edges of the WFG is performed does not alter the final result. Node *init* detects the deadlock if it is not reduced when the deadlock detection algorithm terminates.

In general, WFG reduction can begin at a non-leaf node before recording of the WFG has been completed at that node; this happens when an *ECHO* message arrives and begins reduction at a non-leaf node before all the *FLOOD*s have arrived at it and recorded the complete local WFG at that node. Thus, the activities of recording and reducing the WFG snapshot are done concurrently in a single phase. Unlike the algorithm in [45], no serialization is imposed between the two activities. Since a reduction is done on an incompletely recorded WFG at nodes, the local snapshot at each node has to be carefully manipulated so as to give the effect that WFG reduction is initiated after WFG recording has been completed.

When multiple nodes block concurrently, they may each initiate the deadlock detection algorithm concurrently. Each invocation of the deadlock detection algorithm is treated independently and is identified by the initiator's identity and initiator's timestamp when it blocked. Every node maintains a local snapshot for the latest deadlock detection algorithm initiated by every other node. We will describe only a single instance of the deadlock detection algorithm.

The problem of termination detection

The algorithm requires a termination detection technique so that the initiator can determine that it will not receive any more *ECHO* messages. The algorithm uses a termination detection technique based on weights [20] in conjunction with *SHORT* messages to detect the termination of the algorithm. A weight of 1.0 at the initiator node, when the algorithm is initiated, is distributed among all *FLOOD* messages sent out by the initiator. When the first *FLOOD* is received at a non-leaf node, the weight of the received *FLOOD* is distributed among the *FLOOD*s sent out along outward edges at that node to expand the WFG further. Since any subsequent *FLOOD* arriving at a non-leaf node does not expand the WFG further, its weight is returned to the initiator in a *SHORT* message. When a *FLOOD* is received at a leaf node, its weight is piggybacked to the *ECHO* sent by the leaf node to reduce the WFG. When an *ECHO* arriving at a node unblocks the node, the weight of the *ECHO* is distributed among the *ECHOs* that are sent by that node along the incoming edges in its WFG snapshot. When an *ECHO* arriving at a node does not unblock the node, its weight is sent directly to the initiator in a *SHORT* message.

Note that the following invariant holds in an execution of the algorithm: the sum of the weights in *FLOOD*, *ECHO*, and *SHORT* messages plus the weight at the initiator (received in *SHORT* and *ECHO* messages) is always 1.0. The algorithm terminates when the weight at the initiator becomes 1.0, signifying that all WFG recording and reduction activity has completed.

FLOOD, ECHO, and *SHORT* messages carry weights for termination detection. Variable w, a real number in the range [0, 1], denotes the weight in a message.

10.9.2 The algorithm

A node i stores the local snapshot for snapshots *initiated* by other nodes in a data structure LS_i (local snapshot), which is an array of records:
LS_i: array [1, . . . , n] of record;

A record has several fields to record snapshot related information and is defined in Algorithm 10.4 for an initiator *init*. The deadlock detection algorithm is defined by the following procedures: *SNAPSHOT-INITIATE*, *FLOOD-RECEIVE*, *ECHO-RECEIVE*, and *SHORT-RECEIVE*. They are executed atomically.

Example We now illustrate the operation of the algorithm with the help of an example [26]. Figure 10.3 shows initiation of deadlock detection by node A and Figure 10.4 shows the state after node D is reduced. The notation x/y beside a node in the figures indicates that the node is blocked and needs replies to x out of the y outstanding requests to unblock.

In Figure 10.3, node A sends out *FLOOD* messages to nodes B and C. When node C receives *FLOOD* from node A, it sends *FLOOD*s to nodes D, E, and F. If the node happens to be active when it receives a *FLOOD* message, it initiates reduction of the incoming wait-for edge by returning an *ECHO* message on it. For example, in Figure 10.3, node H returns an *ECHO* to node D in response to a *FLOOD* from it. Note that node can initiate reduction (by sending back an *ECHO* in response to a *FLOOD* along an incoming wait-for edge) even before the states of all other incoming wait-for edges have been recorded in the WFG snapshot at that node. For example, node F

$LS_i[init].out$: set of integers (:= ∅); /*nodes on which i is waiting in the snapshot.*/
$LS_i[init].in$: set of integers (:= ∅); /*nodes waiting on i in the snapshot.*/
$LS_i[init].t$: integer (:= 0); /*time when $init$ initiated snapshot.*/
$LS_i[init].s$: boolean (:= $false$); /*local blocked state as seen by snapshot.*/
$LS_i[init].p$: integer; /*value of p_i as seen in snapshot.*/

SNAPSHOT_INITIATE
/*Executed by node i to detect whether it is deadlocked. */
$init \leftarrow i$;
$w_i \leftarrow 0$;
$LS_i[init].t \leftarrow t_i$;
$LS_i[init].out \leftarrow out(i)$;
$LS_i[init].s \leftarrow true$;
$LS_i[init].in \leftarrow \emptyset$;
$LS_i[init].p \leftarrow p_i$;
send $FLOOD(i, i, t_i, 1/|out(i)|)$ to each j in $out(i)$.

FLOOD_RECEIVE$(j, init, t\_init, w)$
/*Executed by node i on receiving a FLOOD message from j. */
[
$LS_i[init].t < t\_init \wedge j \in in(i) \rightarrow$ /*Valid *FLOOD* for a new snapshot. */
  $LS_i[init].out \leftarrow out(i)$;
  $LS_i[init].in \leftarrow \{j\}$;
  $LS_i[init].t \leftarrow t\_init$;
  $LS_i[init].s \leftarrow wait_i$;
  $wait_i = true \rightarrow$ /* Node is blocked. */
    $LS_i[init].p \leftarrow p_i$;
    send $FLOOD(i, init, t\_init, w/|out(i)|)$ to each $k \in out(i)$;
  $wait_i = false \rightarrow$ /* Node is active. */
    $LS_i[init].p \leftarrow 0$;
    send $ECHO(i, init, t\_init, w)$ to j;
    $LS_i[init].in \leftarrow LS_i[init].in - \{j\}$.
□
$LS_i[init].t < t\_init \wedge j \notin in(i) \rightarrow$ /* Invalid *FLOOD* for a new snapshot. */
  send $ECHO(i, init, t\_init, w)$ to j.
□
$LS_i[init].t = t\_init \wedge j \notin in(i) \rightarrow$ /* Invalid *FLOOD* for current snapshot. */
  send $ECHO(i, init, t\_init, w)$ to j.
□
$LS_i[init].t = t\_init \wedge j \in in(i) \rightarrow$ /*Valid *FLOOD* for current snapshot. */
  $LS_i[init].s = false \rightarrow$
    send $ECHO(i, init, t\_init, w)$ to j;
  $LS_i[init].s = true \rightarrow$

$LS_i[init].in \leftarrow LS_i[init].in \bigcup\{j\}$;
send $SHORT(init, t\_init, w)$ to $init$.

□

$LS_i[init].t > t\_init \rightarrow$ discard the *FLOOD* message. /*Out-dated *FLOOD*. */

]

ECHO_RECEIVE(j, init, t_init, w)

/*Executed by node i on receiving an *ECHO* from j. */

[

/*Echo for out-dated snapshot. */

$LS_i[init].t > t\_init \rightarrow$ discard the *ECHO* message.

□

$LS_i[init].t < t\_init \rightarrow$ cannot happen. /*ECHO for unseen snapshot. */

□

$LS_i[init].t = t\_init \rightarrow$ /*ECHO for current snapshot. */

$LS_i[init].out \leftarrow LS_i[init].out - \{j\}$;
$LS_i[init].s = false \rightarrow$ **send** $SHORT(init, t\_init, w)$ to $init$.
$LS_i[init].s = true \rightarrow$
$LS_i[init].p \leftarrow LS_i[init].p - 1$;
$LS_i[init].p = 0 \rightarrow$ /* getting reduced */
$LS_i[init].s \leftarrow false$;
$init = i \rightarrow$ declare not deadlocked; exit.
send $ECHO(i, init, t\_init, w/|LS_i[init].in|)$ to all k
$\in LS_i[init].in$;
$LS_i[init].p \neq 0 \rightarrow$
send $SHORT(init, t\_init, w)$ to $init$.

]

SHORT_RECEIVE(init, t_init, w)

/*Executed by node i (which is always $init$) on receiving a *SHORT*. */

[

/*SHORT for out-dated snapshot. */

$t\_init < t\_block_i \rightarrow$ discard the message.

□

/*SHORT for uninitiated snapshot. */

$t\_init > t\_block_i \rightarrow$ not possible.

□

/*SHORT for currently initiated snapshot. */

$t\_init = t\_block_i \bigwedge LS_i[init].s = false \rightarrow$ discard. /* $init$ is active. */

$t\_init = t\_block_i \bigwedge LS_i[init].s = true \rightarrow$
$w_i \leftarrow w_i + w$;
$w_i = 1 \rightarrow$ **declare a deadlock**.

]

Algorithm 10.4 Deadlock detection algorithm [26].

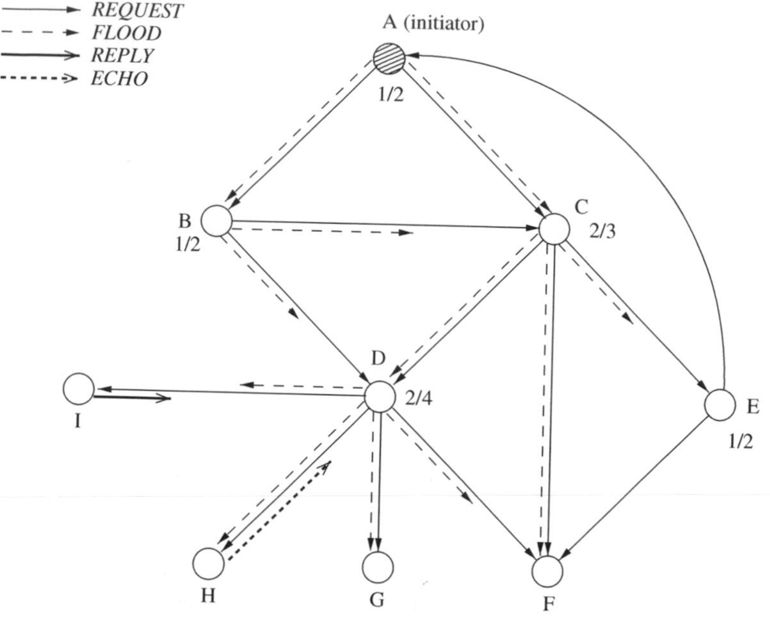

Figure 10.3 An example-run of the algorithm – initiation of deadlock detection by node A [26].

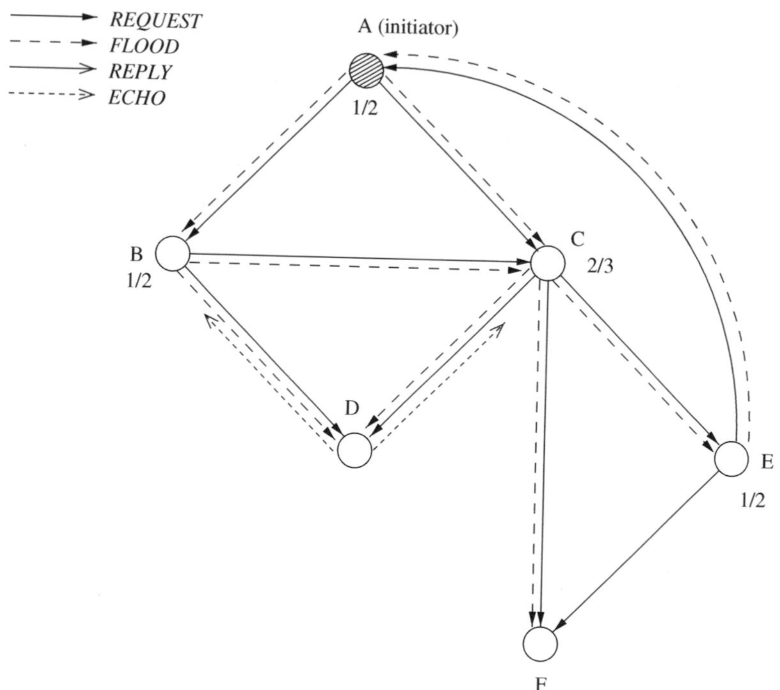

Figure 10.4 An example-run of the algorithm – the state after node D is reduced [26].

in Figure 10.3 starts reduction after receiving a *FLOOD* from C even before it has received *FLOOD*s from D and E.

Note that when a node receives a *FLOOD*, it need not have an incoming wait-for edge from the node that sent the *FLOOD* because it may have already sent back a *REPLY* to the node. In this case, the node returns an *ECHO* in response to the *FLOOD*. For example, in Figure 10.3, when node I receives a *FLOOD* from node D, it returns an *ECHO* to node D.

ECHO messages perform reduction of the nodes and edges in the WFG by simulating the granting of requests in the inward sweep. A node that is waiting a p-out-of-q request gets reduced after it has received *ECHO*s. When a node is reduced, it sends *ECHO*s along all the incoming wait-for edges incident on it in the WFG snapshot to continue the progress of the inward sweep.

In general, WFG reduction can begin at a non-leaf node before recording of the WFG has been completed at that node. This happens when *ECHO*s arrive and begin reduction at a non-leaf node before *FLOOD*s have arrived along all incoming wait-for edges and recorded the complete local WFG at that node. For example, node D in Figure 10.3 starts reduction (by sending an *ECHO* to node C) after it receives ECHOs from H and G, even before *FLOOD* from B has arrived at D. When a *FLOOD* on an incoming wait-for edge arrives at a node which is already reduced, the node simply returns an *ECHO* along that wait-for edge. For example, in Figure 10.4, when a *FLOOD* from node B arrives at node D, node D returns an *ECHO* to B.

In Figure 10.3, node C receives a *FLOOD* from node A followed by a *FLOOD* from node B. When node C receives a *FLOOD* from B, it sends a *SHORT* to the initiator node A. When a *FLOOD* is received at a leaf node, its weight is returned in the *ECHO* message sent by the leaf node to the sender of the *FLOOD*. Note that an *ECHO* is like a reply in the simulated unblocking of processes. When an *ECHO* arriving at a node does not reduce the node, its weight is sent directly to the initiator through a *SHORT* message. For example, in Figure 10.3, when node D receives an *ECHO* from node H, it sends a *SHORT* to the initiator node A. When an *ECHO* that arrives at a node reduces that node, the weight of the *ECHO* is distributed among the *ECHO*s that are sent by that node along the incoming edges in its WFG snapshot. For example, in Figure 10.4, at the time node C gets reduced (after receiving *ECHO*s from nodes D and F), it sends *ECHO*s to nodes A and B. (When node A receives an *ECHO* from node C, it is reduced and it declares no deadlock.) When an *ECHO* arrives at a reduced node, its weight is sent directly to the initiator through a *SHORT* message. For example, in Figure 10.4, when an *ECHO* from node E arrives at node C after node C has been reduced (by receiving *ECHO*s from nodes D and F), node C sends a *SHORT* to initiator node A.

Correctness

Proving the correctness of the algorithm involves showing that it satisfies the following conditions:

1. The execution of the algorithm terminates.
2. The entire WFG reachable from the initiator is recorded in a consistent distributed snapshot in the outward sweep.
3. In the inward sweep, *ECHO* messages correctly reduce the recorded snapshot of the WFG.

The algorithm is initiated within a timeout period after a node blocks on a P-out-of-Q request. On the termination of the algorithm, only all the nodes that are not reduced are deadlocked. For a correctness proof of the algorithm, the readers are referred to the original source [26].

Complexity analysis

The message complexity of the algorithm has been analyzed in [26]. The algorithm has a message complexity of $4e-2n+2l$ and a time complexity[1] of $2d$ hops, where e is the number of edges, n the number of nodes, l the number of leaf nodes, and d the diameter of the WFG. This is better than two-phase algorithms for detecting generalized deadlocks and gives the best time complexity that can be achieved by an algorithm that reduces a distributed WFG to detect generalized deadlocks in distributed systems.

10.10 Chapter summary

Out of the three approaches to handle deadlocks, deadlock detection is the most promising in distributed systems. Detection of deadlocks requires performing two tasks: first, maintaining or constructing whenever needed a WFG; second, searching the WFG for a deadlock condition (cycles or knots).

In distributed deadlock-detection algorithms, every site maintains a portion of the global state graph and every site participates in the detection of a global cycle or knot. Due to lack of globally shared memory, design of distributed deadlock-detection algorithms is difficult because sites may report the existence of a global cycle after seeing its segments at different instants (though all the segments never existed simultaneously).

Distributed deadlock detection algorithms can be divided into four classes: path-pushing, edge-chasing, diffusion computation, and global state detection. In path-pushing algorithms, wait-for dependency information of the global WFG is disseminated in the form of paths (i.e., a sequence of wait-for dependency edges). In edge-chasing algorithms, special messages called probes are

[1] Time complexity denotes the delay in detecting a deadlock after its detection has been initiated.

circulated along the edges of the WFG to detect a cycle. When a blocked process receives a probe, it propagates the probe along its outgoing edges in the WFG. A process declares a deadlock when it receives a probe initiated by it. Diffusion computation type algorithms make use of echo algorithms to detect deadlocks. Deadlock detection messages are successively propagated (i.e, "diffused") through the edges of the WFG. Global state detection-based algorithms detect deadlocks by taking a snapshot of the system and by examining it for the condition of a deadlock.

10.11 Exercises

Exercise 10.1 Consider the following simple approach to handle deadlocks in distributed systems by using "time-outs": a process that has waited for a specified period for a resource declares that it is deadlocked and aborts to resolve the deadlock. What are the shortcomings of using this method?

Exercise 10.2 Suppose all the processes in the system are assigned priorities which can be used to totally order the processes. Modify Chandy *et al.*'s algorithm for the AND model so that when a process detects a deadlock, it also knows the lowest priority deadlocked process.

Exercise 10.3 Show that, in the AND model, false deadlocks can occur due to deadlock resolution in distributed systems [43]. Can something be done about it or they are bound to happen?

Exercise 10.4 Show that in the Kshemkalyani–Singhal algorithm for the P-out-of-Q model, if the weight at the initiator process becomes 1.0, then the intiator is involved in a deadlock.

10.12 Notes on references

Two survey articles on distributed deadlock detection can be found in papers by Knapp [22] and Singhal [43]. The literature is full of distributed deadlock detection algorithms. Path-pushing distributed deadlock detection algorithms can be found in papers by Gligor and Shattuck [11], Menasce and Muntz [33], Ho and Ramamoorthy [18], and Obermarck [38]. Other edge-chasing distributed deadlock detection algorithms can be found in papers by Choudary *et al.* [7], and Kshemkalyani and Singhal [27]. Herman and Chandy [16] discuss detection of deadlocks in the AND/OR model. In [24], Kshemkalyani and Singhal give an optimal algorithm to detect distributed deadlocks under the generalized request model. Other algorithms to detect generalized deadlocks include Bracha and Toueg [2] and Wang *et al.* [45].

In [25], Kshemkalyani and Singhal give a characterization of distributed deadlocks. A rigorous correctness proof of a distributed deadlock detection algorithm is given in Kshemkalyani and Singhal [27]. Brezezinski *et al.* [3] discuss the deadlock models under very generalized blocking conditions. Two knot detection algorithms in distributed systems are given in Chandy and Misra [34] and Manivannan and Singhal [31]. Gray *et al.* [12] present a simple analysis of the probability of deadlocks in database systems. Lee and Kim [30] present a performance analysis of distributed deadlock detection algorithms. Other algorithms for deadlock detection in distributed systems can be found in [1,8–10,13–15,17,21,23,28,32,36,37,39–41,44]. Wu *et al.* [46] present an algorithm to avoid distributed deadlock in the AND model.

References

[1] B. Awerbuch and S. Micali, Dynamic deadlock resolution protocols, in *Proceedings of the Foundations of Computer Science*, Toronto, Canada, 1986, 196–207.

[2] G. Bracha and S. Toueg, Distributed deadlock detection, *Distributed Computing*, **2**(3), 1987, 127–138.

[3] J. Brezezinski, J. M. Helary, M. Raynal, and M. Singhal, Deadlock models and generalized algorithm for distributed deadlock detection, *Journal of Parallel and Distributed Computing*, **31**(2), 1995, 112–125.

[4] K. M. Chandy and L. Lamport, Distributed snapshots: determining global states of distributed systems, *ACM Transactions on Programming Language Systems*, **3**(1), 1985, 63–75.

[5] K. M. Chandy and J. Misra, A distributed algorithm for detecting resource deadlocks in distributed systems, *Proceedings of the ACM Symposium on Principles of Distributed Computing*, Ottawa, Canada, August 1982, 157–164.

[6] K. M. Chandy, J. Misra, and L. M. Haas, Distributed deadlock detection, *ACM Transactions on Computer Systems*, **1**(2), 1983, 144–156.

[7] A. Choudhary, W. Kohler, J. Stankovic, and D. Towsley, A modified priority based probe algorithm for distributed deadlock detection and resolution, *IEEE Transactions on Software Engineering*, **15**(1) 1989, 10–17.

[8] J. R. G. de Mendivil, F. Farina, J. Garitagoitia, C. F. Alastruey, and J. M. Bernabeu-Auban, A distributed deadlock resolution algorithm for the AND model, *IEEE Transactions on Parallel and Distributed Systems*, **10**(5), 1999, 433–447.

[9] A. K. Elmagarmid, N. Soundararajan, and M. T. Liu, A distributed deadlock detection and resolution algorithm and its correctness proof, *IEEE Transactions on Software Engineering*, **14**(10), 1988, 1443–1452.

[10] M. Flatebo and A. K. Datta, Self-stabilizing deadlock detection algorithms, *Proceedings of the 1992 ACM Annual Conference on Communications*, Kansas City, Missouri, March 1992, 117–122.

[11] V. Gligor and S. Shattuck, On deadlock detection in distributed databases, *IEEE Transactions on Software Engineering*, **SE-6**(5), 1980, 435–440.

[12] J. N. Gray, P. Homan, H. F. Korth, and R. L. Obermarck, A straw man analysis of the probability of waiting and deadlock in a database system, *Technical Report RJ 3066*, IBM Research Laboratory, San Jose, CA, 1981.

[13] L. M. Haas, Two approaches to deadlock detection in distributed systems. Ph.D. dissertation, Department of Computer Sciences, University of Texas, Austin, TX, 1981.

[14] L. M. Haas and C. Mohan, A distributed deadlock detection algorithm for a resource-based system, *Research Report RJ 3765*, IBM Research Laboratory, San Jose, CA, 1983.

[15] J. Helary, C. Jard, N. Plouzeau, and M. Raynal, Detection of stable properties in distributed applications, *Proceedings of the ACM Symposium on Principles of Distributed Computing*, Vancouver, Canada, August 1987, 125–136.

[16] T. Herman and K. M. Chandy, *A Distributed Procedure to Detect AND/OR Deadlock*, Technical Report TR LCS-8301, Department of Computer Sciences, University of Texas, Austin, TX, 1983.

[17] B. A. Sanders and P. A. Heuberger, Distributed deadlock detection and resolution with probes, *Proceedings of the 3rd International Workshop on Distributed Algorithms*, September 26–28, 1989, 207–218.

[18] G. S. Ho and C. V. Ramamoorthy, Protocols for deadlock detection in distributed database systems, *IEEE Transactions on Software Engineering*, **8**(6), 1982, 554–557.

[19] R.C. Holt, Some deadlock properties on computer systems, *ACM Computing Surveys*, **4**(3), 1972, 179–196.

[20] S.-T. Huang, Detecting termination of distributed computations by external agents, *ICDCS*, 1989, 79–84.

[21] J. R. Jagannathan and R. Vasudevan, A distributed deadlock detection and resolution scheme: performance study, *Proceedings of the 3rd International Conference on Distributed Computing Systems*, Miami, Florida, 1982, 496–501.

[22] E. Knapp, Deadlock detection in distributed databases, *ACM Computing Surveys*, **19**(4), 1987, 303–328.

[23] M. Krishnamurthi, A. Basavatia, and S. Thallikar, Deadlock detection and resolution in simulation models, Proceedings of the 26th conference on Winter Simulation, p.708-715, December 11-14, 1994, Orlando, Florida.

[24] A. Kshemkalyani and M. Singhal, A one-phase algorithm to detect distributed deadlocks in replicated databases, *IEEE Transactions on Knowledge and Data Engineering*, **11**(6), 1999, 880–895.

[25] A. Kshemkalyani and M. Singhal, On characterization and correctness of distributed deadlocks, *Journal of Parallel and Distributed Computing*, **22**(1), 1994, 44–59.

[26] A. Kshemkalyani and M. Singhal, Efficient detection and resolution of generalized distributed deadlocks, *IEEE Transactions on Software Engineering*, **20**(1), 1994, 43–54.

[27] A. Kshemkalyani and M. Singhal, An invariant-based verification of a probe algorithm for distributed deadlock detection and resolution, *IEEE Transactions on Software Engineering*, **17**(8) 1991, 789–799.

[28] N. Krivokapic, A. Kemper, and E. Gudes, Deadlock detection in distributed database systems: a new algorithm and a comparative performance analysis, *VLDB Journal: Very Large Data Bases*, **8**(2), 1999, 79–100.

[29] L. Lamport, Time, clocks, and the ordering of events in distributed systems, *Communications of the ACM* **21**(7), 1978, 558–565.

[30] S. Lee and J. L. Kim, Performance analysis of distributed deadlock detection algorithms, *IEEE Transactions on Knowledge and Data Engineering*, **13**(4), 2001, 623–636.

[31] D. Manivannan and M. Singhal, An efficient distributed algorithm for detection of knots and cycles in a distributed graph, *IEEE Transactions on Parallel and Distributed Systems*, **14**(10), 2003, 961–972.

[32] J. Mayo and P. Kearns, Distributed deadlock detection and resolution based on hardware clocks, *ICDCS 1999*, 208–215.

[33] D. E. Menasce and R. Muntz, Locking and deadlock detection in distributed databases, *IEEE Transactions on Software Engineering*, **5**(3), 1979, 195–202.

[34] J. Misra and K. M. Chandy, A distributed graph algorithm: knot detection, *ACM Transactions on Programming Language Systems*, **4**(4), 1982, 678–686.

[35] D. P. Mitchell and M. J Merritt, A distributed algorithm for deadlock detection and resolution, *Proceedings of the ACM Symposium on Principles of Distributed Computing*, 1984, 282–284.

[36] N. Natarajan, A distributed scheme for detecting communication deadlock, *IEEE Transactions on Software Engineering* **SE-12**(4), 1986, 531–537.

[37] R. Obermarck, Deadlock detection for all resourse classes, *Research Report RJ2955*, IBM Research Laboratory, San Jose, CA, 1980.

[38] R. Obermarck, Distributed deadlock detection algorithm, *ACM Transactions on Database Systems*, **7**(2), 1982, 187–208.

[39] Y. C. Park, P. Scheuermann, and H. L. Tung, A distributed deadlock detection and resolution algorithm based on a hybrid wait-for graph and probe generation scheme, *Proceedings of the 4th International Conference on Information and Knowledge Management*, Baltimore, MD 1995, 378–386.

[40] Y. C. Park, P. Scheuermann, and S. H. Lee, A periodic deadlock detection and resolution algorithm with a new graph model for sequential transaction processing, *Proceedings of the 8th International Conference on Data Engineering*, 1992, 202–209.

[41] M. Roesler and W. A. Burkhard, Resolution of deadlocks in object-oriented distributed systems, *IEEE Transactions on Computers*, **38**(8), 1989, 1212–1224.

[42] M. K. Sinha and N. Natarajan, A distributed deadlock detection algorithm based on timestamps, *Proceedings of the 4th International Conference on Distributed Computing Systems*, 1984, 546–556.

[43] M. Singhal, Deadlock detection in distributed systems, *IEEE Computer*, November, 1989, 37–48.

[44] J. Villadangos, F. Farina, J. R. G. de Mendivil, J. Garitagoitia, and A. Cordoba, A safe algorithm for resolving OR deadlocks, *IEEE Transactions on Software Engineering*, **29**(7), 2003, 608–622.

[45] J. Wang, S. Huang, and N. Chen, *A distributed algorithm for detecting generalized deadlocks*, Technical Report, Department of Computer Science, National Tsing-Hua University, 1990.

[46] H. Wu, W.-N. Chin, and J. Jaffar, An efficient distributed deadlock avoidance algorithm for the AND model, *IEEE Transactions on Software Engineering*, **28**(1), 2002, 18–29.

CHAPTER

11 Global predicate detection

11.1 Stable and unstable predicates

Specifying predicates on the system state provides an important handle to specify, observe, and detect the behavior of a system. This is useful in formally reasoning about the system behavior. By being able to detect a specified predicate in the execution, we gain the ability to monitor the execution. Predicate specification and detection has uses in distributed debugging, sensor networks used for sensing in various applications, and industrial process control. As an example in the manufacturing process, a system may be monitoring the pressure of Reagent A and the temperature of Reagent B. Only when $\psi_1 = (Pressure_A > 240\,\text{KPa}) \wedge (Temperature_B > 300\,°\text{C})$ should the two reagents be mixed. As another example, consider a distributed execution where variables x, y, and z are local to processes P_i, P_j, and P_k, respectively. An application might be interested in detecting the predicate $\psi_2 = x_i + y_j + z_k < -125$. In a nuclear power plant, sensors at various locations would monitor the relevant parameters such as the radioactivity level and temperature at multiple locations within the reactor.

Observe that the "predicate detection" problem is inherently different from the global snapshot problem. A global snapshot gives one of the possible states that *could have existed* during the period of the snapshot execution. Thus, a snapshot algorithm can observe only one of the predicate values that *could have existed* during the algorithm execution.

Predicates can be either stable or unstable. A *stable* predicate is a predicate that remains true once it becomes true [6]. In traditional systems, a predicate ϕ is stable if $\phi \Longrightarrow \Box\phi$, where "$\Box$" is the "henceforth" operator from temporal logic [21]. In distributed executions, a more precise definition is needed, due to the absence of global time. Formally, a predicate ϕ at a cut C is stable if the following holds:

$$(C \models \phi) \implies (\forall C' \mid C \subseteq C', C' \models \phi).$$

Deadlock in a system is a stable property because the deadlocked processes continue to remain deadlocked (until deadlock resolution is performed). Termination of an execution is another stable property. Specific algorithms to detect termination of the execution, and to detect deadlock were considered in earlier chapters. Here, we look at a general technique to detect a stable predicate.

11.1.1 Stable predicates

Deadlock [13,17]

A deadlock represents a system state where a subset of the processes are blocked on one another, waiting for a reply from the other processes in that subset. The waiting relationship is represented by a wait-for graph (WFG) where an edge from i to j indicates that process i is waiting for a reply from process j. Given a wait-for graph $G = (V, E)$, a *deadlock* is a subgraph $G' = (V', E')$ such that $V' \subseteq V$ and $E' \subseteq E$ and for each process i in V', the process i remains blocked unless it receives a reply from some process(es) in V'. There are two conditions that characterize the deadlock state of the execution:

- (local condition:) each deadlocked process is locally blocked, and
- (global condition:) the deadlocked process will not receive a reply from some process(es) in V'.

Termination [20]

Termination of an execution is another stable property, and is best understood by viewing a process as alternating between two states: *active state* and *passive state*. An *active* process spontaneously becomes *passive* when it has no further work to do; a *passive process* can become *active* only when it receives a message from some other process. If such a message arrives, then the process becomes *active* by doing CPU processing and maybe sending messages as a result of the processing. An execution is *terminated* if each process is *passive*, and will not become active unless it receives more messages. There are two conditions that characterize the termination state of the execution:

- (local condition:) each process is in passive state; and
- (global condition:) there is no message in transit between any pair of processes.

Generalizing from the above two most frequently encountered stable properties, we assume that each stable property can be characterized by a local process state component, and a channel component or a global component. Recall from our discussion of global snapshots [6] that any channel property can be observed by observing the local states at the two endpoints of the channel, in a consistent manner. Thus, any global condition can be observed by observing the local states of the processes.

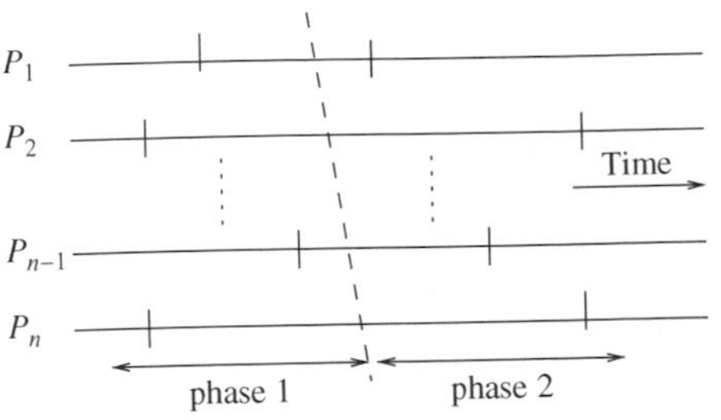

Figure 11.1 Two-phase detection of a stable property. If the values of the relevant local variables that capture the property have not changed between the two phases, then the stable property is true.

We now address the question: "What are the most effective techniques for detecting a stable property?" Clearly, repeatedly or periodically taking a global snapshot will work; if the property is true in some snapshot, then it can be claimed that the property is henceforth true. However, recording a snapshot is expensive; recall that it can require up to $O(n^2)$ control messages without inhibition, or $O(n)$ messages with inhibition. The approach that has been widely adopted is the two-phase approach of observing potentially inconsistent global states. In each state observation, all the local variables necessary for defining the local conditions, as well as the global conditions, are observed. Two potentially inconsistent global states are recorded consecutively, such that the second recording is initiated after the first recording has completed [13,17]. This is illustrated in Figure 11.1. The stable property can be declared to be true if the following holds:

- The variables on which the local conditions as well as the global conditions are defined have not changed in the two observations, as well as between the two observations.

If none of the variables changes between the two observations, it can be claimed that after the termination of the first observation and before the start of the second observation, there is an instant in physical time when the variables still have the same value. Even though the two observations are each inconsistent, if the global property is true at a common physical time, then the stable property will necessarily be true.

The most common ways of taking a pair of consecutive, not necessarily consistent, snapshots using $O(n)$ control messages are as follows:

- Each process randomly records its state variables and sends them to a central process via control messages. When the central process receives this message from each other process, the central process informs each other process to send its (uncoordinated) local state again.
- A token is passed around a ring, and each process appends its local state to the contents of the token. When the token reaches the initiator, it passes the token around for a second time. Each process again appends its local state to the contents of the token.

- On a predefined spanning tree, the root (coordinator) sends a query message in the fan-out sweep of the tree broadcast. In the fan-in sweep of the ensuing tree convergecast, each node collects the local states of the nodes in its subtree rooted at itself and forwards these local states to its parent. When the root gets the local states from all the nodes in its tree, the first phase completes. The second phase, which contains another broadcast followed by a convergecast, is initiated.

11.1.2 Unstable predicates

An *unstable* predicate is a predicate that is not stable and hence may hold only intermittently [8,18]. The following are some of the several challenges in detecting unstable predicates:

- Due to unpredictable message propagation times, and unpredictable scheduling of the various processes on the processors under various load conditions, even for deterministic executions, multiple executions of the same distributed program may pass through different global states. Further, the predicate may be true in some executions and false in others.
- Due to the non-availability of instantaneous time in a distributed system:
  - even if a monitor finds the predicate to be true in a global state, it may not have actually held in the execution;
  - even if a predicate is true for a transient period, it may not be detected by intermittent monitoring.

  Hence, periodic monitoring of the execution is not adequate.

These challenges are faced by snapshot-based algorithms as well as by a central monitor that evaluates data collected from the monitored processes. To address these challenges, we can make two important observations [8,18].

- It seems necessary to examine all the states that arise in the execution, so as not to miss the predicate being true. Hence, it seems useful to define predicates, not on individual states, but on the observation of the entire execution.
- For the same distributed program, even given that it is deterministic, multiple observations may pass through different global states. Further, a predicate may be true in some of the program observations but not in others. Hence it is more useful to define the predicates on all the observations of the distributed program and not just on a single observation of it.

11.2 Modalities on predicates

To address the above complications, predicates are defined, not on global states or on an individual observation of an execution, but on all the possible observations of the distributed execution. The following two modalities on any predicate ϕ are defined [8,18]:

- *Possibly*(ϕ): There exists a consistent observation of the execution such that predicate ϕ holds in a global state of the observation.
- *Definitely*(ϕ): For every consistent observation of the execution, there exists a global state of it in which predicate ϕ holds.

Consider the example in Figure 11.2(a). The execution is run at processes P_1 and P_2. Event e_i^k denotes the kth event at process P_i. Variable a is local to P_1 and variable b is local to P_2. The state lattice for the execution is shown in Figure 11.2(b). Each state is labeled by a tuple (c_1, c_2), where c_1 and c_2 are the event counts at P_1 and P_2, respectively. The execution shown in part (a) goes through the following sequence of global states, and events causing the state transitions between the global states:

$$(0,0), e_2^1, (0,1), e_1^1, (1,1), e_2^2, (1,2), e_1^2, (2,2), e_2^3, (2,3), e_2^4, (2,4), e_1^3, (3,4), e_1^4, (4,4), e_2^5, (4,5), e_1^5, (5,5), e_1^6, (6,5), e_2^6, (6,6), e_2^7, (6,7)$$

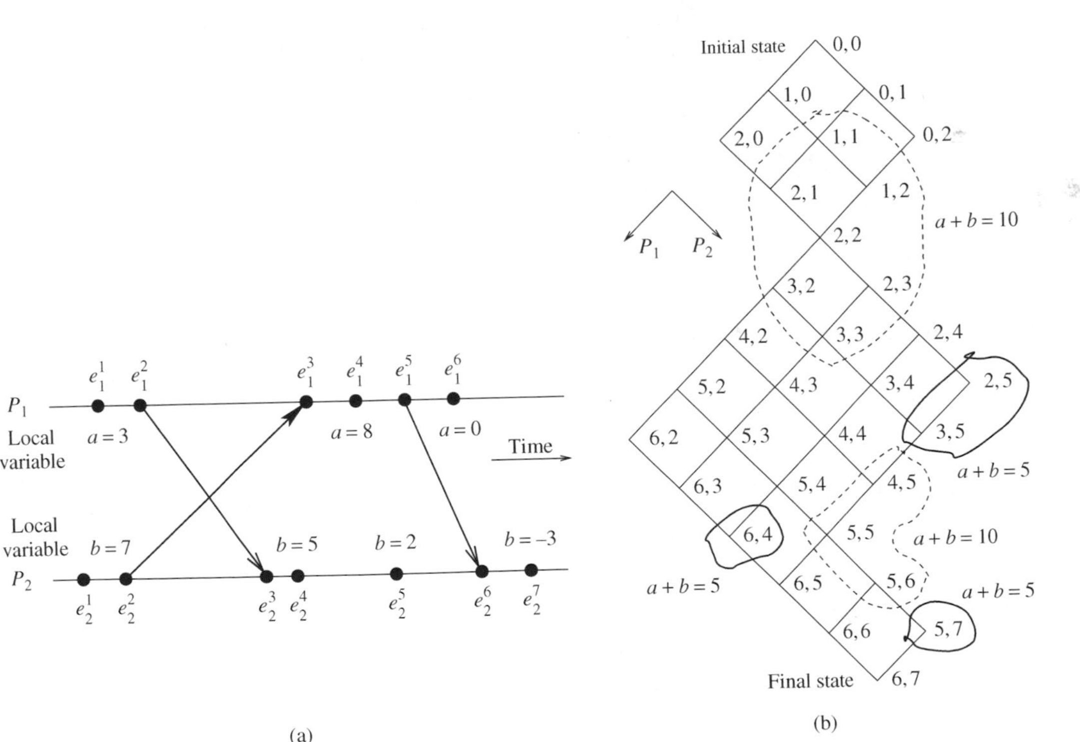

Figure 11.2 Example to illustrate *Possibly*(ϕ) and *Definitely*(ϕ) [16]. (a) The example execution. (b) The state lattice for the execution. Each label in the lattice gives the event count at P_1, P_2.

When the same distributed program is run again, observe that it may not pass through the same intermediate states as the state transitions from the initial state (0, 0) to the final state (6, 7).

- Now observe that $Definitely(a+b=10)$ holds by the following reasoning. When b is assigned 7 at event e_2^1, process P_1's execution may be in any state from the initial state up to the state preceding event e_1^3, in which $a=3$. However, before the value of b changes from 7 to 5 at event e_2^4, and in fact before P_2 executes event e_2^3, P_1 must have executed event e_1^1 at which time $a=3$. This is true for all equivalent executions. Hence, $Definitely(a+b=10)$ holds. With respect to the state lattice in Figure 11.2(b), the states in which $a+b=10$ is true are marked therein. From the state lattice, it can be seen that, in every execution, the state (2, 2) must occur, and in this state, $a+b=10$.
- Observe that $Possibly(a+b=5)$ holds by the following reasoning. The predicate $a+b=5$ can be true only if: (i) $a=3 \wedge b=2$, which is true in states (2, 5) and (3, 5), or (ii) $a=0 \wedge b=5$, which is true in state (6, 4), or (iii) $a=8 \wedge b=-3$, which is true in state (5, 7), in some equivalent execution. State (i) is possible in physical time after the occurrence of event e_2^5 and before the occurrence of e_1^4. In the execution shown, e_2^5 occurs after e_1^4. However, in an equivalent execution, event e_1^4 may be delayed to occur after event e_2^5, in which case b changes to a value other than 2 after a becomes 8. Hence, the predicate is true in this equivalent execution. It so happens that a similar argument also holds for (ii) and (iii).
- Predicate $Definitely(a+b=5)$ is not true in the shown execution because there exists at least one path through the state lattice such that $a+b=5$ is never true in any state along that path.

11.2.1 Complexity of predicate detection

As we suspect from the examples in this section, the predicate detection problem is complex. For n processes and a maximum of m events per process, we need to examine up to an exponential number m^n states. The global predicate detection problem can be readily shown to be NP-complete using a standard reduction from the satisfiability problem (see Exercise 11.4).

11.3 Centralized algorithm for relational predicates

To detect predicates, we first assume that the state lattice is available. A global state $GS=\{s_1^{k_1}, s_2^{k_2}, \ldots, s_n^{k_n}\}$ is abbreviated as $GS^{k_1,k_2,\ldots k_n}$.

- $Possibly(\phi)$: To detect $Possibly(\phi)$, an exhaustive search of the state lattice for any one state that satisfies ϕ needs to be done. The search can terminate as soon as such a state is found. Presumably, there is particular interest in finding the "earliest" state that satisfies ϕ. The level of a global state $\langle s_i^{k_i}(\forall i)\rangle$ is $\sum_{i=1}^{i=n} k_i$. Algorithm 11.1 [8,18] examines the state lattice level-by-level, beginning from the initial state at level 0 and going to the final state. Each level is examined to find a state in which ϕ is true. If such a state is found, the algorithm terminates.

- $Definitely(\phi)$: For $Definitely(\phi)$ to be true, there should exist a set of states satisfying ϕ such that every path through the lattice goes through one of these states. It is sufficient but not necessary that all the states at any particular level in the lattice satisfy ϕ. To see this, consider the execution in Figure 11.3. Here, $Definitely(\phi)$ is true, yet the states satisfying ϕ are at different levels.

  As $Definitely(\phi)$ may be true but there may not exist any level in which all the states satisfy ϕ, Algorithm 11.1 cannot use an approach similar to that used for $Possibly(\phi)$ to detect $Definitely(\phi)$. In particular, replacing the loop condition in line 1a by the following will not work: "(some state in $Reach_\phi$ satisfies $\neg\phi$)." The algorithm examines the state lattice level-by-level but differs in the following two respects:

(variables)
set of global states $Reach_\phi$, $Reach\_Next_\phi \longleftarrow \{GS^{0,0,\ldots 0}\}$
int $lvl \longleftarrow 0$

(1) $Possibly(\phi)$:
(1a) **while** (no state in $Reach_\phi$ satisfies ϕ) **do**
(1b)   **if** ($Reach_\phi = \{final\ \ state\}$) **then return** *false*;
(1c) $lvl \longleftarrow lvl + 1$;
(1d) $Reach_\phi \longleftarrow$ {states at level lvl};
(1e) **return** *true*.

(2) $Definitely(\phi)$:
(2a) remove from $Reach_\phi$ those states that satisfy ϕ
(2b) $lvl \longleftarrow lvl + 1$;
(2c) **while** ($Reach_\phi \neq \emptyset$) **do**
(2d) $Reach\_Next_\phi \longleftarrow$ {states of level lvl reachable from a state in $Reach_\phi$};
(2e)   remove from $Reach\_Next_\phi$ all the states satisfying ϕ;
(2f)   **if** $Reach\_Next_\phi$ = {final state} **then return** *false*;
(2g) $lvl \longleftarrow lvl + 1$;
(2h) $Reach_\phi \longleftarrow Reach\_Next_\phi$;
(2i) **return** *true*.

Algorithm 11.1 Detecting a relational predicate by examining the state lattice (on-line, centralized) [8,18].

1. Rather than track the states (at a level) in which ϕ is true, it tracks the states in which ϕ is *not true*.
2. Additionally, the set of states tracked at a level have to be reachable from the set of those states at the previous level, that are known to satisfy (1) and recursively this same property (2).

The variable $Reach\_Next_{\phi}$ is used to track such states at level lvl, as constructed from the states at the previous level. Thus, $Reach\_Next_{\phi}$ at level lvl contains the set of states at level lvl that are reachable from the initial state *without* passing through any state satisfying ϕ. The algorithm terminates successfully when $Reach\_Next_{\phi}$ becomes the empty set; otherwise it terminates unsuccessfully after examining the final state.

Example Figure 11.3 shows an example execution and the corresponding state lattice. The states belonging to $Reach\_Next_{\phi}$ (line 2d) at any level are either marked by shaded circles or clear circles. The states belonging to $Reach\_Next_{\phi}$ (line 2f) at any level are marked by clear circles. In line 2b, when $lvl = 11$, $Reach_{\phi}$ becomes Ø and the algorithm exits from the loop.

The centralized algorithms in Algorithm 11.1 assumed that the states at any level were readily available. But in an on-line algorithm, these global states need to be assembled from local states, on the fly. How can that be

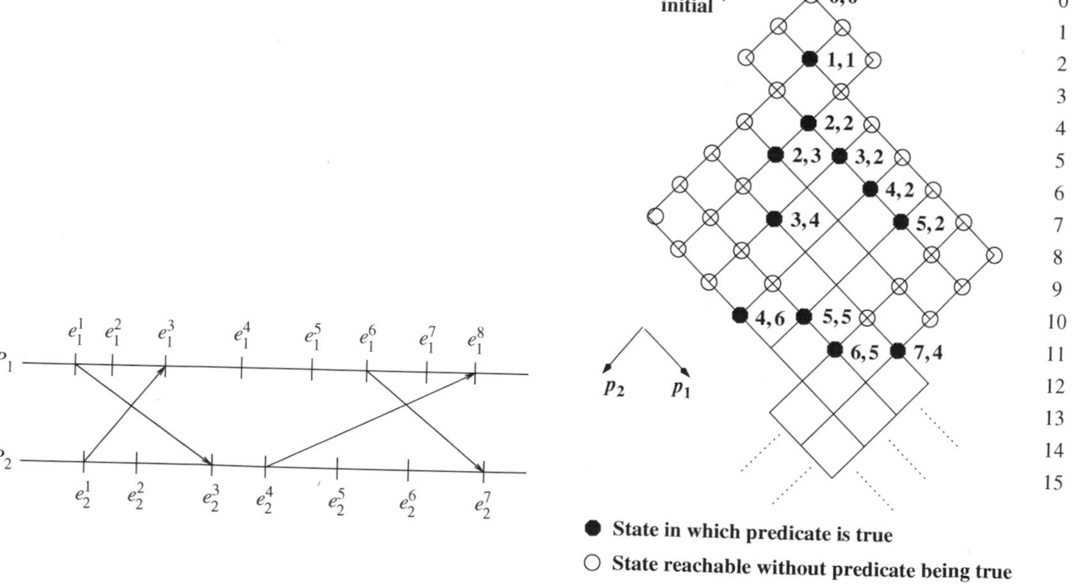

Figure 11.3 Example to show that states in which *Definitely*(ϕ) is satisfied need not be at the same level in the state lattice. (a) Execution. (b) Corresponding state lattice.

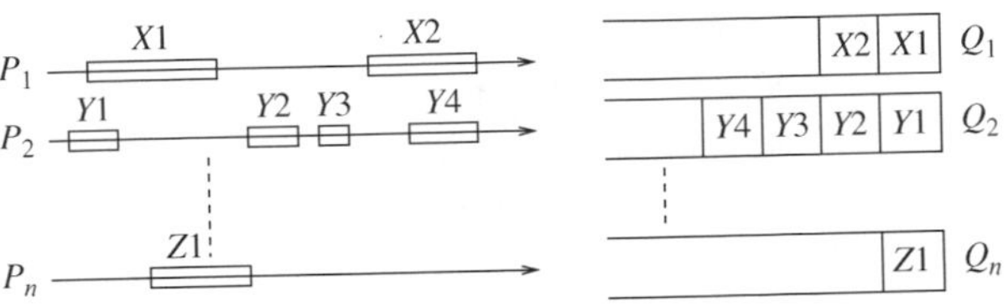

Figure 11.4 Queues Q_1 to Q_n for each of the n processes.

accomplished? Each process P_i can send a local trace of its local states $s_i^{k_i}$, with their vector timestamps, to the central process P_0. P_0 maintains n queues, $Q_1 \ldots Q_n$, for the events of each of the processes, as shown in Figure 11.4. Each local state received from process P_i is enqueued in Q_i. As global state $GS = \{s_1^{k_1}, s_2^{k_2}, \ldots, s_n^{k_n}\}$, also abbreviated as $GS^{k_1,k_2,\ldots k_n}$, is assembled from the corresponding local states, how long does a local state need to be kept in its queue? This is answered by the following two observations, based on the vector clocks VC [9,19]:

- The earliest global state $GS_{min}^{k_1,k_2,\ldots k_n}$ containing $s_i^{k_i}$ is identified as follows. The jth component of $VC(s_i^{k_i})$ is the local value of P_j in its local snapshot state $s_j^{k_j}$. This is expressed as:

$$(\forall j)\ VC(s_j^{k_j})[j] = VC(s_i^{k_i})[j]. \tag{11.1}$$

  It now follows that the lowest level of the state lattice, in which local state $s_i^{k_i}$ (kth local state of P_i) participates, is the sum of the components of $VC(s_i^{k_i})$ – this assumes that in the vector clock operation, the local component is incremented by 1 for each local event.
- The latest global state $GS_{max}^{k_1,k_2,\ldots k_n}$ containing $s_i^{k_i}$ is identified as follows. The ith component of $VC(s_j^{k_j})$ should be the largest possible value but cannot exceed or equal $VC(s_i^{k_i})[i]$ for consistency of the two states $s_i^{k_i}$ and $s_j^{k_j}$. $VC(s_j^{k_j})$ is identified as per Eq. (11.2); but note that the condition on $VC(s_j^{k_j+1})[i]$ is not applicable if $s_j^{k_j}$ is the last state at P_j:

$$(\forall j)\ VC(s_j^{k_j})[i] < VC(s_i^{k_i})[i] \leq VC(s_j^{k_j+1})[i] \tag{11.2}$$

  Hence, the highest level of the state lattice, in which local state $s_i^{k_i}$ participates, is $\sum_{j=1}^{n} VC(s_j^{k_j})[j]$ subject to the above equation.

From Eqs (11.1) and (11.2), we have that $\sum_{j=1}^{n} VC(s_i^{k_i})[j]$ is the lowest level, and $\sum_{j=1}^{n} VC(s_j^{k_j})[j]$, where $s_j^{k_j} \in GS_{max}$, is the highest level, between which local state $s_i^{k_i}$ is useful in constructing a global state.

Given the states of level lvl, the set of states at level $lvl+1$ can be constructed as follows. For each state $GS^{k_1,k_2,\ldots,k_n}$, construct the n global states $GS^{k_1+1,k_2,\ldots,k_n}$, $GS^{k_1,k_2+1,\ldots,k_n}$ $\ldots$ $GS^{k_1,k_2,\ldots,k_n+1}$.

Deterministic versus non-deterministic programs
We need to remember that the entire analysis of predicates and their modalities, and detection algorithms, applies only to deterministic programs. For non-deterministic programs, different executions may have different partial orders.

11.4 Conjunctive predicates

The predicates considered so far are termed *relational predicates* because the predicate can be an arbitrary relation on the variables in the system. A predicate ϕ is a *conjunctive predicate* if and only if ϕ can be expressed as the conjunction $\bigwedge_{i\in N} \phi_i$, where ϕ_i is a predicate local to process i. For a wide range of applications, the predicate of interest can be modeled as a conjunctive predicate. Conjunctive predicates have the following property:

- If ϕ is false in any cut C, then there is at least one process i such that the local state of i in cut C will never form part of any other cut C' such that ϕ is true in C'. More formally, this property of a conjunctive predicate ϕ is defined as the following:

$$C \not\models \phi \Longrightarrow \exists i \in N, \forall C' \in \mathcal{C}uts,\ C' \not\models \phi, \text{ where } C'[i] = C[i].$$

  Here, the state $C[i]$ is a *forbidden state* because it will never form part of any cut that satisfies the predicate. Given a conjunctive predicate, if it is evaluated as false in some cut C, we can advance the local state of at least one process to the next event, and then evaluate the predicate in the resulting cut.

This gives a $O(mn)$ time algorithm, where m is the number of events at any process, to check for a conjunctive predicate, as opposed to an exponential algorithm required by a relational predicate.

Consider the following example on modalities on conjunctive predicates, shown for the same execution considered earlier in Figure 11.2:

- The predicate $Possibly(a=3 \wedge b=2)$ holds by the following reasoning. The predicate can be *true* only if $a=3 \wedge b=2$ simultaneously in the execution. This is possible in physical time after the occurrence of event e_2^5 and before the occurrence of e_1^4. In the execution shown, e_1^4 occurs before e_2^5. However, in an equivalent execution, event e_1^4 may be delayed to occur after event e_2^5, in which case, a changes to a value other than 3 after b becomes 2. Hence, $Possibly(a=3 \wedge b=2)$ is *true*. Note that in Figure 11.2, $a+b=5$ was true in states $(2,5)$, $(3,5)$, $(6,4)$, and $(5,7)$. Among these, $a=3 \wedge b=2$ is true only in $(2,5)$ and $(3,5)$.
- $Definitely(a=3 \wedge b=7)$ holds by the following reasoning. When a is assigned 3 at event e_1^1, process P_2's execution may be at any event from e_2^0

up to but not including e_2^3. However, before the value of a changes from 3 to 8 at event e_1^4, P_2 must have executed event e_2^2 at which time $b = 7$. This is true for all equivalent executions. Note that in Figure 11.2, $a + b = 10$ was true in states (1, 1), (2, 1), (1, 2), (2, 2), (3, 2), (2, 3), (3, 3), (4, 5), (5, 5), and (5, 6). Among these, $a = 3 \wedge b = 7$ is true only in all except (4, 5), (5, 5), and (5, 6).

11.4.1 Interval-based centralized algorithm for conjunctive predicates

Conjunctive predicates are a popular class of predicates. Conjunctive predicates have the advantage that each process can locally determine whether the local component ϕ_i is satisfied; if not, the local state cannot be part of any global state satisfying ϕ. This has the following implication: starting with the initial state, we examine global states. If ϕ is not satisfied, then the local state of at least one process can be advanced and the next global state is examined. Either ϕ is satisfied, or we repeat the step. Within mn steps, we will have examined all necessary global states, giving a $O(mn)$ upper bound on the time complexity.

There are two broad approaches to detecting conjunctive predicates: the *global state*-based approach and the *interval*-based approach [15]. The global state-based approach involves examining the global states, as suggested above and also seen in Section 11.3.

In the interval-based approach, each process identifies alternating time durations when the local predicate alternates between *true* and *false*. This is illustrated in Figure 11.5. Let us consider any two processes P_i and P_j, and let the intervals at these processes when the local predicates ϕ_i and ϕ_j are true be denoted X_i and Y_j, respectively. Let the start and end of an interval X be denoted as $min(X)$ and $max(X)$, respectively. Assume the global predicate is defined on these two processes. We can observe the following definitions of $Definitely(\phi)$ and $Possibly(\phi)$ with the aid of Figure 11.6:

$$Definitely(\phi) : min(X) \prec max(Y) \bigwedge min(Y) \prec max(X), \tag{11.3}$$

$$\overline{Possibly(\phi)} : max(X) \prec min(Y) \bigvee max(Y) \prec min(X). \tag{11.4}$$

When the global predicate is defined on more than two processes, the following results for *Possibly* and *Definitely* are expressible in terms of

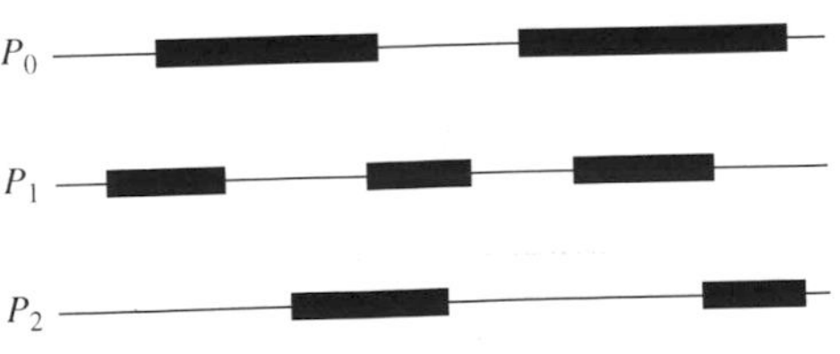

Figure 11.5 For a conjunctive predicate, the shaded durations indicate the periods when the local predicates are true.

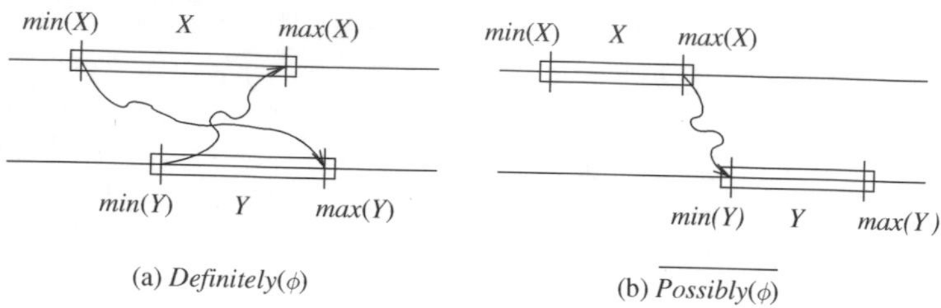

Figure 11.6 Illustrating conditions for *Definitely*(ϕ) and $\neg$*Possibly*(ϕ), for two processes.

Possibly and *Definitely* for pairs of processes [16]. The results can be observed to be true with the help of Figure 11.5.

$$Definitely(\phi) \textbf{ if and only if } \bigwedge_{i,j\in N} Definitely(\phi_i \wedge \phi_j), \tag{11.5}$$

$$Possibly(\phi) \textbf{ if and only if } \bigwedge_{i,j\in N} Possibly(\phi_i \wedge \phi_j). \tag{11.6}$$

Algorithm 11.2 gives an algorithm that is run by a central server P_0 to detect $Possibly(\phi)$ or $Definitely(\phi)$ for a conjunctive predicate ϕ [5,11,12]. Whenever an interval completes, a process could send the vector timestamp of the start and of the end events of that interval as a *Log* entry to the central server process. But observe that, for any two local intervals Y and Y', if there is no send or receive event between the start of the previous interval and the end of the latter interval, then Y and Y' have the exact same relation with respect to all other intervals at all other processes. Hence, an interval needs to be sent to P_0 if there is a send or receive event since the start of the previous interval and the end of this interval. Each execution message thus causes at most four control messages to P_0 – two at the sender and two at the receiver.

The algorithm uses two queues, *updatedQueues* and *newUpdatedQueues*. The *updatedQueues* stores the indices of all the queues whose heads got updated. The latter is a temporary variable for updating *updatedQueues*. A queue gets updated when a new interval potentially becomes the head of the queue; such a new interval becomes a "candidate" interval for the solution. A queue gets updated under two situations: (i) a new interval is enqueued on to an empty queue, or (ii) the current head of a queue gets deleted because it is determined that it cannot possibly be a part of the solution. Each new candidate interval (i.e., the head of some queue) is examined with respect to the heads of all other queues, in accordance with Eqs (11.3) and (11.4), to determine if the desired modality is satisfied. In each comparison, if the desired modality is not satisfied, one of the two intervals examined is marked for deletion (and the corresponding queue is said to be updated).

```
queue of Log: Q1, Q2, ... Qn ←— ⊥
set of int: updatedQueues, newUpdatedQueues ←— {}

On receiving interval from process Pz at P0:
(1)   Enqueue the interval onto queue Qz
(2)   if (number of intervals on Qz is 1) then
(3)       updatedQueues ←— {z}
(4)       while (updatedQueues is not empty)
(5)           newUpdatedQueues ←— {}
(6)           for each i ∈ updatedQueues do
(7)               if (Qi is non-empty) then
(8)                   X ←— head of Qi
(9)                   for j = 1 to n do
(10)                      if (Qj is non-empty) then
(11)                          Y ←— head of Qj
(12)                          if (min(X) ⊀ max(Y)) then                 // Definitely
(13)                              newUpdatedQueues←—{j} ∪ newUpdatedQueues
(14)                          if (min(Y) ⊀ max(X)) then                 // Definitely
(15)                              newUpdatedQueues←—{i} ∪ newUpdatedQueues
(12′)                         if (max(X) ≺ min(Y)) then                 // Possibly
(13′)                             newUpdatedQueues←—{i} ∪ newUpdatedQueues
(14′)                         if (max(Y) ≺ min(X)) then                 // Possibly
(15′)                             newUpdatedQueues←—{j} ∪ newUpdatedQueues
(16)          Delete heads of all Qk where k ∈ newUpdatedQueues
(17)          updatedQueues ←— newUpdatedQueues
(18)      if (all queues are non-empty) then
(19)          solution found. Heads of queues identify intervals solution.
```

Algorithm 11.2 Detecting a conjunctive predicate (centralized, on-line) for *Possibly* or *Definitely* modality. For *Definitely*(ϕ), lines 12–15 are executed. For *Possibly*(ϕ), lines 12′–15′ are executed. To detect both, disjoint data structures are required.

- Specifically, lines 12–15 can be used to check for $Definitely(\phi)$ in accordance with Eq. (11.3).
- Lines 12′–15′ can be used to check for $Possibly(\phi)$ in accordance with Eq. (11.4).

The set *updatedQueues* stores the indices of all the queues whose heads get updated. In each iteration of the **while** loop, the index of each queue whose head is updated is stored in set *newUpdatedQueues* (lines 12–15 *or* 12′–15′). In lines 16 and 17, the heads of all these queues are deleted and indices of the updated queues are stored in the set *updatedQueues*. Thus, an interval gets deleted only if it cannot be part of the solution. Now observe that each interval gets processed unless a solution is found using an interval from each process. From Eqs (11.5) and (11.6), if *every* queue is non-empty and their heads cannot be pruned, then a solution exists and the set of intervals at the head of each queue forms a solution.

Termination

If a solution exists, it is eventually detected by lines 18–19. Otherwise, P_0 waits to receive an interval from some process. The code can be modified to detect the end of the execution at a process, and to notify P_0 about it.

Complexity

Let p be the number of intervals per process, and M be the number of messages sent in the execution.

- **Message complexity** The number of control messages sent by the n processes to P_0 is $min(pn, 4M)$. The first term denotes a message being sent for each interval completed. The second term denotes that at most four control messages get sent for each execution message, in accordance with the observation made earlier. Each control message contains two vector timestamps, which has size $2n$ integers.
- **Space complexity** The space complexity at P_0 is $min(pn, 4M) \cdot 2n$ because all the intervals may have to be queued up among the queues $Q_1, \ldots Q_n$.
- **Time complexity** When an interval is compared with others (loop in lines 6–15), there are $O(n)$ steps. As there are $min(pn, 4M)$ intervals enqueued, the time complexity is $O(n \cdot (min(pn, 4M)))$.

11.4.2 Global state-based centralized algorithm for *Possibly*(ϕ), where ϕ is conjunctive

A more efficient algorithm to detect *Possibly*(ϕ) than the generic algorithm in Algorithm 11.2 can be devised by tailoring an algorithm to this specific modality.

Observe that *Possibly*(ϕ) holds if and only if there is a consistent global state in the execution in which ϕ holds. Thus, detecting *Possibly*(ϕ) is equivalent to identifying a consistent global state in which the local state at each process P_i satisfies ϕ_i. In this consistent global state, for any two local states s_i and s_j at P_i and P_j, respectively, the following must hold:

$$\text{(mutually concurrent)}\ \forall i, \forall j,\ s_i \not\prec s_j \wedge s_j \not\prec s_i. \qquad (11.7)$$

Each process P_i sends the vector timestamp of the local state when ϕ_i becomes true, to the server process P_0. In fact, such a message needs to be sent only each time that the local predicate becomes true for the first time since the previous communication event. This is because internal events that are not separated by communication events are equivalent in terms of consistent global states. Algorithm 11.3 tracks the most recent global state that can potentially satisfy $Possibly(\phi)$ using a two-dimensional array $GS[1 \ldots n, 1 \ldots n]$, where row $GS[i]$ stores the vector timestamp of the local state of process P_i. At P_0, the queuing of the vector timestamps received from P_i into Q_i is not shown explicitly. The algorithm run by P_0 picks any process P_j such that $Valid[j] = 0$ and dequeues the head of Q_j for consideration of consistency with respect

```
integer: GS[1...n, 1...n];   //ith row tracks vector time of P_i
boolean: Valid[1...n];       //Valid[j] = 0 implies P_j state GS[j, ·]
                             //needs to be advanced
queue of array of integer: Q_1, Q_2, ... Q_n ←— ⊥;
                             //Q_i stores timestamp info from P_i

while (∃j | Valid[j] = 0) do  //P_j's state GS[j, ·] is not consistent with
                              //others
    if (Q_j = ⊥ and P_j has terminated) then
        return(0);
    else
        await Q_j becomes non-empty;
        GS[j, 1...n] ←— head(Q_j);  //Consider next state of P_j for
                                    //consistency
        dequeue(head(Q_j));
        Valid[j] ←— 1;
    for k = 1 to n do  //Check P_j's state w.r.t. P_k's state (for every P_k)
        if k ≠ j and Valid[k] = 1 then
            if GS[j, j] ≤ GS[k, j] then  //P_j's state is inconsistent
                                         //with P_k's state
                Valid[j] ←— 0;  //next state of P_j needs to be
                                //considered
            else if GS[k, k] ≤ GS[j, k] then  //P_k's state is inconsistent
                                              //with P_j's state
                Valid[k] ←— 0;  //next state of P_k needs to be
                                //considered
return(1).
```

Algorithm 11.3 Global state-based detection of a conjunctive predicate (centralized, on-line, *Possibly*).

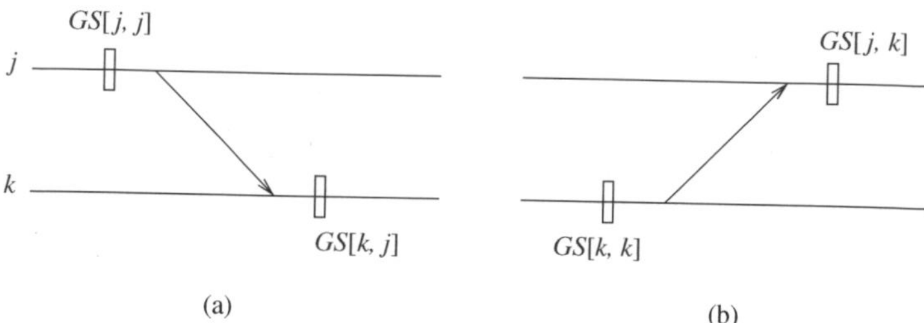

Figure 11.7 In Algorithm 11.3, P_0 tests whether P_j's and P_k's candidate local states are consistent with each other. (a) P_j's old state is invalid. (b) P_k's old state is invalid.

to the current states of all processes (lines 6–8). The main check is in lines 9–14 where P_j's state is checked for mutual consistency with P_k's state, for all k (this check is based on vector clocks [9,19]):

- If P_j's state is old and hence causes inconsistency, it is marked as invalid (lines 11–12). See Figure 11.7(a).
- If P_k's state is old and hence causes inconsistency, it is marked as invalid (lines 13–14). See Figure 11.7(b).

After this main check, the algorithm continues in the main **while** loop and picks another process P_j such that $Valid[j] = 0$. A consistent state is detected when $Valid[j] = 1$ for all j.

Termination

The algorithm terminates successfully (line 15) if $Valid[j] = 1$ for all j, indicating a solution is found. It terminates unsuccessfully (line 3) if some process terminates and its queue is empty.

Complexity

Let m be the number of local states at any process. Let M denote the total number of messages sent in the execution.

- **Time complexity** As there are at most mn local states that are processed by P_0, and for each such local state the **for** loop in line 9 is invoked once and requires $2n$ integer comparisons, the time complexity of the algorithm is $O(n^2m)$.
- **Space complexity** The space complexity at P_0 is $O(n^2m)$ because there are at most mn states, each represented as a vector timestamp, that can be queued among the n queues Q_1 to Q_n.
- **Message complexity** The number of control messages sent by the n processes to P_0 is $2M$, and each message contains the vector timestamp, which has size n integers.

11.5 Distributed algorithms for conjunctive predicates

11.5.1 Distributed state-based token algorithm for *Possibly*(ϕ), where ϕ is conjunctive

Algorithm 11.4 [10] is a distributed version of Algorithm 11.3. Each queue Q_i is maintained locally at P_i. The data structure GS no longer needs to be a $n \times n$ array. Instead, a unique token is passed among the processes serially. The token carries a vector GS corresponding to the vector timestamp of the earliest global state under consideration as a candidate solution.

```
struct token {
        integer: GS[1...n];    //Earliest possible global state as a
                               //candidate solution
        boolean: Valid[1...n]; }Token;   //Valid[j] = 0 indicates
                                         //P_j's state GS[j] is invalid
queue of array of integer: Q_i ←— ⊥;

Initialization. Token is at a randomly chosen process.
On receiving Token at P_i:
    while (Token.Valid[i] = 0) do   //Token.GS[i] is the latest state of P_i
                                    //known to be inconsistent
          await (Q_i to be nonempty);   //with other candidate local
                                        //state of P_j, for some j
          if ((head(Q_i))[i] > Token.GS[i]) then
                Token.GS[i] ←— (head(Q_i))[i];   //earliest possible
                                                 //state of P_i that can be part of
                                                 //solution is written
                Token.Valid[i] ←— 1;   //to Token and its validity is set.
          else dequeue head(Q_i);
    for j = 1 to n (j ≠ i) do   //for each other process P_j: based on P_i's
                                //local state, determine whether
          if j ≠ i and (head(Q_i))[j] ≥ Token.GS[j] then   //P_j's
                                         //candidate local state (in Token)
                                         //is consistent. If not, P_j needs to
                                         //consider a later candidate
                Token.GS[j] ←— (head(Q_i))[j];   // state with a
                                         //timestamp > (head(Q_i))[j]
                Token.Valid[j] ←— 0;
    dequeue (headQ_i);
    if for some k, Token.Valid[k] = 0 then
          send Token to P_k;
    else return(1).
```

Algorithm 11.4 Global state-based detection of a conjunctive predicate (distributed, on-line, *Possibly*) [10]. Code shown is for P_i, $1 \leq i \leq n$.

A process P_i receives a token only when $Token.Valid[i] = 0$. All local states of P_i up to $Token.GS[i]$ will necessarily be *not* consistent with the earliest possible candidate local state of some other process. So P_i has to now consider from its local queue Q_i, the first local state with timestamp greater than $Token.GS[i]$ (lines 3–6). Based on such a state of P_i, now written to $Token.GS[i]$ in line 4, for each j, P_i now determines in line 8 whether P_j's candidate local state $Token.GS[j]$ is consistent with $Token.GS[i]$. This test is illustrated in Figure 11.8.

- If the condition in line 8 is true (Figure 11.8(a)), P_j's state is not consistent. $Token.Valid[j]$ is reset. This implies that the token must visit P_j before termination of the algorithm and P_j needs to find a local state that is mutually consistent with all the other states in $Token.GS$.
- If the condition in line 8 is false (Figure 11.8(b)), P_j's state is consistent.

Termination

The algorithm finds a solution when $Token.Valid[j]$ is 1, for all j (line 14). If a solution is not found, the code hangs in line 2. The code can be modified to terminate unsuccessfully in line 2 by modeling an explicit "process terminated" state in this case.

Complexity

- **Time complexity** Each time a token is received by P_i, at least one local state is examined and deleted. This involves $O(n)$ comparisons in the main loop (lines 7–10). Assuming a total of m states at a process, the time overhead at a process is $O(mn)$. The time overhead across processes is cumulative as the token travels serially. Hence, total time complexity is $O(mn^2)$.
- **Space complexity** In the worst case, all the local states may get queued in Q_i, leading to a space requirement of $O(mn)$. Across all processes the space requirement becomes $O(mn^2)$.
- **Message complexity** The token makes $O(mn)$ hops, and the size of the token is $2n$ integers.

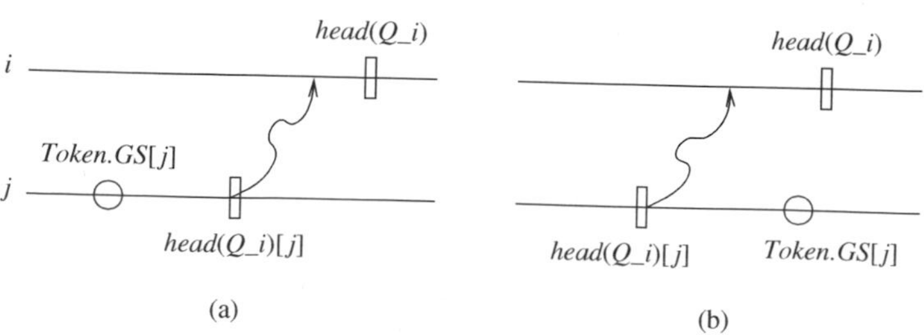

Figure 11.8 In Algorithm 11.4, P_i tests whether P_j's candidate local state $Token.GS[j]$ is consistent with $head(Q_i)[i]$, which is assigned to $Token.GS[i]$. The two possibilities are illustrated. (a) Not consistent. (b) Consistent.

11.5.2 Distributed interval-based token algorithm for *Definitely*(ϕ), where ϕ is conjunctive

We now study an interval-based distributed token-based algorithm to detect $Definitely(\phi)$ based on the tests in Eqs (11.3) and (11.5) [3]. Define $I_i \hookrightarrow I_j$ as: $min(I_i) \prec max(I_j)$.

Problem statement
In a distributed execution, identify a set of intervals $\mathcal{I}$ containing one interval from each process, such that (i) the local predicate ϕ_i is true in $I_i \in \mathcal{I}$, and (ii) for each pair of processes P_i and P_j, $Definitely(\phi_{i,j})$ holds, i.e., $I_i \hookrightarrow I_j$ and $I_j \hookrightarrow I_i$.

The algorithm is given in Algorithm 11.6. The vector timestamps of the start of and of the end of an interval form a data type Log, as shown in Algorithm 11.5. When an interval completes at process P_i, the interval's Log is added to a local queue Q_i *selectively*, as shown in Algorithm 11.5. An interval Y at P_j is deleted if on comparison with some interval X on P_i, $X \not\hookrightarrow Y$,

---

type Log
  start: array$[1 \ldots n]$ of integer;
  end: array$[1 \ldots n]$ of integer;
type Q: queue of Log;
When an interval begins:
$Log_i.start \longleftarrow V_i$.
When an interval ends:
$Log_i.end \longleftarrow V_i$
if (a receive event has occurred since the last time a Log was queued on Q_i) **then**
  Enqueue Log_i on to the local queue Q_i.

Algorithm 11.5 Maintaining intervals for detection of a conjunctive predicate (distributed, on-line, *Definitely*) [3].

i.e., $V_i(min(X))[i] \not\leq V_j(max(Y))[i]$. Thus the interval ($Y$) being deleted or retained depends on its value of $V_j(max(Y))[i]$. The value $V_j(max(Y))[i]$ changes only when a message is received. Hence an interval needs to be stored only if a receive has occurred since the last time a Log of a local interval was queued. Let $V^-(X)$ and $V^+(X)$ denote the vector timestamps of the start of interval X and the end of interval X, respectively.

The token-based algorithm uses three types of messages (see Algorithm 11.6) that are sent among the processes. Request messages of type REQUEST, reply messages of type REPLY, and token messages of type TOKEN, denoted REQ, REP, and T, respectively. Only the token-holder

type *REQUEST* //used by P_i to send a request to each P_j
  start : **integer**; //contains $Log_i.start[i]$ for the interval at the queue //head of P_i
  end : **integer**; //contains $Log_i.end[j]$ for the interval at the queue //head of P_i, when sending to P_j
type *REPLY* //used to send a response to a received request
  updated: **set of integer**; //contains the indices of the updated queues
type *TOKEN* //used to transfer control between two processes
  updatedQueues: **set of integer**; //contains the index of all the updated //queues

(1) Process P_i initializes local state:
(1a) Q_i is empty.

(2) Token initialization:
(2a) A randomly elected process P_i holds the token T.
(2b) $T.updatedQueues \longleftarrow \{1, 2, \ldots, n\}$.

(3) $RcvToken$: When P_i receives a token T:
(3a) Remove index i from $T.updatedQueues$
(3b) **wait until** (Q_i is nonempty)
(3c) $REQ.start \longleftarrow Log_i.start[i]$, where Log_i is the log at head of Q_i
(3d) **for** $j = 1$ to n **do**
(3e) $REQ.end \longleftarrow Log_i.end[j]$
(3f)   Send the request REQ to process P_j
(3g) **wait until** (REP_j is received from each process P_j)
(3h) **for** $j = 1$ to n **do**
(3i) $T.updatedQueues \longleftarrow T.updatedQueues \cup REP_j.updated$
(3j) **if** ($T.updatedQueues$ is empty) **then**
(3k)   Solution detected. Heads of the queues identify intervals that form the solution.
(3l) **else**
(3m)   **if** ($i \in T.updatedQueues$) **then**
(3n)   dequeue the head from Q_i
(3o)   Send token to P_k where k is randomly selected from the set $T.updatedQueues$.

(4) $RcvReq$: When a REQ from P_i is received by P_j:
(4a) **wait until** (Q_j is non-empty)
(4b) $REP.updated \longleftarrow \phi$
(4c) $Y \longleftarrow$ head of local queue Q_j
(4d) $V_i^-(X)[i] \longleftarrow REQ.start$ and $V_i^+(X)[j] \longleftarrow REQ.end$
(4e) Determine $X \hookrightarrow Y$ and $Y \hookrightarrow X$
(4f) **if** ($Y \not\hookrightarrow X$) **then** $REP.updated \longleftarrow REP.updated \cup \{i\}$
(4g) **if** ($X \not\hookrightarrow Y$) **then**
(4h) $REP.updated \longleftarrow REP.updated \cup \{j\}$
(4i)   Dequeue Y from local queue Q_j
(4j) Send reply REP to P_i.

Algorithm 11.6 Interval-based detection of a conjunctive predicate (distributed, on-line, *Definitely*) [3].

process can send *REQ*s and receive *REP*s. The process (P_i) having the token sends *REQ*s to all other processes (line 3f). $Log_i.start[i]$ and $Log_i.end[j]$ for the interval at the head of the queue Q_i are piggybacked on the request *REQ* sent to process P_j (lines 3c–3e). On receiving a *REQ* from P_i, process P_j compares the piggybacked interval X with the interval Y at the head of its queue Q_j (line 4e). The comparisons between intervals on process P_i and P_j can result in these outcomes. (i) $Definitely(\phi_{i,j})$ is satisfied. (ii) $Definitely(\phi_{i,j})$ is not satisfied and interval X can be removed from the queue Q_i. The process index i is stored in $REP.updated$ (line 4f). (iii) $Definitely(\phi_{i,j})$ is not satisfied and interval Y can be removed from the queue Q_j. The interval at the head of Q_j is dequeued and process index j is stored in $REP.updated$ (lines 4g, 4h). Note that outcomes (ii) and (iii) may occur together. After the comparisons, P_j sends REP to P_i. Once the token-holder process P_i receives a REP from all other processes, it stores the indices of all the updated queues in the set $T.updatedQueues$ (lines 3h, 3i). A solution, identified by the set $\mathcal{I}$ formed by the interval I_k at the head of each queue Q_k, is detected if the set $updatedQueues$ is empty. Otherwise, if index i is contained in $T.updatedQueues$, process P_i deletes the interval at the head of its queue Q_i (lines 3m, 3n). If the set $T.updatedQueues$ is non-empty, the token is sent to a process selected randomly from the set (line 3o).

The crux of the correctness of this algorithm is based on Eqs (11.3) and (11.5) for $Definitely(\phi)$. We can make the following observations from the algorithm:

- If $Definitely(\phi_{i,j})$ is not true for a pair of intervals X_i and Y_j, then either i or j is inserted into $T.updatedQueues$.
- An interval is deleted from queue Q_i at process P_i if and only if the index i is inserted into $T.updatedQueues$.
- When a solution $\mathcal{I}$ is detected by the algorithm, the solution is correct, i.e., for each pair $P_i, P_j \in N$, the intervals $I_i = head(Q_i)$ and $I_j = head(Q_j)$ are such that $I_i \hookrightarrow I_j$ and $I_j \hookrightarrow I_i$ (and hence by Eqs (11.3) and (11.5), $Definitely(\phi)$ must be true).
- If a solution $\mathcal{I}$ exists, i.e., for each pair $P_i, P_j \in N$, the intervals I_i, I_j belonging to $\mathcal{I}$ are such that $I_i \hookrightarrow I_j$ and $I_j \hookrightarrow I_i$ (and hence $Definitely(\phi)$ must be true), then the solution is detected by the algorithm.

Complexity

The complexity analysis can be done in terms of two parameters – the maximum number of messages sent per process (m) and the maximum number of intervals per process (p).

- **Space complexity** This is analyzed for each process, and for the entire system.
  - The worst-case space overhead across all the processes is $2mn^2$. The worst-case space overhead at any process is $O(mn^2)$.
  - The total number of *Logs* stored at each process is p because, in the worst case, the *Log* for each interval may need to be stored. As each *Log* has size $2n$, the worst-case overhead is $2np$ integers over all *Logs* per process, and the worst-case space complexity across all processes is $2n^2p = O(n^2p)$.

  As the total number of *Logs* stored on all the processes is $min(np, mn)$, the worst-case space overhead across all the processes is $min(2n^2p, 2n^2m)$. This is equivalent to $min(2np, 2nm)$ per process if the mn message destinations are divided equally among the processes (implying that each process has up to $min(p, m)$ *Logs*). The worst-case space overhead at a process is $min(2np, 2n(n-1)m)$.
- **Time complexity** The two components contributing to time complexity are *RcvReq* and *RcvToken*:

  RcvReq: In the worst case, the number of *REQ*s received by a process is equal to the number of *Logs* on all other processes, because a *REQ* is sent only once for each *Log*. The total number of *Logs* over all the queues is $min(np, mn)$, hence the number of interval pairs compared per process is $min((n-1)p, m(n-1))$. As it takes $O(1)$ time to execute *RcvReq*, the worst-case time complexity per process for *RcvReq* is $O(min(np, mn))$. As the processes execute *RcvReq* in parallel, this is also the total time complexity for *RcvReq*.

  RcvToken: The token makes at most $min(np, mn)$ hops serially and each hop requires $O(n)$ time complexity. Hence the worst-case time complexity for *RcvToken* across all processes is $O(min(pn^2, mn^2))$. In the worst case, a process receives the token each time its queue head is deleted, and this can happen as many times as the number of *Logs* at the process. As the number of *Logs* at a process is $min(p, m(n-1))$, the worst-case time complexity per process is $O(min(pn, mn^2))$.

  The worst-case time complexity across all the processes is $O(min(pn^2, mn^2))$. This is equivalent to $O(min(pn, mn))$ per process if the mn message destinations are divided equally among the processes (implying that each process has up to $min(p, m)$ *Logs*). The worst-case time complexity at a process is $O(min(pn, mn^2))$.
- **Message complexity** For each *Log*, either no messages are sent, or $n-1$ *REQ*s, $n-1$ *REP*s, and one token T are sent.
  - As the total number of *Logs* over all the queues is $min(np, mn)$, hence the worst-case number of messages over all the processes is $O(n\,min(np, mn))$.
  - The size of each T is equal to $O(n)$, while the size of each *REP* and each *REQ* is $O(1)$. Thus for each *Log*, the message space overhead is $O(n)$

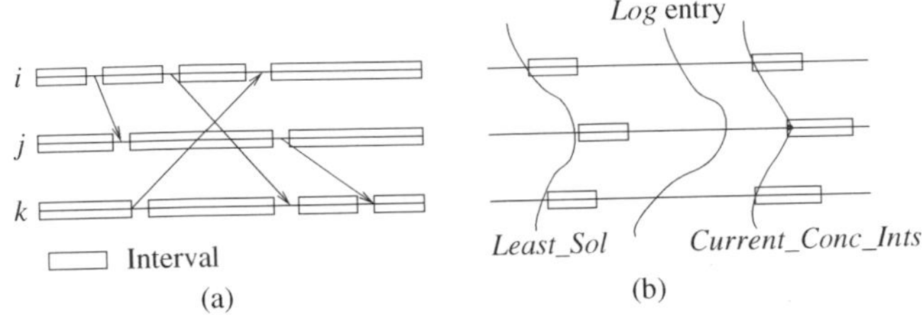

Figure 11.9 Illustrations of definitions used by Algorithm 11.7. (a) Definition of an interval. (b) Definitions of interval vectors *Least_Sol*, *Current_Conc_Ints*, and *Log* entries.

if any messages are sent for that *Log*. Hence the worst-case message space overhead over all the processes is equal to $O(n\, min(np, mn))$.

11.5.3 Distributed interval-based piggybacking algorithm for *Possibly*(ϕ), where ϕ is conjunctive

Unlike the previous algorithm which was a token-based algorithm to detect $Definitely(\phi)$, we now look at the distributed algorithm by Hurfin *et al.* [14] for detecting $Possibly(\phi)$ without using any control messages. Instead, the algorithm piggybacks the necessaryinformation on the application messages. The algorithm therefore illustrates a different design approach.

In Algorithm 11.7, the *semantics of an interval* is that each interval at a process represents the duration between two consecutive communication events at the process (see Figure. 11.9(a)). Intervals are sequentially numbered at any process P_i, as $I_i^0, I_i^1, I_i^2, \ldots$. Two intervals at P_i and P_j are concurrent if $Possibly(\phi)$ is true as determined by Eq. (11.4), and assuming ϕ_i is true in I_i and ϕ_j is true in I_j.

The following variables are used at each process:

- *not_yet_logged*$[1 \ldots n]$, a boolean array, is used to determine whether in the current interval, the "sequence number" of the interval is logged when the predicate first became true in this interval. This variable helps to minimize the number of intervals logged, by ensuring that in any interval, the interval is logged only once in the local log (see below) when the local predicate first becomes true in the interval (lines 1a–1b). Logging just once is important when the predicate may toggle its truth value multiple times within an interval.
- $Current\_Conc\_Ints[1 \ldots n]$, an array of integers, is used to keep track of the latest known concurrent set of intervals (as per Eqs (11.4) and (11.6)). However, it is not necessarily known whether the local predicates at the various processes are true in these intervals, because this array is updated at the start of an interval.
- $Least\_Sol[1 \ldots n]$, an array of integers, is used to track the least possible global state (i.e., set of intervals) that could possibly satisfy $Possibly(\phi)$.

integer: $Least\_Sol[1 \ldots n]$;
boolean: $Valid[1 \ldots n]$;
integer: $Current\_Conc\_Ints[1 \ldots n]$;
queue of $Current\_Conc\_Ints$: $Log \longleftarrow \perp$;
boolean: $not\_yet\_logged_i \longleftarrow 1$;

(1) When local predicate ϕ_i becomes true at P_i:
(1a) **if** $not\_yet\_logged_i$ **then**
(1b) $enqueue(Log_i, Current\_Conc\_Ints_i)$; $not\_yet\_logged_i \longleftarrow false$;
(1c) **if** $Least\_Sol_i[i] = Current\_Conc\_Ints_i[i]$ **then**
(1d) $Valid_i[i] \longleftarrow true$.

(2) P_i sends a message, with $\langle Current\_Conc\_Ints_i, Least\_Sol_i, Valid_i \rangle$ appended:
(2a) $Current\_Conc\_Ints_i[i] \longleftarrow Current\_Conc\_Ints_i[i] + 1$;
(2b) $not\_yet\_logged_i \longleftarrow true$;
(2c) **if** $empty(Log_i)$ **then**
(2d) $Least\_Sol_i[i] \longleftarrow Current\_Conc\_Ints_i[i]$;
(2e) **send** the message with vectors $\langle Current\_Conc\_Ints_i, Least\_Sol_i, Valid_i \rangle$ piggybacked.

(3) When P_i receives a message from P_j with $\langle Current\_Conc\_Ints_j, Least\_Sol_j, Valid_j \rangle$ piggybacked:
(3a) $Current\_Conc\_Ints_i \longleftarrow max(Current\_Conc\_Ints_i, Current\_Conc\_Ints_j)$;
(3b) $(Least\_Sol_i, Valid_i) \longleftarrow Combine\_Maxima((Least\_Sol_i, Valid_i), (Least\_Sol_j, Valid_j))$;
(3c) $Current\_Conc\_Ints_i[i] \longleftarrow Current\_Conc\_Ints_i[i] + 1$;
(3d) $not\_yet\_logged_i \longleftarrow true$;
(3e) **while** $((not\ empty(Log_i))$ and $((head(Log_i))[i] < Least\_Sol_i[i])$ **do**
(3f) $dequeue(Log_i)$;
(3g) **if** $empty(Log_i)$ **then**
(3h) $Least\_Sol_i \longleftarrow Current\_Conc\_Ints_i$; $Valid_i \longleftarrow [0, 0, \ldots, 0]$;
(3i) **else**
(3j) $(Least\_Sol_i, Valid_i) \longleftarrow Combine\_Maxima((Least\_Sol_i, Valid_i), (head(Log_i), [0, 0, \ldots, 0]))$;
(3k) $Valid_i[i] \longleftarrow 1$;
(3l) **if** $Valid_i \longleftarrow [1, 1, \ldots, 1]$ **then**
(3m) $Possibly(\phi)$ is true in global state $Least\_Sol_i$;
(3n) Deliver the message.

(4) function $Combine\_Maxima((C1, A1), (C2, A2))$:
integer: $C[1 \ldots n]$;
boolean: $A[1 \ldots n]$;
(4a) **for** $x = 1$ **to** n **do**
(4b) **case:**
(4c) $C1[x] > C2[x] \longrightarrow (C[x] \longleftarrow C1[x]; A[x] \longleftarrow A1[x])$;
(4d) $C1[x] < C2[x] \longrightarrow (C[x] \longleftarrow C2[x]; A[x] \longleftarrow A2[x])$;
(4e) $C1[x] = C2[x] \longrightarrow (C[x] \longleftarrow C1[x]; A[x] \longleftarrow (A1[x]\ or\ A2[x]))$;
(4f) **return** (C, A).

Algorithm 11.7 Interval-based detection of a conjunctive predicate (distributed, on-line, *Possibly*) [14].

In other words, no interval at any process, that precedes that process's interval in *Least_Sol*, can ever be part of the solution global state.

We have that $(\forall k)$ $Current\_Conc\_Ints[k] \geq Least\_Sol[k]$. See Figure 11.9(b).

- $Valid[1\ldots n]$, a boolean array, tells whether the corresponding interval in *Least_Sol* is valid, i.e., whether the local predicate is ever satisfied in that interval. $Valid[j] = 1$ means ϕ_j is necessarily satisfied in $Least\_Sol[j]$; if 0, it is not yet known whether ϕ_j is satisfied because the interval has not yet completed.
  It follows that $Possibly(\phi)$ is true and ϕ is satisfied in the state identified by *Least_Sol* when all the entries in array *Valid* are true.
- The queue *Log* at each process tracks the various values (vectors) of intervals (one interval per process) that are locally generated, one for each local communication event. In some sense, this tracks the intermediate states *Current_Conc_Ints* as they are generated, between global state *Least_Sol* and the "current" global state *Current_Conc_Ints*.

At the time the local predicate becomes true and *not_yet_logged* is false, (i) *Current_Conc_Ints* is enqueued locally, and (ii) if $Current\_Conc\_Ints[i] = Least\_Sol[i]$, then $Valid[i]$ is set to true (lines 1c–1d).

The array *Current_Conc_Ints*'s local component is always updated for each send and receive event. The array is also piggybacked on each message sent. The receiver takes the maxima of its local array and the sender's array (line 3a). Thus, this global state is always kept up to date.

The array *Least_Sol* plays "catch up" with *Current_Conc_Ints*. At a send event at P_i, $Least\_Sol[i]$ is set to $Current\_Conc\_Int[i]$ if the log Log_i is empty (lines 2c–2d). The arrays *Least_Sol* and *Valid* are also piggybacked on the message sent (line 2e). At a receive event, the receiver P_i takes the more up-to-date information (line 3b) on *Least_Sol* and *Valid* that it has and receives from P_j. (Assuming a solution is not found here, i.e., *Valid* is not all 1, further processing is necessary to advance *Least_Sol*.) In this step, the previous value of $Least\_Sol[i]$ may advance. As a result, entries of *Current_Conc_Ints* in the log Log_i that are older than the new value of $Least\_Sol[i]$ are dequeued and deleted (lines 3e–3f). If Log_i becomes empty, *Least_Sol* catches up completely with *Current_Conc_Ints* and all the entries in the vector *Valid* are reset as we no longer know whether the local predicates were true in the corresponding intervals of *Current_Conc_Ints* (lines 3g–3h). If Log_i is non-empty (line 3i), then the current head of the *Log* represents one of the earlier values of *Current_Conc_Ints*. The information of this queue head and associated validity vector of all 0s, is combined with the value of $(Least\_Sol_i, Valid_i)$ (line 3j) and $Valid[i]$ is set to 1 (line 3k) because the global state from $head(Log_i)$ implies that ϕ_i was true in the local interval in that global state. At this stage, if $Valid_i[k]$ is true for all k, then a solution state is given by *Least_Sol*.

Termination

If $Valid[k] = 1$ for all k, the algorithm finds a set of intervals satisfying $Possibly(\phi)$. Note that for this to happen, some process must have received information about all such intervals and that they were valid. It may happen that such a set of intervals indeed exists but no process is able to see all these intervals under two related conditions: (i) there is not enough communication on which such information can be piggybacked; and (ii) the underlying execution terminates shortly after such a set of intervals come into existence. Exercise 11.8 asks you to analyze this termination condition further.

Complexity

Let M_s and M_c denote the number of messages sent by a process, and the number of communication events at a process, respectively.

- **Time complexity** Each message send and message receive requires $O(n)$ processing. The time complexity at a process is $O(M_c n)$ and, across all processes, this is $O(M_c n^2) = O(M_s n^2)$.
- **Space complexity** The *Log* at a process may have to hold up to M_c intervals, each of size $O(n)$. The other data structures are integer or boolean arrays of size n and require $O(n)$ space locally. Hence, the system space complexity is $O(\sum_{i=1}^{n} M_c n) = O(M_c n^2) = O(M_s n^2)$.
- **Message complexity** On each message sent by the application, $O(3n)$ data is piggybacked. No control messages are used. If a process sends up to M_s messages, the total space overhead is $O(M_s n^2)$.
- **Fault-tolerance** The algorithm is resilient to message losses because it uses piggybacking of control information (See Exercise 11.9).

11.6 Further classification of predicates

We have thus far seen *relational* predicates, *conjunctive* predicates, *local* predicates, and *stable* predicates. Here we formally define local predicates, and then consider two more types of predicates:

- **Local predicate** A local predicate is a predicate whose value is fully controlled by a single process.
- **Disjunctive predicates** If a predicate ϕ can be expressed as the disjunction $\bigvee_{i \in N} \phi_i$, where ϕ_i is a predicate local to process i, then ϕ is a disjunctive predicate. Disjunctive predicates are straightforward to detect; each process monitors the local disjunct, and when it becomes true, informs the other processes. If the disjunct at P_i becomes true after the xth local event, then in the state lattice diagram, ϕ will be true in all global states

having x events at P_i. It is now easy to see that for a disjunctive predicate, $Possibly(\phi) = Definitely(\phi)$.

- **Observer-independent predicates** Different observers may observe different cuts of the execution; an observer can only determine if the predicate ϕ became true in the cuts it can observe. If ϕ is observer-independent, different observers will all agree on whether the predicate ϕ became true.
  We have seen that $Definitely(\phi) \Longrightarrow Possibly(\phi)$. If the predicate ϕ also satisfies the condition $Possibly(\phi) \Longrightarrow Definitely(\phi)$, and thus $Possibly(\phi) = Definitely(\phi)$, then it is an observer-independent predicate. The predicate will be seen to hold or to not hold independent of the observer.

Stable predicates as well as disjunctive predicates are both observer-independent.

The modalities *Possibly* and *Definitely* are coarse-grained. Predicates can also be detected under a rich, fine-grained suite of modalities based on the causality relation [2,15,16].

11.7 Chapter summary

Observing global states is a fundamental problem in asynchronous distributed systems, as studied in Chapter 4. A natural extension of this problem is to detect global states that satisfy a given predicate on the variables of the distributed program. The chapter first considered *stable* predicates, which are predicates that remain true once they become true. Deadlock detection and termination detection are based on stable predicate detection.

Unstable predicates on the program variables are difficult to detect because the values of variables that make the predicate true can change and falsify the predicate. Hence, unstable predicates are defined under modalities: *Possibly* and *Definitely*. Furthermore, a predicate can be broadly classified as *conjunctive* or *relational*. A relational predicate is a predicate using any relation on the distributed variables, whereas a conjuctive predicate is defined to be a conjunct of local predicates.

The chapter studied a centralized algorithm for detecting relational predicates, having exponential complexity. This complexity seems to be inherent for relational predicates. The next centralized algorithms considered for conjunctive predicates were: (i) an interval-based algorithm for detecting both modalities *Possibly* and *Definitely*; and (ii) a global state-based algorithm for detecting under *Possibly* modality.

The chapter then covered three distributed algorithms for conjunctive predicates, all having polynomial complexity. The first was a state-based token-based algorithm for the *Possibly* modality. The second was an interval-based token-based algorithm for the *Definitely* modality. The third was an interval-based piggybacking algorithm for the *Possibly* modality. These representative algorithms illustrate different techniques for conjunctive predicate detection. The chapter concluded by mentioning other more sophisticated predicate modalities.

11.8 Exercises

Exercise 11.1 State whether each of the following is True or False. Justify your answers.

1. $Possibly(\phi) \Longrightarrow \neg Definitely(\phi)$
2. $Possibly(\phi) \Longrightarrow Definitely(\phi)$
3. $Possibly(\phi) \Longrightarrow Definitely(\neg\phi)$
4. $Possibly(\phi) \Longrightarrow \neg Definitely(\neg\phi)$
5. $Definitely(\phi) \Longrightarrow Possibly(\phi)$
6. $Definitely(\phi) \Longrightarrow Possibly(\neg\phi)$
7. $Definitely(\phi) \Longrightarrow \neg Possibly(\phi)$
8. $Definitely(\phi) \Longrightarrow \neg Possibly(\neg\phi)$

Exercise 11.2 A *conjunctive* predicate $\phi = \bigwedge_{i\in N} \phi_i$, where ϕ_i is a predicate defined on variables local to process P_i.

In a distributed execution $(E, \prec)$, let *First_Cut*(ϕ) denote the *earliest* or *smallest* consistent cut in which the global conjunctive predicate ϕ becomes true.

Recall that in different equivalent executions, a different "path" may be traced through the state lattice. Therefore, for different re-executions of this (deterministic) distributed program, is the state *First_Cut*(ϕ) well-defined? That is, is it uniquely identified? In other words, is the set of cuts $C(\phi)$ closed under intersection?

Exercise 11.3 [17] Define all the relevant variables and formulate in detail, a deadlock detection algorithm based on stable predicate detection.

Exercise 11.4 Prove that the predicate detection problem is NP-complete. (Hint: Show a reduction from the satisfiability (SAT) problem.)

Exercise 11.5 If it is known that $Possibly(\phi)$ is true and $Definitely(\phi)$ is false in an execution, then what can be said about ϕ in terms of the paths of the state lattice of that execution?

Exercise 11.6 For Algorithm 11.1, answer the following:

1. When can the algorithm begin constructing the global states of level *lvl*?
2. When are all the global states of level *lvl* constructed?

Exercise 11.7 Can the algorithm for global state-based detection of a conjunctive predicate (centralized, on-line, *Possibly*) of Algorithm 11.3 be modified to detect $Definitely(\phi)$? If yes, give the modified algorithm and show it is correct.

Exercise 11.8 Determine whether the interval-based distributed algorithm (Algorithm 11.7) to detect $Possibly(\phi)$ will always detect $Possibly(\phi)$, even though the algorithm is correct in principle. If it will not, extend the algorithm to ensure that a solution is always detected if it exists. (Hint: Consider the termination of the execution and the *Possibly* modality holding just a little before the termination.)

Exercise 11.9 Analyze the degree to which Algorithm 11.7 is resilient to message losses.

Exercise 11.10 Show the following relationships among the various classes of predicates:

1. The set of stable predicates is a proper subset of the set of observer-independent predicates.
2. The set of disjunctive predicates is a proper subset of the set of observer-independent predicates.

Exercise 11.11 Consider the algorithm for detecting *Possibly*(ϕ) for a conjunctive predicate (Algorithm 11.4). Can the algorithm be modified to delete line 11? How will the correctness of the algorithm be affected?

11.9 Notes on references

The discussion on stable and unstable predicates is based on Chandy and Lamport [6]. Pnueli first introduced a temporal logic for programs with the "henceforth" operator [21]. The discussion on detecting deadlocks is based on Kshemkalyani and Singhal [17] and the discussion on termination detection is based on Mattern [20]. The challenges in detecting unstable predicates, the *Possibly* and *Definitely* modalities, and the notion of the state lattice were formulated by Cooper and Marzullo [8] and Marzullo and Neiger [18]. The centralized algorithms to detect *Possibly* and *Definitely* for relational predicates are based on Cooper and Marzullo [8]. Various techniques to improve the efficiency are given by Alagar and Venkatesan [1]. Conjunctive predicates were discussed by Venkatesan and Dathan [22], Garg and Waldecker [11], and Kshemkalyani [15]. The discussion on the conditions to detect conjunctive predicates is based on Kshemkalyani [15] and Chandra and Kshemkalyani [4]. The centralized algorithm for *Possibly*(ϕ) and *Definitely*(ϕ) where ϕ is conjunctive, in Algorithm 11.2, is adapted from Chandra and Kshemkalyani [2] and Garg and Waldecker [11,12]. The centralized algorithm for *Possibly*(ϕ) where ϕ is conjunctive, in Algorithm 11.3, is based on the test for consistent states using vector clocks of Mattern [19] and Fidge [9]. The distributed state-based algorithm for *Possibly*(ϕ) where ϕ is conjunctive, in Algorithm 11.4, is based on Garg and Chase [10]. The distributed interval-based algorithm for *Definitely*(ϕ) where ϕ is conjunctive, in Algorithm 11.6, is based on Chandra and Kshemkalyani [3]. The distributed interval-based algorithm for *Possibly*(ϕ) where ϕ is conjunctive, in Algorithm 11.7, is based on Hurfin *et al.* [14]. Observer-independent predicates were introduced by Charron-Bost *et al.* [7]. A fine-grained set of modalities was introduced by [15]. Their mapping to the *Possibly*/*Definitely* modalities was proposed in [16]. Algorithms to detect predicates under these fine-grained modalities were given in [2,4,5].

References

[1] S. Alagar and S. Venkatesan, Techniques to tackle state explosion in global predicate detection, *IEEE Transactions Software Engineering*, **27**(8), 2001, 704–714.

[2] P. Chandra and A. D. Kshemkalyani, Algorithms for detecting global predicates under fine-grained modalities, *Proceedings of ASIAN 2003*, December 2003, LNCS, 91–109.

[3] P. Chandra and A. D. Kshemkalyani, Distributed algorithm to detect strong conjunctive predicates, *Information Processing Letters*, **87**(5), 2003, 243–249.

[4] P. Chandra and A. D. Kshemkalyani, Detection of orthogonal interval relations, *Proceedings of the High-Performance Computing Conference*, LNCS 2552, 2002, 323–333.

[5] P. Chandra and A. D. Kshemkalyani, Causality-based predicate detection across space and time, *IEEE Transactions on Computers*, **54**(11), 2005, 1438–1453.

[6] K. M. Chandy and L. Lamport, Distributed snapshots: determining global states of distributed systems, *ACM Transactions on Computer Systems*, **3**(1), 1985, 63–75.

[7] B. Charron-Bost, C. Deloprte-Gallet and H. Fauconnier, Local and temporal predicates in distributed systems, *ACM Transactions on Programming Languages and Systems*, **17**(1), 1995, 157–179.

[8] R. Cooper and K. Marzullo, Consistent detection of global predicates, *Proceedings of the ACM/ONR Workshop on Parallel and Distributed Debugging*, May 1991, 163–173.

[9] C. J. Fidge, Timestamps in message-passing systems that preserve partial ordering, *Australian Computer Science Communications*, **10**(1), 1988, 56–66.

[10] V. K. Garg and C. Chase, Distributed algorithms for detecting conjunctive predicates, *Proceedings of the 15th IEEE International Conference on Distributed Computing Systems*, 1995, 423–430.

[11] V. K. Garg and B. Waldecker, Detection of weak unstable predicates in distributed programs, *IEEE Transactions on Parallel and Distributed Systems*, **5**(3), 1994, 299–307.

[12] V. K. Garg, and B. Waldecker, Detection of strong unstable predicates in distributed programs, *IEEE Transactions on Parallel and Distributed Systems*, **7**(12), 1996, 1323–1333.

[13] G. Ho and C. Ramamoorthy, Protocols for deadlock detection in distributed database systems, *IEEE Transactions on Software Engineering*, **8**(6), 1982, 554–557.

[14] M. Hurfin, M. Mizuno, M. Raynal and M. Singhal, Efficient distributed detection of conjunctions of local predicates, *IEEE Transactions on Software Engineering*, **24**(8), 1998, 664–677.

[15] A. D. Kshemkalyani, Temporal interactions of intervals in distributed systems, *Journal of Computer and System Sciences*, **52**(2), 1996, 287–298.

[16] A. D. Kshemkalyani, A fine-grained modality classification for global predicates, *IEEE Transactions on Parallel and Distributed Systems*, **14**(8), 2003, 807–816.

[17] A. D. Kshemkalyani and M. Singhal, Correct two-phase and one-phase deadlock detection algorithms for distributed systems, *Proceedings of the 2nd IEEE Symposium on Parallel and Distributed Processing*, December 1990, 126–129.

[18] K. Marzullo and G. Neiger, Detection of global state predicates, *Proceedings of the 5th Workshop on Distributed Algorithms*, LNCS 579, October 1991, 254–272.

[19] F. Mattern, Virtual time and global states of distributed systems, *Proceedings of the International Workshop on Parallel and Distributed Algorithms*, October 1998, 215–226.

[20] F. Mattern, Algorithms for distributed termination detection, *Distributed Computing*, **2**, 1987, 161–175.

[21] A. Pnueli, The temporal logic of programs, *Proceedings of the IEEE Symposium on Foundations of Computer Science*, 1977, 46–57.

[22] S. Venkatesan and B. Dathan, Testing and debugging distributed programs using global predicates, *IEEE Transactions on Software Engineering*, **21**(2), 1995, 163–177.

CHAPTER

12 Distributed shared memory

12.1 Abstraction and advantages

Distributed shared memory (DSM) is an abstraction provided to the programmer of a distributed system. It gives the impression of a single monolithic memory, as in traditional von Neumann architecture. Programmers access the data across the network using only *read* and *write* primitives, as they would in a uniprocessor system. Programmers do not have to deal with *send* and *receive* communication primitives and the ensuing complexity of dealing explicitly with synchronization and consistency in the message-passing model. The DSM abstraction is illustrated in Figure 12.1. A part of each computer's memory is earmarked for shared space, and the remainder is private memory. To provide programmers with the illusion of a single shared address space, a memory mapping management layer is required to manage the *shared virtual memory* space.

DSM has the following advantages:

1. Communication across the network is achieved by the read/write abstraction that simplifies the task of programmers.
2. A single address space is provided, thereby providing the possibility of avoiding data movement across multiple address spaces, and simplifying *passing-by-reference* and passing complex data structures containing pointers.
3. If a block of data needs to be moved, the system can exploit locality of reference to reduce the communication overhead.
4. DSM is often cheaper than using dedicated multiprocessor systems, because it uses simpler software interfaces and off-the-shelf hardware.
5. There is no bottleneck presented by a single memory access bus.
6. DSM effectively provides a large (virtual) main memory.
7. DSM provides portability of programs written using DSM. This portability arises due to a common DSM programming interface, which is independent of the operating system and other low-level system characteristics.

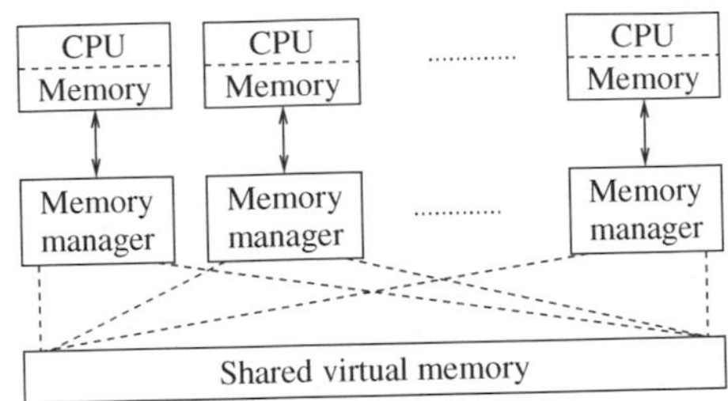

Figure 12.1 Abstract view of DSM.

Although a familiar (i.e., read/write) interface is provided to the programmer (see Figure 12.2) there is a catch to it. Under the covers, there is inherently a distributed system and a network, and the data needs to be shared in some fashion. There is no silver bullet. Moreover, with the possibility of data replication and/or the concurrent access to data, concurrency control needs to be enforced. Specifically, when multiple processors wish to access the same data object, a decision about how to handle concurrent accesses needs to be made. As in traditional databases, if a locking mechanism based on read and write locks for objects is used, concurrency is severely restrained, defeating one of the purposes of having the distributed system. On the other hand, if concurrent access is permitted by different processors to different replicas, the problem of replica consistency (which is a generalization of the problem of cache consistency in computer architecture studies) needs to be addressed. The main point of allowing concurrent access (by different processors) to the same data object is to increase throughput. But in the face of concurrent access, the semantics of what value a read operation returns to the program needs to be specified. Programmers ultimately need to understand this semantics, which may differ from the Von Neumann semantics, because the program logic depends greatly on this semantics. This compromises the assumption that the DSM is transparent to the programmer.

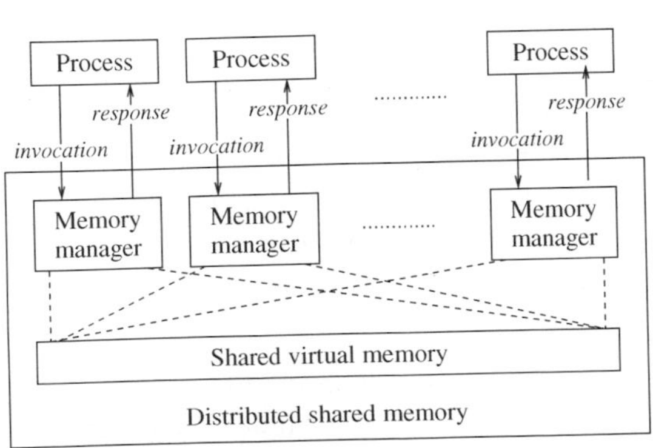

Figure 12.2 Detailed abstraction of DSM and interaction with application processes.

Before examining the challenges in implementing replica coherency in DSM systems, we look at its disadvantages:

1. Programmers are not shielded from having to know about various replica consistency models and from coding their distributed applications according to the semantics of these models.
2. As DSM is implemented under the covers using asynchronous message-passing, the overheads incurred are at least as high as those of a message-passing implementation. As such, DSM implementations cannot be more efficient than asynchronous message-passing implementations. The generality of the DSM software may make it less efficient.
3. By yielding control to the DSM memory management layer, programmers lose the ability to use their own message-passing solutions for accessing shared objects. It is likely that the standard vanilla implementations of DSM have a higher overhead than a programmer-written implementation tailored for a specific application and system.

The main issues in designing a DSM system are the following:

- Determining what semantics to allow for concurrent access to shared objects. The semantics needs to be clearly specified so that the programmer can code his program using an appropriate logic.
- Determining the best way to implement the semantics of concurrent access to shared data. One possibility is to use replication. One decision to be made is the degree of replication – partial replication at some sites, or full replication at all the sites. A further decision then is to decide on whether to use read-replication (replication for the read operations) or write-replication (replication for the write operations) or both.
- Selecting the locations for replication (if full replication is not used), to optimize efficiency from the system's viewpoint.
- Determining the location of remote data that the application needs to access, if full replication is not used.
- Reducing communication delays and the number of messages that are involved under the covers while implementing the semantics of concurrent access to shared data.

There is a wide range of choices on how these issues can be addressed. In part, the solution depends on the system architecture. Recall from Chapter 1 that DSM systems can range from tightly coupled (hardware and software) multicomputers to wide-area distributed systems with heterogenous hardware and software. There are four broad dimensions along which DSM systems can be classified and implemented:

- Whether data is replicated or cached.
- Whether remote access is by hardware or by software.
- Whether the caching/replication is controlled by hardware or software.
- Whether the DSM is controlled by the distributed memory managers, by the operating system, or by the language runtime system.

Table 12.1 Comparison of DSM systems (adapted from [29]).

Type of DSM	Examples	Management	Caching	Remote access
Single-bus multiprocessor	Firefly, Sequent	by MMU	hardware control	by hardware
Switched multiprocessor	Alewife, Dash	by MMU	hardware control	by hardware
NUMA system	Butterfly, CM*	by OS	software control	by hardware
Page-based DSM	Ivy, Mirage	by OS	software control	by software
Shared variable DSM	Midway, Munin	by language runtime system	software control	by software
Shared object DSM	Linda, Orca	by language runtime system	software control	by software

The various options for each of these four dimensions, and their comparison, are shown in Table 12.1, as adapted from [29].

12.2 Memory consistency models

Memory coherence is the ability of the system to execute memory operations correctly. Assume n processes and s_i memory operations per process P_i. Also assume that all the operations issued by a process are executed sequentially (that is, pipelining is disallowed), as shown in Figure 12.3. Observe that there are a total of

$$(s_1 + s_2 + \ldots + s_n)!/(s_1!s_2!\ldots s_n!)$$

possible permutations or interleavings of the operations issued by the processes. The problem of ensuring memory coherence then becomes the problem of identifying which of these interleavings are "correct," which of course requires a clear definition of "correctness." The *memory consistency model* defines the set of allowable memory access orderings. While a traditional definition of correctness says that a correct memory execution is one that returns to each *Read* operation, the value stored by the most recent *Write* operation, the very definition of "most recent" becomes ambigious in the presence of concurrent access and multiple replicas of the data item. Thus, a clear definition of correctness is required in such a system; the objective is to disallow the interleavings that make no semantic sense, while not being overly restrictive so as to permit a high degree of concurrency.

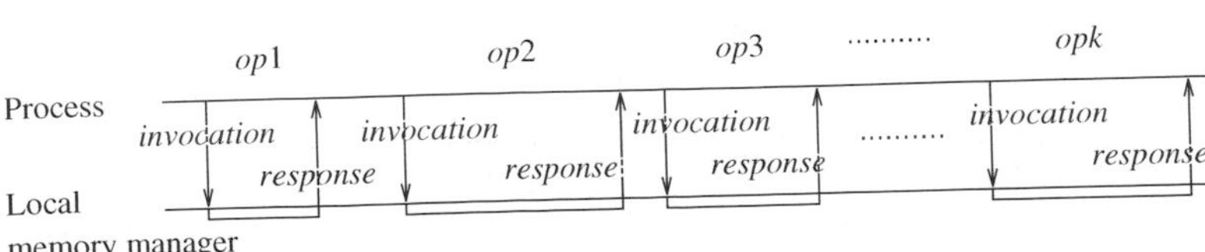

Figure 12.3 Sequential invocations and responses in a DSM system, without any pipelining.

The DSM system enforces a particular memory consistency model; programmers write their programs keeping in mind the allowable interleavings permitted by that specific memory consistency model. A program written for one model may not work correctly on a DSM system that enforces a different model. The model can thus be viewed as a *contract* between the DSM system and the programmer using that system. We now consider six consistency models, which are related as shown in Figure 12.8.

Notation A write of value a to variable x is denoted as $Write(x,a)$. A read of variable x that returns value a is denoted as $Read(x,a)$. A subscript on these operations is sometimes used to denote the processor that issues these operations.

12.2.1 Strict consistency/atomic consistency/linearizability

The strictest model, corresponding to the notion of correctness on the traditional Von Neumann architecture or the uniprocessor machine, requires that any *Read* to a location (variable) should return the value written by the most recent *Write* to that location (variable). Two salient features of such a system are the following: (i) a common global time axis is implicitly available in a uniprocessor system; (ii) each write is immediately visible to all processes. Adapting this correctness model to a DSM system with operations that can be concurrently issued by the various processes gives the *strict consistency model*, also known as the *atomic consistency model*. The model is more formally specified as follows [13,21]:

1. Any *Read* to a location (variable) is required to return the value written by the most recent *Write* to that location (variable) as per a global time reference.

   For operations that do not overlap as per the global time reference, the specification is clear. For operations that overlap as per the global time reference, the following further specifications are necessary.
2. All operations appear to be executed atomically and sequentially.
3. All processors see the same ordering of events, which is equivalent to the global-time occurrence of non-overlapping events.

An alternate way of specifying this consistency model is in terms of the "invocation" and "response" to each *Read* and *Write* operation, as shown in Figure 12.3. Recall that each operation [13] takes a finite time interval and hence different operations by different processors can overlap in time. However, the invocation and the response to each invocation can both be separately viewed as being atomic events. An execution sequence in global time is viewed as a sequence *Seq* of such invocations and responses. Clearly, *Seq* must satisfy the following conditions:

- (Liveness:) Each invocation must have a corresponding response.
- (Correctness:) The projection of *Seq* on any processor i, denoted Seq_i, must be a sequence of alternating invocations and responses if pipelining is disallowed.

Despite the concurrent operations, a linearizable execution needs to generate an equivalent global order on the events that is a permutation of *Seq*, satisfying the semantics of *linearizability*. More formally, a sequence *Seq* of invocations and responses is *linearizable* (LIN) if there is a permutation Seq' of adjacent pairs of corresponding $\langle invoc, resp\rangle$ events satisfying:

1. For every variable v, the projection of Seq' on v, denoted Seq'_v, is such that every *Read* (adjacent $\langle invoc, resp\rangle$ event pair) returns the most recent *Write* (adjacent $\langle invoc, resp\rangle$ event pair) that immediately preceded it.
2. If the response $op1(resp)$ of operation $op1$ occurred before the invocation $op2(invoc)$ of operation $op2$ in *Seq*, then $op1$ (adjacent $\langle invoc, resp\rangle$ event pair) occurs before $op2$ (adjacent $\langle invoc, resp\rangle$ event pair) in Seq'.

Condition 1 specifies that every processor sees a common order Seq' of events, and that in this order, the semantics is that each *Read* returns the most recent completed *Write* value. Condition 2 specifies that the common order Seq' must satisfy the global time order of events, viz., the order of non-overlapping operations in *Seq* must be preserved in Seq'.

Examples Figure 12.4 shows three executions:

- **Figure 12.4(a)** The execution is not linearizable because although the *Read* by P_2 begins after *Write*$(x, 4)$, the *Read* returns the value that existed before the *Write*. Hence, a permutation Seq' satisfying the condition 2 above on global time order does not exist.
- **Figure 12.4(b)** The execution is linearizable. The global order of operations (corresponding to ⟨invocation, response⟩ pairs in Seq'), consistent with the real-time occurrence, is: *Write*$(y, 2)$, *Write*$(x, 4)$, *Read*$(x, 4)$, *Read*$(y, 2)$. This permutation Seq' satisfies conditions 1 and 2.
- **Figure 12.4(c)** The execution is not linearizable. The two dependencies: *Read*$(x, 0)$ before *Write*$(x, 4)$, and *Read*$(y, 0)$ before *Write*$(y, 2)$ cannot both be satisfied in a global order while satisfying the local order of operations at each processor. Hence, there does not exist any permutation Seq' satisfying conditions 1 and 2.

Implementations

Implementing linearizability is expensive because a global time scale needs to be simulated. As all processors need to agree on a common order, the implementation needs to use total order. For simplicity, we assume full replication of each data item at all the processors. Hence, total ordering needs to be combined with a broadcast. Algorithm 12.1 gives the implementation

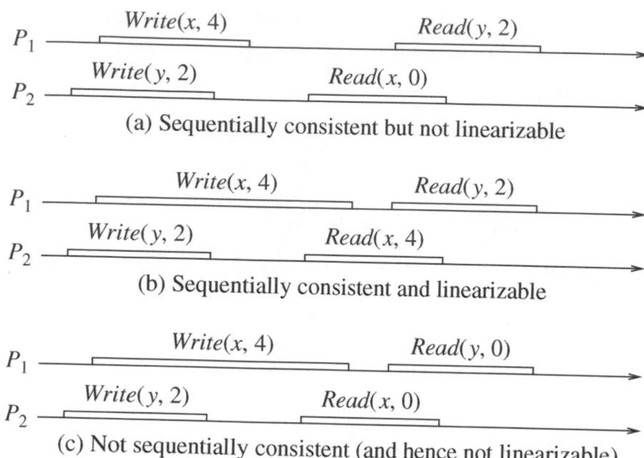

Figure 12.4 Examples to illustrate definitions of linearizability and sequential consistency. The initial values of variables are zero.

assuming the existence of a *total order broadcast* primitive that broadcasts to all processors including the sender. Hence, the memory manager software has to be placed between the application above it and the total order broadcast layer below it.

(shared var)
int: x;

(1) When the memory manager receives a *Read* or *Write* from application:
(1a) **total_order_broadcast** the *Read* or *Write* request to all processors;
(1b) **await** own request that was broadcast;
(1c) **perform** pending response to the application as follows
(1d) **case** *Read*: return value from local replica;
(1e) **case** *Write*: write to local replica and return ack to application.

(2) When the memory manager receives a **total_order_broadcast**(*Write*, *x*, *val*) from network:
(2a) **write** *val* to local replica of x.

(3) When the memory manager receives a **total_order_broadcast**(*Read*, *x*) from network:
(3a) **no operation**.

Algorithm 12.1 Implementing linearizability (LIN) using total order broadcasts [6]. Code shown is for P_i, $1 \leq i \leq n$.

Although Algorithm 12.1 appears simple, it is also subtle. The total order broadcast ensures that all processors see the same order:

- For two non-overlapping operations at different processors, by the very definition of non-overlapping, the response to the former operation precedes the invocation of the latter in global time.
- For two overlapping operations, the total order ensures a common view by all processors.

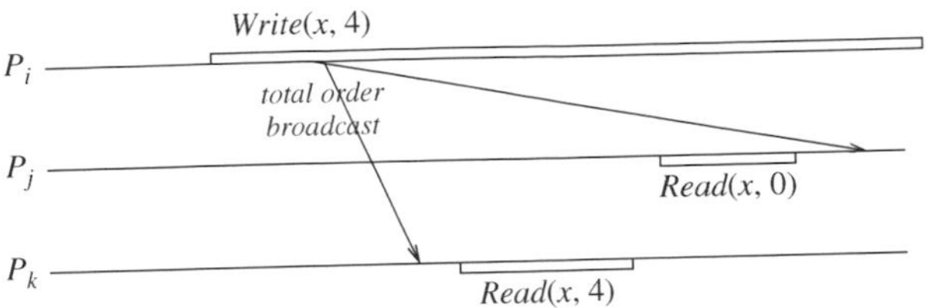

Figure 12.5 A violation of *linearizability* (LIN) if *Read* operations do not participate in the total order broadcast [6].

For a *Read* operation, when the memory managers systemwide receive the total order broadcast, they do not perform any action. Why is the broadcast then necessary? The reason is this. If *Read* operations do not participate in the total order broadcasts, they do not get totally ordered with respect to the *Write* operations as well as with respect to the other *Read* operations. This can lead to a violation of linearizability, as shown in Figure 12.5. The *Read* by P_k returns the value written by P_i. The later *Read* by P_j returns the initial value of 0. As per the global time ordering requirement of linearizability, the *Read* by P_j that occurs after the *Read* by P_k must also return the value 4. However, that is not the case in this example, wherein the *Read* operations do not participate in the total order broadcast.

12.2.2 Sequential consistency

Linearizability or strict/atomic consistency is difficult to implement because the absence of a global time reference in a distributed system necessitates that the time reference has to be simulated. This is very expensive. Programmers can deal with weaker models. The first weaker model, that of *sequential consistency* (SC) was proposed by Lamport [19] and uses logical time reference instead of the global time reference.

Sequential consistency is specified as follows:

- The result of any execution is the same as if all operations of the processors were executed in *some* sequential order.
- The operations of each individual processor appear in this sequence in the local program order.

Although any possible interleaving of the operations from the different processors is possible, all the processors must see *the same* interleaving. In this model, even if two operations from different processors (on the same or different variables) do not overlap in a global time scale, they may appear in reverse order in the *common* sequential order seen by all the processors.

More formally [13], a sequence *Seq* of invocation and response events is sequentially consistent if there is a permutation *Seq'* of adjacent pairs of corresponding $\langle invoc, resp\rangle$ events satisfying:

1. For every variable v, the projection of *Seq'* on v, denoted Seq'_v, is such that every *Read* (adjacent $\langle invoc, resp\rangle$ event pair) returns the most recent *Write* (adjacent $\langle invoc, resp\rangle$ event pair) that immediately preceded it.

2. If the response $op1(resp)$ of operation $op1$ at process P_i occurred before the invocation $op2(invoc)$ of operation $op2$ by process P_i in Seq, then $op1$ (adjacent $\langle invoc, resp\rangle$ event pair) occurs before $op2$ (adjacent $\langle invoc, resp\rangle$ event pair) in Seq'.

Condition 1 is the same as that for linearizability. Condition 2 differs from that for linearizability. It specifies that the common order Seq' must satisfy only the local order of events at each processor, instead of the global order of non-overlapping events. Hence the order of non-overlapping operations issued by different processors in Seq need not be preserved in Seq'.

Examples Three examples are considered in Figure 12.4:

- **Figure 12.4(a)** The execution is sequentially consistent. The global order Seq' is: $Write(y, 2)$, $Read(x, 0)$, $Write(x, 4)$, $Read(y, 2)$.
- **Figure 12.4(b)** As the execution is linearizable (seen in Section 12.2.1), it is also sequentially consistent. The global order of operations (corresponding to ⟨invocation, response⟩ pairs in Seq'), consistent with the real-time occurrence, is: $Write(y, 2)$, $Write(x, 4)$, $Read(x, 4)$, $Read(y, 2)$.
- **Figure 12.4(c)** The execution is not sequentially consistent (and hence not linearizable). The two dependencies: $Read(x, 0)$ before $Write(x, 4)$, and $Read(y, 0)$ before $Write(y, 2)$ cannot both be satisfied in a global order while satisfying the local order of operations at each processor. Hence, there does not exist any permutation Seq' satisfying conditions 1 and 2.

Implementations

As sequential consistency (SC) is less restrictive than linearizability (LIN), it should be easier to implement it. As all processors are required to see the same global order, but global time ordering need not be preserved across processes, it is sufficient to use total order broadcasts for the *Write* operations only. In the simplified algorithm, no total order broadcast is required for *Read* operations, because:

1. all consecutive operations by the same processor are ordered in the same order because pipelining is not used;
2. *Read* operations by different processors are independent of each other and need to be ordered only with respect to the *Write* operations in the execution.

In Exercise 12.1, you will be asked to reason this more thoroughly. Two algorithms for SC by Attiya and Welch [6] that exhibit a trade-off of the inhibition of *Read* versus *Write* operations are given next.

(shared var)
int: x;
(1) When the memory manager at P_i receives a *Read* or *Write* from application:
(1a) **case** *Read*: **return** value from local replica;
(1b) **case** *Write*(x,*val*): **total_order_broadcast**$_i$(*Write*(x, *val*)) to all processors including itself.

(2) When the memory manager at P_i receives a **total_order_broadcast**$_j$ (*Write*, x, *val*) from network:
(2a) **write** *val* to local replica of x;
(2b) **if** $i = j$ **then return** acknowledgement to application.

Algorithm 12.2 Implementing sequential consistency (SC) using local *Read* operations [6]. Code shown is for P_i, $1 \leq i \leq n$.

Local-read algorithm

The first algorithm for SC, given in Algorithm 12.2, is a direct simplification of the algorithm for linearizability, given in Algorithm 12.1. In the algorithm, a *Read* operation completes atomically, whereas a *Write* operation does not. Between the invocation of a *Write* by P_i (line 1b) and its acknowledgement (lines 2a, 2b), there may be multiple *Write* operations initiated by other processors that take effect at P_i (line 2a). Thus, a *Write* issued locally has its completion locally delayed. Such an algorithm is acceptable for *Read*-intensive programs.

Local-write algorithm

Algorithm 12.3 does not delay acknowledgement of *Writes*. For *Write*-intensive programs, it is desirable that a locally issued *Write* gets acknowledged immediately (as in lines 2a–2c), even though the total order broadcast for the *Write*, and the actual update for the *Write* may not go into effect by updating the variable at the same time (line 3a). The algorithm achieves this at the cost of delaying a *Read* operation by a processor until all previously issued local *Write* operations by that same processor have locally gone into effect (i.e., previous *Writes* issued locally have updated their local variables being written to). The variable *counter* is used to track the number of *Write* operations that have been locally initiated but not completed at any time. A *Read* operation completes only if there are no prior locally initiated *Write* operations that have not written to their variables (line 1a), i.e., there are no pending locally initiated *Write* operations to any variable. Otherwise, a *Read* operation is delayed until after all previously initiated *Write* operations have written to their local variables (lines 3b–3d), which happens after the total order broadcasts associated with the *Write* have delivered the broadcast message locally.

```
(shared var)
int: x;
(1)  When the memory manager at P_i receives a Read(x) from application:
(1a) if counter = 0 then
(1b)     return x
(1c) else keep the Read pending.

(2)  When the memory manager at P_i receives a Write(x,val) from
     application:
(2a) counter ⟵ counter + 1;
(2b) total_order_broadcast_i Write(x, val);
(2c) return acknowledgement to the application.

(3)  When the memory manager at P_i receives a total_order_broadcast_j
     (Write, x, val) from network:
(3a) write val to local replica of x;
(3b) if i = j then
(3c)     counter ⟵ counter − 1;
(3d)     if (counter = 0 and any Reads are pending) then
(3e)         perform pending responses for the Reads to the
             application.
```

Algorithm 12.3 Implementing Sequential Consistency (SC) using local *Write* operations [6]. Code shown is for P_i, $1 \leq i \leq n$.

This algorithm performs fast (local) *Write*s and slow *Reads*. The algorithm pipelines all *Write* updates issued by a processor. The *Read* operations have to wait for all *Write* updates issued earlier by that processor to complete (i.e., take effect) locally before the value to be read is returned to the application.

12.2.3 Causal consistency

For the sequential consistency model, it is required that *Write* operations issued by different processors must necessarily be seen in some common order by all processors. This requirement can be relaxed to require only that *Write*s that are *causally related* must be seen in that same order by all processors, whereas "concurrent" *Writes* may be seen by different processors in different orders. The resulting consistency model is the *causal consistency* model, as defined by [4]. We have seen the definition of causal relationships among events in a message-passing system. What does it mean for two *Write* operations to be causally related?

The *causality relation* for shared memory systems is defined as follows:

- **Local order** At a processor, the serial order of the events defines the local causal order.

- **Inter-process order** A *Write* operation causally precedes a *Read* operation issued by another processor if the *Read* returns a value written by the *Write*.
- **Transitive closure** The transitive closure of the above two relations defines the (global) causal order.

Examples The examples in Figure 12.6 illustrate causal consistency:

- **Figure 12.6(a)** The execution is sequentially consistent (and hence causally consistent). Both P_3 and P_4 see the operations at P_1 and P_2 in sequential order and in causal order.
- **Figure 12.6(b)** The execution is not sequentially consistent but it is causally consistent. Both P_3 and P_4 see the operations at P_1 and P_2 in causal order because the lack of a causality relation between the *Writes* by P_1 and by P_2 allows the values written by the two processors to be seen in different orders in the system. The execution is not sequentially consistent because there is no global satisfying the contradictory ordering requirements set by the *Reads* by P_3 and by P_4. What can be said if the two *Read* operations of P_4 returned 7 first and then 4? (See Exercise 12.4.)
- **Figure 12.6(c)** The execution is not causally consistent because the second *Read* by P_4 returns 4 after P_4 has already returned 7 in an earlier *Read*.

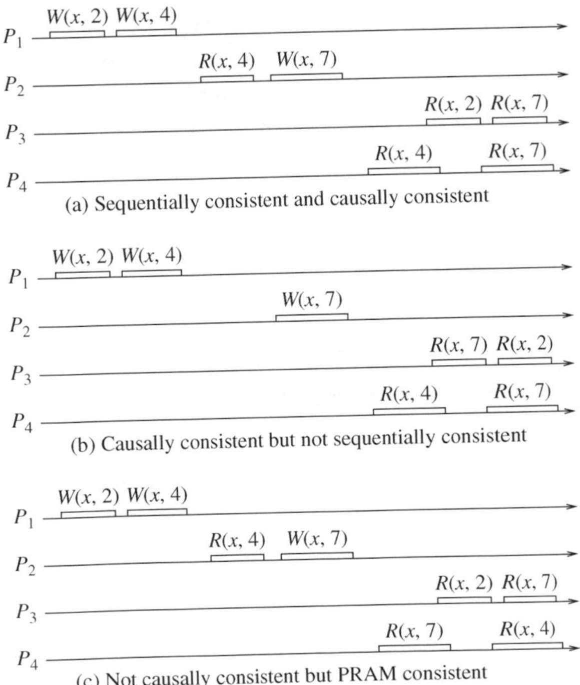

Figure 12.6 Examples to illustrate definitions of sequential consistency (SC), causal consistency (CC), and PRAM consistency. The initial values of variables are zero.

Implementation

We first examine the definition of sequential consistency. Even though all processors only need to see *some* total order of the *Write* operations, observe that if two *Write* operations are related by causality (i.e., the second *Write* begins causally after a *Read* that reads the value written by the first *Write*), then the order of the two *Writes* seen by all the processors also satisfies causal order! In the implementation, even though a total order broadcast primitive is used, observe that it implicitly provides causal ordering on all the *Write* operations. Thus, due to the nature of the definition of causal ordering in shared memory systems, a total order broadcast also provides causal order broadcast, unlike the case for message-passing systems. (Exactly why is it so?)

In contrast to the SC requirement, causal consistency implicitly requires only that causal order be provided. Thus, a causal order broadcast can be used in the implementation. The details of the implementation are left as Exercise 12.5.

12.2.4 PRAM (pipelined RAM) or processor consistency

Causal consistency requires all causally related *Writes* to be seen in the same order by all processors. This may be viewed as being too restrictive for some applications. A weaker form of consistency requires only that *Write* operations issued by the same (any one) processor are seen by all other processors in the same order that they were issued, but *Write* operations issued by different processors may be seen in different orders by different processors. In relation to the "causality" relation between operations, only the local causality relation, as defined by the local order of *Write* operations, needs to be seen by other processors. Hence, this form of consistency is termed *processor consistency*. An equivalent name for this consistency model is *pipelined RAM* (PRAM), to capture the behavior that all operations issued by any processor appear to the other processors in a FIFO pipelined sequence. PRAM consistency was defined by [25].

Examples

- In Figure 12.6(c), the execution is PRAM consistent (even though it is not causally consistent) because (trivially) both P_3 and P_4 see the updates made by P_1 and P_2 in FIFO order along the channels P_1 to P_3 and P_2 to P_3, and along P_1 to P_4 and P_2 to P_4, respectively.
- While PRAM consistency is more permissive than causal consistency, this model must be used with care by the programmer because it can lead to rather unintuitive results. For example, examine the code in Algorithm 12.4, where x and y are shared variables. It is possible that, on a PRAM system, both processes P_1 and P_2 get killed. This can happen as follows: (i) P_1 writes 4 to x in line 1a and P_2 writes 6 to y in line 2a at about the same time; (ii) before these written values propagate to the other

processor, P_1 reads y (as being 0) in line 1b and P_2 reads x (as being 0) in line 2b. Here, a *Read* (e.g., in lines 1b or 2b) can effectively "overtake" a preceding *Write* (of lines 2a or 1a, respectively) if the two accesses by the same processor are to different locations. However, this would not be expected on a conventional machine, where at most one process may get killed, depending on the interleaving of the statements.

- The execution in Figure 12.7 (a) violates PRAM consistency. An explanation is given in Section 12.2.5.

(shared variables)
int: x, y;

Process 1	Process 2
...	...
(1a) $x \longleftarrow 4$;	(2a) $y \longleftarrow 6$;
(1b) **if** $y = 0$ **then kill**(P_2).	(2b) **if** $x = 0$ **then kill**(P_1).

Algorithm 12.4 A counter-intuitive behavior of a PRAM-consistent program. The initial values of variables are zero.

Implementations

PRAM consistency can be implemented using FIFO broadcast. The implementation details are left as Exercise 12.6.

12.2.5 Slow memory

The next weaker consistency model is that of *slow memory* [14]. This model represents a location-relative weakening of the PRAM model. In this model, only all *Write* operations issued by the same processor and to the same memory location must be observed in the same order by all the processors.

Examples The examples in Figure 12.7 illustrate slow memory consistency:

- **Figure 12.7(a)** The updates to each of the variables are seen pipelined separately in a FIFO fashion. The "x" pipeline from P_1 to P_2 is slower than the "y" pipeline from P_1 to P_2. Thus, the overtaking effect is allowed. However, PRAM consistency is violated because the FIFO property is violated over the single common "pipeline" from P_1 to P_2 – the update to y is seen by P_2 but the much older value of $x = 0$ is seen by P_2 later.
- **Figure 12.7(b)** Slow memory consistency is violated because the FIFO property is violated for the pipeline for variable x. "$x = 7$" is seen by P_2 before it sees "$x = 0$" and "$x = 2$" although 7 was written to x after the values of 0 and 2.

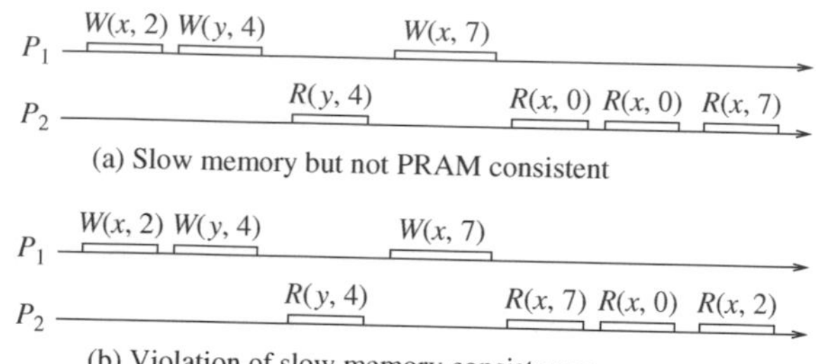

Figure 12.7 Examples to illustrate definitions of PRAM consistency and slow memory. The initial values of variables are zero.

Implementations

Slow memory can be implemented using a broadcast primitive that is weaker than even the FIFO broadcast. What is required is a FIFO broadcast per variable in the system, i.e., the FIFO property should be satisfied only for updates to the same variable. The implementation details are left as Exercise 12.7.

12.2.6 Hierarchy of consistency models

Based on the definitions of the memory consistency models seen so far, there exists a hierarchy among the models, as depicted in Figure 12.8.

12.2.7 Other models based on synchronization instructions

We have seen several popular consistency models. Based on the consistency model, the behavior of the DSM differs, and the programmer's logic therefore depends on the underlying consistency model. It is also possible that newer consistency models may arise in the future.

The consistency models seen so far apply to all the instructions in the distributed program. We now briefly mention some other consistency models that are based on a different principle, namely that the consistency conditions apply only to a set of distinguished "synchronization" or "coordination" instructions. These synchronization instructions are typically from some run-time library. A common example of such a statement is the barrier synchronization. Only

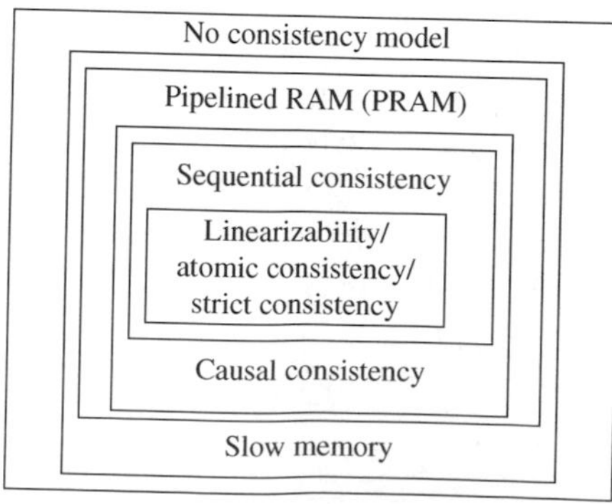

Figure 12.8 A strict hierarchy of the memory consistency models.

the synchronization statements across the various processors must satisfy the consistency conditions; other program statements between synchronization statements may be executed by the different processors without any conditions. Examples of consistency models based on this principle are: *entry consistency*, *weak consistency*, and *release consistency*. The synchronization statements are inserted in the program based on the semantics of the types of accesses. For example, accesses may be conflicting (to the same variable) or non-conflicting (to different variables), conflicting accesses may be competing (a *Read* and a *Write*, or two *Writes*) or non-conflicting (two *Reads*), and so on. We outline the definitions of these consistency models but skip further implementation details of such models.

Weak consistency [11]

Some applications do not require even seeing all *Writes*, let alone seeing them in some order. Consider the case of a process executing a CS, repeatedly reading and writing some variables in a loop. Other processes are not supposed to read or write these variables until the first process has exited its CS. However, if the memory has no way of knowing when a process is in a CS and when it is not, the DSM has to propagate all *Writes* to all memories in the usual way. But by using synchronization variables, processes can deduce whether the CS is occupied.

A synchronization variable in this model has the following *semantics*: it is used to propagate all writes to other processors, *and* to perform local updates with regard to changes to global data that occurred elsewhere in the distributed system. When synchronization occurs, all *Writes* are propagated to other processes, and all *Writes* done by others are brought locally. In an implementation specifically for the CS problem, updates can be propagated in the system only when the synchronization variable is accessed (indicating an entry or exit into the CS).

Weak consistency (defined by [11]) has the following three properties which guarantee that memory is consistent at the synchronization points:

- Accesses to synchronization variables are sequentially consistent.
- No access to a synchronization variable is allowed to be performed until all previous writes have completed everywhere.
- No data access (either *Read* or *Write*) is allowed to be performed until all previous accesses to synchronization variables have been performed.

An access to the synchronization variable forces *Write* operations to complete, and effectively flushes the pipelines. Before reading shared data, a process can perform synchronization to ensure it accesses the most recent data.

Release consistency [12]

The drawback of *weak consistency* is that when a synchronization variable is accessed, the memory does not know whether this is being done because the process is finished writing the shared variables (exiting the CS) or about to begin reading them (entering the CS). Hence, it must take the actions required in both the following cases:

1. Ensuring that all locally initiated *Writes* have been completed, i.e., propagated to all other processes.
2. Ensuring that all *Writes* from other machines have been locally reflected.

If the memory could differentiate between entering the CS and leaving the CS, a more efficient implementation is possible. To provide this information, two kinds of synchronization variables or operations are needed instead of one.

Release consistency provides these two kinds. *Acquire* accesses are used to tell the memory system that a critical region is about to be entered. Hence, the actions for case 2 above need to be performed to ensure that local replicas of variables are made consistent with remote ones. *Release* accesses say that a critical region has just been exited. Hence, the actions for case 1 above need to be performed to ensure that remote replicas of variables are made consistent with the local ones that have been updated. The *Acquire* and *Release* operations can be defined to apply to a subset of the variables. The accesses themselves can be implemented either as ordinary operations on special variables or as special operations.

If the semantics of a CS is not associated with the *Acquire* and *Release* operations, then the operations effectively provide for *barrier synchronization.* Until all processes complete the previous phase, none can enter the next phase.

The following rules are followed by the protected variables in the general case [12]:

- All previously initiated *Acquire* operations must complete successfully before a process can access a protected shared variable.
- All accesses to a protected shared variable must complete before a *Release* operation can be performed.
- The *Acquire* and *Release* operations effectively follow the PRAM consistency model.

A relaxation of the release consistency model is called the *lazy release consistency* model. Rather than propagating the updated values throughout the system as soon as a process leaves a critical region (or enters the next phase in the case of barrier synchronization), the updated values are propagated to the rest of the system only on demand, i.e., only when they are needed. Changes to shared data are only communicated when an *Acquire* access is performed by another process.

Entry consistency [9]

Entry consistency requires the programmer to use *Acquire* and *Release* at the start and end of each CS, respectively. Unlike release consistency, however, entry consistency requires each ordinary shared variable to be associated with some synchronization variable such as a lock or barrier. When an *Acquire* is performed on a synchronization variable, only access to those ordinary shared variables that are guarded by that synchronization variable is regulated.

12.3 Shared memory mutual exclusion

Operating systems have traditionally dealt with multi-process synchronization using algorithms based on first principles (e.g., the well-known bakery algorithm), high-level constructs such as *semaphores* and *monitors*, and special "atomically executed" instructions supported by special-purpose hardware (e.g., *Test&Set*, *Swap*, and *Compare&Swap* [17]). These algorithms are applicable to all shared memory systems. In this section, we will review the bakery algorithm, which requires $O(n)$ accesses in the entry section, irrespective of the level of contention. We will then study *fast mutual exclusion*, which requires $O(1)$ accesses in the entry section in the absence of contention. This algorithm also illustrates an interesting technique in resolving concurrency. As hardware primitives have the in-built atomicity that helps to easily solve the mutual exclusion problem, we will then examine mutual exclusion based on these primitives.

12.3.1 Lamport's bakery algorithm

Lamport proposed the classical *bakery algorithm* for n-process mutual exclusion in shared memory systems [18]. The algorithm is so called because it mimics the actions that customers follow in a bakery store. A process wanting to enter the critical section picks a token number that is one greater than the elements in the array $choosing[1 \ldots n]$. Processes enter the critical section in the increasing order of the token numbers. In case of concurrent accesses to *choosing* by multiple processes, the processes may have the same token number. In this case, a unique *lexicographic order* is defined on the tuple $\langle token, pid \rangle$, and this dictates the order in which processes enter the critical section. The algorithm for process i is given in Algorithm 12.5. The algorithm can be shown to satisfy the three requirements of the critical section problem: (i) mutual exclusion, (ii) bounded waiting, and (iii) progress.

In the entry section, a process chooses a timestamp for itself, and resets it to 0 in the exit section. In lines 1a–1c, each process chooses a timestamp for itself, as the max of the latest timestamps of all processes, plus one. These steps are non-atomic; thus multiple processes could be choosing timestamps in overlapping durations. When process i reaches line 1d, it has to check the status of each other process j, to deal with the effects of any race conditions in selecting timestamps. In lines 1d–1f, process i serially checks the status of each other process j. If j is selecting a timestamp for itself, j's selection interval may have overlapped with that of i, leading to an unknown order of timestamp values. Process i needs to make sure that any other process j ($j < i$) that had begun to execute line 1b concurrently with itself and may still be executing line 1b does not assign itself the same timestamp. Otherwise mutual exclusion could be violated as i would enter the CS, and subsequently, j, having a lower process identifier and hence a lexicographically lower timestamp, would also enter the CS. Hence, i waits for j's timestamp to stabilize, i.e., $choosing[j]$ to be set to *false*. Once j's timestamp is stabilized, i moves from line 1e to line 1f. Either j is not requesting (in which case j's timestamp is 0) or j is requesting. Line 1f determines the relative priority between i and j. The process with a *lexicographically* lower timestamp has higher priority and enters the CS; the other process has to wait (line 1g). Hence, *mutual exclusion* is satisfied.

Bounded waiting is satisfied because each other process j can "overtake" process i at most once after i has completed choosing its timestamp. The second time j chooses a timestamp, the value will necessarily be larger than i's timestamp if i has not yet entered its CS.

Progress is guaranteed because the lexicographic order is a total order and the process with the lowest timestamp at any time in the loop (lines 1d–1g) is guaranteed to enter the CS.

(shared vars)
boolean: $choosing[1 \ldots n]$;
integer: $timestamp[1 \ldots n]$;
repeat
(1) P_i executes the following for the **entry section**:
(1a) $choosing[i] \longleftarrow 1$;
(1b) $timestamp[i] \longleftarrow \max_{k \in [1 \ldots n]}(timestamp[k]) + 1$;
(1c) $choosing[i] \longleftarrow 0$;
(1d) **for** $count = 1$ **to** n **do**
(1e)   **while** $choosing[count]$ **do** no-op;
(1f)   **while** $timestamp[count] \neq 0$ **and** $(timestamp[count], count) < (timestamp[i], i)$ **do**
(1g)   no-op.
(2) P_i executes the **critical section (CS)** after the **entry section**
(3) P_i executes the following **exit section** after the **CS**:
(3a) $timestamp[i] \longleftarrow 0$.
(4) P_i executes the **remainder section** after the **exit section**
until false;

Algorithm 12.5 Lamport's n-process bakery algorithm for shared memory mutual exclusion. Code shown is for process Pi, $1 \leq i \leq n$.

Attempts to improve the bakery algorithm have lead to several important results:

- *Space complexity*: A lower bound of n registers, specifically, the *timestamp* array, has been shown for the shared memory critical section problem [10]. Thus, one cannot hope to have a more space-efficient algorithm for distributed shared memory mutual exclusion.
- *Time complexity*: In many environments, the level of contention may be low. The $O(n)$ overhead of the entry section does not scale well for such environments. This concern is addressed by the field of *fast mutual exclusion* that aims to have $O(1)$ time overhead for the entry and exit sections of the algorithm, in the absence of contention. Although this algorithm guarantees mutual exclusion and progress, unfortunately, this fast algorithm has a price – in the worst case, it does not guarantee bounded delay. Next, we will study Lamport's algorithm for fast mutual exclusion in asynchronous shared memory systems. This algorithm is notable in that it is the first algorithm for fast mutual exclusion, and uses the asynchronous shared memory model. Further, it illustrates an important technique for resolving contention. The worst-case unbounded delay in the presence of persisting contention has been addressed subsequently, by using a timed model of execution, wherein there is an upper bound on the time it takes to execute any step. We will not discuss mutual exclusion under the timed model of execution.

12.3.2 Lamport's WRWR mechanism and fast mutual exclusion

Lamport's *fast mutual exclusion* algorithm [23] is given in Algorithm 12.6. The algorithm illustrates an important technique – the $\langle W - R - W - R\rangle$ sequence that is a necessary and sufficient sequence of operations to check for contention and to ensure safety in the entry section, using only two registers.

Lines 1b, 1c, 1g , and 1h represent a basic $\langle W(x) - R(y) - W(y) - R(x)\rangle$ sequence whose necessity in identifying a minimal sequence of operations for fast mutual exclusion is justified as follows:

1. The first operation needs to be a *Write*, say to variable x. If it were a *Read*, then all contending processes could find the value of the variable even outside the entry section.
2. The second operation cannot be a *Write* to another variable, for that could equally be combined with the first *Write* to a larger variable. The second operation should *not* be a *Read* of x because it follows *Write* of x and if there is no interleaved operation from another process, the *Read* does not provide any new information. So the second operation must be a *Read* of another variable, say y.
3. The sequence must also contain *Read*(x) and *Write*(y) because there is no point in reading a variable that is not written to, or writing a variable that is never read.

```
(shared variables among the processes)
integer: x, y;                                        // shared register initialized
boolean b[1...n];    // flags to indicate interest in critical section
repeat
(1)    P_i (1 ≤ i ≤ n) executes entry section:
(1a)       b[i] ← true;
(1b)       x ← i;
(1c)       if y ≠ 0 then
(1d)               b[i] ← false;
(1e)               await y = 0;
(1f)               goto (1a);
(1g)       y ← i;
(1h)       if x ≠ i then
(1i)               b[i] ← false;
(1j)               for j = 1 to n do
(1k)                       await ¬b[j];
(1l)               if y ≠ i then
(1m)                       await y = 0;
(1n)                       goto (1a);
(2)    P_i (1 ≤ i ≤ n) executes critical section:
(3)    P_i (1 ≤ i ≤ n) executes exit section:
(3a)       y ← 0;
(3b)       b[i] ← false;
forever.
```

Algorithm 12.6 Lamport's deadlock-free fast mutual exclusion solution, using $\Omega(n)$ registers. Code is for process P_i, where $1 \leq i \leq n$.

4. The last operation in the minimal sequence of the entry section must be a *Read*, as it will help determine whether the process can enter CS. So the last operation should be *Read*(x), and the second-last operation should be the *Write*(y).

In the absence of contention, each process writes its own i.d. to x and then reads y. Then finding that y has its initial value, the process writes its own i.d. to y and then reads x. Finding x to still be its own i.d., it enters CS. Correctness needs to be shown in the presence of contention – let us discuss this after considering the structure of the remaining entry and exit section code.

In the exit section, the process must do a *Write* to indicate its completion of the CS. The *Write* cannot be to x, which is also the first variable written in the entry section. So the operation must be *Write*(y).

Now consider the sequence of interleaved operations by processes i, j, and k in the entry section, as shown in Figure 12.9. Process i enters its critical

Figure 12.9 An example showing the need for a boolean vector for fast mutual exclusion.

Process P_i	Process P_j	Process P_k	Variables
	$W_j(x)$		$\langle x = j, y = 0\rangle$
$W_i(x)$			$\langle \mathbf{x} = \mathbf{i}, y = 0\rangle$
$R_i(y)$			$\langle x = i, y = 0\rangle$
	$R_j(y)$		$\langle x = i, y = 0\rangle$
$W_i(y)$			$\langle x = i, \mathbf{y} = \mathbf{i}\rangle$
	$W_j(y)$		$\langle x = i, y = j\rangle$
$R_i(x)$			$\langle x = i, y = j\rangle$
		$W_k(x)$	$\langle x = k, y = j\rangle$
	$R_j(x)$		$\langle x = k, y = j\rangle$

section, but there is no record of its identity or that it had written any variables at all, because the variables it wrote (shown boldfaced above) have been overwritten. In order that other processes can discover when (and who) leaves the CS, there needs to be another variable that is set before the CS and reset after the CS. This is the boolean, $b[i]$. Additionally, y needs to be reset on exiting the CS.

The code in lines 1c–1f has the following use. If a process p finds $y \neq 0$, then another process has executed at least line 1g and not yet executed line 3a. So process p resets its own flag, and before retrying again, it awaits for $y = 0$. If process p finds $y = 0$ in line 1c, it sets $y = p$ in line 1g and checks if $x = p$.

- If $x = p$, then no other process has executed line 1b, and any later process would be blocked in the loop in lines 1c–1f now because $y = p$. Thus, if $x = p$, process p can safely enter the CS.
- If $x \neq p$, then another process, say q, has overwritten x in line 1b and there is a potential race. Two broad cases are possible:
  - Process q finds $y \neq 0$ in line 1c. It resets its flag, and stays in the 1d–1f section at least until p has exited the CS. Process p on the other hand resets its own flag (line 1i) and waits for all other processes such as q to reset their own flags. As process q is trapped in lines 1d–1f, process p will find $y = p$ in line 1l and enter the CS.
  - Process q finds $y = 0$ in line 1c. It sets y to q in line 1g, and enters the race, even closer to process p, which is at line 1h. Of the processes such as p and q that contend at line 1h, there will be a unique winner:
    * If no other process r has since written to x in line 1b, the winner is the process among p and q that executed line 1b last, i.e., wrote its own i.d. to x. That winner will enter the CS directly from line 1h, whereas the losers will reset their own flags, await the winner to exit and reset its flag, and also await other contenders at line 1h and newer contenders to reset their own flags. The losers will compete again from line 1a after the winner has reset y.

* If some other process r has since written its i.d. to x in line 1b, both p and q will enter code in lines 1i–1n. Both p and q reset their flags, await for r, which will be trapped in lines 1d–1f to reset its flag, and then both p and q check the value of y. Between p and q, the process that last wrote to y in line 1g will become the unique winner and enter the CS directly. The loser will then await for the winner to reset y, and then compete again from line 1a.

Thus, mutual exclusion is guaranteed, and progress is also guaranteed. However, a process may be starved, although with decreasing probability, as its number of attempts increases.

12.3.3 Hardware support for mutual exclusion

Hardware support can allow for special instructions that perform two or more operations atomically. Two such instructions, *Test&Set* and *Swap* [17], are defined and implemented as shown in Algorithm 12.7. The atomic execution of two actions (a *Read* and a *Write* operation) can greatly simplify a mutual exclusion algorithm, as seen from the mutual exclusion code in Algorithm 12.8 and Algorithm 12.9, respectively. Algorithm 12.8 can lead to starvation. Algorithm 12.9 is enhanced to guarantee bounded waiting by using a "round-robin" policy to selectively grant permission when releasing the critical section.

(shared variables among the processes accessing each of the different object types)
register: *Reg* ⟵ initial value; // shared register initialized
(local variables)
integer: *old* ⟵ initial value; // value to be returned

(1) *Test&Set*(*Reg*) returns *value*:
(1a) *old* ⟵ *Reg*;
(1b) *Reg* ⟵ 1;
(1c) **return**(*old*).

(2) *Swap*(*Reg*, *new*) returns *value*:
(2a) *old* ⟵ *Reg*;
(2b) *Reg* ⟵ *new*;
(2c) **return**(*old*).

Algorithm 12.7 Definitions of synchronization operations *Test&Set* and *Swap*.

(shared variables)
register: $Reg \longleftarrow false$; // shared register initialized
(local variables)
integer: $blocked \longleftarrow 0$; // variable to be checked before entering CS
repeat
(1) P_i executes the following for the **entry section**:
(1a) $blocked \longleftarrow true$;
(1b) **repeat**
(1c) $blocked \longleftarrow Swap(Reg, blocked)$;
(1d) **until** $blocked = false$;
(2) P_i executes the **critical section (CS)** after the **entry section**
(3) P_i executes the following **exit section** after the **CS**:
(3a) $Reg \longleftarrow false$;
(4) P_i executes the **remainder section** after the **exit section**
until false;

Algorithm 12.8 Mutual exclusion using *Swap*. Code shown is for process *Pi*, $1 \leq i \leq n$.

(shared variables)
register: $Reg \longleftarrow false$; // shared register initialized
boolean: $waiting[1 \ldots n]$;
(local variables)
integer: $blocked \longleftarrow$ initial value; // value to be checked before
// entering CS

repeat
(1) P_i executes the following for the **entry section**:
(1a) $waiting[i] \longleftarrow true$;
(1b) $blocked \longleftarrow true$;
(1c) **while** $waiting[i]$ **and** $blocked$ **do**
(1d) $blocked \longleftarrow Test\&Set(Reg)$;
(1e) $waiting[i] \longleftarrow false$;
(2) P_i executes the **critical section (CS)** after the **entry section**
(3) P_i executes the following **exit section** after the **CS**:
(3a) $next \longleftarrow (i+1) mod\, n$;
(3b) **while** $next \neq i$ **and** $waiting[next] = false$ **do**
(3c) $next \longleftarrow (next+1) mod\, n$;
(3d) **if** $next = i$ **then**
(3e) $Reg \longleftarrow false$;
(3f) **else** $waiting[next] \longleftarrow false$;
(4) P_i executes the **remainder section** after the **exit section**
until false;

Algorithm 12.9 Mutual exclusion with bounded waiting, using *Test&Set*. Code shown is for process *Pi*, $1 \leq i \leq n$.

12.4 Wait-freedom

Processes that interact with each other, whether by message passing or by shared memory, need to synchronize their interactions. Traditional solutions to synchronize asynchronous processes via shared memory objects (also called *concurrent objects*) use solutions based on locking, busy waiting, critical sections, semaphores, or conditional waiting. An arbitrary delay of a process or its crash failure can prevent other processes from completing their operations. This is undesirable.

Wait-freedom is a property that guarantees that any process can complete any synchronization operation in a finite number of lower-level steps, irrespective of the execution speed of other processes [15,24]. More precisely, a wait-free implementation of a concurrent object guarantees that any process can complete an operation on it in a finite number of steps, irrespective of whether other processes crash or encounter unexpected delays. Thus, processes that crash, or encounter unexpected delays (such as delays due to high processor load, swapping out of memory, or CPU schedulng policies) should not delay other processes in a wait-free implementation of a concurrent object.

Not all synchronizations have wait-free solutions. As a trivial example, a producer–consumer synchronization between two processes cannot be implemented in a wait-free manner if the producer process crashes before posting its value – the consumer is necessarily blocked. Nevertheless, the notion of wait-freedom is an important concept in designing fault-tolerant systems and algorithms whenever possible. An alternate view of wait-freedom in terms of fault-tolerance is as follows:

- An f-resilient system is a system in which up to f of the n processes can fail, and the other $n-f$ processes can complete all their operations in a finite number of steps, independent of the states of the f processes that may fail.
- When $f = n-1$, any process is guaranteed to be able to complete its operations in a finite number of steps, independent of all other processes. A process does not depend on other processes, and its execution is therefore said to be *wait-free*. Wait-freedom provides independence from the behavior of other processes, and is therefore a very desirable property.

In the remainder of this chapter, which deals with shared register accesses, only wait-free solutions are considered.

12.5 Register hierarchy and wait-free simulations

Observe from our analysis of DSM consistency models that an underlying assumption was that any memory access takes a finite time interval, and the operation, whether a *Read* or *Write*, takes effect at some point during

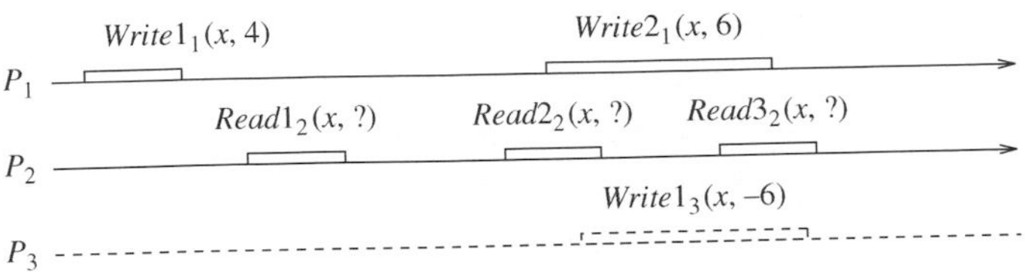

Figure 12.10 Examples to illustrate definitions of *safe*, *regular*, and *atomic* registers. The regular lines assume a SRSW register. If the dashed line is also used, the register is assumed to be SRMW.

this time duration. In the face of concurrent accesses to a memory location, hereafter called a *register*, we cannot predict the outcome. In particular, in the face of a concurrent *Read* and *Write* operation, the value returned by the *Read* is unpredictable. This observation is true even for a simpler multiprocessor memory, without the context of a DSM. This observation led to the research area that tried to define the properties of access orderings for the most elementary memory unit. The access orderings depend on the properties of the register. An implicit assumption is that of the availability of global time. This is a reasonable assumption because we are studying access to a single register. Whether that register value is replicated in the system or not is a lower detail that is not relevant to the level of abstraction of this analysis.

In keeping with the semantics of the *Read* and *Write* operations, the following register types have been identified by Lamport [20–22] to specify the value returned to a *Read* in the face of a concurrent *Write* operation. For the time being, we assume that there is a single reader process and a single writer process.

- **Safe register** A *Read* operation that does not overlap with a *Write* operation returns the most recent value written to that register. A *Read* operation that does overlap with a *Write* operation returns *any one* of the values that the register could possibly contain at any time.
  Consider the example of Figure 12.10, which shows several operations on an integer-valued register. We consider two cases, without and with the *Write* by P_3:
  - **No *Write* by P_3** If the register is *safe*, $Read1_2$ must return the value 4, whereas $Read2_2$ and $Read3_2$ can return any possible integer (up to MAXINT) because these operations overlap with a *Write*, and the value returned is therefore ambiguous.
  - ***Write* by P_3** Same as for the "no *Write*" case.

  If multiple writers are allowed, or if *Write* operations are allowed to be pipelined, then what defines the most recent value of the register in the face of concurrent *Write* operations becomes complicated. We explicitly disallow pipelining in this model and analysis. In the face of *Write* operations from different processors that overlap in time, the notion of a *serialization point* is defined. Observe that each *Write* or *Read* operation has a finite duration between its invocation and its response. In this duration, there is effectively a single time instant at which the operation takes effect. For a *Read* operation, this instant is the one at which the instantaneous value is selected to be returned. For a *Write* operation, this instant is the one at which the value written is first "reflected" in the register. Using this notion of the serialization point, the "most recent" operation is unambiguously defined.

- **Regular register** In addition to being a *safe* register, a *Read* that is concurrent with a *Write* operation returns either the value before the *Write* operation, or the value written by the *Write* operation.
  In the example of Figure 12.10, we consider the two cases, with and without the *Write* by P_3:
  - **No *Write* by P_3** $Read1_2$ must return 4, whereas $Read2_2$ can return either 4 or 6, and $Read3_2$ can also return either 4 or 6.
  - ***Write* by P_3** $Read1_2$ must return 4, whereas $Read2_2$ can return either 4 or −6 or 6, and $Read3_2$ can also return either 4 or −6 or 6.
- **Atomic register** In addition to being a *regular* register, the register is linearizable (defined in Section 12.2.1) to a sequential register.
  In the example of Figure 12.10, we consider the two cases, with and without the *Write* by P_3:
  - **No *Write* by P_3** $Read1_2$ must return 4, whereas $Read2_2$ can return either 4 or 6. If $Read2_2$ returns 4, then $Read3_2$ can return either 4 or 6, but if $Read2_2$ returns 6, then $Read3_2$ must also return 6.
  - ***Write* by P_3** $Read1_2$ must return 4, whereas $Read2_2$ can return either 4 or −6 or 6, depending on the serialization points of the operations.
    1. If $Read2_2$ returns 6 and the serialization point of $Write1_3$ precedes the serialization point of $Write2_1$, then $Read3_2$ must return 6.
    2. If $Read2_2$ returns 6 and the serialization point of $Write2_1$ precedes the serialization point of $Write1_3$, then $Read3_2$ can return +6 or −6.
    3. Cases (3) and (4) where $Read2_2$ returns −6 are similar to cases (1) and (2).

The following properties, summarized in Table 12.2, characterize registers:

- whether the register is single-valued (boolean) or multi-valued
- whether the register is a single-reader (SR) or multi-reader (MR) register
- whether the register is a single-writer (SW) or multi-writer (MW) register
- whether the register is *safe*, *regular*, or *atomic*

The above characteristics lead to a hierarchy of 24 register types, with the most elementary being the boolean SRSW safe register and the most complex being the multi-valued MRMW atomic register.

A study of *register construction* deals with designing the more complex registers using simpler registers. Such constructions allow us to construct

Table 12.2 Classification of registers by type, value, writing access, and reading access. The strength of the register increases down each column.

Type	Value	Writing	Reading
safe	binary	single-writer	single-reader
regular	integer	multi-writer	multi-reader
atomic			

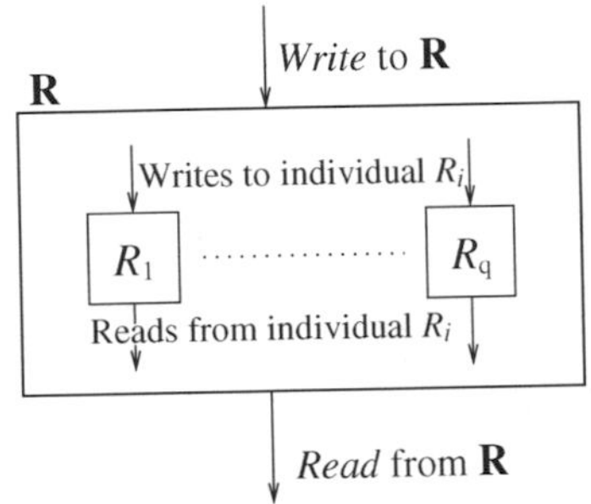

Figure 12.11 Register simulations.

any register type from the most elementary register – the boolean SRSW safe register. We will study such constructions by assuming the following convention: $R_1 \ldots R_q$ are q registers that are used to construct a stronger register R, as shown in Figure 12.11. We assume n processes exist; note that for various constructions, q may be different from n.

Although the traditional memory architecture, based on serialized access via memory ports to a memory location, does not require such an elaborate classification, the bigger picture needs to be kept in mind. In addition to illustrating algorithmic design techniques, this study paves the way for accommodating newer technologies such as quantum computing and DNA computing for constructing system memory.

12.5.1 Construction 1: SRSW *safe* to MRSW *safe*

Algorithm 12.10 gives the construction of a MRSW *safe* register R using only SRSW *safe* registers [21]. Assume the single writer is process P_0 and the n reader processes are P_1 to P_n. Each of the n processes P_i can read only SRSW register R_i. As multiple readers are not allowed to access the same register, in essence, the data needs to be replicated. So in the construction, the writer P_0 writes the same value to the n registers. Register R_i is read by P_i. In Figure 12.11, the value of q would hence be n. When a *Read* by P_i and a *Write* by P_0 do not overlap their access to R_i, the *Read* obtains the correct value. When a *Read* by P_i and a *Write* by P_0 overlap their access to R_i, as R_i is a *safe* register, P_i reads a legitimate value from R_i.

Complexity

This construction has a space complexity of n times the size of a single register, which may be either binary or integer-valued. The time complexity is n steps.

(shared variables)
SRSW safe registers $R_1 \ldots R_n \longleftarrow 0$; // R_i is readable by P_i, writable // by P_0

(1) *Write*(R, val) executed by single writer P_0
(1a) **for all** $i \in \{1 \ldots n\}$ **do**
(1b) $R_i \longleftarrow val$.

(2) *Read*$_i(R, val)$ executed by reader P_i, $1 \leq i \leq n$
(2a) $val \longleftarrow R_i$
(2b) **return**(val).

Algorithm 12.10 Construction 1: SRSW *safe* register to MRSW *safe* register R. This construction can also be used for SRSW *regular* register to MRSW *regular* register R.

12.5.2 Construction 2: SRSW *regular* to MRSW *regular*

This construction is identical to construction 1 (Algorithm 12.10) except that *regular* registers are used instead of *safe* registers [21]. When a *Read* by P_i and a *Write* by P_0 do not overlap their access to R_i, the *Read* obtains the correct value. When a *Read* by P_i and a *Write* by P_0 overlap their access to R_i, as R_i is a *regular* register, P_i reads from R_i either the earlier value or the value being written.

Complexity

This construction has a space complexity of n times the size of a single register, which may be either binary or integer-valued. The time complexity is n steps.

12.5.3 Construction 3: boolean MRSW *safe* to integer-valued MRSW *safe*

Algorithm 12.11 gives the construction of an integer-valued MRSW *safe* register R [21]. Assume the single writer is process P_0 and the n reader processes are P_1 to P_n. The construction can use only boolean MRSW registers – to construct an integer register of size m, at least $log(m)$ boolean registers are necessary. So in the construction, the writer P_0 writes the value in its binary notation to the $log(m)$ registers R_1 to $R_{log(m)}$. Similarly, any reader reads registers R_i to $R_{log(m)}$. When a *Read* by P_i and a *Write* by P_0 do not overlap, the *Read* obtains the correct value. When a *Read* by P_i and a *Write* by P_0

overlap their access to the registers, as the R_i ($i = 1$ to $log(m)$) registers are *safe*, P_i reads a legitimate value.

Complexity
This construction has a space complexity of $O(log(m))$. The time complexity is $O(log(m))$ steps.

(shared variables)
boolean MRSW safe registers $R_1 \ldots R_{log(m)} \longleftarrow 0$; // R_i readable by // all, writable by P_0.

(local variable)
boolean: $Val[1 \ldots log(m)]$;
(1) $Write(R, Val[1 \ldots log\, m])$ executed by single writer P_0
(1a) **for all** $i \in \{1 \ldots log(m)\}$ **do**
(1b) $\quad R_i \longleftarrow Val[i]$.

(2) $Read_i(R, Val[1 \ldots log(m)])$ executed by reader P_i, $1 \leq i \leq n$
(2a) **for all** $j \in \{1 \ldots log\, m\}$ **do** $Val[j] \longleftarrow R_j$
(2b) **return**$(Val[1 \ldots log(m)])$.

Algorithm 12.11 Construction 3: boolean MRSW *safe* register to integer-valued MRSW *safe* register R.

12.5.4 Construction 4: boolean MRSW *safe* to boolean MRSW *regular*

Algorithm 12.12 gives the construction of a boolean MRSW *regular* register R from a MRSW *safe* register [21]. Assume the single writer is process P_0 and the reader processes are P_i ($1 \leq i \leq n$). With respect to Figure 12.11, q has the value 1. P_0 writes R_1 and all n processes read R_1.

When a *Read* by P_i and a *Write* by P_0 do not overlap, the *Read* obtains the correct value. When a *Read* by P_i and a *Write* by P_0 overlap, the *safe* register may not necessarily return the overlapping or the previous value (as required by a *regular* register), but may return a value written much earlier. If the value written before the *Read* begins is α, and the value being written by the concurrent *Write* is also α, the *Read* could return α or $(1-\alpha)$ from the *safe* register, which is a problem for the *regular* register. The solution bypasses this problem by having the *Write* use a local variable *previous* to track the previous value of *val*. If the previous value that was written (line 1b) and stored in *previous* (line 1c) is the same as the new value to be written, then the new value is simply not written. This avoids any concurrent access to R.

(shared variables)
boolean MRSW safe register: $R' \longleftarrow 0$; // R' is readable by all, // writable by P_0.

(local variables)
boolean local to writer P_0: $previous \longleftarrow 0$;

(1) *Write*(R, val) executed by single writer P_0
(1a) **if** $previous \neq val$ **then**
(1b) $R' \longleftarrow val$;
(1c) $previous \longleftarrow val$.

(2) *Read*$_i(R, val)$ executed by process P_i, $1 \leq i \leq n$
(2a) $val \longleftarrow R'$;
(2b) **return**(val).

Algorithm 12.12 Construction 4: boolean MRSW *safe* register to boolean MRSW *regular* register *R*.

Complexity
This construction uses $O(1)$ space and time.

Can the above construction also construct a binary SRSW *atomic* register from a *safe* register? No. Consider P_1 issues a $Write1_1(\alpha)$ that completes; then $Write2_1(1-\alpha)$ begins and overlaps with $Read1_2$ and $Read2_2$ of P_2. With the above construction, $Read1_2$ could return $1-\alpha$ whereas the later $Read2_2$ could return α, thus violating the property of an *atomic* register.

12.5.5 Construction 5: boolean MRSW *regular* to integer-valued MRSW *regular*

Algorithm 12.13 gives the construction of an integer-valued MRSW *regular* register R using boolean MRSW *regular* registers [21]. Assume the single writer is process P_0 and the n reader processes are P_1 to P_n. The construction can use only boolean MRSW registers – to construct an integer register of size m, unary notation is used, so m boolean registers are necessary. In Figure 12.11, $q = m$, and all n processes can read all q registers.

When a *Read* by P_i and a *Write* by P_0 do not overlap, the *Read* obtains the correct value. To deal with a *Read* by P_i and a *Write*(s) by P_0 overlapping their access to the registers, the following approach is used. A reader P_i scans left-to-right looking for a "1" whereas the P_0 writer process writes "1" to the R_{val} location and zeros out the entries right-to-left. The *Read* is guaranteed to see a "1" written by one of the *Write* operations it overlaps with, or the "1" written by the *Write* that completed just before the *Read* began. As each of the bits are regular, its current or previous value is read; if the value is "0," it is guaranteed that a "1" has been written to the right. An implicit assumption here is the integer size, bounded by the

```
(shared variables)
boolean MRSW regular registers R_1 ... R_{m-1} ← 0; R_m ← 1;
                                   // R_i readable by all, writable by P_0.

(local variables)
integer: count;
(1)   Write(R, val) executed by writer P_0
(1a)  R_val ← 1;
(1b)  for count = val − 1 down to 1 do
(1c)        R_count ← 0.

(2)   Read_i(R, val) executed by P_i, 1 ≤ i ≤ n
(2a)  count = 1;
(2b)  while R_count = 0 do
(2c)        count ← count + 1;
(2d)  val ← count;
(2e)  return(val).
```

Algorithm 12.13 Construction 5: boolean MRSW *regular* register to integer-valued MRSW *regular* register R.

number of bits in use. The register is initialized by this largest value. The construction is illustrated in Figure 12.12. In the figure, the reader scans from left to right as marked.

Complexity

This construction uses m binary registers, where m is the largest integer that can be written by the application. The time complexity is $O(m)$.

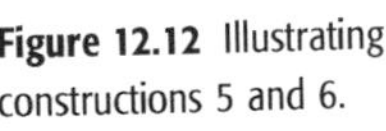

Figure 12.12 Illustrating constructions 5 and 6.

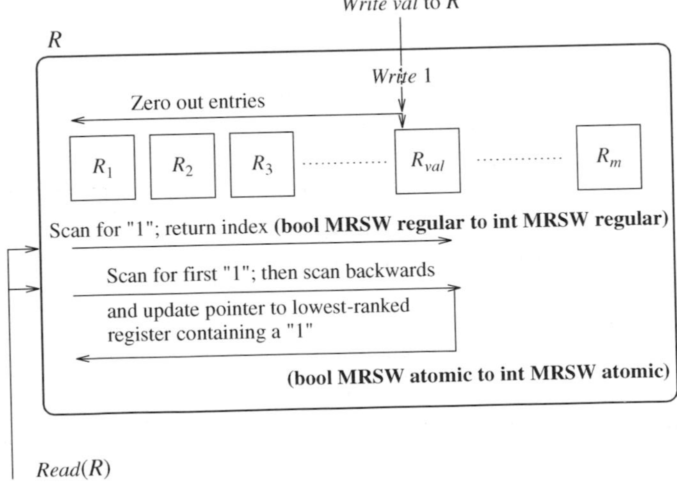

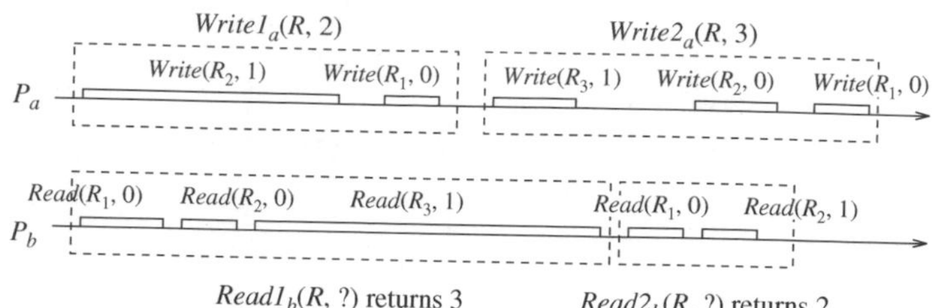

Figure 12.13 Example to illustrate inversion of values read by P_a and P_b.

12.5.6 Construction 6: boolean MRSW *regular* to integer-valued MRSW *atomic*

Can the construction (in Algorithm 12.13) also construct an integer-valued MRSW *atomic* register from boolean MRSW *regular* registers? No. The problem is that when two successive *Read* operations overlap *Write* operations, "inversion" of values returned by the *Read* operations can occur.

Consider the following sequence of operations, depicted in Figure 12.13:

1. $Write1_a(R, 2)$: The low-level operation $Write(R_2, 1)$ begins, i.e., $R_2 \longleftarrow 1$ begins.
2. $Read1_b(R, ?)$: The following low-level operations get executed. $count \longleftarrow 1$; $Read(R_{count}, 0)$; $count \longleftarrow 2$; $Read(R_{count}, 0)$; $count \longleftarrow 3$.
3. $Write1_a(R, 2)$: The low-level operation $Write(R_2, 1)$ from step 1 completes, i.e., the value "1" gets written to R_2; then the left scan to zero out R_1 proceeds by executing $Write(R_1, 0)$.
4. $Write2_a(R, 3)$: The low-level operation $Write(R_3, 1)$ executes, i.e., $R_3 \longleftarrow 1$ begins and ends.
5. $Read1_b(R, ?)$: The low-level operation $Read(R_{count=3}, ?)$ that was to begin after step 2 returns 1; the high-level *Read* completes and returns a value of 3.
6. $Read2_b(R, ?)$: This operation's left-to-right scan for a "1" finds $R_2 = 1$ and returns 2. This is because the low-level operation $Write2(R_2, 0)$ belonging to the high-level operation $Write2_a(R, 3)$ has not yet zeroed out R_2.

Here, $Read2_b(R, 2)$ returns the value written by $Write1_a(R, 2)$; whereas the earlier $Read1_b(R, 3)$ returns the value written by the later $Write2_a(R, 3)$. Hence, this execution is not linearizable.

Algorithm 12.14 gives Vidyasankar's construction [30] of a integer-valued MRSW *atomic* register R by modifying the above solution as follows. The reader makes a right-to-left scan for a "1" after its left-to-right scan completes. If it finds a "1" in a lower index, it updates the value to be returned to this index. The purpose is to make sure that the lowest index (say α) in which a "1" is found in this second "right-to-left" scan is returned by the *Read*. As the writer also zeros out entries "right-to-left," it is not possible that a later *Read* will find a "1" written earlier in a position lower than α, by a *Write* that occurred earlier than the *Write* that wrote α. This allows a linearizable execution. With respect to Figure 12.11, $q = m$, and all n processes can read all q registers. This construction is also illustrated in Figure 12.12, as marked therein.

(shared variables)
boolean MRSW regular registers $R_1 \ldots R_{m-1} \longleftarrow 0$; $R_m \longleftarrow 1$.
// R_i readable by all; writable by P_0.

(local variables)
integer: $count, temp$;

(1) $Write(R, val)$ executed by P_0
(1a) $R_{val} \longleftarrow 1$;
(1b) **for** $count = val - 1$ **down to** 1 **do**
(1c) $\quad R_{count} \longleftarrow 0$.

(2) $Read_i(R, val)$ executed by P_i, $1 \leq i \leq n$
(2a) $count \longleftarrow 1$;
(2b) **while** $R_{count} = 0$ **do**
(2c) $\quad count \longleftarrow count + 1$;
(2d) $val \longleftarrow count$;
(2e) **for** $temp = count$ **down to** 1 **do**
(2f) $\quad$ **if** $R_{temp} = 1$ **then**
(2g) $\quad\quad val \longleftarrow temp$;
(2h) **return**(val).

Algorithm 12.14 Construction 6: boolean MRSW *regular* register to integer-valued MRSW *atomic* register *R*.

A formal argument that this construction is correct needs to show that any execution is linearizable. To do so, it would define the *linearization point* of a *Read* and *Write* operation to capture the notion of the exact instant at which that operation effectively appears to take effect.

- The *value of the MRSW register* at any moment is x, where $R_x = 1$ and $\forall y < x$, $R_y = 0$.
- The linearization point of a $Write(R, x)$ operation is the first instant (line 1a or 1c) when $R_x = 1$ and $\forall y < x$, $R_y = 0$.
- The linearization point of a $Read(R, val)$ that returns (x) is the first instant (line 2d or 2g) when val gets assigned x in the low-level operations.

The following observation can now be made from the construction and the definition of the linearization point of a *Write*:

- The *value of the MRSW register* remains unchanged between the linearization points of any two consecutive *Write* operations.

The *Write* operations are naturally ordered in the linearization sequence. In order to determine a complete linearization of the *Read* operations in addition to the *Write* operations, observe the following:

- A *Read* operation returns the value written by that *Write* operation which has the latest linearization point that precedes the *Read* operation's linearization point.

It naturally follows that a later *Read* will never return the value written by a earlier *Write*, and hence the construction is linearizable.

Complexity

This construction uses m binary registers, where m is the largest integer that is written by the application program. The time complexity is $O(m)$.

12.5.7 Construction 7: integer MRSW *atomic* to integer MRMW *atomic*

We are given MRSW *atomic* registers, i.e., each register has only a single writer. To simulate a MRMW *atomic* register R, the variable has multiple copies, $R_1 \ldots R_n$, one per writer process. Writer P_i can only write to its copy R_i. Reader P_i can read all the registers $R_1 \ldots R_n$. When concurrent updates occur, a global linearization order must be created somehow. The *Read* operations must be able to recognize such a global order, and then return the appropriate version as per the semantics of the atomic register. That is the challenge.

The construction by Vitanyi and Awerbuch [31] is shown in Algorithm 12.15. With respect to Figure 12.11, $q = n$, and all n processes can read all q MRSW registers but only P_i can write to R_i. The idea used is similar to that used by the Bakery algorithm for mutual exclusion (Section 12.3.1). Here each process, when behaving as a writer process, does not compete directly with other writer processes. The competing processes that make concurrent accesses (behaving as the reader processes) then read all the flags and deduce a global order that resolves the contention.

Each register R_i has two fields: $R_i.data$ and $R_i.tag$, where $tag = \langle seq\_no, pid \rangle$. A *lexicographic order* is defined on the tags, using *seq_no* as the primary key, and then *pid* as the secondary key. A common procedure invoked by the readers and writers is *Collect*, which reads all the registers, in no particular order. The reader returns the data corresponding to the (lexicographically) most recent *Write*. A writer chooses a tag greater than the (lexicographically) greatest tag returned by *Collect*, when it writes its new value.

All the *Write* operations are lexicographically totally ordered. Each *Read* is ordered so that it immediately follows that *Write* with the matching tag. Thus, this execution is linearizable.

Complexity

This construction has a space complexity of $O(n)$ integer registers. The time complexity is $O(n)$.

(shared variables)
MRSW atomic registers of type $\langle data, tag\rangle$, where $tag = \langle seq\_no, pid\rangle$:
  $R_1 \ldots R_n$;
(local variables)
MRSW atomic registers of type $\langle data, tag\rangle$, where $tag = \langle seq\_no, pid\rangle$:
  $Reg\_Array[1 \ldots n]$;
integer: seq_no, j, k;

(1) $Write_i(R, val)$ executed by P_i, $1 \le i \le n$
(1a) $Reg\_Array \longleftarrow Collect(R_1, \ldots, R_n)$;
(1b) $seq\_no \longleftarrow \max(Reg\_Array[1].tag.seq\_no, \ldots$
    $Reg\_Array[n].tag.seq\_no) + 1$;
(1c) $R_i \longleftarrow (val, \langle seq\_no, i\rangle)$.

(2) $Read_i(R, val)$ executed by P_i, $1 \le i \le n$
(2a) $Reg\_Array \longleftarrow Collect(R_1, \ldots, R_n)$;
(2b) **identify** j such that for all $k \ne j$, $Reg\_Array[j].tag > Reg\_Array[k].tag$;
(2c) $val \longleftarrow Reg\_Array[j].data$;
(2d) **return**(val).

(3) $Collect(R_1, \ldots, R_n)$ invoked by *Read* and *Write* routines
(3a) **for** $j = 1$ **to** n **do**
(3b) $Reg\_Array[j] \longleftarrow R_j$;
(3c) **return**(Reg_Array).

Algorithm 12.15 Construction 7: integer MRSW *atomic* register to integer MRMW *atomic* register R.

12.5.8 Construction 8: integer SRSW atomic to integer MRSW atomic

We are given SRSW *atomic* registers. To simulate a MRSW *atomic* register R, the variable has multiple copies, $R_1, \ldots R_n$, one per reader process. The single writer can write to all of these registers.

A first attempt at this construction would have the writer write to all the registers $R_1 \ldots R_n$, whereas reader P_i reads R_i. In Figure 12.11, $q = n$, and each R_i is read by P_i and written to by the single writer P_0. However, such a construction does not give a linearizable execution. Consider two reads $Read1_i$ and $Read2_j$ that both overlap a *Write*, and *Read2* begins after *Read1* terminates. It is possible that:

1. $Read1_i$ reads R_i after the *Write* has written to R_i;
2. but $Read2_j$ reads R_j before the writer has had a chance to update R_j.

This results in a non-linearizable execution.

The problem above arose because a reader did not have access to what other readers read; in particular, a reader P_i cannot tell if another *Read* by P_j that completed before this *Read* began got a value that is newer than the value

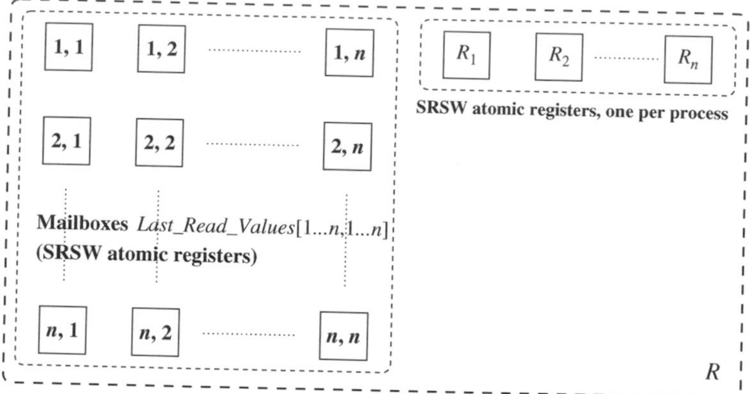

Figure 12.14 Illustrating the data structures for construction 8.

that the writer has written to R_i. In fact, performing multiple reads by the P_i processes, and/or more writes by P_0, and/or using more registers cannot solve this problem.

In the solution by Israeli and Li [16], a reader process P_i must choose the latest of the values that other reader processes have last read, and the value in R_i. As only SRSW registers are available, unfortunately, this requires communication between each pair of reader processes, leading to $O(n^2)$ variables. Thus, a reader process must also write. An array *Last_Read_Values*$[1 \ldots n, 1 \ldots n]$ is used for this purpose. *Last_Read_Values*$[i, j]$ is the value that P_i's last *Read* returned, which P_i has set aside for P_j to know about. Once a reader P_i determines the latest of the values that other readers read (lines 2b–d), and the value written for it by the writer process (line 2a), the reader publishes this value in *Last_Read_Values*$[i, *]$ (lines 2e–2f). As there is a single writer, the format $\langle data, seq\_no \rangle$ for each register value and each *Last_Read_Value* entry is adequate to give a total order on all the values written by it. The construction is shown in Algorithm 12.16 and illustrated in Figure 12.14. Here, $q = n^2 + n$ as there are n^2 SRSW registers that act as personalized mailboxes between the pairs of processes and the n registers that are the mailboxes between writer P_0 and each reader P_i.

Complexity

This construction uses $O(n^2)$ integer registers. The time complexity is $O(n)$.

Achieving linearizability

All the *Write* operations form a total order. A *Read* by P_i returns the value of the latest preceding *Write*, as observed directly from the register R_i, or indirectly from the register R_j and communicated to P_i via *Last_Read_Values*. In a linearized execution, a *Read* is placed after the *Write* whose value it reads. For non-overlapping *Reads*, their relative order represents the order in a linearizable execution, because of the indirect communication among

(shared variables)
SRSW atomic register of type $\langle data, seq\_no\rangle$, **where** $data, seq\_no$ **are integers**: $R_1 \ldots R_n \longleftarrow \langle 0, 0\rangle$;
SRSW atomic register of type $\langle data, seq\_no\rangle$, **where** $data, seq\_no$ **are integers**: $Last\_Read\_Values[1 \ldots n, 1 \ldots n] \longleftarrow \langle 0, 0\rangle$;

(local variables)
type $\langle data, seq\_no\rangle$: $Last\_Read[0 \ldots n]$;
integer: $seq, count$;

(1) $Write(R, val)$ executed by writer P_0
(1a) $seq \longleftarrow seq + 1$;
(1b) **for** $count = 1$ **to** n **do**
(1c) $R_{count} \longleftarrow \langle val, seq\rangle$. // write to each SRSW register

(2) $Read_i(R, val)$ executed by P_i, $1 \leq i \leq n$
(2a) $\langle Last\_Read[0].data, Last\_Read[0].seq\_no\rangle \longleftarrow R_i$;
// $Last\_Read[0]$ stores value of R_i
(2b) **for** $count = 1$ **to** n **do** // read into $Last\_Read[count]$,
// the latest values stored for P_i by P_{count}
(2c) $\langle Last\_Read[count].data, Last\_Read[count].seq\_no\rangle \longleftarrow$
$\langle Last\_Read\_Values[count, i].data,$
$Last\_Read\_Values[count, i].seq\_no\rangle$;
(2d) **identify** j such that for all $k \neq j$, $Last\_Read[j].seq\_no \geq$
$Last\_Read[k].seq\_no$;
(2e) **for** $count = 1$ **to** n **do**
(2f) $\langle Last\_Read\_Values[i, count].data,$
$Last\_Read\_Values[i, count].seq\_no\rangle \longleftarrow$
$\langle Last\_Read[j].data, Last\_Read[j].seq\_no\rangle$;
(2g) $val \longleftarrow Last\_Read[j].data$;
(2h) **return**(val).

Algorithm 12.16 Construction 8: integer SRSW *atomic* register to integer MRSW *atomic* register R.

readers. For overlapping *Reads*, their ordering in a linearized execution is consistent with the *Writes* whose values they read. Hence, the construction is a valid construction.

12.6 Wait-free atomic snapshots of shared objects

Observing the global state of a distributed system is a fundamental problem. For message-passing systems, we have studied how to record global snapshots which represent an instantaneous possible global state that could have occurred in the execution. The snapshot algorithms used message-passing of control messages, and were inherently inhibition-free, although some variants that use fewer control messages do require inhibition.

In this section, we examine the counterpart of the global snapshot problem in a shared-memory system, where only *Read* and *Write* primitives can be used. The problem can be modeled as follows.

Given a set of SWMR atomic registers $R_1 \ldots R_n$, where R_i can be written only by P_i and can be read by all processes, and which together form a compound high-level object, devise a *wait-free* algorithm to observe the state of the object at some instant in time. The following actions are allowed on this high-level object, as also illustrated in Figure 12.15:

- $Scan_i$: This action invoked by P_i returns the atomic snapshot that is an instantaneous view of the object $(R_1, \ldots, R_n)$ at some instant between the invocation and termination of the *Scan*.
- $Update_i(val)$: This action invoked by P_i writes the data *val* to register R_i.

Clearly, any kind of locking mechanism is unacceptable because it is not wait-free. Consider the following attempt at a wait-free solution. The format of each register R_i is assumed to be the tuple: $\langle data, seq\_no \rangle$ in order to uniquely identify each *Write* operation to the register. A scanner would repeatedly scan the high-level object until two consecutive scans, called *double-collect* in the shared memory context, returned identical content. This principle of "double-collect" has been encountered in multiple contexts, such in two-phase deadlock detection and two-phase termination detection algorithms, and essentially embodies the two-phase observation rule (see Chapter 11). However, this solution in not wait-free because between the two observations of each double-collect, an *Update* by another process can prevent the *Scan* from being successful.

A wait-free solution [3,5] is given in Algorithm 12.17. Process P_i can write to its MRSW register R_i and can read all registers $R_1, \ldots R_n$. To design a wait-free solution, it needs to be ensured that a scanner is not indefinitely prevented from getting identical scans in the double-collect, by some writer process periodically making updates. The problem arises because of the imbalance in the roles of the scanner and updater – the updater is inherently more powerful in that it can prevent all scanners from being successful. One elegant solution therefore neutralizes the unfair advantage of the updaters by forcing

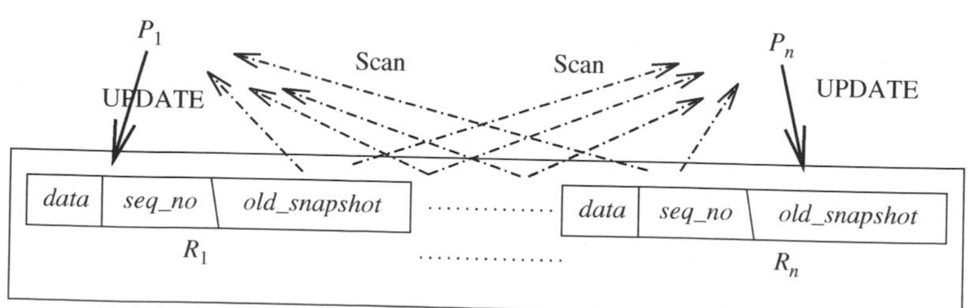

Figure 12.15 Atomic snapshot object, using MRSW atomic registers.

the updaters to follow the same rules as the scanner. Namely, the updaters also have to perform a double-collect, and only after performing a double-collect can an updater write the value it needs to. Additionally, an updater also writes the snapshot it collected in the register, along with the new value of the data item. Now, if a scanner detects that an updater has made an update after the scanner initiated its *Scan*, then the scanner can simply "borrow" the snapshot recorded by the updater in its register. The updater helps the scanner to obtain a consistent value. This is the principle of "helping" that is often used in designing wait-free solutions for various problems.

(shared variables)
MRSW atomic register of type $\langle data, seq\_no, old\_snapshot\rangle$, **where** $data, seq\_no$ **are of type integer, and** $old\_snapshot$ **is array** $[1\ldots n]$ **of integer**: $R_1 \ldots R_n$;

(local variables)
integer: $changed[1\ldots n]$;
type $\langle data, seq\_no, old\_snapshot\rangle$: $v1[1\ldots n], v2[1\ldots n], v[1\ldots n]$;

```
(1)   Update_i(x)
(1a)  v[1...n] ⟵ Scan_i;
(1b)  R_i ⟵ (x, R_i.seq_no + 1, v[1...n]).

(2)   Scan_i
(2a)  for count = 1 to n do
(2b)         changed[count] ⟵ 0;
(2c)  while true do
(2d)         v1[1...n] ⟵ collect();
(2e)         v2[1...n] ⟵ collect();
(2f)         if (∀k, 1 ≤ k ≤ n)(v1[k].seq_no = v2[k].seq_no) then
(2g)                return(v2[1].data, ..., v2[n].data);
(2h)         else
(2i)                for k = 1 to n do
(2j)                       if v1[k].seq_no ≠ v2[k].seq_no then
(2k)                              changed[k] ⟵ changed[k] + 1;
(2l)                              if changed[k] = 2 then
(2m)                                     return(v2[k].old_snapshot).
```

Algorithm 12.17 Wait-free atomic snapshot of a shared MRSW object.

A scanner detects that an updater has made an update after the scanner initiated its *Scan*, by using the local array *changed*. This array is reset to 0 when the *Scan* is invoked. Location $changed[k]$ is incremented (line 2k) if the *Scan* procedure detects (line 2j) that process P_k has changed its *data* and *seq_no* (and implicitly the *old_snapshot*) fields in R_k. Based on the value

Figure 12.16 Nesting of double-collects, in scanning for atomic snapshots of object.

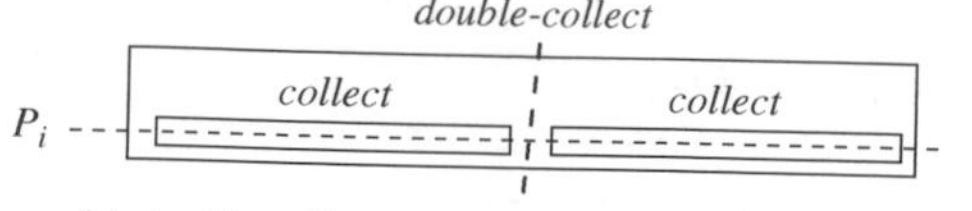

(a) *double-collect* sees identical values in both *collects*

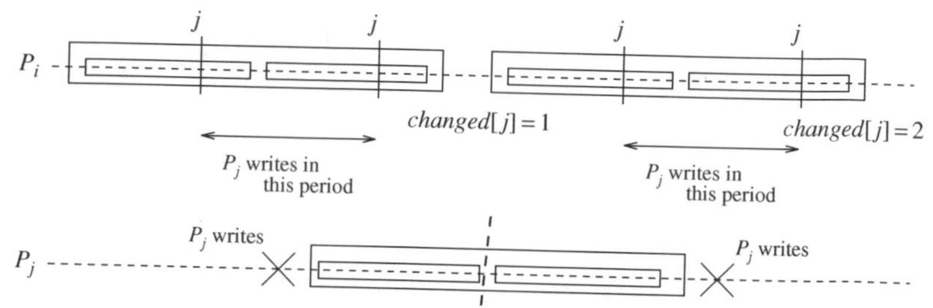

(b) P_j's *double-collect* nested within P_i's SCAN. The *double-collect* is successful, or P_j borrowed snapshot from P_k's *double-collect* nested within P_j's *Scan*. And so on recursively, up to *n* times.

of *changed*[*k*], different inferences can be made, as now explained with the help of Figure 12.16:

- If *changed*[*k*] = 2 (line 2l), then two updates (line 1b) were made by P_k after P_i began its *Scan*. Between the first and the second update, the *Scan* preceding the second update must have completed successfully, and the scanned value was recorded in the *old_snapshot* field. This old snapshot can be safely borrowed by the scanner P_i (line 2m) because it was recorded after P_k finished its first *double-collect*, and hence after the scanner P_i initiated its *Scan*.
- However, if *changed*[*k*] = 1, it cannot be inferred that the *old_snapshot* recorded by P_k was taken after P_i's *Scan* began. When P_k does its *Update* (the first "write" shown in Figure 12.16(b)), the value it writes in *old_snapshot* is only the result of a double-scan that preceded the "write" and may be a value that existed before P_i's *Scan* began.

There are two cases by which a snapshot can be captured, as illustrated using Figure 12.16:

1. A scanner can collect a snapshot (line 2g) if the double-collect (lines 2d–2e) returns identical views (line 2f). (see Figure 12.16(a)). The returned snapshot represents an instantaneous state that existed at all times between the end of the first *collect* (line 2d) and the start of the second *collect* (line 2e).
2. Otherwise the scanner returns a borrowed snapshot (line 2m) from P_k if P_k has been noticed to have made two updates (lines 2l) and therefore P_k has made a *Scan* embedded inside P_i's *Scan*. This borrowed snapshot itself (i) may have been obtained directly via a *double-collect*, or (ii) indirectly been borrowed from another process (line 2l). In case (i), it represents

an instantaneous state in the duration of the *double-collect*. In case (ii), a recursive argument can be applied. Observe that there are n processes, so the recursive argument can hold at most $n-1$ times. The nth time, a *double-collect* must have been successful (see Figure 12.16(b)). Note that between the two *double-collects* of P_i that are shown, there may be up to $(n-2)$ other unsuccessful *double-collects* of P_i. Each of these $(n-2)$ other *double-collects* corresponds to some P_k, $k \neq i, j$, having "changed" once.

The linearization of the *Scan* and *Update* operations follows in a straightforward manner. For example, non-overlapping operations get linearized in the order of their occurrence. An operation by P_i that borrows a snapshot from P_k gets linearized after P_k.

Complexity

The local space complexity is $O(n^2)$ integers. The shared space is $O(n^2)$ corresponding to each of the n registers of size $O(n)$ each. The time complexity is $O(n^2)$. This is because the main *Scan* loop has a complexity of $O(n)$ and the loop may be executed at most n times – the nth time, at least one process P_k must have caused P_i's local $changed[k]$ to reach a value of two, triggering an end to the loop (lines 2k–2l).

12.7 Chapter summary

Distributed shared memory (DSM) is an abstraction whereby distributed programs can communicate with memory operations (*Read* and *Write*) as opposed to using message-passing. The main motivation is to simplify the burden on the programmers. The chapter surveyed this and other motivating factors for DSMs, as well as provided different ways to classify DSMs. The DSM has to be implemented by the middleware layer. Furthermore, in the face of concurrent operations on the shared variables, the expected behavior seen by the programmers should be well-defined. The chapter examined the following consistency models – linearizability, sequential consistency, causal consistency, pipelined RAM (PRAM), and slow memory. Each model is a contract between the programmer and the system provider because the program logic must adhere to the consistency model being provided by the middleware.

The chapter then examined the fundamental problem of mutal exclusion. The well-known bakery algorithm was studied first. Next, Lamport's algorithm for fast mutual exclusion – which gives an $O(1)$ complexity when there are no contentions – was studied. Mutual exclusion using hardware instructions – *Test&Set* and *Swap* – was then examined. Such hardware instructions can perform a *Read* operation and a *Write* operation atomically. Hence, they are powerful, but are also expensive to implement in a machine.

In the context of DSM mutual exclusion, and more generally, DSM synchronization operations, fault-tolerance was then examined. The notion of *wait freedom* is the ability to complete all the operations of a process, irrespective of the behavior of other processes. This makes the system $n-1$ fault tolerant. Next, wait-free register constructions were considered. Registers can be classified as being binary or multi-valued. An orthogonal classification allows single-reader or multiple reader, single-writer or multiple writer registers. Also orthogonally, registers can be *safe*, *regular*, or *atomic*. This allows 24 possible configurations. The chapter considered some of these 24 possible wait-free constructions. The constructions provide insight into how different techniques can be used in the DSM setting. Finally, wait-free atomic snapshots of shared objects was considered. For an object, reading its value atomically in a wait-free manner (without locking) gives an "instantaneous" snapshot of its state. Hence, this is an important problem for DSMs.

12.8 Exercises

Exercise 12.1 Why do the algorithms for sequential consistency (Section 12.2.2) not require the *Read* operations to be broadcast?

Exercise 12.2 Give a formal proof to justify the correctness of Algorithm 12.2 that implements sequential consistency using local *Read* operations.

Exercise 12.3 In the algorithm to implement sequential consistency using local *Write* operations, as given in Algorithm 12.3, why is a single counter *counter* sufficient for the algorithm's correctness?

In other words, why is a separate counter $counter_x$ not required to track the number of updates issued to each variable x, where a *Read* operation on x gets delayed only if $counter_x > 0$? If such a separate counter were used for every variable, what consistency model would be implemented?

Exercise 12.4 1. In Figure 12.6(a), analyze whether the execution is linearizable.
2. In Figure 12.6(b), what forms of memory consistency are satisfied if the two *Read* operations of P_4 return 7 first and then 4?

Exercise 12.5 Give a detailed implementation of causal consistency, and provide a correctness argument for your implementation.

Exercise 12.6 Give a detailed implementation of PRAM consistency, and provide a correctness argument for your implementation.

Exercise 12.7 Give a detailed implementation of slow memory, and provide a correctness argument for your implementation. Is the implementation less expensive than that of PRAM consistency which is a stricter consistency model?

Exercise 12.8 Show that constructions 1 and 2 (Algorithm 12.10) work for binary registers as well as integer-valued registers.

Exercise 12.9 Why are two passes needed by the reader in construction 6, (Algorithm 12.14), for a MRSW atomic register? Why does a single right-to-left pass not suffice?

Exercise 12.10 Assume that the writer does a single pass from left to right in construction 6, (Algorithm 12.14), for a MRSW register. Can the code for the readers be modified to devise a correct algorithm? Justify your answer.

Exercise 12.11 Peterson's mutual exclusion algortihm for two processes is shown in Algorithm 12.18 [26].

1. Show that it satisfies mutual exclusion, progress, and bounded waiting.
2. Use this algorithm as a building block to construct a hierarchical mutual exclusion algorithm for an arbitrary number of processes. (Hint: Use a logarithmic number of steps in the hierarchy.)

(shared variables)
boolean: $turn \longleftarrow false$; // shared register initialized
boolean: $wanting[0, 1]$;

repeat
(1) P_i executes the following for the **entry section**:
(1a) $wanting[i] \longleftarrow true$;
(1b) $turn \longleftarrow 1 - i$;
(1c) **while** $wanting[1 - i]$ **and** $turn = 1 - i$ **do**
(1d) **no-op**;

(2) P_i executes the **critical section (CS)** after the **entry section**

(3) P_i executes the following **exit section** after the **CS**:
(3a) $wanting[i] \longleftarrow false$;

(4) P_i executes the **remainder section** after the **exit section**

until false;

Algorithm 12.18 Peterson's mutual exclusion for two processes $P_i = 0, 1$ [26]. Modulo 2 arithmetic is used.

Exercise 12.12 Determine the average case time complexity of the wait-free atomic snapshot of a shared object, given in Algorithm 12.17.

12.9 Notes on references

A good survey on distributed shared memory systems is given by Protic *et al.* [28] and by Tanenbaum [29]. This includes coverage of the various DSM systems such as Firefly, Sequent, Alewife, Dash, Butterfly, CM*, Ivy, Mirage, Midway, Munin, Linda, and Orca.

The sequential consistency model was defined by Lamport [19]. The linearizability model was formalized by Lamport [21] and developed by Herlihy and Wing [13]. The implementations of linearizability and sequential consistency based on the broadcast primitive and assuming full replication are from Attiya and Welch [6], whereas a similar implementation of sequential consistency is given by Bal *et al.* [8]. The causal consistency model was proposed by [4]. The PRAM model was proposed by Lipton and Sandberg [25]. The slow memory model was proposed by Hutto and Ahamad [14]. Other consistency models such as weak consistency [11], release consistency [12], and entry consistency [9] that apply to selected instructions in the code, were developed mainly in the computer architecture research community, and are discussed in [1,2].

The bakery algorithm for mutual exclusion was presented by Lamport [18]. The fast mutual exclusion algorithm was presented by Lamport [23]. The two-process mutual exclusion algorithm was presented by Peterson [26]. Its modification that is asked as Exercise 12.11 is based on the algorithm by Peterson and Fischer [27].

The notion of wait-freedom was proposed by Lamport [24] and developed by Herlihy [15]. The definition and classification of registers as safe, regular, and atomic were given by Lamport [20–22]. Constructions 1 to 5 were proposed by Lamport [21]. Register construction 6 was proposed by Vidyasankar [30]. Register construction 7 was proposed by Vitanyi and Awerbuch [31]. Register construction 8 was proposed by Israeli and Li [16]. A construction of a MRMW snapshot object using MRSW snapshot objects and MRMW registers was proposed by Anderson [5] and Afek *et al.* [3]. The register constructions are also presented by Attiya and Welch [7].

References

[1] S. Adve and K. Gharachorloo, Shared memory consistency models: a tutorial, *IEEE Computer Magazine*, **29**(12), 1996, 66–76.

[2] S. Adve and M. Hill, A unified formalization of four shared-memory models, *IEEE Transactions on Parallel and Distributed Systems*, **4**(6), 1993, 613–624.

[3] Y. Afek, H. Attiya, D. Dolev, E. Gafni, M. Merritt, and N. Shavit, Atomic snapshots of shared memory, *Journal of the ACM*, **40**(4), 1993, 873–890.

[4] M. Ahamad, G. Neiger, J. Burns, P. Kohli, and P. Hutto, Causal memory: definitions, implementation, and programming, *Distributed Computing*, **9**(1), 1995, 37–49.

[5] J. Anderson, Multi-writer composite registers, *Distributed Computing*, **7**(4), 1994, 175–196.

[6] H. Attiya and J. Welch, Sequential consistency versus linearizability, *ACM Transactions on Computer Systems*, **12**(2), 1994, 91–122.

[7] H. Attiya and J. Welch, *Distributed Computing: Fundamentals, Simulations and Advanced Topics*, Chichester, Wiley, 2004.

[8] H. Bal, F. Kaashoek and A. Tanenbaum, Orca: a language for parallel programming of distributed systems, *IEEE Transactions on Software Engineering*, **18**(3), 1992, 180–205.

[9] B. Bershad, M. Zekauskas, and W. Sawdon, *The Midway Distributed Shared Memory System*, CMU Technical Report CMU-CS-93-119. (Also in *Proceedings of COMPCON 1993*.)

[10] J. Burns and N. Lynch, Bounds on shared memory for mutual exclusion, *Information and Computation*, **107**(2), 1993, 171–184.

[11] M. Dubois and C. Scheurich, Memory access dependencies in shared-memory multiprocessors, *IEEE Transactions on Software Engineering*, **16**(6), 1990, 660–673.

[12] K. Gharachorloo, D. Lenoski, J. Laudon, P. Gibbons, A. Gupta, and J. L. Hennessy, Memory consistency and event ordering in scalable shared-memory multiprocessors, *Proceedings of the 17th International Symposium on Computer Architecture*, Seattle, WA, May 1990, 15–26.

[13] M. Herlihy and J. Wing, Linearizability: a correctness condition for concurrent objects, *ACM Transactions on Programming Languages and Systems*, **12**(3), 1990, 463–492.

[14] P. Hutto and M. Ahamad, Slow memory: weakening consistency to enchance concurrency in distributed shared memories, *Proceedings of the IEEE International Conference on Distributed Computing Systems*, 1990, 302–311.

[15] M. Herlihy, Wait-free synchronization, *ACM Transactions on Programming Languages and Systems*, **13**(1), 1991, 124–149.

[16] A. Israeli and M. Li, Bounded timestamps, *Distributed Computing*, **6**(4), 1993, 205–209.

[17] C. Kruskal, L. Rudolf, and M. Snir, Efficient synchronization of multiprocessors with shared memory, *Proceedings of the ACM Conference on Principles of Distributed Computing*, 1986.

[18] L. Lamport, A new solution of Dijkstra's concurrent programming problem, *Communications of the ACM*, **17**(8), 1974, 453–455.

[19] L. Lamport, How to make a multiprocessor that correctly executes multiprocess programs, *IEEE Transactions on Computers*, **28**(9), 1979, 690–691.

[20] L. Lamport, On interprocess communication, part I: basic formalism, *Distributed Computing*, **1**(2), 1986, 77–85.

[21] L. Lamport, On interprocess communication, part II: algorithms, *Distributed Computing*, **1**(2), 1986, 86–101.

[22] L. Lamport, The mutual exclusion problem, part II: statement and solutions, *Journal of the ACM*, **33**(2), 1986, 327–348.

[23] L. Lamport, A fast mutual exclusion algorithm, *ACM Transactions on Computer Systems*, **5**(1), 1987, 1–11.

[24] L. Lamport, Concurrent reading and writing, *Communications of the ACM*, **20**(11), 1977, 806-811.

[25] R. Lipton and J. Sandberg, *PRAM: a Scalable Shared Memory*, Technical Report CS-TR-180-88, Princeton University, Department of Computer Science, September 1988.

[26] G. L. Peterson, Myths about the mutual exclusion problem, *Information Processing Letters*, **12**, 1981, 115–116.

[27] G. L. Peterson and M. Fischer, Economical solutions for the mutual exclusion problem in a distributed system, *Proceedings of the 9th ACM Symposium on Theory of Computing*, 1977, 91–97.

[28] J. Protic, M. Tomasevic, and V. Milutinovic, Distributed shared memory: concepts and systems, *IEEE Concurrency*, **4**(2), 1996, 63–79.

[29] A. Tanenbaum, *Distributed Operating Systems*, Harlow, Pearson Education, 1995.

[30] K. Vidyasankar, Converting Lamport's regular register to atomic register, *Information Processing Letters*, **28**, 1988, 287–290.

[31] P. Vitanyi and B. Awerbuch, Atomic shared register access by asynchronous hardware, *Proceedings of the 27th IEEE Symposium on Foundations of Computer Science*, 1986, 233–243.

CHAPTER

13 Checkpointing and rollback recovery

13.1 Introduction

Distributed systems today are ubiquitous and enable many applications, including client–server systems, transaction processing, the World Wide Web, and scientific computing, among many others. Distributed systems are not fault-tolerant and the vast computing potential of these systems is often hampered by their susceptibility to failures. Many techniques have been developed to add reliability and high availability to distributed systems. These techniques include transactions, group communication, and rollback recovery. These techniques have different tradeoffs and focus. This chapter covers the rollback recovery protocols, which restore the system back to a consistent state after a failure.

Rollback recovery treats a distributed system application as a collection of processes that communicate over a network. It achieves fault tolerance by periodically saving the state of a process during the failure-free execution, enabling it to restart from a saved state upon a failure to reduce the amount of lost work. The saved state is called a *checkpoint*, and the procedure of restarting from a previously checkpointed state is called *rollback recovery.* A checkpoint can be saved on either the stable storage or the volatile storage depending on the failure scenarios to be tolerated.

In distributed systems, rollback recovery is complicated because messages induce inter-process dependencies during failure-free operation. Upon a failure of one or more processes in a system, these dependencies may force some of the processes that did not fail to roll back, creating what is commonly called a *rollback propagation*. To see why rollback propagation occurs, consider the situation where the sender of a message m rolls back to a state that precedes the sending of m. The receiver of m must also roll back to a state that precedes m's receipt; otherwise, the states of the two processes would be *inconsistent* because they would show that message m was received without being sent, which is impossible in any correct failure-free execution. This phenomenon of cascaded rollback is called the domino effect. In some situations, rollback propagation may extend back to the initial state of the computation, losing all the work performed before the failure.

In a distributed system, if each participating process takes its checkpoints independently, then the system is susceptible to the domino effect. This approach is called *independent* or *uncoordinated checkpointing*. It is obviously desirable to avoid the domino effect and therefore several techniques have been developed to prevent it. One such technique is *coordinated checkpointing* where processes coordinate their checkpoints to form a system-wide consistent state. In case of a process failure, the system state can be restored to such a consistent set of checkpoints, preventing the rollback propagation. Alternatively, *communication-induced checkpointing* forces each process to take checkpoints based on information piggybacked on the application messages it receives from other processes. Checkpoints are taken such that a system-wide consistent state always exists on stable storage, thereby avoiding the domino effect.

The approaches discussed so far implement *checkpoint-based* rollback recovery, which relies only on checkpoints to achieve fault-tolerance. *Log-based* rollback recovery combines checkpointing with logging of non-deterministic events. Log-based rollback recovery relies on the *piecewise deterministic* (PWD) assumption, which postulates that all non-deterministic events that a process executes can be identified and that the information necessary to replay each event during recovery can be logged in the event's *determinant*. By logging and replaying the non-deterministic events in their exact original order, a process can deterministically recreate its pre-failure state even if this state has not been checkpointed. Log-based rollback recovery in general enables a system to recover beyond the most recent set of consistent checkpoints. It is therefore particularly attractive for applications that frequently interact with the *outside world*, which consists of input and output devices that cannot roll back.

13.2 Background and definitions

13.2.1 System model

A distributed system consists of a fixed number of processes, $P_1, P_2, \ldots, P_N$, which communicate only through messages. Processes cooperate to execute a distributed application and interact with the outside world by receiving and sending input and output messages, respectively. Figure 13.1 shows a system consisting of three processes and interactions with the outside world.

Rollback-recovery protocols generally make assumptions about the reliability of the inter-process communication. Some protocols assume that the communication subsystem delivers messages reliably, in first-in-first-out (FIFO) order, while other protocols assume that the communication subsystem can

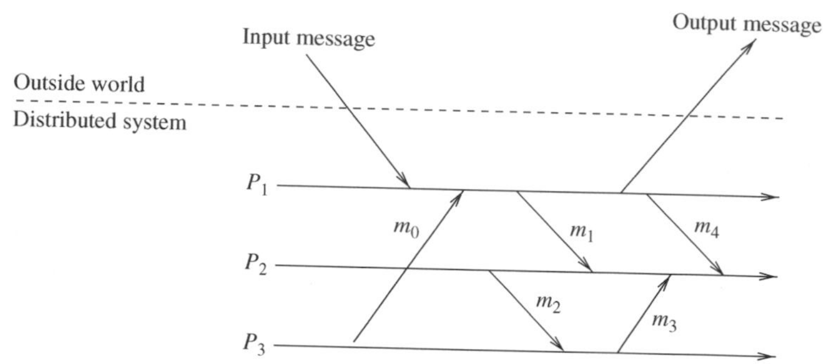

Figure 13.1 An example of a distributed system with three processes.

lose, duplicate, or reorder messages. The choice between these two assumptions usually affects the complexity of checkpointing and failure recovery.

A generic correctness condition for rollback-recovery can be defined as follows [36]: "a system recovers correctly if its internal state is consistent with the observable behavior of the system before the failure." Rollback-recovery protocols therefore must maintain information about the internal interactions among processes and also the external interactions with the outside world.

13.2.2 A local checkpoint

In distributed systems, all processes save their local states at certain instants of time. This saved state is known as a local checkpoint. A local checkpoint is a snapshot of the state of the process at a given instance and the event of recording the state of a process is called local checkpointing. The contents of a checkpoint depend upon the application context and the checkpointing method being used.

Depending upon the checkpointing method used, a process may keep several local checkpoints or just a single checkpoint at any time. We assume that a process stores all local checkpoints on the stable storage so that they are available even if the process crashes. We also assume that a process is able to roll back to any of its existing local checkpoints and thus restore to and restart from the corresponding state.

Let $C_{i,k}$ denote the kth local checkpoint at process P_i. Generally, it is assumed that a process P_i takes a checkpoint $C_{i,0}$ before it starts execution. A local checkpoint is shown in the process-line by the symbol "|".

13.2.3 Consistent system states

A global state of a distributed system is a collection of the individual states of all participating processes and the states of the communication channels. Intuitively, a consistent global state is one that may occur during a failure-free execution of a distributed computation. More precisely, a *consistent system state* is one in which if a process's state reflects a message receipt, then the state of the corresponding sender must reflect the sending of that message [9].

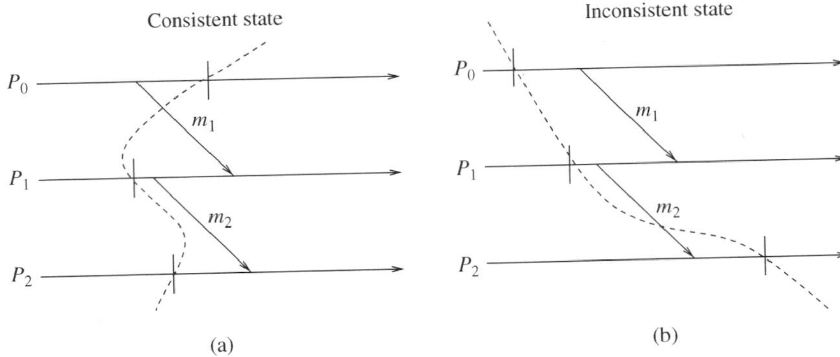

Figure 13.2 Examples of consistent and inconsistent states [13].

For instance, Figure 13.2 shows two examples of global states. The state in Figure 13.2(a) is consistent and the state in Figure 13.2(b) is inconsistent. Note that the consistent state in Figure 13.2(a) shows message m_1 to have been sent but not yet received, but that is alright. The state in Figure 13.2(a) is consistent because it represents a situation in which for every message that has been received, there is a corresponding message send event. The state in Figure 13.2(b) is inconsistent because process P_2 is shown to have received m_2 but the state of process P_1 does not reflect having sent it. Such a state is impossible in any failure-free, correct computation. Inconsistent states occur because of failures. For instance, the situation shown in Figure 13.2(b) may occur if process P_1 fails after sending message m_2 to process P_2 and then restarts at the state shown in Figure 13.2(b).

Thus, a local checkpoint is a snapshot of a local state of a process and a global checkpoint is a set of local checkpoints, one from each process. A consistent global checkpoint is a global checkpoint such that no message is sent by a process after taking its local checkpoint that is received by another process before taking its local checkpoint. The consistency of global checkpoints strongly depends on the flow of messages exchanged by processes and an arbitrary set of local checkpoints at processes may not form a consistent global checkpoint.

The fundamental goal of any rollback-recovery protocol is to bring the system to a consistent state after a failure. The reconstructed consistent state is not necessarily one that occurred before the failure. It is sufficient that the reconstructed state be one that could have occurred before the failure in a failure-free execution, provided that it is consistent with the interactions that the system had with the outside world.

13.2.4 Interactions with the outside world

A distributed application often interacts with the outside world to receive input data or deliver the outcome of a computation. If a failure occurs, the outside world cannot be expected to roll back. For example, a printer cannot roll back the effects of printing a character, and an automatic teller machine cannot recover the money that it dispensed to a customer. To simplify the presentation of how rollback-recovery protocols interact with the outside world, we model the latter as a *special* process that interacts with the rest of the system through message passing. We call this special process the "outside world process" (OWP). It is therefore necessary that the outside world see a consistent behavior of the system despite failures. Thus, before sending output to the OWP, the system must ensure that the state from which the output is sent will be recovered despite any future failure. This is commonly called the output commit problem. Similarly, input messages that a system receives from the OWP may not be reproducible during recovery, because it may not be possible for the outside world to regenerate them. Thus, recovery protocols must arrange to save these input messages so that they can be retrieved when needed for execution replay after a failure. A common approach is to save each input message on the stable storage before allowing the application program to process it.

An interaction with the outside world to deliver the outcome of a computation is shown on the process-line by the symbol "||".

13.2.5 Different types of messages

A process failure and subsequent recovery may leave messages that were perfectly received (and processed) before the failure in abnormal states. This is because a rollback of processes for recovery may have to rollback the send and receive operations of several messages.

In this section, we identify several types of such messages using the example shown in Figure 13.3. Figure 13.3 shows an example consisting of four processes. Process P_1 fails at the point indicated and the whole system recovers to the state indicated by the recovery line; that is, to global state $\{C_{1,8}, C_{2,9}, C_{3,8}, C_{4,8}\}$.

Figure 13.3 Different types of messages [25].

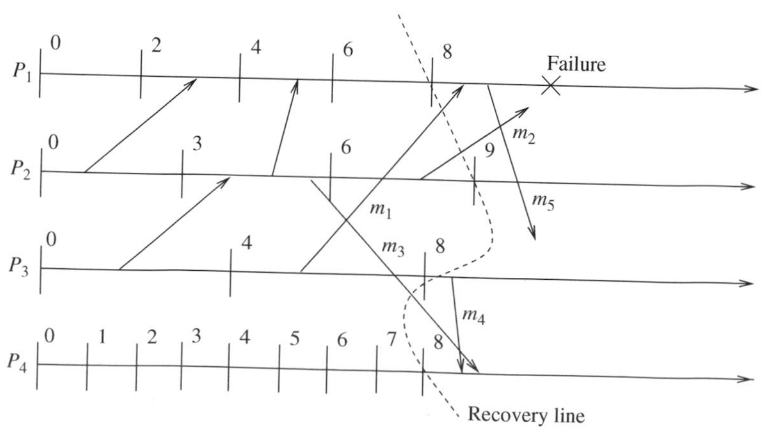

In-transit messages

In Figure 13.3, the global state $\{C_{1,8}, C_{2,9}, C_{3,8}, C_{4,8}\}$ shows that message m_1 has been sent but not yet received. We call such a message an *in-transit* message. Message m_2 is also an in-transit message.

When in-transit messages are part of a global system state, these messages do not cause any inconsistency. However, depending on whether the system model assumes reliable communication channels, rollback-recovery protocols may have to guarantee the delivery of in-transit messages when failures occur. For reliable communication channels, a consistent state must include in-transit messages because they will always be delivered to their destinations in any legal execution of the system. On the other hand, if a system model assumes lossy communication channels, then in-transit messages can be omitted from system state.

Lost messages

Messages whose send is not undone but receive is undone due to rollback are called *lost* messages. This type of messages occurs when the process rolls back to a checkpoint prior to reception of the message while the sender does not rollback beyond the send operation of the message. In Figure 13.3, message m_1 is a lost message.

Delayed messages

Messages whose receive is not recorded because the receiving process was either down or the message arrived after the rollback of the receiving process, are called *delayed* messages. For example, messages m_2 and m_5 in Figure 13.3 are delayed messages.

Orphan messages

Messages with receive recorded but message send not recorded are called *orphan* messages. For example, a rollback might have undone the send of such messages, leaving the receive event intact at the receiving process. Orphan messages do not arise if processes roll back to a consistent global state.

Duplicate messages

Duplicate messages arise due to message logging and replaying during process recovery. For example, in Figure 13.3, message m_4 was sent and received before the rollback. However, due to the rollback of process P_4 to $C_{4,8}$ and process P_3 to $C_{3,8}$, both send and receipt of message m_4 are undone. When process P_3 restarts from $C_{3,8}$, it will resend message m_4. Therefore, P_4 should not replay message m_4 from its log. If P_4 replays message m_4, then message m_4 is called a *duplicate* message.

Message m_5 is an excellent example of a duplicate message. No matter what, the receiver of m_5 will receive a duplicate m_5 message.

13.3 Issues in failure recovery

In a failure recovery, we must not only restore the system to a consistent state, but also appropriately handle messages that are left in an abnormal state due to the failure and recovery [33].

We now describe the issues involved in a failure recovery with the help of a distributed computation shown in Figure 13.4. The computation comprises of three processes P_i, P_j, and P_k, connected through a communication network. The processes communicate solely by exchanging messages over fault-free, FIFO communication channels. Processes P_i, P_j, and P_k have taken checkpoints $\{C_{i,0}, C_{i,1}\}$, $\{C_{j,0}, C_{j,1}, C_{j,2}\}$, and $\{C_{k,0}, C_{k,1}\}$, respectively, and these processes have exchanged messages A to J as shown in Figure 13.4.

Suppose process P_i fails at the instance indicated in the figure. All the contents of the volatile memory of P_i are lost and, after P_i has recovered from the failure, the system needs to be restored to a consistent global state from where the processes can resume their execution. Process P_i's state is restored to a valid state by rolling it back to its most recent checkpoint $C_{i,1}$. To restore the system to a consistent state, the process P_j rolls back to checkpoint $C_{j,1}$ because the rollback of process P_i to checkpoint $C_{i,1}$ created an orphan message H (the receive event of H is recorded at process P_j while the send event of H has been undone at process P_i). Note that process P_j does not roll back to checkpoint $C_{j,2}$ but to checkpoint $C_{j,1}$, because rolling back to checkpoint $C_{j,2}$ does not eliminate the orphan message H. Even this resulting state is not a consistent global state, as an orphan message I is created due to the roll back of process P_j to checkpoint $C_{j,1}$. To eliminate this orphan message, process P_k rolls back to checkpoint $C_{k,1}$. The

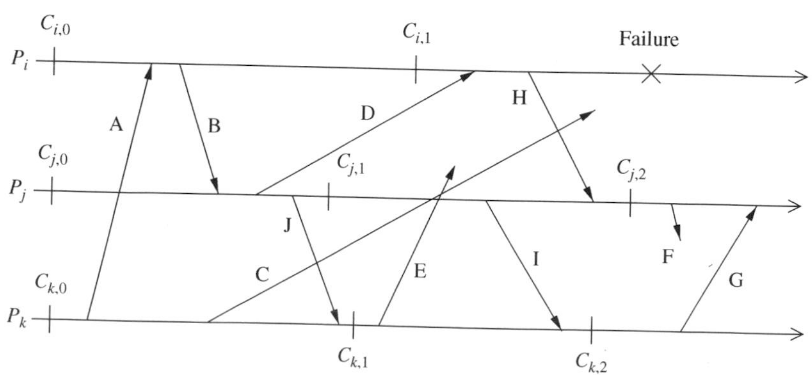

Figure 13.4 Illustration of issues in failure recovery.

restored global state $\{C_{i,1}, C_{j,1}, C_{k,1}\}$ is a consistent state as it is free from orphan messages. Although the system state has been restored to a consistent state, several messages are left in an erroneous state which must be handled correctly.

Messages A, B, D, G, H, I, and J had been received at the points indicated in the figure and messages C, E, and F were in transit when the failure occurred. Restoration of system state to checkpoints $\{C_{i,1}, C_{j,1}, C_{k,1}\}$ automatically handles messages A, B, and J because the send and receive events of messages A, B, and J have been recorded, and both the events for G, H, and I have been completely undone. These messages cause no problem and we call messages A, B, and J normal messages and messages G, H, and I vanished messages [33].

Messages C, D, E, and F are potentially problematic. Message C is in transit during the failure and it is a delayed message. The delayed message C has several possibilities: C might arrive at process P_i before it recovers, it might arrive while P_i is recovering, or it might arrive after P_i has completed recovery. Each of these cases must be dealt with correctly.

Message D is a lost message since the send event for D is recorded in the restored state for process P_j, but the receive event has been undone at process P_i. Process P_j will not resend D without an additional mechanism, since the send D at P_j occurred before the checkpoint and the communication system successfully delivered D.

Messages E and F are delayed orphan messages and pose perhaps the most serious problem of all the messages. When messages E and F arrive at their respective destinations, they must be discarded since their send events have been undone. Processes, after resuming execution from their checkpoints, will generate both of these messages, and recovery techniques must be able to distinguish between messages like C and those like E and F.

Lost messages like D can be handled by having processes keep a message log of all the sent messages. So when a process restores to a checkpoint, it replays the messages from its log to handle the lost message problem. However, message logging and message replaying during recovery can result in duplicate messages. In the example shown in Figure 13.4, when process P_j replays messages from its log, it will regenerate message J. Process P_k, which has already received message J, will receive it again, thereby causing inconsistency in the system state. Therefore, these duplicate messages must be handled properly.

Overlapping failures further complicate the recovery process. A process P_j that begins rollback/recovery in response to the failure of a process P_i can itself fail and develop amnesia with respect process P_i's failure; that is, process P_j can act in a fashion that exhibits ignorance of process P_i's failure. If overlapping failures are to be tolerated, a mechanism must be introduced to deal with amnesia and the resulting inconsistencies.

13.4 Checkpoint-based recovery

In the checkpoint-based recovery approach, the state of each process and the communication channel is checkpointed frequently so that, upon a failure, the system can be restored to a globally consistent set of checkpoints. It does not rely on the PWD assumption, and so does not need to detect, log, or replay non-deterministic events. Checkpoint-based protocols are therefore less restrictive and simpler to implement than log-based rollback recovery. However, checkpoint-based rollback recovery does not guarantee that pre-failure execution can be deterministically regenerated after a rollback. Therefore, checkpoint-based rollback recovery may not be suitable for applications that require frequent interactions with the outside world. Checkpoint-based rollback-recovery techniques can be classified into three categories: *uncoordinated checkpointing*, *coordinated checkpointing*, and *communication-induced checkpointing* [13].

13.4.1 Uncoordinated checkpointing

In uncoordinated checkpointing, each process has autonomy in deciding when to take checkpoints. This eliminates the synchronization overhead as there is no need for coordination between processes and it allows processes to take checkpoints when it is most convenient or efficient. The main advantage is the lower runtime overhead during normal execution, because no coordination among processes is necessary. Autonomy in taking checkpoints also allows each process to select appropriate checkpoints positions. However, uncoordinated checkpointing has several shortcomings [13].

First, there is the possibility of the domino effect during a recovery, which may cause the loss of a large amount of useful work. Second, recovery from a failure is slow because processes need to iterate to find a consistent set of checkpoints. Since no coordination is done at the time the checkpoint is taken, checkpoints taken by a process may be *useless* checkpoints. (A useless checkpoint is never a part of any global consistent state.) Useless checkpoints are undesirable because they incur overhead and do not contribute to advancing the recovery line. Third, uncoordinated checkpointing forces each process to maintain multiple checkpoints, and to periodically invoke a garbage collection algorithm to reclaim the checkpoints that are no longer required. Fourth, it is not suitable for applications with frequent output commits because these require global coordination to compute the recovery line, negating much of the advantage of autonomy.

As each process takes checkpoints independently, we need to determine a consistent global checkpoint to rollback to, when a failure occurs. In order to determine a consistent global checkpoint during recovery, the processes record the dependencies among their checkpoints caused by message exchange

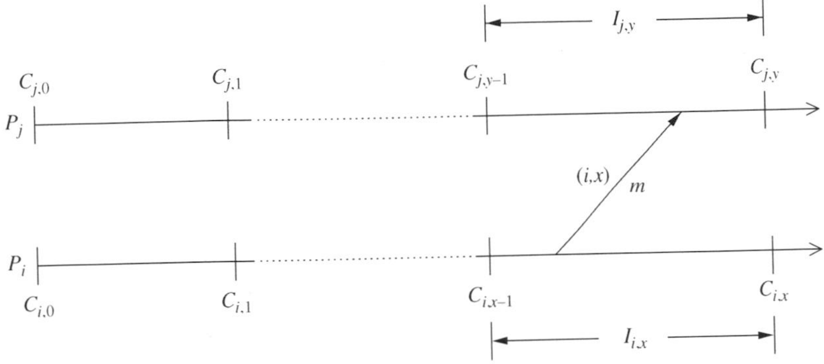

Figure 13.5 Checkpoint index and checkpoint interval [13].

during failure-free operation. The following direct dependency tracking technique is commonly used in uncoordinated checkpointing.

Let $C_{i,x}$ be the xth checkpoint of process P_i, where i is the process i.d. and x is the checkpoint index (we assume each process P_i starts its execution with an initial checkpoint $C_{i,0}$). Let $I_{i,x}$ denote the *checkpoint interval* or simply *interval* between checkpoints $C_{i,x-1}$ and $C_{i,x}$. Consider the example shown in Figure 13.5. When process P_i at interval $I_{i,x}$ sends a message m to P_j, it piggybacks the pair (i, x) on m. When P_j receives m during interval $I_{j,y}$, it records the dependency from $I_{i,x}$ to $I_{j,y}$, which is later saved onto stable storage when P_j takes checkpoint $C_{j,y}$.

When a failure occurs, the recovering process initiates rollback by broadcasting a *dependency request* message to collect all the dependency information maintained by each process. When a process receives this message, it stops its execution and replies with the dependency information saved on the stable storage as well as with the dependency information, if any, which is associated with its current state. The initiator then calculates the recovery line based on the global dependency information and broadcasts a *rollback request* message containing the recovery line. Upon receiving this message, a process whose current state belongs to the recovery line simply resumes execution; otherwise, it rolls back to an earlier checkpoint as indicated by the recovery line.

13.4.2 Coordinated checkpointing

In coordinated checkpointing, processes orchestrate their checkpointing activities so that all local checkpoints form a consistent global state [13]. Coordinated checkpointing simplifies recovery and is not susceptible to the domino effect, since every process always restarts from its most recent checkpoint. Also, coordinated checkpointing requires each process to maintain only one checkpoint on the stable storage, reducing the storage overhead and eliminating the need for garbage collection. The main disadvantage of this method is that large latency is involved in committing output, as a global checkpoint is needed before a message is sent to the OWP. Also, delays and overhead are involved everytime a new global checkpoint is taken.

If perfectly synchronized clocks were available at processes, the following simple method could be used for checkpointing: all processes agree at what instants of time they will take checkpoints, and the clocks at processes trigger the local checkpointing actions at all processes. Since perfectly synchronized clocks are not available, the following approaches are used to guarantee checkpoint consistency: either the sending of messages is blocked for the duration of the protocol, or checkpoint indices are piggybacked to avoid blocking.

Blocking coordinated checkpointing

A straightforward approach to coordinated checkpointing is to block communications while the checkpointing protocol executes. After a process takes a local checkpoint, to prevent orphan messages, it remains blocked until the entire checkpointing activity is complete. The coordinator takes a checkpoint and broadcasts a request message to all processes, asking them to take a checkpoint. When a process receives this message, it stops its execution, flushes all the communication channels, takes a *tentative* checkpoint, and sends an acknowledgment message back to the coordinator. After the coordinator receives acknowledgments from all processes, it broadcasts a commit message that completes the two-phase checkpointing protocol. After receiving the commit message, a process removes the old permanent checkpoint and atomically makes the *tentative* checkpoint permanent and then resumes its execution and exchange of messages with other processes. A problem with this approach is that the computation is blocked during the checkpointing and therefore, non-blocking checkpointing schemes are preferable.

Non-blocking checkpoint coordination

In this approach the processes need not stop their execution while taking checkpoints. A fundamental problem in coordinated checkpointing is to prevent a process from receiving application messages that could make the checkpoint inconsistent. Consider the example in Figure 13.6(a) [13]: message m is sent by P_0 *after* receiving a checkpoint request from the checkpoint coordinator. Assume m reaches P_1 *before* the checkpoint request. This situation results in an inconsistent checkpoint since checkpoint $c_{1,x}$ shows the receipt of message m from P_0, while checkpoint $c_{0,x}$ does not show m being sent from P_0.

If channels are FIFO, this problem can be avoided by preceding the first post-checkpoint message on each channel by a checkpoint request, forcing each process to take a checkpoint before receiving the first post-checkpoint message, as illustrated in Figure 13.6(b). An example of a non-blocking checkpoint coordination protocol using this idea is the snapshot algorithm of

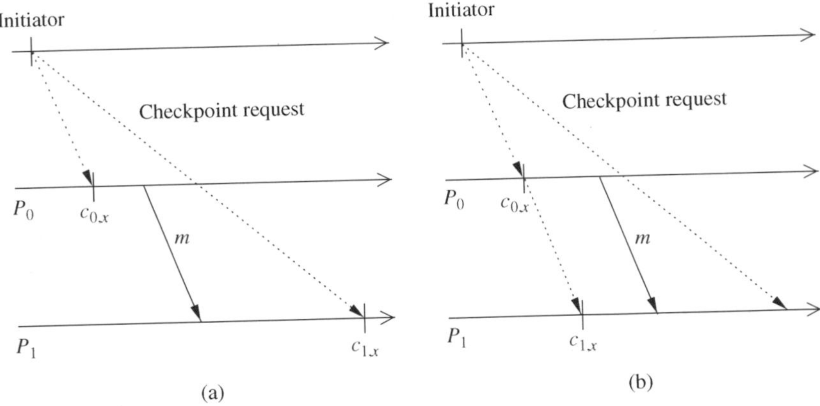

Figure 13.6 Non-blocking coordinated checkpointing: (a) checkpoint inconsistency; (b) a solution with FIFO channels [13].

Chandy and Lamport [9] in which *markers* play the role of the checkpoint-request messages. In this algorithm, the initiator takes a checkpoint and sends a marker (a checkpoint request) on all outgoing channels. Each process takes a checkpoint upon receiving the first marker and sends the marker on all outgoing channels before sending any application message. The protocol works assuming the channels are reliable and FIFO.

If the channels are non-FIFO, the following two approaches can be used: first, the marker can be piggybacked on every post-checkpoint message. When a process receives an application message with a marker, it treats it as if it has received a marker message, followed by the application message. Alternatively, checkpoint indices can serve the same role as markers, where a checkpoint is triggered when the receiver's local checkpoint index is lower than the piggybacked checkpoint index.

Coordinated checkpointing requires all processes to participate in every checkpoint. This requirement generates valid concerns about its scalability. It is desirable to reduce the number of processes involved in a coordinated checkpointing session. This can be done since only those processes that have communicated with the checkpoint initiator either directly or indirectly since the last checkpoint need to take new checkpoints. A two-phase protocol by Koo and Toueg [22] achieves minimal checkpoint coordination.

13.4.3 Impossibility of min-process non-blocking checkpointing

A min-process, non-blocking checkpointing algorithm is one that forces only a minimum number of processes to take a new checkpoint, and at the same time it does not force any process to suspend its computation. Clearly, such checkpointing algorithms will be very attractive. Cao and Singhal [7] showed that it is impossible to design a min-process, non-blocking checkpointing algorithm.

Of course, the following type of min-process checkpointing algorithms are possible. The algorithm consists of two phases. During the first phase, the checkpoint initiator identifies all processes with which it has communicated since the last checkpoint and sends them a request. Upon receiving the request, each process in turn identifies all processes it has communicated with since the last checkpoint and sends them a request, and so on, until no more processes can be identified. During the second phase, all processes identified in the first phase take a checkpoint. The result is a consistent checkpoint that involves only the participating processes. In this protocol, after a process takes a checkpoint, it cannot send any message until the second phase terminates successfully, although receiving a message after the checkpoint has been taken is allowable.

Based on a concept called "Z-dependency," Cao and Singhal proved that there does not exist a non-blocking algorithm that will allow a minimum number of processes to take their checkpoints. Here we give only a sketch of the proof and readers are referred to the original source [7] for a detailed proof.

Z-dependency is defined as follows: if a process P_p sends a message to process P_q during its ith checkpoint interval and process P_q receives the message during its jth checkpoint interval, then P_q Z-depends on P_p during P_p's ith checkpoint interval and P_q's jth checkpoint interval, denoted by $P_p \rightarrow^i{}_j P_q$. If $P_p \rightarrow^i{}_j P_q$ and $P_q \rightarrow {}^j{}_k P_r$, then P_r transitively Z-depends depends on P_p during P_r's kth checkpoint interval and P_p's ith checkpoint interval, and this is denoted as $P_p \ {}^*\!\!\rightarrow^i{}_k P_r$.

A min process algorithm is one that satisfies the following condition: when a process P_p initiates a new checkpoint and takes checkpoint $C_{p,i}$, a process P_q takes a checkpoint $C_{q,j}$associated with $C_{p,i}$ if and only if $P_q \ {}^*\!\!\rightarrow^{j-1}{}_{i-1} P_p$. In a min-process non-blocking algorithm, process P_p initiates a new checkpoint and takes a checkpoint $C_{p,i}$ and if a process P_r sends a message m to P_q after it takes a new checkpoint associated with $C_{p,i}$, then P_q takes a checkpoint $C_{q,i}$ before processing m if and only if $P_q{}^*\!\!\rightarrow {}^{j-1}{}_{i-1} P_p$. According to the min-process definition, P_q takes checkpoint $C_{q,j}$ if and only if $P_q{}^*\!\!\rightarrow^{j-1}{}_{i-1} P_p$, but P_q should take $C_{q,i}$ before processing m. If it takes $C_{q,j}$ after processing m, m becomes an orphan. Therefore, when a process receives a message m, it must know if the initiator of a new checkpoint transitively Z-depends on it during the previous checkpoint interval. But it has been proved that there is not enough information at the receiver of a message to decide whether the initiator of a new checkpoint transitively Z-depends on the receiver. Therefore, no min-process, non-blocking algorithm exists.

13.4.4 Communication-induced checkpointing

Communication-induced checkpointing is another way to avoid the domino effect, while allowing processes to take some of their checkpoints independently. Processes may be forced to take additional checkpoints (over and above their autonomous checkpoints), and thus process independence is constrained to guarantee the eventual progress of the recovery line. Communication-induced checkpointing reduces or completely eliminates the useless checkpoints. In communication-induced checkpointing, processes take two types of checkpoints, namely, autonomous and forced checkpoints. The checkpoints that a process takes independently are called *local* checkpoints, while those that a process is forced to take are called *forced* checkpoints. Communication-induced checkpointing piggybacks protocol-related information on each application message. The receiver of each application message uses the piggybacked information to determine if it has to take a forced checkpoint to advance the global recovery line. The forced checkpoint must be taken before the application may process the contents of the message, possibly incurring some latency and overhead. It is therefore desirable in these systems to minimize the number of forced checkpoints. In contrast with coordinated checkpointing, no special coordination messages are exchanged.

There are two types of communication-induced checkpointing [13]: model-based checkpointing and index-based checkpointing. In *model-based checkpointing*, the system maintains checkpoints and communication structures that prevent the domino effect or achieve some even stronger properties. In *index-based checkpointing*, the system uses an indexing scheme for the local and forced checkpoints, such that the checkpoints of the same index at all processes form a consistent state.

Model-based checkpointing

Model-based checkpointing prevents patterns of communications and checkpoints that could result in inconsistent states among the existing checkpoints. A process detects the potential for inconsistent checkpoints and independently forces local checkpoints to prevent the formation of undesirable patterns. A forced checkpoint is generally used to prevent the undesirable patterns from occurring. No control messages are exchanged among the processes during normal operation. All information necessary to execute the protocol is piggybacked on application messages. The decision to take a forced checkpoint is done locally using the information available.

There are several domino-effect-free checkpoint and communication models. The *MRS* (mark, send, and receive) model of Russell [34] avoids the domino effect by ensuring that within every checkpoint interval all message-receiving events precede all message-sending events. This model can be maintained by taking an additional checkpoint before every message-receiving event that is not separated from its previous message-sending event by a checkpoint. Another way to prevent the domino effect by avoiding rollback propagation completely is by taking a checkpoint immediately after every message-sending event. Recent work has focused on ensuring that every checkpoint can belong to a consistent global checkpoint and therefore is not useless.

Index-based checkpointing

Index-based communication-induced checkpointing assigns monotonically increasing indexes to checkpoints, such that the checkpoints having the same index at different processes form a consistent state. Inconsistency between checkpoints of the same index can be avoided in a lazy fashion if indexes are piggybacked on application messages to help receivers decide when they should take a forced a checkpoint. For instance, the protocol by Briatico *et al.* [5] forces a process to take a checkpoint upon receiving a message with a piggybacked index greater than the local index. More sophisticated protocols piggyback more information on application messages to minimize the number of forced checkpoints.

13.5 Log-based rollback recovery

A log-based rollback recovery makes use of deterministic and non-deterministic events in a computation. So first we discuss these events.

13.5.1 Deterministic and non-deterministic events

Log-based rollback recovery exploits the fact that a process execution can be modeled as a sequence of deterministic state intervals, each starting with the execution of a non-deterministic event. A non-deterministic event can be the receipt of a message from another process or an event internal to the process. Note that a message send event is *not* a non-deterministic event. For example, in Figure 13.7, the execution of process P_0 is a sequence of four deterministic intervals. The first one starts with the creation of the process, while the remaining three start with the receipt of messages m_0, m_3, and m_7, respectively. Send event of message m_2 is uniquely determined by the initial state of P_0 and by the receipt of message m_0, and is therefore not a non-deterministic event.

Log-based rollback recovery assumes that all non-deterministic events can be identified and their corresponding determinants can be logged into the stable storage. During failure-free operation, each process logs the determinants

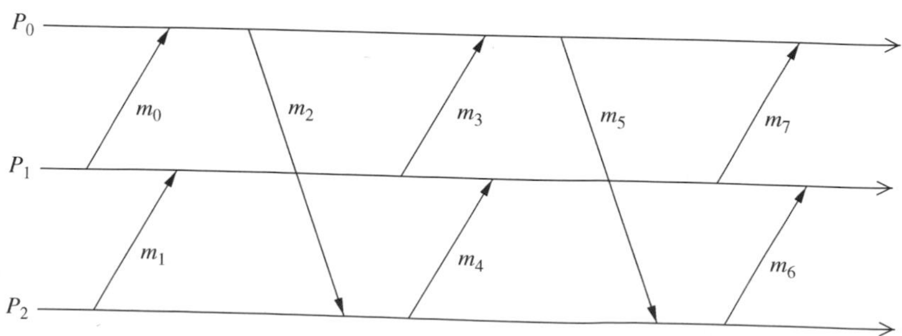

Figure 13.7 Deterministic and non-deterministic events.

of all non-deterministic events that it observes onto the stable storage. Additionally, each process also takes checkpoints to reduce the extent of rollback during recovery. After a failure occurs, the failed processes recover by using the checkpoints and logged determinants to replay the corresponding non-deterministic events precisely as they occurred during the pre-failure execution. Because execution within each deterministic interval depends only on the sequence of non-deterministic events that preceded the interval's beginning, the pre-failure execution of a failed process can be reconstructed during recovery up to the first non-deterministic event whose determinant is not logged.

The no-orphans consistency condition

Let e be a non-deterministic event that occurs at process p. We define the following [13]:

- *Depend*(e): the set of processes that are affected by a non-deterministic event e. This set consists of p, and any process whose state depends on the event e according to Lamport's *happened before* relation [23].
- *Log*(e): the set of processes that have logged a copy of e's determinant in their volatile memory.
- *Stable*(e): a predicate that is true if e's determinant is logged on the stable storage.

Suppose a set of processes Ψ crashes. A process p in Ψ becomes an orphan when p itself does not fail and p's state depends on the execution of a non-deterministic event e whose determinant cannot be recovered from the stable storage or from the volatile memory of a surviving process. Formally, it can be stated as follows [13]:

$$\forall(e) : \neg Stable(e) \Longrightarrow Depend(e) \subseteq Log(e).$$

This property is called the *always-no-orphans* condition [13]. It states that if any surviving process depends on an event e, then either event e is logged on the stable storage, or the process has a copy of the determinant of event e. If neither condition is true, then the process is an orphan because it depends on an event e that cannot be generated during recovery since its determinant is lost.

Log-based rollback-recovery protocols guarantee that upon recovery of all failed processes, the system does not contain any orphan process, i.e., a process whose state depends on a non-deterministic event that cannot be reproduced during recovery. Log-based rollback-recovery protocols are of three types: pessimistic logging, optimistic logging, and causal logging protocols. They differ in their failure-free performance overhead, latency of output commit, simplicity of recovery and garbage collection, and the potential for rolling back surviving processes.

13.5.2 Pessimistic logging

Pessimistic logging protocols assume that a failure can occur after any non-deterministic event in the computation. This assumption is "pessimistic" since in reality failures are rare. In their most straightforward form, pessimistic protocols log to the stable storage the determinant of each non-deterministic event before the event affects the computation. Pessimistic protocols implement the following property, often referred to as *synchronous logging*, which is stronger than the always-no-orphans condition [13]:

$$\forall e : \neg Stable(e) \Longrightarrow |Depend(e)| = 0.$$

That is, if an event has not been logged on the stable storage, then no process can depend on it. In addition to logging determinants, processes also take periodic checkpoints to minimize the amount of work that has to be repeated during recovery. When a process fails, the process is restarted from the most recent checkpoint and the logged determinants are used to recreate the pre-failure execution. Consider the example in Figure 13.8. During failure-free operation the logs of processes P_0, P_1, and P_2 contain the determinants needed to replay messages m_0, m_4, m_7, m_1, m_3, m_6, and m_2, m_5, respectively. Suppose processes P_1 and P_2 fail as shown, restart from checkpoints B and C, and roll forward using their determinant logs to deliver again the same sequence of messages as in the pre-failure execution. This guarantees that P_1 and P_2 will repeat exactly their pre-failure execution and re-send the same messages. Hence, once the recovery is complete, both processes will be consistent with the state of P_0 that includes the receipt of message m_7 from P_1. In a pessimistic logging system, the observable state of each process is always recoverable.

The price paid for these advantages is a performance penalty incurred by synchronous logging. Synchronous logging can potentially result in a high performance overhead. Implementations of pessimistic logging must use special techniques to reduce the effects of synchronous logging on the performance. This overhead can be lowered using special hardware. For

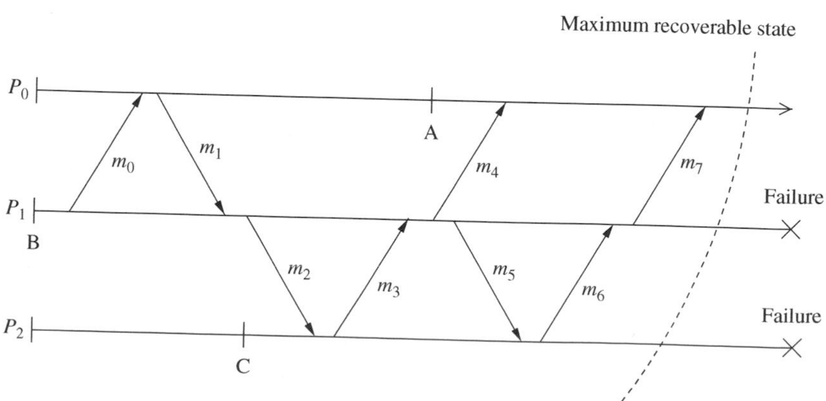

Figure 13.8 Pessimistic logging [13].

example, fast non-volatile semiconductor memory can be used to implement the stable storage. Another approach is to limit the number of failures that can be tolerated. The overhead of pessimistic logging is reduced by delivering a message or executing an event and deferring its logging until the process communicates with another process or with the outside world.

Synchronous logging in such an implementation is orders of magnitude cheaper than with a traditional implementation of stable storage that uses magnetic disk devices. Another form of hardware support uses a special bus to guarantee atomic logging of all messages exchanged in the system. Such hardware support ensures that the log of one machine is automatically stored on a designated backup without blocking the execution of the application program. This scheme, however, requires that all non-deterministic events be converted into *external* messages.

Some pessimistic logging systems reduce the overhead of synchronous logging without relying on hardware. For example, the *sender-based message logging* (SBML) protocol keeps the determinants corresponding to the delivery of each message m in the volatile memory of its sender. The determinant of m, which consists of its content and the order in which it was delivered, is logged in two steps. First, before sending m, the sender logs its content in volatile memory. Then, when the receiver of m responds with an acknowledgment that includes the order in which the message was delivered, the sender adds to the determinant the ordering information. SBML avoids the overhead of accessing stable storage but tolerates only one failure and cannot handle non-deterministic events internal to a process. Extensions to this technique can tolerate more than one failure in special network topologies.

13.5.3 Optimistic logging

In optimistic logging protocols, processes log determinants *asynchronously* to the stable storage [13]. These protocols optimistically assume that logging will be complete before a failure occurs. Determinants are kept in a volatile log, and are periodically flushed to the stable storage. Thus, optimistic logging does not require the application to block waiting for the determinants to be written to the stable storage, and therefore incurs much less overhead during failure-free execution. However, the price paid is more complicated recovery, garbage collection, and slower output commit. If a process fails, the determinants in its volatile log are lost, and the state intervals that were started by the non-deterministic events corresponding to these determinants cannot be recovered. Furthermore, if the failed process sent a message during any of the state intervals that cannot be recovered, the receiver of the message becomes an orphan process and must roll back to undo the effects of receiving the message.

Optimistic logging protocols do not implement the *always-no-orphans* condition. The protocols allow the temporary creation of orphan processes which

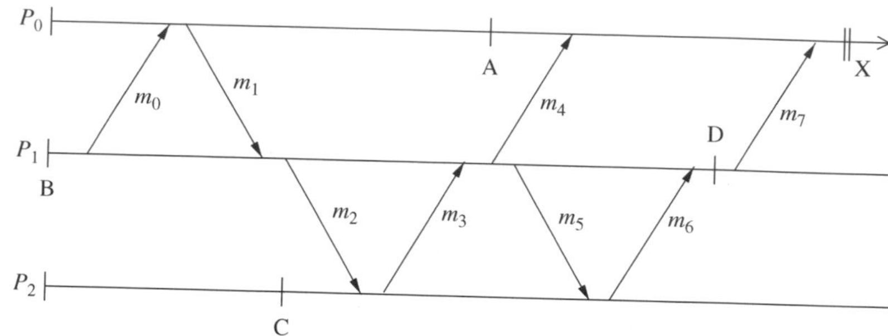

Figure 13.9 Optimistic logging [13].

are eventually eliminated. The *always-no-orphans* condition holds after the recovery is complete. This is achieved by rolling back orphan processes until their states do not depend on any message whose determinant has been lost.

Consider the example shown in Figure 13.9. Suppose process P_2 fails before the determinant for m_5 is logged to the stable storage. Process P_1 then becomes an orphan process and must roll back to undo the effects of receiving the orphan message m_6. The rollback of P_1 further forces P_0 to roll back to undo the effects of receiving message m_7.

To perform rollbacks correctly, optimistic logging protocols track causal dependencies during failure free execution. Upon a failure, the dependency information is used to calculate and recover the latest global state of the pre-failure execution in which no process is in an orphan. Optimistic logging protocols require a non-trivial garbage collection scheme. Also note that pessimistic protocols need only keep the most recent checkpoint of each process, whereas optimistic protocols may need to keep multiple checkpoints for each process.

Since determinants are logged asynchronously, output commit in optimistic logging protocols requires a guarantee that no failure scenario can revoke the output. For example, if process P_0 needs to commit output at state X, it must log messages m_4 and m_7 to the stable storage and ask P_2 to log m_2 and m_5. In this case, if any process fails, the computation can be reconstructed up to state X.

13.5.4 Causal logging

Causal logging combines the advantages of both pessimistic and optimistic logging at the expense of a more complex recovery protocol [13]. Like optimistic logging, it does not require synchronous access to the stable storage except during output commit. Like pessimistic logging, it allows each process to commit output independently and never creates orphans, thus isolating processes from the effects of failures at other processes. Moreover, causal logging limits the rollback of any failed process to the most recent checkpoint on the stable storage, thus minimizing the storage overhead and the amount of lost work.

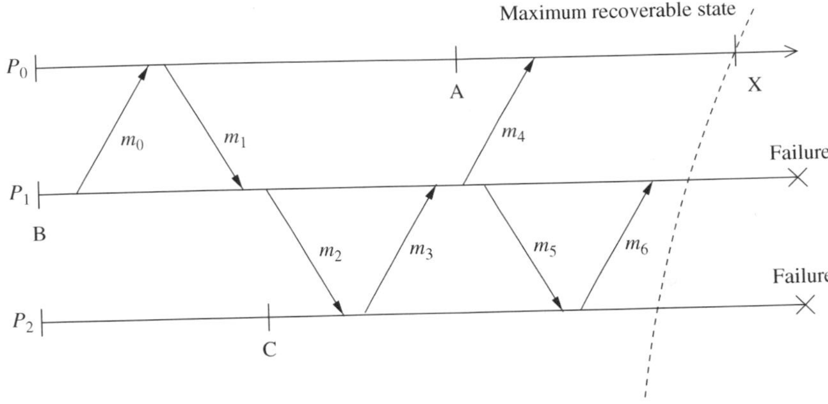

Figure 13.10 Causal logging [13].

Causal logging protocols make sure that the *always-no-orphans property* holds by ensuring that the determinant of each non-deterministic event that causally precedes the state of a process is either stable or it is available locally to that process. Consider the example in Figure 13.10. Messages m_5 and m_6 are likely to be lost on the failures of P_1 and P_2 at the indicated instants. Process P_0 at state X will have logged the determinants of the non-deterministic events that causally precede its state according to Lamport's *happened-before* relation. These events consist of the delivery of messages m_0, m_1, m_2, m_3, and m_4. The determinant of each of these non-deterministic events is either logged on the stable storage or is available in the volatile log of process P_0. The determinant of each of these events contains the order in which its original receiver delivered the corresponding message. The message sender, as in sender-based message logging, logs the message content. Thus, process P_0 will be able to "guide" the recovery of P_1 and P_2 since it knows the order in which P_1 should replay messages m_1 and m_3 to reach the state from which P_1 sent message m_4. Similarly, P_0 has the order in which P_2 should replay message m_2 to be consistent with both P_0 and P_1. The content of these messages is obtained from the sender log of P_0 or regenerated deterministically during the recovery of P_1 and P_2. Note that information about messages m_5 and m_6 is lost due to failures. These messages may be resent after recovery possibly in a different order. However, since they did not causally affect the surviving process or the outside world, the resulting state is consistent.

Each process maintains information about all the events that have causally affected its state. This information protects it from the failures of other processes and also allows the process to make its state recoverable by simply logging the information available locally. Thus, a process does not need to run a multi-host protocol to commit output. It can commit output independently.

13.6 Koo–Toueg coordinated checkpointing algorithm

Koo and Toueg's [22] coordinated checkpointing and recovery technique takes a consistent set of checkpoints and avoids the domino effect and livelock problems during the recovery. Processes coordinate their local checkpointing actions such that the set of all checkpoints in the system is consistent [9].

13.6.1 The checkpointing algorithm

The checkpoint algorithm makes the following assumptions about the distributed system: processes communicate by exchanging messages through communication channels. Communication channels are FIFO. It is assumed that end-to-end protocols (such as the sliding window protocol) exist to cope with message loss due to rollback recovery and communication failure. Communication failures do not partition the network.

The checkpoint algorithm takes two kinds of checkpoints on the stable storage: permanent and tentative. A permanent checkpoint is a local checkpoint at a process and is a part of a consistent global checkpoint. A tentative checkpoint is a temporary checkpoint that is made a permanent checkpoint on the successful termination of the checkpoint algorithm. In case of a failure, processes roll back only to their permanent checkpoints for recovery.

The checkpointing algorithm assumes that a single process invokes the algorithm at any time to take permanent checkpoints. The algorithm also assumes that no process fails during the execution of the algorithm.

The algorithm consists of two phases.

First phase

An initiating process P_i takes a tentative checkpoint and requests all other processes to take tentative checkpoints. Each process informs P_i whether it succeeded in taking a tentative checkpoint. A process says "no" to a request if it fails to take a tentative checkpoint, which could be due to several reasons, depending upon the underlying application. If P_i learns that all the processes have successfully taken tentative checkpoints, P_i decides that all tentative checkpoints should be made permanent; otherwise, P_i decides that all the tentative checkpoints should be discarded.

Second phase

P_i informs all the processes of the decision it reached at the end of the first phase. A process, on receiving the message from P_i, will act accordingly. Therefore, either all or none of the processes advance the checkpoint by taking permanent checkpoints.

The algorithm requires that after a process has taken a tentative checkpoint, it cannot send messages related to the underlying computation until it is informed of P_i's decision.

Correctness

A set of permanent checkpoints taken by this algorithm is consistent because of the following two reasons: first, either all or none of the processes take permanent checkpoints; second, no process sends a message after taking a tentative checkpoint until the receipt of the initiating process's decision, as by then all processes would have taken checkpoints. Thus, a situation will not arise where there is a record of a message being received but there is no record of sending it. Thus, a set of checkpoints taken will always be consistent.

An optimization

Note that the above protocol may cause a process to take a checkpoint even when it is not necessary for consistency. Since taking a checkpoint is an expensive operation, we would like to avoid taking checkpoints if it is not necessary.

Consider the example shown in Figure 13.11. The set $\{x_1, y_1, z_1\}$ is a consistent set of checkpoints. Suppose process X decides to initiate the checkpointing algorithm after receiving message m. It takes a tentative checkpoint x_2 and sends "take tentative checkpoint" messages to processes Y and Z, causing Y and Z to take checkpoints y_2 and z_2, respectively. Clearly, $\{x_2, y_2, z_2\}$ forms a consistent set of checkpoints. Note, however, that $\{x_2, y_2, z_1\}$ also forms a consistent set of checkpoints. In this example, there is no need for process Z to take checkpoint z_2 because Z has not sent any message since its last checkpoint. However, process Y must take a checkpoint since it has sent messages since its last checkpoint.

13.6.2 The rollback recovery algorithm

The rollback recovery algorithm restores the system state to a consistent state after a failure. The rollback recovery algorithm assumes that a single process

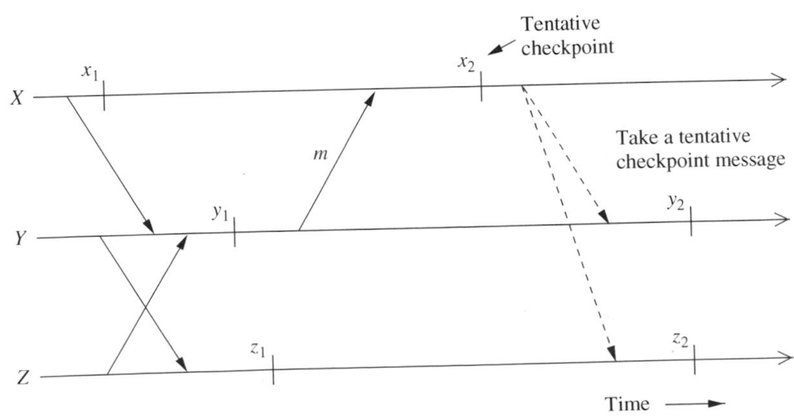

Figure 13.11 Example of checkpoints taken unnecessarily.

invokes the algorithm. It also assumes that the checkpoint and the rollback recovery algorithms are not invoked concurrently. The rollback recovery algorithm has two phases.

First phase
An initiating process P_i sends a message to all other processes to check if they all are willing to restart from their previous checkpoints. A process may reply "no" to a restart request due to any reason (e.g., it is already participating in a checkpoint or recovery process initiated by some other process). If P_i learns that all processes are willing to restart from their previous checkpoints, P_i decides that all processes should roll back to their previous checkpoints. Otherwise, P_i aborts the rollback attempt and it may attempt a recovery at a later time.

Second phase
P_i propagates its decision to all the processes. On receiving P_i's decision, a process acts accordingly.

During the execution of the recovery algorithm, a process cannot send messages related to the underlying computation while it is waiting for P_i's decision.

Correctness

All processes restart from an appropriate state because, if they decide to restart, they resume execution from a consistent state (the checkpointing algorithm takes a consistent set of checkpoints).

An optimization

The above recovery protocol causes all processes to roll back irrespective of whether a process needs to roll back or not. Consider the example shown in Figure 13.12. In the event of failure of process X, the above protocol will require processes X, Y, and Z to restart from checkpoints x_2, y_2, and z_2, respectively. However, note that process Z need not roll back because there has been no interaction between process Z and the other two processes since the last checkpoint at Z.

13.7 Juang–Venkatesan algorithm for asynchronous checkpointing and recovery

We now describe the algorithm of Juang and Venkatesan [18] for recovery in a system that employs asynchronous checkpointing.

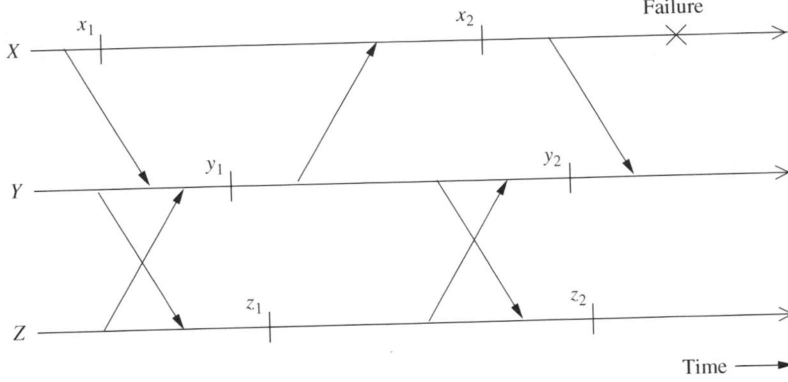

Figure 13.12 Example of an unnecessary rollback.

13.7.1 System model and assumptions

The algorithm makes the following assumptions about the underlying system: the communication channels are reliable, deliver the messages in FIFO order, and have infinite buffers. The message transmission delay is arbitrary, but finite. The processors directly connected to a processor via communication channels are called its neighbors.

The underlying computation or application is assumed to be event-driven: a processor P waits until a message m is received, it processes the message m, changes its state from s to s', and sends zero or more messages to some of its neighbors. Then the processor remains idle until the receipt of the next message. The new state s' and the contents of messages sent to its neighbors depend on state s and the contents of message m. The events at a processor are identified by unique monotonically increasing numbers, e_{x0}, e_{x1}, e_{x2}, ... (see Figure 13.13).

To facilitate recovery after a process failure and restore the system to a consistent state, two types of log storage are maintained, volatile log and stable log. Accessing the volatile log takes less time than accessing the stable

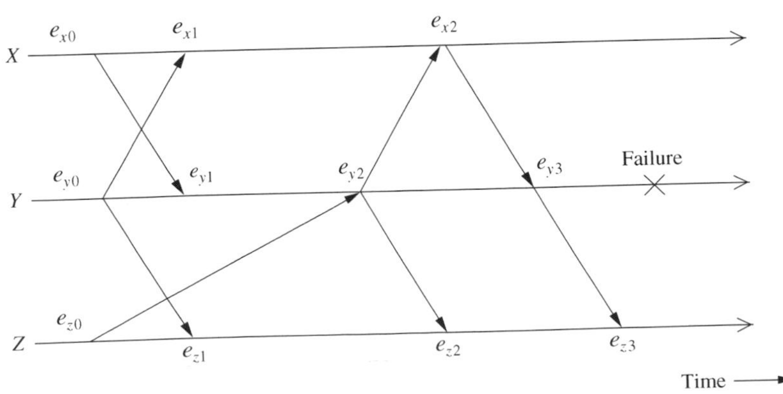

Figure 13.13 An event-driven computation.

log, but the contents of the volatile log are lost if the corresponding processor fails. The contents of the volatile log are periodically flushed to the stable storage.

13.7.2 Asynchronous checkpointing

After executing an event, a processor records a triplet $\{s, m, msgs\_sent\}$ in its volatile storage, where s is the state of the processor before the event, m is the message (including the identity of the sender of m, denoted as $m.sender$) whose arrival caused the event, and $msgs\_sent$ is the set of messages that were sent by the processor during the event. Therefore, a local checkpoint at a processor consists of the record of an event occurring at the processor and it is taken without any synchronization with other processors. Periodically, a processor independently saves the contents of the volatile log in the stable storage and clears the volatile log. This operation is equivalent to taking a local checkpoint.

13.7.3 The recovery algorithm

Notation and data structure

The following notation and data structure are used by the algorithm:

- $RCVD_{i \leftarrow j}(CkPt_i)$ represents the number of messages received by processor p_i from processor p_j, from the beginning of the computation until the checkpoint $CkPt_i$.
- $SENT_{i \rightarrow j}(CkPt_i)$ represents the number of messages sent by processor p_i to processor p_j, from the beginning of the computation until the checkpoint $CkPt_i$.

Basic idea

Since the algorithm is based on asynchronous checkpointing, the main issue in the recovery is to find a consistent set of checkpoints to which the system can be restored. The recovery algorithm achieves this by making each processor keep track of both the number of messages it has sent to other processors as well as the number of messages it has received from other processors. Recovery may involve several iterations of roll backs by processors. Whenever a processor rolls back, it is necessary for all other processors to find out if any message sent by the rolled back processor has become an orphan message. Orphan messages are discovered by comparing the number of messages sent to and received from neighboring processors. For example, if $RCVD_{i \leftarrow j}(CkPt_i) > SENT_{j \rightarrow i}(CkPt_j)$ (that is, the number of messages received by processor p_i from processor p_j is greater than the number of messages sent by processor p_j to processor p_i, according to the current states of the processors), then one or more messages at processor p_j are orphan messages. In this case, processor

p_j must roll back to a state where the number of messages received agrees with the number of messages sent.

Consider an example shown in Figure 13.13. Suppose processor Y crashes at the point indicated and rolls back to a state corresponding to checkpoint e_{y1}. According to this state, Y has sent only one message to X; however, according to X's current state (e_{x2}), X has received two messages from Y. Therefore, X must roll back to a state preceding e_{x2} to be consistent with Y's state. We note that if X rolls back to checkpoint e_{x1}, then it will be consistent with Y's state, e_{y1}. Likewise, processor Z must roll back to checkpoint e_{z1} to be consistent with Y's state, e_{y1}. Note that similarly processors X and Z will have to resolve any such mutual inconsistencies (provided they are neighbors).

Description of the algorithm

When a processor restarts after a failure, it broadcasts a *ROLLBACK* message that it has failed.[1] The recovery algorithm at a processor is initiated when it restarts after a failure or when it learns of a failure at another processor. Because of the broadcast of *ROLLBACK* messages, the recovery algorithm is initiated at all processors. The algorithm is shown in Algorithm 13.1.

The rollback starts at the failed processor and slowly diffuses into the entire system through *ROLLBACK* messages. Note that the procedure has N iterations. During the kth iteration ($k \neq 1$), a processor p_i does the following: (i) based on the state $CkPt_i$ it was rolled back to in the $(k-1)$th iteration, it computes $SENT_{i \to j}(CkPt_i)$ for each neighbor p_j and sends this value in a *ROLLBACK* message to that neighbor; and (ii) p_i waits for and processes *ROLLBACK* messages that it receives from its neighbors in the kth iteration and determines a new recovery point $CkPt_i$ for p_i based on information in these messages. At the end of each iteration, at least one processor will rollback to its final recovery point, unless the current recovery points are already consistent.

Example Consider an example shown in Figure 13.14 consisting of three processors. Suppose processor Y fails and restarts. If event e_{y2} is the latest checkpointed event at Y, then Y will restart from the state corresponding to e_{y2}. Because of the broadcast nature of *ROLLBACK* messages, the recovery algorithm is also initiated at processors X and Z. Initially, X, Y, and Z set $CkPt_X \leftarrow e_{x3}$, $CkPt_Y \leftarrow e_{y2}$ and $CkPt_Z \leftarrow e_{z2}$, respectively, and X, Y, and Z send the following messages during the first iteration: Y sends $ROLLBACK(Y, 2)$ to X and $ROLLBACK(Y, 1)$ to Z; X sends $ROLLBACK(X, 2)$ to Y and $ROLLBACK(X, 0)$ to Z; and Z sends $ROLLBACK(Z, 0)$ to X and $ROLLBACK(Z, 1)$ to Y.

[1] Such a broadcast can be done using only $O(|E|)$ messages where $|E|$ is the total number of communication links.

Procedure RollBack_Recovery: processor p_i executes the following:

STEP (a)

if processor p_i is recovering after a failure **then**

$CkPt_i$:= latest event logged in the stable storage

else

$CkPt_i$:= latest event that took place in p_i {The latest event at p_i can be either in stable or in volatile storage.}

end if

STEP (b)

for $k = 1$ to N {N is the number of processors in the system} **do**

for each neighboring processor p_j **do**

compute $SENT_{i\rightarrow j}(CkPt_i)$

send a $ROLLBACK(i, SENT_{i\rightarrow j}(CkPt_i))$ message to p_j

end for

for every $ROLLBACK(j, c)$ message received from a neighbor j **do**

if $RCVD_{i\leftarrow j}(CkPt_i) > c$ {Implies the presence of orphan messages} **then**

find the latest event e such that $RCVD_{i\leftarrow j}(e) = c$ {Such an event e may be in the volatile storage or stable storage.}

$CkPt_i := e$

end if

end for

end for{for k}

Algorithm 13.1 Juang–Venkatesan algorithm

Since $RCVD_{X\leftarrow Y}(CkPt_X) = 3 > 2$ (2 is the value received in the $ROLLBACK(Y, 2)$ message from Y), X will set $CkPt_X$ to e_{x2} satisfying $RCVD_{X\leftarrow Y}(e_{x2}) = 2 \leq 2$. Since $RCVD_{Z\leftarrow Y}(CkPt_Z) = 2 > 1$, Z will set $CkPt_Z$ to e_{z1} satisfying $RCVD_{Z\leftarrow Y}(e_{z1}) = 1 \leq 1$. At Y, $RCVD_{Y\leftarrow X}(CkPt_Y) = 1 < 2$

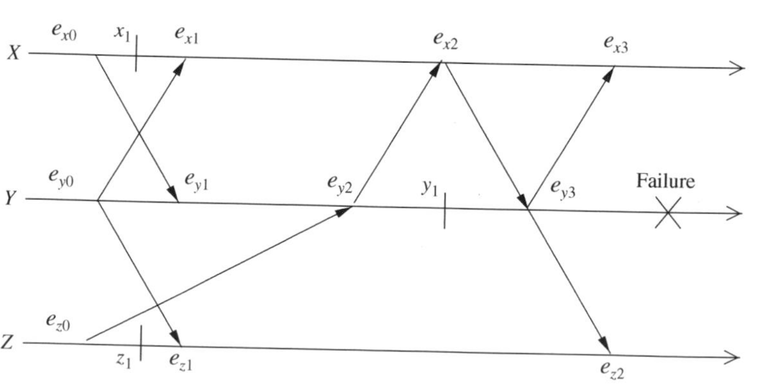

Figure 13.14 An example execution of the Juang–Venkatesan algorithm.

and $RCVD_{Y\leftarrow Z}(CkPt_Y) = 1 = SENT_{Z\leftarrow Y}(CkPt_Z)$. Hence, Y need not roll back further. In the second iteration, Y sends $ROLLBACK(Y, 2)$ to X and $ROLLBACK(Y, 1)$ to Z; Z sends $ROLLBACK(Z, 1)$ to Y and $ROLLBACK(Z, 0)$ to X; X sends $ROLLBACK(X, 0)$ to Z and $ROLLBACK(X, 1)$ to Y. Note that if Y rolls back beyond e_{y3} and loses the message from X that caused e_{y3}, X can resend this message to Y because e_{x2} is logged at X and this message is available in the log. The second and third iteration will progress in the same manner. Note that the set of recovery points chosen at the end of the first iteration, $\{e_{x2}, e_{y2}, e_{z1}\}$, is consistent, and no further rollback occurs.

13.8 Manivannan–Singhal quasi-synchronous checkpointing algorithm

When processes independently take their local checkpoints, there is a possiblity that some local checkpoints can never be included in any consistent global checkpoint. (Recall that such local checkpoints are called the useless checkpoints.) In the worst case, no consistent checkpoint can ever be formed.

The Manivannan–Singhal quasi-synchronous checkpointing algorithm [25] improves the performance by eliminating useless checkpoints. The algorithm is based on communication-induced checkpointing, where each process takes basic checkpoints asynchronously and independently, and in addition, to prevent useless checkpoints, processes take forced checkpoints upon the reception of messages with a control variable.

The Manivannan–Singhal quasi-synchronous checkpointing algorithm combines coordinated and uncoordinated checkpointing approaches to get the best of both:

- It allows processes to take checkpoints asynchronously.
- It uses communication-induced checkpointing to eliminates the "useless" checkpoints.
- Since every checkpoint lies on a consistent checkpoint, determination of the recovery line during a rollback recovery is simple and fast.

Each checkpoint at a process is assigned a unique sequence number. The sequence numbers assigned to basic checkpoints are picked from the local counters, which are incremented periodically.

When a process P_i sends a message, it appends the sequence number of its latest checkpoint to the message. When a process P_j receives a message, if the sequence number received in the message is greater than the sequence number of the latest checkpoint of P_j, then, before processing the message, P_j takes a (forced) checkpoint and assigns the sequence number received in the message as the sequence number of the checkpoint taken. When it is time for a process to take a basic checkpoint, it skips taking a basic checkpoint if its latest checkpoint has a sequence number greater than or equal to the current value of its counter. This strategy helps to reduce the checkpointing overhead, i.e., the number of checkpoints taken. An alternative approach to reduce the number of checkpoints is to allow a process to delay processing a received message until the sequence number of its latest checkpoint is greater than or equal to the sequence number received in the message.

13.8.1 Checkpointing algorithm

Now, we present the quasi-synchronous checkpointing algorithm formally (Algorithm 13.2). The variable $next_i$ of process P_i represents its local counter. It keeps track of the current number of checkpoint intervals at process P_i. The value of the variable sn_i represents the sequence number of the latest checkpoint of P_i at any time. So, whenever a new checkpoint is taken, the checkpoint is assigned a sequence number and sn_i is updated accordingly. $C.sn$ denotes the sequence number assigned to checkpoint C and $M.sn$ denotes the sequence number piggybacked to message M.

Properties

When processes take checkpoints in this manner, checkpoints satisfy the following interesting properties ($C_{i,k}$ denotes a checkpoint with sequence number k at process P_i):

1. Checkpoint $C_{i,m}$ of process P_i is concurrent with checkpoints $C_{*,m}$ of all other processes. For example, in Figure 13.15, checkpoint $C_{2,3}$ is concurrent with checkpoints $C_{1,3}$ and $C_{3,3}$.
2. Checkpoints $C_{*,m}$ of all processes form a consistent global checkpoint. For example, in Figure 13.15, checkpoints $\{C_{1,4}, C_{2,4}, C_{3,4}\}$ form a consistent global checkpoint. An interesting application of this result is that if process P_3 crashes and restarts from checkpoint $C_{3,5}$ (in Figure 13.15), then P_1 will need to take a checkpoint $C_{1,5}$ (without rolling back) and the set of checkpoints $\{C_{1,5}, C_{2,5}, C_{3,5}\}$ will form a consistent global checkpoint. Since there may be gaps in the sequence numbers assigned to checkpoints at a process, we have the following result:
3. The checkpoint $C_{i,m}$ of process P_i is concurrent with the earliest checkpoint $C_{j,n}$ at process P_j such that $m \leq n$. For example, in Figure 13.15, checkpoints $\{C_{1,3}, C_{2,2}, C_{3,2}\}$ form a consistent global checkpoint.

The following corollary gives a sufficient condition for a set of local checkpoints to be a part of a global checkpoint.

Corollary 13.1 *Let* $S = \{C_{i_1,m_{i_1}}, C_{i_2,m_{i_2}}, \ldots, C_{i_k,m_{i_k}}\}$ *be a set of local checkpoints from distinct processes. Let* $m = min\{m_{i_1}, m_{i_2}, \ldots, m_{i_k}\}$. *Then, S can be extended to a global checkpoint if* $\forall\ l\ (1 \leq l \leq k)$, $C_{i_l,m_{i_l}}$ *is the earliest checkpoint of* P_{i_l} *such that* $m_{i_l} \geq m$.

The following corollary gives a sufficient condition for a global checkpoint to be consistent.

Data Structures at Process P_i:

$sn_i := 0;$ {Sequence number of the current checkpoint, initialized to 0. This is updated every time a new checkpoint is taken.}

$next_i := 1;$ {Sequence number to be assigned to the next basic checkpoint; initialized to 1.}

When it is time for process P_i to increment $next_i$:

$next_i := next_i + 1;$ {$next_i$ is incremented at periodic time intervals of X time units}

When process P_i sends a message M:

$M.sn := sn_i;$ {sequence number of the current checkpoint is appended to M}

send (M);

Process P_j receives a message from process P_i:

```
if sn_j < M.sn then               {if sequence number of the current checkpoint
      Take checkpoint C;           is less than checkpoint number received in the
      C.sn := M.sn;                message, then take a new checkpoint before
      sn_j := M.sn;                processing the message}
Process the message.
```

When it is time for process P_i to take a basic checkpoint:

```
if next_i > sn_i then             {skips taking a basic checkpoint if next_i ≤ sn_i
      Take checkpoint C;           (i.e., if it already took a forced checkpoint
      sn_i := next_i;              with sequence number ≥ next_i)}
      C.sn := sn_i;
```

Algorithm 13.2 Manivannan–Singhal quasi-synchronous checkpointing algorithm [25].

Corollary 13.2 *Let $S = \{C_{1,m_1}, C_{2,m_2}, \ldots, C_{N,m_N}\}$ be a set of local checkpoints one for each process. Let $m = min\{m_1, m_2, \ldots, m_N\}$. Then, S is a global checkpoint if $\forall\ i\ (1 \le i \le N)$, C_{i,m_i} is the earliest checkpoint of P_i such that $m_i \ge m$.*

These properties have a strong implication on the failure recovery. The task of finding a consistent global checkpoint after a failure is considerably simplified. If the failed process rolls back to a checkpoint with sequence number m, then all other processes simply need to roll back to the earliest local checkpoint $C_{*,n}$ such that $m \le n$.

Example We illustrate the basic idea behind the checkpoints algorithm using an example.

Consider a system consisting of three processes P_1, P_2, and P_3 shown in Figure 13.15. The basic checkpoints are shown in the figure as "|" and forced checkpoints are shown as "|*". The sequence numbers assigned to checkpoints are also shown in the figure. Each process P_i increments its variable $next_i$ every x time units. Process P_3 takes a basic checkpoint every x time units, P_2 takes a basic checkpoint every $2x$ time units, and P_1 takes a basic checkpoint every $3x$ time units. Message M_0 forces P_3 to take a forced checkpoint with sequence number 2 before processing message M_0. As a result P_3 skips taking a basic checkpoint with sequence number 2. Message M_1 forces process P_2 to take a forced checkpoint with sequence number 3 before processing M_1 because $M_{1.sn} > sn_2$ while receiving the message. Similarly message M_2 forces P_1 to take a checkpoint before processing the message and M_4 forces P_2 to take a checkpoint before processing the message. However, M_3 does not force process P_3 to take a checkpoint before processing it. Note that there may be gaps in the sequence numbers assigned to checkpoints at a process.

13.8.2 Recovery algorithm

The recovery process is asynchronous; that is, when a process fails, it just rolls back to its latest checkpoint and broadcasts a rollback request message to every other process and continues its processing without waiting for any reply message from them. The recovery is based on the assumption that if a process P_i fails, then no other process fails until the system is restored to a consistent state. In addition to the variables defined in the checkpoint algorithm, the processes also maintains two other variables: inc_i and rec_line_i. The inc_i is the incarnation number for process P_i. It is incremented every time a process fails and restarts from its latest checkpoint. The rec_line_i is the recovery line number. These variables are stored in the stable storage, so that they are made available for recovery. Initially, $\forall i$, $inc_i = 0$ and $rec\_line_i = 0$. With each message M, the current values of the three variables inc_i, sn_i, and rec_line_i are piggybacked. The values of these variables piggybacked to M are denoted by $M._{inc}$, $M._{sn}$, and $M._{rec\_line}$, respectively. $C._{sn}$ denotes the sequence number of checkpoint C. We present the basic recovery algorithm formally in Algorithm 13.3.

An explanation

When process P_i fails, it rolls back to its latest checkpoint and broadcasts a $rollback(inc_i, rec\_line_i)$ message to all other processes and continues its normal execution. Upon receiving this rollback message, a process P_j rolls back to its earliest checkpoint whose sequence number $\geq rec\_line_i$, and continues normal execution. If process P_j does not have such a checkpoint, it takes a checkpoint with the sequence number equal to rec_line_i, and continues

---

Data structures at process P_i:
integer $sn_i = 0$;
integer $next_i = 1$;
integer $inc_i = 0$;
integer $rec\_line_i = 0$;

Checkpointing algorithm:
When it is time for process P_i to increment $next_i$
    $next_i := next_i + 1$;

When it is time for process P_i to take a basic checkpoint
  If $(next_i > sn_i)$ {
    Take checkpoint C;
    $C._{sn} := next_i$;
    $sn_i := C._{sn}$;
  }

When process P_i sends a message M:
  $M_{.sn} := sn_i$;
  $M._{rec\_line} := rec\_line_i$;
  $M._{inc} := inc_i$;
  $send(M)$;

When process P_j receives a message M:
  If $(M._{inc} > inc_j)$ {
    $rec\_line_j := M._{rec\_line}$;
    $inc_j := M._{inc}$;
    $Roll\_Back(P_j)$;
  }
  If $(M._{sn} > sn_j)$ {
    Take checkpoint C;
    $C._{sn} := M._{sn}$;
    $sn_j := C._{sn}$;
  }
  Process the message;

Basic recovery algorithm (BRA):
Recovery initiated by process P_i after failure:
  Restore the latest checkpoint;
  $inc_i := inc_i + 1$;
  $rec\_line_i := sn_i$;
  send $rollback(inc_i, rec\_line_i)$ to all other processes;
  resume normal execution;

Process P_j upon receiving $Roll\_Back(inc_i, rec\_line_i)$ from P_i:
  If $(inc_i > inc_j)$ {

```
            inc_j := inc_i;
            rec_lin_j := rec_line_i;
            Roll_Back(P_j);
            continue as normal;
    }
    else
            Ignore the rollback message;

Procedure Roll_Back(P_j):
    If (rec_line_j > sn_j) {
            Take checkpoint C;
            C.sn := rec_line_j;
            sn_j := C.sn;
    }
    else
    {
            Find the earliest checkpoint C with C.sn ≥ rec_line_j;
            sn_j := C.sn;
            Restore checkpoint C;
            Delete all checkpoints after C;
    }
```

Algorithm 13.3 Manivannan–Singhal quasi-synchronous recovery algorithm [25].

normally. Due to message delays, the broadcast message might be delayed and a process P_j may come to know about a rollback indirectly through some other process that has already seen the rollback message. Since every message is piggybacked with $M._{inc}$, $M._{sn}$, and $M._{rec\_line}$, the indirect application message that P_j receives indicates a rollback incarnation by some other process. If process P_j receives such a message M, and $M._{inc} > inc_j$, then P_j infers that some failed process had initiated a rollback with incarnation number $M._{inc}$ and P_j rolls back to its earliest checkpoint whose sequence number $\geq M._{rec\_line}$; if P_j later receives a rollback message corresponding to this incarnation, it ignores it. Thus, after knowing directly or indirectly about the failure of a process P_i, all other processes rollback to their earliest checkpoint whose sequence number is greater than equal to rec_line_i. If any process does not have such a checkpoint, it takes a checkpoint and adds it to the *rec_line* and proceeds normally. Note that not all processes need to perform a rollback to their earliest checkpoints.

Example We illustrate the basic recovery using the example in Figure 13.15. Suppose process P_3 fails at the instant shown. When P_3 recovers, it increments inc_3 to 1, sets rec_line_3 to $sn_3(=5)$, rolls back to its latest checkpoint $C_{3,5}$ and sends a *rollback*(1, 5) message to all other processes. Upon receiving

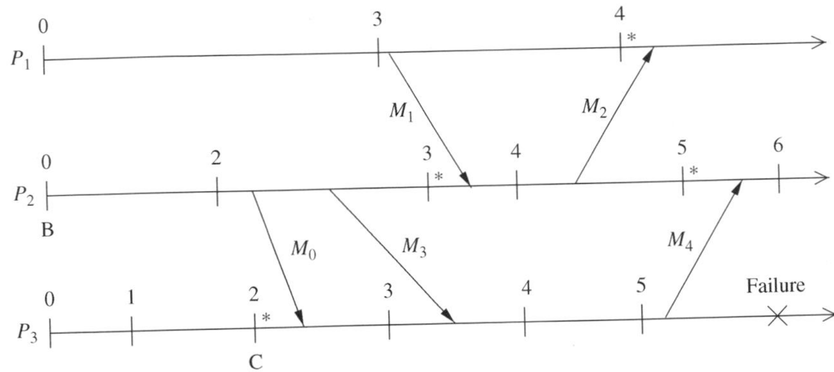

Figure 13.15 An example illustrating the Manivannan–Singhal algorithm [25].

this message, P_2 will rollback to checkpoint $C_{2,5}$ since $C_{2,5}$ is the earliest checkpoint at P_2 with sequence number ≥ 5. However, since P_1 does not have a checkpoint with sequence number greater than or equal to 5, it takes a local checkpoint and assigns 5 as its sequence number. Thus, $\{C_{1,5}, C_{2,5}, C_{3,5}\}$ is the recovery line corresponding this failure.

Thus, the recovery is simple. The failed process (on recovering from the failure) rolls back to its latest checkpoint and requests other processes to rollback to a consistent checkpoint which they can easily determine solely based on the local information. There is no domino effect and the recovery is fast and efficient.

In this example, we find that the sequence number of all checkpoints in the recovery line is the same, but it need not always be the case.

13.8.3 Comprehensive message handling

Rollback to a recovery line that is consistent may result in lost, delayed, orphan, or even duplicated messages. Existence of these types of message may lead the system to an inconsistent state. Next, we discuss on how to modify the BRA to handle these messages.

Handling the replay of messages

Not all messages stored in the stable storage need to be replayed. The BRA has to be modified so that it can decide which messages need to be replayed. In Figure 13.16, we assume that process P_1 fails at the point marked X and initiates a recovery with a new incarnation. After failure it rolls back to its latest checkpoint, $C_{1,10}$, then increments the incarnation inc_1 to 1 and sets the rec_line_1 to 10, and sends a $rollback(1, 10)$ message to all other processes. Upon receiving the rollback message from P_1, process P_2 rolls back to its checkpoint $C_{2,12}$. Consequently, all other processes roll back to an appropriate checkpoint following the BRA approach. After all the processes have rolled back to a set of consistent checkpoints, these checkpoints form a recovery

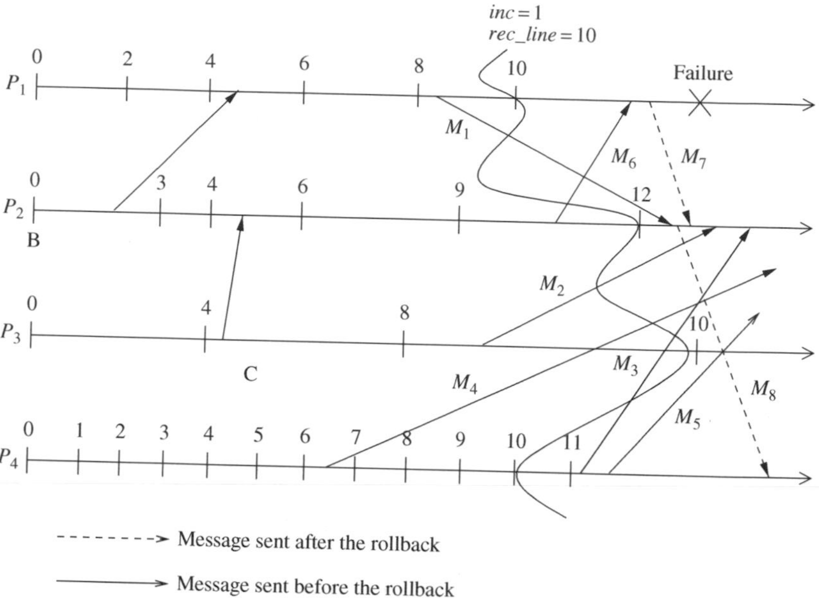

Figure 13.16 Handling of messages during the recovery [25].

line with number 10. The messages sent to the left of the recovery line carry incarnation number 0 and messages sent to the right of the recovery line carry incarnation 1.

To avoid lost messages, when a process rolls back it must replay all messages from its log whose receive was undone and whose send was not undone. In other words, a process must replay only those messages that originated from the left of the recovery line and were delivered to the right of the recovery line. In the example, after the rollback process P_2 must replay messages M_1 and M_2 from its log but must not replay M_3, because the send of M_1 and M_2 were not undone but the send of M_3 was. It is easy to determine the origin of the send of a message M by looking at the sequence number $(M._{sn})$ piggybacked. Therefore, we can state a rule for replaying messages as follows:

Message replay rule: After a process P_j rolls back to checkpoint C, it replays a message M only if it was received after C and if $M._{sn}$ < recovery line number.

Handling of received messages

This section discusses how a process handles received messages. Suppose process P_j receives a message M from process P_i. At the time of receiving the message, if P_j is replaying messages from the message log, then P_j will buffer the message M and will process it only after it finishes with the replaying of messages from the message log. If P_j does not do this then the following three cases may occur.

Case 1: M is a delayed message

A delayed message with respect to a recovery line carries an incarnation number less than the incarnation number of the receiving process. The process P_i that sent such a message M was not aware of the recovery process at the time of sending of M. Therefore, the piggybacked incarnation number of P_i is less than the latest incarnation number of P_j, the receiving process. In such a situation, if $M._{sn} < rec\_line_j$, then M is first logged in the message log and then processed; otherwise, it is discarded because P_i will eventually rollback and resend the message. In the figure, M_4 is logged and then processed by P_2 so that P_2 might have to replay M_4 due to a failure that may occur later, whereas M_5 is discarded by P_2. P_2 discards M_5 because $M._{sn} > rec\_line_2$ $(11 > 10)$ and $M._{inc}(= 0)$ is less than $inc_2(= 1)$. Therefore, we have the following rule for handling delayed messages:

Rule for determining and handling delayed messages: A delayed message M received by process P_j has $M._{inc}$ less than inc_j. Such a delayed message is processed by process P_j only if $M._{sn} < rec\_line_j$; otherwise, it is discarded.

Case 2: M was sent in the current incarnation

Suppose P_j receives a message M such that $inc_j = M._{inc}$. In this case, if $M._{sn}$ $< sn_j$, then P_j must log M before processing it. This is done because P_j might need to replay M due to a future failure. For example, in Figure 13.16, message M_7 is sent by process P_1 to process P_2 after P_1's recovery and after P_2's rollback during the same incarnation. In this case, $M._{inc} = inc_2 = 1$ and $M._{sn}(= 10) < sn_2(= 12)$, and M_7 must be logged before being processed because P_2 might have to roll back to checkpoint $C_{2,12}$ in case of a failure. In that case, P_2 will need to replay message M_7. Therefore, the rule for message logging in this case is stated as follows:

Message logging rule: A message received by process P_j is logged into the message log if $M._{inc} < inc_j$ and $M._{sn} < rec\_line_j$ or $M._{inc} = inc_j$ and $M._{sn} < sn_j$.

Case 3: Message M was sent in a future incarnation

In this case, $M._{inc} > inc_j$ and P_j handles it as follows: P_j sets rec_line_j to $M._{rec\_line}$ and inc_j to $M._{inc}$, and then rolls back to the earliest checkpoint with sequence number $\geq rec\_line_j$. After the roll back, message M is handled as in case 2, because $M._{inc} = inc_j$.

Features

The Manivannan–Singhal quasi-synchronous checkpointing algorithm has several interesting features:

- Communication-induced checkpointing intelligently guides the checkpointing activities to eliminates "useless" checkpoints. Thus, every checkpoint lies on consistent checkpoint.
- There is no extra message overhead involved in checkpointing. Only a scalar is piggybacked on application messages.

- It ensures the existence of a recovery line consistent with the latest checkpoint of any process all the time. This helps bound the depth of rollback during a rollback recovery.
- A failed process rolls back to its latest checkpoint and requests other processes to rollback to a consistent checkpoint (no domino-effect).
- Helps in garbage collection. After a process has established a recovery line, all checkpoints preceding the line can be deleted.
- The algorithm achieves the best of both worlds:
  - it has ease and low overhead of uncoordinated checkpointing;
  - it has recovery time advantages of coordinated checkpointing.

13.9 Peterson–Kearns algorithm based on vector time

The Peterson–Kearns [28] checkpointing and recovery protocol is based on the optimistic rollback. Vector time is used to capture causality to identify events and messages that become orphans when a failed process rolls back.

13.9.1 System model

We assume that there are N processors in the system, which are logically configured as a ring. Each processor knows its successor on the ring and this knowledge is stored in its stable storage since it is critical that it be recoverable after a failure. We assume a single process is executing on each processor. These N processes are denoted as $P_0, P_1, P_2 \ldots P_{N-1}$. We assume that $P_{(i+1)}$ mod N is the successor of P_i for $0 \le i < N$.

Each process P_i has a vector clock $V_i[j]$, $0 \le j \le N-1$. $V_i(e_i)$ denotes the clock value of an event e_i which occurred at P_i. The ith component of the vector is incremented before each event at process P_i and the current timestamp vector is sent on each message to update the receiving process's clock. $V_i(p_i)$ denotes the current vector clock time of process P_i and e_i denotes the most recent event in P_i. Thus $V_i(p_i) = V_i(e_i)$. Each send and receive event increments the vector time. The processes take periodic checkpoints of process state and also maintain a message log on the stable storage. The receipt of incoming messages is also logged periodically. The current vector clock value is considered a part of the process state and is logged to the stable storage when a checkpoint is taken.

Notation

The following notation is used to explain the algorithm:

- e^i_j: The ith event on P_j. We use e' and e'' to refer to generic events of P_j.
- s: A send event of the underlying computation.
- $\sigma(s)$: The process where send event s occurs.

- $\rho(s)$: The process where the receive event matched with send event s occurs.
- f_j^i: The ith failure on P_j.
- ck_j^i: The ith state checkpoint on P_j. The checkpoint resides on the stable storage.
- rs_j^i: The ith restart event on P_j.
- rb_j^i: The ith rollback event on P_j.
- $e_i^j \mapsto e_i^l$ iff $e_i^j \rightarrow e_i^l \wedge \nexists e_i^k$ such that $e_i^j \rightarrow e_i^k \rightarrow e_i^l$ where e_i^j and e_i^l are in the same process, p_i.
- $LastEvent(f_j^i) = e'$ iff $e' \mapsto rs_j^i$.

In a rollback protocol, every process must be contacted at least once to indicate that a failure has occurred and to send it the information necessary for recovery. This process is characterized as a series of one or more polling waves which are typified by the arrival of a polling message which transmits information necessary for rollback and a response by the polled process. We define two new event types:

- $C_{i,k}(m)$: The arrival of the final polling wave message for rollback from failure f_i^m at process P_k.
- $w_{i,k}(m)$: The response to this final polling wave by P_k. If no response is required, $w_{i,k}(m) = C_{i,k}(m)$

The final polling wave for recovery from failure f_i^m is defined as:

$$PW_i(m) = \bigcup_{k=0}^{N-1} w_{i,k}(m) \cup \bigcup_{k=0}^{N-1} C_{i,k}(m).$$

13.9.2 Informal description of the algorithm

When a process P_i restarts after failure f_i^m, it retrieves its latest checkpoint, including its vector clock value $V_i(Latest.ck(f_i^m))$, from the stable storage and rolls back to it. The message log is replayed until it is exhausted. Since the vector time of each message is logged with the message, when the messages are replayed, the clock value of the recovering process is appropriately updated. After the logged messages have been replayed, the recovering process executes a restart event, rs_i^m, to begin the global rollback protocol, originates a *token* message containing the vector timestamp of $rs_i{}^m$ and sends the token to its successor process. The token associated with failure f_i^m and restart event $rs_i{}^m$ is denoted by $tk(i,m)$. The timestamp of this token is denoted as $tk(i, m).ts$. Process P_i buffers all incoming application messages until the return of the token. When this occurs, P_i resumes normal execution.

The token is circulated through all the processes on the ring. When the token arrives at process P_j, the timestamp in the token is used to determine whether the process P_j must roll back. If $tk(i, m).ts < V_j(p_j)$, then an orphan event has occurred at P_j and P_j must roll back to an earlier state. This is accomplished by restoring P_j to the state of ck'_j , where ck'_j is the latest checkpoint at P_j for which $V_j(ck'_j) < tk(i, m).ts$, and then replaying logged messages as long as the timestamp of the message is less than $tk(i, m).ts$.

It is possible that an orphan event in P_j is the receipt of a message originating in a non-orphaned send event in process P_i. Since the send event corresponding to such a receipt does not causally succeed any lost event in P_i, the recovery of P_i will not result in the replay of such messages. Therefore, these messages are lost unless some special actions are taken. To make sure that these messages are not lost, P_j must request their retransmission during the rollback.

During the rollback, P_j must also retransmit any message that it sent to P_i that was lost due to failure. Process P_j can determine whether the messages it had sent have been received by the failed process P_i by comparing the vector timestamps of the messages to the timestamp in the token. If $V_j(s)[j] > V_i(rs_i^m(j))$, where s is the message that was sent to P_i, then it is possible that the failed process has lost the message and it must be resent. It is also possible that the message is not lost, but is still in transit; thus P_i must discard any duplicate messages. Because channels are FIFO, P_i can identify any duplicate message from its timestamp.

After the logged messages have been replayed and retransmissions of the required messages are done, P_j instigates a rollback event, $rb^k{}_j$, to indicate that rollback at it is complete. Vector time is not incremented for this event so $V(_j rb^k{}_j)= V_j(e'_j)$, where e'_j is the last event replayed. Any logged event whose vector time exceeds $tk(i, m).ts$ is discarded.

If $tk(i, m).ts \not< V_j(P_j)$ when the token arrives, the state of P_j is not changed. For consistency, however, a rollback event is instigated to indicate that rollback is complete at P_j and to allow the token to be propagated.

Note that, after the rollback is complete, $V_j(P_j) \not> V_j(rs_i^m)$, that is, every event in P_j either happens before the restart event $rs_i{}^m$ or is concurrent to it. The property of vector time that $e'_i \rightarrow e''_j$ iff $V_i(e'_i) < V_j(e''_j)$ allows us to make this claim.

The token is propagated from process P_i to process $P_{(i+1)modN}$. As the token propagates, it rolls back orphan events at every process. When the token returns to the originating process, the roll back recovery is complete.

Handling in-transit orphan messages

It is possible for orphan messages to be in transit during the rollback process. If these messages are received and processed during or after the rollback procedure, an inconsistent global state will result. To identify these orphan messages and discard them on arrival, it is necessary to include an incarnation number with each message and with the token. inc_i denotes the current incarnation number of process P_i, and $Inc(e_i)$ denotes the incarnation number of event e_i. The value returned for an event equals the current incarnation number of the process in which the event occurred. The incarnation number in the token is denoted by $tk(i, m).inc$.

When P_i initiates the rollback process, it increments its current incarnation number by one and attaches it to the token. A process receiving the token saves both the vector timestamp of the token and the incarnation number in the stable storage. Because there is no bound on message transmission time, the vector timestamps and associated incarnation numbers that have arrived in the token must be accumulated in a set denoted as $OrVect_i$. The set $OrVect_i$ is composed of ordered pairs of token timestamps and incarnation numbers received by process P_i.

When an application message is received by process P_i, the vector timestamp of the message is compared to the vector timestamps stored in $OrVect_i$. If the vector timestamp of the message is found to be greater than a timestamp in $OrVect_i$, then the incarnation number of the message is compared to the incarnation number corresponding to the timestamp in $OrVect_i$. If the message incarnation number is smaller, then the message is discarded. Clearly, this is an orphan message that was in transit during the rollback process. In all other cases, the message is accepted and processed. Upon the receipt of a token, the receiving process sets its incarnation number to that in the token.

13.9.3 Formal description of the rollback protocol

The causal rollback protocol is described as set of six rules, CRB1 to CRB6. For each rule, we first present its formal description and then give a verbal explanation of the rule.

The rollback protocol

CRB1 $w_{i,i}(m)$ occurs iff there exists f_i^m, $rs_i{}^m$ such that $f_i^m \mapsto rs_i^m \rightarrow w_{i,i}(m)$.

A formerly failed process creates and propagates a token, event $w_{i,i}(m)$, only after restoring the state from the latest checkpoint and executing the message log from the stable storage.

CRB2 The occurrence of $w_{i,i}(m)$ implies that
$tk.(i, m).ts = V_i(rs_i{}^m) \wedge$

$tk.(i, m).inc = Inc(Latest.ck(f_i^m)) + 1 \wedge$
$Inc_i = Inc(Latest.ck(f_i^m)) + 1$

The restart event increments the incarnation number at the recovering process, and the token carries the vector timestamp of the restart event and the newly incremented incarnation number.

CRB3 $w_{i,j}(m), i \neq j$ occurs iff
$\exists\, rb_i^k$ such that $c_{i,j}(m) \rightarrow rb_i^k \rightarrow w_{i,j}(m) \wedge$
$\forall\, e_j^{'}$ such that $V_j(e_j^{'}) > tk(i, m).ts, \neg\, Recorded(e_j^{'})$

A non-failed process will propagate the token only after it has rolled back.

CRB4 The occurrence of $w_{i,j}(m)$ implies that
$Inc_i = tk(i, m).inc \wedge\ (tk(i, m).ts, tk(i, m).inc) \in OrVect_j$

A non-failed process will propagate the token only after it has incremented its incarnation number and has stored the vector timestamp of the token and the incarnation number of the token in its *OrVect* set.

CRB5 Polling wave $PW_i(m)$ is complete when $C_{i,j}(m)$ occurs.

When the process that failed, recovered, and initiated the token, receives its token back, the rollback is complete.

CRB6 Any message received by event, $n(s)$, is discarded iff $\exists\ m \in OrVect_{(p(s))}$ such that $Inc(s) < Inc(m) \wedge V(m) < V(s)$.

Messages that were in transit and which were orphaned by the failure and subsequent restart and recovery must be discarded.

Example Consider an example consisting of three processes shown in Figure 13.17. The processes have taken checkpoints C_0^1, C_1^1, C_2^1. Each event on a process time line is tagged with the vector time (x, y, z) of its occurrence. Each message is tagged with $[i](x, y, z)$, where i is the incarnation number associated with the message send event, and (x, y, z) is the vector time of the send event. Process P_0 fails just after sending message m_5, which increments its vector clock to (5, 4, 0).

Upon restart of P_0, the checkpoint C_0^1 is restored, and the restart event, rs_0^1 is performed by the protocol. We assume that message m_4 was not logged into the stable storage at P_0, hence it cannot be replayed during the recovery. A token, [1](4, 0, 0), is created and propagated to P_1. This is shown in the figure by a dotted vertical arrow. Upon the receipt of the token, P_1 rolls back to a point such that its vector time is not greater than (3, 0, 0), the time in the token. Hence P_1 rolls back to its state at time (1, 4, 0). P_1 then records the token in its *OrVect* set and sends the token to P_2. P_2 takes a similar action and rolls back to message send event with time (1, 4, 4). The token is then returned to P_0 and recovery is complete.

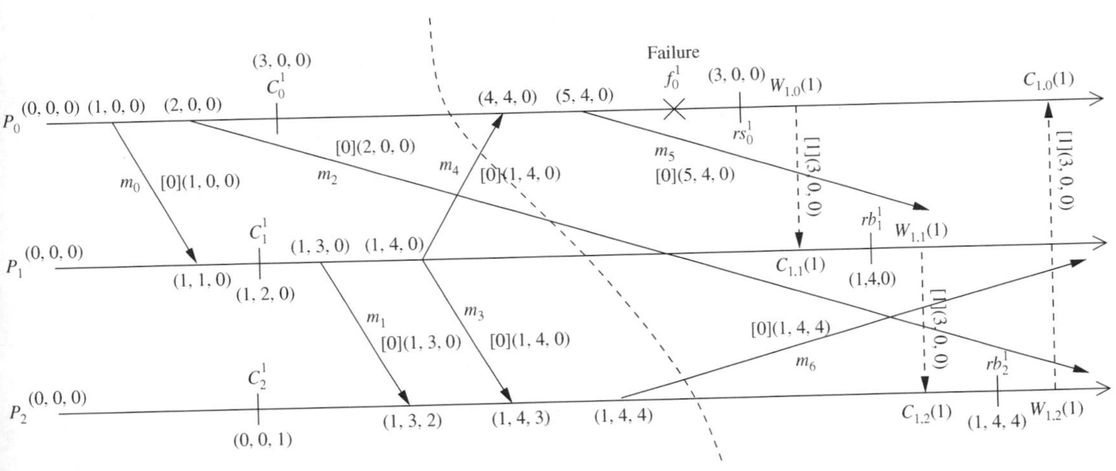

Figure 13.17 An example of rollback recovery in the Peterson–Kearns algorithm.

Three messages are in transit while the polling wave is executing. The message m_2 from P_0 to P_2 with label [0](2, 0, 0) will be accepted when it arrives. Likewise, message m_6 from P_2 will be accepted by P_1 when it arrives. However, application of rule CRB6 will result in message m_5 with label [0](5, 4, 0) being discarded when it arrives at P_1. The net effect of the recovery process is that the application is rolled back to a consistent global state indicated by the dotted line, and all processes have sufficient information to discard messages sent from orphan events on their arrival.

13.9.4 Correctness proof

First we show that all orphaned events are detected and eliminated [28].

Theorem 13.1 *The completion of a wave in casual rollback protocol insures that every event orphaned by failure f_i^m is eliminated before the final polling wave.*

Proof We prove that when the initiator process receives the token back, all the orphan events have been detected and eliminated. That is, for an event $w_{i,j}(m)$, as specified in the causal rollback protocol,

$$\neg Orphan(w_{i,j}(m), f_i^m).$$

First we prove that the token, as constructed during the restoration of a failed process, contains necessary information to determine if any event is orphaned by a failure. If there exists any orphan event e_i' due to failure f_m^j, then the vector timestamp in the token will be less than the vector time of the event, i.e., $tk(j, m).ts < V_i(e_i')$. By CRB2, the vector timestamp in the token, $tk(j, m).ts$ must equal to $V_j(rs_j^m)$, and $V_j(rs_j^m) = V_j(LastEvent(f_j^m))$. In other words, the timestamp in the token must be equal to the vector time of the restart event rs_j^m at process P_j denoted as $V_j\ (rs_j^m)$, and the vector time of the restart event at P_j will be one more than the vector time of the latest event before failure f_m^j. Since rs_j^m occupies the same position in causal partial order as e_j' and $LastEvent(f_j^m) \mapsto e_j'$, the following must hold: $V_j(rs_j^m) \leq V_j(e_j')$. If there exists an orphan e_i', then there exist e_j' such that $LastEvent(f_j^m) \rightarrow e_j' \rightarrow e_i'$.

Therefore, $V_j(e_j') < V_i(e_i')$ and $V_j(rs_j^m) < V_i(e_i')$, which proves that when

$$tk(j, m).ts < V_j(e_i'), \tag{13.1}$$

there exists an orphan event e_i'.

We use the above result to prove that there exists no orphan event at the end of the final polling wave:

$$\neg Orphan(w_{i,j}(m), f_i^m). \tag{13.2}$$

The proof is by contradiction. Let us assume that there exists a polling event $w_{i,j}(m)$ for which $Orphan(w_{i,j}(m), f_i^m)$ is true. Then there exists an event e_i' such that $LastEvent(f_i^m) \rightarrow e_i' \rightarrow w_{i,j}(m)$. Then there must exist e_j' such that $e_i' \rightarrow e_j' \rightarrow w_{i,j}(m)$. This implies $Orphan(e_j', f_i^m)$. But according to Eq. (13.1), $tk(i, m).ts < Vj(e_j')$, which contradicts CRB3: $w_{i,j}(m)$ occurs iff there exists rb_j^k such that $c_{i,j}(m) \rightarrow rb_j^k \rightarrow w_{i,j}(m)$ and for every e_j' such that $V_j(e_j') > tk(i, m).ts$, $\neg\ recorded(e_j')$.

Therefore, every event orphaned by a failure f_i^m is eliminated before the final polling wave is completed. □

Now we show that only all orphaned messages are discarded [28].

Theorem 13.2 *All orphaned messages are discarded and all non-orphaned messages are eventually delivered.*

Proof Let us consider a send event s, which is not orphaned by the failure f_i^m. In this case, $n(s) \rightarrow w_{i,p(s)}(m) \vee w_{i,p(s)}(m) \rightarrow n(s)$.

Given reliable channels, the message will eventually arrive. The receipt of a message can only disappear from the causal order if it is lost by a failed process, rolled back by the protocol, or discarded upon arrival.

The first possibility is that process P_i lost the message due to its failure. In this case the receiving process $p(s)$ is i. During the rollback at $P_{\sigma(s)}$ (the process where the send event occurred), this message will be retransmitted as the occurrence of the rb event associated with $w_{i,\sigma(s)}(m)$ guarantees this. Therefore $w_{i,i} \rightarrow n(s)$.

The second possibility is that $n(s) \rightarrow w_{i,p(s)}$ and $n(s)$ has rolled back because $n(s)$ was orphaned by the failure f_i^m. However, if event s is not orphaned by f_i^m, $P_{p(s)}$ (the receiving process) will request retransmission before the occurrence of the rollback event rb, and $w_{i,p(s)} \rightarrow n(s)$.

The final possibility is that $n(s)$ occurs after the wave but is discarded upon arrival. By CRB6, $n(s)$ will be discarded if and only if $V(s) > tk(i, m).ts$ and $inc(s) < tk(i, m).inc$. If $s \rightarrow w_{i,\sigma(s)}$ and $Orphan(s, f_m^i)$, then $V(s) \not> tk(i, m).ts$. If $w_{i,\sigma(s)} \rightarrow s$, then $Inc(s) \not< tk(i, m).inc$. Therefore, $n(s)$ will not be discarded and $w_{i,p(s)} \rightarrow n(s)$.

We now prove the converse:

$$\text{If } n(s) \rightarrow w_{i,p(s)}(m) \vee w_{i,p(s)} \rightarrow n(s) \text{ then } \neg Orphan(s, f_i^m).$$

Assume $n(s) \rightarrow w_{i,p(s)}$. From Eq. (13.2), we know $\neg\ Orphan(w_{i,p(s)}, f_i^m)$. Therefore $\neg\ Orphan(n(s), f_i^m)$ and $\neg\ Orphan(s, f_i^m)$.

Assume $w_{i,p(s)} \rightarrow n(s)$ and $Orphan(s, f_i^m)$. By Eq. (13.1), this implies $tk(i, m).ts < V_{\sigma(s)}$. Rule CRB2 of the protocol guarantees that if $Orphan(s, f_i^m)$ is true, then $Inc(s) < tk(i, m).inc$. Rule CRB4 requires that $tk(i, m).ts$ and $tk(i, m).inc$ are stored in $OrVect_j$ before $w_{i,j}(m)$ occurs. Therefore, there exists $z \in OrVect_j$ such that $V(z) < V(s)$ and $Inc(z) > Inc(s)$. CRB6 requires such a message must be discarded, contradicting our assumption that $w_{i,p(s)} \rightarrow n(s)$. □

13.10 Helary–Mostefaoui–Netzer–Raynal communication-induced protocol

The Helary–Mostefaoui–Netzer–Raynal [15,16] communication-induced checkpointing protocol prevents useless checkpoints and does it efficiently. To prevent useless checkpoints, some coordination is required in taking local checkpoints. Coordinated checkpointing protocols use additional control messages to synchronize their checkpointing activities, but these result in reduced process autonomy and degraded performance of the underlying application. Communication-induced checkpointing protocols achieve this coordination by piggybacking control information on application messages. No control messages are needed and no synchronization is added to the application. More precisely, processes take local checkpoints independently, called basic checkpoints, and the protocol directs them to take additional local checkpoints, called forced checkpoints. A process takes a forced checkpoint when it receives a message and a predicate at it becomes true. This predicate is based on local control variables of the receiving process and on the control values carried by the message. The values of the local control variables at the process are based on causal dependencies appearing in its past.

The Helary–Mostefaoui–Netzer–Raynal communication-induced checkpointing protocol ensures that no local checkpoint is useless and it takes as few forced checkpoints as possible. The protocol is based on Z-path

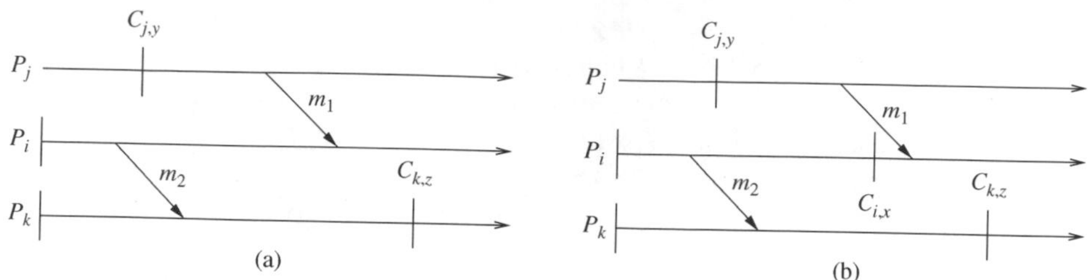

Figure 13.18 To checkpoint or not to checkpoint [16]?

and Z-cycle theory introduced by Netzer and Xu [27] who showed that a useless checkpoint exactly corresponds to the existence of a Z-cycle in the distributed computation. At the model level, the protocol prevents Z-cycles. At the operational level, each message is piggybacked with an integer (Lamport's clock value), a vector of integers (checkpoint sequence number), and two boolean vectors (the size of each vector is n, the number of processes). An interesting feature of this protocol is that for any checkpoint C, it is very easy to determine a consistent global checkpoint to which C belongs.

13.10.1 Design principles

With each checkpoint C, let us associate a timestamp denoted by $C.t$. The protocol depends on the following result:

For any pair of checkpoints $C_{j,y}$ and $C_{k,z}$, such that there is a Z-path from $C_{j,y}$ to $C_{k,z}$, $C_{j,y}.t < C_{k,z}.t$ implies that there is no Z-cycle.

Thus, if we can manage the timestamps and take checkpoints in such a way that the timestamps always increase along any Z-path, then no Z-cycles will form, and no checkpoints will be useless. Each process P_i has a logical clock lc_i managed in the following way:

1. Before a process P_i takes a (basic or forced) checkpoint, it increases its clock by 1 and associates the new clock value with the checkpoint.
2. Every message m is timestamped with the value of its sender clock (let $m.t$ denote the timestamp associated with message m).
3. When a process P_i receives a message m, it updates its local clock $lc_i = max(lc_i, m.t)$.

It follows from this mechanism that, if there is a causal Z-path from $C_{j,y}$ to $C_{k,z}$, then we have $C_{j,y}.t < C_{k,z}.t$.

To checkpoint or not to checkpoint?

Let us consider the computation depicted in Figure 13.18, where $C_{j,y}$ is a local checkpoint taken by P_j before sending m_1 and $C_{k,z}$ is the first checkpoint of P_k taken after the delivery of m_2. As the sending of m_2 and the delivery of

m_1 belong to the same interval of P_i, messages m_1 and m_2 constitute a Z-path from $C_{j,y}$ to $C_{k,z}$. When P_i receives m_1, two cases can occur:

1. $m_1.t \leq m_2.t$. In this case, $C_{j,y}.t < m_1.t < m_2.t < C_{k,z}.t$. Hence, the Z-path due to messages m_1 and m_2 in Figure 13.18(a) is in accordance with the above result.
2. $m_1.t > m_2.t$. In this case, a safe strategy to prevent Z-cycle formation is to direct P_i to take a forced checkpoint $C_{i,x}$ before delivering m_1 (as shown in Figure 13.18(b)). This "breaks" the m_1, m_2 Z-path, so it is no longer a Z-pattern.

This strategy can be implemented in the following way. Each process P_i maintains a boolean array $sent\_to_i[1 \ldots n]$ to determine whether the reception of a message creates a Z-pattern. The value of $sent\_to_i[k]$ is true iff P_i has sent a message to P_k since its last checkpoint. Each process P_i also maintains an array of integers $min\_to_i[1 \ldots n]$, where $min\_to_i[k]$ keeps the timestamp of the first message P_i sent to P_k since its last checkpoint.

The condition $m_1.t > m_2.t$ is then expressed as:

$$C \equiv (\exists\, k{:}\ sent\_to_i[k] \wedge m_1.t > min\_to_i[k]).$$

Therefore, P_i takes a forced checkpoint if C is true. The predicate C is true if there exists a message from P_i to P_k since its last checkpoint and the timestamp of m_1 is greater than that of the first message P_i sent to P_k since its last checkpoint.

Reducing the number of forced checkpoints

Each process P_i maintains the local clock values of other processes. For each $k(1 \leq k \leq n)$, let $cl_i(k)$ denote the value of P_k's local clock as perceived by P_i. If $k = i$, obviously $cl_i(i) = lc_i$. However, if $k \neq i$, the perception of P_k's local clock by P_i is only an approximation such that $cl_i(k) \leq lc_k$. Consider again the situation in Figure 13.18. If the following property holds:

$$(m_1.t < m_2.t) \vee P, \text{ where } P \equiv (C_{j,y}.t \leq m_1.t \leq cl_i(k) < C_{k,z}.t),$$

then the Z-path due to messages m_1 and m_2 is in accordance with the above result. Let us consider the property P in the case where $m_1.t > m_2.t$. Since $m_1.t$ carries the value lc_j when m_1 is sent, the first relation $C_{j,y}.t \leq m_1.t$ necessarily holds when m_1 is received. So, the property P can be violated only if, when m_1 is received, $m_1.t > cl_i(k)$ or if $cl_i(k) \geq C_{k,z}.t$.

Therefore, to prevent the formation of a Z-path due to messages m_1 and m_2 that would violate property P, the protocol requires process P_i to take a forced checkpoint before delivering m_1 if $m_1.t > cl_i(k)$ or if $cl_i(k) \geq C_{k,z}.t$.

Now we have to determine which value of cl_k, the approximation $cl_i(k)$ refers to. Let us consider the following two possible cases:

1. The value of $cl_i(k)$ has been brought to P_i by a causal Z-path that started from P_k and ended before $C_{k,z}$. This situation is illustrated in Figure 13.19. The value of $cl_i(k)$ is brought to P_i by m' in Figure 13.19(a) and by m'' and m_1 in Figure 13.19(b). In this case, we have $cl_i(k) < C_{k,z}.t$ and, consequently, P_i has to take a forced checkpoint only if $m_1.t > cl_i(k)$.
2. The value of $cl_i(k)$ has been brought to P_i by a causal Z-path that started from P_k and ended after $C_{k,z}$. This situation is illustrated in Figure 13.20. Here the relevant causal Z-path is m' in Figure 13.20(a) and m'' and m_1 in Figure 13.20(b). Both these figures can be redrawn so that they correspond to the pattern in Figure 13.21. In one case, m' brings the last value of P_k's local clock to P_i, and in the other case it is $m''.m_1$. In this case, we have $cl_i(k) \geq C_{k,z}.t$ and P_i has to recognize this pattern and take a forced checkpoint if it occurs. Let C_1 be a predicate describing this pattern occurrence.

The previous condition C can be redefined as C' as follows:

$$C' \equiv (\exists k{:}\ sent\_to_i[k] \wedge (m_1.t > min\_to_i[k]) \wedge (m_1.t > cl_i(k) \vee C_1)).$$

The predicate C' has two parts. The first part is the previous condition C and the second part is a predicate C_1. The second part will be true if the timestamp of message m_1 is greater than P_k's local clock value as perceived by P_i or if predicate C_1 is true.

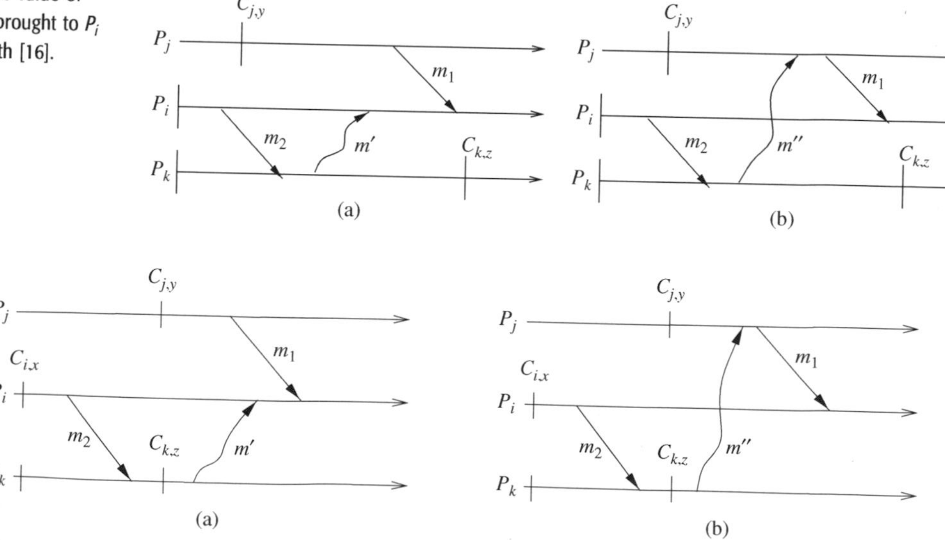

Figure 13.19 The value of $cl_i(k)$ has been brought to P_i by a causal Z-path [16].

Figure 13.20 The value of $cl_i(k)$ has been brought to P_i by a causal Z-path [16].

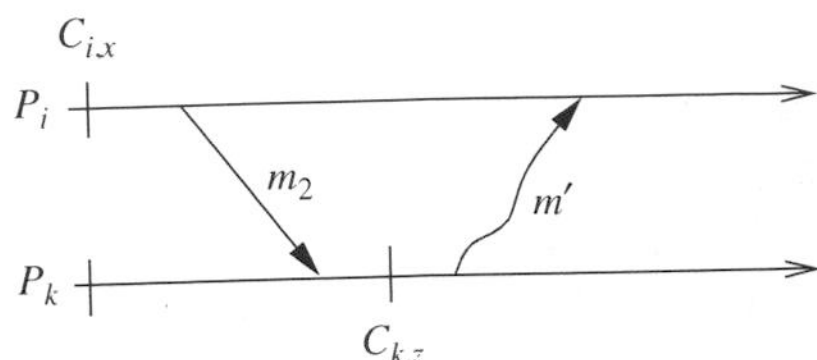

Figure 13.21 An example of a Z-cycle [16].

To evaluate the predicate C_1, each process maintains two additional arrays:

1. Array $ckpt_i$ is a vector that counts the number of checkpoints taken by each process. So, $ckpt_i[k]$ denoted the number of checkpoints taken by P_k to P_i's knowledge. Let *m.ckpt* be the value appended to m by its sender P_i, which is the value of the array $ckpt_i$ at the time of sending of message m.
2. A boolean array $taken_i$ is used in conjunction with $ckpt_i$ to evaluate C_1. The value of $taken_i[k]$ is true iff there is a causal Z-path from the last checkpoint of P_k known by P_i to the next checkpoint of P_i and this causal Z-path includes a checkpoint.

The array $taken_i$ is updated in the following way:

- When a process P_i takes a checkpoint, it sets to true all entries of $taken_i$ except $taken_i[i]$, which always remains false: $\forall k \neq i$: $taken_i\,[k]$ = true.
- When process P_i sends a message, P_i appends its current value of $taken_i$ to the message.
- When process P_i receives m, P_i updates $taken_i$ in the following way:

$\forall k \neq i$ do case
$m.ckpt[k] < ckpt_i[k] \rightarrow$ skip
$m.ckpt[k] > ckpt_i[k] \rightarrow taken_i[k] := m.taken[k]$
$m.ckpt[k] = ckpt_i[k] \rightarrow taken_i[k] := (taken_i[k] \vee m.taken[k])$
end do case

With these data structures, the predicate C_1 can be expressed as follows:

$$C_1 \equiv (m_1.ckpt[i] = ckpt_i[i]) \wedge m_1.taken[i].$$

Consider the example shown in Figure 13.21. The first part of the condition C_1 states that there is a causal Z-path starting from $C_{i,x}$ and arriving at P_i before $C_{i,x+1}$, while the second part indicates that some process has taken a checkpoint along this causal Z-path.

13.10.2 The checkpointing protocol

Next (see Algorithm 13.4) we describe the Helary–Mostefaoui–Netzer–Raynal communication-induced checkpointing protocol, which takes as few forced checkpoints as possible and also ensures that no local checkpoint is useless.

Procedure take-checkpoint:

$\forall k$ do $sent\_to_i[k]$:= false end do;
$\forall k$ do $min\_to_i[k] := +\infty$ end do;
$\forall k \neq i$ do $taken_i[k]$:= true end do;
$clock_i[i] := clock_i[i] + 1$;
Save the current local state with a copy of $clock_i[i]$;
/* let $C_{i,x}$ denote this checkpoint. We have $C_{i,x}.t = clock_i[i]$ */
$ckpt_i[i] := ckpt_i[i] + 1$;

($S0$) initialization:

$\forall k$ do $clock_i[k] := 0$; $ckpt_i[k] := 0$ end do;
$taken_i[i] := false$;
$take\_checkpoint$;

($S1$) When P_i sends a message to P_k:

if $\neg$ $sent\_to_i[k]$ then $sent\_to_i[k] := true$; $min\_to_i[k]$:= clock$_i[i]$
end if;
Send (m, $clock_i$, $ckpt_i$, $taken_i$) to P_k;

($S2$) When P_i receives (m, $clock_i$ i, $ckpt_i$, $taken_i$) *from* P_j:

/* $m.clock[j]$ is the Lamport's timestamp of m (i.e., $m.t$) */
if ($\exists k$: $sent\_to_i[k] \wedge (m.clock[j] > min\_to_i[k]) \wedge$
$((m.clock[j] > max(clock_i[k], m.clock[k])) \vee$
$(m.ckpt[i] = ckpt_i[i] \wedge m.taken[i])))$
then $take\_checkpoint$ /*forced checkpoint */
end if;
$clock_i[i] := max(clock_i[i], m.clock[j])$; /* update of the scalar clock $lc_i \equiv clock_i[i]$ */
$\forall k \neq i$ do
$clock_i[k] := max(clock_i[k], m.clock[k])$;
case
$m.ckpt[k] < ckpt_i[k] \rightarrow$ skip
$m.ckpt[k] > ckpt_i[k] \rightarrow ckpt_i[k] := m.ckpt[k]$; $taken_i[k] := m.taken[k]$
$m.ckpt[k] < ckpt_i[k] \rightarrow taken_i[k] := taken_i[k] \vee m.taken[k]$
end case
end do
deliver (m);

Algorithm 13.4 The Helary–Mostefaoui–Netzer–Raynal communication-induced checkpointing protocol [16].

The protocol is executed by each process P_i. $S0$, $S1$, and $S2$ describe the initialization, the statements executed by P_i when it sends a message, and statements it executes when it receives a message, respectively. The procedure *take-checkpoint* is called each time P_i takes a checkpoint (basic or forced).

The protocol uses the following additional data structure: every process P_i maintains an array $clock_i[1 \ldots n]$, where $clock_i[j]$ denotes the highest value of lc_j known to P_i. $clock_i[1 \ldots n]$ is initialized to (0, 0,…,0) and is updated as follows:

- When a process P_i takes a (basic or forced) checkpoint, it increases $clock_i[i]$ by 1.
- When P_i sends a message m, the current value of $clock_i$ is sent on the message. Let $m.clock$ be the timestamp associated with a message m.
- When a process P_i receives a message m from P_j, it updates its clock as follows:
    - $clock_i[i] := max(clock_i[i], m.clock[j])$
    - $\forall k \neq i : clock_i[k] := max(clock_i[k], m.clock[k])$

Note that $clock_i[i]$ is lc_i, so we do not need to keep lc_i.

Helary *et al.* [15] showed that given a local checkpoint $C_{i,x}$ with timestamp a, the checkpoint can be associated with the consistent global checkpoint it belongs to using the following result:

Theorem 13.5 *Let a be a Lamport timestamp and C_a be a global checkpoint, $\{C_{1,x1}, C_{2,x2}, \ldots, C_{n,xn}\}$. If $\forall$ k, $C_{k,xk}$ is the last checkpoint of P_k such that $C_{k,xk}.t \leq a$, then C_a is a consistent global checkpoint.*

Proof For a proof, the readers are referred to the original source [15,16]. This result implies that given a local checkpoint at a process, it is easy to determine what local checkpoints at other processes form a consistent global checkpoint with it. This result has a strong implication on the recovery from a failure. □

13.11 Chapter summary

Rollback recovery achieves fault tolerance by periodically saving the state of a process during the failure-free execution, and restarting from a saved state on a failure to reduce the amount of lost computation.

There are three basic approaches for checkpointing and failure recovery: uncoordinated checkpointing, coordinated checkpointing, and communication-induced checkpointing. In uncoordinated checkpointing, each participating process takes its checkpoints independently; during a failure recovery, all processes communicate to find a consistent global checkpoint. In coordinated checkpointing, processes coordinate their checkpointing activities to form a system-wide consistent state. In case of a process failure, the system state can be restored to such a consistent set of checkpoints, preventing the rollback propagation. This techniques has additional overhead at run time but it avoids the domino effect at recovery time. Communication-induced checkpointing forces each process to take checkpoints based on information piggybacked on the application messages it receives from other processes. Checkpoints are taken such that construction of a consistent checkpoint at recovery is simple, efficient, and fast and the domino effect is avoided.

Message logging can help with the handling of various types of abnormal-messages and with the recovery after a failure. There are three types of message logging: pessimistic logging, optimistic logging, and causal logging protocols.

Over the last two decades, checkpointing and failure recovery has been a very active area of research and several checkpointing and failure recovery algorithms have been proposed. In this chapter, we described a set of representative checkpointing and recovery algorithms. Lately, useless checkpoints and techniques to eliminate useless checkpoints have been the main focus of attention.

13.12 Exercises

Exercise 13.1 Consider the following simple checkpointing algorithm. A process takes a local checkpoint right after sending a message. Show that the last checkpoint at all processes will always be consistent. What are the trade-offs with this method?

Exercise 13.2 Show by example that, in the Koo–Toueg checkpointing algrithm, if processes do not block after taking a tentative checkpoint, then the global checkpoint taken by all processes may not be consistent.

Exercise 13.3 Show that, in the Manivannan–Singhal algorithm, every checkpoint taken is useful.

Exercise 13.4 Design a checkpointing and recovery algorithm that uses vector clocks, and does not assume any underlying topology (like ring or tree).

Exercise 13.5 Give a rigorous proof of the impossibility of a min-process, non-blocking checkpointing algorithm.

13.13 Notes on references

Checkpointing and failure recovery is a well-studied topic and a large number of checkpointing and failure-recovery algorithms exist. A classical paper on fault tolerance is by Randell [32]. Classical failure-recovery algorithms are Leu–Bhargava [4], Sistla–Welch [35], Kim [19–21], and Strom–Yemini [36]. Other checkpointing and failure recovery algorithms can be found in [3,8,11,12,14,15,24,30,31,38–41] [15].

An excellent review paper on the topic is by Elnozahy et al. [13]. Richard and Singhal give a comprehensive recovery protocol using vector timestamp [33].

An impossibility proof of min-process non-blocking in coordinated checkpointing is given in [7]. Cao and Singhal introduced the concept of mutable checkpointing [6,8] to improve the performance. Alvisi and Marzullo discuss various message logging techniques [1]. Netzer and Xu discuss necessary and sufficient conditions for consistent global snapshots in distributed systems [27]. Manivannan *et al.* [26] and Wang [41] discuss how to construct consistent global checkpoints that contain a given set of local checkpoints. Prakash and Singhal discuss how to take maximal global snapshot with concurrent initiators [29]. Other communication-induced checkpointing algorithms can be found in papers by Baldoni *et al.* [2,3,17]. Tong *et al.* [37] present rollback recovery using loosely synchronized clocks.

References

[1] L. Alvisi and K. Marzullo, Message logging: pessimistic, optimistic, causal, and optimal, *IEEE Transactions on Software Engineering*, **24**(2), 1998, 149–159.

[2] Roberto Baldoni, Jean-Michel Helary and Michel Raynal, Consistent records in asynchronous computations, *Acta Informatica*, Vol 35, No 6, June 1998, pp. 441–455.

[3] R. Baldoni, A communication-induced checkpointing protocol that ensures rollback-dependency trackability, *Proceedings of the 27th International Symposium on Fault-Tolerant Computing (FTCS'97)*, June 25–27, 1997, p. 68.

[4] B. Bhargava and P. Leu, Concurrent robust checkpointing and recovery in distributed systems, *Proceedings of the IEEE International Conference on Data Engineering*, February 1988, 154–163.

[5] D. Briatico, A. Ciuffoletti, and L. Simoncini, A distributed domino-effect free recovery algorithm, *Proceedings of the Symposium on Reliability in Distributed Software and Database Systems*, Silver Spring, MD, October 1984, 207–215.

[6] G. Cao and M. Singhal, Mutable checkpoints: a new checkpointing approach for mobile computing systems, *IEEE Transactions on Parallel and Distributed Systems*, **12**(2), 2001, 157–172. Available online at: www.cse.psu.edu/ gcao/paper/gcao/TPDS01.pdf).

[7] G. Cao and M. Singhal, On the impossibility of min-process non-blocking checkpointing and an efficient checkpointing algorithm for mobile computing systems, *Proceedings of the 1998 International Conference on Parallel Processing*, 1998, 37–44.

[8] G. Cao and M. Singhal, Checkpointing with mutable checkpoints, *Theoretical Computer Science*, **290**(2), 2003, 1127–1148. Available online at: www.cse.psu.edu/ gcao/paper/gcao/TCS03.pdf.

[9] K. M. Chandy and L. Lamport, Distributed snapshots: determining global states of distributed systems, *ACM Transactions on Computer Systems* **3**(1), 1985, 63–75.

[10] M. Chandy and C. V. Ramamoorthy, Rollback and recovery strategies for computer programs, *IEEE Transactions on Computers* **21**(6), 1972, 546–556.

[11] O.P. Damani, Yi-Min Wang, and V. K. Garg, Distributed recovery with K-optimistic logging, *Journal of Parallel and Distributed Computing*, **63**(12), 2003, 1193–1218.

[12] E.N. Elnozahy and W. Zwaenepoel, Manetho: transparent rollback-recovery with low overhead, limited rollback, and fast output commit, available online at: www.cs.utexas.edu/users/mootaz/cs372/Projects/paper2.pdf.

[13] E. N. Elnozahy, L. Alvisi, Y.-M. Wang, and D. B. Johnson, A survey of rollback-recovery protocols in message-passing systems, *ACM Computing Surveys*, **34**(3), 2002, 375–408. Available online at: www.cs.utexas.edu/users/lorenzo/papers/SurveyFinal.pdf.

[14] E. N. Elnozahy and J. S. Plank, Checkpointing for peta-scale systems: a look into the future of practical rollback-recovery, *IEEE Transactions on Dependable and Secure Computing*, **1**(2), 2004, 97–108.

[15] J. M. Helary, A. MosteFaul, R. H. Netzer, and M. Raynal, Communication-based prevention of useless checkpoints in distributed computations, *Distributed Computing*, **13**(1), 2000, 183–190.

[16] J.-M. Helary and A. Mostefaoui, and M. Raynal, Preventing useless checkpoints in distributed computations, *Proceedings of the 16th Symposium on Reliable Distributed Systems* (SRDS'97), October 22–24, 1997, 183–190.

[17] J.-M. Helary, A. Mostefaoui, and M. Raynal, Communication-induced determination of consistent snapshots, *IEEE Transactions on Parallel and Distributed Systems*, **10**(9), 1999, 865–877.

[18] T. T.-Y. Juang and S. Venkatesan, Crash recovery with little overhead, *Proceedings of the 11th International Conference on Distributed Computer Systems*, 1991, 454–461.

[19] K. H. Kim, Programmer-transparent coordination of recovering concurrent processes: philosophy and rules for efficient implementation, *IEEE Transactions on Software Engineering*, **14**(6), 1988, 810–821.

[20] K. H. Kim, Approach to mechanization of the conversation scheme based on monitor, *IEEE Transactions on Software Engineering*, **8**(3), 1982, 189–197.

[21] K. H. Kim, Software fault tolerance, in C. R. Vick and C. V. Ramamoorthy (eds), *Handbook of Software Engineering*, New York, Van Nostrand Reinhold, 1984.

[22] R. Koo and S. Toueg, Checkpointing and rollback-recovery for distributed systems, *IEEE Transactions on Software Engineering*, **13**(1) 1987, 23–31.

[23] L. Lamport, Time, clocks, and the ordering of events in a distributed system, *Communications of the ACM*, **21**(7), 1978, 558–565.

[24] D. Manivannan and M. Singhal, Asynchronous recovery without using vector timestamps, *Journal of Parallel and Distributed Computing*, **62**(12), 2002, 1695–1728.

[25] D. Manivannan and M. Singhal, A low overhead recovery technique using quasi-synchronous checkpointing, *Proceedings of the 16th International Conference on Distributed Computing Systems*, 1996, 100–107.

[26] D. Manivannan, R. H. B. Netzer, and M. Singhal, Finding consistent global checkpoints in a distributed computation, *IEEE Transactions on Parallel and Distributed Systems*, **8**(6), 1997, 623–627.

[27] R. H. B. Netzer and J. Xu, Necessary and sufficient conditions for consistent global snapshots, *IEEE Transactions on Parallel and Distributed Systems*, **6**(2), 1995, 165–169.

[28] S. L. Peterson and P. Kearns, Rollback based on vector time, *Proceedings of the Symposium on Reliable Distributed Systems*, 1993, 68–77.

[29] R. Prakash and M. Singhal, Maximal global snapshot with concurrent initiators, *Proceedings of the 6th IEEE Symposium on Parallel and Distributed Processing*, Dallas, TX, 1994, 344–351.

[30] R. Prakash and M. Singhal, Low-cost checkpointing and failure recovery in mobile computing systems, *IEEE Transactions on Parallel and Distributed Systems*, **7**(10), 1996, 1035–1048.

[31] P. Ramanathan and K. G. Shin, Use of common time base for checkpointing and rollback recovery in a distributed system, *IEEE Transactions on Software Engineering*, **19**(6), 1993, 571–583.

[32] B. Randell, System structure for software fault tolerance, *IEEE Transactions on Software Engineering*, **1**(2), 1975, 220–232.

[33] G. G. Richard III and M. Singhal, Complete process recovery: using vector time to handle multiple failures in distributed systems, *IEEE Parallel and Distributed Technology: Systems and Technology*, **5**(2), 1997, 50–59.

[34] D. L. Russell, State restoration in systems of communicating processes, *IEEE Transactions of Software Engineering*, **6**(2), 1980, 183–194.

[35] A. P. Sistla and J. L. Welch, Efficient distributed recovery using message logging, *Proceedings of the 8th Annual ACM Symposium on Principles of Distributed Computing*, Edmonton, Alberta, Canada, 1989, 223–238.

[36] R. Strom and S. Yemini, Optimistic recovery in distributed systems, *ACM Transactions on Computer Systems*, **3**(3), 1985, 204–226.

[37] Z. Tong, R. Y. Kain, and W. T. Tsai, Rollback recovery in distributed systems using loosely synchronized clocks, *IEEE Transactions on Parallel and Distributed Systems*, **3**(2), 1992, 246–251.

[38] S. Venkatesan, T. T.-Y. Juang and S. Alagar, Optimistic crash recovery without changing application messages, *IEEE Transactions on Parallel and Distributed Systems*, **8**(3), 1997, 263–271.

[39] Y.-M. Wang and W. K. Fuchs, Lazy checkpoint coordination for bounding rollback propagation, *IEEE Symposium on Reliable Distributed Systems*, 1993.

[40] Y.-M. Wang, P.-Y. Chung, I.-J. Lin, and W. K. Fuchs, Checkpoint space reclamation for uncoordinated checkpointing in message-passing systems, *IEEE Transactions on Parallel and Distributed Systems*, **6**(5), 1995, 546–554.

[41] Y.-M. Wang, Consistent global checkpoints that contain a given set of local checkpoints, *IEEE Transactions on Computers*, **46**(4), 1997, 456–468.

CHAPTER

14 Consensus and agreement algorithms

14.1 Problem definition

Agreement among the processes in a distributed system is a fundamental requirement for a wide range of applications. Many forms of coordination require the processes to exchange information to negotiate with one another and eventually reach a common understanding or agreement, before taking application-specific actions. A classical example is that of the *commit* decision in database systems, wherein the processes collectively decide whether to *commit* or *abort* a transaction that they participate in. In this chapter, we study the feasibility of designing algorithms to reach agreement under various system models and failure models, and, where possible, examine some representative algorithms to reach agreement.

We first state some assumptions underlying our study of agreement algorithms:

- **Failure models** Among the n processes in the system, at most f processes can be faulty. A faulty process can behave in any manner allowed by the failure model assumed. The various failure models – fail-stop, send omission and receive omission, and Byzantine failures – were discussed in Chapter 5. Recall that in the fail-stop model, a process may crash in the middle of a step, which could be the execution of a local operation or processing of a message for a send or receive event. In particular, it may send a message to only a subset of the destination set before crashing. In the Byzantine failure model, a process may behave arbitrarily. The choice of the failure model determines the feasibility and complexity of solving consensus.
- **Synchronous/asynchronous communication** If a failure-prone process chooses to send a message to process P_i but fails, then P_i cannot detect the non-arrival of the message in an asynchronous system because this scenario is indistinguishable from the scenario in which the message takes a very long time in transit. We will see this argument again when we consider

the impossibility of reaching agreement in asynchronous systems in any failure model. In a synchronous system, however, the scenario in which a message has not been sent can be recognized by the intended recipient, at the end of the round. The intended recipient can deal with the non-arrival of the expected message by assuming the arrival of a message containing some default data, and then proceeding with the next round of the algorithm.

- **Network connectivity** The system has full logical connectivity, i.e., each process can communicate with any other by direct message passing.
- **Sender identification** A process that receives a message always knows the identity of the sender process. This assumption is important – because even with Byzantine behavior, even though the payload of the message can contain fictitious data sent by a malicious sender, the underlying network layer protocols can reveal the true identity of the sender process.

  When multiple messages are expected from the same sender in a single round, we implicitly assume a scheduling algorithm that sends these messages in sub-rounds, so that each message sent within the round can be uniquely identified.
- **Channel reliability** The channels are reliable, and only the processes may fail (under one of various failure models). This is a simplifying assumption in our study. As we will see even with this simplifying assumption, the agreement problem is either unsolvable, or solvable in a complex manner.
- **Authenticated vs. non-authenticated messages** In our study, we will be dealing only with *unauthenticated* messages. With unauthenticated messages, when a faulty process relays a message to other processes, (i) it can forge the message and claim that it was received from another process, and (ii) it can also tamper with the contents of a received message before relaying it. When a process receives a message, it has no way to verify its authenticity. An unauthenticated message is also called an *oral* message or an *unsigned* message.

  Using authentication via techniques such as digital signatures, it is easier to solve the agreement problem because, if some process forges a message or tampers with the contents of a received message before relaying it, the recipient can detect the forgery or tampering. Thus, faulty processes can inflict less damage.
- **Agreement variable** The agreement variable may be boolean or multivalued, and need not be an integer. When studying some of the more complex algorithms, we will use a boolean variable. This simplifying assumption does not affect the results for other data types, but helps in the abstraction while presenting the algorithms.

Consider the difficulty of reaching agreement using the following example, that is inspired by the long wars fought by the Byzantine Empire in the Middle

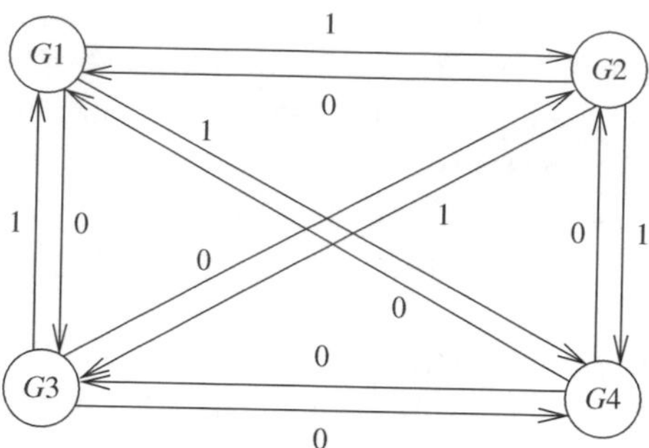

Figure 14.1 Byzantine generals sending confusing messages.

Ages. Four camps of the attacking army, each commanded by a general, are camped around the fort of Byzantium.[1] They can succeed in attacking only if they attack simultaneously. Hence, they need to reach agreement on the time of attack. The only way they can communicate is to send messengers among themselves. The messengers model the messages. An asynchronous system is modeled by messengers taking an unbounded time to travel between two camps. A lost message is modeled by a messenger being captured by the enemy. A Byzantine process is modeled by a general being a traitor. The traitor will attempt to subvert the agreement-reaching mechanism, by giving misleading information to the other generals. For example, a traitor may inform one general to attack at 10 a.m., and inform the other generals to attack at noon. Or he may not send a message at all to some general. Likewise, he may tamper with the messages he gets from other generals, before relaying those messages.

A simple example of Byzantine behavior is shown in Figure 14.1. Four generals are shown, and a *consensus* decision is to be reached about a boolean value. The various generals are conveying potentially misleading values of the decision variable to the other generals, which results in confusion. In the face of such Byzantine behavior, the challenge is to determine whether it is possible to reach agreement, and if so under what conditions. If agreement is reachable, then protocols to reach it need to be devised.

14.1.1 The Byzantine agreement and other problems

The Byzantine agreement problem

Before studying algorithms to solve the agreement problem, we first define the problem formally [20,25]. The *Byzantine agreement* problem requires a designated process, called the *source process*, with an *initial value*, to reach

[1] Byzantium was the name of present-day Istanbul; Byzantium also had the name of Constantinople.

agreement with the other processes about its initial value, subject to the following conditions:

- **Agreement** All non-faulty processes must agree on the same value.
- **Validity** If the source process is non-faulty, then the agreed upon value by all the non-faulty processes must be the same as the initial value of the source.
- **Termination** Each non-faulty process must eventually decide on a value.

The validity condition rules out trivial solutions, such as one in which the agreed upon value is a constant. It also ensures that the agreed upon value is correlated with the source value. If the source process is faulty, then the correct processes can agree upon any value. It is irrelevant what the faulty processes agree upon – or whether they terminate and agree upon anything at all.

There are two other popular flavors of the Byzantine agreement problem – the *consensus* problem, and the *interactive consistency* problem.

The consensus problem

The consensus problem differs from the Byzantine agreement problem in that each process has an initial value and all the correct processes must agree on a single value [20,25]. Formally:

- **Agreement** All non-faulty processes must agree on the same (single) value.
- **Validity** If all the non-faulty processes have the same initial value, then the agreed upon value by all the non-faulty processes must be that same value.
- **Termination** Each non-faulty process must eventually decide on a value.

The interactive consistency problem

The interactive consistency problem differs from the Byzantine agreement problem in that each process has an initial value, and all the correct processes must agree upon a set of values, with one value for each process [20,25]. The formal specification is as follows:

- **Agreement** All non-faulty processes must agree on the same array of values $A[v_1 \ldots v_n]$.
- **Validity** If process i is non-faulty and its initial value is v_i, then all non-faulty processes agree on v_i as the ith element of the array A. If process j is faulty, then the non-faulty processes can agree on any value for $A[j]$.
- **Termination** Each non-faulty process must eventually decide on the array A.

14.1.2 Equivalence of the problems and notation

The three problems defined above are equivalent in the sense that a solution to any one of them can be used as a solution to the other two problems [9]. This equivalence can be shown using a reduction of each problem to the other two problems. If problem A is reduced to problem B, then a solution to problem B can be used as a solution to problem A in conjunction with the reduction. Exercise 14.1 asks the reader to show these reductions.

Formally, the difference between the *agreement problem* and the *consensus problem* is that, in the agreement problem, a single process has the initial value, whereas in the consensus problem, all processes have an initial value. However, the two terms are used interchangably in much of the literature and hence we shall also use the terms interchangably.

14.2 Overview of results

Table 14.1 gives an overview of the results and lower bounds on solving the consensus problem under different assumptions.

It is worth understanding the relation between the consensus problem and the problem of attaining common knowledge of the agreement value. For the "no failure" case, consensus is attainable. Further, in a synchronous system, common knowledge of the consensus value is also attainable, whereas in the asynchronous case, concurrent common knowledge of the consensus value is attainable.

Consensus is not solvable in asynchronous systems even if one process can fail by crashing. To circumvent this impossibility result, weaker variants

Table 14.1 Overview of results on agreement. f denotes number of failure-prone processes. n is the total number of processes.

Failure mode	Synchronous system (message-passing and shared memory)	Asynchronous system (message-passing and shared memory)
No failure	Agreement attainable Common knowledge also attainable	Agreement attainable Concurrent common knowledge attainable
Crash failure	Agreement attainable $f < n$ processes $\Omega(f+1)$ rounds	Agreement not attainable
Byzantine failure	Agreement attainable $f \leq \lfloor (n-1)/3 \rfloor$ Byzantine processes $\Omega(f+1)$ rounds	Agreement not attainable

Table 14.2 Some solvable variants of the agreement problem in an asynchronous system. The overhead bounds are for the given algorithms, and are not necessarily tight bounds for the problem.

Solvable variants	Failure model and overhead	Definition
Reliable broadcast	Crash failures, $n > f$ (MP)	Validity, agreement, integrity conditions (Section 14.5.7)
k-set consensus	Crash failures, $f < k < n$ (MP and SM)	Size of the set of values agreed upon must be at most k (Section 14.5.4)
ϵ-agreement	Crash failures, $n \geq 5f + 1$ (MP)	Values agreed upon are within ϵ of each other (Section 14.5.5)
Renaming	Up to f fail-stop processes, $n \geq 2f + 1$ (MP) Crash failures, $f \leq n - 1$ (SM)	Select a unique name from a set of names (Section 14.5.6)

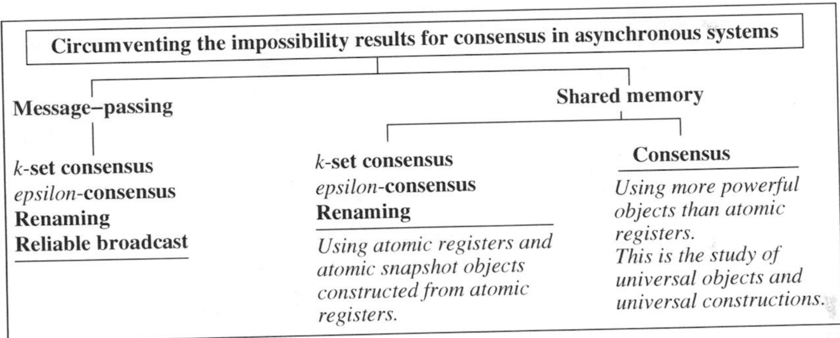

Figure 14.2 Circumventing the impossibility result for consensus in asynchronous systems.

of the consensus problem are defined in Table 14.2. The overheads given in this table are for the algorithms described. Figure 14.2 shows further how asynchronous message-passing systems and shared memory systems deal with trying to solve consensus.

14.3 Agreement in a failure-free system (synchronous or asynchronous)

In a failure-free system, consensus can be reached by collecting information from the different processes, arriving at a "decision," and distributing this decision in the system. A distributed mechanism would have each process broadcast its values to others, and each process computes the same function on the values received. The decision can be reached by using an application-specific function – some simple examples being the *majority*, *max*, and *min* functions. Algorithms to collect the initial values and then distribute the decision may be based on the token circulation on a logical ring, or the three-phase tree-based broadcast–convergecast–broadcast, or direct communication with all nodes.

- In a synchronous system, this can be done simply in a constant number of rounds (depending on the specific logical topology and algorithm used). Further, common knowledge of the decision value can be obtained using an additional round (see Chapter 8).
- In an asynchronous system, consensus can similarly be reached in a constant number of message hops. Further, concurrent common knowledge of the consensus value can also be attained, using any of the algorithms in Chapter 8.

Reaching agreement is straightforward in a failure-free system. Hence, we focus on failure-prone systems.

14.4 Agreement in (message-passing) synchronous systems with failures

14.4.1 Consensus algorithm for crash failures (synchronous system)

Algorithm 14.1 gives a consensus algorithm for n processes, where up to f processes, where $f < n$, may fail in the fail-stop model [8]. Here, the consensus variable x is integer-valued. Each process has an initial value x_i. If up to f failures are to be tolerated, then the algorithm has $f+1$ rounds. In each round, a process i sends the value of its variable x_i to all other processes if that value has not been sent before. Of all the values received within the round and its own value x_i at the start of the round, the process takes the minimum, and updates x_i. After $f+1$ rounds, the local value x_i is guaranteed to be the consensus value.

(global constants)
integer: f; // maximum number of crash failures tolerated
(local variables)
integer: $x \longleftarrow$ local value;

(1) Process P_i ($1 \le i \le n$) executes the consensus algorithm for up to f crash failures:
(1a) **for** *round* **from** 1 **to** $f+1$ **do**
(1b) **if** the current value of x has not been broadcast **then**
(1c) **broadcast**(x);
(1d) $y_j \longleftarrow$ value (if any) received from process j in this round;
(1e) $x \longleftarrow min_{\forall j}(x, y_j)$;
(1f) **output** x as the consensus value.

Algorithm 14.1 Consensus with up to f fail-stop processes in a system of n processes, $n > f$ [8]. Code shown is for process P_i, $1 \le i \le n$.

- The *agreement condition* is satisfied because in the $f+1$ rounds, there must be at least one round in which no process failed. In this round, say round r, all the processes that have not failed so far succeed in broadcasting their values, and all these processes take the minimum of the values broadcast and received in that round. Thus, the local values at the end of the round are the same, say x_i^r for all non-failed processes. In further rounds, only this value may be sent by each process at most once, and no process i will update its value x_i^r.
- The *validity condition* is satisfied because processes do not send fictitious values in this failure model. (Thus, a process that crashes has sent only correct values until the crash.) For all i, if the initial value is identical, then the only value sent by any process is the value that has been agreed upon as per the *agreement condition*.
- The *termination condition* is seen to be satisfied.

Complexity

There are $f+1$ rounds, where $f < n$. The number of messages is at most $O(n^2)$ in each round, and each message has one integer. Hence the total number of messages is $O((f+1) \cdot n^2)$. The worst-case scenario is as follows. Assume that the minimum value is with a single process initially. In the first round, the process manages to send its value to just one other process before failing. In subsequent rounds, the single process having this minimum value also manages to send that value to just one other process before failing.

Algorithm 14.1 requires $f+1$ rounds, independent of the actual number of processes that fail. An *early-stopping* consensus algorithm terminates sooner; if there are f' actual failures, where $f' < f$, then the early-stopping algorithm terminates in $f'+1$ rounds. Exercise 14.2 asks you to design an early-stopping algorithm for consensus under crash failures, and to prove its correctness.

A lower bound on the number of rounds [8]

At least $f+1$ rounds are required, where $f < n$. The idea behind this lower bound is that in the worst-case scenario, one process may fail in each round; with $f+1$ rounds, there is at least one round in which no process fails. In that guaranteed failure-free round, all messages broadcast can be delivered reliably, and all processes that have not failed can compute the common function of the received values to reach an agreement value.

14.4.2 Consensus algorithms for Byzantine failures (synchronous system)

14.4.3 Upper bound on Byzantine processes

In a system of n processes, the Byzantine agreement problem (as also the other variants of the agreement problem) can be solved in a synchronous

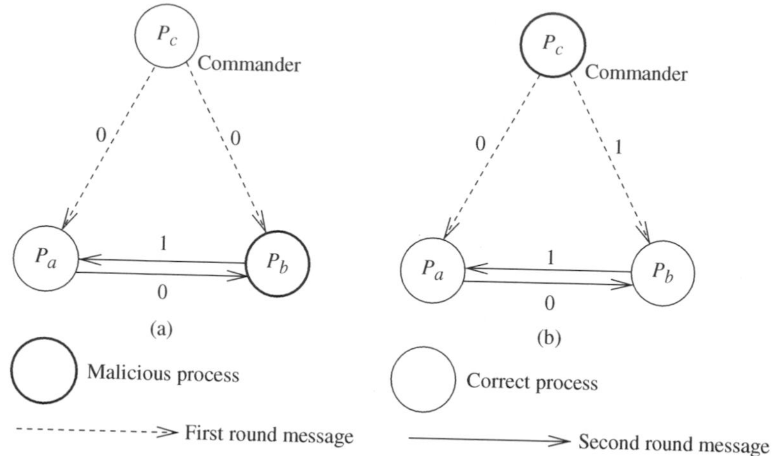

Figure 14.3 Impossibility of achieving Byzantine agreement with $n = 3$ processes and $f = 1$ malicious process.

system only if the number of Byzantine processes f is such that $f \leq \lfloor \frac{n-1}{3} \rfloor$ [20,25].

We informally justify this result using two steps:

- With $n = 3$ processes, the Byzantine agreement problem cannot be solved if the number of Byzantine processes $f = 1$. The argument uses the illustration in Figure 14.3, which shows a commander P_c and two lieutenant processes P_a and P_b. The malicious process is the lieutenant P_b in the first scenario (Figure 14.3(a)) and hence P_a should agree on the value of the loyal commander P_c, which is 0. But note the second scenario (Figure 14.3(b)) in which P_a receives identical values from P_b and P_c, but now P_c is the disloyal commander whereas P_b is a loyal lieutenant. In this case, P_a needs to agree with P_b. However, P_a cannot distinguish between the two scenarios and any further message exchange does not help because each process has already conveyed what it knows from the third process.

  In both scenarios, P_a gets different values from the other two processes. In the first scenario, it needs to agree on a 0, and if that is the default value, the decision is correct, but then if it is in the second indistinguishable scenario, it agrees on an incorrect value. A similar argument shows that if 1 is the default value, then in the first scenario, P_a makes an incorrect decision. This shows the impossibility of agreement when $n = 3$ and $f = 1$.
- With n processes and $f \geq n/3$ processes, the Byzantine agreement problem cannot be solved. The correctness argument of this result can be shown using reduction. Let $Z(3, 1)$ denote the Byzantine agreement problem for parameters $n = 3$ and $f = 1$. Let $Z(n \leq 3f, f)$ denote the Byzantine agreement problem for parameters $n(\leq 3f)$ and f. A reduction from $Z(3, 1)$ to $Z(n \leq 3f, f)$ needs to be shown, i.e., if $Z(n \leq 3f, f)$ is solvable, then $Z(3, 1)$ is also solvable. After showing this reduction, we can argue that as $Z(3, 1)$ is not solvable, $Z(n \leq 3f, f)$ is also not solvable.

The main idea of the reduction argument is as follows. In $Z(n \leq 3f, f)$, partition the n processes into three sets S_1, S_2, S_3, each of size $\leq n/3$. In $Z(3, 1)$, each of the three processes P_1, P_2, P_3 simulates the actions of the corresponding set S_1, S_2, S_3 in $Z(n \leq 3f, f)$. If one process is faulty in $Z(3, 1)$, then at most f, where $f \leq n/3$, processes are faulty in $Z(n, f)$. In the simulation, a correct process in $Z(3, 1)$ simulates a group of up to $n/3$ correct processes in $Z(n, f)$. It simulates the actions (send events, receive events, intra-set communication, and inter-set communication) of each of the processes in the set that it is simulating.

With this reduction in place, if there exists an algorithm to solve $Z(n \leq 3f, f)$, i.e., to satisfy the validity, agreement, and termination conditions, then there also exists an algorithm to solve $Z(3, 1)$, which has been seen to be unsolvable. Hence, there cannot exist an algorithm to solve $Z(n \leq 3f, f)$.

Byzantine agreement tree algorithm: exponential (synchronous system)

Recursive formulation

We begin with an informal description of how agreement can be achieved with $n = 4$ and $f = 1$ processes [20,25], as depicted in Figure 14.4. In the first round, the commander P_c sends its value to the other three lieutenants, as shown by dotted arrows. In the second round, each lieutenant relays to the other two lieutenants, the value it received from the commander in the first round. At the end of the second round, a lieutenant takes the majority of the values it received (i) directly from the commander in the first round, and (ii) from the other two lieutenants in the second round. The majority gives a correct estimate of the "commander's" value. Consider Figure 14.4(a) where the commander is a traitor. The values that get transmitted in the two rounds are as

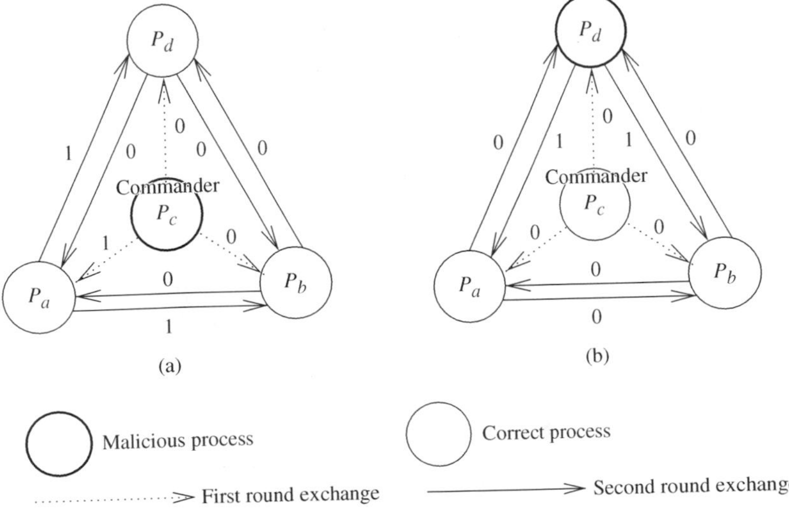

Figure 14.4 Achieving Byzantine agreement when $n = 4$ processes and $f = 1$ malicious process.

shown. All three lieutenants take the majority of $(1, 0, 0)$ which is "0," the agreement value. In Figure 14.4(b), lieutenant P_d is malicious. Despite its behavior as shown, lieutenants P_a and P_b agree on "0," the value of the commander.

(variables)

boolean: $v \longleftarrow$ initial value;

integer: $f \longleftarrow$ maximum number of malicious processes, $\leq \lfloor (n-1)/3 \rfloor$;

(message type)

OM(v, *Dests*, *List*, *faulty*), where

v is a boolean,

Dests is a set of destination process i.d.s to which the message is sent,

List is a list of process i.d.s traversed by this message, ordered from most recent to earliest,

faulty is an integer indicating the number of malicious processes to be tolerated.

Oral_Msg (f), where $f > 0$:

(1) The algorithm is initiated by the commander, who sends his source value v to all other processes using a $OM(v, N, \langle i \rangle, f)$ message. The commander returns his own value v and terminates.

(2) **[Recursion unfolding:]** For each message of the form $OM(v_j$, *Dests*, *List*, $f')$ received in this round from some process j, the process i uses the value v_j it receives from the source j, and using that value, acts as a *new* source. (If no value is received, a default value is assumed.)

To act as a new source, the process i initiates *Oral_Msg* $(f' - 1)$, wherein it sends

$OM(v_j, Dests - \{i\}, concat(\langle i \rangle, L), (f' - 1))$

to destinations not in $concat(\langle i \rangle, L)$

in the next round.

(3) **[Recursion folding:]** For each message of the form $OM(v_j$, *Dests*, *List*, $f')$ received in step 2, each process i has computed the agreement value v_k, for each k not in *List* and $k \neq i$, corresponding to the value received from P_k after traversing the nodes in *List*, at one level lower in the recursion. If it receives no value in this round, it uses a default value. Process i then uses the value $majority_{k \notin List, k \neq i}(v_j, v_k)$ as the agreement value and returns it to the next higher level in the recursive invocation.

Oral_Msg(0):

(1) **[Recursion unfolding:]** Process acts as a source and sends its value to each other process.

(2) **[Recursion folding:]** Each process uses the value it receives from the other sources, and uses that value as the agreement value. If no value is received, a default value is assumed.

Algorithm 14.2 Byzantine generals algorithm – exponential number of unsigned messages, $n > 3f$. Recursive formulation.

Table 14.3 Relationships between messages and rounds in the oral messages algorithm for the Byzantine agreement.

Round number	A message has already visited	Aims to tolerate these many failures	Each message gets sent to	Total number of messages in round
1	1	f	$n-1$	$n-1$
2	2	$f-1$	$n-2$	$(n-1)\cdot(n-2)$
...	...	...	...	...
x	x	$(f+1)-x$	$n-x$	$(n-1)(n-2)\ldots(n-x)$
$x+1$	$x+1$	$(f+1)-x-1$	$n-x-1$	$(n-1)(n-2)\ldots(n-x-1)$
...	...	...	...	...
$f+1$	$f+1$	0	$n-f-1$	$(n-1)(n-2)\ldots(n-f-1)$

The first algorithm for solving Byzantine agreement was proposed by Lamport *et al.* [20]. We present two versions of the algorithm.

The recursive version of the algorithm is given in Algorithm 14.2. Each message has the following parameters: a consensus estimate value (v); a set of destinations ($Dests$); a list of nodes traversed by the message, from most recent to least recent ($List$); and the number of Byzantine processes that the algorithm still needs to tolerate ($faulty$). The list $L = \langle P_i, P_{k_1} \ldots P_{k_{f+1-faulty}} \rangle$ represents the sequence of processes (subscripts) in the knowledge expression $K_i(K_{k_1}(K_{k_2} \ldots K_{k_{f+1-faulty}}(v_0)\ldots))$. This knowledge is what $P_{k_{f+1-faulty}}$ conveyed to $P_{k_{f-faulty}}$ conveyed to ... P_{k_1} conveyed to P_i who is conveying to the receiver of this message, the value of the commander ($P_{k_{f+1-faulty}}$)'s initial value.

The commander invokes the algorithm with parameter $faulty$ set to f, the maximum number of malicious processes to be tolerated. The algorithm uses $f+1$ synchronous rounds. Each message (having this parameter $faulty = k$) received by a process invokes several other instances of the algorithm with parameter $faulty = k-1$. The terminating case of the recursion is when the parameter $faulty$ is 0. As the recursion folds, each process progressively computes the majority function over the values it used as a source for that level of invocation in the unfolding, and the values it has just computed as consensus values using the majority function for the lower level of invocations.

There are an exponential number of messages $O(n^f)$ used by this algorithm. Table 14.3 shows the number of messages used in each round of the algorithm, and relates that number to the number of processes already visited by any message as well as the number of destinations of that message.

As multiple messages are received in any one round from each of the other processes, they can be distinguished using the $List$, or by using a scheduling algorithm within each round. A detailed iterative version of the high-level recursive algorithm is given in Algorithm 14.3. Lines 2a–2e correspond to the unfolding actions of the recursive pseudo-code, and lines 2f–2h correspond to the folding of the recursive pesudo-code. Two operations are defined in the list L: $head(L)$ is the first member of the list L, whereas $tail(L)$

(variables)
boolean: $v \longleftarrow$ initial value;
integer: $f \longleftarrow$ maximum number of malicious processes, $\leq \lfloor (n-1)/3 \rfloor$;
tree of boolean:

- level 0 root is v^L_{init}, where $L = \langle\rangle$;
- level h ($f \geq h > 0$) nodes: for each v^L_j at level $h-1 = sizeof(L)$, its $n-2-sizeof(L)$ descendants at level h are $v_k^{concat(\langle j\rangle,L)}$, $\forall k$ such that $k \neq j, i$ and k is not a member of list L.

(message type)
$OM(v, Dests, List, faulty)$, where the parameters are as in the recursive formulation.

(1) Initiator (i.e., commander) initiates the oral Byzantine agreement:
(1a) **send** $OM(v, N-\{i\}, \langle P_i\rangle, f)$ to $N-\{i\}$;
(1b) **return**(v).

(2) (Non-initiator, i.e., lieutenant) receives the oral message (OM):
(2a) **for** $rnd = 0$ **to** f **do**
(2b) **for** each message OM that arrives in this round, **do**
(2c) **receive** $OM(v, Dests, L = \langle P_{k_1} \dots P_{k_{f+1-faulty}}\rangle, faulty)$ from P_{k_1}; // $faulty + rnd = f$; $|Dests| + sizeof(L) = n$
(2d) $v_{head(L)}^{tail(L)} \longleftarrow v$; // $sizeof(L) + faulty = f+1$. fill in estimate.
(2e) **send** $OM(v, Dests-\{i\}, \langle P_i, P_{k_1} \dots P_{k_{f+1-faulty}}\rangle, faulty-1)$ to $Dests-\{i\}$ **if** $rnd < f$;
(2f) **for** $level = f-1$ **down to** 0 **do**
(2g) **for** each of the $1\cdot(n-2)\cdot\ldots(n-(level+1))$ nodes v^L_x in level $level$, **do**
(2h) $v^L_x(x \neq i, x \notin L) = majority_{y \notin concat(\langle x\rangle,L);\, y\neq i}(v^L_x, v_y^{concat(\langle x\rangle,L)})$;

Algorithm 14.3 Byzantine generals algorithm – exponential number of unsigned messages, $n > 3f$. Iterative formulation. Code for process P_i.

is the list L after removing its first member. Each process maintains a tree of boolean variables. The tree data structure at a non-initiator is used as follows:

- There are $f+1$ levels from level 0 through level f.
- Level 0 has one root node, $v^{\langle\rangle}_{init}$, after round 1.

- Level h, $0 < h \leq f$ has $1 \cdot (n-2) \cdot (n-3) \cdots (n-h) \cdot (n-(h+1))$ nodes after round $h+1$. Each node at level $(h-1)$ has $(n-(h+1))$ child nodes.
- Node v_k^L denotes the command received from the node $head(L)$ by node k which forwards it to node i. The command was relayed to $head(L)$ by $head(tail(L))$, which received it from $head(tail(tail(L)))$, and so on. The very last element of L is the commander, denoted P_{init}.
- In the $f+1$ rounds of the algorithm (lines 2a–2e of the iterative version), each level k, $0 \leq k \leq f$, of the tree is successively filled to remember the values received at the end of round $k+1$, and with which the process sends the multiple instances of the *OM* message with the fourth parameter as $f-(k+1)$ for round $k+2$ (other than the final terminating round).
- For each message that arrives in a round (lines 2b–2c of the iterative version), a process sets $v_{head(L)}^{tail(L)}$ (line 2d). It then removes itself from *Dests*, prepends itself to L, decrements *faulty*, and forwards the value v to the updated *Dests* (line 2e).
- Once the entire tree is filled from root to leaves, the actions in the folding of the recursion are simulated in lines 2f–2h of the iterative version, proceeding from the leaves up to the root of the tree. These actions are crucial – they entail taking the majority of the values at each level of the tree. The final value of the root is the agreement value, which will be the same at all processes.

Example Figure 14.5 shows the tree at a lieutenant node P_3, for $n = 10$ processes P_0 through P_9 and $f = 3$ processes. The commander is P_0. Only one branch of the tree is shown for simplicity. The reader is urged to work through all the steps to ensure a thorough understanding. Some key steps from P_3's perspective are outlined next, with respect to the iterative formulation of the algorithm.

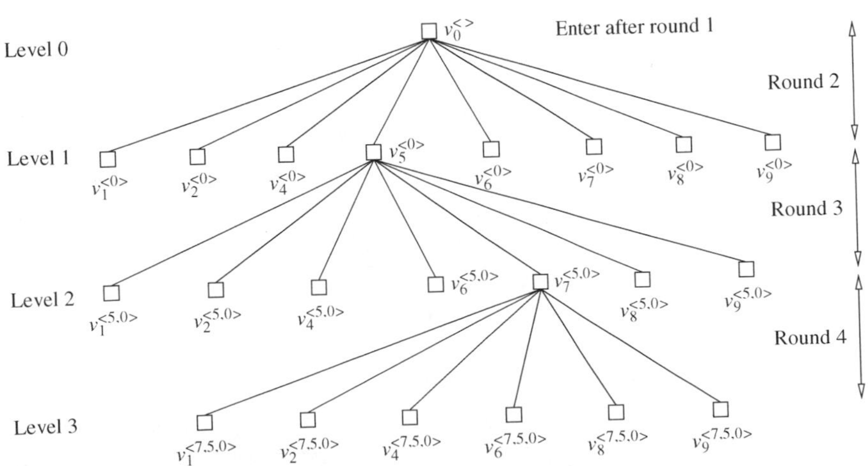

Figure 14.5 Local tree at P_3 for solving the Byzantine agreement, for $n = 10$ and $f = 3$. Only one branch of the tree is shown for simplicity.

- **Round 1** P_0 sends its value to all other nodes. This corresponds to invoking *Oral_Msg* (3) in the recursive formulation. At the end of the round, P_3 stores the received value in $v_0^{\langle\rangle}$.
- **Round 2** P_3 acts as a source for this value and sends this value to all nodes except itself and P_0. This corresponds to invoking *Oral_Msg* (2) in the recursive formulation. Thus, P_3 sends 8 messages. It will receive a similar message from all other nodes except P_0 and itself; the value received from P_k is stored in $v_k^{\langle 0\rangle}$.
- **Round 3** For each of the 8 values received in round 2, P_3 acts as a source and sends the values to all nodes except (i) itself, (ii) nodes visited previously by the corresponding value, as remembered in the superscript list, and (iii) the direct sender of the received message, as indicated by the subscript. This corresponds to invoking *Oral_Msg* (1) in the recursive formulation. Thus, P_3 sends 7 messages for each of these 8 values, giving a total of 56 messages it sends in this round. Likewise it receives 56 messages from other nodes; the values are stored in level 2 of the tree.
- **Round 4** For each of the 56 messages received in round 3, P_3 acts a source and sends the values to all nodes except (i) itself, (ii) nodes visited previously by the corresponding value, as remembered in the superscript list, and (iii) the direct sender of the received message, as indicated by the subscript. This corresponds to invoking *Oral_Msg* (0) in the recursive formulation. Thus, P_3 sends 6 messages for each of these 56 values, giving a total of 336 messages it sends in this round. Likewise, it receives 336 messages, and the values are stored at level 3 of the tree. As this round is *Oral_Msg* (0), the received values are used as estimates for computing the majority function in the folding of the recursion.

An example of the majority computation is as follows:

- P_3 revises its estimate of $v_7^{\langle 5,0\rangle}$ by taking *majority* $(v_7^{\langle 5,0\rangle}, v_1^{\langle 7,5,0\rangle}, v_2^{\langle 7,5,0\rangle}, v_4^{\langle 7,5,0\rangle}, v_6^{\langle 7,5,0\rangle}, v_8^{\langle 7,5,0\rangle}, v_9^{\langle 7,5,0\rangle})$. Similarly for the other nodes at level 2 of the tree.
- P_3 revises its estimate of $v_5^{\langle 0\rangle}$ by taking *majority* $(v_5^{\langle 0\rangle}, v_1^{\langle 5,0\rangle}, v_2^{\langle 5,0\rangle}, v_4^{\langle 5,0\rangle}, v_6^{\langle 5,0\rangle}, v_7^{\langle 5,0\rangle}, v_8^{\langle 5,0\rangle}, v_9^{\langle 5,0\rangle})$. Similarly for the other nodes at level 1 of the tree.
- P_3 revises its estimate of $v_0^{\langle\rangle}$ by taking *majority*$(v_0^{\langle\rangle}, v_1^{\langle 0\rangle}, v_2^{\langle 0\rangle}, v_4^{\langle 0\rangle}, v_5^{\langle 0\rangle}, v_6^{\langle 0\rangle}, v_7^{\langle 0\rangle}, v_8^{\langle 0\rangle}, v_9^{\langle 0\rangle})$. This is the consensus value.

Correctness

The correctness of the Byzantine agreement algorithm (Algorithm 14.3) can be observed from the following two informal inductive arguments. Here we assume that the *Oral_Msg* algorithm is invoked with parameter x, and that there are a total of f malicious processes. There are two cases depending on whether the commander is malicious. A malicious commander causes more chaos than an honest commander.

Loyal commander

Given f and x, if the commander process is loyal, then *Oral_Msg* (x) is correct if there are at least $2f+x$ processes.

This can easily be seen by induction on x:

- For $x=0$, *Oral_Msg* (0) is executed, and the processes simply use the (loyal) commander's value as the consensus value.
- Now assume the above induction hypothesis for any x.
- Then for *Oral_Msg* $(x+1)$, there are $2f+x+1$ processes including the commander. Each loyal process invokes *Oral_Msg* (x) to broadcast the (loyal) commander's value v_0 – here it acts as a commander for this invocation it makes. As there are $2f+x$ processes for each such invocation, by the induction hypothesis, there is agreement on this value (at all the honest processes) – this would be at level 1 in the local tree in the folding of the recursion. In the last step, each loyal process takes the majority of the direct order received from the commander (level 0 entry of the tree), and its estimate of the commander's order conveyed to other processes as computed in the level 1 entries of the tree. Among the $2f+x$ values taken in the majority calculation (this includes the commanders's value but not its own), the majority is loyal because $x>0$. Hence, taking the majority works.

No assumption about commander

Given f, *Oral_Msg* (x) is correct if $x \geq f$ and there are a total of $3x+1$ or more processes.

This case accounts for both possibilities – the commander being malicious or honest. An inductive argument is again useful.

- For $x=0$, *Oral_Msg* (0) is executed, and as there are no malicious processes ($0 \geq f$) the processes simply use the (loyal) commander's value as the consensus value. Hence the algorithm is correct.
- Now assume the above induction hypothesis for any x.
- Then for *Oral_Msg* $(x+1)$, there are at least $3x+4$ processes including the commander and at most $x+1$ are malicious.
  - (Loyal commander:) If the commander is loyal, then we can apply the argument used for the "loyal commander" case above, because there will be more than $(2(f+1)+(x+1))$ total processes.
  - (Malicious commander:) There are now at most x other malicious processes and $3x+3$ total processes (excluding the commander). From the induction hypothesis, each loyal process can compute the consensus value using the majority function in the protocol.

Illustration of arguments

In Figure 14.6(a), the commander who invokes *Oral_Msg* (x) is loyal, so all the loyal processes have the same estimate. Although the subsystem of $3x$ processes has x malicious processes, all the loyal processes have the same view to begin with. Even if this case repeats for each nested invocation of *Oral_Msg*, even after x rounds, among the processes, the loyal processes are in a simple majority, so the majority function works in having them maintain the same common view of the loyal commander's value. (Of course, had we known the commander was loyal, then we could have terminated after a single round, and neither would we be restricted by the $n > 3x$ bound.) In Figure 14.6(b), the commander who invokes *Oral_Msg* (x) may be malicious and can send conflicting values to the loyal processes. The subsystem of $3x$ processes has $x-1$ malicious processes, but all the loyal processes do not have the same view to begin with.

Complexity

The algorithm requires $f+1$ rounds, an exponential amount of local memory, and

$$(n-1)+(n-1)(n-2)+\cdots+[(n-1)(n-2)\cdots(n-f-1)] \text{ messages.}$$

Phase-king algorithm for consensus: polynomial (synchronous system)

The Lamport–Shostak–Pease algorithm [21] requires $f+1$ rounds and can tolerate up to $f \leq \lfloor \frac{n-1}{3} \rfloor$ malicious processes, but requires an exponential number of messages. The *phase-king* algorithm proposed by Berman and Garay [4] solves the consensus problem under the same model, requiring $f+1$ phases, and a polynomial number of messages (which is a huge saving),

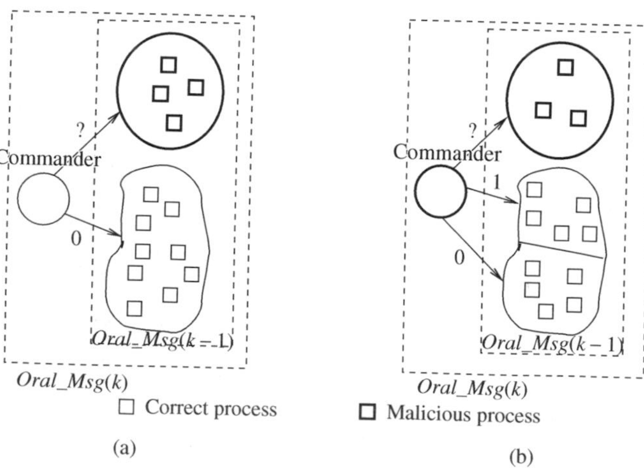

Figure 14.6 The effects of a loyal or a disloyal commander in a system with $n = 14$ and $f = 4$. The subsystems that need to tolerate k and $k-1$ traitors are shown for two cases. (a) Loyal commander. (b) No assumptions about commander.

but can tolerate only $f < \lceil n/4 \rceil$ malicious processes. The algorithm is so called because it operates in $f+1$ phases, each with two rounds, and a unique process plays an asymmetrical role as a leader in each round.

The phase-king algorithm is given in Algorithm 14.4, and assumes a binary decision variable. The message pattern is illustrated in Figure 14.7.

(variables)

boolean: $v \longleftarrow$ initial value;

integer: $f \longleftarrow$ maximum number of malicious processes, $f < \lceil n/4 \rceil$;

(1) Each process executes the following $f+1$ phases, where $f < n/4$:

(1a) **for** *phase* = 1 **to** $f+1$ **do**

(1b) Execute the following round 1 actions:

(1c) **broadcast** v to all processes;

(1d) **await** value v_j from each process P_j;

(1e) *majority* $\longleftarrow$ the value among the v_j that occurs $> n/2$ times (default value if no majority);

(1f) *mult* $\longleftarrow$ number of times that *majority* occurs;

(1g) Execute the following round 2 actions:

(1h) **if** $i = phase$ **then**

(1i) **broadcast** *majority* to all processes;

(1j) **receive** *tiebreaker* from P_{phase} (default value if nothing is received);

(1k) **if** $mult > n/2 + f$ **then**

(1l) $v \longleftarrow majority$;

(1m) **else** $v \longleftarrow tiebreaker$;

(1n) **if** $phase = f+1$ **then**

(1o) output decision value v.

Algorithm 14.4 Phase-king algorithm [4] – polynomial number of unsigned messages, $n > 4f$. Code is for process P_i, $1 \le i \le n$.

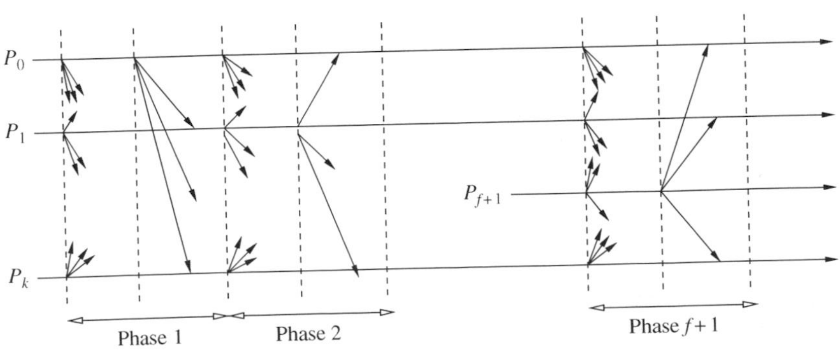

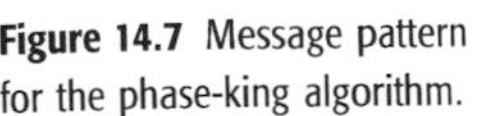
Figure 14.7 Message pattern for the phase-king algorithm.

- **Round 1** In the first round (lines 1b–1f) of each phase, each process broadcasts its estimate of the consensus value to all other processes, and likewise awaits the values broadcast by others. At the end of the round, it counts the number of "1" votes and the number of "0" votes. If either number is greater than $n/2$, then it sets its *majority* variable to that consensus value, and sets *mult* to the number of votes received for the majority value. If neither number is greater than $n/2$, which may happen when the malicious processes do not respond, and the correct processes are split among themselves, then a default value is used for the *majority* variable.
- **Round 2** In the second round (lines 1g–1o) of each phase, the phase king initiates processing – the phase king for phase k is the process with identifier P_k, where $k \in \{1 \ldots n\}$. The phase king broadcasts its majority value *majority*, which serves the role of a tie-breaker vote for those other processes that have a value of *mult* of less than $n/2+f$. Thus, when a process receives the tie-breaker from the phase king, it updates its estimate of the decision variable v to the value sent by the phase king if its own *mult* variable $< n/2+f$. The reason for this is that among the votes for its own *majority* value, f votes could be bogus and hence it does not have a clear majority of votes (i.e., $> n/2$) from the non-malicious processes. Hence, it adopts the value of the phase king. However, if *mult* $> n/2+f$ (lines 1k–1l), then it has received a clear majority of votes from the non-malicious processes, and hence it updates its estimate of the consensus variable v to its own majority value, irrespective of what tie-breaker value the phase king has sent in the second round.

At the end of $f+1$ phases, it is guaranteed that the estimate v of all the processes is the correct consensus value.

Correctness

The correctness reasoning is in three steps:

1. Among the $f+1$ phases, the phase king of some phase k is non-malicious because there are at most f malicious processes.
2. As the phase king of phase k is non-malicious, all non-malicious processes can be seen to have the same estimate value v at the end of phase k. Specifically, observe that any two non-malicious processes P_i and P_j can set their estimate v in three ways:

   (a) Both P_i and P_j use their own *majority* values. Assume P_i's *majority* value is x, which implies that P_i's *mult* $> n/2+f$, and of these voters, at least $n/2$ are non-malicious. This implies that P_j must also have received at least $n/2$ votes for x, implying that its majority value *majority* must also be x.

(b) Both P_i and P_j use the phase king's tie-breaker value. As P_k is non-malicious it must have sent the same tie-breaker value to both P_i and P_j.
(c) P_i uses its majority value as the new estimate and P_j uses the phase king's tie-breaker as the new estimate. Assume P_i's *majority* value is x, which implies that P_i's $mult > n/2+f$, and of these voters, at least $n/2$ are non-malicious. This implies that phase king P_k must also have received at least $n/2$ votes for x, implying that its majority value *majority* that it sends as tie-breaker must also be x.

For all three possibilities, any two non-malicious processes P_i and P_j agree on the consensus estimate at the end of phase k, where the phase king P_k is non-malicious.

3. All non-malicious processes have the same consensus estimate x at the start of phase $k+1$ and they continue to have the same estimate at the end of phase $k+1$. This is self-evident because we have that $n > 4f$ and each non-malicious process receives at least $n-f > n/2+f$ votes for x from the other non-malicious processes in the first round of phase $k+1$. Hence, all the non-malicious processes retain their estimate v of the consensus value as x at the end of phase $k+1$.
The same logic holds for all subsequent phases. Hence, the consensus value is correct.

Complexity
The algorithm requires $f+1$ phases with two sub-rounds in each phase, and $(f+1)[(n-1)(n+1)]$ messages.

14.5 Agreement in asynchronous message-passing systems with failures

14.5.1 Impossibility result for the consensus problem

Fischer *et al.* [12] showed a fundamental result on the impossibility of reaching agreement in an asynchronous (message-passing) system, even if a single process is allowed to have a crash failure. This result, popularly known as the FLP impossibility result, has a significant impact on the field of designing distributed algorithms in a failure-susceptible system. The correctness proof of this result also introduced the important notion of *valency* of global states.

For any global state GS, let $v(GS)$ denote the set of possible values that can be agreed upon in some global state reachable from GS. $|v(GS)|$ is defined as the *valency* of global state GS. For a boolean decision value, a global state can be *bivalent*, i.e., have a valency of two, or *monovalent*, i.e., having a valency of one. A monovalent state GS is *1-valent* if $v(GS) = \{1\}$ and it is *0-valent* if $v(GS) = \{0\}$. Bivalency of a global state captures the idea of uncertainty in the decision, as either a 0-valent or a 1-valent state may be reachable from this bivalent state.

In an (asynchronous) failure-free system, Section 14.3 showed how to design protocols that can reach consensus. Observe that the consensus value can be solely determined by the inputs. Hence, the initial state is monovalent.

In the face of failures, it can be shown that a consensus protocol necessarily has a bivalent initial state (assuming each process can have an arbitrary initial value from $\{0, 1\}$, to rule out trivial solutions). This argument is by contradiction. Clearly, the initial state where inputs are all 0 is 0-valent and the initial state where inputs are all 1 is 1-valent. Transforming the input assignments from the all-0 case to the all-1 case, observe that there must exist input assignments $\vec{I_a}$ and $\vec{I_b}$ that are 0-valent and 1-valent, respectively, and that they differ in the input value of only one process, say P_i. If a 1-failure tolerant consensus protocol exists, then:

1. Starting from $\vec{I_a}$, if P_i fails immediately, the other processes must agree on 0 due to the termination condition.
2. Starting from $\vec{I_b}$, if P_i fails immediately, the other processes must agree on 1 due to the termination condition.

However, execution 2 looks identical to execution 1, to all processes, and must end with a consensus value of 0, a contradiction. Hence, there must exist at least one bivalent initial state.

Observe that reaching consensus requires some form of exchange of the intial values (either by message-passing or shared memory, depending on the model). Hence, a running process cannot make a unilateral decision on the consensus value. The key idea of the impossibility result is that, in the face of a potential process crash, it is not possible to distinguish between a crashed process and a process or link that is extremely slow. Hence, from a bivalent state, it is not possible to transition to a monovalent state. More specifically, the argument runs as follows. For a protocol to transition from a bivalent global state to a monovalent global state, and using the global time interleaved model for reasoning in the proof, there must exist a *critical step* execution that changes the valency by making a decision on the consensus value. There are two possibilities:

- The *critical step* is an event that occurs at a single process. However, other processes cannot tell apart the two scenarios in which this process has crashed, and in which this process is extremely slow. In both scenarios, the other processes can continue to wait forever and hence the processes may not reach a consensus value, remaining in bivalent state.
- The *critical step* occurs at two or more independent (i.e., not send–receive related) events at different processes. However, as independent events at different processes can occur in any permutation, the *critical step* is not well-defined and hence this possibility is not admissible.

Thus, starting from a bivalent state, it is not possible to transition to a monovalent state. This is the key to the impossibility result for reaching consensus in asynchronous systems.

The impossibility result is significant because it implies that all problems to which the agreement problem can be reduced are also not solvable in any asynchronous system in which crash failures may occur. As all real systems are prone to crash failures, this result has practical significance. We can show that all the problems, such as the following, requiring consensus are not solvable in the face of even a single crash failure:

- The leader election problem.
- The computation of a network-side global function using broadcast–convergecast flows.
- Terminating reliable broadcast.
- Atomic broadcast.

The common strategy is to use a reduction mapping from the consensus problem to the problem X under consideration. We need to show that, by using an algorithm to solve X, we can solve consensus. But as consensus is unsolvable, so must be problem X.

14.5.2 Terminating reliable broadcast

As an example, consider the terminating reliable broadcast problem, which states that a correct process always gets a message even if the sender crashes while sending. If the sender crashes while sending the message, the message may be a null message but it must be delivered to each correct process. The formal specification of reliable broadcast was studied in Chapter 6; here we have an additional termination condition, which states that each correct process must eventually deliver some message.

- **Validity** If the sender of a broadcast message m is non-faulty, then all correct processes eventually deliver m.
- **Agreement** If a correct process delivers a message m, then all correct processes deliver m.
- **Integrity** Each correct process delivers a message at most once. Further, if it delivers a message different from the null message, then the sender must have broadcast m.
- **Termination** Every correct process eventually delivers some message.

The reduction from consensus to terminating reliable broadcast is as follows. A commander process broadcasts its input value using the terminating reliable broadcast. A process decides on a "0" or "1" depending on whether it receives "0" or "1" in the message from this process. However, if it receives the null message, it decides on a default value. As the broadcast is done using the terminating reliable broadcast, it can be seen that the conditions of the consensus problem (Section 14.1.1) are satisfied. But as consensus is not solvable, an algorithm to implement terminating reliable broadcast cannot exist.

14.5.3 Distributed transaction commit

Database transactions require the *commit* operation to preserve the ACID properties (atomicity, consistency, integrity, durability) of transactional semantics. The *commit* operation requires polling all participants whether the transaction should be committed or rolled back. Even a single rollback vote requires the transaction to be rolled back. Whatever the decision, it is conveyed to all the participants in the transaction. Clearly, this can be seen to be a consensus problem. Exercise 14.5 asks you to formally prove that distributed commit is not solvable under a crash failure.

Despite the unsolvability of the distributed *commit* problem under crash failure, the (blocking) two-phase commit and the non-blocking three-phase commit protocols do solve the problem. This is because the protocols use a somewhat different model in practice, than that used for our theoretical analysis of the consensus problem. The two-phase protocol waits indefinitely for a reply, and it is assumed that a crashed node eventually recovers and sends in its vote. Optimizations such as *presumed abort* and *presumed commit* are pessimistic and optimistic solutions that are not guaranteed to be correct under all circumstances. Similarly, the three-phase commit protocol uses timeouts to default to the "abort" decision when the coordinator does not get a reply from all the participants within the timeout period.

14.5.4 *k*-set consensus

Although consensus is not solvable in an asynchronous system under crash failures, a weaker version, known as the *k-set consensus* problem [6], is solvable as long as the number of crash failures f is less than the parameter k. The parameter k indicates that the nonfaulty processes agree on different values, as long as the size of the set of values agreed upon is bounded by k.

Assuming that the consensus value is from a multi-valued domain, the problem specification is as follows:

- ***k*-agreement** All non-faulty processes must make a decision, and the set of values that the processes decide on can contain up to k values.
- **Validity** If a non-faulty process decides on some value, then that value must have been proposed by some process.
- **Termination** Each non-faulty process must eventually decide on a value.

The k-agreement condition is new, the validity condition is different from that for regular consensus, and the termination condition is unchanged from that for regular consensus. The protocol in Algorithm 14.5 can be seen to solve k-set consensus in a straightforward manner, as long as the number of crash failures f is less than k. Let $n = 10, f = 2, k = 3$ and let each process propose a unique value from $\{1, 2 \ldots 10\}$. Then the 3-set is $\{8, 9, 10\}$.

(variables)
integer: $v \longleftarrow$ initial value;
(1) A process P_i, $1 \leq i \leq n$, executes k-set consensus:
(1a) **broadcast** v to all processes;
(1b) **await** values from $|N| - f$ processes and add them to set V;
(1c) **decide** on $max(V)$.

Algorithm 14.5 Protocol for k-set consensus [6]. Code shown is for process P_i, $1 \leq i \leq n$.

14.5.5 Approximate agreement

Another weaker version of consensus that is solvable in an asynchronous system under crash failures is known as the *approximate consensus* problem. Like k-set consensus, approximate agreement also assumes the consensus value is from a multi-valued domain. However, rather than restricting the set of consensus values to a set of size k, ϵ-approximate agreement requires that the agreed upon values by the non-faulty processes be within ϵ of each other. The problem specification is as follows.

- **ϵ-agreement** All non-faulty processes must make a decision and the values decided upon by any two non-faulty processes must be within ϵ range of each other.
- **Validity** If a non-faulty process P_i decides on some value v_i, then that value must be within the range of values initially proposed by the processes.
- **Termination** Each non-faulty process must eventually decide on a value.

Algorithm outline

The Dolev *et al.* [7] algorithm to solve approximate agreement in the message-passing model is studied next. The algorithm for the message-passing model assumes $n \geq 5f + 1$, although the problem is solvable for $n > 3f + 1$.

The asynchronous approximate agreement algorithm simulates synchronous communication by operating in rounds (Algorithm 14.6). Lines 1a–1c perform the initialization computation to decide the number of synchronous rounds to be simulated. We will examine this logic after examining the rest of the algorithm. The main loop, in lines 1d–1f, performs an all-to-all message exchange asynchronously for the determined number of rounds. In each round (simulated by *Asynchronous_Exchange*), a process broadcasts its estimate of the agreement value, and awaits $n - f$ such messages from other processes before moving to the next round. After each round, each process revises its estimate of the consensus value. The estimate is revised in such a way that the choices of the different processes are guaranteed to converge at a certain rate.

(variables)
real: $v \longleftarrow$ input value; //initial value
multiset of real V;
integer $r \longleftarrow 0$; // number of rounds to execute

(1) Execution at process P_i, $1 \leq i \leq n$:
(1a) $V \longleftarrow Asynchronous\_Exchange(v, 0)$;
(1b) $v \longleftarrow$ any element in$(reduce^{2f}(V))$;
(1c) $r \longleftarrow \lceil log_c(diff(V))/\epsilon \rceil$, where $c = c(n-3f, 2f)$.
(1d) **for** *round* **from** 1 **to** r **do**
(1e) $\quad V \longleftarrow Asynchronous\_Exchange(v, round)$;
(1f) $\quad v \longleftarrow new_{2f,f}(V)$;
(1g) **broadcast** $(\langle v,$ **halt**$\rangle, r+1)$;
(1h) **output** v as decision value.

(2) *Asynchronous_Exchange(v,h)* returns V:
(2a) **broadcast** (v, h) to all processes;
(2b) **await** $n-f$ responses belonging to round h and add to V;
(2c) for each process P_k that sent $\langle x,$ **halt**$\rangle$ as value, use x as its input henceforth;
(2d) **return** the multiset V.

Algorithm 14.6 Asynchronous approximation agreement algorithm [7]. Here, $n \geq 5f+1$.

Consider any sorted collection U. The new estimate of a process is chosen by computing $new_{k,f}(U)$, which is parameterized by k and f, and defined as $mean(select_k(reduce^f(U)))$:

- $reduce^f(U)$ removes the f largest and f smallest members of U.
- $select_k(U)$ selects every kth member of U, beginning with the first. If U has m members, $select_k(U)$ has $c(m,k) = \lfloor (m-1)/k \rfloor + 1$ members. This constant c represents a *convergence factor* towards the final agreement value, i.e., if x is the range of possible values held by correct processes before a round, then x/c is the possible range of estimate values held by those processes after that round.

Illustration of definitions

Figure 14.8 shows the $select_k(reduce^f(U))$ operation, with $k=5$ and $f=4$. The mean of the selected members is the new estimate $new_{5,4}(U)$.

The algorithm uses $m = n-3f$ and $k = 2f$. So $c(n-3f, 2f)$ will represent the *convergence factor* towards reaching approximate agreement and $new_{2f,f}$ is the new estimate after each round. The choice of these parameters will be justified.

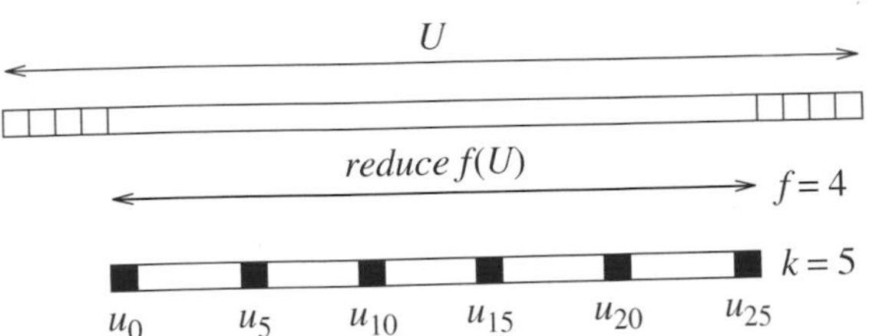

Figure 14.8 Illustrating $select_k(reduce^f(U))$, with $k = 5$ and $f = 4$. $reduce^4(U)$ has 26 members, hence $c(26, 5) = 6$ members are selected.

Notation

The algorithm uses *multisets*, which are sets with repeating elements included. Union, intersection, and set difference operations on multisets are natural extensions of the counterparts for regular sets. $mean(U)$ is the arithmetic mean of U, calculated by considering each instance in the multiset. $min(U)$ and $max(U)$ are defined as for sets. $range(U)$ is the interval $[min(U), max(U)]$. $diff(U)$ is $max(U) - min(U)$.

Some essential combinatorial results are first proved. Let $|U| = m$, and let the m elements $u_0 \ldots u_{m-1}$ of multiset U be in non-decreasing order. The following properties on non-empty multisets U, V, and W can easily be seen:

- **Property 1** The number of the elements in multisets U and V is reduced by at most 1 when the smallest element is removed from both. Similarly for the largest element.
- **Property 2** The number of elements common to U and V before and after j reductions differ by at most $2j$. Thus, for $j \geq 0$ and $|V|, |W| \geq 2j$, $|V \cap W| - |reduce^j(V) \cap reduce^j(W)| \leq 2j$.
- **Property 3** Let V contain at most j values not in U, i.e., $|V - U| \leq j$, and let size of V be at least $2j$. Then by removing the j low and j high elements from V, it is easy to see that remaining elements in V must belong to the range of U, see Figure 14.9. Thus,

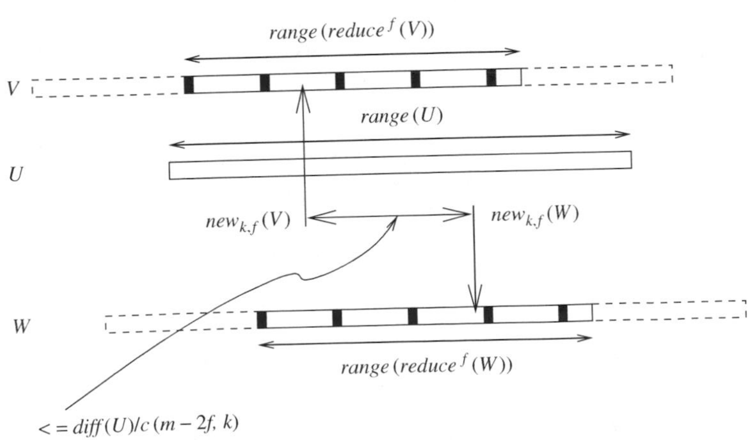

Figure 14.9 Illustrating property 3 for the ϵ-agreement problem. $|V| = |W| = m$, $|V - W| = |W - V| = k$, and $|V - U|, |W - U| \leq f$. Note that the horizontal spacing in the figure shows only the relative positioning of elements in the sorted multisets and need not be to scale.

- each value in $reduce^j(V)$ is in the range of U, i.e., $range(reduce^j(V)) \subseteq range(U)$;
- $new_{k,j}(V) \in range(U)$.

Convergence rate of approximation

Let U be the multiset of estimates, one estimate per correct process, at the start of a round. Let V and W be the multisets received at two arbitrary correct proceses in that round. The processes use the approximation function to choose their values for the next round. The new estimates chosen by any two arbitrary correct processes, using the approximation function $new_{k,f}$, are guaranteed to be within $range(U)/c(m,k)$ of each other, when (i) $|V| = |W| = m$, (ii) $|W-V|, |V-W| \leq k$, and (iii) $|V-U|, |W-U| \leq f$.

Convergence rate

Let $k > 0$, $f \geq 0$, and $m > 2f$. For the multisets received, $|V| = |W| = m$. Let the multisets received differ from U in at most f elements ($|V-U|$, $|W-U| \leq f$), and let the multisets received differ from each other in at most k elements ($|W-V|, |V-W| \leq k$). Then

$$|new_{k,f}(V) - new_{k,f}(W)| \leq diff(U)/c(m-2f, k). \qquad (14.1)$$

The proof of this relationship is outlined next. There are exactly $m-2f$ members in each of $M = reduce^f(V)$ and $N = reduce^f(W)$. Hence, $select_k(M) = \{m_0, m_1 \ldots m_{c-1}\}$ and $select_k(N) = \{n_0, n_1 \ldots n_{c-1}\}$, where $select_k(M)$ and $select_k(N)$ each have $c = c(m-2f, k)$ members. Observe that (i) at least $ki+1$ members of M are less than or equal to any m_i (likewise for N). Also, (ii) at most ki members of M are less than m_i (likewise for N). The following can be shown using the earlier properties and definitions:

$$max(m_i, n_i) \leq min(m_{i+1}, n_{i+1}), \text{ where } 0 \leq i \leq c-2. \qquad (14.2)$$

This directly follows if $m_i \leq n_{i+1}$ and $n_i \leq m_{i+1}$ can be shown.

Assume to the contrary that $m_i > n_{i+1}$. From (i), at least $k(i+1)+1$ elements of N are less than or equal to n_{i+1}, and hence less than m_i. But from (ii), at most ki elements of M are less than m_i. Hence, at least $k+1$ elements in N are not in M, i.e., $|N-M| \geq k+1$.

Observe that $|W-V| \leq k$ and $|W \cap V| \geq m-k$. Using property 2, this implies that $|N \cap M| \geq m-k-2f$ and hence $|N-M| \leq (m-2f) - (m-k-2f) \leq k$. This contradicts the conclusion of the assumption about $m_i > n_{i+1}$. Hence, $m_i \leq n_{i+1}$. Symmetrically, $n_i \leq m_{i+1}$ can be shown. Therefore, Eq. (14.2) holds:

$$|new_{k,f}(V) - new_{k,f}(W)| = \frac{1}{c}|\sum_{i=0}^{c-1}(m_i - n_i)| \leq \frac{1}{c}\sum_{i=0}^{c-1}|m_i - n_i|$$

$$= \frac{1}{c}\sum_{i=0}^{c-1}(max(m_i, n_i) - min(m_i, n_i))$$

Using Eq. (14.2) in the R.H.S., expanding terms, and simplifying:

$$|new_{k,f}(V) - new_{k,f}(W)| \leq \frac{1}{c}(max(m_{c-1}, n_{c-1}) - min(m_0, n_0)).$$

Using property 3, $max(m_{c-1}, n_{c-1}) - min(m_0, n_0) \leq range(U)$ and Eq. (14.1) follows.

Correctness

Let T, the set of correct processes, be such that $|T| \geq n-f$. Let U and U' be the multiset of estimates (one estimate from each process) before and after some round h. $|V| = |W| = n-f$. Also, $|V-U|, |W-U| \leq f$ because at most f processes are faulty. $|V \cap W| \geq n-3f$ because both p and q would have received the same values from the correct processes from which both received messages. Hence, the difference between V and W, $|V-W| = |W-V| = |V| - |V \cap W| \leq 2f$ (the upper bound on this was denoted as k in Eq. (14.1)). Then, we have the following:

- **ϵ-agreement** $|new_{2f,f}(V) - new_{2f,f}(W)| \leq diff(U)/c(n-3f, 2f)$. This immediately follows by observing that the multisets U, V, and W satisfy Eq. (14.1) when m is set to $n-f$ and k is set to $2f$, and hence $c(m-2f, k)$ becomes $c(n-3f, 2f)$.
  This inequality implies that the range of the multiset of estimates chosen by all processes in T reduces by a factor of $c(n-3f, 2f)$. This ≥ 2 as the algorithm assumes that $n \geq 5f+1$. Hence, after a logarithmic number of iterations (determined in lines 1a–1c and described below), this range reduces to below ϵ.
- **Validity** $range(U') \subseteq range(U)$. As the multisets U and V satisfy Property 3, we have that $new_{2f,f}(V) \in range(U)$. For each round, it can be seen that the value of each correct process is within the range of the values of the correct processes at the start of the first round.

Initialization (lines 1a–1c)

The upper bound on the number of iterations is determined in the initialization phase, in lines 1a–1c. Let the multisets of estimates received by two arbitrary correct processes P_p and P_q after line 1a be V_p and V_q. $|V_p|, |V_q| > 4f$ because $n \geq 5f+1$; and $|V_p - V_q|, |V_q - V_p| \leq 2f$ (shown above). We can apply property 2 to both V_p and W_q with respect to each other (and by setting $j = 2f$) – to get that $range(reduce^{2f}(V_p)) \subseteq range(V_q)$ and $range(reduce^{2f}(V_q)) \subseteq range(V_p)$.

It follows that $v_p \in range(V_q)$ and $v_q \in range(V_p)$ after line 1b. This guarantees that each correct process P_q knows at the end of the initialization round that its range $range(V_q)$ contains *all* the values v_p of all correct processes P_p at the end of this initialization round. Knowing ϵ and the convergence rate c, $\epsilon \geq \lceil diff(V)/c^{round} \rceil$ and hence it is adequate to execute $round = \lceil log_c(diff(V)/\epsilon) \rceil$ rounds. Hence, the number of rounds computed in line 1c is an upper bound on the number of iterations in which every two correct processes are guaranteed to converge to within ϵ.

Termination (lines 1g–1h)
Observe that each process may determine a different number of rounds to execute at line 1c. When a process finishes the required number of rounds, it executes lines 1g–1h wherein it sends a special symbol "**halt**" and terminates itself. When some process P_q receives such a message from P_p, it should use the value of P_p for this and all of its subsequent rounds until it finishes its own precomputed number of rounds. This detail is left out of the pseudo-code for simplicity.

Complexity

- **Time complexity** $\lceil log_c(diff(V)/\epsilon)\rceil + 1$ rounds.
- **Message complexity** $n \times [\lceil log_c(diff(V)/\epsilon)\rceil + 1]$ messages of size $O(1)$ each.

14.5.6 Renaming problem

Problem definition
The consensus problem which was a problem about agreement required the processes to agree on a single value, or a small set of values (k-set consensus), or a set of values close to one another (approximate agreement), or reach agreement with high probability (probabilistic or randomized agreement). A different agreement problem introduced by Attiya et al. [1] requires the processes to agree on necessarily distinct values. This problem is termed as the *renaming* problem. The renaming problem assigns to each process P_i, a name m_i from a domain M, and is formally specified as follows:

- **Agreement** For non-faulty processes P_i and P_j, $m_i \neq m_j$.
- **Termination** Each nonfaulty process is eventually assigned a name m_i.
- **Validity** The name m_i belongs to M.
- **Anonymity** The code executed by any process must not depend on its initial identifier.

The renaming problem is useful for name space transformation. A specific example where this problem arises is when processes from different domains need to collaborate, but must first assign themselves distinct names from a small domain. A second example of the use of renaming is when processes need to use their names as "tags" to simply mark their presence, as in a priority queue. A third example is when the name space has to be condensed. This can occur when, for a system consisting of a large number of processes, k-mutual exclusion has to be enforced. Of the large pool of processes, only k can be in the mutual exclusion at any time to use the k copies of a replicated resource. Each resource can be viewed as holding a permit, 1 through k. For a process to gain access to the resource, it has to gain a permit.

The assumptions about the renaming problem are as follows:

- The n processes $P_1 \ldots P_n$ have their identifiers in the old name space. P_i knows only its identifier, and the total number of processes, n. The names of other processes are not known to a process.
- The n processes take on new identifiers $m_1 \ldots m_n$, respectively, from the name space M.
- Due to asynchrony, each process that chooses its new name must continue to cooperate with the others until they have chosen their new names.

The above formulation of the renaming problem is called the *one-time renaming* problem. If processes continually acquire and release names from a common pool, then the formulation becomes the *long-lived renaming* problem. *Long-lived renaming* is a resource acquisition problem.

Algorithm

Attiya *et al.* [1] give a algorithm for one-time renaming when $n \geq 2f+1$, and up to f processes may fail in a fail-stop manner. The size of the transformed name space M is $n+f$.

The high-level functioning of the algorithm is given in Figure 14.10. Each process has a list *View* in which it tracks the latest proposed name by each process, as and when it learns of it. Its own proposed name is tracked in *View*[1]. In more details, the view of a name has four components, as described in the *View* data structure in Algorithm 14.7. *View* is a list of up to n

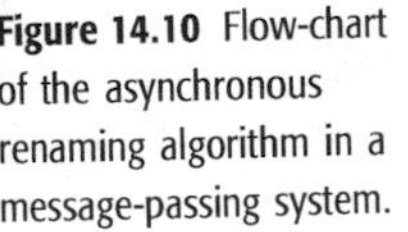
Figure 14.10 Flow-chart of the asynchronous renaming algorithm in a message-passing system.

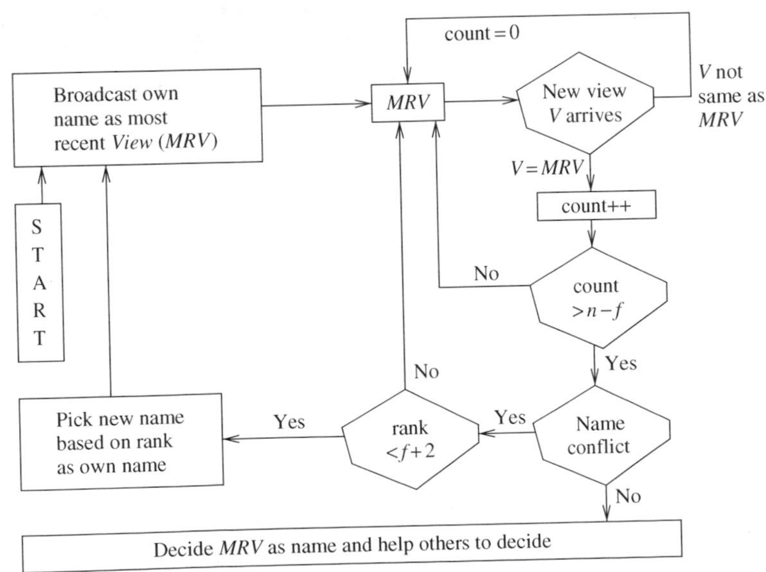

objects of type *bid*. Various views are ordered by the $\leq$ relation, defined as follows:

$View \leq View'$ if and only if for each process P_i such that $View[k].P = P_i$, we also have that for some k', $View'[k'].P = P_i$ and $View[k].attempt \leq View'[k'].attempt$.

If $View' \not\leq View$ (line 1n), then $View$ is updated using $View'$ (line 1o) by:

1. including all process entries from $View'$ that are missing in $View$ (i.e., $View'[k'].P$ is not equal to $View[k].P$, for all k), so such entries $View'[k']$ are added to $View$.
2. replacing older entries for the same process with more recent ones, (i.e., if $View'[k'].P = P_i = View[k].P$ and $View'[k'].attempt > View[k].attempt$, replace $View[k]$ by $View'[k']$).

Any new information learnt is broadcast to all processes (lines 1c, 1v), and a process uses a counter *count* to track the number of other processes that have broadcast the exact same view as the latest view of this process (line 1k). If the view in a received message contains information that is not in the current view (line 1n), the current view is updated (line 1o). Note that this is similar to taking the pairwise maximum of vector clocks. However, a crucial difference is that the ordering of the components is not predetermined, as each process may order the other processes differently. When *count* reaches $n-f$ (line 1l), no more messages may arrive because the other f processes may have failed. Such a view for which $n-f$ affirmations were received is said be a *stable view*.

Once a process determines a view to be stable (lines 1m, 1q), the process checks if there is a conflict with its choice of a new name and the choices of other processes (lines 1r, 1s). If there is no conflict, it finalizes its choice of the new name (lines 1t, 1u) and goes to the loop (lines 1G-1K) wherein it helps other processes to gain stable views and finalize their new name

---

(local variables)
struct *bid*:
  integer P;   // old name of process
  integer x;   // new name being bid by the process
  integer *attempt*;   // the number of bids so far, including this current // bid
  boolean *decide*;   // whether new name x is finalized
list of bid: $View[1 \ldots n] \longleftarrow \langle\langle i, 0, 0, false\rangle\rangle$;   // initialize list with an // entry for P_i
integer *count*;   // number of copies of the latest local view, received from // others
boolean: *restart*, *stable*, *no_choose*;   // loop control variables

```
A process P_i, 1 ≤ i ≤ n, participates in renaming:
repeat
  restart ⟵ false;
  broadcast message(View);
  count ⟵ 1;
  repeat
        no_choose ⟵ 0;
        repeat
              await message(View′);
              stable ⟵ false;
              if View′ = View then
                    count ⟵ count + 1;
                    if count ≥ n − f then
                          stable ⟵ true;
              else if View′ ≰ View then
                    update View using View′ by taking latest
                      information for each process;
                    restart ⟵ true;
        until (stable = true or restart = true);    // n − f copies
                                        // received, or new view obtained
        if restart = false then                     // View[1] has
                                                  // information about P_i
              if View[1].x ≠ 0 and View[1].x ≠ View[j].x
                                        for any j then
                    decide View[1].x;
                    View[1].decide ⟵ true;
                    broadcast message(View);
              else
                    let r be the rank of P_i in UNDECIDED(View);
                    if r ≤ f + 1 then
                          View[1].x ⟵ FREE(View)(r), the rth
                            free name in View;
                          View[1].attempt ⟵ View[1].attempt + 1;
                          restart ⟵ 1;
                    else
                          no_choose ⟵ 1;
     until no_choose = 0;
  until restart = 0;
  repeat
    on receiving message(View′)
          update View with View′ if necessary;
          broadcast message(View);
  until false.
```

Algorithm 14.7 Asynchronous renaming in the message-passing model [1]. Code shown is for process P_i, $1 \le i \le n$.

choices. If there is a conflict (lines 1w–1F), a new name must be chosen once again and competed with other processes. There are two cases here, depending on the *rank* of the process among all the processes that have not yet finalized their new names (i.e., among all processes except those for which $View[j].decide = 1$). Let the set of such processes be denoted as $UNDECIDED(View)$. Clearly, as the new names of such processes are not finalized, the rank is determined based on the old names (line 1x).

- If the *rank* r is less than $f+2$ (line 1y), the process chooses the rth free name from $FREE(View)$, the "free" names from M that have not been finalized by the processes (which have their *decide* component set to 1 in *View*). The process has to restart the bidding process, by going back to step (1a), broadcasting its updated view (line 1c), and so on.
- If the *rank* r exceeds $f+1$ (lines 1C,ID), the process goes to line (1e) and then waits for some other process to send its updated views. The logic here is that at least one correct process will have a rank up to $f+1$ among *UNDECIDED*, and will pick and stabilize its new name before processes with rank greater than $f+1$ begin to compete for a new name.

Some definitions and properties are now given:

- **P1** An algorithm is *locally proper* if for each run and each process, the sequence of the *View* list is totally ordered by $\leq$. Algorithm 14.7 is seen to be locally proper, from lines 1j–1o.
- **P2** A view is *stable with respect to a process* if the process has received $n-f-1$ messages containing identical information in the accompanying view. (Along with its own identical view, there are $n-f$ affirmations.) A view is *stable in a run* if it is stable with respect to some process.
- **P3** If an algorithm is locally proper, then in any run, the set of stable views is totally ordered.

  This is seen as follows. Let views *View* and *View*′ be stable with respect to processes i and j, respectively. Then $n-f$ processes (say, set A_i) agree on *View*, and $n-f$ processes (say, set A_j) agree on *View*′.

  If *View* and *View*′ are not totally ordered, $A_i \cap A_j = \emptyset$. Disjointness implies size of A_j is at most $n-(n-f)=f$. Thus, $n-f \leq f$, implying, $n \leq 2f$. This contradicts the assumption that $n \geq 2f+1$, hence, at least one process must have sent both *View* and *View*′. So *View* and *View*′ must be totally ordered.
- **P4** As Algorithm 14.7 is locally proper, its set of stable views is totally ordered.

Correctness

Safety

A process finalizes a new name once it has a stable view. P_i and P_j cannot finalize the same name because the stable views are totally ordered. Without loss of generality, assume that P_i's stable view $\leq$ P_j's stable view when they respectively finalize their names. Then P_j's stable view must include the name finalized by P_i, and P_j will not pick the same name.

Liveness/termination

Observe that when a process picks a new name (line 1z), there are at most $n-1$ names used by others, so $f+1$ names are available. To show that all processes eventually finalize a name, let $FREE(View)$ be the set of free names from M as per $View$. Let $DECIDED$ be the set of processes that finalize their new names (i.e., for which $bid.decide$ is true). Then $N - DECIDED$ is $UNDECIDED$, the set of processes which cannot finalize a new name. We now argue using contradiction that $UNDECIDED$ is empty.

- Consider the execution after the time that all processes in $DECIDED$ have decided their new names, and at least one bid sent by every other correct process has been received by each correct process, implying that $|View| \geq n-f$. As no correct process blocks, this point in time will occur. Let $View_{min}$ be the smallest stable view after this point in time. By P4, all the views are totally ordered and hence $View_{min}$ is uniquely defined. Let the set of free names at this time be denoted as $FREE(View_{min})$ and the set of undecided processes at this time be denoted as $UNDECIDED(View_{min})$.
- Among the processes in $UNDECIDED(View_{min})$, consider the process P_{min} with the smallest *rank*, based on the old names. The rank is at most $f+1$, and hence the process will select a new name (lines 1y, 1z, 1A). As $rank$ is unique, no other process in $UNDECIDED(View_{min})$ will now or henceforth choose this name chosen by P_{min}.
- P_{min} updates and broadcasts its view. When other processes receive this view, they update their local views with this new information, and will also broadcast their updated views:
  - either in the loop (lines 1G–1K); or
  - via execution of lines (1C–1D), then lines (1n–1o), and then (1b–1c).

  P_{min} and all other correct processes receive at least $n-f$ confirmations, making the view containing P_{min}'s choice of a new name a stable view. Hence, P_{min} can decide a new name, leading to a contradiction that $UNDECIDED(View_{min})$ is empty.

Complexity

Each time a process bids with a new name for itself, a broadcast is sent ($n-1$ messages) and each recipient of the broadcast, seeing a new view, also does a broadcast ($n-1$ messages). This leads to $O(n^2)$ messages per new name bid. Let the final stable view be denoted by $View_{final}$. The total number of messages is $\Sigma_{i=1}^{n} View_{final}.attempt_i \times n^2$. Exercise 14.9 asks you to analyze the bound on the number of attempts made by the processes.

14.5.7 Reliable broadcast

Although reliable terminating broadcast (RTB) is not solvable under failures (recall that we showed a reduction from consensus to that problem in Section 14.5.2), a weaker version of RTB, namely reliable broadcast, in which the termination condition is dropped, is solvable under crash failures. The protocol is shown in Algorithm 14.8. This protocol uses up to $O(n^2)$ messages to broadcast message M and works in the face of any number of failures. The key difference between RTB and reliable broadcast is that RTB requires eventual delivery of some message – even if the sender fails just when about to broadcast. In this case, a null message must get sent, whereas this null message need not be sent under reliable broadcast. Thus, RTB requires the recognition of the failure (as described above) as opposed to no message getting sent. This reduces to the ability of being able to distinguish between a slow process and a failed process, which was the crux in solving the consensus problem under crash failure.

---

(1) Process P_0 initiates reliable broadcast:
(1a) **broadcast** message M to all processes.

(2) A process P_i, $1 \le i \le n$, receives message M:
(2a) **if** M was not received earlier **then**
(2b) **broadcast** M to all processes;
(2c) deliver M to the application.

Algorithm 14.8 Protocol for reliable broadcast.

14.6 Wait-free shared memory consensus in asynchronous systems

14.6.1 Impossibility result

The impossibility of achieving consensus in asynchronous message-passing systems in a system prone to crash failures (discussed in Section 14.5.1) also extends to asynchronous shared memory systems. A shared memory system can be emulated by a message-passing system – if consensus could be reached in a shared memory system, it could also be reached in a message-passing system, leading to a contradiction. Thus, consensus cannot be reached in an asynchronous shared memory system in the crash failure model. The intuition behind the impossibility result in shared memory systems is similar – in the face of a potential process crash, it is not possible to distinguish between a crashed process and a process that is extremely slow in doing its *Read* or *Write* operation. The FLP argument using 0-valent and 1-valent states and the *critical step* used earlier for asynchronous message-passing systems can also be used here for asynchronous shared memory systems. The reasoning to show that consensus cannot be achieved even if a single process fails runs informally along the following lines [21,22].

Assume there exists a protocol in which consensus can be reached even if a single process fails. Recall from Section 14.5.1 that there exists a bivalent initial state. Due to the termination requirement of the problem, there must exist some process i that makes a transition from a bivalent state to an univalent state even if there are no failures. (For a wait-free consensus, this is also true.) So there must be some execution prefix X that is bivalent, but from which a step by i makes it 0-valent, whereas a step by i after an extension Y of X leads to a 1-valent state. (See Figure 14.11.) If there are multiple events between X and Y, then there must be a prefix Z such that a step by i leads to 0-valence but a step by another process j ($j \neq i$ as processes are assumed to be deterministic) followed by a step by i leads to 1-valence.

The argument now uses a simple case analysis based on the actions of i and j after Z, to show that the configuration of Z as shown in Figure 14.11 is impossible, showing the impossibility of a 1-failure consensus protocol. The notation $extend(Z, i \circ j)$ denotes the state after processes i and j take steps in that order, after execution Z.

- **Process i's event is a *Read* (see Figure 14.12(a))** Then $extend(Z, i \circ j)$ and $extend(Z, j \circ i)$ are identical to all processes except i. If i does not take any step after $extend(Z, i \circ j)$, then all process must eventually terminate with consensus on 0 while executing a suffix, say δ. But if the same suffix is executed after $extend(Z, j \circ i)$, they must reach a consensus on 1. As $extend(Z, i \circ j)$ and $extend(Z, j \circ i)$ are isomorphic to all processes except the stopped process i, we have a contradiction.
- **Process j's event is a *Read*** The states after $extend(Z, i)$ and $extend(Z, j \circ i)$ are identical to all processes except j. The same logic as for the previous case, this time letting j stop instead of i, leads to a similar contradiction.
- **Processes i and j execute *Write* on different variables (see Figure 14.12(b))** The system state after $extend(Z, i \circ j)$, which is 0-valent, is the same as the system state after $extend(Z, j \circ i)$, which is 1-valent. There now arises a contradiction, irrespective of whether all processes decide on 0 or on 1.

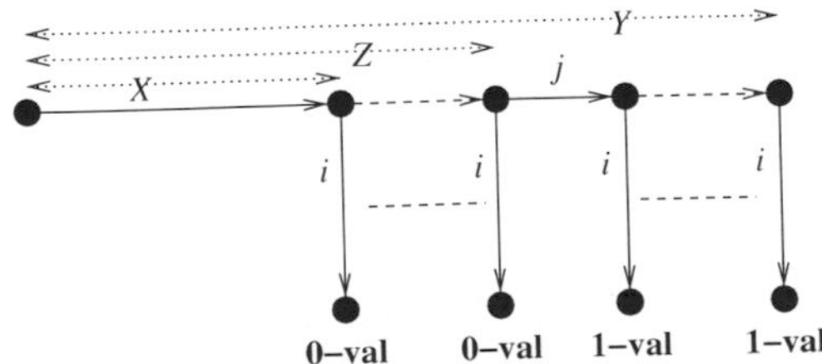

Figure 14.11 Execution prefix used to show impossibility of 1-failure tolerant consensus [21,22].

- **Processes i and j execute *Write* on the same variable (see Figure 14.12(c))** The system state after $extend(Z, i)$ and $extend(Z, j \circ i)$ are identical to all processes except j. If all processes except j run after $extend(Z, i)$, the consensus value must be 0. If all processes except j run after $extend(Z, j \circ i)$, the consensus value must be 1. As $extend(Z, j \circ i)$ and $extend(Z, i)$ are isomorphic to all processes except the stopped process j, we have a contradiction.

Hence, there cannot exist any bivalent state that allows any process to go a univalent state.

The key reason why this result for the 1-failure case is different from that for the failure-free case is that the 1-failure case allows for a bivalent initial state, whereas the initial state for a failure-free execution is univalent.

Between the time a process reads various registers and (deciding on a consensus value) writes its consensus value, the values of the other registers read can get updated by other processes. Herein lies the difficulty for shared memory systems – the reads and the writes are not together guaranteed to be an atomic action – and hence taking action about deciding a consensus value, independent of processes that are "suspected" to have failed, can lead to an erroneous decision on consensus. Hence, from a bivalent state, it is not possible to transition to a univalent state. This leads to the following two results – the second one follows trivially from the first:

- It is not possible to reach consensus in an asynchronous shared memory system using Read/Write atomic registers, even if a single process can fail by crashing.
- There is no wait-free consensus algorithm for reaching consensus in an asynchronous shared memory system using Read/Write atomic registers.

There are two ways of overcoming the impossibility result:

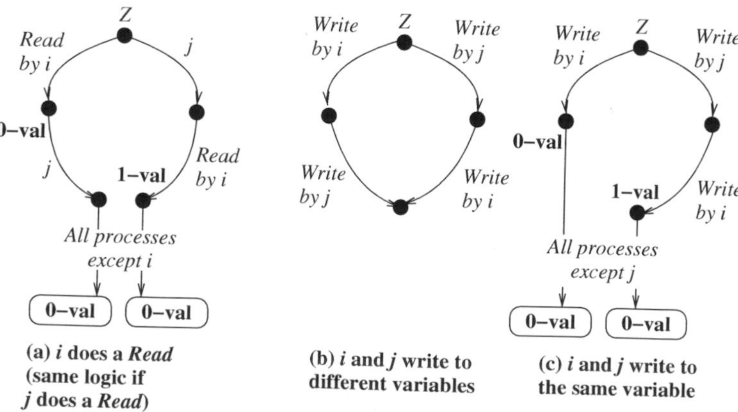

Figure 14.12 Various cases to show impossibility of 1-failure tolerant consensus in the asynchronous message-passing model [21,22].

- Weakening the consensus problem, as was done for message-passing systems. This area covers the design of asynchronous algorithms for k-set consensus, approximate consensus, and renaming using atomic registers and atomic snapshot objects which are built from atomic registers, studied in Chapter 12. These algorithms are studied in Sections 14.6.4–14.6.6.
- Using memory that is stronger than atomic Read/Write memory to design wait-free consensus algorithms. Such a memory would have corresponding access primitives.

  Recall that a *wait-free* algorithm in a system of n processes is a $(n-1)$-crash resilient algorithm. Thus, any process should be able to perform its execution, independent of any other processes. The above results lead to the question:

  - Are there objects (with supporting operations) for which there is a wait-free algorithm for reaching consensus in a n-process system?

In the remainder of this section, we assume only the crash failure model, and also require the solutions to be *wait-free*.

As it turns out, the answer is Yes [14]. Objects/primitives such as *Test&Set*, *Swap*, *Compare&Swap*, and *Memory Move*, which were designed in the context of efficient computer architectures, do indeed allow consensus to be reached in a wait-free manner. Such objects are stronger than the safe, regular, or atomic Read/Write registers. The notion of *consensus number* provides a metric to measure the degree to which these various primitives allow consensus to be reached. This study of these more complex objects also extends our study of the register simulations of Chapter 12, wherein stronger register types were simulated from weaker register types.

14.6.2 Consensus numbers and consensus hierarchy [14]

Definition 14.1 *An object of type X has consensus number k, denoted as CN(X) = k, if k is the largest number for which the object X can solve wait-free k-process consensus in an asynchronous system subject to k − 1 crash failures, using only objects of type X and read/write objects.*

The consensus numbers of some well-known objects are shown in Table 14.4. Algorithm 14.9 gives the definitions of some of these objects. Definitions of *Swap* and *Fetch&Increment* were seen in Algorithm 12.7. As seen from Definition 14.1, there is an infinite hierarchy – called the *consensus hierarchy* – that gets defined, according to the power of the objects to solve wait-free consensus under crash failures.

Table 14.4 Consensus numbers of some object types [14]. Some of these objects are described in Algorithm 14.9.

Object	Consensus number
Read/Write objects	1
Test&Set, stack, FIFO queue, *Fetch&Inc*	2
Augmented queue with *peek* – size k	k
Compare&Swap, augmented queue, memory–memory move, memory-memory swap, *Fetch&Cons*, store-conditional	∞

(shared variables among the processes accessing each of the different object types)
register: $Reg \longleftarrow$ initial value; // shared register initialized
(local variables)
integer: $old \longleftarrow$ initial value; // value to be returned
integer: $key \longleftarrow$ comparison value for conditional update;

(1) *RMW*(*Reg*, function f) returns *value*:
(1a) $old \longleftarrow Reg$;
(1b) $Reg \longleftarrow f(Reg)$;
(1c) **return**(*old*).

(2) *Compare&Swap*(*Reg*, *key*, *new*) returns *value*:
(2a) $old \longleftarrow Reg$;
(2b) **if** $key = old$ **then**
(2c) $\quad Reg \longleftarrow new$;
(2d) **return**(*old*).

(3) *Fetch&Inc*(*Reg*) returns *value*:
(3a) $old \longleftarrow Reg$;
(3b) $Reg \longleftarrow old + 1$;
(3c) **return**(*old*).

Algorithm 14.9 Definitions of synchronization operations *RMW*, *Compare&Swap*, *Fetch&Inc* [17].

A natural consequence of the definition of *consensus number* is the following result.

Theorem 14.1 *For objects X and Y such that $CN(X) < CN(Y)$, there is no wait-free simulation of object Y using X and read/write registers (whose consensus number is* 1*) in a system with more than $CN(X)$ processes.*

If such a simulation did exist, then by Definition 14.1, $CN(X) = CN(Y)$, leading to a contradiction. Note that if there are up to $CN(X)$ processes, it is possible (as shown in Section 14.6.3) for X and read/write registers to wait-free simulate Y because the full power of reaching consensus among more than $CN(X)$ processes is never required to be exercised.

A corollary of this result is that there is no wait-free simulation of any object with consensus number more than one, using only read/write atomic registers. This corollary is important because it implies that objects with stronger properties than the read/write atomic register are needed. The ability to read and write, perhaps conditionally, in an atomic manner was earlier found to be useful in designing semaphores in operating systems, and certain primitives in computer architecture and design. Several of the objects in Algorithm 14.9 were first designed in hardware in these specialized contexts [17]. We will now see two examples of achieving wait-free consensus – one using the FIFO queue, and another using the *Compare&Swap* instruction.

FIFO queue

Algorithm 14.10 shows how 2-consensus is achieved using a FIFO queue [14]. The queue operations are **enqueue** and **dequeue**. The queue is initialized with a single value, 0. Both processes try to dequeue from the queue. However, due to the atomicity of the **dequeue** operation, access is always serialized. The first process that dequeues the "0" element uses its own initial value (local x) as the consensus value and outputs it. The other process, on completing its **dequeue** operation, gets $\perp$, and learns that the first process has dequeued first, and therefore borrows the value set aside by the first process in $Choice[1-i]$.

(shared variables)
queue: $Q \longleftarrow \langle 0 \rangle$; // queue Q initialized
integer: $Choice[0, 1] \longleftarrow [\perp, \perp]$; // preferred value of // each process

(local variables)
integer: $temp \longleftarrow 0$;
integer: $x \longleftarrow$ initial choice;

(1) Process P_i, $0 \le i \le 1$, executes this for 2-process consensus using a FIFO queue:
(1a) $Choice[i] \longleftarrow x$;
(1b) $temp \longleftarrow dequeue(Q)$;
(1c) **if** $temp = 0$ **then**
(1d) **output**(x)
(1e) **else output**$(Choice[1-i])$.

Algorithm 14.10 Protocol for 2-process wait-free consensus using a FIFO queue [14]. Code for P_i, $0 \le i \le 1$.

Thus, both processes agree on the same value and hence 2-process consensus is achieved. The operations of any process can be seen to be wait-free. The same logic cannot be extended to three processes because of the following informal reasoning. Some one process will dequeue the "0." When the other two processes dequeue and get a $\perp$, they know that one of the other two processes' value is the consensus value, but do not know which of the other two processes it is. This is because the queue object does not atomically allow the first process to leave behind (i.e., write) its identifier as an imprint for the second and third processes to learn about when they issue their **dequeue**. Therefore, $CN(queue) = 2$.

Compare&Swap

Algorithm 14.11 shows how wait-free consensus is achieved among any number of processes using the *Compare&Swap* operation (see Algorithm 14.9) on a shared register *Reg* [14]. The *Compare&Swap* performs all actions of an invocation atomically, thus serializing all concurrent accesses. Each process executes *Compare&Swap*$(Reg, \perp, x)$. The value of the object *Reg* is read into local variable *val*, and if this value *val* equals the key $\perp$, then the process's preference x gets written to *Reg* atomically. Due to the serialization of the operations, some process always gets serialized first, even if accesses are concurrent. There are thus two cases.

(shared variables)
integer: $Reg \longleftarrow \perp$; // shared register *Reg* initialized
(local variables)
integer: $temp \longleftarrow 0$; // temp variable to read value of *Reg*
integer: $x \longleftarrow$ initial choice; // initial preference of process

(1) Process P_i, $(\forall i \geq 1)$, executes this for consensus using *Compare&Swap*
(1a) $temp \longleftarrow$ *Compare&Swap*$(Reg, \perp, x)$;
(1b) **if** $temp = \perp$ **then**
(1c) **output**(x)
(1d) **else output**$(temp)$.

Algorithm 14.11 Protocol for wait-free consensus using *Compare&Swap*, for any number of processes [14]. Code for P_i, $1 \leq i \leq \infty$.

- Consider the process that gets serialized first. The value of *Reg* read via *Compare&Swap*$(Reg, \perp, x)$ equals the key $\perp$, and the preference x of this process gets written to *Reg*. The process returns its x as the consensus value.
- Any other process executing *Compare&Swap*$(Reg, \perp, x)$ will find that the value of *Reg* (which is the value x set by the first process) does not match the key $\perp$. Hence it leaves *Reg* unmodified and returns the value of *Reg* as the consensus value. The implication is that another process has earlier found $Reg = \perp$ and set its own preference as the value of *Reg*. So this process borrows the value set by the earlier process in *Reg* as the consensus value.

Due to the atomicity of the *Compare&Swap* operation and the fact that this logic works for any number of processes, the code for consensus is wait-free and can tolerate up to $n-1$ failures, for all n. Hence, $CN(Compare\&Swap)$ is ∞.

Read–modify–write abstraction

Read–modify–write (abbreviated as *RMW*) abstracts several objects wherein a register can be read and modified using an arbitrary function f atomically [17]. Such objects include *Fetch&Inc*, *Swap*, and *Test&Set*. The *RMW* object has a consensus number of at least 2 because the first process to read the object can atomically modify its value to leave an imprint that the object has been accessed at least once (e.g., as in the FIFO queue). If the imprint can also include the identity of the first process to read, or of the choice of the first process, processes that subsequently access the object can by pointed to the choice made by the first process, and the consensus number may then be more than 2.

The various *RMW* objects differ in their function f. A function is termed as *interfering* if for all process pairs i and j, and for all legal values v of the register, (i) $f_i(f_j(v)) = f_j(f_i(v))$, i.e., function is commutative, or (ii) the function is not write-preserving, i.e., $f_i(f_j(v)) = f_i(v)$ or vica-versa with the roles of i and j interchanged.

Examples The *Fetch&Inc* commutes even though it is write-preserving. The *Test&Set* commutes and is not write-preserving. The *Swap* does not commute but it is not write-preserving. Hence, all three objects uses functions that are *interfering*.

Algorithm 14.12 shows how wait-free consensus is achieved among two processes using the *RMW* operation (defined in Algorithm 14.9) on a shared register *Reg* [14]. The *RMW* performs all actions of an invocation atomically, thus serializing all concurrent accesses. Each process executes *RMW*(*Reg*, *f*), where x is the initial choice of the process. The shared data structures are shown in Figure 14.13. *Reg* has an initial distinguished value $\perp$, known to all processes. The assumption here is that the function f is non-trivial, meaning it is not the identity function.

Although any non-trivial RMW operation has a consensus number of at least 2, it can be seen that a nontrivial *interfering* RMW operation has a consensus number of exactly 2, i.e., there is no algorithm to reach consensus with three processes. An informal argument to see this is as follows. Consider the third process to access the object. If the RMW operation is commutative, the third process cannot tell which of the other two processes accessed the object first, and hence does not know what consensus value to use. If the RMW operation is not write-preserving, the third process cannot tell if it is the second or the third process to access the object; and hence does not know what consensus value to use. Operations such as *Compare&Swap* are noninterfering operations, and hence have consensus numbers higher than 2.

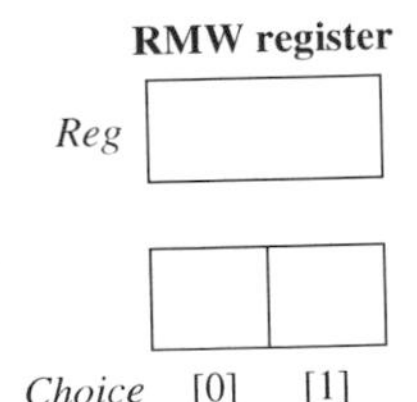

Figure 14.13. Shared data structures for solving 2-process wait-free consensus using the *RMW* operation.

(shared variables)
integer: $Reg \longleftarrow \perp$; // shared register Reg initialized
integer: $Choice[0, 1] \longleftarrow [\perp, \perp]$; // data structure
(local variables)
integer: $x \longleftarrow$ initial choice; // initial preference of process

(1) Process P_i, $(0 \leq i \leq 1)$, executes this for consensus using RMW:
(1a) $Choice[i] \longleftarrow x$;
(1b) $val \longleftarrow RMW\ (Reg, f)$;
(1c) **if** $val = \perp$ **then**
(1d) **output**$(Choice[i])$
(1e) **else output**$(Choice[1-i])$.

Algorithm 14.12 Protocol for wait-free consensus for two processes using *RMW* [14]. Code is for P_i, $0 \leq i \leq 1$.

14.6.3 Universality of consensus objects [14]

In Chapter 12, we studied the wait-free simulations of various types of registers using weaker forms of registers. We now build on this notion of wait-free simulation of one object type using another object type, in the context of consensus under crash failures. An object is defined to be *universal* if that object along with read/write registers can simulate any other object in a wait-free manner [14]. The main result of this section is that in any system containing up to k processes, an object X such that $CN(X) = k$ is *universal*, i.e., it can simulate any other object. The condition on the number of processes in the system is essential; because X does not and cannot manifest the greater power that is required when the number of objects exceeds $CN(X)$. If the condition were removed, then an object X would truly wait-free simulate another object with a greater consensus number in a system with more than $CN(X)$ processes, leading to a violation of the definition of consensus number.

For any system with up to k processes, the universality of objects X with consensus number k is shown by giving a *universal* algorithm to wait-free simulate *any* object using only objects of type X and read/write registers. This is shown in two steps:

1. A *universal* algorithm to wait-free simulate *any* object whatsoever using read/write registers and arbitrary k-processor consensus objects is given. This is the main step.
2. Then, the arbitrary k-process consensus objects are simulated with objects of type X, also having consensus number k. This trivially follows after the first step.

Hence, any object X with consensus number k is universal in a system with $n \leq k$ processes. In the rest of this subsection, we study a universal algorithm to wait-free simulate any object whatsoever using read/write registers and arbitrary k-processor consensus objects (step 1). The following two concepts are useful:

- An arbitrary consensus object X allows a single operation, $Decide(X, v_{in})$ and returns a value v_{out}, where both v_{in} and v_{out} have to assume a legal value from known domains V_{in} and V_{out}, respectively. For the correctness of this shared object version of the consensus problem, all v_{out} values returned to each invoking process must equal the v_{in} of some process.
- A *non-blocking* operation, in the context of shared memory operations, is an operation that may not complete itself but is guaranteed to complete (i.e., provide a *response indication* (see Chapter 12) to) at least one of the pending operations in a finite number of steps. This operation is a weaker version of a wait-free operation.

We will first study a universal algorithm that does a non-blocking simulation of any object, and then refine this algorithm to get a wait-free algorithm.

A non-blocking universal algorithm

Algorithm 14.13 uses a linked list (with the initial record termed *anchor_record*) to store the linearized sequence of operations and resulting states on an arbitrary object Z. The data structure *op* defines the format of one such element in this linked list. The linked list and data structure format are illustrated in Figure 14.14. Operations to the arbitrary object Z are simulated in a non-blocking way using only an arbitrary consensus object (namely, the field *op−>next* in each record) which is accessed via the *Decide* call. We are not concerned with how the consensus object itself or *Decide* is implemented.

When an object Z being simulated is invoked using *invoc*, a record pointed to by *my_new_record* is allocated and the record's *operation* field is set to the invoked operation (lines 1a–1b). The main challenge in simulating Z is to linearize all the operations being invoked on it concurrently by the various processes – there is competition among the processes to apply their own operation next, i.e., to thread their own operation next to the tail of the linked list. This is where the consensus object comes in useful – with respect to the current most recent operation that has been linearized, the consensus object "decides" on the next operation that is to be linearized.

Before a process competes, it first needs to identify the tail of the linked list which is dynamically changing. Array *Head* stores pointers to the tail of the linked list; $Head[i]$ is P_i's best estimate of the pointer that points to the tail record. In loop (1c)-(1e), P_i selects the most up to date estimate of the tail pointer. However, observe that this may still be hopelessly out of date

```
(shared variables)
record op
      integer: seq ←− 0;             // sequence number of serialized operation
      operation ←−⊥;                 // operation, with associated parameters
      state ←− initial state;        // the state of the object after the operation
      result ←−⊥;   // the result of the operation, to be returned to invoker
      op *next ←−⊥;                                 // pointer to the next record
op *Head[1...k] ←− &(anchor_record);
(local variables)
op *my_new_record, *winner;

(1)   Process P_i, 1 ≤ i ≤ k performs operation invoc on an arbitrary
      consensus object:
(1a)  my_new_record ←− malloc(op);
(1b)  my_new_record−>operation ←− invoc;
(1c)  for count = 1 to k do
(1d)         if Head[i]−>seq < Head[count]−>seq then
(1e)                Head[i] ←− Head[count];
(1f)  repeat
(1g)         winner ←− Decide(Head[i]−>next, my_new_record);
(1h)         winner−>seq ←− Head[i]−>seq + 1;
(1i)         winner−>state, winner−>result ←− apply(winner−>
                operation, Head[i]−>state);
(1j)         Head[i] ←− winner;
(1k)  until winner = my_new_record;
(1l)  enable the response to invoc, that is stored at winner−>result.
```

Algorithm 14.13 Non-blocking universal algorithm to simulate an arbitrary object using any consensus object [2,14]. Code for P_i, $1 \leq i \leq k$.

due to the nonatomic nature of scanning the array *Head*. Still, *Head*$[i]$ is P_i's best estimate of the pointer to the record that is at the tail of the linked list. In the main loop (lines 1f–1k), P_i competes on the consensus object *Head*$[i]$*−>next* to thread itself next to the list (line 1g). The following possibilities arise:

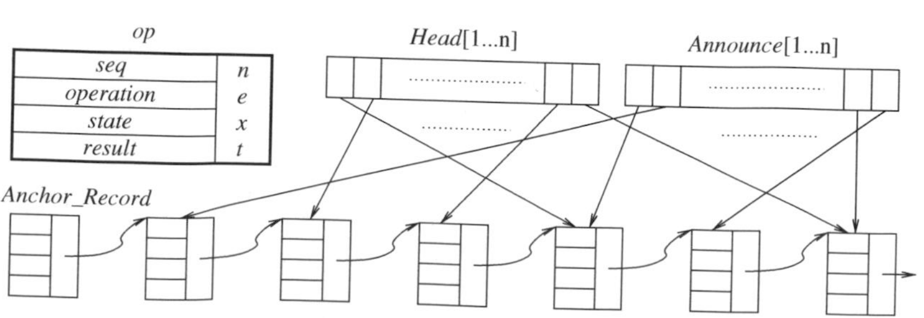

Figure 14.14 Wait-free simulation of a universal consensus object [2]. For a non-blocking simulation of the object, the array *Announce* is not used.

1. *Head*[*i*] points to the correct tail of the list. The process P_i invokes *Decide* on the consensus object which is the *next* field of the record pointed to by *Head*[*i*] – to learn if it succeeds in threading its operation next. But there may be concurrent calls to *Decide*. The winner of the "race" is pointed to by *winner*. (We do not yet know if P_i won.) The fields of the record – its new state, new sequence number, new result – are computed and stored as per lines 1h–1i. *Head*[*i*] is updated to *winner* (line 1j).
   (a) If *winner* is the same as *my_new_record* (line 1k), then P_i won the race and succeeded in threading its operation after the *Head*[*i*] record before the current iteration of the **repeat** loop. The process exits after returning the value stored in the *result* field (line 1l).
   (b) If *winner* is not the same as *my_new_record*, then P_i lost the race. The pointer to the record of the true winner of the race was returned in *winner* by the consensus object. The record of the true winner got filled in again by P_i in lines 1h–1j. But now *Head*[*i*] is pointing to the next record, i.e., the record with sequence number one more than in the previous iteration. The process competes again by going through the next iteration of the **repeat** loop.
2. *Head*[*i*] points to an old tail of the list. The process executes the **repeat** loop (see case 1(b), which repeats itself) until *Head*[*i*] points to the record that is the most recent tail. It then competes to thread its own operation pointed to by *my_new_record* as in step 1.

We make some notes that give an insight into the design of this algorithm:

- We cannot use a single consensus object because consensus has to be reached on-line with respect to the current most recent operation, on the next operation to be linearized. A consensus object always returns the same decision value. Thus the algorithm uses as many consensus objects (the *next* fields of the records) as there are records on whose order to reach consensus.
- A single pointer in a read/write object cannot be used instead of the array *Head* to point to the latest operation record. This is because reading the pointer to contend, and updating it after contention is over and threaded to the list, cannot be done atomically in a wait-free manner.
- The linearization of the operations is given by the sequence numbers. The sequence numbers increase monotonically along the linked list.
- A process may never succeed in threading its own operation to the list. It continues the **repeat** loop forever. This may happen if it loses the contention every time to another process trying to thread concurrently. This can be used to observe that the algorithm is not wait-free but the algorithm is non-blocking.
- The estimate of the tail of the list in lines 1c–1e may be very out of date due to the way it is computed. This is a drawback as the process has to

iterate through the **repeat** loop at least as many times as the number of operations by which the estimate is out of date.

Complexity

The worst-case time complexity to thread a specific operation is not bounded due to the non-blocking nature of the algorithm. Exercise 14.14 asks you to perform an average-case analysis.

A wait-free universal algorithm

The non-blocking algorithm in the previous section is enhanced to make it wait-free. To ensure that a process does not happen to continually lose the contention, a round-robin approach of "helping" is used. If a process P_j determines that the next operation is to be assigned sequence number x, then it first checks whether the process P_i such that $i = x\,(mod\,n)$ is contending for threading its operation. If so, then P_j tries to thread P_i's operation instead of its own.

The algorithm is shown in Algorithm 14.14. The implementation of the round-robin "helping" is done using the array $Announce[1 \ldots n]$. When a process P_i wants to thread its operation, it first announces it by making $Announce[i]$ point to the record where the operation is stored (lines 1a–1b). It then proceeds as before to estimate the latest tail of the list, using the *Head* array (lines 1c–1e). Each process is required to determine whether it should try to thread the record of the rightful process (lines 1g–1h), as determined by the modulo function, or its own (line 1j). Only if the "rightful" process is not interested in threading its own operation does a process try to thread its own operation (line 1j).

We argue using contradiction that within n iterations of the **while** loop, process P_i will have succeeded in having its operation threaded to the linked list, and exit the loop. Assume by way of contradiction that P_i's record is not threaded by P_i's $(n+1)$th iteration of the **while** loop. After the $Announce[i]$ having been set in lines 1a–1b, n other records initiated by other processes must have been threaded to the linked list. But of these n sequence numbers, one of them modulo n must have equalled i and the other processes would have threaded P_i's record instead of their own (see lines 1g–1i).

Complexity

Each process completes its operation within n iterations of the main **while** loop, irrespective of the other processes.

14.6.4 Shared memory *k*-set consensus

The message-passing version of k-set consensus for the crash failure model and $k > f$ was presented in Section 14.5.4. Here, its counterpart for the shared memory model assuming an atomic snapshot object is given in

```
(shared variables)
record op
    integer: seq ⟵ 0;               // sequence number of serialized operation
    operation ⟵⊥;                    // operation, with associated parameters
    state ⟵ initial state;           // state of the object after the operation
    result ⟵⊥;            // result of the operation, to be returned to invoker
    op *next ⟵⊥;                               // pointer to the next record
op *Head[1...k], *Announce[1...k] ⟵ &(anchor_record);
(local variables)
op *winner; *my_new_record;
(1)  Process P_i, 1 ≤ i ≤ k performs operation invoc on an arbitrary
                consensus object:
(1a) Announce[i] ⟵ malloc(op);
(1b) Announce[i]−>operation ⟵ invoc; Announce[i]−>seq ⟵ 0;
(1c) for count = 1 to k do
(1d)      if Head[i]−>seq < Head[count]−>seq then
(1e)           Head[i] ⟵ Head[count];
(1f) while Announce[i]−>seq = 0 do
(1g)      turn ⟵ (Head[i]−>seq + 1)mod (k);
(1h)      if Announce[turn]−>seq = 0 then
(1i)           my_new_record ⟵ Announce[turn];
(1j)      else my_new_record ⟵ Announce[i];
(1k)      winner ⟵ Decide(Head[i]−>next, my_new_record);
(1l)      winner−>seq ⟵ Head[i]−>seq + 1;
(1m)      winner−>state, winner−>result ⟵ apply(winner−>
               operation, Head[i]−>state);
(1n)      Head[i] ⟵ winner;
(1o) enable the response to invoc, that is stored at winner−>result.
```

Algorithm 14.14 Wait-free universal algorithm to simulate an arbitrary object using any consensus object [2,14]. Code for P_i, $1 \leq i \leq k$.

Algorithm 14.15. The algorithm can be easily derived from the message-passing algorithm. A process P_i writes its initial value v to its component within the shared object $Obj[i]$, and repeatedly scans the shared object until $n-f$ processes have written to the object. It then takes the maximum of the values scanned.

14.6.5 Shared memory renaming

The renaming problem was introduced in Section 14.5.6 and an algorithm to solve renaming in the message passing model was given. An asynchronous algorithm for wait-free renaming for the shared memory model under crash failures is given in Algorithm 14.16. The algorithm assumes an atomic snapshot object Obj, which has the nice property that it linearizes all asynchronous operations to it. Each P_i can write to its component in *Obj* and read all components atomically (see Section 12.6). We assume P_i does not have a unique index from $[1 \ldots n]$ to access *Obj*. Each process begins by bidding a new name of "1" for itself (line 1a). The process then repeats the following loop. It writes its latest bid to its component of *Obj* (line 1c); it reads the entire object using a **scan** into its

---

(local variables)

integer: $v \longleftarrow$ initial value;

array of integer $local\_array \longleftarrow \overline{\perp}$;

(shared variables)

atomic snapshot object $Obj[1 \ldots n] \longleftarrow \overline{\perp}$;

(1) A process P_i, $1 \leq i \leq n$, executes k-set consensus:

(1a) **update**$_i$($Obj[i]$) with v;

(1b) **repeat**

(1c) $\quad local\_array \longleftarrow$ **scan**$_i$(Obj);

(1d) **until** there are at least $|N| - f$ non-null values in Obj;

(1e) $v \longleftarrow$ maximum of the values in $local\_array$.

Algorithm 14.15 Asynchronous protocol for k-set consensus in the shared memory model using an atomic snapshot object. Code shown is for process P_i, $1 \leq i \leq n$.

(local variables)

integer: $m_i \longleftarrow 0$;

integer: $P_i \longleftarrow$ name from old domain space;

list of integer tuples $local\_array \longleftarrow \overline{\langle \perp, \perp \rangle}$;

(shared variables)

atomic snapshot object $Obj \longleftarrow \overline{\langle \perp, \perp \rangle}$; // n components

(1) A process P_i, $1 \leq i \leq n$, participates in wait-free renaming:

(1a) $m_i \longleftarrow 1$;

(1b) **repeat**

(1c) $\quad$ **update**$_i$(Obj, $\langle P_i, m_i \rangle$); // update own component with bid m_i

(1d) $\quad local\_array(\langle P_1, m_1 \rangle, \ldots \langle P_n, m_n \rangle) \longleftarrow$ **scan**$_i$(Obj);

(1e) $\quad$ **if** $m_i = m_j$ for some $j \neq i$ **then**

(1f) $\qquad$ Determine rank $rank_i$ of P_i in $\{P_j \mid P_j \neq \perp \wedge j \in [1, n]\}$;

(1g) $\qquad m_i \longleftarrow rank_i$th smallest integer not in $\{m_j \mid m_j \neq \perp \wedge j \in [1, n] \wedge j \neq i\}$;

(1h) $\quad$ **else**

(1i) $\qquad$ **decide**(m_i); **exit**;

(1j) **until** *false*.

Algorithm 14.16 Asynchronous wait-free renaming using an atomic snapshot object in the shared memory model [2]. Code shown is for process P_i, $1 \leq i \leq n$.

local array (line 1d). P_i examines the local array for a possible conflict with its proposed new name (line 1e).

- If P_i detects a conflict with its proposed name m_i (line 1e) it determines its rank *rank* among the *old* names (line 1f); and selects the *rank*th smallest integer among the names that have not been proposed in the view of the object just read (line 1g). This will be used as P_i's bid for a new name in the next iteration.
- If P_i detects no conflict with its proposed name m_i (line 1e), it selects this name and exits (line 1i).

We now consider the following properties of this algorithm:

- **Correctness** If two processes were to choose the same new name, then the Scans returned to them in their final iteration must have indicated that the name they bid was unique. However, due to the linearizability property of the atomic snapshot object *Obj*, the Scan that was returned to the "later" process could not have indicated that the name it bid was unique. Hence, no two processes can choose the same name when they terminate.
- **Size of name space** At any time, there are at most $n-1$ names that are bid by other processes, and the rank of a process is at most n. Hence, a process will never bid a name greater than $2n-1$. The name space is confined to $[1, 2n-1]$.
- **Termination** Assume there is a subset $\overline{T} \subseteq N$ of processes that never terminate. Let $min(\overline{T})$ be the process in $\overline{T}$ with the lowest ranked process identifier (old name). Let $rank(min(\overline{T}))$ be the rank of this process among *all* the processes $P_1 \ldots P_n$. Once every process in $\overline{T}$ has done at least one **update**, and once all the processes in T have terminated, we have the following:
  - The set of names of the terminated processes, say M_T, remains fixed.
  - The process $min(\overline{T})$ will choose a name not in M_T, that is ranked $rank(min(\overline{T}))$. As $rank(min(\overline{T}))$ is unique, no other process in $\overline{T}$ will ever choose this name.
  - Hence, $min(\overline{T})$ will not detect any conflict with $rank(min(\overline{T}))$ and will terminate.

  As $min(\overline{T})$ cannot exist, the set $\overline{T} = \emptyset$.
- **Wait-freedom** A process can choose its new name independent of the actions of the other processes.

Complexity

Exercise 14.17 asks you to perform a time complexity analysis of this algorithm, and show the following lower bounds.

Lower bounds

Let M be the new name space. For crash-failures, the following lower bounds can be seen to exist:

- For wait-free renaming, wherein all other $n-1$ processes may fail, the name space must be of size $2n-1$.
- To tolerate up to f failures, the name space must be of size $n+f$.

14.6.6 Shared memory renaming using splitters

Moir and Anderson [23] presented a very elegant wait-free renaming algorithm using the *splitter* concurrent object defined as follows [19]. When n ($n \geq 1$) processes invoke the *splitter*, each is returned a value from the set $\{stop, down, right\}$ subject to the following constraints:

- At most one process is returned *stop*.
- At most $n-1$ processes are returned *down*.
- At most $n-1$ processes are returned *right*.

```
(shared variables)
integer X ⟵ ⊥;
boolean Y ⟵ false;

(1)   splitter(), executed by process P_i, 1 ≤ i ≤ n:
(1a)  X ⟵ i;
(1b)  if Y then
(1c)       return(right);
(1d)  else
(1e)       Y ⟵ true;
(1f)       if X = i then return(stop);
(1g)       else return(down).
```

Algorithm 14.17 A wait-free implementation of a splitter [23]. Code shown is for process P_i, $1 \leq i \leq n$.

Figure 14.15 shows a schematic definition of a splitter. Algorithm 14.17 shows a wait-free implementation of a splitter using MRMW atomic variables.

- The first time that some process P_i finds X equal to its own identifier in line 1f, Y must be true, and hence all other processes must get the value *right* (unless they fail) while P_i must get value *stop*. Hence, at most one process is returned *stop*.
- Let P_i be the last process to execute line 1a. Unless P_i crashes, it will either get value *right* if another process executed line 1e, or if no other process executed line 1e yet, P_i will execute line 1f and return *stop*. Hence at most $n-1$ processes are returned *down*.

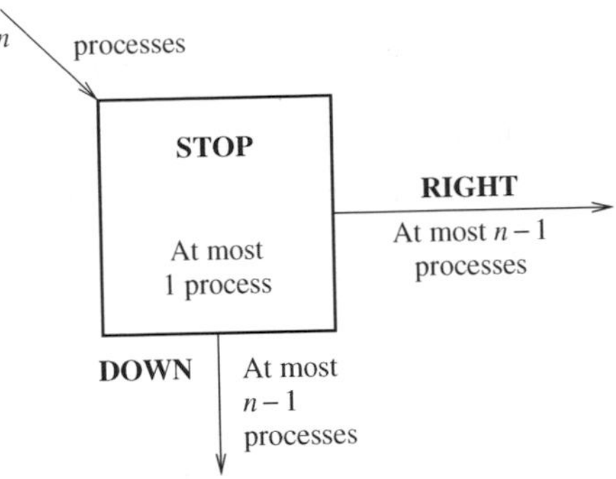

Figure 14.15 The structure for a splitter [23].

- The first process that reads Y in line 1b cannot get value *right* because Y is initialized to false. Hence, not all processes can are returned *right*.

The renaming algorithm is now constructed using $n(n+1)/2$ splitters arranged as shown in Figure 14.16. Each splitter is labelled by coordinates r, d. Observe that each process is guaranteed to get a *stop* value from one of the $n(n+1)/2$ splitters, and no two processes will stop at the same splitter. So the coordinates of the splitter where a process stops can serve as the new label. The code is shown in Algorithm 14.18. The array of shared variables corresponding to the grid of splitter is not shown.

Complexity

The new name space is $n(n+1)/2$ when the number of processes is n. Each process takes $O(n)$ steps to select its new name. The algorithm is clearly wait-free.

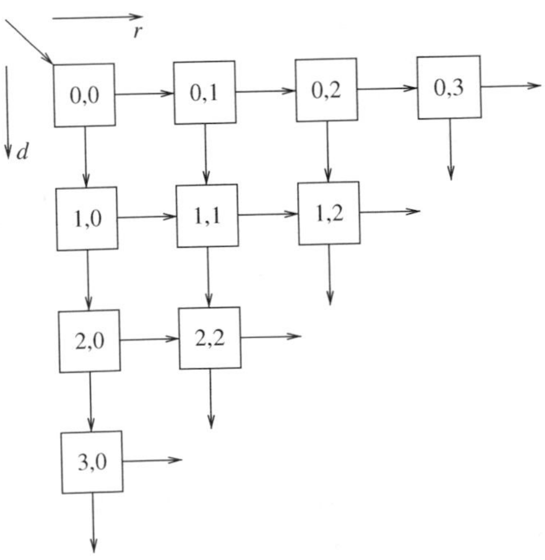

Figure 14.16 The Moir–Anderson wait-free renaming algorithm using splitters [23]. Code shown is for P_i, $1 \leq i \leq n$.

```
(local variables)
next, r, d, new_name ⟵ 0;
Process P_i, 1 ≤ i ≤ n, participates in wait-free renaming:
r, d ⟵ 0;
while next ≠ stop do
        next ⟵ splitter(r, d);
        case
              next = right then r ⟵ r + 1;
              next = down then d ⟵ d + 1;
              next = stop then break();
return(new_name = n · d − d(d − 1)/2 + r).
```

Algorithm 14.18 Moir and Anderson's asynchronous wait-free renaming using splitters [23]. Code shown is for process P_i, $1 \leq i \leq n$.

14.7 Chapter summary

Consensus problems are fundamental aspects of distributed computing because they require inherently distributed processes to reach agreement. This chapter first covers different forms of the consensus problem, which are shown to be equivalent to one another. Consensus is attainable in fault-free systems. The chapter then gives an overview of what forms of consensus are solvable under different failure models and different assumptions on the synchrony/asynchrony.

The chapter covers agreement in the following categories. (i) Synchronous message-passing systems with failures. Here, different fault models are considered – the fail-stop model and the Byzantine model. Lower bounds on the number of failure-prone processes are given. Also, representative algorithms under different assumptions and fault models are given. (ii) Asynchronous message-passing systems with failures. The first result here is that it is impossible to reach consensus in this model. Hence, several weaker versions of the consensus problem, such as k-set consensus, approximate consensus, the renaming problem, and reliable broadcast are considered. Algorithms to solve the weakened forms of consensus in these models are then given. (iii) Wait-free shared memory consensus in asynchronous systems. Here, the first result is the impossibility result, analogous to that for message-passing systems. The chapter then solves consensus using registers (or objects) that are stronger than the atomic read/write registers. The consensus hierarchy that naturally emerges for stronger consensus objects is then studied. Algorithms for shared memory renaming and k-set consensus are also covered.

14.8 Exercises

Exercise 14.1 For each of the six ordered pairs of problems among: the *Byzantine agreement* problem, the *Consensus* problem, and the *Interactive consistency* problem, demonstrate a reduction from the former to the latter.

Exercise 14.2 Modify Algorithm 14.1 to design an *early-stopping* algorithm for consensus under failstop failures, that terminates within $f'+1$ rounds, where f', the actual number of stop-failures, is less than f. Prove the correctness of your algorithm.
(Hint: A process can be required to send a mesage in each round, even if the value was sent in the earlier round. Processes should also track the other processes that failed, which is detectable by identifying the processes from which no message was received.)

Exercise 14.3 Modify the iterative Byzantine Agreement algorithm and the tree data structure specification given in Algorithm 14.3, as well as the example in Figure 14.5, to now solve the *consensus problem.*

Exercise 14.4 Examine the phase-king algorithm for consensus in the face of Byzantine failures, as given in Algorithm 14.4. This algorithm works when $n > 4f$. Presumably, the algorithm will fail for $4f \geq n > 3f$, even though this condition is a sufficient condition for the existence of a solution to the consensus problem in a synchronous message-passing system.

1. Why will the algorithm fail for $4f \geq n > 3f$?
2. Even though the algorithm is not correct for $4f \geq n > 3f$, under some circumstance(s), the correct processors will end up with the same value. Characterize one such circumstance, independent of the behavior of the malicious processes.
3. To derive a correct solution for $4f > n > 3f$, change line 1k to read:

$$if\ mult > 2f.$$

   Will this solution work?
4. To derive another correct solution for $4f \geq n > 3f$, run the algorithm for $4(f+1)$ rounds instead of for $2(f+1)$ rounds of the original algorithm. Will this solution work?

Exercise 14.5 Prove that the *distributed commit* problem is not solvable under a crash failure.
(Hint: Show a reduction from the consensus problem to the distributed commit problem.)

Exercise 14.6 Prove that the *leader election* problem is not solvable under a crash failure.

Exercise 14.7 In the ϵ-agreement problem, can a correct process halt if it receives $f+1$ halting tags from other processes, even before it has completed its precomputed number of rounds? Justify your answer.

Exercise 14.8 How can the algorithm for ϵ-agreement, given in Algorithm 14.6, be simplified if a synchronous system is available? Identify all the changes to the various parameter values. Can a better value be obtained for the convergence rate?

Exercise 14.9 Analyze the number of bids for a new name made by each process in the asynchronous renaming algorithm given in Algorithm 14.7.

Exercise 14.10 How can the algorithm for asynchronous renaming, given in Algorithm 14.7, be simplified if a synchronous system is available?

Exercise 14.11 Examine the *Test&Set* instruction in Algorithm 12.7. What is the consensus number x of this register object? Give an algorithm to achieve consensus for this consensus number.

Exercise 14.12 (k-Write instruction)

1. Consider the *2-Write* instruction that can write two locations atomically. Show how the *2-Write* instruction can be used to implement a wait-free 2-consensus protocol. (Hint: structure the solution using a structure similar to that of the protocols for *RMW* and *Swap*.)
2. Consider the *k-Write* instruction. Can this *k-Write* instruction be used to implement a wait-free consensus protocol for k processes? Justify your answer.

Exercise 14.13 Examine the standard stack object, having its standard *push* and *pop* operations. What is the consensus number x of the stack? Give the code for achieving 2-process consensus using the stack.

Exercise 14.14 Perform an average-case time complexity analysis of the non-blocking universal algorithm for consensus objects given in Algorithm 14.13.

Exercise 14.15 Simplify the non-blocking universal algorithm for consensus objects (Algorithm 14.14) by using the specific *Compare&Swap* object, but also eliminating the *Head* array.

Exercise 14.16 Adapt the asynchronous message-passing approximate agreement algorithm given in Section 14.5.5 for a shared memory system.

Exercise 14.17 Perform a time complexity analysis of the wait-free renaming algorithm using the atomic snapshot object in asynchronous systems, given in Algorithm 14.16. Also prove the lower bounds on the size of the name space, as indicated in Section 14.6.5.

Exercise 14.18 Show how the number of splitters used in the renaming algorithm of Section 14.6.6 can be reduced to $n(n-1)/2$.

14.9 Notes on references

The Byzantine agreement and the consensus problems were defined by Lamport *et al.* [20,25]. The exponential messages algorithm for solving consensus in the face of Bzyantine failures and the $3f+1$ lower bound were given in these papers. A later proof of the exponential algorithm was given by Bar-Noy *et al.* [3], and a later proof of the $3f+1$ lower bound was given by Fischer *et al.* [11]. The polynomial-message phase-king algorithm to solve consensus in the same Byzantine failure model was given by Berman and Garey [4]. A polynomial-message algorithm requiring $t+1$ rounds and $n > 3t$ processes has been given by Garey and Moses [13]. The result on the impossibility of reaching consensus in an asynchronous message-passing system was given by Fischer *et al.* [12]. The same impossbility result for an asynchronous shared memory system was given by Loui and Abu-Amara [21]. Fischer and Lynch [10] and Dolev and Strong [8] proved the lower bound of $f+1$ rounds for reaching consensus in the Byzantine failure and crash failure models, respectively.

The k-set consensus problem was defined by Chaudhuri [6]. This work also presented the first algorithm for solving k-set consensus under f faults, where $f < k$. The lower bound of $f < k$ crash-failure processes for solving this problem was shown by Borowski and Gafni [5], Herlihy and Shavit [15], and Saks and Zaharoglou [27]. The approximate agreement problem was proposed, and solved for crash failure and Byzantine failures in the message-passing model by Dolev *et al.* [7]. The wait-free shared memory solution to this problem was proposed by Moran [24].

Wait-free synchronization was introduced by Lamport [18] and developed by Peterson [26]. The theory of wait-free synchronization, consensus hierarchy, and the universal constructions for arbitrary consensus objects was given by Herlihy [14]. The discussion of RMW operations and the analysis of the consensus number of RMW objects with interfering operations is given by Kruskal *et al.* [17]. The renaming problem was proposed and solved for the message-passing model by Attiya *et al.* [1]. They also showed that at least $n+1$ new names are needed if f crash failures are to be tolerated. This lower bound was tightened to $n+f$ by Herlihy and Shavit [15]. This lower bound, as well as the lower bound for k-set consensus are derived from a theorem that characterizes the solvable problems by a f-resilient algorithm using only *Read* and *Write* operations, as shown by Herlihy and Shavit [16]. The wait-free k-set consensus for shared memory is adapted from Attiya and Welch [2]. The wait-free renaming algorithm for the shared memory pardigm is adapted from [1] and Attiya and Welch [2]. The wait-free shared memory renaming algorithm using splitters was proposed by Moir and Anderson [23]. The abstraction of wait-free splitters was proposed and implemented by Lamport [19].

References

[1] H. Attiya, A. Bar-Noy, D. Dolev, D. Peleg, and R. Reischuk, Renaming in an asynchronous environment, *Journal of the ACM*, **41**(1), 1990, 524–548.

[2] H. Attiya and J. Welch, *Distributed Computing: Fundamentals, Simulations, and Advanced Topics*, 2nd edn, Hoboken, NJ, Wiley Interscience, 2004.

[3] A. Bar-Noy, D. Dolev, C. Dwork, and H. R. Strong, Shifting gears: changing algorithms on the fly to expedite Byzantine agreement, *Information and Computation*, **92**(2), 1992, 205–233.

[4] P. Berman and J. Garay, Closure votes: $n/4$-resilient distributed consensus in $(t+1)$ rounds, *Mathematical Systems Theory*, **26**(1), 1993, 3–19.

[5] E. Borowski and E. Gafni, Generalized FLP impossibility result for t-resilient asynchronous computations, *Proceedings of the 25th IEEE Symposium on Theory of Computing*, 1993, 91–100.

[6] S. Chaudhuri, More choices allow more faults: set consensus problems in totally asynchronous systems, *Information and Computation*, **105**(1), 1993, 132–158.

[7] D. Dolev, N. Lynch, S. Pinter, E. Stark, and W. Weihl, Reaching approximate agreements in the presence of faults, *Journal of the ACM*, **33**(3), 1986, 499–516.

[8] D. Dolev and H. R. Strong, Authenticated algorithms for Byzantine agreement, *SIAM Journal of Computing*, **12**(4), 1983, 656–666.

[9] M. Fischer, The consensus problem in unreliable distributed systems (a brief survey), *1983 Conference on Fault-Tolerant Computing*, 1983.

[10] M. Fischer and N. Lynch, A lower bound for the time to assure interactive consistency, *Information Processing Letters*, **14**(4), 1982, 183–186.

[11] M. Fischer, N. Lynch, and M. Merritt, Easy impossibility proofs for distributed consensus problems, *Distributed Computing*, **1**(1), 1986, 26–39.

[12] M. Fischer, N. Lynch, and M. Paterson, Impossibility of distributed consensus with one faulty processor, *Journal of the ACM*, **32**(2), 1985, 374–382.

[13] J. Garey and Y. Moses, Fully polynomial Byzantine agreement for $n > 3t$ processors in $t+1$ rounds, *SIAM Journal of Computing*, **27**(1), 1998, 247–290.

[14] M. Herlihy, Wait-free synchronization, *ACM Transactions on Programming Languages and Systems*, **11**(1), 1991, 124–149.

[15] M. Herlihy and N. Shavit, The asynchronous computability theorem for t-resilient tasks, *Proceedings of the 25th IEEE Symposium on Theory of Computing*, 1993, 111–120.

[16] M. Herlihy and N. Shavit, The topological structure of asynchronous computability, *Journal of the ACM*, **46**(6), 1999, 858–923.

[17] C. Kruskal, L. Rudolph, and M. Snir, Efficient synchronization of multiprocessors with shared memory, *Proceedings of ACM Principles of Distributed Computing*, August 1986, 218–228.

[18] L. Lamport, Concurrent reading and writing, *Communications of the ACM*, **20**(11), 1977, 806–811.

[19] L. Lamport, A fast mutual exclusion algorithm, *ACM Transactions on Computer Systems*, **5**(1), 1987, 1–11.

[20] L. Lamport, R. Shostak, and M. Pease, The Byzantine generals problem, *ACM Transactions on Programming Languages and Systems*, **4**(3), 1982, 382–401.

[21] M. C. Loui and H. H. Abu-Amara, Memory requirements for agreement among unreliable asynchronous processes, *Advances in Computing Research, Vol. 4: Parallel and Distributed Computing*, Greenwich, CT, JAI Press, 1987, 163–183.

[22] N. Lynch, *Distributed Algorithms*, MIT Press and Morgan Kaufmann, 1996

[23] M. Moir and J. Anderson, Wait-free algorithms for fast long-lived renaming, *Science of Computer Programming*, **25**(1), 1995, 1–39.

[24] S. Moran, Using approximate agreement to obtain complete disagreement: the output structure of input free asynchronous computations, *Proceedings of the 3rd Israeli Symposium on Theory of Computing and Systems*, 1995, 251–257.

[25] M. Pease, R. Shostak, and L. Lamport, Reaching agreement in the presence of faults, *Journal of the ACM*, **27**(2), 1980, 228–234.

[26] G. Peterson, Concurrent reading while writing, *ACM Transactions on Programming Languages and Systems*, **5**(1), 1983, 46–55.

[27] M. Saks and F. Zaharoglou, Wait-free k-set agreement is impossible: the topology of public knowledge, *Proceedings of the 25th IEEE Symposium on Theory of Computing*, 1993, 101–110.

CHAPTER

15 Failure detectors

15.1 Introduction

This chapter deals with the design of fault-tolerant distributed systems. It is widely known that the design and verification of fault-tolerent distributed systems is a difficult problem. Consensus and atomic broadcast are two important paradigms in the design of fault-tolerent distributed systems and they find wide applications. Consensus allows a set of processes to reach a common decision or value that depends upon the initial values at the processes, regardless of failures. In atomic broadcast, processes reliably broadcast messages such that they agree on the set of messages delivered and the order of message deliveries.

This chapter focuses on solutions to consensus and atomic broadcast problems in asynchronous distributed systems. In asynchronous distributed systems, there is no bound on the time it takes for a process to execute a computation step or for a message to go from its sender to its receiver. In an asynchronous distributed system, there is no upper bound on the relative processor speeds, execution times, clock drifts, and delay during the transmission of messages, although they are finite. This is mainly caused by unpredictable loads on the system that causes asynchrony in the system and one cannot make any timing assumptions of any types. On the other hand, synchronous systems are characterized by strict bounds on the execution times and message transmission delays.

The asynchronous model of a distributed system has simpler semantics when compared to synchronous model. Applications based on the asynchronous model are easily portable because there are no strict timing assumptions to take care of. The asynchronous model of distributed systems is very popular and has attracted lot of attention due to these reasons. Inspite of the attractiveness of asynchronous distributed systems, it is well known that consensus, atomic broadcast, and several other reliable broadcast problems cannot be solved deterministically even for a single process failure due to the unbounded timing characteristics. The main cause of this impossibility result is that it is very difficult to determine in asynchronous systems whether a process has failed or is simply taking a long time for execution; so it is difficult to deal with failures in these systems. On the other hand, in synchronous systems due to strict timing constraints, failures can easily be detected.

The asynchronous model of distributed systems is widely used, and such systems are prone to failures. Thus, detection and/or prevention of failures in these systems is of vital importance. The detection of process failures is a crucial task in the design of fault tolerant distributed systems. Detection of crashed processes is especially difficult in asynchronous systems as it is impossible to determine whether a process has really crashed or is very slow (as there are no timing constraints present).

In this chapter, we discuss the concept of unreliable failure detectors to deal with the impossibility results in asynchronous distributed systems with crash failures. Basically, the asynchronous model of computation is extended with a failure detection mechanism that is prone to errors in the sense that a process can brand another process as crashed even though the process is running. We study failure detectors in asynchronous distributed systems. We investigate two major problems faced in asynchronous distributed environments, namely, consensus and atomic broadcast. We study several solutions for these problems.

15.2 Unreliable failure detectors

Chandra and Toueg [3] introduced the concept of unreliable failure detectors and showed how unreliable failure detectors can be used to solve two fundamental paradigms of asynchronous distributed systems with crash failures, namely, consensus and atomic broadcast.

15.2.1 The system model

We consider asynchronous distributed systems in which there is no bound on message delay, clock drift, or the time taken to execute a step. The system consists of a finite set of n processes, $Q = \{p_1, p_2, ..., p_n\}$. Each pair of processes is connected by a reliable communication channel. A process can fail by crashing only, i.e., by prematurely halting. A process behaves correctly (i.e., according to its specification) until it crashes.

A discrete global clock is assumed, and the range of the clock's ticks, Φ, is the set of natural numbers. The global clock is used for the sake of simplicity of presentation and reasoning and is not accessible to the processes.

A process p_i is said to crash at time t if p_i does not perform any action after time t. Process failures are permanent; once a process crashes, it does not recover. A *correct* process is a process that has not crashed.

Informally, a run is an infinite execution of the system. Given any run σ, *Crashed*(t, σ) is the set of processes that have crashed by time t and $Up(t, \sigma)$ is the set of processes that are correct (i.e., have not crashed) by time t, that is, $Up(t, \sigma) = Q - Crashed(t, \sigma)$. $Crashed(\sigma)$ is the set of processes that have crashed in a run σ and is equal to $\bigcup_t Crashed(t, \sigma)$. $Up(\sigma)$ is the set of processes that are correct in a run σ and is equal to $Q - Crashed(\sigma)$. If a process $p \in Crashed(\sigma)$, we say that p is a faulty process in σ. If a process $p \in Up(\sigma)$, we say that p is a correct process in σ. We consider only runs where at least one process is correct.

Failure patterns and environments

A failure pattern is a function F from Φ to 2^Q, where $F(t)$ denotes the set of processes that have crashed through time t. An environment E is a set of failure patterns. Environments describe the crashes that can occur in a system. In general, we consider the environments that contain all possible failure patterns, i.e., there is no bound on the number of processes that crash.

Each process p_i has a local failure detector module of D, denoted by D_i. Associated with each failure detector D is a range R_D of values output by the failure detector. A failure detector history H is a function from $QX\Phi$ to 2^Q. $D(F)$ denotes the set of possible failure detector histories permitted for the failure pattern F, i.e., each history represents a possible behavior of D for the failure pattern F. For any failure detector D, any failure pattern F, and any history H in $D(F)$, $H(p_i, t)$ is the set of processes suspected by process p_i at time t.

15.2.2 Failure detectors

A failure detector D is a distributed oracle that gives hints about failure patterns. Each process p_i in the distributed environment has its own local failure detector D_i, which monitors all other processes and maintains a list of processes, currently p_i suspects to have crashed. The suspicion is based on relative timeouts of other processes at p_i.

Thus, a failure detector D as the vector $D = \langle D_{p1}, D_{p2}, ...D_{pn} \rangle$, where D_i is the failure detector module at process p_i, that outputs the set of processes that it currently suspects to have crashed. Formally, a failure detector is a function "from time and the set of all runs" to 2^Q. $D_p(t, \sigma)$ is the set of processes that are suspected to have crashed by p's failure detector module at time t in run σ. If $q \in D_p(t, \sigma)$, we say that p suspects q at time t in run σ. After a process crashes, it is immaterial what its failure detector module indicates. We formalize this by assuming that if $p \in Crashed(t, \sigma)$, then $D_p(t, \sigma) = \phi$.

The failure detectors can make mistakes, i.e., a correct process may be added to the list of suspects and can later be removed if the failure detector realizes that it was a mistake. Thus, a failure detector may continually add and remove processes from its list of suspects. Processes can be added and removed from the list of suspects by each failure detector module any number of times. At any time, failure detector modules at two processes may have different lists of suspects.

It should be noted that the addition of a correct process to the list of suspects by any other process or by all other processes should not prevent this process from behaving correctly, according to its specifications.

15.2.3 Completeness and accuracy properties

Chandra and Toueg [3] classified failure detectors in terms of their completeness and accuracy properties. Informally, completeness requires that a failure detector eventually suspects all processes that have crashed and accuracy restricts the mistakes a failure detector can make (i.e., a correct process suspects another correct process). They define two types of completeness and four types of accuracy properties, giving rise to eight classes of failure detectors.

Chandra and Toueg [3] introduced the concept of reducibility among failure detectors. Informally, a failure detector D is *reducible* into another failure detector D' if there exists a distributed algorithm that can transform D into D'. In this case, any problem that can be solved using D' can also be solved using D. If two failure detectors are reducible to each other, they are said to be *equivalent*.

Chandra and Toueg [3] put failure detectors into eight classes and ordered them into a hierarchy according to the reducibility relationship. In this hierarchy, some failure detectors can solve the consensus problem with any number of process failures, while others require a certain number of correct processes to solve the consensus problem. This requirement and the boundary where this requirement becomes necessary have been clearly specified.

We now define completeness and accuracy properties of a failure detector.

Completeness

Definition 15.1 (Completeness) There is a time after which every process that has crashed is permanently suspected by a correct process.

Completeness can be of two types:

- **Strong completeness** Eventually every process that crashes is permanently suspected by every correct process. Notationally,

$$\forall \sigma, \forall p \in Crashed(\sigma), \forall q \in Up(\sigma), \exists t \text{ such that } \forall t' \geq t : p \in D_q(t', \sigma).$$

- **Weak completeness** Eventually every process that crashes is permanently suspected by some correct process. Notationally,

$$\forall \sigma, \forall p \in Crashed(\sigma), \exists q \in Up(\sigma), \exists t \text{ such that } \forall t' \geq t : p \in D_q(t', \sigma).$$

Note that completeness by itself may not be of much use. For example, a failure detector may satisfy the strong completeness property by having every process permanently suspect all other processes. Such a failure detector is useless because it provides no information about actual failures. Thus, a failure detector must satisfy some accuracy property to be useful. We define this property next.

Accuracy

Definition 15.2 (Accuracy) There is a time after which a correct process is never suspected by any correct process.

There are two types of accuracy property:

- **Strong accuracy** Correct processes are never suspected by any correct process. Formally,

$$\forall \sigma, \forall t, \forall p, q \in Up(t, \sigma) : p \notin D_q(t, \sigma).$$

Since in any practical system it is extremely difficult to achieve accuracy, we weaken it as follows:

- **Weak accuracy** Some correct process is never suspected by any correct process. Formally,

$$\forall \sigma, \exists p \in Up(\sigma), \forall t, \forall q \in Up(t, \sigma) : p \notin D_q(t, \sigma).$$

We collectively refer to strong accuracy and weak accuracy as the *perpetual accuracy* properties because these properties hold all the time. Note that even weak accuracy is difficult to achieve, because a failure detector (even at a correct process) may suspect a correct process and then later correct its mistake. The weak accuracy property does not permit this. Thus, we further weaken the accuracy requirement and allow failure detectors that may suspect a correct process at some points in the run, but they *eventually* satisfy the strong and weak accuracy properties.

Eventual accuracy

Definition 15.3 (Eventual accuracy) We need not require accuracy property to be satisfied by each process at all the time. Instead, we require the accuracy property to be eventually satisfied.

There are two types of eventual accuracy:

- **Eventual strong accuracy** There is a time after which correct processes are not suspected by any correct process. Formally,

$$\forall \sigma, \exists t, \forall t' \geq t, \forall p, q \in Up(t', \sigma) : p \notin D_q(t', \sigma).$$

- **Eventual weak accuracy** There is a time after which some correct process is not suspected by any correct process. Formally,

$$\forall \sigma, \exists t, \forall t' \geq t, \exists p \in Up(\sigma), \forall q \in Up(\sigma) : p \notin D_q(t', \sigma).$$

We collectively refer to eventual strong accuracy and eventual weak accuracy as the *eventual accuracy* properties because these properties hold eventually.

15.2.4 Types of failure detectors

Based on types of accuracies and completeness defined above, failure detectors can be classified into the following categories:

- **Perfect failure detectors** (P) Failure detectors that satisfy the strong completeness and the strong accuracy properties are called perfect failure detectors.
- **Eventually perfect failure detectors** ($\diamond P$) Failure detectors that satisfy the strong completeness and the eventual strong accuracy properties are called eventually perfect failure detectors.
- **Strong failure detectors** (S) Failure detectors that satisfy the strong completeness and the weak accuracy properties are called strong failure detectors.
- **Eventually strong failure detectors** ($\diamond S$) Failure detectors that satisfy the strong completeness and the eventual weak accuracy properties are called eventually strong failure detectors.
- **Weak failure detectors** (W) Failure detectors that satisfy the weak completeness and the weak accuracy properties are called weak failure detectors.
- **Eventually weak failure detectors** ($\diamond W$) Failure detectors that satisfy the weak completeness and the eventual weak accuracy properties are called eventually weak failure detectors.
- Another class of failure detector is the one that satisfies weak completeness and strong accuracy properties. This class is denoted by ϑ.
- The last class is the set of failure detectors that satisfy weak completeness and eventually strong accuracy properties. This class is denoted by $\diamond\vartheta$.

15.2.5 Reducibility of failure detectors

A failure detector D is reducible to another failure detector D' if there is an algorithm $T_{D \rightarrow D'}$ that transforms a failure detector D into another failure detector D'. A natural question is: what does it mean that an algorithm transforms D into D'? Algorithm $T_{D \rightarrow D'}$ requires D' to maintain a variable $output_p$ at every process p. This variable is a part of the local state of p and emulates the output of D' at p. An algorithm $T_{D \rightarrow D'}$ transforms a failure detector D into another failure detector D' if and only if for every run R of $T_{D \rightarrow D'}$ under a failure pattern F

using D, $output^R \in D'(F)$, where $output^R$ is the output of run R using failure detector D and $D'(F)$ denotes the set of histories of failure detector D' for failure pattern F. Thus, $T_{D \rightarrow D'}$ can emulate D' using D. $T_{D \rightarrow D'}$ need not emulate all failure detector histories of D'; however, all failure detector histories it emulates must be histories of D'. Algorithm $T_{D \rightarrow D'}$ is called the *reduction algorithm.*

Given a reduction algorithm $T_{D \rightarrow E}$, any problem that can be solved using E, can also be solved using D. We illustrate this with an example: suppose a given algorithm A requires failure detector E, but only failure detector D is available. We can execute A using failure detector D as follows. Concurrently with A, processes run $T_{D \rightarrow E}$ to transform D to E. Algorithm A is modified at process p as follows: whenever A requires that p queries its failure detector module, p reads the current value of $output_p$, which is concurrently maintained by $T_{D \rightarrow E}$.

Since $T_{D \rightarrow E}$ is able to use D to emulate E, D must provide at least as much information about process failures as E does. Thus, if there is an algorithm $T_{D \rightarrow E}$ that transforms D into E, we say that E is weaker than D and denote it by $D \sqsubseteq E$. Note that $\sqsubseteq$ is a transitive relation. If $D \sqsubseteq E$ and $E \sqsubseteq D$, then we say that D and E are *equivalent* and denote it by $D \equiv E$.

If D and ε are two classes of failure detectors and there exists an algorithm $T_{D' \rightarrow E}$ that can transform every failure detector $D' \in D$ into a failure detector $E \in \varepsilon$, then we say that the class of failure detectors D is reducible to the class of failure detectors ε and this is denoted by $D \sqsubseteq \varepsilon$. In this case, ε is weaker than D. If $D \sqsubseteq \varepsilon$ and $\varepsilon \sqsubseteq D$, then D and ε are equivalent and this is denoted by $D \equiv \varepsilon$.

From a trivial reduction algorithm, where each process p periodically writes the current output of its failure detector module into $output_p$, the following relations between the classes of failure detectors are obvious:

Observation 15.1 $P \sqsubseteq \vartheta$, $S \sqsubseteq W$, $\Diamond P \sqsubseteq \Diamond\vartheta$, $\Diamond S \sqsubseteq \Diamond W$.

15.2.6 Reducing weak failure detector *W* to a strong failure detector *S*

In Algorithm 15.1, we give a reduction algorithm $T_{D \rightarrow D'}$ (due to Chandra and Toueg [3]) that transforms any given failure detector D that satisfies weak completeness, into a failure detector D' that satisfies strong completeness. D' satisfies the same accuracy property that D satisfies. Thus, this algorithm strenghtens the completeness while preserving the accuracy.

Informally, the conversion of any weak failure detector W to a strong failure detector S is as follows: initially, for every process p, $output_p$ is set to null. (Recall that $output_p$ is the variable emulating the output of the failure detector module D'_p.) Every process p periodically sends $(p, suspects_p)$ to every process, where $suspects_p$ denotes the set of processes that p suspects according to its failure detector module D_p. When a process p recieves a message $(q, suspects_q)$ from a process q, process p adds the suspect list of process q, $suspects_q$, to its output, $output_p$, and removes the process q from its output as it is a correct process.

Every process p executes the following:
$output_p \leftarrow \phi$
cobegin
||Task 1: repeat forever
$suspects_p \leftarrow D_p$ {p queries its local failure detector module D_p}
$send(p, suspects_p)$ to all other processes.
||Task 2: when receive $(q, suspects_q)$ from a process q
$output_p \leftarrow (output_p \cup suspects_q) - \{q\}$ {$output_p$ emulates E_p}
coend

Algorithm 15.1 Transforming weak completeness to strong completeness [3].

A correctness argument

The correctness proof of the algorithm involves showing the following three properties:

1. It transforms weak completeness into strong completeness.
2. It preserves the perpetual accuracy.
3. It preserves the eventual accuracy.

We show these properties in the following three lemmas.

Lemma 15.1 *Let p be any process that crashes. If eventually some correct process permanently suspects p in H_D, then eventually all correct processes permanently suspect p in $output^R$, where H_D is the history of failure detector D and $output^R$ is the output of an arbitrary run R using failure detector D.*

Since process p crashes, there is a time t' after which no process recieves a message from p. Suppose there is a correct process q that permanently suspects p in H_D after time t. Consider the execution of *task* 1 by process q after time $t_p = max(t, t')$. Process q sends a message $(q, suspects_q)$ such that $p \in suspects_q$ to all processes. Eventually, every correct process recieves $(q, suspects_q)$ and adds p to output (in *task* 2). Since no correct process recieves any messages from p after time t' and $t_p \geq t'$, no correct process removes p from its *output* after t_p. Thus, there is a time after which every correct process permanently suspects p in $output^R$.

Lemma 15.2 *Let p be any process. If no process suspects p in H_D before time t, then no process suspects p in $output^R$ before time t.*

Suppose there is a time t before which no process suspects process p in H_D. Thus, no process sends a message of type $(-, suspects)$ such that $p \in suspects$ before time t. Thus, no process q adds p to $output_q$ before time t.

Lemma 15.3 *Let p be a correct process. If there is a time after which no correct process suspects p in H_D, then there is a time after which no correct process suspects p in $output^R$.*

Suppose there is a time t after which no correct process suspects p in H_D. Thus, all processes that suspect p after time t eventually crash. Thus, there is time after which no process will send messages of type (–, *suspects*) such that $p \in suspects$. Thus, there is a time t' after which no correct process recieves a message of type (–, *suspects*) such that $p \in suspects$. Let q be a correct process. We need to show that there is a time after which q does not suspect p *in* $output^R$. Consider the execution of *task* 1 by process p after time t'. Process p sends the message $(p, suspects_p)$ to q. When q receives this message, it removes p from $output_q$ if p is present in $output_q$ (*task* 2). Note that q does not receive any messages of type (–, *suspects*) such that $p \in suspects$ after time t'; therefore, q does not add p to $output_q$ after time t'. Thus, there is a time after which q does not suspect p in $output^R$.

Theorem 15.1 $\vartheta \sqsubseteq P, W \sqsubseteq S, \diamond\vartheta \sqsubseteq \diamond P$ *and* $\diamond W \sqsubseteq \diamond S$.

Proof Let D be any failure detector in ϑ, W, $\diamond\vartheta$, or $\diamond$ W. We show that $T_{D \to E}$ transforms D into a failure detector E in P, S, $\diamond P$, or $\diamond S$. Since D satisfies weak completeness, E satisfies strong completeness (from Lemma 15.1). We now argue that D and E have the same accuracy properties. If D is in ϑ or W, then D and E have the same accuracy property (from Lemma 15.2). If D is in $\diamond\vartheta$ or $\diamond W$, then D and E have the same accuracy property (from Lemma 15.3).

Thus, we have:

$$\vartheta \sqsubseteq P, W \sqsubseteq S, \diamond\vartheta \sqsubseteq \diamond P \text{ and } \diamond W \sqsubseteq \diamond S. \qquad \square$$

From Theorem 15.1 and Observation 15.1, we have the following result:

$$P \equiv \vartheta, S \equiv W, \diamond P \equiv \diamond\vartheta, \text{ and } \diamond S \equiv \diamond W.$$

A significance of this result is that if we solve a problem for the four failure detectors with strong completeness, the problem is automatically solved for the remaining four failure detectors.

15.2.7 Reducing an eventually weak failure detector $\diamond W$ to an eventually strong failure detector $\diamond S$

Algorithm 15.2 gives an algorithm that converts any eventually weak failure detector $D \in \diamond W$ into an eventually strong failure detector $E \in \diamond S$. Q is the set of all processes.

At process p, variable $suspected_p(r, q)$ denotes how many times process q has suspected process r and variable $refuted_p(r, q)$ denotes how many times process r has refuted process q. Both variables are initialized to zero. S_p denotes the suspect list of process p.

Process p runs the following:

for all $q, r \in Q$
{Number of times q suspected r according to p}
$suspected_p(r, q) \leftarrow 0$
{Number of times r refuted q according to p}
$refuted_p(r, q) \leftarrow 0$

cobegin
||Task 1: repeat forever
if ($r \in D_p$ and $refuted_p(r, p) \leq suspected_p(r, p)$) then
  p rbcasts (p, *suspects*, r, $refuted_p(r, p) + 1$)

||Task 2: when p rbdelivers (q, *suspects*, r, k)
  $suspected_p(r, q) \leftarrow k$
  if $p = r$ then p rbcasts (p, *refutes*, q, k)

||Task 3: when p redelivers (r, *refutes*, q, k)
  $refuted_p(r, q) \leftarrow k$

||Task 4: repeat forever
  for all processes r
    if $\exists\, q : suspected_p(r, q) > refuted_p(r, q)$
      then $S_p \leftarrow S_p \cup \{r\}$
    else $S_p \leftarrow S_p - \{r\}$
coend

Algorithm 15.2 An algorithm to reduce an eventually weak failure detector into an eventually strong failure detector [3].

An explanation of the algorithm

The algorithm consists of four tasks.

In task 1, a process p continuously performs the following for every process r that it suspects according to its failure detector module D_p: if the number of times process r is suspected by p is greater than the number of times r has refuted p, then p broadcasts a suspect message that contains the incremented refuted value.

In task 2, when process p receives a suspect message (q, *suspects*, r, k) from a process q, it updates $suspected_p(r, q)$ to k. If process p discovers that it was erroneously suspected by process q, p broadcasts an appropriate refutation, refuting the suspicion of process q.

In task 3, when process p receives a refutation message (r, *refutes*, q, k) from process r, it updates $refuted_p(r, q)$ to k.

In task 4, the following is repeatedly done for every process r: if there exists a process q such that the number of times q suspects process r is greater than the number of times the process r refutes q according to p, then process r is added to the suspect list of process p. Otherwise, r is removed from the suspect list of process p.

Correctness argument

A correctness argument of the algorithm is as follows. When a process q receives a suspect message accusing process p, process q may add p to its list of suspects S_q. However, upon receiving p's refutation, process q will remove p from its list of suspects S_q. However, p can be suspected again and added to S_q a second time. However, a further refutation from p will cause p to be again removed from S_q. Thus, a possibly infinite sequence of suspicions followed by corresponding refutations may occur, resulting in p being repeatedly added to and removed from S_q. However, from the eventual weak accuracy property of D, there is a time after which some correct process is not suspected. That is, there is a process p such that there is a time after which no correct process receives a message of type ($*$, *suspects*, p, k), suspecting p. Thus, after a time no correct process adds process p to its suspect list. Together with the refutation mechanism, this ensures the eventual weak accuracy property of the constructed E.

Now let us see why E satisfies the strong completeness property. Since D satisfies the weak completeness property, eventually every process that crashes is permanently suspected by some correct process, say p. Thus, eventually process p will repeatedly broadcast (p, *suspects*, $*$, k) messages for these crashed processes and since these processed have crashed, no one will send refute messages for them. Thus all crashed processes will eventually belong to the suspect list of all correct processes. Thus, due to the broadcast of suspect messages and weak completeness property of D, E satisfies the strong completeness property. Thus E satisfies strong completeness and weak accuracy.

15.3 The consensus problem

In the consensus problem, each correct process proposes a value and all processes must reach a unanimous and irrevocable decision on a value that is related to the proposed values [9]. The consensus problem is defined in terms of the following properties:

- **Termination** Every correct process eventually decides some value.
- **Uniform integrity** Every process decides at most once.
- **Agreement** No two correct processes decide differently.
- **Uniform validity** If a process decides a value v, then some process proposed v.

It is widely known that the consensus cannot be solved in asynchronous systems in the presence of even a single crash failure. This is primarily because one cannot distinguish between a process that has crashed and a process that is responding very slow (may be due to the slow network).

15.3.1 Solutions to the consensus problem

Chandra and Toueg [3] showed how to solve the consensus problem using unreliable failure detectors for each of the eight classes of failure detectors. From the following property, the classes of failure detectors P, S, $\Diamond P$, $\Diamond S$ are, respectively, equivalent to failure detectors ϑ, W, $\Diamond\vartheta$, $\Diamond W$. Notationally,

$$P \equiv \vartheta,\ S \equiv W,\ \Diamond P \equiv \Diamond\vartheta, \text{ and } \Diamond S \equiv \Diamond W.$$

So the problem of solving the consensus problem using unreliable failure detectors reduces to solving it for four classes of failure detectors that satisfy strong completeness (i.e., P, S, $\Diamond P$ and $\Diamond S$), instead of solving it for all eight classes. Since P is reducible to S and $\Diamond P$ is reducible to $\Diamond S$ (i.e., $P \sqsubseteq S$ and $\Diamond$ P $\sqsubseteq \Diamond$ S), the algorithms for solving consensus using S also solve the consensus using P and the algorithms for solving consensus using $\Diamond S$ also solve the consensus using $\Diamond P$.

Next, we present algorithms that solve consensus using S and $\Diamond S$. The consensus algorithm using S can tolerate any number of process failures. However, the consensus algorithm using $\Diamond S$ requires a majority of the processes to be operational.

15.3.2 A solution using strong failure detector *S*

Algorithm 15.3 solves the consensus problem in an asynchronous system using a failure detector D that satisfies strong completeness and weak accuracy (i.e., $D \in S$). This algorithm tolerates any number of process failures (up to $n-1$ faulty processes among a total of n processes).

The following notation will be used:

- i_p is the value proposed by process p.
- $\perp$ is the null value.
- $V_p[q]$ is the process p's estimate of process q's proposed value.
- V_p is process p's estimate of the proposed values by all other processes.
- Δ_p contains all the values of V_p.
- r_p is the current round number of process p.
- $msgs_p(r_p)$ is the set of messages that p recieves from other processes about the proposed values in round r_p.
- $lastmsgs_p$ contains the recieved V_q for all processes q by process p.

Every process p executes the following:
procedure propose(i_p)
$V_p \leftarrow \langle \bot, \bot, \ldots, \bot \rangle$ {p's estimate of the proposed values}
$V_p[p] \leftarrow i_p$;
$\Delta_p \leftarrow V_p$

Phase 1: {Execute round r_p, $1 \leq r_p \leq n-1$}
for $r_p \leftarrow 1$ to $n-1$
    p sends (r_p, Δ_p, p) to all other processes
    wait until [$\forall q$: *received* (r_p, Δ_q, q) or $q \in D_p$] {query the failure detector}

    $msgs_p[r_p] \leftarrow \{(r_p, \Delta_q, q) \mid received(r_p, \Delta_q, q)\}$
    $\Delta_p \leftarrow \langle \bot, \bot, \ldots, \bot \rangle$
    for $k \leftarrow 1$ to n
        if ($V_p[k] = \bot$ and $\exists (r_p, \Delta_q, q) \in msgs_p(r_p)$ with $\Delta_q[k] \neq \bot$)
        then $V_p[k] \leftarrow \Delta_q[k]$
            $\Delta_p[k] \leftarrow \Delta_q[k]$

Phase 2: p sends V_p to all processes
wait until [$\forall q$: *received* V_q or $q \in D_p$] {query the failure detector}
$lastmsgs_p \leftarrow \{V_q \mid$ *received* $V_q\}$
for $k \leftarrow 1$ to n
    if $\exists V_q \in lastmsgs_p$ with $V_q[k] = \bot$
        then $V_p[k] \leftarrow \bot$

Phase 3: decide on the first non-$\bot$ element of V_p

Algorithm 15.3 An algorithm to solve the consensus problem using a strong failure detector $D \in S$ [3].

An explanation of the algorithm

This algorithm has three phases. Initially, V_p is set to null and $V_p[p]$ contains the value, i_p, proposed by process p.

In the first phase, each process executes $n-1$ asynchronous rounds. In each round, processes broadcast and relay their proposed values. Then, each process p waits until it receives a round r message from every process that is not in D_p, before proceeding to round $r+1$. While p is waiting for a message from a process q in round r, it is possible that q is added to D_p. If this is the case, p does not wait for q's message before it proceeds to round $r+1$. All messages recieved by p in round r_p are stored in $msgs_p(r_p)$. If p's estimate of some process k's proposed value is null and it has recieved a message of the form (r_p, Δ_q, q) such that q's estimate of process k's proposed value is not null, then p updates its estimate of k's proposed value to q's estimate of process k's proposed value.

In the second phase, a process p broadcasts its estimate of the proposed values of the processes and waits until it receives the estimate from every process that is not in D_p. While p is waiting for an estimate from q, it is possible that q is added to D_p. If this occurs, p stops waiting for q's estimate. By the end of the second phase, correct processes agree on a vector based on the proposed values of all processes. The ith element of this vector either contains the proposed value of process p_i or $\perp$. If any of the correct processes does not agree with the proposed value of a process, say p_i, then the ith element in the vector is set to null and consensus is not reached on the proposed value. It has been shown that this vector contains the proposed value of at least one process.

In the third phase, all correct processes decide the first non-trivial component of this vector.

This solution for the consensus problem using strong failure detectors, even one having weak accuracy property, has an excellent fault tolerance capacity; the solution tolerates any number of process failures.

Also, since a weak failure detector W is reducible to a strong failure detector S using the algorithm given above, this algorithm also solves the consensus using a weak failure detector W.

15.3.3 A solution using eventually strong failure detector $\diamond S$

The previous solution to the consensus problem used failure detectors with *weak accuracy*: some correct process is never suspected. We now present a solution to the consensus problem using a failure detector that satisfies *eventual weak accuracy*: all processes may be erroneously added to the lists of suspects at one time or another, but there is a time after which a correct process p is permanently removed from the list of suspects. However, at any given time t, processes cannot determine if a particular process is correct, or whether a correct process will never be suspected after time t.

Algorithm 15.4 presents a solution to the consensus problem using an eventually strong failure detector $D \in \diamond S$. Such failure detectors satisfy strong

---

Every process p executes the following:
$estimate_p \leftarrow i_p$ {p's estimate of the decision value}
$state_p \leftarrow$ undecided
$r_p \leftarrow 0$ {r_p denotes the current round number}
$ts_p \leftarrow 0$ {the round in which $estimate_p$ was last updated, initially 0}
cobegin
||Task 1: {Rotate through coordinators until a decision is reached}
while $state_p =$ undecided
  $r_p \leftarrow r_p + 1$
  $c_p \leftarrow (r_p \; mod \; n) + 1$ {c_p is the current coordinator}

Phase 1: {All processes p send $estimate_p$ to the current coordinator}
p sends (p, r_p, $estimate_p$, ts_p) to c_p

Phase 2: {The current coordinator gathers $\lceil (n+1)/2 \rceil$ estimates and proposes a new estimate}
if $p = c_p$ then
wait until [for $\lceil (n+1)/2 \rceil$ processes q: $received(q, r_p, estimate_q, ts_q)$ from q]
$msgs_p[r_p] \leftarrow \{(q, r_p, estimate_q, ts_q) | \ p \ received(q, r_p, estimate_q, ts_q)$ from $q\}$
$t \leftarrow$ largest ts_q such that $(q, r_p, estimate_q, ts_q) \in msgs_p[r_p]$
$estimate_p \leftarrow$ select one $estimate_q$ such that $(q, r_p, estimate_q, t) \in msgs_p[r_p]$
p sends (p, r_p, $estimate_p$) to all processes

Phase 3: {All processes wait for the new estimate proposed by the current coordinator}
wait until [$received(c_p, r_p, estimate_{c_p})$ from c_p or $c_p \in D_p$] {Query the failure detector}
if [$received(c_p, r_p, estimate_{c_p})$ from c_p] then {p received $estimate_{c_p}$ from c_p}
$estimate_p \leftarrow estimate_{c_p}$
$ts_p \leftarrow r_p$
p sends (p, r_p, ack) to c_p
else
p sends (p, r_p, $nack$) to c_p {p suspects that c_p crashed}

Phase 4: {The current coordinator waits for $\lceil (n+1)/2 \rceil$ replies. If these replies indicate that $\lceil (n+1)/2 \rceil$ processes adopted its estimate, the coordinator broadcasts a request to decide.}
if $p = c_p$ then
wait until [for $\lceil (n+1)/2 \rceil$ processes q: $received(q, r_p, ack)$ or $(q, r_p, nack)$]
if [for $\lceil (n+1)/2 \rceil$ processes q: $received(q, r_p, ack)$]
then p R-broadcasts (p, r_p, $estimate_p$, $decide$)

||Task 2: {When p receives a *decide* message, it decides}
when p R-delivers (q, r_q, $estimate_q$, $decide$) for some q
if $state_p = undecided$ then
decide on $estimate_q$
$state_p \leftarrow decided$
coend

Algorithm 15.4 An algorithm to solve the consensus problem using an eventually strong failure detector $D \in \diamond S$ [3].

completeness and eventual weak accuracy. The algorithm requires that a majority of the processes are always up. If f is the maximum number of processes that may crash at any time, this algorithm requires that $f < \lceil n/2 \rceil$, that is, at least $(n+1)/2$ processes are correct at all times.

An explanation of the algorithm

This algorithm proceeds in asynchronous rounds and makes use of the *rotating coordinator* paradigm until a decision is reached. All processes know that during round r, the coordinator is process $c = (r \bmod n) + 1$. All messages are either to or from the "current" coordinator. The "current" coordinator tries to determine a consistent decision value. If the current coordinator is correct and is not suspected by any surviving process, then it succeeds and broadcasts the decision value.

The algorithm goes through three asynchronous stages where each stage can contain several asynchronous rounds. In the first stage, several decision values are proposed. In second stage, a value gets locked: no other decision value is possible. In the third and final stage, the processes decide on the locked value and consensus is reached.

Initially, the state of a process p is "undecided" and its estimate of the decision value is i_p. A timestamp ts_p is associated with every process p that contains the round number when its estimate was last updated.

Each round of task 1 consists of four asynchronous phases. In phase 1, every process p sends its current estimate of the decision value to the current coordinator c_p. It also sends the round number (ts_p) in which it adopted this estimate.

In phase 2, the coordinator c_p gathers $\lceil (n+1)/2 \rceil$ such estimates and proposes a new estimate. The current coordinator waits until it receives estimates from $\lceil (n+1)/2 \rceil$ processes. It stores all these estimates in the array $msgs_p[r_p]$, selects one with the largest timestamp, and sends it to all the processes as the new estimate, $estimate_p$.

In phase 3, all processes wait for the new estimate proposed by the current coordinator. For each process p, there are two possibilities:

- Process p receives $estimate_{c_p}$ from the coordinator c_p: in this case, p updates its timestamp to the current round number and sends an *ack* to c_p to indicate that it adopted $estimate_{c_p}$ as its own estimate.
- Process p does not receive an $estimate_{c_p}$ from the coordinator c_p and, upon consulting its failure detector module D_p, suspects that the coordinator c_p has crashed: in this case, p sends a *nack* to c_p.

In phase 4, the coordinator c_p waits for $\lceil (n+1)/2 \rceil$ replies (*acks* or *nacks*). If all replies are *acks*, then c_p knows that a majority of processes changed their estimates to $estimate_{c_p}$ and thus $estimate_p$ is locked and c_p broadcasts a request to decide value $estimate_p$.

In task 2, at any time, if a process receives such a request, it decides accordingly, i.e., when a process p receives a message of the form (q, r_q, $estimate_q$, *decide*) from a process q, then p decides on the estimate of q provided it has not already decided. In this case, process p changes its state to "decided."

For correctness of the algorithm, we have to show that the algorithm satisfies termination, uniform validity, agreement, and uniform integrity properties. The readers are referred to the original source for a correctness proof.

This algorithm requires that $f < \lceil n/2 \rceil$, i.e., at least $\lceil n/2 \rceil$ processes are correct, and assumes that processes have a priori knowledge of the list of potential coordinators.

15.4 Atomic broadcast

Atomic broadcast is one of the fundamental problems in fault-tolerant distributed computing. It is a powerful paradigm in the design of fault-tolerant distributed computing systems. Chandra and Toueg [3] showed that the results of consensus can be applied to solve the problem of atomic broadcast. Informally, atomic broadcast requires that all correct processes deliver the same set of messages in the same order (i.e., deliver the same sequence of messages). Formally, *atomic broadcast* can be defined as a reliable broadcast with the total order property.

The total order property If two correct processes p and q deliver two messages m and m', then p delivers m before m' if and only if q delivers m before m'.

The total order and agreement properties of atomic broadcast ensure that all correct processes deliver the same sequence of messages.

In asyncronous sytems with crash failures, consensus and atomic broadcast are equivalent and this can be shown by reducing one to the another. Consensus can be reduced to atomic broadcast as follows: In the consensus problem, to propose a value, a process atomically broadcasts it. To decide a value, a process picks the value of the first message that it atomically delivers. The total order property of atomic broadcast ensures that all correct processes deliver the same first message. Hence, all correct processes choose the same value and the agreement property of the consensus is satisfied. In the next section, we show how to reduce atomic broadcast to consensus.

A consequence of this equivalence is that a solution for one can be used to solve the other. In addition, it implies the following for solving atomic broadcast in asynchronous systems:

- Since consensus has no deterministic solution in aynchronous systems, even if we assume that at most one process may fail by crashing, atomic broadcast cannot be solved by a deterministic algorithm even if at most one process may fail by crashing.
- As consensus is solvable using randomization or unreliable failure detectors in asynchronous systems, atomic broadcast can be solved using these techniques.

15.5 A solution to atomic broadcast

Algorithm 15.5 presents a solution (due to Chandra and Toueg [3]) to the atomic broadcast problem using the consensus problem in asynchronous systems. This algorithm shows how to transform any consensus algorithm into an atomic broadcast algorithm in asynchronous systems. This atomic broadcast algorithm tolerates as many faulty processes as the consensus algorithm does.

This atomic broadcast algorithm uses repeated executions of consensus. The kth execution of consensus is used to decide on the kth batch of messages to be atomically delivered. Processes distinguish between these executions by tagging all the messages pertaining to the kth execution of consensus with the counter k.

The atomic broadcast algorithm uses $R\_broadcast(m)$ and $R\_deliver(m)$ primitives of reliable broadcast. To avoid any confusion, note that the primitives $A\_broadcast(m)$ and $A\_deliver(m)$ respectively refer to a broadcast and a delivery in atomic broadcast, while primitives $R\_broadcast(m)$ and $R\_deliver(m)$ respectively refer to a broadcast and a delivery associated with reliable broadcast. $propose(k, -)$ and $decide(k, -)$ are the propose and decide primitives corresponding to the kth execution of consensus.

An explanation of the algorithm

The algorithm consists of three tasks such that: (i) a task that is enabled is eventually executed and (ii) a task i can execute concurrently with another task j provided $i \neq j$.

In task 1, when a process p wants to A-broadcast a message m, it R_broadcasts m. In task 2, a message m is added to set $R\_delivered_p$ when process p R_delivers it.

In task 3, when a process p A_delivers a message m, it adds m to set $A\_delivered_p$. $A\_undelivered_p$ (defined as $R\_delivered_p - A\_delivered_p$) is the set of messages that p has R_delivered but has not A_delivered yet. Process p periodically checks whether $A\_undelivered_p$ contains messages. If $A\_undelivered_p$ contains messages, p enters its next execution of consensus, say the kth one, and proposes $A\_undelivered_p$ as the next batch of messages to be A_delivered. Process p then waits for the kth consensus decision, which is denoted by $msgSet^k$. $msgSet^k$ contains messages that are R_delivered but are

Every process p executes the following:

Initialization:

$R\_delivered \leftarrow \emptyset$
$A\_delivered \leftarrow \emptyset$
$k \leftarrow 0$

To execute $A\_broadcast(m)$: {Task 1}
$R\_broadcast(m)$

$A\_deliver(-)$ occurs as follows:
when $R\_deliver(m)$ {Task 2}
  $R\_delivered \leftarrow R\_delivered \bigcup \{m\}$
when $R\_delivered - A\_delivered \neq \emptyset$ {Task 3}
  $k \leftarrow k+1$
  $A\_undelivered \leftarrow R\_delivered - A\_delivered$
  $propose(k,\ A\_undelivered)$
  wait until $decide(k,\ msgSet^k)$
  $A\_deliver^k \leftarrow msgSet^k - A\_delivered$
  atomically deliver all messages in $A\_deliver^k$ in some determinisic order
  $A\_delivered \leftarrow A\_delivered \bigcup A\_deliver^k$

Algorithm 15.5 A solution to atomic broadcast using consensus [3].

yet to be A_delivered. Finally, p A_delivers all the messages in $msgSet^k$ except those already A_delivered by it (i.e, all the messages in the set $A\_deliver_p^k = msgSet^k - A\_delivered_p$) in some deterministic order that was agreed *a priori* by all processes.

For a correctness proof of the algorithm, the readers should refer to the original source.

15.6 The weakest failure detectors to solve fundamental agreement problems

Delporte-Gallet *et al.* [5] showed that, if we exclude unrealistic failure detectors,[1] then in an environment where we do not bound the number of faulty processes, the class of perfect failure detectors P is the weakest to solve fundamental agreement problems like uniform consensus, atomic broadcast, and terminating reliable broadcast (also called the Byzantine Generals).

[1] Unrealistic failure detectors are failure detectors that can guess the future and thus cannot be implemented even in a perfectly synchronous systems.

Delporte-Gallet *et al.* [5] collapsed the Chandra–Toueg failure detector hierarchy in this environment, and showed that P is the only useful class to solve these agreement problems. This explains why most known reliable distributed systems rely on a group membership service that precisely aims at emulating a perfect failure detector P, that is, when a process is suspected due to a timeout, it is excluded from the group. Thus, every suspicion is taken as being accurate.

Uniform consensus

In consensus, the agreement property allows the bad processes to decide differently from good processes. This fact can be sometimes undesirable as it does not prevent a bad process from propagating a different decision in the system before crashing. In the uniform consensus, the uniform-agreement property allows no two processes (good or bad) to decide differently, which enforces the same decision on any process that decides.

Terminating reliable broadcast

Solving the consensus problem is equivalent to solving the atomic broadcast problem, in any system with reliable channels (i.e., where only a finite number of messages can be lost). Atomic broadcast entails delivering messages to processes in a reliable and totally ordered manner. Terminating reliable broadcast is a stronger form of atomic broadcast. In *terminating reliable broadcast*, the processes deliver messages in the same sequence as atomic broadcast does, but, in addition, processes should deliver a specific nil value for every message that was broadcast by a faulty process but was not delivered by any correct process. This problem is a rephrasing of the famous Byzantine Generals problem in the fail-stop model.

Delporte-Gallet *et al.* [5] showed that, in environments where the number of faulty processes is not bounded, uniform consensus is strictly harder than consensus, and uniform consensus and atomic broadcast are strictly weaker than terminating reliable broadcast.

In environments where the number of faulty processes is not bounded, the exact information about failures needed to solve consensus (hence atomic broadcast) and terminating reliable broadcast, is captured by P. Thus, in the failure detector hierarchy, P is the only useful class to solve the agreement problems.

15.6.1 Realistic failure detectors

Note that a failure detector has been defined as *any* function of the failure pattern and this function may be able to provide information about the future failures. Such a failure detector does not factor out synchrony assumptions of the system and cannot be implemented even in a perfectly synchronous system.

Delporte-Gallet *et al.* [5] restricted the scope of failure detectors as functions of the "past" failure patterns and defined the class of realistic failure detectors R, which cannot guess the future.

A failure detector is realistic if it cannot guess the future, i.e., there is no time t and no failure pattern F at which the failure detector can provide exact information about crashes that will hold after t in F.

Formally, the class of realistic failure detectors R is the set of failure detectors D that satisfy the following property: (An environment E is a set of all possible failure patterns and Ω denotes the set of all the processes.)

$$\forall (F, F') \in E, \forall t \in \phi \text{ such that } \forall t_1 \leq t;\ F(t_1) = F'(t_1).$$

We have:

$$\forall H \in D(F), \exists H' \in D(F') \text{ such that } \forall t_1 \leq t;\ \forall p_i \in \Omega\text{: } H(p_i, t_1) = H'(p_i, t_1).$$

That is, a failure detector D is realistic if for any pair of failure patterns F and F' that are similar up to a given time t, whenever D outputs some information at a time $t-k$ in F, D could output the very same information at $t-k$ in F'. Thus, a realistic failure detector cannot distinguish two failure patterns according to what will happen in the future. In other words, the output of a realistic failure detector depends only upon the past. For a realistic failure detector D, for any failure pattern F, the output of D at time t is a function of F up to time t.

Example We now present two failure detector examples to illustrate the concept. The first failure detector is realistic and the second is non-realistic.

1. **Scribe** (C) A scribe, "C," is a failure detector that sees what is happening at all processes in real time and outputs a list of processes based on what it sees. For any failure pattern F, failure detector C outputs, at any time t, the list of values of F up to time t, denoted by $F[t]$. For each failure pattern F, $C(F)$ is the singleton set that contains the failure detector history H, such that:

$$\forall t \in \phi,\ \forall p_i \in \Omega,\ H(p_i, t) = F[t].$$

   Therefore, C is an example of a realistic failure detector.

2. **The Marabout** (M) Failure detector M (Marabout) outputs a list of processes. For any failure pattern F and at any process p_i, the output of the failure detector M is constant and is the list of faulty processes in F. Thus, M outputs the list of processes that have crashed or will crash in F. This is an example of an unrealistic failure detector.

   To better understand why M is an unrealistic failure detector, consider the failure patterns F and F' such that (i) all processes are correct in F except p_1, which crashes at time $t = 10$, (ii) all processes are correct in F', and (iii) F and F' are same up to time $t = 9$.

Consider any history H of $M(F)$ and any history H' of $M(F')$. By the definition of M:

- the output at any process and at any time of H' is ϕ, and
- for any history $H \in M(F)$, for any process p_i, and any time $t \in \Phi$, the output, $H(p_i, t)$, is $\{p_1\}$.

However, if M was realistic, its failure detector histories H in $M(F)$ and H' in $M(F')$ should be such that H' and H are identical up to time $t = 9$. Thus, M is unrealistic because it is accurate about the future.

15.6.2 The weakest failure detector for consensus

Recall that, in the consensus problem, every process proposes an initial value and all processes must agree on one of these values such that termination, agreement, and validity properties are satisfied. Delporte-Gallet *et al.* [5] showed that, if the number of faulty processes is not restricted, then P is the weakest "realistic" failure detector class to solve consensus. Precisely, they showed that if the number of faulty processes is not restricted, any realistic failure detector that solves consensus can be tranformed into a failure detector of class P. We next give an intuitive proof of this lower bound, which includes the following two parts:

- First, we show that "any consensus algorithm is total," that is, the causal chain of any decision event contains a message from every process that has not crashed at the time of the decision.

  We argue that a consensus decision cannot be reached by any process without having consulted every other correct process. If this is not true, a situation is possible where, after the decision, all the consulted processes crash except the one that is not consulted and this process later decides differently. If all the processes are consulted before every decision, we call such an algorithm "total".
- Second part of the proof entails showing that "if a realistic failure detector D implements a total consensus algorithm, then D can be transformed into a perfect failure detector P."

  This proof uses the fact that D is realistic and the algorithm is total. Therefore, for accurate tracking of process failures, no decision is taken without consulting every correct process. A process is suspected to have crashed in a sequence of consensus instances, if and only if a decision is reached and the process was not consulted in the decision.

15.6.3 The weakest failure detector for terminating reliable broadcast

Terminating reliable broadcast is a strong form of reliable broadcast in which processes must deliver a specific value *nil* if the sender process has crashed; else, the processes must deliver the message m, broadcast by $sender(m)$.

A general variant of the problem is considered where every process is a potential initiator of the broadcast. The kth instance of the broadcast initiated by process p_i is denoted by (i, k). Instance $(i, *)$ is defined by the following properties:

- **Validity** If a correct process p_i broadcasts a message m, then p_i eventually delivers m.
- **Agreement** If a process delivers a message m, then every correct process delivers m.
- **Integrity** If a process delivers a message m and p_i is correct, then $sender(m) = p_i$.

If we do not bound the number of processes that can crash, then among realistic failure detectors, the weakest class to solve terminating reliable broadcast is P. A sketch of the proof is as follows.

Sufficient condition
Terminating reliable broadcast problem can be solved by any perfect failure detector, including realistic failure detectors. When instance (k, k') of the terminating reliable broadcast is executed, each process waits until it receives the value from p_k or it suspects p_k. In the former case, it proposes the received value to consensus, and in the latter case, it proposes value *nil*. The value delivered is the consensus value.

Necessary condition
Suppose A is any terminating reliable broadcast algorithm using a failure detector D. We can emulate the output of D, a failure detector of class P, using terminating reliable broadcast algorithm A in a distributed variable $output(P)$ in the following way: whenever a process p_j delivers *nil* for an instnace $(i, *)$ of the algorithm, p_j adds p_i to $output(P)_j$. Any process that crashes will eventually be permanently added to $output(P)$ at every correct process. Thus, strong completeness will be ensured. A process p_i is added to $output(P)_j$ at some time t only if p_i is faulty. Since D is assumed to be realistic, p_i must have crashed by time t.

15.7 An implementation of a failure detector

Now we present an algorithm to implement a failure detector. The algorithm is a timeout-based implementation of eventually perfect failure detector $D \in \Diamond P$

in partially synchronous models. The concept of partial synchrony in a distributed system lies between the cases of a synchronous system and an asynchronous system. In partial synchrony, the system is asynchronous initially but after an unknown time t, the system becomes synchronous. This assumption captures the fact that the system does not always behave as synchronous. Generally distributed systems are synchronous most of the time, and then they experience bounded asynchrony periods. We expect from partial synchrony a period of synchrony long enough to terminate the distributed algorithm.

Each process p maintains a default timeout interval for every other process in the system. A process sets a timeout based on the worst-case round trip of a message exchange. To measure the elapsed time, each process p maintains a local clock, say, by counting the number of steps that it takes.

The following variables are used in the algorithm:

- $Output_p$ (called the suspect list of p) is a set to hold all the suspected processes by process p. This set is initially empty. This set is local to process p, which is executing the algorithm.
- q is the loop variable used to identify each process in the system.
- Π is a set of all processes in the system.
- $\Delta_p(q)$ is the duration of p's timeout interval for q.

The algorithm is presented in Algorithm 15.6.

Every process p executes the following:
$Output_p \leftarrow \emptyset$ {Initializes output set to empty}
for all $q \in \prod$
$\Delta_p(q) \leftarrow$ default time-out interval {Set the timeout interval}
cobegin

||Task 1: **repeat periodically**
  send "p-is-alive" to all

||Task 2: **repeat periodically**
  for all $q \in \prod$
  if $q \neq Output_p$ and
  p did not receive "q-is-alive" during the last $\Delta_p(q)$ ticks of p's clock
    $Output_p \leftarrow Output_p \cup \{q\}$ {p times-out on q and starts suspecting that q has crashed}

||Task 3: when receive "q-is-alive" for some q
  If $q \in Output_p$ {p knows that it prematurely timed-out on q}
    $Output_p \leftarrow Output_p - \{q\}$ {p repents on q}
    $\Delta_p(q) \leftarrow \Delta_p(q) + 1$ {p increases its time-out period for q}
coend

Algorithm 15.6 A timeout-based implementation of $D \in \diamond P$ in the partial synchrony model [3].

Explanation of the algorithm

- **Task 1** Each process p periodically sends a "p-is-alive" message to all other processes. This is like a heart-beat message that informs other processes that process p is alive.
- **Task 2** If a process p does not receive a "q-is-alive" message from a process q within $\Delta_p(q)$ time units on its clock, then p adds q to its set of suspects if q is not already in the suspect list of p.
- **Task 3** When a process delivers a message from a suspected process, it corrects its error about the suspected process and increases its timeout for that process. If process p receives "q-is-alive" message from a process q that it currently suspects, p knows that its previous timeout on q was premature – p removes q from its set of suspects and increases its timeout period for process q, $\Delta_p(q)$.

Correctness of the algorithm

The algorithm insures the properties of an eventually perfect failure detector as discussed below:

- **Strong completeness** If a process p crashes, it will stop sending "p-is-alive" messages. Eventually every process that crashes is permanently detected by every correct process. Therefore, a crashed process will be suspected by any correct process and no process will revise the judgement.
- **Eventual strong accuracy** After time t, the system becomes synchronous, i.e., after time t, a message sent by a correct process p to another process q will be delivered within a bounded time. If p was wrongly suspected by q, then q will revise its suspicious. Eventually, no correct process is ever suspected.

15.8 An adaptive failure detection protocol

In this section, we discuss an adaptive failure detection protocol that allows a process to monitor other processes and eventually detects its crash [8]. The protocol relies as much as possible on application messages to do this monitoring and uses control messages only when no application message is sent by the monitoring process to the observed process. More precisely, the proposed protocol allows a process to monitor another process using the application messages it is exchanging to communicate with the other process, saving failure detection messages. A failure detector (thus, failure detection messages) are used when the processes are not communicating. The cost associated with the implementation of a failure detector incurs only when the failure detector is used (hence, it is called a lazy failure detector). When the underlying system satisfies the partial synchrony assumption, the protocol implements an eventually perfect failure detector $D \in \diamond P$. Recall that an eventually perfect failure detector makes no mistake (i.e., the list of suspects at a process includes all crashed processes, but no correct process) after a finite, but unknown time.

For any failure detector in $\diamond P$, after it becomes perfect, if the average observed transmission delay is finite and the upper layer application terminates within a bounded number of steps, then it terminates correctly when run with the proposed protocol. These properties make the protocol attractive: it is inexpensive, implementable, and powerful.

The basic failure detection protocol (denoted by FD_L) ensures that if a process queries another process that has crashed, then it will definitely suspect it. Thus, completeness of the detection is satisfied. The failure detection protocol is plugged into two particular contexts. The first context is defined by the properties to be satisfied by the lower layer, namely, partial synchrony. When the failure detection protocol is plugged in such a system, the protocol provides a failure detector of the class $\diamond P$. The second context is defined by a property assumed to be satisfied by the upper layer, i.e., the application and some weaker properties to be satisfied by the lower level. The first context is defined by partial synchrony.

The second context defines a property (called $\diamond P$-terminating) that the application has to satisfy. A failure detector-based application (the failure detector it uses belongs to $\diamond P$) is $\diamond P$-terminating if it terminates correctly within at most some l steps after the failure detector becomes perfect. When run with a $\diamond P$-terminating application, the protocol provides the application with the same properties as $\diamond P$ if the average observed transmission delay is finite. Interestingly, unlike the first context, the second context does not require an upper bound on message transfer delays. These two contexts show that this failure detection protocol is inexpensive, implementable, and powerful.

15.8.1 Lazy failure detection protocol (FD_L)

Assumptions

The basic system consists of a finite set of processes $P = \{p_1, p_2, \ldots p_n\}$. Each process p_i has a local hardware clock hc_i that strictly monotonically increases. The local clocks are not required to be synchronized, and there is no assumption on their possible drift. The behavior of a process can be modeled by a finite state automaton. Each step of a process is triggered by a message. An event is the execution of a communication statement by a process. The history h_i of a process p_i is the sequence of communication events it produces.

Every pair of processes is connected by a channel and they communicate by sending and receiving messages through channels. Channels are not required to be FIFO. They are only assumed to be reliable in the following sense: they do not create, duplicate, alter or loose messages, i.e., if a process p_j is correct, message sent by a process p_i to p_j is eventually received by p_j.

Primitives provided

The protocol provides the following primitives to each upper layer application process p_i:

- *SEND M* to p_j: used by p_i to send an application message M to p_j.
- *RECEIVE M*: used by p_i to receive an application message M.
- *QUERY*(j): used to know whether p_j is suspected to have crashed. This primitive returns an answer, namely, the value suspect or no_suspect.

At an operational level, the protocol uses three types of messages: "*appl*," "*ack*," and "*ping*." To send an application message M to p_j, a process p_i invokes "*sendappl*(m) to p_j" where the protocol message m includes M plus some control information. When it receives such a message, p_j systematically acknowledges it by sending back $ack(m)$. When it receives $ack(m)$, p_i computes the round trip delay of the pair $appl(m) + ack(m)$. For each destination process p_j, p_i aditionally computes the maximum round trip delay for the messages that have been acknowledged by p_j.

The answer provided by *QUERY*(j) when it is invoked by the upper layer depends on the existence of a "pending" message, i.e., a message m such that $appl(m)$ has been sent to p_j but the corresponding $ack(m)$ has not yet been received by p_i: (i) if there is no such message, the answer is *no_suspect*, but p_i sends a ping message to p_j in order to verify its answer, (ii) if there are such "pending" messages, the answer depends on the maximum round trip delay already experienced.

The protocol FD_L

The protocol manages two arrays of local variables for each process p_i: (i) $pending\_msg\_st_i[j]$ – this set is initially empty and it contains the sending times of the messages sent by p_i to p_j, whose acknowledgements have not yet been received by p_i; (ii) $max\_rtd_i[j]$ – this contains the biggest round trip time of the messages that p_i sent to p_j and that have been acknowledged. Initially, this variable has the value zero. If the value of $max\_rtd_i[j]$ from the previous execution is known, then $max\_rtd_i[j]$ can be initialized to this value.

A call to *SEND M* is interpreted as a message reception from the upper layer. Similarly, *RECEIVE M* is interpreted as a message sent to the upper layer. A protocol message m has a type $(appl/ack/ping)$. In addition to a content $(m.content)$, a message m also carries the local send time $(m.st)$.

More precisely, $appl(m)$ and $ping(m)$ carry their local send time and $ack(m)$ carries the send time of the $appl(m)$ or $ping(m)$ message it is associated with.

The protocol is described for process p_i in Algorithm 15.7 and works as follows:

```
when SEND M to p_j is invoked:
m.content ← M; m.st ← hc_i;
pending_msg_st_i[j] ← pending_msg_st_i[j] ∪ {m.st}
send appl(m) to p_j

when type(m) is recieved from p_j:
 case
   type = appl then transmit M = m.content to upper layer,
           {* RECEIVE M *} send ack(m) to p_j {* m.st keeps its value *}
   type = ack then rt ← hc_i;
     max_rtd_i[j] ← max(max_rtd_i[j], rt-m.st);
     pending_msg_st_i[j] ← pending_msg_st_i[j] – {m.st}
   type = ping then send ack(m) to p_j {* m.st keeps its value *}
 endcase

when QUERY(j) is invoked:
if pending_msg_st_i[j] = Ø then create a control message m;
   m.content ← null; m.st ← hc_i
   send ping(m) to p_j;
   pending_msg_st_i[j] ← {m.st};
   return(no_suspect)
else
   rt ← hc_i;
   if rt-min(pending_msg_st_i[j]) > max_rtd_i[j]
     then return (suspect)
     else return (no_suspect)
   endif
endif
```

Algorithm 15.7 Lazy failure detection protocol for process p_i [8].

When *SEND M* to p_j is invoked by p_i, *m.content* is initialized to the application message M and *m.st* is initialized to the local hardware clock time. Since the acknowledgement of this message is not yet received by p_i, *m.st* is added to the set $pending\_msg\_st_i[j]$. Now, p_i sends the application message $appl(m)$ to p_j.

When p_i receives a message from p_j, it acts as follows: if the message received by process p_i is of type *appl*, then the message *content*(*m.content*) is transmitted to the upper layer and an acknowledgement message, $ack(m)$

is sent to p_j. If the message is an acknowledgement, *ack*, then the maximum round trip delay time of the messages sent to p_j by process p_i is updated to the maximum of the previous and current round trip delay times. Since this is an acknowledgement message, its sending time is deleted from the pending time set. When the message of type *ping* is received by p_i, it sends an acknowledgement message $ack(m)$ to p_j.

When $QUERY(j)$ is invoked by the process p_i, the following two conditions arise: (i) if $pending\_msg\_st_i[j]$ is empty, then a control message m is created and is used to ping process p_j. A control message is used as there is no communication between the processes. The ping message send time is added to the pending time set and a value *no_suspect* is returned. (ii) When $pending\_msg\_st_i[j]$ is non-empty, if the time taken to receive an acknowledgement from process p_j is greater than the $max\_rtd_i[j]$, then the process p_j is suspected to be crashed and a value *suspect* is returned, else *no_suspect* is returned.

Properties of FD_L

If from some time t, a process p_i obtains the answer *suspect* each time it invokes $QUERY(j)$, we say that from that time it "permanently suspects p_j" from t.

Completeness property

Let us assume that p_i is correct, while p_j is faulty (i.e., it has crashed). Then, FD_L ensures that eventually p_i permanently suspects p_j to have crashed.

The protocol in partially synchronous systems[2]

If the underlying system is partially synchronous, there is a time t after which FD_L ensures that no correct process is suspected by a correct process.

$\diamond P$ terminating protocol

If the upper layer protocol is $\diamond P$-terminating, then it terminates with probability 1 when, instead of using a failure detector of $\diamond P$, it uses FD_L.

Message cost

Each *appl*() or *ping*() message generates at most one *ack*() message. Both *appl*() and *ping*() are due to the application layer: *appl*() when it sends an application message and *ping*() when it invokes *QUERY*().

The cost of invocation of $QUERY(j)$ by a process p_i after p_j has crashed: According to the current state of $pending\_msg\_st_i[j]$, p_i can be forced to send $ping(m)$ message to p_j. But from now, the condition $pending\_msg\_st_i[j] \neq \emptyset$ remains permanently true. Consequently, the next invocations of $QUERY(j)$ do not send messages, and their communication cost is zero.

[2] This means that there is a time after which there are upper bounds on messages transfer delays and associated processing times.

15.9 Exercises

Exercise 15.1 It is well known fact that consensus and atomic broadcast problems cannot be solved deterministically in asynchronous distributed systems even for a single process failure. Then how do failure detectors solve these problems?

15.10 Notes on references

The area of failure detectors was initiated by Chandra and Toueg [3] and a large number of researchers followed it. An excellent short review paper on the topic is by Raynal [22]. Delporte-Gallet *et al.* [5] present a realistic failure detector. An adaptive failure detector can be found in Fetzer *et al.* [8].

Implementations of failure detectors can be found in [17]–[20]. Garg and Mitchell [11] describe implementable failure detectors. Gupta *et al.* [14] discuss scalable failure detectors. Hurfin *et al.* [15,16] present a family of consensus protocols based on failure detectors. Schiper [23] discusses early consensus using weak failure detectors. Chandra *et al.* [4] discuss the weakest failure detector to solve consensus. Guerraoui [12] presents non-blocking atomic commit using failure detectors. Delporte-Gallet *et al.* [6] discuss how to achieve mutual exclusion in asynchronous distributed systems with failure detectors. Other work on failure detectors can be found in [1,2,7,10,12,13,21,24].

References

[1] M. Aguilera, W. Chen, and S. Toueg, Heartbeat: a timeout-free failure detector for quiescent reliable communication, *Workshop on Distributed Algorithms*, 1997, 126–140.

[2] M. Aguilera, W. Chen, and S. Toueg, Using the heartbeat failure detector for quiescent reliable communication and consensus in partitionable networks, *Theoretical Computer Science*, **220**(1), 1999, 3–30.

[3] T. D. Chandra and S. Toueg, Unreliable failure detectors for reliable distributed systems, *Journal of the ACM*, **43**(2), 1996, 225–267. (First version published in *Proceedings of the 10th ACM Symposium on Principles of Distributed Computing*, 1991.)

[4] T. D. Chandra, V. Hadzilacos, and S. Toueg, The weakest failure detector for solving consensus, *Journal of the ACM*, **43**(4), 1996, 685–722.

[5] C. Delporte-Gallet, H. Fauconnier, and R. Guerraoui, A realistic look at failure detectors, *Proceedings IEEE International Conference on Dependable Systems and Networks (DSN'02)*, Washington DC, 2002, 345–352.

[6] C. Delporte-Gallet, H. Fauconnier, R. Guerraoui, and P. Kouznetsov, Mutual exclusion in asynchronous systems with failure detectors, *Journal of Parallel and Distributed Computing*, **65**(4), 2005, 492–505.

[7] C. Delporte-Gallet, H. Fauconnier, and R. Guerraoui, Failure detection lower bounds on registers and consensus, *Proceedings of the 16th Symposium on Distributed Computing (DISC'02)*, 2002, 237–251.

[8] C. Fetzer, M. Raynal, and F. Tronel, An adaptive failure detection protocol, *Proceedings of the 8th IEEE Pacific Rim International Symposium on Dependable Computing (PRDC'01)*, Seoul, Korea, 2001, 146–153.

[9] M. J. Fischer, N. A. Lynch, and M. S. Paterson. Impossibility of distributed consensus with one faulty process, *Journal of the ACM*, **32**(3), 1985, 374–382.

[10] R. Friedman, A. Mostefaoui, and M. Raynal, A weakest failure detector based asynchronous consensus protocol for $f < n$, *Information Processing Letters*, **90**(1), 2004, 39–46.

[11] V. K. Garg and J. R. Mitchell, Implementable Failure Detectors in Asynchronous Systems, *Proceedings of the 18th Conference on Foundations of Software Technology and Theoretical Computer Science*, Chennai, India, December 17–19, 1998, Berlin/Heidelberg, Springer, 1998, 158–170.

[12] R. Guerraoui, Nonblocking atomic commit in asynchronous distributed systems with failure detectors, *Distributed Computing*, **15**, 2002, 17–25.

[13] R. Guerraoui, Indulgent algorithms, *Proceedings of the 19th ACM Symposium on Principles of Distributed Computing, (PODC'00)*, Portland, OR, 2000, 289–298.

[14] I. Gupta, T. D. Chandra, and G. S. Goldszmidt, On scalable and efficient distributed failure detectors, *Proceedings of the 20th Annual ACM Symposium on Principles of Distributed Computing*, Newport, RI, August 2001, 170–179.

[15] M. Hurfin, A. Mostefaoui, and M. Raynal, A versatile family of consensus protocols based on Chandra–Toueg's unreliable failure detectors, *IEEE Transactions on Computers*, **51**(4), 2002, 395–408.

[16] M. Hurfin and M. Raynal, A simple and fast asynchronous consensus protocol based on a weak failure detector, *Distributed Computing*, **12**(4), 1999, 209–223.

[17] M. Larrea, S. Arevalo, and A. Fernandez, Efficient algorithms to implement unreliable failure detectors in partially synchronous systems, *Proceedings of the 13th International Symposium on Distributed Computing*, September 27–29, 1999, 34–48.

[18] M. Larrea, A. Fernandez, and S. Arevalo, Optimal implementation of the weakest failure detector for solving consensus, *Proceedings of the 19th IEEE Symposium on Reliable Distributed Systems (SRDS'00)*, October 16–18, 2000, 52.

[19] G. Le Lann and U. Schmid, *How to Implement a Time-Free Perfect Failure Detector in Partially Synchronous Systems*, Technical University of Vienna, Institute for Technische Informatik, Research Report, Number 28/2005, 2005.

[20] A. Mostefaoui, E. Mourgaya, and M. Raynal, Asynchronous implementation of failure detectors, *Proceedings of the International IEEE Conference on Dependable Systems and Networks (DSN'03)*, San Francisco, CA, 2003, 351–360.

[21] A. Mostefaoui, E. Mourgaya, and M. Raynal, An introduction to oracles for asynchronous distributed systems, *Future Generation Computer Systems*, **18**(6), 2002, 757–767.

[22] M. Raynal, A short introduction to failure detectors for asynchronous distributed systems, *ACM SIGACT News*, **36**(1), 2005, 53–70.

[23] A. Schiper, Early consensus in an asynchronous system with a weak failure detector, *Distributed Computing*, **10**(3), 1997, 149–157.

[24] L. Temal and D. Conan, Failure, connectivity and disconnection detectors, *Proceedings of the 1st French-speaking Conference on Mobility and Ubiquity Computing*, Nice, France, June 1–3, 2004.

CHAPTER

16 Authentication in distributed systems

16.1 Introduction

A fundamental concern in building a secure distributed system is the authentication of local and remote entities in the system [41]. In a distributed system, the hosts communicate by sending and receiving messages over the network. Various resources (such as files and printers) distributed among the hosts are shared across the network in the form of network services provided by servers. The entities in a distributed system, such as users, clients, servers, and processes, are collectively referred to as principals. A distributed system is susceptible to a variety of threats mounted by intruders as well as legitimate users of the system.

In an environment where a principal can impersonate another principal, principals must adopt a mutually suspicious attitude toward one another and authentication becomes an important requirement. Authentication is a process by which one principal verifies the identity of another principal. For example, in a client–server system, the server may need to authenticate the client. Likewise, the client may want to authenticate the server so that it is assured that it is talking to the right entity. Authentication is needed for both authorization and accounting functions. In one-way authentication, only one principal verifies the identity of the other principal, while in mutual authentication both communicating principals verify each other's identity. A user gains access to a distributed system by logging on to a host in the system. In an open access environment where hosts are scattered across unrestricted areas, a host can be arbitrarily compromised, necessitating mutual authentication between the user and host. In a distributed system, authentication is carried out using a protocol involving message exchanges and these protocols are termed *authentication protocols* [41].

16.2 Background and definitions

In simple terms, authentication is identification plus verification. *Identification* [41] is the procedure whereby an entity claims a certain identity, while *verification* is the procedure whereby that claim is checked. Authentication is a process of verifying that the principal's identity is as claimed. The *correctness* of authentication relies heavily on the verification procedure employed.

A successful identity authentication results in a belief held by the authenticating principal (the *verifier*) that the authenticated principal (the *claimant*) possesses the claimed identity. The other types of authentication include message origin authentication and message content authentication. In this chapter, we restrict our attention to identity authentication only.

Authentication in distributed systems is carried out using protocols. A protocol is a precisely defined sequence of communication and computation steps. A communication step transfers messages from one principal (the sender) to another (the receiver), while a computation step updates a principal's internal state. Two distinct states can be identified upon the termination of the protocol: one signifying successful authentication and the other failure.

Although the goal of any authentication is to verify the claimed identity of a principal, specific success and failure states are highly protocol dependent. For example, the success of an authentication during the connection establishment phase of a communication protocol is usually indicated by the distribution of a fresh session key between two mutually authenticated peer processes. On the other hand, in a user login authentication, success usually results in the creation of a login process on behalf of the user.

16.2.1 Basis of authentication

Authentication generally is based on the possession of some secret information, like password, known only to the entities participating in the authentication. When an entity wants to authenticate another entity, the former will verify if the latter possesses the knowledge of the secret. If the entity demonstrates the knowledge of the right secret information, the authentication succeeds, else authentication fails. Examples of secret information for the purpose of authentication include the following: something known (e.g., a shared key), something possessed (e.g., a smartcard), or something inherent (e.g., biometrics). However, the verification process should not allow an attacker to reuse an authentication exchange to impersonate an entity. The verification process must provide the verifier with enough confidence that an attacker is not trying to impersonate an entity.

16.2.2 Types of principals

In a distributed system, the entities that require identification are hosts, users, and processes [26]. They thus are the principals involved in an authentication.

- **Hosts** These are addressable entities at the network level. A host is usually identified by its name (for example, a fully qualified domain name) or its network address (for example, an IP address).
- **Users** These entities are ultimately responsible for all system activities. Users initiate and are accountable for all system activities. Most access control and accounting functions are based on users. Typical users include humans, as well as accounts maintained in the user database. Users are considered to be outside the system boundary.
- **Processes** The system creates processes within the system boundary to represent users. A process requests and consumes resources on the behalf of its user.

Processes fall into two classes: client and server. Client processes are consumers who obtain services from server processes, who are service providers. A particular process can act as both a client and a server.

16.2.3 A simple classification of authentication protocols

Authentication protocols can be categorized based on the following criteria [28]: type of cryptography (symmetric vs. asymmetric), reciprocity of authentication (mutual vs. one-way), key exchange, real-time involvement of a third party (on-line vs. off-line), nature of trust required from a third party, nature of security guarantees, and storage of secrets.

In this chapter, we classify authentication protocols [41] primarily based on the cryptographic technique used. There are two basic types of cryptographic techniques: symmetric ("private key") and asymmetric ("public key"). Symmetric cryptography uses a single private key to both encrypt and decrypt data. Any party that has the key can use it to encrypt and decrypt data. Symmetric cryptography algorithms are typically fast and are suitable for processing large streams of data. Asymmetric cryptography, also called public-key cryptography, uses a secret key that must be kept from unauthorized users and a public key that is made public. Both the public key and the private key are mathematically linked: data encrypted with the public key can be decrypted only by the corresponding private key, and data signed with the private key can only be verified with the corresponding public key. Both keys are unique to a communication session.

16.2.4 Notation

We specify authentication protocols [39] with precise syntax and semantics and define a system model that characterizes protocol executions. We assume a given set of constant symbols, which denote the names of principals, nonces, and keys. In symmetric key cryptography, let $\{X\}_k$ denote the encryption of X using a symmetric key k and $\{Y\}_{k^{-1}}$ denote the decryption of Y using a symmetric key k. In asymmetric key cryptography, for a principal x, K_x and K_x^{-1} denote its public and private keys, respectively.

We present authentication protocols using the following format. A communication step whereby P sends a message M to Q is represented as $P \rightarrow Q$: M, whereas a computation step of P is written as P: ..., where "..." is a specification of the computation step.

For example, a typical login protocol between a host H and a user U is given in Algorithm 16.1 (f denotes a *one-way* function, that is, given y, it is computationally infeasible to find an x such that $f(x) = y$).

$U \rightarrow H:$	U
$H \rightarrow U:$	"Please enter password"
$U \rightarrow H:$	p
$H:$	compute $y = f(p)$
:	Retrieve user record $(U, f(\text{password}_U))$ from the database
:	If y $= f(\text{password}_U)$, then accept; otherwise reject

Algorithm 16.1 A login protocol.

Since authentication protocols for distributed systems directly use cryptosystems, their basic design principles also follow the type of cryptosystem used. Specifically, we identify two basic categories of authentication: one based on symmetric cryptosystems and other on asymmetric cryptosystems. Protocols presented in this chapter are intended to illustrate basic design principles and a realistic protocol is certainly a refinement of these basic protocols.

16.2.5 Design principles for cryptographic protocols

Abadi and Needham set out a set of principles [2] to denote prudent engineering practices for cryptographic protocols design [2,4]. They are not meant to apply to every protocol in every instance, but they do provide rules of thumb that should be considered when designing a cryptographic protocol.

We next present these principles and briefly comment on them [2,4].

- **Principle 1** Every message should say what it means: the interpretation of the message should depend only on its content. It should be possible to write down a straightforward English sentence describing the content – though if there is a suitable formalism available, which is good, too.
- **Principle 2** The conditions for a message to be acted upon should be clearly set out so that someone reviewing the design may see whether they are acceptable or not.

- **Principle 3** If the identity of a principal is essential to the meaning of a message, it is prudent to mention the principal's name explicitly in the message.
- **Principle 4** Be clear as to why encryption is being done. Encryption is not wholly cheap, and not asking precisely why it is being done can lead to redundancy. Encryption is not synonymous with security, and its improper use can lead to errors.
- **Principle 5** When a principal signs material that has already been encrypted, it should not be inferred that the principal knows the content of the message. On the other hand, it is proper to infer that the principal that signs a message and then encrypts it for privacy knows the content of the message.
- **Principle 6** Be clear about what properties you are assuming about nonces. What may do for ensuring temporal succession may not do for ensuring association – and perhaps association is best established by other means.
- **Principle 7** The use of a predictable quantity (such as the value of a counter) can serve in guaranteeing newness, through a challenge–response exchange. But if a predictable quantity is to be effective, it should be protected so that an intruder cannot simulate a challenge and later replay a response.
- **Principle 8** If timestamps are used as freshness guarantees by reference to absolute time, then the difference between local clocks at various machines must be much less than the allowable age of a message deemed to be valid. Furthermore, the time maintenance mechanism everywhere becomes part of the trusted computing base.
- **Principle 9** A key may have been used recently, for example, to encrypt a nonce, yet be quite old, and possibly compromised. Recent use does not make the key look any better than it would otherwise.
- **Principle 10** If an encoding is used to present the meaning of a message, then it should be possible to tell which encoding is being used. In the common case where the encoding is protocol dependent, it should be possible to deduce that the message belongs to this protocol, and in fact to a particular run of the protocol, and to know its number in the protocol.
- **Principle 11** The protocol designer should know which trust relations his protocol depends on, and why the dependence is necessary. The reasons for particular trust relations being acceptable should be explicit though they will be founded on judgment and policy rather than on logic.

16.3 Protocols based on symmetric cryptosystems

In a symmetric cryptosystem, knowing the shared key lets a principal encrypt and decrypt arbitrary messages [41]. Without such knowledge, a principal cannot create the encrypted version of a message, or decrypt an encrypted message. Hence, authentication protocols can be designed using the following principle:

> If a principal can correctly encrypt a message using a key that the verifier believes is known only to a principal with the claimed identity (outside of the verifier), this act constitutes sufficient proof of identity.

Thus, the principle embodies the fact that a principal's knowledge is indirectly demonstrated through its ability to encrypt or decrypt.

16.3.1 Basic protocol

Using the above principle, we immediately obtain the basic protocol (shown in Algorithm 16.2) where principal P is authenticating itself to principal Q. "k" denotes a secret key that is shared between only P and Q [41].

P :	Create a message $m =$ "I am P."
:	Compute $m' = \{m, Q\}_k$
$P \rightarrow Q$:	m, m'
Q :	verify $\{m, Q\}_k = m'$
:	if equal then accept; otherwise the authentication fails

Algorithm 16.2 Basic protocol.

In this protocol, the principal P prepares a message m, and encrypts the message and identity of Q using the symmetric key k and sends to Q both the plain text and encrypted messages. Principal Q, on receiving the message, encrypts the plain text message and its identity to get the encrypted message. If it is equal to the encrypted message sent by P, then Q has authenticated P, else the authentication fails.

Weaknesses

Clearly, this method is sound only if the underlying cryptosystem is strong (one cannot create the encrypted version of a message without knowing the key) and the key is secret (it is shared only between the real principal and the verifier). Note that this protocol performs only one-way authentication; mutual authentication can be achieved by reversing the roles of P and Q.

One major weakness of the protocol is its vulnerability to replays. More precisely, an adversary could masquerade as P by recording the message m, m' and later replaying it to Q. As mentioned, replay attacks can be countered by using nonces or timestamps. Since both plain text message m and its encrypted version m' are sent together by P to Q, this method is vulnerable to known plain text attacks. Thus the cryptosystem must be able to withstand known plain text attacks.

16.3.2 Modified protocol with nonce

To prevent replay attacks, we modify the protocol by adding a challenge-and-response step using a nonce (shown in Algorithm 16.3). A nonce is a large random or pseudo-random number that is drawn from a large space so that it is difficult to guess by an intruder. This property of a nonce helps ensure that old communications cannot be reused in replay attacks.

$P \rightarrow Q:$ "I am P."
$Q:$ generate nonce n
$Q \rightarrow P:$ n
$P:$ compute $m' = \{P, Q, n\}_k$
$P \rightarrow Q:$ m'
$Q:$ verify $\{P, Q, n\}_k = m'$
: if equal then accept; otherwise the authentication fails

Algorithm 16.3 Challenge-and-response protocol using a nonce.

In the modified version of the protocol [41], the principal P wants to authenticate itself to Q. Q generates a nonce and sends this nonce to P. P then encrypts Q, the nonce, and its own identity with the secret key and sends this encrypted message to Q. Q verifies this encrypted message by encrypting its identity, P's identity, and the nonce with the key k. Q authenticates P if the encrypted information equals that sent by P, else the authentication fails.

Replay is foiled by the freshness of nonce n and because n is drawn from a large space. Therefore, it is highly unlikely that the nonce n generated by Q in the current session is the same as one used in a previous session. Thus an attacker cannot use a message of type m' from a previous session to mount a replay attack. In addition, even if an eavesdropper has monitored all previous authentication conversations between P and Q, it is impossible to produce the message m' because it does not know the secret key k. The challenge-and-response step can be repeated any number of times until the desired level of confidence is reached by Q.

Weaknesses

This protocol has scalability problems because each principal must store the secret key for every other principal it would ever want to authenticate [41]. This presents major initialization (the predistribution of secret keys) and storage problems. Moreover, the compromise of one principal can potentially compromise the entire system. Note that this protocol is also vulnerable to known plain text attacks.

16.3.3 Wide-mouth frog protocol

The above raised problems can be significantly reduced by postulating a centralized server S. The wide-mouth frog protocol [28] uses a similar approach where a principal A authenticates itself to principal B using a Server S. The protocol works as follows:

$$A \rightarrow S: \quad A, \{T_A, K_{AB}, B\}_{K_{AS}}$$

$$S \rightarrow B: \quad \{T_S, K_{AB}, A\}_{K_{BS}}$$

A decides that it wants to set up communication with B. A sends to S its identity and a packet encrypted with the key, K_{AS}, it shares with S. The packet contains the current timestamp, A's desired communication partner, and a randomly generated key K_{AB}, for communication between A and B. S decrypts the packet to obtain K_{AB} and then forwards this key to B in an encrypted packet that also contains the current timestamp and A's identity. B decrypts this message with the key it shares with S and retrieves the identity of the other party and the key, K_{AB}. Any principal receiving a message with an out-of-date timestamp during this protocol discards it to prevent replay attacks. This protocol achieve two objectives: first, it securely establishes a secret key between two principals A and B; and second, A authenticates itself to B with the help of the server S. This is because only the server S could have constructed the message $\{T_S, K_{AB}, A\}_{K_{BS}}$ in step 2 only after receiving a message from A in step 1.

A weakness of the protocol is that a global clock is required and the protocol will fail if the server S is compromised.

16.3.4 A protocol based on an authentication server

Another approach to solve the problem is by using a centralized *authentication server* S that shares a secret key K_{XS} with every principal X in the system [41]. The basic authentication protocol is shown in Algorithm 16.4.

In the protocol using an authentication server, the principal P sends its identity to Q. Q generates a nonce and sends this nonce to P. P then encrypts P, Q and n with the key K_{PS} and sends this encrypted value x to Q. Q then encrypts P, Q and x with K_{QS} and sends this encrypted value y to authentication server S. Since S knows both the secret keys, it decrypts y with K_{QS}, recovers x, decrypts x with K_{PS} and recovers P, Q and n. Server S then encrypts P, Q and n with key K_{QS} and sends the encrypted value m to Q. Q then computes P, Q and $n_{K_{QS}}$ and verifies if this value is equal to the value received from S. If both values are equal, then authentication succeeds, else it fails.

Thus Q's verification step is preceded by a *key-translation* step by S. Since P and Q do not share a secret key, the authentication server S does the

$$
\begin{aligned}
P \rightarrow Q &: \text{ "I am } P\text{."} \\
Q &: \text{ generate nonce } n \\
Q \rightarrow P &: \ n \\
P &: \text{ compute } x = \{P, Q, n\}_{K_{PS}} \\
P \rightarrow Q &: \ x \\
Q &: \text{ compute } y = \{P, Q, x\}_{K_{QS}} \\
Q \rightarrow A &: \ y \\
A &: \text{ recover } P, Q, x \text{ from } y \text{ by decrypting } y \text{ with } K_{QS} \\
&: \text{ recover } P, Q, n \text{ from } y \text{ by decrypting } x \text{ with } K_{PS} \\
&: \text{ compute } m = \{P, Q, n\}_{K_{QS}} \\
A \rightarrow Q &: \ m \\
Q &: \text{ independently compute } \{P, Q, n\}_{K_{QS}} \text{ and verify } \{P, Q, n\}_{K_{QS}} = m \\
&: \text{ if equal, then accept; otherwise, the authentication fails}
\end{aligned}
$$

Algorithm 16.4 A protocol using an authentication server.

key translation because it shares a secret key with both principals P and Q. Q sends the message (encrypted with K_{PS} that it received from P) to S. S does the key translation by decrypting it with K_{PS}, encrypting P, Q and n with K_{QS} and sending the message encrypted with K_{QS} to Q. This is termed as the key-translation step [41].

The basis of this protocol is a challenge for Q to P if P can encrypt the nonce n with the secret key that it shares with server S. The protocol correctness rests on S's trustworthiness – that S will properly decrypt using P's key and reencrypt using Q's key. The initialization and storage problems are greatly alleviated because each principal needs to keep only one key. The risk of compromise is mostly shifted to S, whose security can be guaranteed by various measures, such as encrypting stored keys using a master key and putting S in a physically secure room.

16.3.5 One-time password scheme

In the one-time password scheme [24], a password can only be used once. A one-time password system generates a list of passwords and secretly communicates this list to the client and the server. The client uses the passwords in the list to log on to a server. Once a password has been used, it cannot be used again. To log on again, the client must use the next password in the list. The server always expects the next password in the list at the next logon. Therefore, even if a password is disclosed, the possibility of replay attacks is eliminated because the system expects the next password in the subsequent logon. This protocol is best suited for distributed systems where authentication mainly takes place between client and server.

Protocol description

The protocol consists of two steps:

1. **The Registration stage**, where the client registers with the server and gets a list of passwords.
2. **The Login and authentication stage**, where the server authenticates the client.

Step 1: registration

1. Every client shares a pre-shared secret key, represented as $SEED$ with the server. It is a large random number secretly communicated by the server to the client.
2. The server generates a session key (SK) with the help of a random number D and a timestamp T, i.e., $SK = D||T$. The server computes and sends $SEED \oplus SK$ to the client. When the client receives $SEED \oplus SK$, it computes the value of SK as follows:

$$SK := SEED \oplus (SEED \oplus SK).$$

   The client then generates an initial key IK with the help of a randomly generated secret key K,

$$IK := K \oplus SEED.$$

   The client then decides the number of times (N) it wants to login to the server and sends the generated initial key (IK) to the server. To do this, the client performs $IK \oplus SK$ and $N \oplus SK$ and sends these values to the server.
3. When the server receives $IK \oplus SK$ and $N \oplus SK$, it retrieves IK and N from the received values and computes

$$p_0 := H^N(IK) \text{ for the user, where } H \text{ is a hash function,}$$

   and performs $p_0 := p_0 \oplus SK$, stores p_0 and N in its database, and sends $p_0 \oplus SK$ back to the client as a response. It also computes p_1 and p_2 as follows:

$$p_1 := H^{N-1}(IK), \text{ and}$$
$$p_2 := H^{N-2}(IK).$$

   The server then sends $p_0 \oplus SK$, $p_1 \oplus SK$, and $p_2 \oplus SK$ to the client.
4. On receiving $p_0 \oplus SK$, $p_1 \oplus SK$, and $p_2 \oplus SK$ from the server, the client performs the XOR operation on SK and $p_0 \oplus SK$, $p_1 \oplus SK$, and $p_2 \oplus SK$ separately, to obtain p_0, p_1, and p_2, respectively. The client hashes IK for N times and then compares it with p_0. If both values are equal, the client

is sure of the authenticity of the server and that it is not communicating with an intruder.

It then saves the values of p_0, p_1, p_2, and N for future communication with the server. This marks the end of the registration stage.

The above steps are described in Algorithm 16.5.

If N is 50, the user can log in to the server 50 times and $p_0 = H^{50}(IK)$. After 50 logins, the user must repeat the steps in the registration.

Server → Client : $SEED$
Server → Client : $SEED \oplus SK$
Client → Server : $IK \oplus SK$ and $N \oplus SK$
Server → Client : $p_0 \oplus SK,\ p_1 \oplus SK,\ p_2 \oplus SK$

Algorithm 16.5 The registration stage.

Step 2: Login and authentication

Once the client is registered, every time it needs to access a service provided by the server, the client needs to get authenticated. Authentication requires the following steps:

1. If the client is logging in for the tth time, the server generates a new session key (SK):

   $$SK := D||T, \text{ where } T \text{ is the timestamp and } D \text{ is a random number.}$$

   The server also computes $p_{t-1} = H^{C+1}(IK)$ where $C = N - t$. It then performs $p_{t-1} \oplus SK$ and $SK \oplus SEED$ ($SEED$ is stored in the database) and sends these values to the client.
2. On the receipt of the values from the server, the client computes SK as follows:

   $$SK := p_{t-1} \oplus (p_{t-1} \oplus SK).$$

   Then the client checks the timestamp T of the session key SK. If the timestamp is valid, the client computes $SEED := SK \oplus (SK \oplus SEED)$ and checks the value of $SEED$ with the one saved to make sure of the server's identity. If they match, the server's authenticity is verified.
3. Now the client proves its identity to the server as follows: it sends $SK \oplus p_t$ to the server. The client uses the p_t saved in the previous login in this EX-OR operation.

   The server calculates p_t from $SK \oplus p_t$ received from the client as follows:

   $$p_t := SK \oplus (SK \oplus p_t).$$

From the received p_t value, it calculates $p_{t-1} := H(p_t)$ and compares it with p_{t-1} obtained in step 1. If both match, the identity of the client is verified.

Finally, the server updates N with C, where $C = N - t$, and computes p_{t+1} using p_0 and sends $p_{t+1} \oplus SK$ to the client.

4. The client computes value of p_{t+1} as $p_{t+1} = SK \oplus (SK \oplus p_{t+1})$ and stores it for its next login.
   For example, if $t = 10$ and $N = 100$, then $p_{t-1} := H^{91}(IK)$, $p_t := H^{90}(IK)$, and $p_{t+1} := H^{89}(IK)$.

The above steps are described in Algorithm 16.6.

Server → Client :	$p_{t-1} \oplus SK, SEED \oplus SK$
Client → Server :	$p_t \oplus SK$
Server → Client :	$p_{t+1} \oplus SK$

Algorithm 16.6 The login and authentication stage.

In this protocol, the client and the server communicate with each other by passing parameters that are encrypted, i.e., exclusive ORed with either SK or $SEED$. SK is the session key of a particular session and $SEED$ is the pre-shared secret key. Since these two values are known only to the client and server, eavesdropping of the connection does not have any effect. Since SK is obtained by using the timestamp, replay of previous session does not work and thus the scheme is robust against replay attacks. The use of the hash function makes dictionary attacks impossible.

Weaknesses

One-time passwords that are not time-synchronized are vulnerable to phishing. Phishing usually occurs when a fraudster sends an email that contains a link to a fraudulent website where the users are asked to provide personal account information. The email and website are usually disguised to appear to recipients as though they are from a bank or another well-known brand. In late 2005, customers of a Swedish bank were tricked into giving up their passwords.

16.3.6 Otway–Rees protocol

The Otway–Rees protocol [28] is a server-based protocol that provides authenticated key transport only in four messages without requiring timestamps. It provides key authentication and key freshness assurances. It does not, however, provide entity authentication or key confirmation.

The notations used in the protocol are as follows: K_{AB} is a session key that the sever S generates for users A and B to share. N_A and N_B are nonces chosen by A and B, respectively, to allow verification of key freshness (thereby detecting replay attacks). M is another nonce chosen by A which serves as a transaction identifier. S shares symmetric keys K_{AS} and K_{BS} with A and B, respectively. This protocol is shown in Algorithm 16.7.

$$\begin{array}{ll}(1)\ A \rightarrow B: & M, A, B, (N_A, M, A, B)_{K_{AS}} \\ (2)\ B \rightarrow S: & M, A, B, (N_A, M, A, B)_{K_{AS}}, (N_B, M, A, B)_{K_{BS}} \\ (3)\ S \rightarrow B: & (N_A, K_{AB})_{K_{AS}}, (N_B, K_{AB})_{K_{BS}} \\ (4)\ B \rightarrow A: & M, (N_A, K_{AB})_{K_{AS}}\end{array}$$

Algorithm 16.7 Otway–Rees protocol.

In step 1, user A encrypts two nonces, N_A and M, and the identities of itself and the identity of the party B to whom it wishes to communicate, with the key K_{AS} and sends this to B along with M, A, and B in plain text. On the receipt of this message, user B creates its own nonce N_B and an analogous encrypted message, $(N_B, M, A, B)_{K_{BS}}$, in step 2 and sends this along with A's message to server S. When the server S receives this message, it uses the clear (plain) text identifiers in the message to retrieve K_{AS} and K_{BS}, then verifies if the clear text (M, A, B) matches that recovered upon decrypting both parts of the message in step 2. Verifying M in particular confirms that the encrypted parts are linked. If so, S decides on a new key K_{AB} for communication between A and B, prepares two distinct messages $(N_A, K_{AB})_{K_{AS}}$, $(N_B, K_{AB})_{K_{BS}}$ for A and B, respectively, and sends both to B in step 3. When B receives this message, it decrypts the second part of the message received in step 3 and checks if N_B matches that sent in step 2. If so, it sends the first part to A in step 4. When A receives this message, it decrypts message received in step 4 and checks if N_A matches that sent in step 1.

If all checks pass, A and B are assured that K_{AB} is fresh (due to their respective nonces), and trust that $(N_A, K_{AB})_{K_{AS}}$ and $(N_B, K_{AB})_{K_{BS}}$ have been constructed by the server S. A knows that B is active as verification of step 4 implies that B sent the message in step 2 recently; B, however, has no assurance that A is active until subsequent use of K_{AB} by A, since B cannot determine if the message in step 1 is fresh.

Weaknesses

One problem with this protocol is that a malicious intruder can arrange for A and B to end up with different keys as follows: A and B execute the first three messages; at this point, B has received the key K_{AB}. The intruder intercepts the fourth message. He/she replays step 2, which results in S generating a new key K'_{AB} and sending it to B in step 3. The intruder intercepts this message, too, but sends to A the part of it that B would have sent to A. So A has finally received the expected fourth message, but with K'_{AB} instead of K_{AB}. Another problem is that, although the server tells B that A used a nonce, B doesn't know if this was a replay of an old message.

16.3.7 Kerberos authentication service

Kerberos [20,30,32] primarily addresses client–server authentication using a symmetric cryptosystem. Kerberos is an authentication system designed for MIT's Project Athena [1]. The goal of Project Athena was to create an educational computing environment based on high-performance workstations, high-speed networking, and servers of various types. Researchers envisioned a large-scale (10 000 workstations to 1000 servers) open network computing environment in which individual workstations could be privately owned and operated. Therefore, a workstation cannot be trusted to identify its users correctly to network services. Kerberos is not a complete authentication service required for secure distributed computing in general; it only addresses issues of client–server interactions.

In this section, we describe the Kerberos authentication protocol. Kerberos' design is based on the use of a symmetric cryptosystem together with trusted third-party authentication servers. The basic components include authentication servers (*Kerberos servers*) and *ticket-granting servers* (TGSs).

Initial registration

Every client/user registers with the Kerberos server by providing its user i.d., U, and a password, $password_u$. The Kerberos server computes a key $k_u = f(password_u)$ using a one-way function f and stores this key in a database. Note that k_u is a secret key that depends on the password of the user and is shared by client U and the Kerberos server only.

The authentication protocol

Authentication in Kerberos proceeds in three steps:

1. **Initial authentication at login** The Kerberos server authenticates user login at a host and installs a ticket for the ticket-granting server, TGS, at the login host.
2. **Obtain a ticket for the server** Using the ticket for the ticket-granting server, the client requests the ticket-granting server, TGS, for a ticket for the server.
3. **Requesting service from the server** The client uses the server ticket obtained from the TGS to request services from the server.

These steps are shown in Figure 16.1. Next, we explain these steps in detail.

Step 1: Initial authentication at login

Initial authentication at login uses a Kerberos server and is shown in Algorithm 16.8. Let U be a user who is attempting to log into a host H.

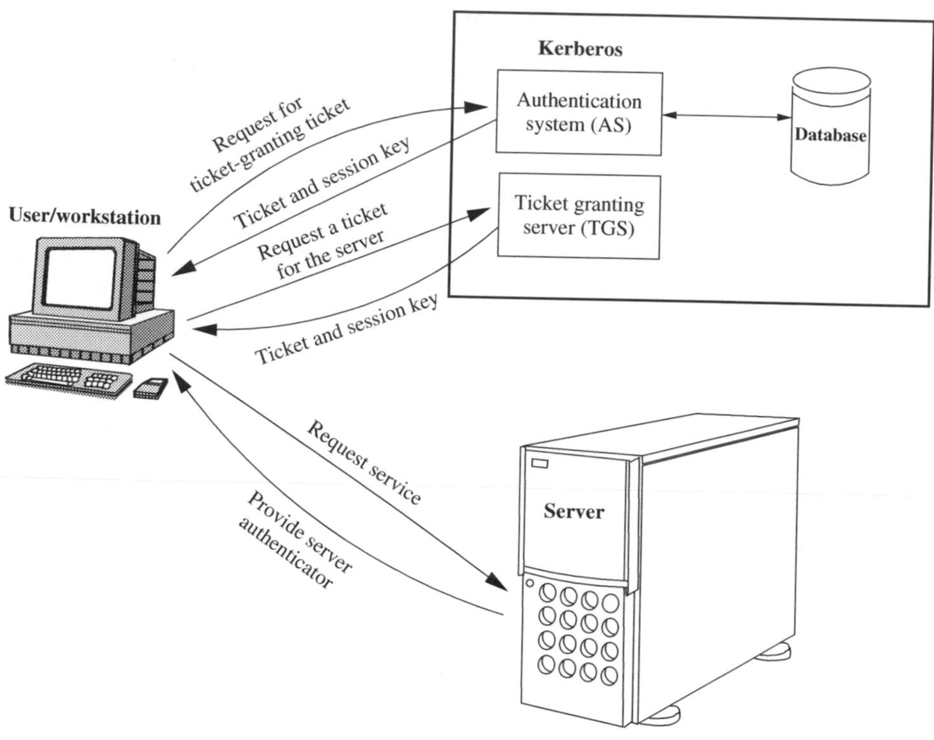

Figure 16.1 Steps in authentication in Kerberos.

(1)	$U \rightarrow H$	:	U
(2)	$H \rightarrow$ Kerberos	:	U, TGS
(3)	Kerberos	:	retrieve k_U and k_{TGS} from database
		:	generate new session key k
		:	create a ticket-granting ticket
		:	$tick_{TGS} = \{U, TGS, k, T, L\}_{K_{TGS}}$
(4)	Kerberos $\rightarrow H$	:	$\{TGS, k, T, L, tick_{TGS}\}_{k_U}$
(5)	$H \rightarrow U$	:	"Password?"
(6)	$U \rightarrow H$	:	*password*
(7)	H	:	compute $k'_U = f(password)$
		:	recover k, $tick_{TGS}$ by decrypting $\{TGS, k, T, L, tick_{TGS}\}_{k_U}$ with k'_U
		:	if decryption fails, abort login, otherwise, retain $tick_{TGS}$ and k
		:	erase *password* from the memory

Algorithm 16.8 Initial authentication at login.

In step 1, user U initiates login by entering his/her username. In step 2, the login host H forwards the login request and the i.d. of the TGS to a Kerberos server. In step 3, the Kerberos server retrieves k_U and k_{TGS} from the database, generates a new session key k and creates a *ticket-granting* ticket $tick_{TGS} = \{U, TGS, k, T, L\}_{K_{TGS}}$, where U is the identity of the user who wishes to communicate with the server, TGS is the identity of the ticket-granting server, k is the session key, T is a timestamp, L is the ticket's lifetime and k_{TGS} is the key shared between the TGS and the Kerberos server. In step 4, the Kerberos server encrypts the ticket $tick_{TGS}$, the identity of the TGS, the session key, the timestamp, and lifetime with k_U and sends it to host H.

In step 5, on receiving this message from the Kerberos server, host H prompts the user for his/her password, which the user supplies in step 6. In step 7, host H computes the key, k'_U, corresponding to the password using the one-way function f. The host recovers the session key k by decrypting $\{TGS, k, T, L, tick_{TGS}\}_{k_U}$ with k'_U. If the password supplied by the user is not the valid password of U, k'_U would not be identical to k_U, and the authentication will fail. Thus, the user is authenticated if the host is able to decrypt the message for the Kerberos server. Upon successful authentication, the host saves the new session key k and the ticket-granting ticket, $tick_{TGS}$, for further use and erases the user password from the memory. The ticket-granting ticket is used to request server tickets from the TGS. Note that $tick_{TGS}$ is encrypted with k_{TGS}, the key shared between the TGS and the Kerberos server.

Step 2: Obtain a ticket for the server

The client executes the steps shown in Algorithm 16.9 to request a ticket for the server from the TGS. Basically, the client sends the ticket $tick_{TGS}$ to the TGS, requesting a ticket for the server S. (T_1 and T_2 are timestamps.)

Because a ticket is susceptible to interception and replay, it does not by itself constitute sufficient proof of identity. For authentication, a principal presenting a ticket must also demonstrate the knowledge of the session key k named in the ticket. An *authenticator*, $\{C, T\}_k$, where C is the client identity, T is the timestamp, and k is the session key, provides the demonstration. Unlike the ticket, which is reusable, an authenticator can be used only once and has a very short lifetime. The ticket proves the client's identity and also distributes the key; however, it is susceptible to replay attacks. The authenticator is used to counter this attack. Because an authenticator can be used only once and has a very short lifetime, the threat of an opponent stealing the ticket for a replay attack is countered.

$$
\begin{array}{lrl}
(1) & C \rightarrow TGS: & S, tick_{TGS}, \{C, T_1\}_k \\
(2) & TGS: & \text{recover } k \text{ from } tick_{TGS} \text{ by decrypting with } k_{TGS}, \\
 & : & \text{recover } T_1 \text{ from } \{C, T_1\}_k \text{ by decrypting with } k \\
 & : & \text{check timelines of } T_1 \text{ with respect to local clock} \\
 & : & \text{generate new session key } k \\
 & : & \text{Create server ticket } tick_S = \{C, S, k, T, L\}_{k_S} \\
(3) & TGS \rightarrow C: & \{S, k, T, L, tick_S\}_k \\
(4) & C: & \text{recover } k, tick_S \text{ by decrypting the message with } k
\end{array}
$$

Algorithm 16.9 Obtain a ticket for the server.

In step 1, to request a ticket for server S, client C presents its ticket-granting ticket $tick_{TGS}$ along with the authenticator to the TGS. C's knowledge of k is demonstrated using the authenticator $\{C, T_1\}_k$. In step 2, the TGS decrypts $tick_{TGS}$ with k_{TGS} to recover k, verifies the authenticity of the authenticator by decrypting $\{C, T_1\}_k$ with k, and checks the timeliness of T_1 in the authenticator and T in $tick_{TGS}$. If both decryptions in step 2 are successful and T_1 is timely, the TGS is convinced of the authenticity of the ticket, and creates a ticket $tick_S = \{C, S, k, T, L_{k_S}\}$ for server S, where C is the identity of the client, S is the server identity, k is the new session key, T is the timestamp of the TGS, L is the lifetime of the ticket, and k_S is the key shared between the TGS and server S. This ticket is returned to C in step 3. In step 4, C recovers k and $tick_S$ from $\{S, k, T, L, tick_S\}_k$ by decrypting it with k.

Step 3: requesting service from the server

Client C sends the ticket and the authenticator to server. The server decrypts $tick_S$ and recovers k. It then uses k to decrypt the authenticator $\{C, T_2\}_k$ and checks if the timestamp is current and the client identifier matches with that in the $tick_S$ before granting service to the client. If mutual authentication is required, the server returns an authenticator (Algorithm 16.10).

(1) $C \rightarrow S:$ $tick_S, \{C, T_2\}_k$
(2) $S:$ recover k from $tick_S$ by decrypting it with k_S
$:$ recover T_2 from $\{C, T_2\}_k$ by decrypting with k
$:$ check timeliness of T_2 with respect to the local clock
(3) $S \rightarrow C:$ $\{T_2 + 1\}_k$

Algorithm 16.10 Requesting service from the server.

In step 1, C presents S with $tick_S$ and a new authenticator. In step 2, S recovers k from $tick_S$ by decrypting it with k_S and uses k obtained to decrypt $\{C, T_2\}_k$. If both decryptions are successful and T_2 is timely, then S is assured of the authenticity of the Client. Finally, step 3 assures C of the server's identity.

Weaknesses

Kerberos [5,21] makes no provisions for host security; it assumes that it is running on trusted hosts with an untrusted network. If host security is compromised, then Kerberos is compromised as well. Kerberos uses a principal's password (encryption key) as the fundamental proof of identity. If a user's Kerberos password is stolen by an attacker, then the attacker can impersonate that user with impunity. Since the Kerberos' password database holds all the passwords for all of the principals in a realm, if the host security on the database is compromised, then the entire realm is compromised. In Kerberos version 4, authenticators are valid for a particular time. If an attacker sniffs the network for authenticators, they have a small time window in which they can re-use it and gain access to the same service. Kerberos version 5 introduced a replay cache that prevents any authenticator from being used more than once. Since anybody can request a ticket-granting ticket for any user, and that ticket is encrypted with the user's secret key (password), it is simple to perform an offline attack on this ticket by trying to decrypt it, say using the dictionary attack. Kerberos version 5 introduced pre-authentication to solve this problem.

16.4 Protocols based on asymmetric cryptosystems

In an asymmetric cryptosystem [41], each principal P publishes its public key k_p and keeps secret its private key k_p^{-1}. Thus only P can generate $\{m\}_{k_p^{-1}}$ for any message m by signing it using k_p^{-1}. The signed message $\{m\}_{k_p^{-1}}$ can be verified by any principal with the knowledge of k_p (assuming a commutative asymmetric cryptosystem). Asymmetric authentication protocols can be constructed using a design principle called ASYM, which is as follows:

> If a principal can correctly sign a message using the private key of the claimed identity, this act constitutes a sufficient proof of identity.

This ASYM principle follows the proof-by-knowledge principle for authentication, in that a principal's knowledge is indirectly demonstrated through its signing capability.

16.4.1 The basic protocol

Using ASYM, we obtain a basic protocol as shown in Algorithm 16.11 [41].

$P \rightarrow Q$: "I am P."
Q : generate nonce n
$Q \rightarrow P$: n
P : compute $m = \{P, Q, n\}_{k_p^{-1}}$
$P \rightarrow Q$: m
Q : verify $(P, Q, n) = \{m\}_{k_p}$
: if equal, then accept; otherwise, the authentication fails

Algorithm 16.11 Basic protocol.

In this protocol, Q sends a random number n to P and challenges it to encrypt with its private key. P encrypts (P, Q, n) with its private key k_p^{-1} and sends it to Q. Q verifies the received message by decrypting it with P's public key k_p and checking with the identity of P, Q, and n. This protocol depends on the guarantee that $\{P, Q, n\}_{k_p^{-1}}$ cannot be produced without the knowledge of k_p^{-1} and the correctness of k_p as published by P and kept by Q.

16.4.2 A modified protocol with a certification authority

The basic protocol requires that Q has the knowledge of P's public key. A problem arises if Q does not know P's public key. This problem is alleviated by postulating a centralized *certification authority* (CA) that maintains a database of all published public keys [41]. If a user A does not have the public key of another user B, A can request B's public key from the CA.

The basic protocol can be modified as shown in Algorithm 16.12 to address this issue.

$P \rightarrow Q$:	"I am P."
Q :	generate nonce n
$Q \rightarrow P$:	n
P :	compute $m = \{P, Q, n\}_{k_p^{-1}}$
$P \rightarrow Q$:	m
$Q \rightarrow CA$:	"I need P's public key."
CA :	retrieve public key k_P of P from key database
	Create certificate $c = \{P, k_P\}_{k_{CA}^{-1}}$
$CA \rightarrow Q$:	P, c
Q :	recover P, k_P from c by decrypting with k_{CA}
	verify $(P, Q, n) = \{m\}_{k_P}$
:	if equal, then accept; otherwise, the authentication fails

Algorithm 16.12 A modified protocol with a certification authority, *CA*.

This protocol is similar to the basic protocol described above but a certification authority, CA, is involved. When Q receives a message encrypted with P's private key from P, it requests the authentication server for P's public key. CA retrieves the public key of P from the key database and provides Q with a certificate for P's public key. The certificate, $\{P, k_P\}_{k_{CA}^{-1}}$ contains P's identity and its public key, encrypted with the private key of the certification authority. Q retrieves the public key of P by decrypting the certificate with the public key of CA. Then it decrypts the message m it received from P using the public key k_P and checks if $\{m\}_{k_P}$ equals $\{P, Q, n\}$. If both are equal, authentication succeeds, else it fails.

Note that c, called a *public key certificate*, represents a certified statement by CA that P's public key is k_p. Other information such as an expiration date and the classification of principal P can also be included in the certificate. However, each principal in the system must know the public key k_{CA} of CA.

In this protocol, CA is an example of an *on-line* certification authority. It supports interactive queries and is actively involved in authentication exchanges. A certification authority can also operate *off-line*. In this case, a public key certificate is issued to a principal when it first registered. The certificate is kept by the principal and is forwarded during an authentication exchange, thus eliminating the need to make a separate query to a CA. Forgery is impossible, since a certificate is signed by the certification authority.

16.4.3 Needham and Schroeder protocol

The Needham–Schroeder public key protocol [29] uses a trusted key server that issues certificates containing the public key of a user. The protocol is described in Algorithm 16.13. In this protocol, the initiator A seeks to establish a session with responder B with the help of trusted key server S. (Recall that, for a principal x, K_x and K_x^{-1}denote its public and private keys, respectively.)

$$
\begin{array}{lll}
(1) & A \rightarrow S: & A, B \\
(2) & S \rightarrow A: & \{K_b, B\}_{K_s^{-1}} \\
(3) & A \rightarrow B: & \{N_a, A\}_{K_b} \\
(4) & B \rightarrow S: & B, A \\
(5) & S \rightarrow B: & \{K_a, A\}_{K_s^{-1}} \\
(6) & B \rightarrow A: & \{N_a, N_b\}_{K_a} \\
(7) & A \rightarrow B: & \{N_b\}_{K_b}
\end{array}
$$

Algorithm 16.13 The Needham–Schroeder protocol.

In step 1, A sends a message to the server S, requesting B's public key. S responds by returning B's public key K_b along with B's identity (to prevent attacks based upon diverting key deliveries), encrypted using S's secret key (to assure A that this message originated from S). A then seeks to establish a connection with B by selecting a nonce N_a, and sending it along with its identity to B (message 3) encrypted using B's public key. When B receives this message, it decrypts the message to obtain the nonce N_a and to learn that user A is trying to communicate with it. It then requests the public key of A from server S (message 4), which the server sends to B in message 5. B then returns nonce N_a, along with a new nonce N_b, to A, encrypted with A's public key (message 6). When A receives this message, it decrypts it with its private key and is assured that it is talking to B, since only B could have decrypted the message in step 3 to obtain N_a. A then returns nonce N_b to B, encrypted with B's key. When B receives this message, it is assured that it is talking to A, since only A could have decrypted the message in step 6 to obtain N_b. Thus, after step 7, A and B have mutually authenticated themselves.

This protocol can be considered as the interleaving of two logically disjoint protocols: messages 1, 2, 4, and 5 are concerned with obtaining public keys, whereas messages 3, 6, and 7 are concerned with the authentication of A and B.

Weaknesses

This protocol provides no guarantee that the public keys obtained are current and not replays of old, possibly compromised keys. This problem can be overcome in various ways. For example, one way is that the server S includes timestamps in messages 2 and 5; however, this requires synchronized clocks at processes. Another method is that A sends a nonce in message 1 and S returns the same nonce in message 2.

An impersonation attack on the protocol

We now show how an intruder can mount an impersonation attack on this protocol [25]. We assume that the intruder I is a user of the computer network, and so is able to set up standard sessions with other users, and other users may try to set up sessions with I. We assume that the intruder can intercept any messages in the system and introduce new messages. However, we make some assumptions about what sort of messages the intruder may introduce. We assume that the intruder cannot guess the value of nonces being passed in encrypted messages, unless those messages are encrypted with his own key. Thus the intruder can only produce new messages using nonces that it invented itself, or that it has previously seen and understood. It can also replay complete encrypted messages, even if it is unable to understand the contents.

The attack shown in Algorithm 16.14, starts with a user A trying to establish a session with I.

The attack on the protocol allows an intruder I to impersonate the user A to set up a false session with a user B. The attack involves two simultaneous runs of the protocol: in run 1, A establishes a valid session with I; in run 2, I impersonates A to establish a fake session with B. In Algorithm 16.14, 1.3 represents message 3 in run 1 and $I(A)$ represents the intruder I impersonating A.

(1.3) $A \rightarrow I: \{N_a, A\}_{K_i}$
(2.3) $I(A) \rightarrow B: \{N_a, A\}_{K_b}$
(2.6) $B \rightarrow I(A): \{N_a, N_b\}_{K_a}$
(1.6) $I \rightarrow A: \{N_a, N_b\}_{K_a}$
(1.7) $A \rightarrow I: \{N_b\}_{K_i}$
(2.7) $I(A) \rightarrow B: \{N_b\}_{K_b}$

Algorithm 16.14 An impersonation attack on the Needham–Schroeder protocol.

In step 1.3, A starts to establish a session with I, sending it a nonce N_a. In step 2.3, the intruder impersonates A to try to establish a false session with B sending it the nonce N_a obtained in the previous message from A. B responds in step 2.6 by selecting a new nonce N_b and returning it, along with N_a, to A. The intruder intercepts this message, but cannot decrypt it because it is encrypted with A's public key. The intruder uses A as an oracle, by forwarding the message to A in step 1.6; note that this message is of the form expected by A in run 1 of the protocol. A decrypts the message to obtain N_b and returns this to I in step 1.7. I decrypts this message to obtain N_b and returns it to B in step 2.7, thus completing run 2 of the protocol. After B receives the message in step 2.7, B is led to believe that A has correctly established a session with it.

A solution to the attack

The main cause of this attack is that step 6 does not contain the identity of the responder. If we include the responder's identity in step 6 of the protocol:

$$(6)\ B \rightarrow A: \quad \{B, N_a, N_b\}_{k_a},$$

then step 2.6 of the attack would become:

$$(2.6)\ B \rightarrow I(A): \quad \{B, N_a, N_b\}_{k_a},$$

and the intruder I cannot successfully replay this message in step 1.6 because A is expecting a message containing I's identity.

16.4.4 SSL protocol

The secure sockets layer (SSL) protocol [37] was developed by Netscape and is the standard Internet protocol for secure communications. The secure hypertext transfer protocol (HTTPS) is a communications protocol designed to transfer encrypted information between computers over the World Wide Web. HTTPS is http using a secure socket layer (SSL). SSL resides between TCP/IP and upper-layer applications, requiring no changes to the application layer. SSL is used typically between server and client to secure the connection. One advantage of SSL is that it is application protocol independent. A higher-level protocol can layer on top of the SSL protocol transparently.

SSL protocol allows client–server applications to communicate in a way so that eavesdropping, tampering, and message forgery are prevented. The SSL protocol, in general, provides the following features:

- **End point authentication** The server is the "real" party that a client wants to talk to, not someone faking the identity.
- **Message integrity** If the data exchanged with the server has been modified along the way, it can be easily detected.
- **Confidentiality** Data is encrypted. A hacker cannot read your information by simply looking at the packets on the network.

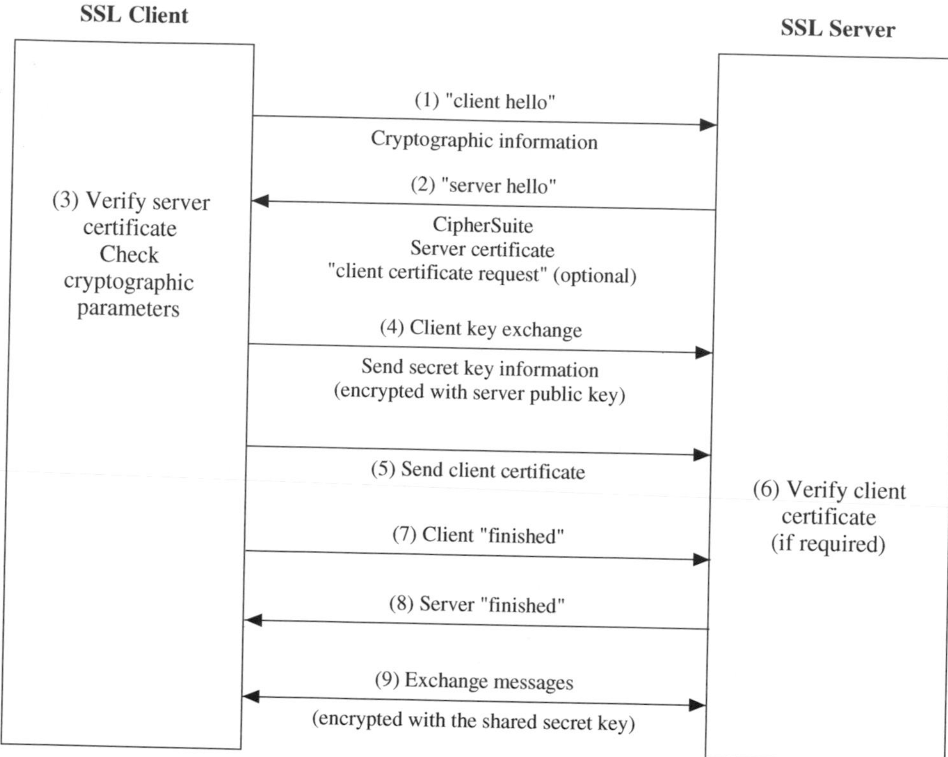

Figure 16.2 SSL handshake protocol and data exchange.

SSL record protocol

The record protocol takes an application message to be transmitted, fragments the data into manageable blocks, optionally compresses the data, applies MAC, encrypts, adds a header, and transmits the resulting unit into a TCP segment. Received data are decrypted, verified, decompressed, and reassembled, and then delivered to high-level users.

SSL handshake protocol

The SSL handshake protocol [37] allows the server and client to authenticate each other and to negotiate an encryption algorithm and cryptographic keys before the application protocol transmits or receives its first byte of data.

The following steps, shown in Figure 16.2, are involved in the SSL handshake:

1. The SSL client sends a "client hello" message that lists cryptographic information such as the SSL version and, in the client's order of preference, the CipherSuites supported by the client. The message also contains a random byte string that is used in subsequent computations.

2. The SSL server responds with a "server hello" message that contains the CipherSuite chosen by the server from the list provided by the SSL client, the session ID, and another random byte string. The SSL server also sends its digital certificate. If the server requires a digital certificate for client authentication, the server sends a "client certificate request" that includes a list of the types of certificates supported and the distinguished names of acceptable certification authorities (CAs).
3. The SSL client verifies the digital signature on the SSL server's digital certificate and checks that the CipherSuite chosen by the server is acceptable.
4. The SSL client, using all data generated in the handshake so far, creates a premaster secret for the session that enables both the client and the server to compute the secret key to be used for encrypting subsequent message data. The premaster secret itself is encrypted with the server's public key.
5. If the SSL server sent a "client certificate request," the SSL client sends another signed piece of data which is unique to this handshake and known only to the client and server, along with the encrypted premaster secret and the client's digital certificate, or a "no digital certificate alert." This alert is only a warning, but with some implementations the handshake fails if client authentication is mandatory.
6. The SSL server verifies the signature on the client certificate.
7. The SSL client sends the SSL server a "finished" message, which is encrypted with the secret key, indicating that the client part of the handshake is complete.
8. The SSL server sends the SSL client a "finished" message, which is encrypted with the secret key, indicating that the server part of the handshake is complete.
9. For the duration of the SSL session, the SSL server and SSL client can now exchange messages that are encrypted with the shared symmetric secret key.

How SSL provides authentication

During both client and server authentication, there is a step that requires data to be encrypted with one of the keys in an asymmetric key pair and is decrypted with the other key of the pair [37].

For server authentication, the client uses the server's public key to encrypt the data that is used to compute the secret key. The server can generate the secret key only if it can decrypt that data with the correct private key.

For client authentication, the server uses the public key in the client certificate to decrypt the data the client sends during step 5 of the handshake. The exchange of finished messages that are encrypted with the secret key (steps 7 and 8 in the overview) confirms that authentication is complete.

If any of the authentication steps fails, the handshake fails and the session terminates.

The exchange of digital certificates during the SSL handshake is a part of the authentication process. The certificates required are as follows, where CA *X* issues the certificate to the SSL client, and CA *Y* issues the certificate to the SSL server:

For server authentication only, the SSL server needs the following:

- The personal certificate issued to the server by CA *Y*.
- The server's private key.

The SSL client needs:

- The CA certificate for CA *Y* or the personal certificate issued to the server by CA *Y*.

If the SSL server requires client authentication, the server verifies the client's identity by verifying the client's digital certificate with the public key for the CA that issued the personal certificate to the client, in this case CA *X*. For both server and client authentication, the SSL server needs:

- The personal certificate issued to the server by CA *Y*.
- The server's private key.
- The CA certificate for CA *X* or the personal certificate issued to the client by CA *X*.

The SSL client needs:

- The personal certificate issued to the client by CA *X*.
- The client's private key.
- The CA certificate for CA *Y* or the personal certificate issued to the server by CA *Y*.

Both the SSL server and the SSL client might need other CA certificates to form a certificate chain to the root CA certificate.

16.5 Password-based authentication

The use of passwords is a highly popular technique to achieve authentication because of low cost and convenience. This section is concerned with authentication techniques that are based on passwords.

A problem with passwords is that people tend to pick a password that is convenient, i.e., short and easy to remember. Such passwords are vulnerable to a password-guessing attack, which works as follows: an adversary builds a database of possible passwords, called a dictionary. The adversary picks a password from the dictionary and checks if it works. This may amount to generating a response to a challenge or decrypting a message using the password or a function of the password. After every failed attempt, the adversary picks a different password from the dictionary and repeats the process. This non-interactive form of attack is known as the *off-line dictionary attack*.

Preventing off-line dictionary attacks

Thus, a major problem is that users tend to choose weak passwords, which are chosen from a sample space small enough to be enumerated by an adversary. Hence, protocols that are stronger than simple challenge–response protocols are needed to use these cryptographically weak passwords to securely authenticate entities. A password-based authentication protocol aims at preventing off-line dictionary attacks by producing a cryptographically strong shared secret key, called the session key, after a successful run of the protocol. This session key can be used by both entities to encrypt subsequest messages for a seceret session.

In this section, we focus on protocols designed to prevent off-line dictionary attacks on password-based authentication. Next, we present two password-based authentication protocols.

16.5.1 Encrypted key exchange (EKE) protocol

The first attempt to protect a password protocol against off-line dictionary attacks was made by Bellovin and Merritt [6] who developed a password-based encrypted key exchange (EKE) protocol using a combination of symmetric and asymmetric cryptography. Algorithm 16.15 describes the EKE protocol that works as follows: suppose users A and B are participating in a run of the protocol. (Recall that $\{X\}_k$ denotes the encryption of X using a symmetric key k and $\{Y\}_{k^{-1}}$ denotes the decryption of Y using a symmetric key k.)

In step 1, user A generates a public/private key pair (E_A, D_A) and also derives a secret key K_{pwd} from his/her password pwd. In step 2, A encrypts his/her public key E_A with K_{pwd} and sends it to B. In steps 3 and 4, B decrypts the message and uses E_A together with K_{pwd} to encrypt a session key K_{AB} and sends it to A. In steps 5 and 6, A uses this session key to encrypt a unique challenge C_A and sends the encrypted challenge to B. In step 7, B decrypts the message to obtain the challenge and generates a unique challenge C_B. In step 8, B then encrypts $\{C_A, C_B\}$ with the session key K_{AB} and sends it to A. In step 9, A decrypts this message to obtain C_A and C_B and compares the former with the challenge it had sent to B. If they match, the correctness of B's response is verified (i.e., B is authenticated). In step 10, A encrypts B's challenge C_B with the session key K_{AB} and sends it to B. When B receives this message, it decrypts the message to obtain C_B and uses it verify the correctness of A's response and to authenticate A. Note that the protocol results in a session key (stronger than the shared password) which the users can later use to encrypt sensitive data.

(1) A: (E_A, D_A), $K_{pwd} = f(pwd)$. {* f is a function. *}
(2) $A \rightarrow B: A, \{K_{pwd}\}_{E_A}$.
(3) B: Compute $E_A = \{\{E_A\}_{K_{pwd}}\}_{K^{-1}_{pwd}}$ and generate a random secret key K_{AB}
(4) $B \rightarrow A$: $\{\{K_{AB}\}_{E_A}\}_{K_{pwd}}$.
(5) A: $K_{AB} = \{\{\{\{K_{AB}\}_{E_A}\}_{\{K_{pwd}\}}\}_{K^{-1}_{pwd}}\}_{D_A}$. Generate a unique challenge C_A.
(6) $A \rightarrow B: \{C_A\}_{K_{AB}}$.
(7) B: Compute $C_A = \{\{C_A\}_{K_{AB}}\}_{K^{-1}_{AB}}$ and generate a unique challenge C_B.
(8) $B \rightarrow A$: $\{C_A, C_B\}_{K_{AB}}$.
(9) A: Decrypt message sent by B to obtain C_A and C_B. Compare the former with own challenge. If they match, go to the next step, else abort.
(10) $A \rightarrow B$: $\{C_B\}_{K_{AB}}$.

Algorithm 16.15 Encrypted key exchange protocol.

The EKE protocol suffers from the plain-text equivalence, which means that the user and the host have access to the same secret password or hash of the password.

16.5.2 Secure remote password (SRP) protocol

Wu [44] combined the technique of zero-knowledge proof with asymmetric key exchange protocols to develop a verifier-based protocol, called the secure remote password (SRP) protocol. The SRP protocol eliminates plain-text equivalence.

All computations in SRP are carried out on the finite field $\mathcal{F}_n$, where n is a large prime. Let g be a generator of $\mathcal{F}_n$. Let A be a user and B be a server. Before initiating the SRP protocol, A and B do the following:

1. A and B agree on the underlying field.
2. A picks a password pwd, a random salt s and computes the verifier $v = g^x$, where $x = H(s, pwd)$ is the long-term private-key and H is a cryptographic hash function. A *salt* is a random string of data used to modify a password hash.
3. B stores the verifier v and the salt s.

Now, A and B can engage in the SRP protocol (shown in Algorithm 16.16). The SRP protocol works as follows. In step 1, A sends its username “A” to server B. In step 2, B looks-up A’s verifier v and salt s and sends A the salt. In steps 3 and 4, A computes its long-term private-key $x = H(s, pwd)$, generates an ephemeral public-key $K_A = g^a$, where a is randomly chosen from the interval $1 < a < n$, and sends K_A to B. In steps 5 and 6, B computes ephemeral public-key $K_B = v + g^b$, where b is randomly chosen from the interval $1 < a < n$, and sends K_B and a random number r to A. In step 7, A computes $S = (K_B - g^x)^{a+rx} = g^{ab+brx}$ and B computes $S = (K_A v^r)^b = g^{ab+brx}$. The values of S computed by A and B will match if the password A entered in step 3 matches the one that A used to calculate the verifier v which is stored at B. In step 8, both A and B use a cryptographically strong hash function to compute a session key $K_{AB} = H(S)$. In step 9, A computes $C_A = H(K_A, K_B, K_{AB})$ and sends it to B as an evidence that it has the session key. C_A also serves as a challenge. In step 10, B computes C_A itself and matches it with A's message. B also computes $C_B = H(K_A, C_A, K_{AB})$. In step 11, B sends C_B to A as an evidence that it has the same session key as A. In step 12, A verifies C_B, accepts if the verification passes and aborts otherwise.

---

(1) $A \to B: \quad A.$
(2) $B \to A: \quad s.$
(3) $A: \quad x := H(s, pwd); K_A := g^a.$
(4) $A \to B: \quad K_A.$
(5) $B: \quad K_B := v + g^b.$
(6) $B \to A: \quad K_B, r.$
(7) $A: \quad S := (K_B - g^x)^{a+rx}$ and $B: \quad S := (K_A v^r)^b.$
(8) $A, B: \quad K_{AB} := H(S).$
(9) $A \to B: \quad C_A := H(K_A, K_B, K_{AB}).$
(10) B verifies C_A and computes $C_B := H(K_A, C_A, K_{AB})$.
(11) $B \to A: \quad C_B.$
(12) A verifies C_B. Accept if verification passes; abort otherwise.

Algorithm 16.16 Secure remote password (SRP) protocol.

Note that unlike EKE, none of the protocol messages are encrypted in the SRP protocol. Since neither the user nor the server has access to the same secret password or hash of the password, SRP eliminates plaintext equivalence. SRP was unique in its swapped-secret approach in building a verifier-based, zero-knowledge protocol, resisting off-line dictionary attacks.

16.6 Authentication protocol failures

Despite the apparent simplicity of the basic design principles, realistic authentication protocols [11,29] are notoriously difficult to design [39]. There are several reasons for this:

- First, most realistic cryptosystems satisfy algebraic additional identities. These extra properties may generate undesirable effects when combined with a protocol logic.
- Second, even assuming that the underlying cryptosystem is perfect, unexpected interactions among the protocol steps can lead to subtle logical flaws.

- Third, assumptions regarding the environment and the capabilities of an adversary are not explicitly specified, making it extremely difficult to determine when a protocol is applicable and what final states are achieved.

We illustrate the difficulty by showing an authentication protocol proposed, with a subtle weakness. Consider the authentication protocol shown in Algorithm 16.17 (k_p and k_q are symmetric keys shared between P and A, and Q and A, respectively, where A is an authentication server and k is a session key).

(1)	$P \rightarrow A:$	P, Q, n_p
(2)	$A \rightarrow P:$	$\{n_p, Q, k, \{k, P\}_{k_Q}\}_{k_p}$
(3)	$P \rightarrow Q:$	$\{k, P\}_{k_Q}$
(4)	$Q \rightarrow P:$	$\{n_Q\}_K$
(5)	$P \rightarrow Q:$	$\{n_Q+1\}_K$

Algorithm 16.17 Authentication protocol with a subtle weakness.

The message $\{k, P\}_{k_Q}$ in step 3 can only be decrypted by Q and hence can only be understood by Q. Step 4 reflects Q's knowledge of k, while step 5 assures Q of P's knowledge of k; hence the authentication handshake is based entirely on the knowledge of k.

The subtle weakness in the protocol arises from the fact that the message $\{k, P\}_{k_Q}$ sent in step 3 contains no information for Q to verify its freshness. This is the first message sent to Q about P's intention to establish a secure connection. An adversary who has compromised an old session key k' can impersonate P by replaying the recorded message $\{k', P\}_{k_Q}$ in step 3 and subsequently executing steps 4 and 5 using k'.

To avoid protocol failures, formal methods may be employed in the design and verification of authentication protocols. A formal design method should embody the basic design principles. For example, informal reasoning such as "if you believe that only you and Bob know k, then you should believe any message you receive encrypted with k was originally sent by Bob" should be formalized by a verification method.

16.7 Chapter summary

Authentication is a process by which one principal verifies the identity of the other principal. For example, in a client–server system, the client and the server may need to verify each other's identity to assure that each is talking to the right entity. Generally, authentication is based on the possession of a secret information, like password, that is known only to the entities participating in the authentication. For a successful authentication, the entity must demonstrate the knowledge of the right secret information.

In this chapter, we described several user authentication protocols based on symmetric and asymmetric cryptosystems. We also discussed authentication techniques that are based on passwords and which mitigate dictionary attacks. Authentication protocols are vulnerable to several attacks.

16.8 Exercises

Exercise 16.1 List three attacks/threats that are associated with user authentication on the Internet.

Exercise 16.2 What is a nonce? What security problem does it solve?

Exercise 16.3 Consider the following simple method to handle attacks on the password based authentication. If a user fails to login in three successive attempts, the system locks his account suspecting an attack/intrusion. What major problem do you see with this method?

Exercise 16.4 Choose two principles given by Needham and Abadi for designing cryptographic protocols. For each, give an example where their principle applies and results in an improved protocol.

Exercise 16.5 Consider the following protocol for authentication/key distribution (X and Y are two principals, A is a certificate authority or a key distribution center, R_X is a randon number, and E_X means encrypted with the secret key of X).

$$\begin{array}{lll} (1) & X \rightarrow A: & X, Y, R_X \\ (2) & A \rightarrow X: & E_X(R_X, Y, K, E_Y(K, X)) \\ (3) & X \rightarrow Y: & E_Y(K, X) \\ (4) & Y \rightarrow X: & E_K(R_Y) \\ (5) & X \rightarrow Y: & E_K(R_Y - 1) \end{array}$$

1. What does the presence of R_X in message 2 assure?
2. What problem will be created if an attacker were to break an old K (and the attacker has also copied messages for that session)? Explain your answer.
3. Suggest a method to solve this problem.

Exercise 16.6 Discuss two biometric based methods for authentication. What are pros and cons of biometric based methods for authentication?

16.9 Notes on references

Authentication in distributed systems is a well studied topic and a large number of authentication protocols exist. An excellent survey on the topic is by Woo and Lam [39]. Burrows *et al.* discuss the logic of authentication [7]. A classical paper on the topic is by Needham and Schroeier [29]. Syverson and Cervesato [36] discuss the logic of authentication protocols. Two relevant books on the topic are by Schneier [33] and Stallings [35]. Lampson *et al.* [23] discuss the theory and practice of authentication.

A review paper on password based authentication is by Chakrabarti and Singhal [8]. Conklin *et al.* [10] give a system's perspective of password-based authentication. Biometric authentication has been very popular recently. Information on this topic can be found in [15–17,31,34]. King and Dos Santos [22] discuss AI-based methods for human authentication. Kaminsky *et al.* [18] discuss user authentication in a global file system. A list of papers on authentication can be found at: www.passwordresearch.com/papers/pubindex.html. Other relevant work on authentication in distributed systems can be found in [3,9,12–14,19,27,38,40,42,43].

References

[1] J.M. Arfman and P. Roden, Project Athena: Supporting distributed computing at MIT, *IBM Systems Journal*, **31**(3), 1992, 550–563.

[2] M. Abadi and R. Needham, Prudent engineering practices for cryptographic protocols, *Proceedings of the IEEE Computer Society Symposium on Research in Security and Privacy*, May 1994, 122–136.

[3] M. Abadi, M. Burrows, C. Kaufman, and B.W. Lampson, Authentication and delegation with smart-cards, *Science of Computer Programming*, **21**(2), 1993, 93–113.

[4] R. Anderson and R. Needham, Robustness principles for public key protocols, in D. Coppersmith (ed.), *Advances in Cryptology – CRYPTO'95*, New York, Springer-Verlag, 1995, 236–247.

[5] S.M. Bellovin and M. Merritt, Limitations of the Kerberos authentication system, *Proceedings of USENIX Winter Conference*, Dallas, TX, January 1991, 253–267.

[6] S.M. Bellovin and M. Merritt, Encrypted key exchange: password-based protocol secure against dictionary attacks, *Proceedings of the IEEE Symposium on Security and Privacy*, Oakland, CA, 1992, 72–84.

[7] M. Burrows, M. Abadi, and R.M. Needham, A logic of authentication, *ACM Transactions on Computer Systems*, **8**(1), 1990, 18–36.

[8] S. Chakrabarti and M. Singhal, Password-based authentication: preventing dictionary attacks, *IEEE Computer*, **40**(6), 2007, 68–74.

[9] CCITT Recommendation X.509, *The Directory – Authentication Framework*, 1988. See also ISO/IEC 9594-8, 1989.

[10] A. Conklin, G. Dietrich, and D. Walz, Password-based authentication: a system perspective, *Proceedings of the 37th Hawaii International Conference on System Sciences*, January 2004.

[11] D.E. Denning, *Cryptography and Data Security*, Addison-Wesley, 1982.

[12] D. Dolev and A.C. Yao, On the security of public key protocols, *IEEE Transactions on Information Theory*, **IT-29**(2), 1983, 198–208.

[13] M. Gasser, A. Goldstein, C. Kaufman, and B.W. Lampson, The Digital distributed system security architecture, *Proceedings of the 12th National Computer Security Conference*, Baltimore, MD, October 1989, 305–319.

[14] M. Gasser and E. McDermott, An architecture for practical delegation in a distributed system, *Proceedings of the 11th IEEE Symposium on Research in Security and Privacy*, Oakland, CA, May 7–9, 1990, 20–30.

[15] L. O'Gorman, Practical systems for personal fingerprint authentication, *IEEE Computer*, **33**(2), 2000, 58–60.

[16] A. Jain, L. Hong, and S. Pankanti, Biometrics identification, *Communications of the ACM*, **43**(2), 2000, 91–98.

[17] M. Indovina, U. Uludag, R. Snelick, A. Mink, and A. Jain, Multimodal biometric authentication methods: a COTS approach, *Proceedings of MMUA 2003, Workshop on Multimodal User Authentication*, Santa Barbara, CA, December 11–12, 2003, 99–106.

[18] M. Kaminsky, G. Saviddes, D. Mazieres, and M. F. Kaashoek, Decentralized user authentication in a global file system, *Symposium on Operating System Principles*, 2003, 60–73.

[19] C. Kaufman, *DASS Distributed Authentication Security Service*, September 1993, RFC 1507.

[20] J. T. Kohl, B. C. Neuman, and T. Y. Tso, The evolution of the Kerberos authentication system, in F. Brazier and D. Johansen (eds), *Distributed Open Systems*, New York, IEEE Computer Society Press, 1994, 78–94.

[21] Kerberos Frequently Asked Questions, available online at: www.nrl.navy.mil/CCS/people/kenh/kerberos-faq.html.

[22] J. King and A. dos Santos, A user-friendly approach to human authentication of messages, *Proceedings of the 9th International Conference on Financial Cryptography and Data Security*, Roseau, Dominica, 2005, 225–239.

[23] B. Lampson, M. Abadi, and M. Burrows, Authentication in distributed systems: theory and practice, *ACM Transactions on Computer Systems*, (10):265–310, November 1992.

[24] M.-H. Lin and C.-C. Chang, A secure one-time password authentication scheme with low-computation for mobile communications, *ACM SIGOPS Operating Systems Review*, **38**(2), 2004, 76–84.

[25] G. Lowe, An attack on the Needham–Schroeder public-key authentication protocol, *Information Processing Letters*, **56**(3) 1995, 131–133.

[26] J. Linn, Practical authentication for distributed computing, *Proceedings of the 11th IEEE Symposium on Research in Security and Privacy*, Oakland, CA, May 7–9 1990, 31–40.

[27] C.-C. Lee, M.-S. Hwang, and L.-H. Li, A new key authentication scheme based on discrete logarithms, *Applied Mathematics and Computation*, **139**(2–3), 2003.

[28] A. Menezes, P. van Oorschot, and S. Vanstone, Key establishment protocols, Chapter 12 in *Handbook of Applied Crytography*, New York, CRC Press, 1996.

[29] R. M. Needham and M. D. Schroeder, Using encryption for authentication in large networks of computers, *Communications of the ACM*, **21**(12), 1978, 993–999.

[30] B. C. Neuman and T. Y. Ts'o, An authentication service for computer networks, *IEEE Communications Magazine*, **32**(9), 1994, 33–38.

[31] N. K. Ratha, J. H. Connell, and R. M. Bolle, Enhancing security and privacy in biometrics-based authentication systems, *IBM Systems Journal*, **40**(3), 2001, 614–634.

[32] J. G. Steiner, C. Neuman, and J. I. Schiller, Kerberos: an authentication service for open network systems, *Proceedings of USENIX Winter Conference*, Dallas, TX, February 1988, 191–202.

[33] B. Schneier, *Applied Cryptography*, New York, John Wiley & Sons, Inc., 1996.

[34] B. Schneier, The uses and abuses of biometrics, *Communications of the ACM*, **42**(8), 1999, 136.

[35] W. Stallings, *Cryptography and Network Security: Principles and Practice*, 4th edn., Englewood Cliffs, NJ, Prentice-Hall, 2005.

[36] P. Syverson and I. Cervesato, *The Logic of Authentication Protocols*, New York, Springer-Velag, 2001.

[37] The Secure Socket Layer, available online at: http://publib.boulder.ibm.com/infocenter/wmqv6/v6r0/index.jsp?topic=/com.ibm.mq.csqzas.doc/cssauthentication.htm.

[38] J. J. Tardo and K. Alagappan, SPX: global authentication using public key certificates, in *Proceedings of the 12th IEEE Symposium on Research in Security and Privacy*, Oakland, CA, May 20–22, 1991, 232–244.

[39] T. Y. C. Woo and S. S. Lam, Authentication for distributed systems, *Computer*, **25**(1), 1992, 39–52. See also: Authentication revisited, *Computer*, **25**(3), 1992, 10.

[40] T. Y. C. Woo and S. S. Lam, A lesson on authentication protocol design, *ACM Operating Systems Review*, **28**(3), 1994, 24–37.

[41] T. Y. C. Woo and S. Lam, Authentication for distributed systems, D. Denning and P. Denning (eds), in *Internet Besieged: Countering Cyberspace Scofflaws*, Addison–Wesley and ACM Press, 1998.

[42] T. Y. C.Woo and S. S. Lam, Design, verification, and implementation of an authentication protocol, *Proceedings of International Conference on Network Protocols*, Boston, MA, October 25–28, 1994. (Also available online at: www.cs.utexas.edu/users/lam/NRL/.)

[43] T. Y. C. Woo, R. Bindignavle, S. Su, and S. S. Lam, SNP: an interface for secure network programming, *Proceedings of USENIX Summer Technical Conference*, Boston, MA, June 6–10, 1994. (Also available online at: www.cs.utexas.edu/users/lam/NRL/.)

[44] T. D. Wu, The secure remote password protocol, *Proceedings of the Network and Distributed Systems Security, NDSS 1998*, San Diego, CA, 1998, 97–111.

CHAPTER

17 Self-stabilization

17.1 Introduction

The idea of self-stabilization in distributed computing was first proposed by Dijkstra in 1974 [34]. The concept of self-stabilization is that, regardless of its initial state, the system is guaranteed to converge to a legitimate state in a bounded amount of time by itself without any outside intervention. A non-self-stabilizing system may never reach a legitimate state or it may reach a legitimate state only temporarily. The main complication in designing a self-stabilizing distributed system is that nodes do not have a global memory that they can access instantaneously. Each node must make decisions based on the local knowledge available to it and actions of all nodes must achieve a global ojective.

The definition of legitimate and illegitimate states depends on the particular application. Generally, all illegitimate states are defined to be those states which are not legitimate states. Dijkstra also gave an example of the concept of self-stabilization using a self-stabilizing token ring system. For any given token ring when there are multiple tokens or there is no token, then such global states are known as illegitimate states. When we consider a distributed system where a large number of systems are widely distributed and communicate with each other using message passing or shared memory approach, there is a possibility for these systems to go into an illegitimate state, for example, if a message is lost. The concept of self-stabilization can help us recover from such situations in distributed system.

Let us explain the concept of self-stabilization using an example. Let us take a group of children and ask them to stand in a circle. After few minutes, you will get an almost perfect circle without having to take any further action. In addition, you will discover that the shape of this circle is stable, at least until you ask the children to disperse. If you force one of the children out of position, the others will move accordingly, moving the entire circle in another position, but keeping its shape unchanged.

In this example, the group of children build a self-stabilizing circle: if some thing goes wrong with the circle, they are able to rebuild the circle by themselves, without any external intervention. The time required for stabilization varies from experiment to experiment, depending on the (random) initial position. However, if the field size is limited, this time will be bounded. The algorithm does not define the position of the circle in the field and so it will not always be the same. The position of each child relative to each other will also vary.

The self-stabilization principle applies to any system built on a significant number of components which are evolving independently from one another, but which are cooperating or competing to achieve common goals. This applies, in particular, to large distributed systems which tend to result from the integration of many subsystems and components developed separately at earlier times or by different people.

In this chapter, we first present the system model of a distributed system and present definitions of self-stabilization. Next, we discuss Dijkstra's seminal work and use it to motivate the topic. We discuss the issues arising from the Dijkstra's original presentation as well as several related issues in the design of self-stabilizing algorithms and systems. After that, we discuss three important themes that have recently emerged. In particular, we discuss the methods that have been used to design complex self-stabilizing systems, we discuss the role of compilers in designing self-stabilization, and we enumerate factors that have been found to interfere with self-stabilization. We also discuss self-stabilizing protocols for construction of spanning trees and present a self-stabilizing algorithm for 1-maximal independent set. We conclude the chapter with limitations of self-stabilization.

17.2 System model

The term distributed system is used to describe a set of computers that communicate over a network. Variants of distributed systems have similar fundamental coordination requirements among the communicating entries, whether they are computers, processors or processes. Thus an abstract model that ignores the specific settings and captures the important characteristics of a distributed system is usually used.

In a distributed system, each computer runs a program composed of executable statements. Each execution changes the content of the computer's logical memory. An abstract way to model a computer that executes a program is to use the state machine model. A distributed system model comprises of a set of n state machines called processors that communicate with each other. We usually denote the ith processor in the system by P_i. Neighbors of a processor are processors that are directly connected to it. A processor can directly communicate with its neighbors. A distributed system can be conveniently represented by a graph in which each processor is represented by a node and every pair of neighboring nodes are connected by a link.

The communication between neighboring processors can be carried out either by message passing or shared memory. Communication by writing in and reading from the shared memory usually fits systems with processors that are geographically close together, such as multiprocessor computers. A message-passing distributed model fits both processors that are located close to each other as well as that are widely distributed over a network.

In the message-passing model, neighbors communicate by sending and receiving messages. In asynchronous distributed systems, the speed of processors and message transmission can vary. First-in first-out (FIFO) queues are used to model asynchronous delivery of messages. A communication link is either unidirectional or bidirectional. A unidirectional communication link from processor P_i to P_j transfers messages only from P_i to P_j. The abstraction used for such a unidirectional link is a first-in first-out (FIFO) queue $Q_{i,j}$ that contains all messages sent by a processor P_i to its neighbor P_j that have not yet been received. Whenever P_i sends a message m to P_j, the message is enqueued (added to the tail of the queue). The bidirectional communication link between processors P_i and P_j is modeled by two FIFO queues, one from P_i to P_j and the other from P_j to P_i.

It is convenient to identify the state of a computer or a distributed system at a given time, so that no additional information about the past of the computation is needed in order to predict the future behavior (state transitions) of the computer or the distributed system. A full description of a message passing distributed system at a particular time consists of the state of every processor and the content of every queue (messages traveling in the communication links). The term system configuration (or configuration) is used for such a description. A configuration is denoted by $c = (s_1, s_2, \ldots, s_n, q_{1,2}, q_{1,3}, \ldots, q_{i,j}, \ldots, q_{n,n-1})$, where s_i, $1 \leq i \leq n$ is the state of P_i and $q_{i,j}$, $i \neq j$ is the state of queue $Q_{i,j}$, that is, messages sent by P_i to P_j but not yet received. The behavior of a system consists of a set of states, a transition relation between those states, and a set of fairness criteria on the transition relation [79].

The system is usually modeled as a graph of processing elements (modeled as state machines), where edges between these elements model unidirectional or bidirectional communication links. Let N be an upper bound on n (the number of nodes in the system). The communication network is usually restricted to the neighbors of a particular node. Let δ denote the diameter of the network (i.e., the length of the longest unique path between two nodes) and let Δ denote the upper bound on δ. A network is static if the communication topology remains fixed. It is dynamic if links and network nodes can go down and recover later. In the context of dynamic systems, self-stabilization refers to the time after the "final" link or node failure. The term "final failure" is typical in the literature on self-stabilization. Since stabilization is only guaranteed eventually, the assumption that faults eventually stop to occur implies that there are no faults in the system for "sufficiently long period" for the system to stabilize. In any case, it is assumed that the topology remains connected, i.e., there exists a path between any two nodes.

In the shared memory model, processors communicate using shared communication registers (hereafter, called registers). Processors may write in a set of registers and may read from a possibly different set of registers. Two neighboring nodes have access to a common data structure, variable or register which can store a certain amount of information. These variables can be distinguished between input and output variables (depending on which process can modify them). When executing a step, a process may read all its input variables, perform a state transition and write all its output variables in a single atomic operation. This is called composite atomicity. A weaker notion of a step (called read/write atomicity) also exists where a process can only either read or write its communication variables in one atomic step.

The configuration of a system with n processors and m communication registers is denoted by $c = (s_1, s_2, s_3, \ldots, s_n, r_1, r_2, \ldots, r_m)$, where s_i, $1 \leq i \leq n$, is the state of P_i and r_j, $1 \leq j \leq m$, is the contents of a communication register.

Algorithms are modeled as state machines performing a sequence of steps. A step consists of reading input and the local state, then performing a state transition and writing output. Communication can be by exchanging messages over the communication channels. An algorithm may be randomized, i.e., have access to a source of randomness (a random number generator or a random coin flip). If an algorithm is not randomized, we will call it deterministic. A related characteristic of a system model is its execution semantics. In self-stabilization, this has been encapsulated within the notion of a scheduler or daemon (also demon). Under a central daemon, at most one processing element is allowed to take a step at the same time.

17.3 Definition of self-stabilization

We have seen an informal definition of self-stabilization at the beginning. Formally, we define self-stabilization for a system S with respect to a predicate P over its set of global states, where P is intended to identify its correct execution [79]. States satisfying P are called legitimate states and those not satisfying P are called illegitimate states. We use the terms safe and unsafe interchangeably with legitimate and illegitimate, respectively.

A system S is self-stabilizing with respect to predicate P if it satisfies the following two properties:

- **Closure** P is closed under the execution of S. That is, once P is established in S, it cannot be falsified.
- **Convergence** Starting from an arbitrary global state, S is guaranteed to reach a global state satisfying P within a finite number of state transitions.

Arora and Gouda [12] introduced a more generalized definition of self-stabilization, called *stabilization*, which is defined as follows. We define stabilization for a system S with respect to two predicates P and Q, over its set of global states. Predicate Q denotes a restricted start condition. S satisfies $Q \to P$ (read as Q stabilizes to P) if it satisfies the following two properties:

- **Closure** P is closed under the execution of S. That is, once P is established in S, it cannot be falsified.
- **Convergence** If S starts from any global state that satisfies Q, then S is guaranteed to reach a global state satisfying P within a finite number of state transitions.

Note that self-stabilization is a special case of stabilization where Q is always *true*, that is, if S is self-stabilizing with respect to P, then this may be restated as TRUE $\to P$ in S.

Next, we define two terms relevant to the discussion of self-stabilization.

Reachable Set

Often when a programmer writes a program, he/she does not have a particular definition of safe and unsafe states in mind but develops the program to function from a particular set of start states. In such situations, it is reasonable to define as safe those states that are reachable under normal program execution from the set of legitimate start states. These states are referred to as the *reachable set*. So, when we say that a program is self-stabilizing without mentioning a predicate, we mean with respect to the reachable set. By definition, the reachable set is closed under program execution, and it corresponds to a predicate over the set of states [79].

We use the transient failure model in the discussion.

Transient failure

A transient failure is temporary (short lived) and it does not persist. A transient failure may be caused by corruption of local state of processes or by corruption of channels or shared memory. A transient failure may change the state of the system, but not its behavior.

17.3.1 Randomized and probabilistic self-stabilization

Randomized methods for self-stabilization are useful in achieving self-stabilization under process symmetry (i.e., all processes are identical). Depending on the stabilization time, self-stabilization can be classified as randomized and probabilistic self-stabilization:

- **Randomized self-stabilization** A system is said to be a *randomized self-stabilizing system*, if and only if it is self-stabilizing and the expected number of rounds needed to reach a correct state (legal state) is bounded by some constant k.

- **Probabilistic self-stabilization** A system S is said to be *probabilistically self stabilizing* with respect to a predicate P if it satisfies the following two properties:
  - **Closure** P is closed under the execution of S. That is, once P is established in S, it cannot be falsified.
  - **Convergence** There exists a function f from natural numbers to [0,1] satisfying $f_{lim\ k\rightarrow\infty}(k) = 0$, such that the probability of reaching a state satisfying P, starting from an arbitrary global state within k state transitions, is $1 - f(k)$.

A *pseudo-stabilizing system* is one that, if started in an arbitrary state, is guaranteed to reach a state after which it *does not* deviate from its intended specification. A *stabilizing system* is one that, if started at an arbitrary state, is guaranteed to reach a state after which it *cannot* deviate from its intended specification. Thus, the difference between the two notions comes down to the difference between cannot and does not – a difference that hardly matters in many practical situations. The stronger requirement of self-stabilization is advantageous over pseudo-stabilization in finite-state systems, since self-stabilization property implies a bounded convergence span while the pseudo stabilization does not. Algorithms have been proposed for probabilistic orientation of an asynchronous bi-directional ring, as well as for a synchronous ring with odd number of processes and one token.

In the next section, we discuss the issues in the design of self-stabilization algorithms.

17.4 Issues in the design of self-stabilization algorithms

A distributed system comprises of many individual units and many issues arise in the design of self-stabilization algorithms in distributed system. Some of the main issues are as follows:

- Number of states in each of the individual units in a distributed system.
- Uniform and non-uniform algorithms in distributed systems.
- Central and distributed demon.
- Reducing the number of states in a token ring.
- Shared memory models.
- Mutual exclusion.
- Costs of self-stabilization.

Dijkstra's self-stabilizing token ring system

We explain the above issues with the help of Dijkstra's landmark self-stabilizing token ring system [34]. His system consisted of a set of n finite-state machines connected in the form a ring. He defines a privilege of a machine to be the ability to change its current state. This ability is based on a Boolean predicate that consists of its current state and the states of its neighbors. When a machine has a privilege, it is able to change its current state, which is referred to as a move. Furthermore, when multiple machines enjoy a privilege at the same time, the choice of the machine that is entitled to make a move is made by a central demon, which arbitrarily decides which privileged machine will make the next move.

A legitimate state must satisfy the following constraints:

- There must be at least one privilege in the system (liveness or no deadlock).
- Every move from a legal state must again put the system into a legal state (closure).
- During an infinite execution, each machine should enjoy a privilege an infinite number of times (no starvation).
- Given any two legal states, there is a series of moves that change one legal state to the other (reachability).

Dijkstra [34] considered a legitimate (or legal) state as one in which exactly one machine enjoys the privilege. This corresponds to a form of mutual exclusion, because the privileged process is the only process that is allowed in its critical section. Once the process leaves the critical section, it passes the privilege to one of its neighbors.

With this background, let us see how the above issues affect the design of a self-stabilization algorithm.

17.4.1 The number of states in each of the individual units

The number of states that each machine must have for the self-stabilization is an important issue. Dijkstra offered three solutions for a directed ring with n machines, 0, 1, ..., $n-1$, each having K states, (i) $K \geq n$, (ii) $K = 4$, (iii) $K = 3$. It was later proven by Ghosh [49] that a minimum of three states is required in a self-stabilizing ring. In all three algorithms, Dijkstra assumed the existence of at least one exceptional machine that behaved differently from the others.

The first solution ($K \geq n$) is described below.

First solution

For any machine, we use the symbols S, L, and R to denote its own state, the state of the left neighbor and the state of the right neighbor on the ring, respectively.

The exceptional machine:

If $L = S$ then
$S := (S + 1) \bmod K$
End If;

The other machines:

If $L \neq S$ then
$S := L$
End If;

In this algorithm, except the exceptional machine (machine 0), all other machines follow the same algorithm. In the ring topology, each machine compares its state with the state of the anti-clockwise neighbor and if they are not same, it updates its state to be the same as that of its anti-clockwise neighbor.

So, if there are n machines and each of them is initially at a random state $r \in K$, then all the machines (except the exceptional machine, machine 0) whose states are not the same as their anti-clockwise neighbor are said to be privileged and there is a central demon that decides which of these privileged machines will make the move.

Suppose machine 6 (assume $n \gg 6$) makes the first move. It is obvious that its state is not the same as that of machine 5 and hence it had the privilege to make the move and finally sets its state to be the same as that of machine 5. Now machine 6 loses its privilege as its state is same as that of its anti-clockwise neighbor (machine 5). Next, suppose machine 7, whose state is different from the state of machine 6, is given the privilege. It results in making the state of machine 7 the same as that of machine 6. Now machines 5, 6, and 7 are in the same state. Eventually, all the machines will be in the same state in the similar manner. At this point, only the exceptional machine (machine 0) will be privileged as its condition $L = S$ is satisfied, i.e., its state is the same as that of its anti-clockwise neighbor. Now there exists only one privilege or token in the system (at machine 0). Machine 0 makes a move and changes its state from S to $(S+1)$ mod K. This will make the next machine, machine 1, privileged as its state is not the same as its anti-clockwise neighbor, i.e., machine 0. Thus, it can be interpreted as the token is currently with machine 1. Machine 1, as per the algorithm, changes its state to the same state as that of machine 0. This will move the token to machine 2 as its state is now not same as that of machine 1. Likewise, the token keeps circulating around the ring and the system is stable.

This is a simple algorithm, but it requires a number of states, which depends on the size of the ring, which may be awkward for some applications.

Second solution

The second solution uses only three-state machines and is presented in Algorithm 17.1. The state of each machine is in $\{0, 1, 2\}$.

In the first algorithm, there is only one exceptional machine, machine 0. In the second solution, there are two such machines, machine 0, referred to as the bottom machine, and machine $n-1$, referred to as the top machine.

The bottom machine, machine 0:

If $(S+1) \bmod 3 = R$ then
  $S := (S-1) \bmod 3$

The top machine, machine $n-1$:

If $L = R$ and $(L+1) \bmod 3 \neq S$ then
  $S := (L+1) \bmod 3$

The other machines:

If $(S+1) \bmod 3 = L$ then
  $S := L$
If $(S+1) \bmod 3 = R$ then
  $S := R$

Algorithm 17.1 The second solution.

In this algorithm, the bottom machine, machine 0, behaves as follows:

If $(S+1) \bmod 3 = R$ then
  $S := (S-1) \bmod 3$

Thus, the state of the bottom machine depends upon its current state and the state of its right neighbor.

The condition $(s+1) \bmod 3$ covers the three possible states; for $s = 0$, 1, 2, we have $(s+1) \bmod 3 = 1$, 2, 0. These result in the following three possibilities:

1. If $s = 0$ and $r = 1$, then the state of s is changed to 2.
2. If $s = 1$ and $r = 2$, then the state of s is changed to 0.
3. If $s = 2$ and $r = 0$, then the state of s is changed to 1.

The top machine, machine $n-1$, behaves as follows:

---

if $L = R$ and $(L+1) \bmod 3 \neq S$ then
  $S := (L+1) \bmod 3$

The state of the top machine depends upon both its left and right neighbors (the bottom machine). The condition specifies that the left neighbor (L) and the right neighbor (R) should be in the same state and $(L+1) \bmod 3$ should not be equal to S. (Note that $(L+1) \bmod 3$ is 1, 2, 0 when L is 0, 1, 2, respectively) Thus, the state of the top machine is as follows:

1. 1, when its left neighbor is 0.
2. 2, when its left neighbor is 1.
3. 0 when its left neighbor is 2.

Table 17.1 An example execution of Dijkstra's three-state algorithm [42].

State of machine 0	State of machine 1	State of machine 2	State of machine 3	Privileged machines	Machine to make move
0	1	0	2	0, 2, 3	0
2	1	0	2	1, 2	1
2	2	0	2	1	1
2	0	0	2	0	0
1	0	0	2	1	1
1	1	0	2	2	2
1	1	1	2	2	2
1	1	2	2	1	1
1	2	2	2	0	0
0	2	2	2	1	1
0	0	2	2	2	2
0	0	0	2	3	3
0	0	0	1	2	2

All other machines behave as follows:

```
If (S+1) mod 3 = L then
        S := L
If (S+1) mod 3 = R then
        S := R
```

While finding out the state of the other machines (machines 1 and 2 in the example below), we first compare the state of a machine with its left neighbor:

1. If $s = 0$ and $L = 1$, then $s = 0$.
2. If $s = 1$ and $L = 2$, then $s = 2$.
3. If $s = 2$ and $L = 0$, then $s = 1$.

If the above conditions are not satisfied, then the machine compares its state with its right neighbor.

A sample execution of Dijkstra's three-state algorithm for a ring of four processes (0, 1, 2, 3) is shown in Table 17.1 [42]. Machine 0 is the bottom machine and machine 3 is the top machine. The last column in the table gives the number of the machine chosen to make the next move. Initially, three privileges exist in the system. The number of privileges decreases until only one privilege is left in the system.

We make the following observations:

- There are no deadlocks in any state (at least one privilege is present).
- The closure property is satisfied (the system moves from a legal state to a legal state).

- No starvation (each machine has a chance of making more than 1 move).
- Reachability (there are always a series of moves to reach from one legal state to other).

All four constraints for a legitimate state (given at the start of Section 17.4) are satisfied. So the system is stabilized.

Special networks

In the above two algorithms, each processor needs K states and three states, respectively. Ghosh [47] found that there are special networks, where the number of states required by each processor is two.

Ghosh's solution

All nodes (machines) in the network shown in Figure 17.1 require only two states [47]. However, a node needs to use information from all of its neighbors. Let *s[i]* denote the state of machine i. There are two possible states for each machine, 0 and 1. In the algorithm [47], let b denote an arbitrary state (0 or 1) and $\tilde{b}$ denote the complementary state of b.

For machine 0:
  If $(s[0], s[1]) = (\tilde{b}, b)$ then $s[0] := b$

For machine $2n-1$:
  If $(s[2n-1], s[2n-2]) = (b, b)$ then $s[2n-1] := \tilde{b}$

For even numbered machines:
  If $(s[2i-2], s[2i-1], s[2i], s[2i+1]) = (b, b, \tilde{b}, b)$ then
  $s[2i] := b$

For odd numbered machines:
  If $(s[2i-2], s[2i-1], s[2i], s[2i+1]) = (b, b, b, \tilde{b})$ then
  $s[2i-1] := \tilde{b}$

Each machine must examine the states of all its neighbors. In this algorithm, a large atomicity is assumed because each machine must be able to examine the states of all its neighbors in one atomic step. In addition, the algorithm requires that the number of machines in the network must be even and at least six. However, the algorithm shows that self-stabilizing algorithms requiring two states are possible.

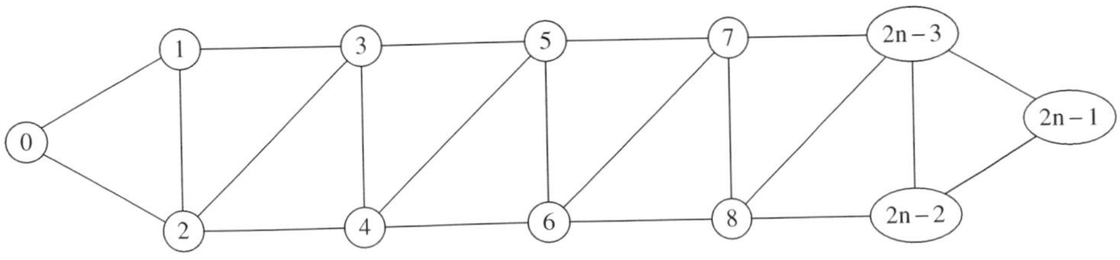

Figure 17.1 A special network needing only binary state machines [47].

Dolev et al.'s solution

For a system with an odd number of machines in a ring, the solution for self-stabilization described by Dolev *et al.* [40] is as follows: each node has two states, 0 and 1. Given a global state, the nodes make moves according to the following rules:

- If the local state is different from its left neighbor's state, then the state is changed to be the same as its left neighbor.
- If the local state is the same as its left neighbor's state, the state is chosen randomly from 0 and 1.

Nodes make moves in synchronization in each step. A node has a privilege if its state is the same as that of its left neighbor. Using a probabilistic argument, it has been shown that eventually only one privilege exists in the system. This algorithm shows that the number of states required for each node may be reduced using a probabilistic algorithm if nodes operate synchronously.

17.4.2 Uniform vs. non-uniform networks

Whether processes (or machines) are uniform or not is an important issue in self-stabilization. In a distributed system, it is desirable and also possible to have each machine use the same algorithm. To design self-stabilizing systems, however, it is often necessary to have non-uniformity among machines. From the examples of the preceding section, we notice that at least one of the machines (known as the exceptional machines) had a privilege and executed steps that were different from other machines.

The individual processes can be anonymous, meaning they are indistinguishable and all run the same algorithm. Often, anonymous networks are called uniform networks. A network is semi-uniform if there is one process (the root) which executes a different algorithm. While there is no way to distinguish nodes, in uniform or semi-uniform algorithms nodes usually have a means of distinguishing their neighbors by ordering the incoming communication links. In the most general case it is assumed that processes have globally unique identifiers.

Self-stabilization algorithms for distributed systems should be uniform, but this is not always possible. As a simple example, consider the ring of four processors shown in Figure 17.2 [47].

Figure 17.2 A ring of four processors [47].

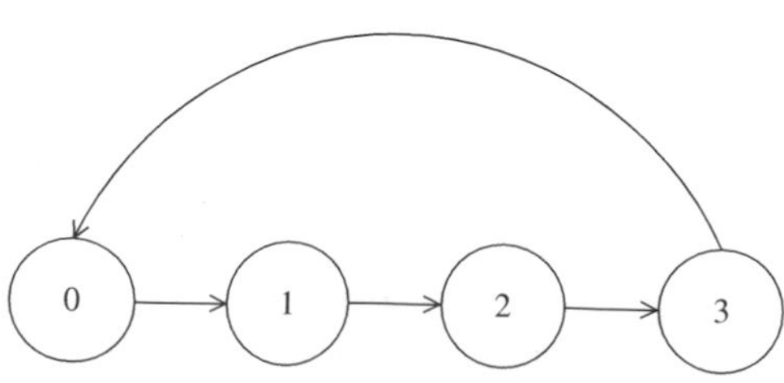

Assume there is a uniform self-stabilizing algorithm for this ring. In a distributed system, the state of a machine/process is changed depending on the state of its neighbors. In this example, if all processors have the same state when started, all must have privileges because there must be at least one privilege in the system (property 1 of a legal state).

Note that 0 and 2 make a move (because if one makes a move, it does not affect the neighbors of the other), and change their states. In this example, 0 and 2 make an independent set. After the transition, 0 and 2 are in the same state and so are 1 and 3.

The system is partitioned into two sets: {0,2} and {1,3}. At least two machines must have a privilege because 0 and 2 have the same states and also their neighbors 1 and 3 have the same states. Thus once again, machines 0 and 2 can make moves and leave the network in a similar situation. The scenario with 1 and 3 is also the same, they both are in the same state and their neighbors 0 and 2 are in the same state. So, if 1 has privilege, then 3 will also have privilege and both machines can make moves and leave the network in a similar situation. So, in either case, there will be two privileged machines at any time in the network.

Even though uniformity is a desirable property, most self-stabilizing algorithms are non-uniform (i.e., they use at least one exceptional machine). However, uniformity is sometimes attainable. For example, Burns and Pachl [21] developed a uniform self-stabilizing algorithm for a ring of n processors, where n is prime. However, it was observed that for a ring of composite size, the algorithm failed only because it could deadlock. Thus if deadlocks can be tolerated or can be corrected easily, then the algorithm may be useful. Thus the uniformity may be achieved if we are willing to sacrifice the property of self-stabilization.

17.4.3 Central and distributed demons

Generally, the presence of a central demon is assumed in self-stabilizing algorithms. For example, Dijkstra assumed a central demon to decide which machine with a privilege will make the next move. However, the presence of a central demon is an undesirable constraint. In a self-stabilizing algorithm with a distributed demon, each privileged machine makes its own decision whether to make a move. Clearly, a distributed demon is more desirable in distributed systems. In a self-stabilizing system without a central demon, each machine makes a decision unilaterally and decisions of machines eventually take the system towards a global goal. When this global goal is achieved, the system is self-stabilized.

Interestingly, many early algorithms (e.g., Dijkstra's three-, four-, and K-state algorithms) were developed assuming the presence of a central demon and they did not deal with the possibility of having a distributed demon, yet these algorithms also work with distributed demons.

Even though a central demon is not a desirable feature to have, the presence of a central demon considerably simplifies the verification of a weak correctness criterion of a self-stabilizing algorithm. Consequently, self-stabilizing systems are often developed and verified for the weak correctness assuming the presence of a central demon and after the weak correctness is verified, the system is examined to see if it is still self-stabilizing when the assumption of a central demon is removed.

Burns *et al.* [22,24] examine the extensibility of some algorithms. They showed that letting all machines operate simultaneously will not affect the correctness of some algorithms. Such interleaving assumption is very useful in the verification of self-stabilizing systems. As an example, Burns *et al.* [24] verified that Dijkstra's algorithms are correct even in the presence of a distributed demon. Dijkstra's algorithms were originally proven to be correct in the presence of a central demon. Burns *et al.* [24] showed that the central demon assumption is not necessary for the three and the four state algorithms. The K-state solution is valid for a distributed demon only if $K > n$ (n is the number of machines), because a cycle of illegal global states occurs if $K = n$.

Burns *et al.* [24] also developed results, which can be used to show that an algorithm that is correct in the presence of a central demon is also correct when the central demon assumption is removed. This is useful in the verification process because once the algorithm is verified in the presence of a central demon, the algorithm may be correct even when the central demon assumption is lifted without any modification to the algorithm. This of course may not be the case for all algorithms, but these results can be helpful in the process of verification.

17.4.4 Reducing the number of states in a token ring

A natural question is: what is the number of states of a machine to achieve self-stabilization in various configurations? Clearly, the objective is to minimize the number of states of a machine for efficient implementation.

It has been shown that if self-stabilization is not a requirement, then there exists an asymmetric token ring with two states per machine. In a self-stabilizing token ring with a central demon and deterministic execution, Ghosh [49] showed that a minimum of three states per machine is required. However, for a non-ring topology, the number of states can be reduced to two per machine. There exists a non-trivial self-stabilizing system with two states per machine [47]. It requires a high degree of atomicity in each action. Each non-exceptional process reads from three of its neighbors. Thus, obviously, the topology is non-ring.

Herman [59] presented a unidirectional and symmetric solution, with only two states, which showed that, for a "probabilistically" self-stabilizing synchronous token ring with randomized actions, a solution requiring two states per machine exists. Flatebo and Datta [41] developed a two-state, unidirectional and asymmetric solution for a "probabilistically" self-stabilizing token ring with randomized actions under the assumption of a randomized central demon. With a randomized central demon, a demon is chosen randomly among privileged machines, and this minimizes the problem of malicious scheduling on the part of the demon.

Thus, it appears that to obtain self-stabilizing systems with two states per machine, we must either relax the objective to "probabilistic self-stabilization" using randomized actions, or use a non-ring topology with higher atomicity in the actions.

17.4.5 Shared memory models

Self-stabilizing algorithms have also been developed for distributed systems with shared memory where processes communicate with each other by reading and writing to shared registers. In this type of model, no processor has direct access to the state of its neighbors, and the only way to determine the state is by passing information through shared registers. If two processors, P_i and P_j, are neighbors, then there are two registers, i and j, between the two nodes. To communicate, P_i writes to i and reads from j and P_j writes to j and reads from i. It is convenient to represent a distributed system by a graph in which each processor is represented by a node and the neighboring nodes are connected by a link that shows the communication between a node and its neighbors.

The self-stabilizing algorithms work for an arbitrarily connected graph. They also work if the system is dynamic and the graph changes during execution (due to a node failure, etc.). In a self-stabilization algorithm, eventually only one process can change a register at any instance, and this happens when the system is stabilized. It is assumed that all read/write operations on the registers are atomic. Later in this chapter, we study a dynamic self-stabilizing algorithm.

Dolev *et al.* [40] present a dynamic self-stabilizing algorithm for mutual exclusion. The algorithm only requires that all nodes be connected (that is, the network should not be partitioned). Node failures may cause an illegal global state, but the system again converges to a legal state. Thus, the protocol is dynamic and self-stabilizing. If a node is restarted, an illegal global state may again occur, but the system will automatically correct itself. The size of the registers are on the order of log (n), where n is the number of processors. The only assumption made is that the read/write operations on the registers are atomic, which is a weak assumption but makes the implementation of the algorithm feasible.

17.4.6 Mutual exclusion

In previous sections, we discussed self-stabilizing systems where there is only one action (e.g., changing a state or the contents of a register) being done after a finite amount of time. In a mutual exclusion algorithm, each process has a critical section of code, and only one process can enter its critical section at any time, and every process that wants to enter its critical section, must be able to enter its critical section in finite time. If a process has a privilege, it can enter its critical section, and once it is finished (execuing the critical section), it passes the privilege to the neighbor. If the process does not want to enter its critical section, it simply passes the privilege to its neighbor. Since the self-stabilizing algorithms mentioned adhere to the four properties discussed previously, mutual exclusion is also satisfied. Since eventually, there is only one privilege in the system and each process enjoys a privilege an infinite number of times, a process is guaranteed to enter its critical section in finite time.

A self-stabilizing mutual exclusion system can also be described in terms of a token system [42], which has the processes circulating tokens. If a process has one of these tokens, it is allowed to enter its critical section. Brown *et al.* [20] used this system to develop self-stabilizing mutual exclusion systems. Initially, there may be more than one token in the system, but after a finite time, only one token exists in the system which is circulated among the processes. Such systems are easier to implement in circuits, and Brown *et al.* showed how the implementation is done using flip-fops. All of the models, token systems, privileges, and shared memory are forms of mutual exclusion, and the algorithms also tolerate node failures and restarts or a bad initialization. So these algorithms are more tolerant of errors than other mutual exclusion algorithms [42].

17.4.7 Costs of self-stabilization

The definition of self-stabilization does not put any upper bound on the number of transitions required by the system to reach a safe state starting from an unsafe one. Thus, the system might remain in an unsafe state for a considerable amount of time before reaching a safe state. A study and assessment of these cost factor is very important in any practical implementation.

Gouda and Evangelist [53] introduced the following two concepts related to the cost of self-stabilization:

- **Convergence span** The maximum number of transitions that can be executed in a system, starting from an arbitrary state, before it reaches a safe state.
- **Response span** The maximum number of transitions that can be executed in a system to reach a specified target state, starting from some initial state. The choice of initial state and target state depends upon the application.

Clearly, the aim of the designer of a self-stabilizing algorithm is to reduce the convergence span and response span.

Time-complexity measure for self-stabilizing algorithms is the number of rounds. In synchronous models, algorithms execute in rounds, i.e., processors execute steps at the same time and at a constant rate. Rounds can be defined in asynchronous models too, where the first round ends in a computation when every processor has executed at least one step. In general, the ith round ends when every processor has executed at least i steps. Generally, communication between any two processors in a particular system takes at least d rounds. This is because it normally takes at least one round to propagate information between two adjacent processors.

17.5 Methodologies for designing self-stabilizing systems

Having seen the issues in the design of self-stabilizing system, let us now discuss the methodologies for designing self-stabilizing systems.

Self-stabilization is characterized in terms of a "malicious adversary" whose objective is to disrupt the normal operation of the system. This adversary (e.g., a virus or a hardware problem) may destroy some portions of the system, or disrupt the operation of one or more portions. Furthermore, it might not be possible for a system to detect that it has been "attacked," as soon as the attack appears. To be called self-stabilizing, a system must have the capability to recover normal operation when exposed to such attacks. If the system (or parts of it) is destroyed completely, so that it is no longer possible for the system to operate, then no self-stabilizing system can work. The adversary succeeds in achieving his goals. However, if enough components are left for the system to operate, then a self-stabilizing system will slowly resume normal operation after the attack. It is up to the designer to decide under what conditions the system may be termed "completely destroyed" or "still capable of operating."

17.5.1 Layering and modularization

The most commonly used techniques for building self-stabilizing systems are layering and modularization. The basic idea is to divide the system into smaller components, make each component self-stabilizing independently, and then integrate them to compose the system.

Self-stabilization is amenable to layering because the self-stabilization relation is transitive, i.e. if $P \rightarrow Q$ (P stabilizes Q) and $Q \rightarrow R$, then $P \rightarrow R$. Thus, different layers of self-stabilizing programs (each by itself self-stabilizing) can be composed. First step is to build a self-stabilizing "platform" and any program written on that platform automatically becomes self-stabilizing. The basic idea behind a self-stabilizing platform is to provide primitives that can be used to write other programs.

To develop self-stabilizing systems using the technique of layering, we require primitives to provide structures on which algorithms may be built.

There are two basic structuring mechanism primitives: common clock primitives and topology-based primitives.

Common clock primitives

Unison is the process of maintaining time through the use of local clocks in shared memory systems. The properties required here are the safety property and the progress property. For a synchronous shared memory system, the safety and progress properties for unison are as follows:

- **Safety** All clocks have the same value.
- **Progress** At each step, each clock is incremented by the same amount.

For asynchronous systems with shared memory, the safety and progress properties for unison are as follows:

- **Safety** Clocks of two neighboring nodes can differ by at most 1.
- **Progress** A clock is incremented to $i+1$ when clocks at all neighboring nodes have value i or $i+1$.

Topology-based primitives

Leader election is perhaps the most basic primitive with respect to an arbitrary dynamic topology. Once a leader has been found, a spanning tree might be constructed. Algorithms for mutual exclusion and distributed reset can be easily developed on top of self-stabilizing spanning-tree algorithms for arbitrarily connected graphs.

We now discuss two examples of self-stabilizing programs, namely, mutual exclusion and reset, developed using the concept of layering.

Example A two-layered self-stabilizing algorithm for mutual exclusion [40]. The first layer creates a spanning tree from an arbitrarily connected graph, whose topology might change dynamically with the exception of a distinguished process (the root). The self-stabilizing spanning tree protocol is based on a breadth-first search of the graph, rooted at the distinguished node. The distinguished node is needed to break symmetry and all other nodes execute an identical program.

The second layer achieves mutual exclusion on a dynamic tree structured system. It is a token-based system. When a node receives the token/privilege, it executes its critical section (if it wants to) and then it passes to the token to its children in left-to-right order. Thus, the token traverses the tree in a depth-first manner.

Finally, the two protocols are superposed to obtain a single self-stabilizing protocol for mutual exclusion on an arbitrarily connected graph.

Example A self-stabilizing reset algorithm for an asynchronous shared memory system [12].

Arora and Gouda [12] used a layering technique to develop a self-stabilizing reset algorithm for asynchronous shared-memory systems. The algorithm allows dynamic topology as long as the underlying graph remains connected. There is no distinguished process, however, each process has a unique identifier.

The algorithm consists of three layers. In the first layer, a root is elected forming a spanning tree. In the second layer, the root initiates a diffusing computation in which reset requests are propagated to the leaf nodes and are reflected back to the root node. The reset request passes through every node, detecting any anomaly in the global state. When the reset returns to the root, the reset is complete.

A self-stabilizing "platform" resets the system upon encountering an illegitimate state. The platform writes to variables of the original program only if an illegitimate state is detected. The platform does not affect the original program under normal execution.

17.6 Communication protocols

A communication protocol is a collection of processes that exchange messages over communication links in a network. A protocol may be adversely affected for several reasons:

- Initialization to an illegal state.
- A change in the mode of operation. Not all processes get the request for the change at the same time, so an illegal global state may occur.
- Transmission errors because of message loss or corruption.
- Process failure and recovery.
- A local memory crash which changes the local state of a process.

Previously, these five types of errors have been treated separately. However, if a protocol is self-stabilizing, they will all be corrected in a finite number of steps, regardless of the reason for the loss of coordination.

A communication protocol is stabilizing if and only if starting from any unsafe state (i.e., one that violates the intended invariant of the protocol), the protocol is guaranteed to converge to a safe state within a finite number of state transitions. Stabilization allows the processes in a protocol to reestablish coordination between one another whenever coordination is lost due to some failure.

Gouda and Multari [55] showed that a communication protocol must satisfy the following three properties to be self-stabilizing:

- It must be non-terminating.
- There are an infinite number of safe states.
- There are timeout actions in a non-empty subset of processes.

Self-stabilizing systems can automatically recover from arbitrary state perturbations in finite time. They are therefore well-suited for dynamic, failure-prone environments. The spanning-tree construction in distributed systems is a fundamental operation that forms the basis for many other network algorithms (like token circulation or routing). Next, we discuss some self-stabilizing algorithms that construct a spanning tree within a network of processing entities.

Let us consider an arbitrary distributed algorithm, e.g., for termination detection, and start it in a state where one of its variables has been set to a random value from its domain. Usually, the behavior is not predictable: either the algorithm will output garbage (e.g., declare a computation as finished although it is still running), or (most probably) it will deadlock (e.g., it will fail to output anything at all). It may be argued that changing the value of a variable is unfair: no algorithm can tolerate such manipulations since algorithms have to rely on proper initialization. This argument, however, is not true.

Self-stabilizing algorithms are guaranteed to recover from an arbitrary perturbation of their local state in a finite number of execution steps. This means that the variables of such algorithms do not need to be initialized properly. If we assign each variable an arbitrary value from its domain, the algorithm will eventually start to behave as expected. Arbitrary state perturbations can also happen without curious users playing around with their algorithm: cosmic rays in spacecraft, for example, can arbitrarily change the contents of memory cells in random access memory. Self-stabilizing algorithms have the desirable property of being able to recover from such faults automatically.

17.7 Self-stabilizing distributed spanning trees

In distributed systems, a spanning tree is the basis for many complex distributed protocols. To define a spanning tree, the network is modeled as a graph $G = (V, E)$, where V is the set of network nodes (vertices) and E is the set of communication links (edges) between network nodes (formally, it is a subset of $E \times E$). A spanning tree $T = (V, E')$ of G is a graph consisting of the same set of nodes V, but only a subset E' of edges E such that there exists exactly one path between every pair of network nodes. Basically, this means that the graph is connected and does not contain cycles. A basic theorem of spanning trees states that, in a network of n nodes, the tree contains exactly $n-1$ communication links. A spanning tree of a graph is in general not unique (even if the root node is fixed). Figure 17.3 shows an example of a network of five nodes and a spanning tree of the network [43].

A spanning tree in a network is often a prerequisite for more involved network protocols like routing or token circulation. It generally increases the efficiency of network protocols. For example, consider the problem of broadcasting messages in the network. There are algorithms that flood the network,

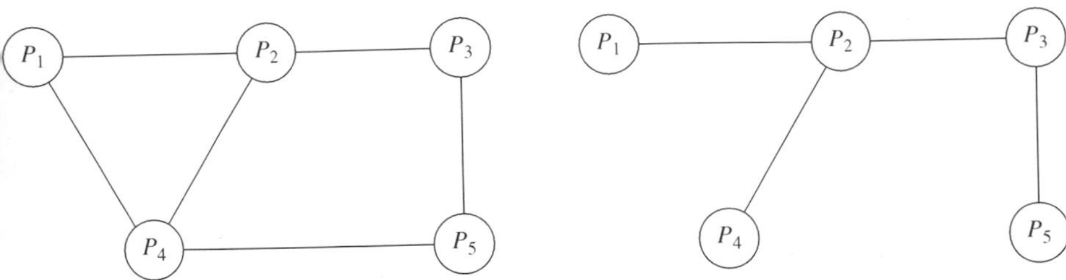

Figure 17.3 A network and its spanning tree [43].

i.e., the broadcast message is recursively sent to all neighbors. Consequently, the message crosses all communication links before the protocol terminates. However, if a spanning tree of the network is available, the message only needs to be sent along all the edges of the spanning tree. Instead of crossing all E links, the message just crosses $n-1$ links. Since $|E|$ is usually significantly larger than $n-1$, a spanning tree can considerably reduce the message complexity of the broadcast algorithm.

Two kinds of spanning trees may be distinguished: breadth-first search (BFS) trees result from a breadth-first traversal of the underlying network topology. Similarly, depth-first search (DFS) trees are obtained from a depth-first traversal. A notion underlying both DFS and BFS trees is that of a rooted tree. A rooted spanning tree is a spanning tree of the network where the tree edges are consistently directed with respect to a particular node (the root). Edges can be directed towards the root or "away from" the root. Rooted spanning trees have a notion of "parent" and naturally result from the execution of semi-uniform algorithms. In fact, since almost all algorithms use a single pointer (to a neighbor or the parent) to store the structures of the tree, all these algorithms implicitly construct a rooted spanning tree.

In spanning-tree construction, it is impossible to deterministically construct a spanning tree in uniform networks. Intuitively, this is caused by problems of symmetry, and so at least a semi-uniform setting (e.g., a distinguished root processor) or a source of randomization is needed.

The time-complexity of self-stabilizing algorithms is often measured by the number of rounds. In self-stabilizing spanning-tree construction, an arbitrary initial state may make it necessary to propagate information through the entire network. Therefore, a general lower bound of d rounds can be assumed for self-stabilizing spanning-tree algorithms. By combining the algorithm with a hierarchical structure and sacrificing true distribution, this bound can be lowered.

Space complexity measures the amount of state necessary to perform self-stabilizing spanning-tree construction. Dolev *et al.* [38] derived the following result on lower bounds regarding the space complexity: self-stabilizing spanning-tree construction needs at least $\log n$ bits per processor if the algorithm is silent (i.e., if the contents of the communication registers eventually stop changing). If the algorithm is not required to be silent, Johnen [66] showed that it is possible to construct an algorithm using only $O(1)$ bits per edge in a uniform rooted network with a central demon.

17.8 Self-stabilizing algorithms for spanning-tree construction

In this section, we discuss a set of representative self-stabilizing algorithms for constructing spanning trees [43].

17.8.1 Dolev, Israeli, and Moran algorithm

Dolev *et al.* [40] developed a self-stabilizing BFS spanning-tree construction algorithm for semi-uniform systems with a central demon under read/write atomicity. In the algorithm, every node maintains two variables: (i) a pointer to one if its incoming edges (this information is kept in a bit associated with each communication register), and (ii) an integer measuring the distance in hops to the root of the tree. The distinguished node in the network acts as the root.

The algorithm works as follows: the network nodes periodically exchange their distance value (current distance from the root node) with each other. After reading the distance values of all neighbors, a network node chooses the neighbor with minimum distance *dist* as its new parent. It then writes its own distance into its output registers, which is *dist* + 1. The distinguished root node does not read the distance values of its neighbors and always sends a value of 0.

The algorithm stabilizes starting from the root process. After sufficient activations of the root, it has written 0 values into all of its output variables. These values will not change anymore. Note that without a distinguished root process the distance values in all nodes would grow without bound. More specifically, after reading all neighbors values for k times, the distance value of a process is at least $k+1$. This means that, after the root has written its output registers, the direct neighbors of the root – after inspecting their input variables – will see that the root node has the minimum distance of all other nodes (the other nodes have distance at least 1). Hence, all direct neighbors of the root will select the root as their parent and update their distance correctly to 1. This line of reasoning can be continued incrementally for all other distances from the root. That is, after all nodes at distance d from the root have computed their distance from the root correctly and written it in their registers, their registers no longer change and nodes at distance $d+1$ from the root are ready to compute their distance from the root. After $O(\delta)$ update cycles, the entire tree will have stabilized.

```
Variables:

no_neighbors = number of processor's neighbors
i = the writing processor
m = for whom the data is written
lr_ji (local register r_ji) = the last value of r_ji read by P_i

Root Node:
{do forever}
while TRUE do
    for m := 1 to no_neighbors do
        write lr_im := ⟨0, 0⟩
    end
end

Other Nodes:
{do forever}
while TRUE do
    for m := 1 to no_neighbors do
        lr_mi := read(lr_mi)
        FirstFound := false
        dist := 1 + min(lr_mi.dist) ∀ m: 1 ≤ m ≤ no_neighbors
        for m := 1 to no_neighbors do
            if not FirstFound and lr_mi.dis = dist − 1 then
            write r_im := ⟨1, dist⟩
            FirstFound := true
            else write r_im := ⟨0, dist⟩
        end
    end
end
```

Algorithm 17.2 Dolev *et al.*'s spanning-tree construction algorithm for P_i [40].

Dolev *et al.*'s self-stabilizing algorithm for constructing spanning trees is shown in Algorithm 17.2. Two neighbors P_i and P_j communicate with each other by reading from and writing to two shared registers, r_{ij} and r_{ji}. To communicate, P_i writes to r_{ij} and reads from r_{ji} and P_j writes to r_{ji} and reads from r_{ij}.

The root node repeatedly writes values $\langle 0, 0\rangle$ in the registers of all of its neighbors. All other processors repeatedly perform the following steps: in each iteration, the processor reads the registers of all of its neighbors and computes the a value for variable *dist* as follows: it chooses the minimum distance of their neighbors, sets its *dist* variable to the minimum distance plus 1, and

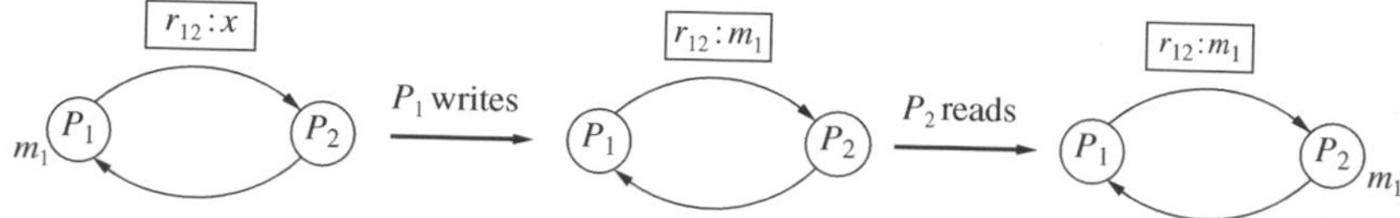

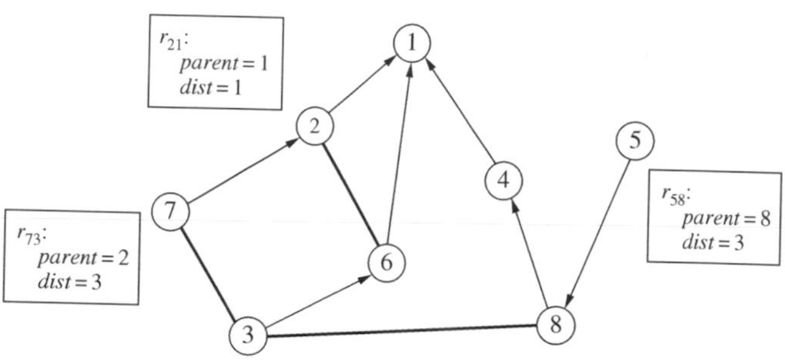

Figure 17.4 An example of Dolev *et al.*'s algorithm.

updates the registers of its neighbors. The internal variable corresponding to register r_{ij} is denoted by lr_{ij}. It stores the last value of r_{ji} that is read by P_i.

A snapshot of the system state in Dolev *et al.*'s self-stabilizing algorithm is given in Figure 17.4. This algorithm has been used as the basis for a topology update algorithm in dynamic networks. Based on a similar idea, Collin and Dolev [31] present a semi-uniform spanning-tree algorithm under a central demon and read/write atomicity that constructs a DFS tree (instead of a BFS tree). A similar algorithm, which also constructs a DFS tree but uses composite atomicity, was developed by Herman. In this algorithm, the outgoing links at every process are ordered, and the DFS tree is defined as the tree resulting from a DFS graph traversal always selecting the smallest outgoing edge. Instead of writing its current level into the output registers, it writes a representation of its current estimate of the path (the sequence of outgoing link identifiers) to the root. The root repeatedly writes the "empty path" ($\perp$) to its output registers. If a node has k neighbors, there are k alternative paths to choose from. From these, the node chooses the path that is minimal according to a lexicographic order that prefers smaller link identifiers. For example, $(\perp) < (\perp, 1) < (\perp, 1, 1) < (\perp, 2) < (1)$. Thus, a node does not choose the shortest path to the root but a path along the smallest link identifiers.

The memory requirement for the DFS algorithm is $O(n \log K)$ bits, where K is an upper bound on the maximum degree of a node. The time complexity is $O(\delta n K)$ rounds.

17.8.2 Afek, Kutten, and Yung algorithm for spanning-tree construction

The algorithm by Afek *et al.* [3] constructs a BFS spanning-tree in the read/write atomicity model. However, this algorithm does not make the assumption of a distinguished root process. Instead, it assumes that all nodes have globally unique identifiers that can be totally ordered. The node with the largest identifier will eventually become the root of the tree.

The idea of the algorithm is as follows: every node maintains a parent pointer and a distance variable as in the algorithm by Dolev, Israeli, and Moran algorithm. In addition, it stores the identifier of the root of the tree in which it thinks it is present. Periodically, nodes exchange this information. If a node notices that it has the maximum identifier in its neighborhood, it makes itself the root of its own tree. If a node learns that there is a tree with a larger root identifier nearby, it joins this tree by sending a "join request" to the root of that tree and receiving a "grant" back from the root. Local consistency checks ensure that cycles and fake root identifiers are eventually detected and removed.

The algorithm stabilizes in $O(n^2)$ asynchronous rounds and needs $O(\log n)$ space per edge to store the process identifier. Afek *et al.* argued that this is optimal since message communication buffers need to communicate "at least" the identifier.

17.8.3 Arora and Gouda algorithm for spanning-tree construction

Arora and Gouda [12] developed a self-stabilizing BFS spanning-tree algorithm for the composite atomicity model under the assumption of a central daemon. Like Afek *et al.*, they also assume unique identifiers and the node with maximum identifier eventually becomes the root of the system. However, the algorithm needs a bound N on the number n of nodes in the network to work correctly. The bound on the number of nodes is necessary because the algorithm uses a different technique to detect and remove cycles.

Every node maintains variables for distance, parent, and root identifier. Periodically, every node compares its own distance and root identifier values with the values stored in the node pointed to by the parent variable. In the final spanning tree, the root identifiers should be identical and the distance should be the distance of the parent plus one. If this is not the case, the root identifier is copied from the parent and the distance is set to the parent's distance plus one. A node also continuously monitors the root identifier and distance settings of its neighbors. If a neighbor has a larger root identifier or the same identifier with smaller distance, the node adjusts its values accordingly.

Cycles are detected in the following manner: if there is a cycle in the tree (or the graph to be precise), say, due to improper initialization, the distance values are incremented along this cycle without bound. Hence, a cycle is detected when the distance value exceedes the bound N. The first node to detect this makes itself the root of a new tree.

A bound on the number of nodes in the network, N, allows the Arora and Gouda algorithm to be simpler than the one by Afek *et al.*. However, the stabilization time in Arora and Gouda algorithm is $O(N^2)$, which can be much larger than that of Afek *et al.*, $O(n^2)$. In dynamic networks where network nodes may go down, a stabilization time in the order of the actual number of nodes is preferable.

17.8.4 Huang *et al.* algorithms for spanning-tree construction

Chen *et al.* [29] developed a self-stabilizing spanning tree algorithm for semi-uniform systems with composite atomicity. It is based on the same idea of cycle breaking (bumping up the distance counter). The fact that there is a distinguished root makes the algorithm even simpler than the one by Arora and Gouda [12]. However, the algorithm does not necessarily stabilize to a BFS tree since the choice of a new parent after a cycle is broken is non-deterministic and is governed by the scheduler.

This algorithm was later improved by Huang and Chen [63] to yield an algorithm which constructs a BFS tree using knowledge of the size n of the network.

17.8.5 Afek and Bremler algorithm for spanning-tree construction

Afek and Bremler [6] gave a self-stabilizing algorithm for constructing spanning trees for systems with unidirectional, bounded capacity communication links. They assumed node have unique identifiers and adopted the algorithm for the synchronous and the asynchronous networks. The network node with the smallest identifier eventually becomes the root of the spanning tree.

The algorithm is based on a new idea called "power supply." The power supply method exploits the fact that self-stabilizing algorithms must continuously check their own state. Nodes that are part of a spanning tree expect to receive "power" from the root of the tree. Power, like electric current, means a continuous flow of certain messages, one per round. The basic idea is that only legal roots may be the source of power and nodes attached to fake roots eventually fail to receive power and subsequently make themselves the root of a new tree.

Whenever a node receives power from a neighbor with a smaller identifier, it attaches itself to its tree. In the asynchronous case, the power supply idea is implemented using different types of messages: weak messages are exchanged periodically between the nodes to synchronize their states, while strong messages carry power.

The idea of power supply imparts the algorithm several interesting features. For example, the algorithm stabilizes in $O(n)$ rounds without processes to have the knowledge of n. Afek and Bremler gave a generic power supply algorithm which can be instantiated to a leader election algorithm, or an algorithm to construct DFS or BFS spanning trees.

The spanning-tree algorithms discussed in this section have been applied in many different settings in practice. For example, a variant of the algorithm by Dolev *et al.* [40] was used to implement a reliable data storage subsystem for the self-stabilizing file system developed at the Ben Gurion University [68].

As another example [43], consider the protocol to eliminate redundant paths in switched Ethernets [30]. If a network segment becomes unreachable or network parameters are changed, the protocol automatically reconfigures the spanning-tree topology by activating a standby path. The protocol can be briefly described as follows: initially, switches believe they are the root of the spanning tree but they do not forward any packets. Based by a timer, they regularly exchange status information. The status information contains (i) the identifier of the transmitting switch (usually a MAC address), (ii) the identifier of the switch which is believed to be the root of the tree, and (iii) the "cost" of the path towards the root. A switch uses this information to choose the shortest path towards the root. If there are multiple possible roots, it selects the root with the smallest identifier (lowest MAC address). Links that are not included in the spanning tree are placed in blocking mode and do not forward packets, but still transport status information.

17.9 An anonymous self-stabilizing algorithm for 1-maximal independent set in trees

In a distributed system, an independent set is defined as a large subset of nodes that are pair-wise non-adjacent. A *maximal independent set* is a set of nodes such that every node not in the set is adjacent to a node in the set. Maximal independent sets are important in several distributed network applications and several parallel or distributed algorithms have been developed for this task [73]. A 1-maximal independent set is a maximal independent set provided one cannot increase the cardinality of the independent set by removing one node and adding more nodes.

In this section, we discuss Shi *et al.*'s [81] self-stabilizing algorithm for finding a 1-maximal independent set in a tree. The algorithm uses a constant space at each node. The algorithm is somewhat unusual in that it stabilizes on all graphs, but it is only guaranteed to be correct on some graphs.

A distributed system is modeled as a connected, undirected graph G with node set V and edge set E. Two nodes joined by an edge are said to be *neighbors* and $N(i)$ is used to denote the set of neighbors of node i. A self-stabilizing algorithm is presented as a set of rules, each with a Boolean predicate and an action. A node is said to be *privileged* if the predicate is true. If a node becomes privileged, it may execute the corresponding action called a *move*. The assumption is that there exists a *central demon*, which at each time-step selects one of the privileged nodes to move (and thus two nodes never move at the same time). When no further move is possible, the system is said to be in a *stable configuration*. While the definition of self-stabilizing is normally more general, since this is a graph algorithm we say that an algorithm is self-stabilizing if from any initial configuration it always terminates in a legitimate stable configuration after a finite number of moves no matter the selections of the daemon.

Description of algorithm

In Shi *et al.*'s algorithm [81] for finding a 1-maximal independent set in a tree, each node is in one of a finite number of distinct states. Those nodes in the state called 0 will end up being in the desired set: let us call this set M. A node with no neighbor in state 0 will change to state 0 and a node in state 0 with a neighbor in state 0 will change to something else. This idea readily produces a maximal independent set.

To achieve 1-maximality, however, a node must be able to leave set M when that would allow two of its neighbors to enter M. Available neighbors are those which have no other neighbor in state 0: these will be in the state called 1. In order to allow this *interchange*, we implement a hand-shaking process: the node offers to leave M by changing to state $0'$, the relevant neighbors agree to enter M by changing to state $1'$, the node leaves by changing to state $2'$, and then the relevant neighbors go in by changing to state 0.

Specifically, the set of states is $0, 0', 1, 1', 2, 2'$. The states with a prime are *transition states*, used in the hand-shaking process described above. These transition states will be absent when the algorithm terminates if the network satisfies certain properties.

For the purpose of the algorithm, the nodes with state $0'$ are also considered to be in M. The states 0, 1, and 2 are used to indicate that a node has zero, one, or at least two neighbors in M, respectively.

Actions of a node in the algorithm can be summarized as follows:

1. If not involved in a transition process, then set state to the number of neighbors in state 0 or state $0'$. (The value 2 is used to indicate two or more such neighbors.)
2. If in state 0 and adjacent to at least two 1s, change state to $0'$.
3. If in state 1 and adjacent to a $0'$, change state to $1'$.
4. If in state $0'$ and adjacent to at least two $1'$s, change state to $2'$.
5. If in state $1'$ and adjacent to no $0'$, change state to 0.
6. If in state $2'$ and adjacent to no $1'$, change state to 2.

The complexity of the actual algorithm arises from invalid initial states and from two interchanges affecting one another.

For a state y, we use the notation $S(y)$ to represent the set of nodes in state y. Furthermore, we use the notation $S(y1/y2/y3/\ldots)$ to denote $S(y1) \cup S(y2) \cup S(y3) \ldots$.

For example, the notation $S(0)$ denotes the nodes in state 0. The state of a node is stored in a local variable denoted by x. The states with a prime are transition states. We will also identify the prime with a virtual flag – we will say the flag is set when the node is in a transition state, and clearing the flag will mean changing from state i' to state i.

To define the rules of the algorithm, we define the following function f. Let i be a node and t a state. Then we define $f_i(t) = \min\{2, |N(i) \cap S(t)|\}$.

The function f_i gives the number of neighbors of node i in a specified state. We further use the notation: $f_i(x/y/z/\ldots) = \min\{2, f_i(x) + f_i(y) + f_i(z) + \ldots\}$ When the node i is clear from the context, we will drop the subscript from f_i. We also utilize the concept of *bad edge*, which is defined next. The rules will be such that a bad edge can only occur as a result of faulty initialization. A bad edge is an edge connecting two nodes in the following list of pairs of states: 0–0, 0–0′, 0′–0′, 0′–2′, 1′–1′, and 2′–2′.

The complete algorithm for finding a 1-maximal independent set is given in Algorithm 17.3.

{* All moves are tried in the listed order *}

V1: **if** flag is set **and** node is incident on bad edge
  and after clearing flag node would not be incident on any bad edge
  then clear flag

V2: **if** flag is clear **and** $x' = f(0/0')$ **and** $(f(0/0') \geq 1$ **or** $f(1'/2') = 0)$
  then set $x = f(0/0')$

C1: **if** $x = 0$ **and** $f(1) = 2$ **and** $f(0/0'/2') = 0$
  then set $x = 0'$

C2: **if** $x = 0'$ **and** $(f(1/1') \leq 1$ **or** $f(0/0') \geq 1)$
  then set $x = f(0/0')$

C3: **if** $x = 0'$ **and** $f(1') = 2$ **and** $f(0'/2') = 0$
  then set $x = 2'$

C4: **if** $x = 2'$ **and** $f(1') = 0$
  then set $x = f(0/0')$

C5: **if** $x = 1$ **and** $f(0') = 1$ **and** $f(0/1'/2') = 0$
  then set $x = 1'$

C6: **if** $x = 1'$ **and** $(f(0') \neq 1$ **or** $f(0/1'/2') \geq 1)$
  then set $x = f(0/0')$

Algorithm 17.3 Shi *et al.*'s algorithm for finding a 1-maximal independent set [81].

The algorithm converges in $O(mn)$ time, where m is the number of edges and n is the number of nodes in the network. The algorithm stabilizes to a 1-maximal independent set in $O(n^2)$ steps in an arbitrary tree.

Having seen two regular self-stabilizing algorithms, let us now discuss a probabilistic self-stabilizing algorithm.

17.10 A probabilistic self-stabilizing leader election algorithm

We now discuss a probabilistic self-stabilizing leader election algorithm by Dolev *et al.* [35]. The distributed system consist of n stations (sites) and they need to choose a leader among themselves by using a leader election algorithm. The following three possibilities arise: during a time unit, stations can detect either silence, success, or collision.

Silence in the system implies that no station tried to transmit a message. Success implies that only one station used the channel to transmit a message, and finally, a collision implies that at least two stations attempted to transmit messages.

The leader election algorithm is shown in Algorithm 17.4.

{**Termination condition**}
If $n = 1$ then Stop.
{**Randomized selection process**}
If $n \geq 2$ then randomly divide n into $(n_1, n - n_1)$.
If $n_1 \neq 0$ then Apply $d(n_1)$.
Else Apply $d(n)$.

Algorithm 17.4 Dolev *et al.*'s leader election algorithm $d(n)$.

If the number of stations is greater than or equal to two, then, in the first time unit, all the stations send their messages via the channel and, as a result, a collision occurs. The first time unit is nothing but the first instance of a time unit.

If we take a station S into consideration, in the next time unit, there are two possibilities:

Case 1 S tries to send the message again, or
Case 2 S does not try to send the message again.

There are two possibilities for the first case (S tries to send the message again): success or collision. If result is a success, then S is the leader, else (a collision occurs) S flips a coin (send/not send).

For the second case (S does not participate), there are three possibilities: silence, success, or collision. If silence occurs, then the station S flips the coin (send/not send) again, if success occurs, then station S detects the leader, and if a collision occurs, S is eliminated.

Thus, the algorithm can be written as shown in Algorithm 17.5.

---

n stations, $n \geq 2$
First time unit: All stations send their messages via the channel;
  Collision → Each station flips a coin (send or not send) again.
Next time unit:
  Case 1: Station S tries again
    Success: S is the leader
    Collision: S flips a coin again
  Case 2: Station S isn't participating
    Silence: S flips again the coin
    Success: S detects the leader
    Collision: S is eliminated

Algorithm 17.5 Modified leader election algorithm.

Example We now illustrate the algorithm using the example shown in Figure 17.5. In Figure 17.5 initially (in the first time unit), ABCD try to send messages and a collision occurs. In the next time unit, A and B send message again, while C and D do not participate. Since both A and B try to send their messages, there is a collision again. As a result, C and D are eliminated. Now in the next time unit, both A and B do not participate and the result is a Silence. So both of them flip a coin and B decides to send a message again and A decides not to participate. Since B is the only one sending a message, the result is Success and B is the leader and the algorithm terminates.

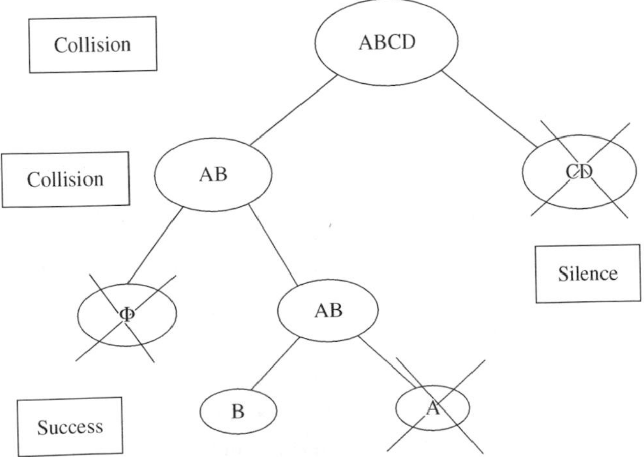

Figure 17.5 An example of the leader election algorithm.

17.11 The role of compilers in self-stabilization

A compiler converts a program written in a language into an equivalent program in another language. Typically, the latter is an object program that is to be run on a particular architecture. Formally, a compiler is a homomorphism $f: A \rightarrow B$, where A and B are two classes of architecture or systems. Then, for each $m \in A$, $f(M)$ mimics the actions of M in some well-defined fashion [79].

When a source program is self-stabilizing, we expect the compiler to produce an object program that is self-stabilizing on the target architecture. It would be highly desirable to have a "self-stabilizing compiler" that will convert a non-self-stabilizing source program into self-stabilizing object code.

It is very important for the compiler to preserve the properties of the source program that are important to the designers:

- In a sequential paradigm under termination, it is important that both the programs compute the same results (quantitative).
- In a distributed or parallel paradigm, preservation of qualitative properties due to the need for control and coordination among the processes is important.

Dijkstra [34], in his seminal work, implied that there doesn't exist a compiler from asymmetric rings to symmetric rings that forces or preserves self-stabilization. However, if self-stabilization is not required, we can compile an asymmetric ring into a symmetric ring [79].

Gouda *et al.* [53] showed that self-stabilization across architectures is in principle unstable. They also demonstrated that the ability to force or preserve self-stabilization is very much dependent on how certain properties, such as termination, fairness, and concurrency, are required to be preserved when compiling from one system to other.

Next we discuss compilers that force self-stabilization in sequential programs, asynchronous distributed systems, and shared memory systems [79].

17.11.1 Compilers for sequential programs

The main focus of research on self-stabilization has been in the domains of concurrent and distributed systems, where the goals of algorithms are both qualitative and quantitative. Achieving self-stabilization in sequential programs becomes much more difficult due to the termination requirement.

Browne *et al.* [19] and Schneider [77] suggested a rule-based program model. A rule-based program consists of an initialization section and a finite set of rules. A *rule* is a multiple assignment statement with an enabling condition, called a *guard*, which is a predicate over the variables of the program. If the guard of a rule is true for a state, then the rule is said to be *enabled*. A *computation* is a sequence of rule firings, where at each step an enabled rule is non-deterministically selected for execution. A program is said to have *terminated* when a *fixed point* is reached. A *fixed point* is a state in which the values of the variables can no longer change. A *partial fixed point* is defined as a state from which the values of a subset of variables do not change.

For a terminating program to be self-stabilizing, the relation it computes should be verifiable in one step, else it might terminate in an unsafe state. Browne *et al.* [19] showed that a class of programs exists for which there is a compiler that forces self-stabilization while preserving termination. Object programs have runtime and size within a constant factor of the source program. However, it is assumed that inputs are incorruptible. These programs must satisfy the following properties [79]:

- The data dependency graphs of these programs are acyclic.
- Each rule in the program assigns only one variable.
- For any pair of enabled rules with the same target variable, both rules will assign the same value to the variable.

For arbitrary programs, one cannot obtain the same result as for acyclic programs. Consider the class of programs restricted to boolean variables. Schneider [77] showed that if there exists a compiler that forces self-stabilization onto boolean programs, while preserving termination, then PSPACE = NP, which is not a very likely result. Further, if we require that the source and target to have the same set of variables, then PSPACE = P, which is even less likely result.

However, if we waive the requirement that the object program reach a fixed point (i.e., the condition of termination), life becomes simpler. Schneider [77,78] introduced the notion of partial fixed-points (where termination is not required) and showed that one can produce, in quadratic time, an equivalent self-stabilizing program with time complexity and size within a constant factor of the original.

17.11.2 Compilers for asynchronous message passing systems

Such a compiler should convert a non-self-stabilizing program into a self-stabilizing version in an asynchronous message passing system [69]. This is accomplished through a self-stabilizing platform, which when interleaved with a non-self-stabilizing program, yields a self-stabilizing program. The resulting program is called an extension of the original program.

The algorithm consists of three components [69]:

- A self-stabilizing version of Chandy–Lamport's global snapshot algorithm [28].
- A self-stabilizing reset algorithm that is superposed on it.
- A non-self-stabilizing program on which the former two are superposed to obtain a self-stabilizing program.

The algorithm works as follows: a distinguished initiator repeatedly takes global snapshots. After the distinguished initiator has obtained a snapshot, it evaluates a predicate[1] on the collected state. If an illegitimate global state is detected, then the initiator initiates the execution of the reset algorithm, which resets the global state of the source program to an initial state. In this methodology, the compiler takes the program and the predicate (specifying the set of safe states) as input and produces a self-stabilizing version of the program.

An extension

Informally, a program Q is an extension of program P if the subset of Q corresponding to P behaves exactly like P, except that the same state may repeat. If P terminates, its extension Q needs to repeat the final state of P forever, changing only the variables not present in P, in order to achieve self-stabilization. If Q terminates, it cannot be a self-stabilizing extension of P because it could terminate in an illegal state.

The Katz and Perry methodology [69] has the following two drawbacks: first, it might not be always possible to find a predicate that can distinguish between legitimate and illegitimate states. Legitimate states could be defined in terms of the reachable states. However, computing the reachable set might become intractable. Second, a global snapshot algorithm does not produce the current state. It captures a possible successor to the state it was initited in. If the original program stabilizes by itself, we might end up doing a reset from a legitimate state.

17.11.3 Compilers for asynchronous shared memory systems

In shared-memory systems, we can write the snapshot and reset algorithms in a way very similar to message-passing systems. A self-stabilizing synchronous shared memory system might be compiled into an asynchronous self-stabilizing shared memory system as follows [79]: assume that a process can read and write in one atomic action and that each shared variable is written by only one process (called the owner of the variable). The steps of the synchronous system are simulated using a self-stabilizing asynchronous unison algorithm. One step of the synchronous algorithm is executed each time the clock is incremented. For each shared variable, two copies are maintained: one to store the current value and another to store the previous value. This allows a process to access the previous value of a shared variable even if it has been updated by another process. When the local clock of a process ticks from i to $i+1$, it concurrently executes one step of the synchronous system and updates current and previous values of all variables that it owns.

[1] It is assumed that there exists a decidable predicate that can detect whether a global state is legitimate.

17.12 Self-stabilization as a solution to fault tolerance

Self-stabilization is the property of a system, component, process, or object to correct itself no matter how severely its state variables, including memory, message buffers, and registers, are corrupted. Self-stabilization is most interesting for distributed and concurrent systems because local detection of a faulty condition is difficult.

Self stabilization has risen beyond the theory and has served as a guiding principle in many network protocols (in fact, a number of Internet and LAN protocols are self-stabilizing or very close to it). Recent applied research has succeeded in demonstrated self-stabilizing file systems and in implementing protocols for routing, reprogramming, and synchronizing nodes in sensor networks. These examples show that the principles of self-stabilization can be used to implement lightweight solutions to the problems of fault tolerance in real-life systems.

Fault tolerance

Fault tolerance is defined as tolerance to transient failures, in which the state of a component changes spontaneously, but the component remains correct.

Fault-tolerance or graceful degradation is the property that enables a system to continue operating properly in the event of the failure of some of its components. The quality of operation may decrease in proportion to the severity of the failure, while in a naively designed system, even a small failure can cause the total system breakdown.

In a system, fault tolerance is generally achieved by anticipating exceptional conditions and designing the system to cope with them. The concept of self-stabilization has emerged as a complementary paradigm to fault tolerance in distributed computing. A system is said to be self-stabilizing if, starting from any state, it automatically recovers to a specified set of legal states in finite time. The arbitrary state from which the system starts may be a faulty state due to a transient failure within the system. Such a fault could be the corruption of local memory, loss of a message, or reception of a corrupted message. During the recovery process, the user may experience a partial loss of services and performance, but guarantee is given that correct system operation will eventually resume.

Self-stabilizing systems meet a stronger notion of correctness under failures. If a transient error pushes the system into an inconsistent or incorrect state, then regardless of the origin and type of the failure, the system eventually coverges to a correct state without any outside assistance. The fact that the type of fault is not specified further contains the striking power of the paradigm: the ability to mask the effect of faults is traded for the ability to tolerate any kind and any number of faults. Thus, self-stabilizing systems offer a degree of fault tolerance that goes beyond the shortcomings of traditional approaches for designing fault-tolerant systems.

Robustness is one of the most important requirements of modern distributed systems and a practical distributed system should be able to recover from any transient faults of the processors and communication links. Ideally, the recovery process should automatically start as soon as a fault is detected and must not rely on the assumption that it is possible to start the system from a well-defined state. It is not reasonable to assume that the code executed by every processor is not altered by transient faults. This code may be stored in a read-only memory or may be reloaded from a non-volatile memory after a transient fault. A distributed self-stabilizing system is a system that can start from any possible initial state and reach a legitimate state in finite time.

Self-stabilization is a different way of looking at distributed system fault tolerance; it provides a "built-in-safeguard" against "transient failures" that might corrupt the data in a distributed system; self-stabilization enables systems to recover from failures automatically without any intervention by any external agency. Stabilizing algorithms are optimistic in the sense that the distributed system may temporarily behave inconsistently but a return to correct system behavior is guaranteed in a finite time, while traditional robust distributed algorithms follow a pessimistic approach in that it protects against the worst possible scenario which demands an assumption of the upper bound on the number of faults.

Self-stabilization provides a unified approach to transient failures by formally incorporating them into the design model. The following transient faults can be handled by a self-stabilizing system [79]:

- **Inconsistent initialization** Different processes in the program may be initialized to local states that are inconsistent with one another.
- **Mode of change** There can be different modes of execution of a system. In changing the mode of operation, it is impossible for all of the processes to effect the change at the same time. The program is bound to reach a global state in which some processes have changed while others have not.
- **Transmission errors** These errors include loss, corruption, or reordering of messages and can cause inconsistency between the states of the sender and receiver.
- **Process failure and recovery** If a process goes down and recovers later, its local state may be inconsistent with the rest of the system/program.
- **Memory crash** A memory crash may cause the loss of local state, making it inconsistent with the rest of the system/program.

Traditional approaches to fault tolerance have addressed each of these issues separately. Self-stabilization provides a unified approach to fault tolerance by handling all these issues single-handedly.

Global initialization is not necessary; each component can be started separately in an arbitrary state. Self-stabilization does not rely on particular initial state as other distributed algorithms do. There is no need for proper and consistent initialization. A self-stabilizing distributed system eventually reaches a legitimate system state, regardless of its initial state. Because of this property, a self-stabilizing distributed system is extremely robust against failures; it tolerates any finite number of transient failures.

Self-stabilization can be applied to topology preservation/control. After a topological change the system converges to a new feasible configuration. The self-stabilization principle applies to any system built on significant number of components which are evolving independently from one another, but which are cooperating or competing to achieve some common goals. This applies, in particular to large distributed systems which tend to result from the integration of many subsystems and components developed separately earlier by different people.

The investigation and use of self-stabilization as an approach to fault-tolerance has been undergoing a renaissance. Dijkstra's notion of self-stabilization, which originally had a very narrow scope of application, is proving to encompass a formal and unified approach to fault tolerance under a model of transient failures for distributed systems. Self-stabilization has most obvious application to the network protocols area since communication protocols should be especially tolerant to temporary faults.

17.13 Factors preventing self-stabilization

In this section, we discuss some of the factors that prevent self-stabilization. The factors preventing self-stabilization include the following:

- Symmetry
- Termination
- Isolation
- Look-alike configuration.

Symmetry

Self-stabilization requires that all processes should not be identical/symmetric because a self-stabilization solution generally relies on a distiguished process. Asymmetry must be maintained in systems where processes may synchronize with one another such as mutual exclusion, dining philosophers, drinking philosophers, and resource allocation systems under deterministic rules.

A system can be asymmetric by state or asymmetric by identity. A system is asymmetric by state when all processes are identical; however, they start from different initial local states. A system is asymmetric by identity when not all of the processes are identical. In general, a system asymmetric only by state cannot be self-stabilizing, while a system asymmetric by identity can be self-stabilizing.

Termination

Self-stabilization is generally incompatible with termination. If any unsafe global state is a final state, then a system will not be able to stabilize. Though self-stabilization is generally incompatible with termination, there is one exceptional case where self-stabilization can be achieved in the presence of termination. That is, in the case of finite-state sequential programs, since the number of states is finite, a compiler can remove all the unsafe states [79].

While the property of termination is very natural when dealing with algorithms whose goal is to compute a function (i.e., quantitative), it is unnatural in the domain of distributed systems, where computations are non-terminating by design and have qualitative goals such as coordination and control.

One form of termination that occurs within distributed systems is deadlock where one or more processes wait for an event that will never occur [69]. In a distributed message-passing system, processes will be waiting for messages to come from other processes. A process sends a message and then waits for a response. By way of a malicious adversary, control of a local process could be placed at a point just after a send instruction without a message actually having been sent. Thus at any local process state that follows the sending of a message, it is impossible for that process to know whether a message has in fact been sent.

This situation can lead to deadlock where one or more processes wait for messages that will never come. The problem of deadlock is not seen in a shared memory system. Because a process can test the value of shared memory when required, there is no waiting for messages and thus no deadlock.

Isolation

Isolation occurs within a system when the local state and computation of each process is consistent with some safe global state and computation; however, the resulting global state and computation is not safe. In such a situation, the system is unable to stabilize due to inadequate communication and coordination between its processes [53,79].

Look-alike configurations

Look-alike configurations result when the same computation (sequence of actions) is enabled in two different states with no way to differentiate between them [53,79]. If one of the two states is unsafe, then the system cannot guarantee convergence from the unsafe state.

17.14 Limitations of self-stabilization

The problem in self-stabilizing systems is the time it takes for a system to correct itself when started in an illegal state or there is an error causing it to go in an illegal state. If a system cannot tolerate this initial unknown period, then self-stabilization does not help. Even if the initial unknown can be tolerated for a brief period of time, the system may not converge to a legal state quickly enough.

Need for an exceptional machine

Almost all self-stabilization algorithms rely on the fact that there is at least one exceptional machine in the system. This may be difficult to achieve in some systems, but it is not a major drawback in most distributed systems.

Convergence–response tradeoffs

The convergence span denotes the maximum number of critical transitions made before the system reaches a legal state and the response span denotes the maximum number of transitions to get from some starting state to some goal state. Critical transitions are similar to errors occurring in the system due to a move. For example, in a mutual exclusion system, if one process is in its critical section and another process makes a move and enters its critical section, an error has occurred because more than one process has been allowed to enter its critical section.

Several self-stabilizing termination detection algorithms, each having different properties, have been developed. For a ring of n processes, if one has comparative convergence and response spans, one has a fast convergent span and a slow response span, and another shows the relationship between the two spans. If the convergence span is decreased by a factor of k $(1 \leq k \leq n)$, the response span is increased by the same factor. So, the convergence span is of the order of n/k while the response span is $n*k$. This relationship exists in all the other classes of self-stabilizing systems.

This relationship is reasonable because the more checks that are made, the longer it will take to converge, while there will be a fewer number of errors made. This relationship is very useful in the design of self-stabilizing systems because the system can be modified according to the goal of the system. Depending on the requirements of the system, one can have fast convergence with many errors or slower convergence with fewer errors or something in between.

Pseudo-stabilization

It is sometimes expensive to design self-stabilizing systems. Lessening the requirements of the system can reduce some of the cost. A system is said to stabilize if and only if every computation has some state in it such that any computation starting from this state will be in the set of legal computations. On the other hand, in order for a system to pseudo-stabilize, every computation only needs to have some state such that the suffix of the computation beginning at this state is in the set of legal computations [23]. The property of pseudo-stabilization is obviously weaker than the requirement of stabilization, although, it is less expensive to implement.

Verification of self-stabilizing systems

When designing self-stabilizing systems, verifying the correctness of these algorithms may be difficult, but there has been some work in this area. A convergence stair method has been developed where the legal states are built up step by step. Proof that the algorithm stabilizes in each step, verifies the correctness of the entire algorithm. The interleaving assumptions can be relaxed to make it easier to verify the correctness of the algorithm. Algorithms that are pseudo-stabilizing [23] are usually good enough for many systems, and these are easier to implement, easier to verify, and more efficient to run.

17.15 Chapter summary

Self-stabilization has been used in many areas and the areas of study continue to grow. Algorithms have been developed using central or distributed demons and uniform and non-uniform networks. The algorithms that assume a central demon can usually be easily extended to support distributed demon, so these algorithms are still useful when applied to distributed systems.

Extensions of communication protocols that are self-stabilizing have also been developed, such as the sliding window protocol, the two-way handshake, and the alternating-bit protocol [2]. The major drawback of self-stabilizing systems is the initial illegal configurations. The system must converge quickly in order to make the illegal configurations less serious. Verification of the systems can be difficult, but there are ways to make it easier. Relaxing interleaving assumptions and usage of a convergence stair are two of the ways. Some of the assumptions made while designing the systems make it nearly impossible to implement the systems. For example, self-stabilizing protocols require a timeout action that needs to examine the contents of the communication link and also needs to know the values of some non-local variables. Global timeout actions are usually avoided, which makes these algorithms not easy to implement.

These requirements may not be necessary in some cases. The alternating-bit protocol [2], for example, does not need unbounded sequence numbers, nor does it need expensive global timeout actions. Therefore, this protocol can be implemented relatively easily, and even though the algorithm is pseudo-stabilizing and not exactly stabilizing, this does not affect the usefulness of the algorithm in most situations.

17.16 Exercises

Exercise 17.1 When self-stabilization claims to solve so many problems in fault tolerance in a unified manner, why are people still studying and investigating each of those problems individually?

Exercise 17.2 Describe the self-stabilizing alternating-bit protocol.

Exercise 17.3 Give a psuedo-stabilization algorithm. Discuss how it reduces the cost compared to stabilization.

Exercise 17.4 What is "superstabilization"? What type of guarantees do superstabilization provide?

Exercise 17.5 What are the trade-offs in a self-stabilizing system/algorithm?

Exercise 17.6 Fault containment is a problem with self-stablizing algorithms. What are fault-containing self-stablizing algorithms [48]? Describe how they solve the problem.

Exercise 17.7 Describe a self-stabilizing mutual exclusion algorithm.

Exercise 17.8 One weakness of self-stabilization is that it is a global property. A failure that is local to a machine may spread and lead to corrective actions across the entire system. Discuss how this problem can be addressed by local detection and correction of failures [4,15].

17.17 Notes on references

The idea of self-stabilization was first proposed by Dijkstra in a seminal paper in 1974 [34]. Since then considerable volume of work has been done on this topic. The most extensive work in self-stabilization has been done in the area of mutual exclusion [24,64,67,74]. The reason for this is mainly due to Dijkstra's original self-stabilizing model, where a legal state is defined as a state in which only one privilege exists in the system.

An excellent survey on the topic is due to Schneider [79]. An excellent review article on the topic is due to Flatebo *et al.* [42]. An excellent monograph on the topic is Dolev [37]. Gartner [43] presents a survey of algorithms for construction of self-stabilizing spanning trees. Aggarwal [6,7] presents a time optimal self-stabilizing algorithm for spanning trees. Antonioiu and Srimani [8]–[10] discuss self-stabilizing algorithms to construct minimum spanning trees. More details on distributed reset can be found in [12]. Chang *et al.* [27] discuss the cost of self-stabilization. Self-stabilization has been used to design more robust distributed mechanisms, such a synchronization [13]. Other interesting papers on the topic include [1,11,14,16–18,26,32,33,36,38,39,42,44–46,50–52,56–58,60–62,65,70–72,75,78,82,83].

References

[1] M. Abadi and M. G. Gouda, The stabilizing computer, *Proceedings of the 1992 International Conference on Parallel and Distributed Systems*, December, Taiwan, 1992, 90–96.

[2] Y. Afek and G. Brown, Self-stabilization of the alternating-bit protocol, *Proceedings of the 8th Symposium on Reliable Distributed Systems*, 1989, 80–83.

[3] Y. Afek, S. Kutten, and M. Yung, Memory efficient self-stabilizing protocols for general networks, *Proceedings of the 4th International Workshop on Distributed Algorithms*, 1991, 15–28.

[4] Y. Afek, S. Kutten, and M. Yung, The local detection paradigm and its applications to self-stabilization, *Theoretical Computer Science*, **186**(1–2), 1997, 199–229.

[5] Y. Afek and A. Bremler, Self-stabilizing unidirectional network algorithms by power supply, *Chicago Journal of Theoretical Computer Science*, **1998**(3), 1998.

[6] S. Aggarwal, *Time Optimal Self-stabilizing Spanning Tree Algorithms*, Technical Report MIT-LCS/MIT/LCS/TR-632, Massachusetts Institute of Technology, Laboratory for Computer Science, August 1994.

[7] S. Aggarwal and S. Kutten, Time optimal self-stabilizing spanning tree algorithms, in R. K. Shyamasundar (ed.). *Proceedings of Foundations of Software Technology and Theoretical Computer Science*, Berlin, Germany, December 1993, 400–410.

[8] G. Antonoiu and P. K. Srimani, A self-stabilizing distributed algorithm to construct an arbitrary spanning tree of a connected graph, *Computers and Mathematics with Applications*, **30**, 1995, 1–7.

[9] G. Antonoiu and P. K. Srimani, Distributed self-stabilizing algorithm for minimum spanning tree construction, *Proceedings of Euro-Par '97 Parallel Processing*, 1997, 480–487.

[10] G. Antonoiu and P. K. Srimani, A self-stabilizing distributed algorithm for minimal spanning tree problem in a symmetric graph, *Computers and Mathematics with Applications*, **35**(10), 1998, 15–23.

[11] A. Arora and M. G. Gouda, Closure and convergence: a foundation for fault-tolerant computing, *Proceedings of the 22nd International Conference on Fault-Tolerant Computing Systems*, 1992, 396–403.

[12] A. Arora and M. G. Gouda, Distributed reset, *IEEE Transactions on Computers*, **43**(9), 1994, 1026–1038.

[13] B. Awerbuch, S. Kutten, Y. Mansour, B. Patt-Shamir, and G. Varghese, Time optimal self-stabilizing synchronization, *Proceedings of the 25th Annual ACM Symposium on the Theory of Computing*, San Diego, CA, May 16–18, 1993, 652–661.

[14] B. Awerbuch and R. Ostrovsky, Memory-efficient and self-stabilizing network reset, *Symposium on Principles of Distributed Computing (PODC '94)*, New York, August 1994, 254–263.

[15] B. Awerbuch, B. Patt-Shamir, and G. Varghese, Self-stabilization by local checking and correction, *Proceedings of the 31st Annual IEEE Symposium on Foundations of Computer Science, FOCS91*, 1991, 268–277.

[16] B. Awerbuch and G. Varghese, Distributed program checking: a paradigm for building self-stabilizing distributed protocols, *Proceedings of the 32nd IEEE Symposium on Foundations of Computer Science*, October 1991, 268–277.

[17] P. Awerbuch and G. Varghese, Self-stabilization by local checking and correction, *Proceedings of the 32nd IEEE Symposium on Foundations of Computer Science*, October 1991.

[18] Y. Bastani and Y. Zhao, On self-stabilization, non-determinism, and inherent fault tolerance, *Proceedings of the MCC Workshop on Self-Stabilizing Systems*, MCC Technical Report STP-379-89, 1989.

[19] J. C. Browne, A. Emerson, M. Gouda, D. Miranker, A. Mok, and L. Rosier, Bounded time fault-tolerant rule-based systems, *Telematics Information*, **7**(3/4), 1990, 441–454.

[20] G. M. Brown, M. G. Gouda, and C.-L. Wu, Token systems that self stabilize, *IEEE Transactions on Computers*, **38**(6), 1989, 845–852.

[21] J. E. Burns and J. K. Pachl, Uniform self-stabilizing rings, *ACM Transactions on Programming Languages and Systems (TOPLAS)*, **11**(2), 1989, 330–344.

[22] J. E. Burns, *Self-stabilizing Rings Without Demons*, Technical Report GITICS-87/36, Georgia Institute of Technology, 1987.

[23] J. E. Burns, M. G. Gouda, and R. E. Miller, Stabilization and pseudo-stabilization, *Distributed Computing archive*, 7(1), 1993.

[24] J. E. Burns, M. G. Gouda, and R. E. Miller, On relaxing interleaving assumptions, *Proceedings of the MCC Workshop on Self-Stabilizing Systems*, MCC Technical Report STP-379-89, 1989.

[25] R. W. Buskens and R. P. Bianchini, Jr., Self-stabilizing mutual exclusion in the presence of faulty nodes, *Proceedings of the 25th International Symposium on Fault Tolerant Computing* 1995, 144–153.

[26] F. Butelle, C. Lavault, and M. Bui, A uniform self-stabilizing minimum diameter tree algorithm (extended abstract), in J.-M. Ĥelary and M. Raynal, (eds), *Distributed Algorithms, 9th International Workshop, WDAG '95*, Le Mont-Saint-Michel, France, September 13–15, 1995, 25–272.

[27] E. J. H. Chang, G. H. Gonnet, and D. Rotem, On the costs of self-stabilization, *Information Processing Letters*, **24**(5), 1987, 311–316.

[28] K. M. Chandy and L. Lamport, Distributed snapshots: determining global states of distributed systems, *ACM Transactions on Computer Systems*, 1985, 63–75.

[29] N. S. Chen, H.-P. Yu, and S.-T. Huang, A self-stabilizing algorithm for constructing spanning trees, *Information Processing Letters*, **39**, 1991, 147–151.

[30] Cisco Systems Inc, *Using Vlan Director*, system documentation, 1998, available online at: www.cisco.com/univercd/cc/td/doc/product/rtrmgmt/swntman/cwsimain/cwsi2/cwsiug2/vlan2/index.htm.

[31] Z. Collin and S. Dolev, Self-stabilizing depth first search, *Information Processing Letters*, **49**, 1994, 297–301.

[32] J.-M., Couvreur, N. Francez, and M. G. Gouda, Asynchronous unison, *Proceedings of the 12th International Conference on Distributed Computing Systems*, Yokohama, Japan, June 1992, 486–493.

[33] E. W. Dijkstra, A belated proof of self-stabilization, *Distributed Computing*, **1**, 1986, 5–6.

[34] E. W. Dijkstra, Self stabilizing systems in spite of distributed control, *Communications of the ACM*, **17**(11), 1974, 643–644.

[35] S. Dolev, A. Israeli, and S. Moran, Uniform dynamic self-stabilizing leader election, *IEEE Transactions on Parallel and Distributed Systems*, **8**(4), 1997, 424–440.

[36] S. Dolev, Optimal time self-stabilization in dynamic systems (preliminary version), in André Schiper (ed.), *Proceedings of the 7th International Workshop on Distributed Algorithms (WDAG93)*, Lausanne, Switzerland, September 27–29, 1993, 160–173.

[37] S. Dolev, *Self-Stabilization*, MIT Press, 2000.

[38] S. Dolev, M. G. Gouda, and M. Schneider, Memory requirements for silent stabilization, *Acta Informatica*, **36**(6), 1999, 447–462.

[39] S. Dolev, A. Israeli, and S. Moran, Self stabilization of dynamic systems, *Proceedings of the MCC Workshop on Self-Stabilizing Systems*, MCC Technical Report STP-379-89, 1989.

[40] S. Dolev, A. Israeli, and S. Moran, Self-stabilization of dynamic systems assuming only read/write atomicity, *Proceedings of the 9th Annual ACM Symposium on Principles of Distributed Computing*, Quebec City, Canada, August 22–24, 1990, 103–117.

[41] M. Flatebo and A. Datta, Two-state self-stabilizing algorithms, *Proceedings of the 6th International Parallel Processing Symposium*, Beverly Hills, CA, March 1992, 198–203.

[42] M. Flatebo, A. K. Datta, and S. Ghosh, Self-stabilization in distributed systems, in T. L. Casavant and M. Singhal (eds), *Readings in Distributed Computer Systems*, New York, IEEE Computer Society Press, 1994, 100–114.

[43] F. C. Gartner, *A Survey of Self-Stabilizing Spanning-Tree Construction Algorithms*, EPFL Technical Report, 2003. Available online at: icwww.epfl.ch/publications/documents/IC_TECH_REPORT_200338.pdf.

[44] F. C. Gartner and H. Pagnia, Time-efficient self-stabilizing algorithms through hierarchical structures, *Proceedings of the 6th Symposium on Self-Stabilizing Systems*, San Francisco, June 2003, Springer-Verlag.

[45] C. Genolini and S. Tixeuil, A lower bound on dynamic k-stabilization in asynchronous systems, *Proc. of the 21st Symposium on Reliable Distributed Systems*, 2002, 211–221.

[46] S. Ghosh, Binary self-stabilization in distributed systems, *Information Processing Letters*, **40**(3), 1991, 153–159.

[47] S. Ghosh, Self-stabilizing distributed systems with binary machines, *Proceedings of the 28th Annual Allerton Conference*, 1990, 988–997.

[48] S. Ghosh, A. Gupta, T. Herman, and S. V. Pemmaraju, Fault-containing self-stabilizing algorithms, *Proceedings of the 15th Annual ACM Symposium on Principles of distributed computing*, Philadelphia, 1996, 45–54.

[49] S. Ghosh, *Understanding Self-stabilization in Distributed Systems*, Technical Report TR-90-02, Department of Computer Science, University of Iowa, 1990.

[50] S. Ghosh, A. Gupta, and S. Pemmaraju, A fault-containing self-stabilizing algorithm for spanning trees, *Journal of Computing and Information*, **2**, 1996, 322–338.

[51] S. Ghosh, A. Gupta, M. Karaata, and S. Pemmaraju, Self-stabilizing dynamic programming algorithms on trees, *Proceedings of the 2nd Workshop on Self-Stabilizing Systems*, 1995, 11.1–11.15.

[52] M. G. Gouda, *The Stabilizing Philosopher: Asymmetry by Memory and by Action*, Technical Report TR-87-12, Department of Computer Sciences, University of Texas at Austin, 1987.

[53] M. G. Gouda and M. Evangelist, Convergence/response tradeoffs in concurrent systems, *Proceedings of the 2nd IEEE Symposium on Parallel and Distributed Processing*, December 1990, 288–292.

[54] M. G. Gouda and T. Herman, Stabilizing unison, *Information Processing Letters*, **35**, 1990, 171–175.

[55] M. G. Gouda and N. Multari, Stabilizing communication protocols, *IEEE Transactions on Computers*, **40**(4), 1991, 448–458.

[56] M. G. Gouda, R. R. Howell, and L. E. Rosier, The instability of self-stabilization, *Acta Informatica*, **27**, 1990, 697–724.

[57] F. F. Haddix, *Stabilization of Bounded Token Rings*, Technical Report ARL-TR-91-31, Applied Research Laboratory, University of Texas at Austin, 1991.

[58] S. M. Hedetniemi, S. T. Hedetniemi, D. P. Jacobs, and P. K. Srimani, Self-stabilizing algorithms for minimal dominating sets and maximal independent sets, *Computers, Mathematics and Applications*, **46**(5–6), 2003, 805–811.

[59] T. Herman, Probabilistic self-stabilization, *Information Processing Letters*, **35**(2), 1990, 63–67.

[60] T. Herman, Self-stabilization: randomness to reduce space, *Distributed Computing*, **6**, 1992, 95–98.

[61] L. Higham and Z. Liang, Self-stabilizing minimum spanning tree construction on message-passing networks, *Proceedings of the 15th International Symposium on Distributed Computing (DISC)*, Lisbon, Portugal, October 2001.

[62] S.-C. Hsu and S.-T. Huang, A self-stabilizing algorithm for maximal matching, *Information Processing Letters*, **43**(2), 1992, 77–81.

[63] S. Huang and N. Chen, A self-stabilizing algorithm for constructing breadth-first trees, *Information Processing Letters*, **41**, 1992, 109–117.

[64] A. Israeli and M. Jaflon, Token management schemes and random walks yield self-stabilizing mutual exclusion, *Proceedings of the 9th Annual ACM Symposium on Principles of Distributed Computing*, Quebec City, Quebec, Canada, August, 22–24, 1990, 119–131.

[65] G. Itkis and L. Levin, Fast and lean self-stabilizing asynchronous protocols, *Proceedings of the 35th Annual Symposium on Foundations of Computer Science*, Santa Fe, New Mexico, November 20–22, 1994, 226–239.

[66] C. Johnen, Memory efficient, self-stabilizing algorithm to construct BFS spanning trees, *Proceedings of the 16th Annual ACM Symposium on Principles of Distributed Computing (PODC '97)*, August 1997, 288.

[67] H. Kakugawa and M. Yamashita, *A Universal Self-Stabilizing Mutual Exclusion Algorithm*, Dagstuhl Seminar 00431: SelfStabilization, Dagstuhl, Germany, 2000. Available online at http://citeseer.ist.psu.edu/yamashita00universal.html.

[68] R. Kat, Self-stabilizing replication file system, September 2002, Available online at: www.cs.bgu.ac.il/srfs/.

[69] S. Katz and K. J. Perry, Self-stabilizing extensions for message-passing systems, *Proceedings of the 9th Annual ACM Symposium on Principles of Distributed Computing*, Quebec City, Canada, August 1990, 22–24, 91–101.

[70] H. S. M. Kruijer, Self-stabilization (in spite of distributed control) in tree-structured systems, *Information Processing Letters*, **8**(2), 1979, 2–79.

[71] D. Lehman and M. Rabin, On the advantages of free choice: a symmetric and fully distributed solution of the dining philosopher's problem, *Proceedings of the 8th Annual ACM Symposium on Principles of Programming Languages*, 1981.

[72] X. Lin and S. Ghosh, Self-stabilizing maxima finding, *Proceedings of the 28th Annual Allerton Conference*, 1991, 662–671.

[73] M. Luby, A simple parallel algorithm for the maximal independent set problem, *SIAM Journal on Computing*, **15**(4), 1986, 1036–1055.

[74] M. Mizuno, M. Nesterenko, and H. Kakugawa, Lock based self-stabilizing distributed mutual exclusion algorithms, *International Conference on Distributed Computing Systems*, 1996, 708–716.

[75] R.-C. Pan, J.-Z. Wang, and L. R. Chow. A self-stabilizing distributed spanning tree construction algorithm with a distributed demon, *Tamsui Oxford Journal of Mathematical Sciences*, **15**, 1999, 23–32.

[76] M. Schneider, *Self-Stabilization – A Unified Approach to Fault Tolerance in the Face Transient Errors*, TechReport TR-91-18, Department of Computer Science, University of Texas at Austin, TX, 1991.

[77] M. Schneider, *Compiling Self-Stabilization into Sequential Programs*, Department of Computer Science, University of Texas at Austin, TX, 1992.

[78] M. Schneider, *Lecture notes on Self-Stabilization*, The University of Texas at Austin, available online at: www.cs.utexas.edu/users/plaxton/c/395t/slides/Schneider.pdf.

[79] M. Schneider, Self stabilization, *ACM Computing Surveys*, **25**(1), 1993.

[80] S. Shukla, D. Rosenkrantz, and S. Ravi, Observations on self-stabilizing graph algorithms for anonymous networks, *Proceedings of the Second Workshop on Self-Stabilizing Systems*, 1995, 7.1–7.15.

[81] Z. Shi, W. Goddard, and S. T. Hedetniemi, An anonymous self-stabilizing algorithm for 1-maximal independent set in trees, *Information Processing Letters*, **91**(2), 2004, 77–83.

[82] S. Sur and P. K. Srimani, A self-stabilizing distributed algorithm to construct BFS spanning trees of a symmetric graph, *Parallel Processing Letters*, **2**(2–3), 1992, 171–179.

[83] M.-S. Tsai and S.-T. Huang, A self-stabilizing algorithm for the shortest paths problem with a fully distributed demon, *Parallel Processing Letters*, **4**(1–2), 1994, 65–72.

CHAPTER

18 Peer-to-peer computing and overlay graphs

18.1 Introduction

Peer-to-peer (P2P) network systems use an application-level organization of the network overlay for flexibly sharing resources (e.g., files and multimedia documents) stored across network-wide computers. In contrast to the client–server model, any node in a P2P network can act as a server to others and, at the same time, act as a client. Communication and exchange of information is performed directly between the participating peers and the relationships between the nodes in the network are equal. Thus, P2P networks differ from other Internet applications in that they tend to share data from a large number of end users rather than from the more central machines and Web servers. Several well known P2P networks that allow P2P file-sharing include Napster [25], Gnutella [16,17], Freenet [10], Pastry [30], Chord [32], and CAN [27].

Traditional distributed systems used DNS (domain name service) to provide a lookup from host names (logical names) to IP addresses. Special DNS servers are required, and manual configuration of the routing information is necessary to allow requesting client nodes to navigate the DNS hierarchy. Further, DNS is confined to locating hosts or services (not data objects that have to be a priori associated with specific computers), and host names need to be structured as per administrative boundary regulations. P2P networks overcome these drawbacks, and, more importantly, allow the location of arbitrary data objects.

An important characteristic of P2P networks is their ability to provide a large combined storage, CPU power, and other resources while imposing a low cost for scalability, and for entry into and exit from the network. The ongoing entry and exit of various nodes, as well as dynamic insertion and deletion of objects is termed as *churn*. The impact of churn should be as transparent as possible. P2P networks exhibit a high level of self-organization and are able to operate efficiently despite the lack of any prior infrastructure or authority. The philosophy of this model requires that if a node wants to

Table 18.1 Desirable characteristics and performance features of P2P systems.

Features	Performance
Self-organizing	Large combined storage, CPU power, and resources
Distributed control	Fast search for machines and data objects
Role symmetry for nodes	Scalable
Anonymity	Efficient management of churn
Naming mechanism	Selection of geographically close servers
Security, authentication, trust	Redundancy in storage and paths

enjoy the services which other nodes provide, that node should provide service to other nodes. Some desirable features of P2P systems are summarized in Table 18.1.

18.1.1 Napster

One of the earliest popular P2P systems, Napster [25], used a server-mediated central index architecture organized around clusters of servers that store direct indices of the files in the system. The central server maintains a table with the following information of each registered client: (i) the client's address (IP) and port, and offered bandwidth, and (ii) information about the files that the client can allow to share. The basic steps of operation to search for content and to determine a node from which to download the content are the following:

1. A client connects to a meta-server that assigns a lightly loaded server from one of the close-by clusters of servers to process the client's query.
2. The client connects to the assigned server and forwards its query along with its own identity.
3. The server responds to the client with information about the users connected to it and the files they are sharing.
4. On receiving the response from the server, the client chooses one of the users from whom to download a desired file. The address to enable the P2P connection between the client and the selected user is provided by the server to the client.

Users are generally anonymous to each other. The directory serves to provide the mapping from a particular host that contains the required content, to the IP address needed to download from it.

18.1.2 Application layer overlays

A core mechanism in P2P networks is searching for data, and this mechanism depends on how (i) the data, and (ii) the network, are organized. Search algorithms for P2P networks tend to be data-centric, as opposed to the host-centric algorithms for traditional networks. P2P search uses the *P2P overlay*, which is a logical graph among the peers that is used for the object search and object storage and management algorithms. Note that above the P2P overlay is the application layer overlay, where communication between peers is point-to-pont (representing a logical all-to-all connectivity) once a connection is established.

The P2P overlay can be *structured* (e.g., hypercubes, meshes, butterfly networks, de Bruijn graphs) or *unstructured*, i.e., no particular graph structure is used. Structured overlays use some rigid organizational principles based on the properties of the P2P overlay graph structure, for the object storage algorithms and the object search algorithms. Unstructured overlays use very loose guidelines for object storage. As there is no definite structure to the overlay graph, the search mechanisms are more "ad-hoc," and typicaly use some forms of *flooding* or *random walk* strategies. Thus, object storage and search strategies are intricately linked to the overlay structure as well as to the data organization mechanisms.

18.2 Data indexing and overlays

The data in a P2P network is identified by using indexing. Data indexing allows the physical data independence from the applications. Indexing mechanisms can be classified as being *centralized*, *local*, or *distributed*:

- **Centralized indexing** entails the use of one or a few central servers to store references (indexes) to the data on many peers. The DNS lookup as well as the lookup by some early P2P networks such as Napster used a central directory lookup.
- **Distributed indexing** involves the indexes to the objects at various peers being scattered across other peers throughout the P2P network. In order to access the indexes, a structure is used in the P2P overlay to access the indexes. Distributed indexing is the most challenging of the indexing schemes, and many novel mechanisms have been proposed, most notably the *distributed hash table (DHT)*. Various DHT schemes differ in the hash mapping, search algorithms, diameter for lookup, search diameter, fault-tolerance, and resilience to churn.

  A typical DHT uses a flat key space to associate the mapping between network nodes and data objects/files/values. Specifically, the node address is mapped to a logical identifier in the key space using a consistent hash function. The data object/file/value is also mapped to the same key space using hashing. These mappings are illustrated in Figure 18.1.
- **Local indexing** requires each peer to index only the local data objects and remote objects need to be searched for. This form of indexing is typically used in unstructured overlays in conjunction with flooding search or random walk search. Gnutella uses local indexing.

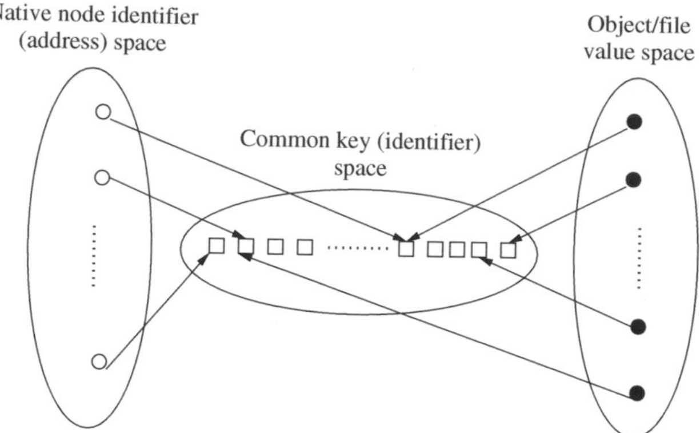

Figure 18.1 The mappings from node address space and object space in a typical DHT scheme, e.g., Chord, CAN, Tapestry.

An alternate way to classify indexing mechanisms is as being a *semantic index mechanism* or a *semantic-free* index mechanism. A semantic index is human readable, for example, a document name, a keyword, or a database key. A semantic-free index is not human readable and typically corresponds to the index obtained by a hash mechanism, e.g., the DHT schemes. A semantic index mechanism supports keyword searches, range searches, and approximate searches, whereas these searches are not supported by semantic-free index mechanisms.

18.2.1 Distributed indexing

Structured overlays

The P2P network topology has a definite structure, and the placement of files or data in this network is highly deterministic as per some algorithmic mapping. (The placement of files can sometimes be "loose," as in some earlier P2P systems like Freenet, where "hints" are used.) The objective of such a deterministic mapping is to allow a very fast and deterministic lookup to satisfy queries for the data. These systems are termed as *lookup systems* and typically use a hash table interface for the mapping. The hash function, which efficiently maps *keys* to *values*, in conjunction with the regular structure of the overlay, allows fast search for the location of the file.

An implicit characteristic of such a deterministic mapping of a file to a location is that the mapping can be based on a single characteristic of the file (such as its name, its length, or more generally some *predetermined* function computed on the file). A disadvantage of such a mapping is that arbitrary queries, such as range queries, attribute queries and exact keyword queries cannot be handled directly.

Another implicit effect of the tight coupling of the regular overlay structure and the rigid mapping function to enable fast access is that file insertions and deletions incur some overhead which may be nontrivial under churn.

Unstructured overlays

The P2P network topology does not have any particular controlled structure, nor is there any control over where files/data is placed. Each peer typically indexes only its local data objects, hence, *local indexing* is used. Node joins and departures are easy – the local overlay is simply adjusted. File placement is not governed by the topology. Search for a file may entail high message overhead and high delays. However, complex queries are supported because the search criteria can be arbitrary.

Although the P2P network topology does not have any controlled structure, some topologies naturally emerge. The following topologies are common and will be studied in later sections:

- **Power law random graph** (PLRG) This is a random graph where the node degrees follow the power law. Here, if the nodes are ranked in terms of their degree, then the ith node has c/i^{α} neighbors, where c is a constant.
- **Normal random graph** This is a normal random graph where the nodes typically have a uniform degree.

We study search in unstructured overlay networks in the next section.

18.3 Unstructured overlays

18.3.1 Unstructured overlays: properties

Unstructured overlays have the serious disadvantage that queries may take a long time to find a file or may be unsuccessful even if the queried object exists. The message overhead of a query search may also be high.

The following are the main advantages of unstructured overlays such as the one used by Gnutella:

- Exact keyword queries, range queries, attribute-based queries, and other complex queries can be supported because the search query can capture the semantics of the data being sought; and the indexing of the files and data is not bound to any non-semantic structure.
- Unstructured overlays can accommodate high churn, i.e., the rapid joining and departure of many nodes without affecting performance.

The following are advantages of unstructured overlays if certain conditions are satisfied:

- Unstructured overlays are efficient when there is some degree of data replication in the network.
- Users are satisfied with a best-effort search.
- The network is not so large as to lead to scalability problems during the search process.

18.3.2 Gnutella

Gnutella uses a fully decentralized architecture [16,17]. In Gnutella logical overlays, nodes index only their local content. The actual overlay topology can be arbitrary as nodes join and leave randomly. A node joins the Gnutella network by forming a connection to some nodes found in standard Gnutella directory-like databases. (Note that the function of joining the network cannot be said to be fully decentralized.) Users communicate with each other, performing the role of both *serv*er and cli*ent*, termed as *servent*. The following are the main message types used by Gnutella:

- *Ping* messages are used to discover hosts, and allow a new host to announce itself.
- *Pong* messages are the responses to *Pings*. The *Pong* messages indicate the port and (IP) address of the responder, and some information about the amount of data (the number and size of files) that node can make available.
- *Query* messages. The search strategy used is flooding. *Query* messages contain a search string and the minimum download speed required of the potential responder, and are flooded in the network.
- *QueryHit* messages are sent as responses if a node receiving a *Query* detects a local match in response to a query. A *QueryHit* contains the port and address (IP), speed, the number of files found, and related information. The path traced by a *Query* is recorded in the message, so the *QueryHit* follows the same path in reverse.

18.3.3 Search in Gnutella and unstructured overlays

Consider a system with n nodes and m objects. Let q_i be the popularity of object i, as measured by the fraction of all queries that are queries for object i. All objects may be equally popular, or more realistically, a Zipf-like power law distribution of popularity exists. Thus [23],

$$\sum_{i=1}^{m} q_i = 1, \tag{18.1}$$

$$\text{uniform: } q_i = 1/m; \qquad \text{Zipf-like: } q_i \propto i^{-\alpha}. \tag{18.2}$$

Let r_i be the number of replicas of object i, and let p_i be the fraction of all objects that are replicas of i. Three static replication strategies are: uniform, proportional, and square root. Thus,

$$\sum_{i=1}^{m} r_i = R; \qquad p_i = r_i/R, \tag{18.3}$$

$$\text{uniform: } r_i = R/m; \quad \text{proportional: } r_i \propto q_i; \quad \text{square-root: } r_i \propto \sqrt{q_i}. \tag{18.4}$$

Under uniform replication, all objects have an equal number of replicas and hence the performance for all query rates is the same. With a uniform query rate, proportional and square-root replication schemes reduce to the uniform replication scheme.

For an object search, some of the more popular metrics of efficiency are:

- the probability of success of finding the queried object;
- delay or the number of hops in finding an object;
- the number of messages processed by each node in a search;
- node coverage, the fraction of (distinct) nodes visited;
- *message duplication*, which is (#messages − #nodes visited)/#messages;
- maximum number of messages at a node;
- *recall*, the number of objects found satisfying the desired search criteria. This metric is useful for keyword, inexact, and range queries;
- *message efficiency*, which is the recall per message used.

Guided versus unguided search

In unguided or blind search, there is no history of earlier searches, and hence, each search is inherently independent. In guided search, nodes store some history of past searches to aid future searches. Various mechanisms for caching hints to guide and narrow down future searches are used. In this chapter, we focus on unguided searches in the context of unstructured overlays.

Search strategies

Flooding [23]

- In order to curtail the high message overhead that flooding introduces, the initial strategy was to use *checking*. Here, a node checks back with the query originator before forwarding a query. Unfortunately, this cause heavy load on the originator, in addition to excessive delays, and hence is not practical.
- The next approach is to use the *time to live* (TTL) field or the hop count. However, this does not guarantee that a match can be found for the query even if the object exists in the network, and requires a high value of TTL to have a high degree of success.
- A refinement that allows more control is the *expanding ring* strategy. A node first floods with a small TTL. If the search is not successful, it starts another flood with a larger TTL, and so on. This strategy is more successful when objects are replicated.
  The expanding ring approach is significantly more successful than the TTL approach, for all replication strategies, and all query distributions, and the cost is only a relatively small increase in delay.

Although the expanding ring is superior to TTL, both are flooding-based strategies and suffer from message duplication.

Random walk

Another strategy to use is that of *random walking*. Here, a query is randomly forwarded by a node when it is received. Random walk greatly reduces the message overhead but it increases the search latency. Hence, k *random walkers* can be used. To terminate the k random walkers, a "checking-cum-TTL" strategy is effective. Here, each walker periodically (after a certain number of hops) checks with the query originator whether to terminate; the TTL is used to prevent looping, and is usually set to a large value.

Performance

The performance of searches in unstructured overlays has been studied via simulations and by experiments. The following are some of the relationships of interest, for both flooding and for k-random walk (for various values of k) for various graph topologies such as the random graph and the PLRG:

- The success rate as a function of the number of message hops, or TTL.
- The number of messages as a function of the number of message hops, or TTL.
- The above metrics as the replication ratio and the replication strategy changes.
- The node coverage, recall, and message efficiency, as a function of the number of hops, or TTL; and as a function of various replication ratios and replication strategies.

Guidelines

- Adaptively determining the termination condition is important. Checking is adaptive whereas TTL is not.
- Message duplication must be minimized, as it represents wasted resources.
- At each step in the search, the number of messages (or number of nodes visited) should not increase by a large amount.

Overall, k-random walk performs much better than flooding and is more scalable, for various replication and query distributions, and various graph topologies.

18.3.4 Replication strategies

Cohen and Shenker [12] studied the degree of replication for blind or *unguided* search in random overlay graphs. The various parameters used to study replication are defined in Table 18.2. Random search is modeled by the following process. A node is repeatedly drawn at random from a bin, examined for a match with the copy of the object, and replaced in the bin, until the object is found. The metric then is the number of nodes drawn (or equivalently, the

Table 18.2 Parameters to study replication.

n	number of nodes in the system
m	number of objects in the system
q_i	normalized query rate, where $\sum_{i=1}^{m} q_i = 1$
r_i	number of replicas of object i
ρ	capacity (measured as number of objects) per node
R	$n\rho = \sum_{i=1}^{m} r_i$, the total capacity in the system
p_i	r_i/R, the population fraction of object i replicas

number of hops of a random walker) until success. The probability that the object is found on the kth drawing is:

$$Pr_i(k) = \frac{r_i}{n}(1 - \frac{r_i}{n})^{k-1}.$$

The average search size for i, denoted as A_i, is:

$$A_i = E_{over\ all\ k}(Pr_i(k)) = \sum_{k=1}^{n} \left[k\frac{r_i}{n}\left(1 - \frac{r_i}{n}\right)^{k-1} \right] \sim \frac{n}{r_i}, \text{ for large } n. \quad (18.5)$$

Across the system, the average search size A is:

$$\text{average search size } A = \sum_{i=1}^{m} q_i A_i = n \sum_i \frac{q_i}{r_i}. \quad (18.6)$$

Setting r_i to n maximizes A, but requires full replication. As resources are constrained, assume that average number of replicas per node is $\rho = R/n < m$. (It is easy to see that $R \geq m \geq \rho$.) Substituting for n with R/ρ in the equation above, we have:

$$\text{average search size } A = \frac{R}{\rho} \sum_i \frac{q_i}{r_i} = \frac{1}{\rho} \sum_i \frac{q_i}{p_i}. \quad (18.7)$$

The *utilization rate* u_i of a replica of object i is the average rate of requests serviced by a replica of i. With random search, $u_i = q_i/p_i = R(q_i/r_i)$. Over all replicas of object i, the utilization is simply $= Rq_i$. The average utilization rate over (all copies of) all objects is $u = \sum_{i=1}^{m} r_i(u_i/R) = \sum_{i=1}^{m} p_i(q_i/p_i) = 1$. This average is a constant, and independent of the replication scheme. It is desirable to have a low maximum utilization rate in order to distribute the load more uniformly.

The replication problem is formulated as the optimization solution for Eq. (18.7). We assume that all objects are of uniform size. To simplify analysis, we also assume that each object that is queried exists in the system and a search continues until the object is found, i.e., all searches are eventually successful. (In practice, there is a parameter L – such as TTL – that controls the maximum search size. Search on insoluble queries continues until this parameter is exceeded. The cost of such queries is $f_s A + (1 - f_s)L$, where f_s is the fraction of queries that are soluble.)

Two natural replication strategies are *uniform* and *proportional*:

- **Uniform** $r_i = R/m$, which implies $p_i = r_i/R = 1/m$.

  The average search size for object i is $A_i = n/r_i$. This equals $R/(\rho r_i) = R/(\rho R/m_i) = m/\rho$ and is the same for all objects.

  From Eq. (18.7), the average search size $A_{uniform} = (1/\rho)\sum_i (q_i/p_i) = \frac{1}{\rho}\sum_i mq_i = (m/\rho)$.

  The utilization of a replica of i is $u_i = q_i/p_i$, which is proportional to the query rate as p_i is the same for all objects.

  The maximum utilization of a replica of i is $max_i u_i = max_i(q_i/p_i) = R(q_i/r_i)$, which can vary significantly.
- **Proportional** $r_i = Rq_i$, which implies $p_i = q_i$.

  The average search size for object i is $A_i = n/r_i = n/Rp_i = n/(Rq_i) = 1/(\rho q_i)$, which is inversely proportional to the query rate.

  From Eq. (18.7), the average search size $A_{proportional} = (1/\rho)\sum_i (q_i/p_i) = (1/\rho)\sum_{i=1}^{m} 1 = m/\rho$.

  The utilization of a replica of i is $u_i = q_i/p_i = 1$, a constant for all replicas of all objects.

  The maximum utilization of a replica of i is $max_i u_i = max_i(q_i/p_i) = max_i(q_i/q_i) = 1$ for all i.

Both uniform and proportional replication have the same average search size, which is independent of the query distribution. However, objects whose query rates are below the average have lower overhead with uniform replication, while those with query rates larger than the average have lower overhead with proportional replication.

- **Square root** The optimal replication strategy that minimizes the average search size is the square-root replication, which is defined as having $p_i = r_i/R \propto \sqrt{q_i}/\sum_j \sqrt{q_j}$, assuming that $1/R \le \sqrt{q_i}/\sum_j \sqrt{q_j} \le n/R$ for all i.

  The optimality of square-root replication can be seen as follows. Substituting $1 - \sum_{i=1}^{m-1} p_i$ for p_m in the cost function of Eq. (18.7), we have:

$$\text{search size } A_{sq-rt} = \frac{1}{\rho}\sum_i q_i/p_i = \frac{1}{\rho}\left[\sum_{i=1}^{m-1} q_i/p_i + q_m \Bigg/ \left(1 - \sum_{i=1}^{m-1} p_i\right)\right].$$

By solving $ds/dp_i = 0$, the value of p_i that minimizes A_{sq-rt} is seen to be $p_m\sqrt{q_i/q_m}$.

Analogous to uniform and proportional replications, the values of A, A_i, and u_i for square-root replication can be dervied. Exercise 18.1 asks you to show the derivations. The results are summarized in Table 18.3. It can be seen that to minimize A, $r_i = R\sqrt{q_i}/\sum_j \sqrt{q_j}$.

Table 18.3 Comparison of uniform, proportional, and square-root replication [23].

	r_i	A	$A_i = n/r_i$	$u_i = Rq_i/r_i$
Uniform	constant, R/m	m/ρ	m/ρ	$q_i m$
Proportional	$q_i R$	m/ρ	$1/(\rho q_i)$	1
Square-root	$R\sqrt{q_i}/\sum_j \sqrt{q_j}$	$(\sum_i \sqrt{q_i})^2/\rho$	$\frac{\sum_j \sqrt{q_j}/\sqrt{q_i}}{\rho}$	$\sqrt{q_i}\sum_j \sqrt{q_j}$

The square-root replication rate ($\propto \sqrt{q_i}$) is more than that of uniform ($\propto 1$), but less than that of proportional ($\propto q_i$). It has been shown that:

- any allocation rate "in between" that of uniform and of proportional has a lower average search size A than that of uniform and proportional;
- any allocation rate either less than that of uniform, or greater than that of proportional has a higher average search size A than that of uniform and proportional.

18.3.5 Implementing replication strategies

Proportional and uniform can be trivially implemented. For proportional, each query creates a copy; for uniform, a fixed number of copies are made when an object is created [12].

The simple "path replication" scheme, wherein the number of copies made is proportional to the length of the (successful) search path, implements square-root replication. Here object i is replicated $c(n/r_i)$ times per query, where c is some constant. Then r_i can be captured by the following equation: $dr_i/dt = q_i c(n/r_i)$. Let $a = ln(r_i/r_j)$, then:

$$\frac{da}{dt} = cn\left(\frac{q_j}{r_j^2} - \frac{q_i}{r_i^2}\right) = \frac{1}{r_j}\frac{dr_j}{dt} - \frac{1}{r_i}\frac{dr_i}{dt}.$$

Square-root replication, wherein $r_i = (R/\sum \sqrt{q_i})\sqrt{q_i}$, is a fixed-point solution of this equation. Therefore, path replication implements square-root replication.

The analysis implicitly assumes that replicas also get deleted, in a way that is independent of their object identity or query rate, and the lifetime of a replica is a non-decreasing function of its age. (Policies such as random and FIFO satisfy this condition, but LRU and LFU do not.) Then, during steady state, the creation rate can equal the deletion rate.

An alternate way of analyzing replication schemes is as follows. Let C be the number of replicas created on a successful query; $\overline{C}$ is its average. Then, in steady state,

$$\frac{p_i}{p_j} = \frac{q_i\overline{C_i}}{q_j\overline{C_j}}. \tag{18.8}$$

To implement distributed algorithms for various replication policies, it is necessary to determine C_i locally without knowing p_i or q_i:

- For proportional replication, C is the same for all objects.
- For square-root replication, if $\overline{C_i} \propto 1/\sqrt{q_i}$ then $p_i/p_j = \sqrt{q_i/q_j}$, by substituting in Eq. (18.8).

  As $A_i \propto nR/p_i$ and $p_i \propto q_i\overline{C_i}$, therefore $A_i \propto 1/(q_i\overline{C_i})$.

  With *path replication*, $C_i \propto A_i$, hence $C_i \propto A_i \propto 1/(q_i\overline{C_i})$.

  In steady state, A_i and $\overline{C_i}$ are equal. Solving $C_i \propto 1/(q_i\overline{C_i})$ for the fixed point, $C_i \propto 1/\sqrt{q_i}$. As $p_i \propto q_i C_i$ when C_i is steady, this gives $p_i \propto \sqrt{q_i}$. In a practical implementation, it needs to be ensured that convergence occurs once steady state sets in.

18.4 Chord distributed hash table

18.4.1 Overview

The Chord protocol, proposed by Stoica *et al.* [32], uses a flat key space to associate the mapping between network nodes and data objects/files/values. The node address as well as the data object/file/value is mapped to a logical identifier in the common key space using a consistent hash function. These mappings are illustrated in Figure 18.1. Both these mappings should ensure that the keys are distributed roughly equally among the nodes. This also insures that with high probability, the overhead of key management when nodes join or leave the P2P network is low. Specifically, when a node joins or leaves the network having n nodes, only $O(1/n)$ keys need to be moved from one location to another.

The Chord key space is flat, thus giving applications flexibility in mapping their files/data to keys. Chord supports a single operation, *lookup*(x), which maps a given key x to a network node. Specifically, Chord stores a file/object/value at the node to which the file/object/value's key maps. Two steps are involved:

1. Map the object/file/value to its key in the common address space.
2. Map the key to the node in its native address space using *lookup*. The design of *lookup* is the main challenge.

In Chord, a node's IP address is hashed to an m-bit identifier that serves as the node identifier in the common key (identifier) space. Similarly, the file/data key is hashed to an m-bit identifier that serves as the key identifier. m is sufficiently large so that the probability of collisions during the hash is negligible. The Chord overlay uses a logical ring of size 2^m. The identifier space is ordered on the logical ring modulo 2^m. Henceforth in this section, we will assume modulo 2^m arithmetic. A key k gets assigned to the first node such that its node identifier equals or follows the key identifier of k in the

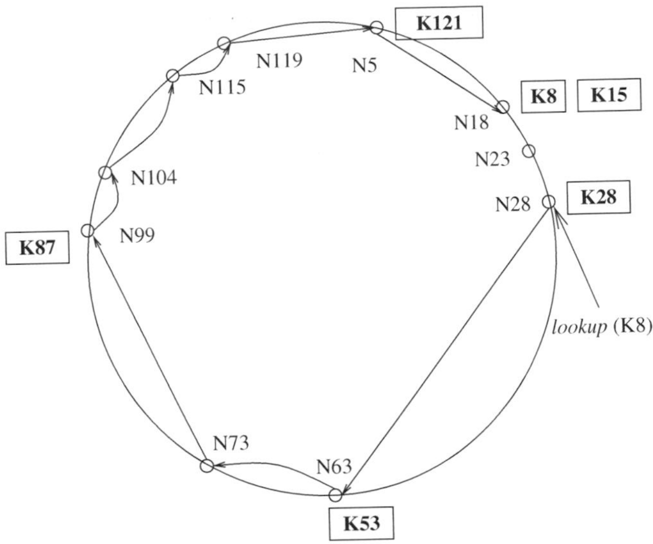

Figure 18.2 An example Chord ring with $m = 7$, showing mappings to the Chord address space, and a query lookup using a simple scheme [32].

common identifier space. The node is the successor of k, denoted $succ(k)$. A Chord ring for $m = 7$ is depicted in Figure 18.2. Nodes N5, N18, N23, N28, N63, N73, N99, N104, N115, and N119 are shown. Six keys, K8, K15, K28, K53, K87, and K121, are stored among these nodes as follows: $succ(8) = 18$, $succ(15) = 18$, $succ(28) = 28$, $succ(53) = 63$, $succ(87) = 99$, and $succ(121) = 5$.

18.4.2 Simple lookup

A simple key lookup algorithm that requires each node to store only 1 entry in its routing table works as follows. Each node tracks its successor on the ring, in the variable *successor*; a query for key x is forwarded to the successors of nodes until it reaches the first node such that that node's identifier y is greater than the key x, modulo 2^m. The result, which includes the IP address of the node with key y, is returned to the querying node along the reverse of the path that was followed by the query. This mechanism requires $O(1)$ local space but $O(n)$ hops, where n is the number of nodes in the P2P network. The pseudo-code for this simple lookup is given in Algorithm 18.1. The following convention is assumed. Notation $(x, y]$ represents the left-open right-closed segment of the Chord logical ring modulo m. Notation $x.Proc(\cdot)$ is a RPC to execute *Proc* on node x while $x.var$ is a RPC to read the variable *var* at process x.

Example The steps for the query: lookup(K8) initiated at node 28, are shown in Figure 18.2 using arrows.

(variables)
integer: *successor* $\longleftarrow$ initial value;

(1) *i.Locate_Successor(key)*, where $key \neq i$:
(1a) **if** $key \in (i, successor]$ **then**
(1b) **return**(*successor*)
(1c) **else return** (*successor.Locate_Successor(key)*).

Algorithm 18.1 A simple object location algorithm in Chord at node *i* [32].

18.4.3 Scalable lookup

A scalable lookup algorithm that uses $O(log\, n)$ message hops at the cost of $O(m)$ space in the local routing tables, uses the following idea. Each node *i* maintains a routing table, called the *finger table*, with at most $O(log\, n)$ distinct entries, such that the *x*th entry ($1 \leq x \leq m$) is the node identifier of the node $succ(i+2^{x-1})$. This is denoted by $i.finger[x] = succ(i+2^{x-1})$. This is the first node whose key is greater than the key of node *i* by at least $2^{x-1}\ mod\ 2^m$. Note that each finger table entry would have to contain the IP address and port number in addition to the node identifier, in order that *i* can communicate with $i.finger[x]$; henceforth we will assume this implicitly without showing these entries.

The size of the finger table is bounded by *m* entries. Due to the logarithmic structure, the finger table has more information about nodes closer ahead of it in the Chord overlay, than about nodes further away. Given any key whose node is to be located, the highly scalable logarithmic search shown in Algorithm 18.2 is used. For a query on key *key* at node *i*, if *key* lies between

(variables)
integer: *successor* $\longleftarrow$ initial value;
integer: *predecessor* $\longleftarrow$ initial value;
integer *finger*$[1 \ldots m]$;

(1) *i.Locate_Successor(key)*, where $key \neq i$:
(1a) **if** $key \in (i, successor]$ **then**
(1b) **return**(*successor*)
(1c) **else**
(1d) $j \longleftarrow$ *Closest_Preceding_Node(key)*;
(1e) **return** (*j.Locate_Successor(key)*).

(2) *i.Closest_Preceding_Node(key)*, where $key \neq i$:
(2a) **for** $count = m$ **down to** 1 **do**
(2b) **if** $finger[count] \in (i, key]$ **then**
(2c) **break**();
(2d) **return**(*finger*[*count*]).

Algorithm 18.2 A scalable object location algorithm in Chord at node *i* [32].

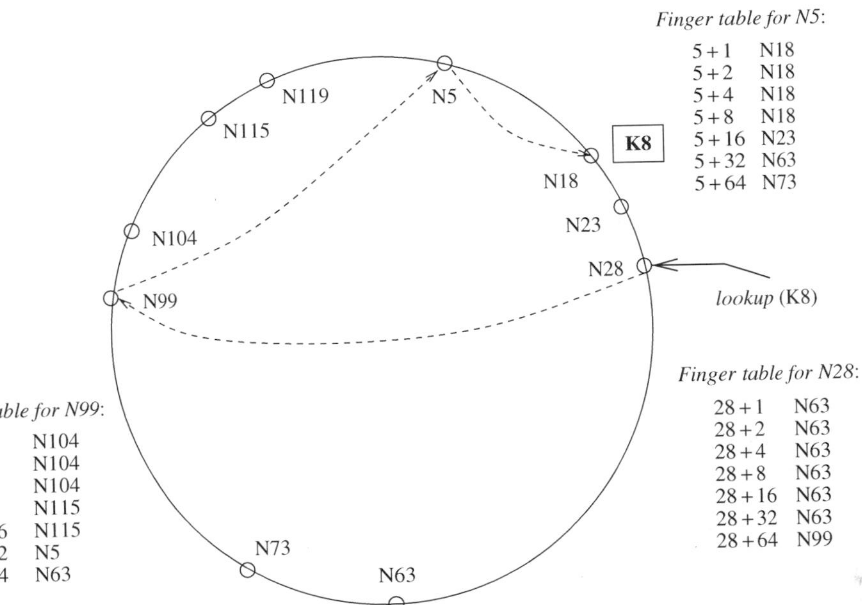

Figure 18.3 An example showing a query lookup using the logarithmically-structured finger tables [32].

i and its successor, the *key* would reside at the successor and the successor's address is returned. If *key* lies beyond the successor, then node i searches through the m entries in its finger table to identify the node j such that j most immediately precedes *key*, among all the entries in the finger table. As j is the closest known node that precedes *key*, j is most likely to have the most information on locating *key*, i.e., locating the immediate successor node to which *key* has been mapped.

Example The use of the finger tables in answering the query lookup(K8) at node N28 is illustrated in Figure 18.3. The finger tables of N28, N99, and N5 that are used are shown.

18.4.4 Managing Churn

The code to manage dynamic node joins, departures, and failures is given in Algorithm 18.3.

Node joins

To create a new ring, a node i executes *Create_New_Ring* which creates a ring with the singleton node. To join a ring that contains some node j, node i invokes *Join_Ring*(j). Node j locates i's successor on the logical ring and informs i of its successor. Before i can participate in the P2P exchanges, several actions need to happen: i's successor needs to update its predecessor

```
(variables)
integer: successor ⟵ initial value;
integer: predecessor ⟵ initial value;
integer finger[1...m];
integer: next_finger ⟵ 1;

(1)   i.Create_New_Ring():
(1a)  predecessor ⟵⊥;
(1b)  successor ⟵ i.

(2)   i.Join_Ring(j), where j is any node on the ring to be joined:
(2a)  predecessor ⟵⊥;
(2b)  successor ⟵ j.Locate_Successor(i).

(3)   i.Stabilize():  // executed periodically to verify and inform successor
(3a)  x ⟵ successor.predecessor;
(3b)  if x ∈ (i, successor) then
(3c)       successor ⟵ x;
(3d)  successor.Notify(i).

(4)   i.Notify(j):  // j believes it is predecessor of i
(4a)  if predecessor =⊥ or j ∈ (predecessor, i)) then
(4b)       transfer keys in the range (predecessor, j] to j;
(4c)       predecessor ⟵ j.

(5)   i.Fix_Fingers():  // executed periodically to update the finger table
(5a)  next_finger ⟵ next_finger + 1;
(5b)  if next_finger > m then
(5c)       next_finger ⟵ 1;
(5d)  finger[next_finger] ⟵ Locate_Successor(i + 2^(next_finger−1)).

(6)   i.Check_Predecessor():  // executed periodically to verify whether
                              // predecessor still exists
(6a)  if predecessor has failed then
(6b)       predecessor ⟵⊥.
```

Algorithm 18.3 Managing churn in Chord. Code shown is for node i [32].

entry to i, i's predecessor needs to revise its successor field to i, i needs to identify its predecessor, the finger table at i needs to be built, and the finger tables of all nodes need to be updated to account for i's presence. This is achieved by procedures *Stabilize*(), *Fix_Fingers*(), and *Check_Predecessor*() that are periodically invoked by each node.

Figure 18.4 illustrates the main steps of the joining process. A recent joiner node i that has executed *Join_Ring*($\cdot$) gets integrated into the ring by the following sequence:

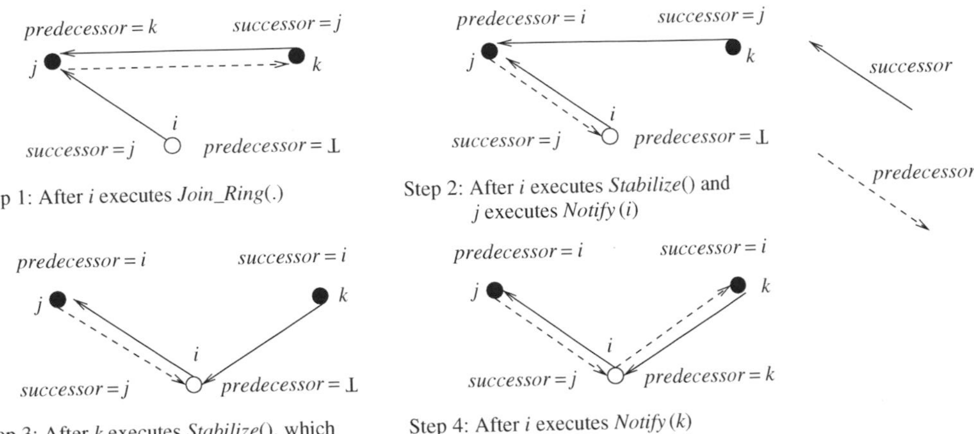

Figure 18.4 Steps in the integration of node i in the ring, where $j > i > k$ [32].

1. The configuration after a recent joiner node i has executed $Join\_Ring(\cdot)$.
2. Node i executes $Stabilize()$, which allows its successor j to adjust j's variable $predecessor$ to i. Specifically, when node i invokes $Stabilize()$, it identifies the successor's predecessor k. If $k \in (i, successor)$, then i updates its $successor$ to k. In either case, i notifies its successor of itself via $successor.Notify(i)$, so the successor has a chance to adjust its $predecessor$ variable to i.
3. The earlier predecessor k of j (i.e., the predecessor in Step 1) executes $Stabilize()$ and adjusts its $successor$ pointer from j to i.
4. Node i executes $Fix\_Fingers()$ to build its finger table, and other nodes also execute the procedure to update their finger tables if necessary.

Once all the successor variables and finger tables have stabilized, a call by any node to $Locate\_Successor(\cdot)$ will reflect the new joiner i. Until then, a call to $Locate\_Successor(\cdot)$ may result in the $Locate\_Successor(\cdot)$ call performing a conservative scan. The loop in $Closest\_Preceding\_Node$ that scans the finger table will result in a search traversal using smaller hops rather than truly logarithmic hops, resulting in some inefficiency. Still, the node i will be located although via more hops.

Showing the correctness of the Chord protocol in the face of concurrent join operations and stablize operations in which pointers are being rewired is non-trivial. It can be shown that for any set of concurrent join operations, at some point after the last join operation completes, all the pointers and finger tables will be correct. However, in the transient period before the Chord ring stabilizes, an object search can result in three outcomes:

- The finger tables used in a search are up to date and the correct successor of the key is sought in $O(log\, n)$ hops.

- The finger tables are not up to date but the successor pointers are correct. The sought key will be located but may take more steps as the full advantage of a logarithmic search space pruning cannot be used.
- If the successor pointers are incorrect, or the key transfer to the new joiners in procedure *Notify* has not completed, the search may fail. This is during a transient duration, and the source has the choice of reissuing the query.

Node failures and departures

When a node j fails abruptly, its successor i on the ring will discover the failure when the successor i executes *Check_Predecessor*() periodically. Process i gets a chance to update its *predecessor* field when another node k causes i to execute *Notify*(k). But that can happen only if k's *successor* variable is i. This requires the predecessor of the failed node to recognize that its successor has failed, and get a new functioning successor. In fact, the successor pointers are required for object search; the predecessor variables are required only to accommodate new joiners. Note from Algorithm 18.2 that knowing that the successor is functional, and that the nodes pointed to by the finger pointers are functional, is essential.

Example In Figure 18.3, assume that node N63 fails. The closest successor that node N28 can find via the finger table is N99. N73 cannot be detected, and keys K64 through K73 will effectively be lost.

A solution such as introducing a *Check_Successor*() procedure analogous to *Check_Predecessor* procedure will not solve the problem because it does not help to identify the functional successor. The Chord protocol proposes that, rather than maintain a single successor, each node maintains a list of α successors, which are the node's first α successors. If the first successor does not respond, the node can try the next successor from the list, and so on. Only the simultaneous failure of all the α successors can then cause the protocol to fail. Maintaining a list of successors requires some changes to the code in Algorithm 18.3. Exercise 18.2 asks you to adapt this code to the changes required for maintaining successor lists.

The provision for a successor list at each node provides a natural mechanism for the application to manage replicated objects. The replicas get placed at the node corresponding to the object key, as well as at the nodes in the successor list of that node. As Chord is able to update its successor list as the successor list changes, Chord can also interface with the application to let it track the locations of the replicas.

A voluntary departure from the ring can be treated as a failure. However, a failed node causes all the data (keys) stored at that node to be lost until corrective action is taken. When a node departs voluntarily, it should first transfer all the keys it is responsible for to its successor. The departing node should also inform its successor and predecessor. This will enable the successor to update its predecessor to the predecessor of the departing node.

The predecessor will also be able to update its successor list by deleting the departing node and adding the last successor of the departing node's successor list to its own successor list.

18.4.5 Complexity

The following results on the complexity have a non-trivial correctness proof and interested readers should consult the Chord papers for the proofs.

1. For a Chord network with n nodes, each node is responsible for at most $(1+\epsilon)K/n$ keys, with "high probability," where K is the total number of keys.
   Using consistent hashing, ϵ can be shown to be bounded by $O(log\, n)$. The "high probability" clause is required because the validity of the result depends on the randomness and conflict-free mappings of the hash function used.
2. The search for a successor in *Locate_Successor* in a Chord network with n nodes requires time complexity $O(log\, n)$ with high probability.
   This result is based on the observation that assuming completely random distributions of the key mappings and node mappings, after $2\, log\, n$ hops, the distance between the key being searched for and the present node that the query has reached is at most $1/n$.
3. The size of the finger table is $log(n) \leq m$.
4. The average lookup time is $1/2\, log(n)$.

Exercises 18.2 and 18.3, based on the Chord papers, ask you to prove further results about the complexity under churn conditions.

18.5 Content addressible networks (CAN)

18.5.1 Overview

A content-addressible network (CAN) is essentially an indexing mechanism that maps objects to their locations in the network. The CAN project originated from the observation that the bottleneck to designing a scalable P2P network is this indexing mechanism. An efficient and scalable CAN is useful not only for object location in P2P networks, but also for large-scale storage management systems and wide-area name resolution services that decouple name resolution and the naming scheme. All these applications inherently require efficient and scalable addition of and location of objects using arbitrary location-independent names or keys for the objects.

A CAN supports three basic operations: insertion, search, and deletion of (*key, value*) tuples. (A "value" is an object in the context of a CAN.) A good CAN design is distributed, fault-tolerant, scalable, independent of the naming structure, implementable at the application layer, and *autonomic*, i.e., self-organizing and self-healing. Although CAN is a generic phrase, it also specifically denotes the particular design of a CAN proposed by Ratnasamy *et al.* [27]. We now study this particular CAN design.

CAN is a *logical d-dimensional Cartesian coordinate space organized as a d-torus logical topology*, i.e., a virtual overlay d-dimensional mesh with wrap-around. A two-dimensional torus was shown in Figure 1.5(a) in Chapter 1. The entire space is partitioned *dynamically* among all the nodes present, so that each node i is assigned a disjoint region $r(i)$ of the space. As nodes arrive, depart, or fail, the set of participating nodes, as well as the assignment of regions to nodes, change.

For any object v, its key $k(v)$ is mapped using a deterministic hash function to a point $\vec{p}$ in the Cartesian coordinate space. The (k, v) pair is stored at the node that is presently assigned the region that contains the point $\vec{p}$. In other words, the (k, v) pair is stored at node i if presently the point $\vec{p}$ corresponding to (k, v) lies in region $r(i)$. Analogously, to retrieve object v, the same hash function is used to map its key k to the same point $\vec{p}$. The node that is presently assigned the region that contains $\vec{p}$ is accessed (using a CAN routing algorithm) to retrieve v. The three core components of a CAN design are the following:

1. Setting up the CAN virtual coordinate space, and partitioning it among the nodes as they join the CAN.
2. Routing in the virtual coordinate space to locate the node that is assigned the region containing $\vec{p}$.
3. Maintaining the CAN due to node departures and failures.

18.5.2 CAN initialization

1. Each CAN is assumed to have a unique DNS name that maps to the IP address of one or a few bootstrap nodes of that CAN. A bootstrap node is responsible for tracking a partial list of the nodes that it believes are currently participating in the CAN. These are reasonable assumptions, and perhaps the most "non-distributed" portions of the CAN design.
2. To join a CAN, the joiner node queries a bootstrap node via a DNS lookup, and the bootstrap node replies with the IP addresses of some randomly chosen nodes that it believes are participating in the CAN.
3. The joiner chooses a random point $\vec{p}$ in the coordinate space. The joiner sends a request to one of the nodes in the CAN, of which it learnt in step 2, asking to be assigned a region containing $\vec{p}$. The recipient of the request routes the request to the owner $old\_owner(\vec{p})$ of the region containing $\vec{p}$, using the CAN routing algorithm.
4. The $old\_owner(\vec{p})$ node splits its region in half and assigns one half to the joiner. The region splitting is done using an a priori ordering of all

the dimensions, so as to decide which dimension to split along. This also helps to methodically merge regions, if necessary. The (k, v) tuples for which the key k now maps to the zone to be transferred to the joiner, are also transferred to the joiner.

5. The joiner learns the IP addresses of its neighbors from $old\_owner(\vec{p})$. The neighbors are $old\_owner(\vec{p})$ and a subset of the neighbors of $old\_owner(\vec{p})$. $old\_owner(\vec{p})$ also updates its set of neighbors. The new joiner as well as $old\_owner(\vec{p})$ inform their neighbors of the changes to the space allocation, so that they have correct information about their neighborhood and can route correctly. In fact, each node has to send an immediate update of its assigned region, followed by periodic HEARTBEAT refresh messages, to all its neighbors.

When a node joins a CAN, only the neighboring nodes in the coordinate space are required to participate in the joining process. The overhead is thus of the order of the number of neighbors, which is $O(d)$ and independent of n, the number of nodes in the CAN.

18.5.3 CAN routing

CAN routing uses the straight-line path from the source to the destination in the logical Euclidean space. This routing is realized as follows. Each node maintains a routing table that tracks its neighbor nodes in the logical coordinate space. In d-dimensional space, nodes x and y are neighbors if the coordinate ranges of their regions overlap in $d-1$ dimensions, and abut in one dimension. All the regions are *convex* and can be characterized as follows. Let $region(x) = [[x^1_{min}, x^1_{max}], \ldots, [x^d_{min}, x^d_{max}]]$. Let $region(y) = [[y^1_{min}, y^1_{max}], \ldots, [y^d_{min}, y^d_{max}]]$. Nodes x and y are neighbors if there is some dimension j such that $x^j_{max} = y^j_{min}$ and for all other dimensions i, $[x^i_{min}, x^i_{max}]$ and $[y^i_{min}, y^i_{max}]$ overlap. An example of neighbouring nodes in two-dimensional space is shown in Figure 18.5.

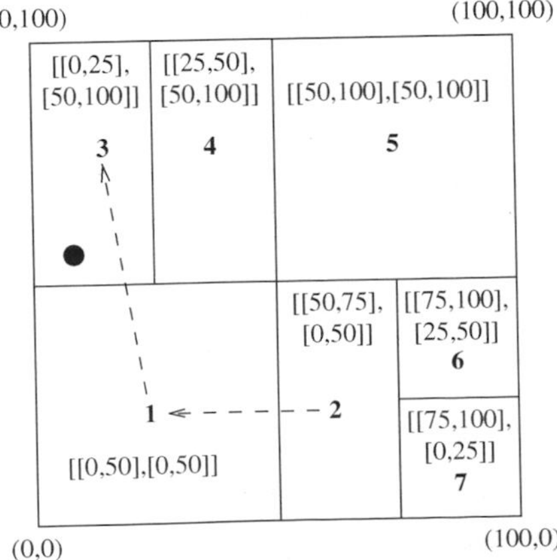

Figure 18.5 Two-dimensional CAN space. Seven regions are shown. The dashed arrows show the routing from node 2 to the coordinate $\vec{p}$ shown by the shaded circle [27].

The routing table at each node tracks the IP address and the virtual coordinate region of each neighbor. To locate value v, its key $k(v)$ is mapped to a point $\vec{p}$ whose coordinates are used in the message header. Knowing the neighbors' region coordinates, each node follows simple greedy routing by forwarding the message to that neighbor having coordinates that are closest to the destination's coordinates. To implement greedy routing to a destination node x, the present node routes a message to that neighbor among the neighbors $k \in Neighbors$, given by

$$argmin_{k \in Neighbors}[\min |\vec{x} - \vec{k}|].$$

Here, $\vec{x}$ and $\vec{k}$ are the coordinates of nodes x and k.

Assuming equal-sized zones in d-dimensional space, the average number of neighbors for a node is $O(d)$. The average path length is $(d/4) \cdot n^{1/d}$. The implication on scaling is that each node has about the same number of neighbors and needs to maintain about the same amount of state information, irrespective of the total number of nodes participating in the CAN. In this respect, the CAN structure is superior to that of Chord. Also note that unlike in Chord, there are typically many paths for any given source-destination pair. This greatly helps for fault-tolerance. Average path length in CAN scales as $O(n^{1/d})$ as opposed to $log\, n$ for Chord.

18.5.4 CAN maintainence

When a node voluntarily departs from CAN, it hands over its region and the associated database of $(key, value)$ tuples to one of its neighbors. The neighbor is chosen as follows. If the node's region can be merged with that of one of its neighbors to form a valid convex region, then such a neighbor is chosen. Otherwise the node's region is handed over to the neighbor whose region has the smallest volume or load – the regions are not merged and the neighbor handles both zones temporarily until a periodic background region reassignment process runs to integrate the regions and prevent further fragmentation.

CAN requires each node to periodically send a HEARTBEAT update message to each neighbor, giving its assigned region coordinates, the list of its neighbors, and their assigned region coordinates. When a node dies, the neighbors suspect its death and initiate a TAKEOVER protocol to decide who will take over the crashed node's region. Despite this TAKEOVER protocol, the $(key, value)$ tuples in the crashed node's database remain lost until the primary sources of those tuples refresh the tuples. Requiring the primary sources to periodically issue such refreshes also serves the dual purpose of updating stale (dirty) objects in the CAN.

The TAKEOVER protocol is as follows. When a node suspects that a neighbor has died, it starts a timer in proportion to its region's volume.

On timeout, it sends a TAKEOVER message, with its region volume piggybacked on the message, to all the neighbors of the suspected failed node. When a TAKEOVER message is received, a node cancels its bid to take over the failed node's region if the received TAKEOVER message contains a smaller region volume than that of the recipient's region. This protocol thus helps in load balancing by choosing the neighbor whose region volume is the smallest, to take over the failed node's region. As all nodes initiate the TAKEOVER protocol, the node taking over also discovers its neighbors and vica versa. In the case of multiple concurrent node failures in one vicinity of the Cartesian space (this is rare), a more complex protocol using a expanding ring search for the TAKEOVER messages can be used.

A graceful departure as well as a failure can result in a neighbor holding more than one region if its region cannot be merged with that of the departed or failed node. To prevent the resulting fragmentation and restore the $1 \rightarrow 1$ node to region assignment, there is a background reassignment algorithm that is run periodically. Conceptually, consider a binary tree whose root represents the entire space. An internal node represents a region that existed earlier but is now split into regions represented by its children nodes. A leaf represents a currently existing region, and (overloading the semantics and the notation), also the node that represents that region.

When a leaf node x fails or departs, there are two cases:

1. If its sibling node y is also a leaf, then the regions of x and y are merged and assigned to y. The region corresponding to the parent of x and y becomes a leaf and it is assigned to node y.
2. If the sibling node y is not a leaf, run a depth-first search in the subtree rooted at y until a pair of sibling leaves (say, $z1$ and $z2$) is found. Merge the regions of $z1$ and $z2$, making their parent z a leaf node, assign the merged region to node $z2$, and the region of x is assigned to node $z1$.

Figure 18.6 illustrates this reassignment. If node 2 fails, its region is assigned to node 3. If node 7 fails, regions 5 and 6 get merged and assigned to node 5 whereas node 6 is assigned the region of the failed node 7.

A distributed version of the above depth-first centralized tree traversal can be performed by the neighbors of a departed node. The distributed traversal leverages the fact that when a region is split, it is done in accordance to a

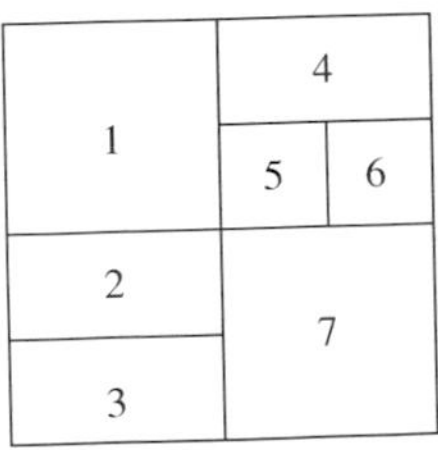

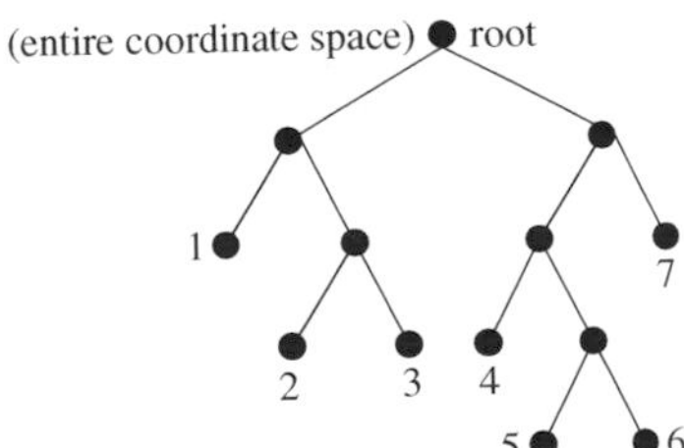

Figure 18.6 Example showing region reassignment in a CAN [27].

particular ordering on the dimensions. Node i performs its part of the depth-first traversal (initiated by the node to which the region of the departed node x is assigned in the TAKEOVER protocol) as follows:

1. Identify the highest ordered dimension dim_a that has the shortest coordinate range $[i_{min}^{dim_a}, i_{max}^{dim_a}]$. Node i's region was last halved along dimension dim_a.
2. Identify neighbor j such that j is assigned the region that was split off from i's region in the last partition along dimension dim_a. Node j's region abuts i's region along dimension dim_a.
3. If j's region volume equals i's region volume, the two nodes are siblings and the regions can be combined. This is the terminating case of the depth-first tree search for siblings. Node j is assigned the combined region, and node i takes over the region of the departed node x. This takeover by node i is done by returning the recursive search request to the originator node, and communicating i's identity on the replies.
4. Otherwise, j's region volume must be smaller than i's region volume. Node i forwards a recursive depth-first search request to j.

18.5.5 CAN optimizations

The following design techniques aim to improve one or more of the performance factors: the per-hop latency, the path length, fault tolerance, availability, and load balancing. These techniques typically demonstrate a trade-off.

- **Multiple dimensions** As the path length is $O(d \cdot n^{1/d})$, increasing the number of dimensions decreases the path length and increases routing fault tolerance at the expense of larger state space per node.
- **Multiple realities** A coordinate space is termed as a *reality*. The use of multiple independent realities assigns to each node a different region in each different reality. This implies that in each reality, the same node will store different (k, v) tuples belonging to the region assigned to it in that reality, and will also have a different neighbor set. The data contents (k, v) get replicated in each reality, leading to higher data availability. Furthermore, the multiple copies of each (k, v) tuple, one in each reality, offer a choice – the closest copy can be accessed. Routing fault tolerance also improves because each reality offers a set of different paths to the same (k, v) tuple. All these advantages come at the cost of more storage – for state information for the neighbors in each reality, as well as for the (k, v) tuples mapped to the region allocated to a node in each reality.
- **Delay latency** Rather than using just the Cartesian distance as a metric to make routing decisions, the delay latency (measured using the round-trip time (RTT)) on each of the candidate logical links can also be used in making the routing decision.
- **Overloading coordinate regions** Each region can be shared by multiple nodes, up to some upper limit. This offers several advantages. First, the

path length and path latency get reduced because overloading is equivalent to having fewer nodes in the CAN. Second, the fault tolerance improves because a region becomes empty only if all the nodes assigned to it depart or fail concurrently. Third, the per-hop latency decreases because a node can select the closest node from the neighboring region to forward a message towards the destination. The cost of gaining these advantages is that many of the aspects of the basic CAN protocol need to be reengineered to accommodate overloading of coordinate regions (see Exercise 18.5).

- **Multiple hash functions** The use of multiple hash functions maps each key to different points in the coordinate space. This replicates each (k, v) pair for each hash function used. The effect is similar to that of using multiple realities.
- **Topologically sensitive overlay** The CAN overlay described so far has no correlation to the physical proximity or to the IP addresses of domains. Logical neighbors in the overlay may be geographically far apart, and logically distant nodes may be physical neighbors. By constructing an overlay that accounts for physical proximity in determining logical neighbors, the average query latency can be significantly reduced.

18.5.6 CAN complexity

The time overhead for a new joiner is $O(d)$ for updating the new neighbors in the CAN, and $O(d/4 \cdot log(n))$ for routing to the appropriate location in the coordinate space. This is also the overhead in terms of the number of messages. The time overhead and the overhead in terms of the number of messages for a node departure is $O(d^2)$, because the TAKEOVER protocol uses a message exchange between each pair of neighbors of the departed node. Exercise 18.4 asks you to compute the complexity of the distributed region reassignment protocol.

18.6 Tapestry

18.6.1 Overview

The Tapestry P2P overlay network provides efficient scalable location-independent routing to locate objects distributed across the Tapestry nodes [20,21,30,36]. Much of the design is adapted from an earlier design of Plaxton trees [26]. The notable enhancements of Tapestry include dealing with node churn as well as dynamic addition and deletion of objects. As in Chord, nodes as well as objects are assigned identifiers obtained by mapping from their native name spaces to a common large identifier space using a uniformly distributed hash function such as SHA-1. The hashed node identifiers are termed VIDs (the acronym for *virtual* i.d.s) and the hashed object identifiers are termed as GUIDs (acronym for *globally unique* i.d.s). For brevity, a specific node v's virtual identifier is denoted v_{id} and a specific object O's GUID is denoted O_G.

18.6.2 Overlay and routing

Root and surrogate root

Tapestry uses a common identifier space specified using m bit values. This identifier is typically expressed in hexadecimal notation, i.e., base $b = 16$, and presently Tapestry recommends $m = 160$. Each identifier O_G in this common overlay space is mapped to a set of *unique* nodes that exists in the network, termed as the identifier's root set denoted $\mathcal{O}_{G_R}$. Typically, $|\mathcal{O}_{G_R}|$ is a small constant, and the main purpose of having $|\mathcal{O}_{G_R}| > 1$ is to increase fault-tolerance. In our discussion, we assume $|\mathcal{O}_{G_R}| = 1$, and refer to a root node of O_G as O_{G_R}.

If there exists a node v such that $v_{id} = O_{G_R}$, then v is the root of identifier O_G. If such a node does not exist, then a globally known deterministic rule is used to identify another unique node sharing the largest common prefix with O_G, that acts as the *surrogate* root. To access object O, the goal is to reach the root O_{G_R} (whether real or surrogate). Routing to O_{G_R} is done using distributed routing tables that are constructed using *prefix routing* information. Prefix routing in Tapestry is somewhat analogous to prefix routing within the telephone network, or to address allocation in the Internet using classless interdomain routing (CIDR). Unlike the telephone numbers or CIDR-assigned IP addresses, Tapestry's VIDs are in a virtual space without correlation to topology, however, topological information can be used to select nodes that are "close" as per some metric.

Prefix routing

Prefix routing at any node to select the next hop is done by increasing the prefix match of the next hop's VID with the destination O_{G_R}. Thus, a message destined for $O_{G_R} = 62C35$ could be routed along nodes with VIDs 6****, then 62***, then 62C**, then 62C3*, and then to 62C35. Let $M = 2^m$. The routing table at node v_{id} contains $b \cdot log_b M$ entries, organized in $log_b M$ levels $i = 1, \ldots, log_b M$. Each entry is of the form $\langle w_{id}, IP\,address \rangle$. In level i, there are b entries with the following property:

- Each entry denotes some "neighbor" node VIDs with an $(i-1)$-digit prefix match with v_{id} – thus, the entry's w_{id} matches v_{id} in the $(i-1)$-digit prefix. Further, in level i, for each digit j in the chosen base (e.g., $0, 1, \ldots, E, F$ when $b = 16$), there is an entry for which the ith digit position is j. Specifically, the jth entry (counting from 0) in level i has value j for digit position i. Let an i digit prefix of v_{id} be denoted as $prefix(v_{id}, i)$. Then the jth entry (counting from 0) in level i begins with an i-digit prefix $prefix(v_{id}, i-1) \circ j$. For example, the fifth entry in level 2 at node 9F248 will be 94***, thus having a two-digit prefix "94."

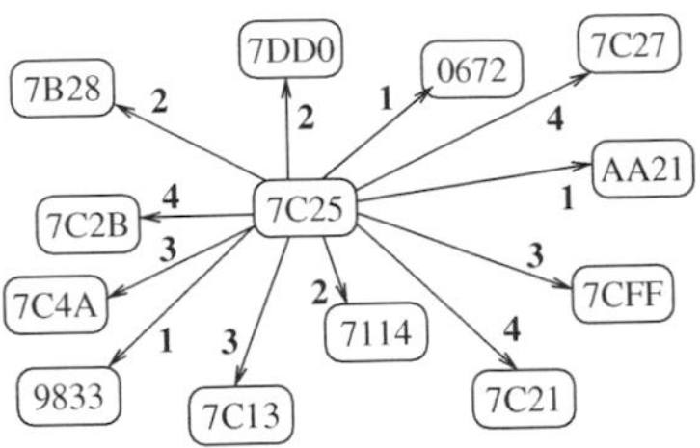

Figure 18.7 Some example links of the Tapestry routing mesh at node with identifier "7C25"[35]. Three links from each level 1 through 4 are labeled by the level.

Router Table

The nodes in the router table at v_{id} are the *neighbors* in the overlay, and these are exactly the nodes with which v_{id} communicates. A part of the routing mesh at one node is shown in Figure 18.7. For each *forward pointer* from node v to v', there is a *backward pointer* from v' to v. Observe the following regarding the router table construction:

- There is a choice of which entry to add in the router table. For example, the jth entry in level i can be the VID of any node whose i-digit prefix is determined; the $(m-i)$-digit suffix can vary. The flexibility is useful to select a node that is "close", as defined by some metric space (e.g., round-trip time). In fact, this choice also allows a more fault-tolerant strategy for routing. Multiple VIDs can be stored in the routing table, as follows. For each prefix β of a node v's identifier and for each digit $j \in \{0, \ldots, b-1\}$ in the alphabet, define the *neighbor set* $\mathcal{N}^{v}_{\beta,j}$ as the set of all nodes whose identifiers share prefix $\beta \circ j$. The nodes in this neighbor set are also referred to as (β, j) neighbors of v. The b sets, one for each value of j, form the routing table of level $|\beta|+1$. $|\mathcal{N}^{v}_{\beta,j}|$ grows exponentially as $|\beta|$ decreases, so the size of this set can be limited by a predetermined parameter c. The closest node in each set is the primary neighbor. Thus the size of the routing table is: $c \cdot b \cdot log_b M$.

  The route from v^0_{id} (source) to destination $j_1 \circ j_2 \cdots \circ j_{log\,M}$, is via nodes $v^1, v^2, \ldots, v^{log\,M}$, where $v^1 \in \mathcal{N}^{v^0}_{\perp,j_1}$ (first hop), $v^2 \in \mathcal{N}^{v^1}_{j_1,j_2}$ (second hop), $v^1 \in \mathcal{N}^{v^2}_{j_1 \circ j_2, j_3}$ (third hop), and so on. The primary neighbor is chosen at each hop. Observe that this provides *location-independent* routing, i.e., irrespective of the source, the same unique root node is reached.

- The jth entry in level i may not exist because no node meets the criterion. This is a *hole* in the routing table. Stated more generally, $|\mathcal{N}^{v}_{\beta,j}|$ may be 0, signifying a hole for digit j at level $|\beta|+1$.

  Surrogate routing can be used to route around holes. If the jth entry in level i should be chosen but is missing, route to the next non-empty entry in level i, using wraparound if needed. All the levels from 1 to $log_b 2^m$ need to be considered in routing, thus requiring $log_b 2^m$ hops. The code for determining the next hop using $NEXT\_HOP(i, O_G)$ is shown in Algorithm 18.4. This is invoked as $NEXT\_HOP(1, O_G)$ at the source node. To determine hop i of the route, the node v that executes the function has a prefix at least $i-1$ digits in common with O_G.

(variables)

integer $Table[1 \ldots log_b 2^m, 1 \ldots b]$; // routing table

(1) $NEXT\_HOP(i, O_G = d_1 \circ d_2 \ldots \circ d_{log_b M})$ executed at node v_{id} to route to O_G:

// i is (1 + length of longest common prefix), also level of the table

(1a) **while** $Table[i, d_i] = \perp$ **do** // d_j is jth digit of destination

(1b) $d_i \longleftarrow (d_i + 1) \bmod b$;

(1c) **if** $Table[i, d_i] = v$ **then** // node v also acts as next hop // (special case)

(1d) **return** $(NEXT\_HOP(i+1, O_G))$ // locally examine next digit of // destination

(1e) **else return**$(Table[i, d_i])$. // node $Table[i, d_i]$ is next hop

---

Algorithm 18.4 Routing in Tapestry [35]. The logic for determining the next hop at a node with node identifier v, $1 \leq v \leq n$, based on the ith digit of O_G, i.e., based on the digit in the ith most significant position in O_G.

Example An example of routing is shown in Figure 18.8.

Property 1 Surrogate routing leads to a unique root. If the routing were to lead to different nodes A and B, let the most significant position in which the digits of A and B differ be i. This implies level i routing caused the routing at some nodes X and Y along different digits. However, the first i digits do not change henceforth, and, assuming synchronized routing tables, the holes would be consistent in the tables at X and Y. Hence both should route to the same ith digit, which is a contradiction. It can now be seen that:

Property 2 For each identifier v_{id}, the routing algorithm identifies a unique spanning tree rooted at v_{id}.

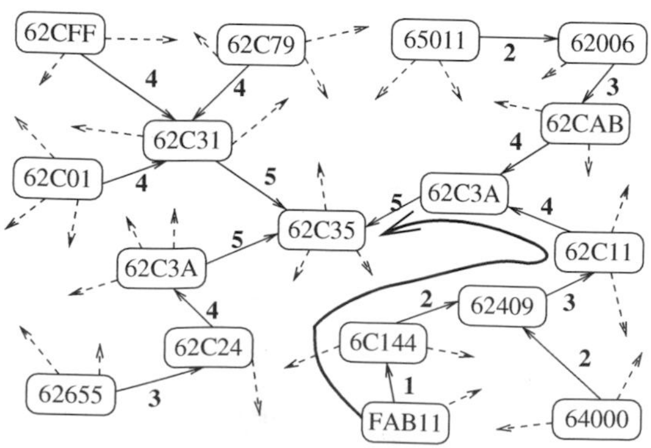

Figure 18.8 An example of routing from FAB11 to 62C35 [35]. The numbers on the arrows show the level of the routing table used. The dashed arrows show some unused links.

18.6.3 Object publication and object search

The unique spanning tree used to route to v_{id} is used to publish and locate an object whose unique root identifier O_{G_R} is v_{id}. A server S that stores object O having GUID O_G and root O_{G_R} periodically publishes the object by routing a *publish* message from S towards O_{G_R}. At each hop and including the root node O_{G_R}, the *publish* message creates a pointer to the object. Ideally, "each node between O and O_{G_R} must maintain a pointer to O despite churn." (Note that the publishing is done by each server at which a replica of the object resides, as well as for each GUID of the object. Recall that an object can be assigned multiple GUIDs, each mapping to a different root node, and giving rise to the set of root nodes $\mathcal{O}_{G_R}$.) If a node lies on the path from two or more servers storing replicas, that node will store a pointer to each replica, sorted in terms of a distance metric (such as latency from itself). This is the directory information for objects, and is maintained as a *soft-state*, i.e., it requires periodic updates from the server, to deal with changes and to provide fault-tolerance.

Example An example showing publishing of an object with O_G = 72EA1 by two replicas, at 1F329 and C2B40 is shown in Figure 18.9.

To search for an object O with GUID O_G, a client sends a query destined for the root O_{G_R}. Along the $log_b\, 2^m$ hops, if a node finds a pointer to the object residing on server S, the node redirects the query directly to S. Otherwise, it forwards the query towards the root O_{G_R} which is guaranteed to have the pointer for the location mapping. A query gets redirected directly to the object as soon as the query path overlaps the publish path towards the same root. Each hop towards the root reduces the choice of the selection of its next node by a factor of b; hence, the more likely by a factor of b that a query path

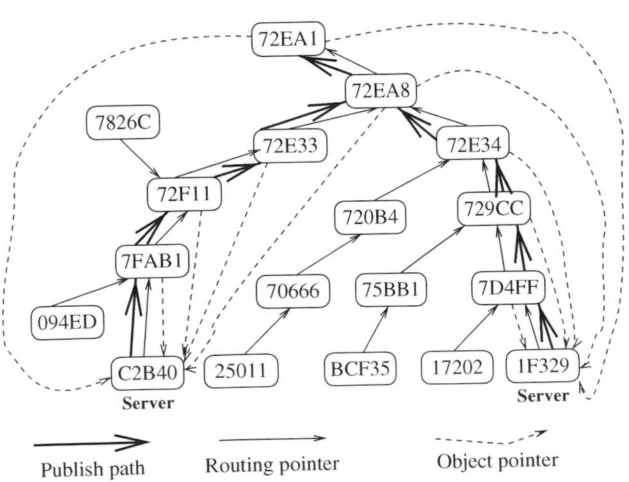

Figure 18.9 An example showing publishing of object with identifier 72EA1 at two replicas 1F329 and C2B40 [35].

and a publish path will meet. Furthermore, as the next hop is chosen based on the network distance metric whenever there is a choice, we also observe that the closer the client is to the server in terms of the distance metric, the more likely that their paths to the object root will meet sooner, and the faster the query will be redirected to the object.

Example Consider the object O_G which has identifier 72EA1 and two replicas at 1F329 and C2B40, as shown in Figure 18.9. A query for the object from 094ED will find the object pointer at 7FAB1. A query from 7826C will find the object pointer at 72F11. A query from BCF35 will find the object pointer at 729CC.

18.6.4 Node insertion

When nodes join the network, the result should be the same as though the network and the routing tables had been initialized with the nodes as part of the network. The procedure for the insertion of node X should maintain the following property of Tapestry:

Property 3 For any node Y on the path between a publisher of object O and the root G_{O_R}, node Y should have a pointer to O.

More generally, the insertion should satisfy the following properties:

- Nodes that have a hole in their routing table should be notified if the insertion of node X can fill that hole.
- If X becomes the new root of existing objects, references to those objects should now lead to X.
- The routing table for node X must be constructed.
- The nodes near X should include X in their routing tables to perform more efficient routing.

The main steps in node insertion are as follows:

1. Node X uses some gateway node into the Tapestry network to route a message to itself. This leads to its "surrogate," i.e., the root node with identifier closest to that of itself (which is X_{id}). The surrogate Z identifies the length α of the longest common prefix that Z_{id} shares with X_{id}.
2. Node Z initiates a MULTICAST-CONVERGECAST on behalf of X by essentially creating a logical spanning tree as follows. Acting as a root, Z contacts all the (α, j) nodes, for all $j \in \{0, 1, \ldots, b-1\}$ (tree level 1). These are the nodes with prefix α followed by digit j. Each such (level 1) node $Z1$ contacts all the $(prefix(Z1, |\alpha|+1), j)$ nodes, for all $j \in \{0, 1, \ldots, b-1\}$ (tree level 2). This continues up to level $log_b 2^m - |\alpha|$ and completes the MULTICAST. The nodes at this level are the leaves

of the tree, and initiate the CONVERGECAST, which also helps to detect the termination of this phase.

All the nodes contacted fill in any holes in their routing table and, if necessary, transfer any references of pointers that are rooted locally. All these nodes also contact X with their information, so that X can build its routing table from level $|\alpha|+1$ up to $log_b 2^m$. All these nodes that contact X have a common prefix of α.

To construct the rest of its routing table from levels 1 through $|\alpha|$, node X procures similar lists for successively smaller prefixes until it gets closest b nodes matching the empty prefix. Node X begins with the list of nodes for level α, corresponding to the level l of its routing table which is already filled. To construct the level $l-1$ list, node X contacts all the nodes in the level l list to find out all the level $l-1$ nodes they know about by asking for both forward pointers and backward pointers. Level $l-1$ of the routing table is filled in using the k closest nodes from the level $l-1$ list, for each of the digits $0, \ldots, b-1$. In this manner, X completes its routing table, and all the nodes contacted in the process can optimize their routing tables by using X if it helps.

The insertion protocols are fairly complex and deal with concurrent insertions.

18.6.5 Node deletion

When a node A leaves the Tapestry overlay, the following actions are performed:

1. Node A informs the nodes to which it has (routing) backpointers. It also provides them with replacement entries for each level from its routing table. This is to prevent holes in their routing tables. (The notified neighbors can periodically run the nearest neighbor algorithm to fine-tune their tables.)
2. The servers to which A has object pointers are also notified. The notified servers send object republish messages.
3. During the above steps, node A routes messages to objects rooted at itself to their new roots. On completion of the above steps, node A informs the nodes reachable via its backpointers and forward pointers that it is leaving, and then leaves.

Node failures are handled by using the redundancy that is built in to the routing tables and object location pointers. For example, each routing table entry has up to c neighbors in the neighbor set $\mathcal{N}^{v}_{\beta,j}$. A node X detects a failure of another node A by using soft-state beacons or when a node sends a message but does not get a response. Node X updates its routing table entry for A with a suitable substitute node, running the nearest neighbor algorithm if necessary. If A's failure leaves a hole in the routing table of X, then X contacts the suggorate of A in an effort to identify a node to fill the hole. The details of the protocol can be found in the Tapestry papers.

In addition to repairing the routing mesh, the object location pointers also have to be adjusted. Objects rooted at the failed node may be inaccessible until the object is republished. The protocols for doing so essentially have to (i) maintain path availability, and (ii) optionally collect garbage/dangling pointers that would otherwise persist until the next soft-state refresh and timeout.

Overall, experiments have shown that Tapestry continues to perform well with high probability, despite dynamic node insertions and failures.

Complexity

- A search for an object is expected to take $(log_b 2^m)$ hops. However, the routing tables are optimized to identify nearest neighbor hops (as per the space metric). Thus, the latency for each hop is expected to be small, compared to that for CAN and Chord protocols.
- The size of the routing table at each node is $c \cdot b \cdot log_b 2^m$, where c is the constant that limits the size of the neighbor set that is maintained for fault-tolerance.

The larger the Tapestry network, the more efficient is the performance. Hence, it is better that different applications share the same overlay.

18.7 Some other challenges in P2P system design

18.7.1 Fairness: a game theory application

P2P systems depend on all the nodes cooperating to store objects and allowing other nodes to download from them. However, nodes tend to be selfish in nature; thus there is a tendancy to download files without reciprocating by allowing others to download the locally available files. This behavior, termed as *leaching* or *free-riding*, leads to a degradation of the overall P2P system performance. Hence, penalties and incentives should be built in the system to encourage sharing and maximize the benefit to all nodes.

We now examine the classical problem, termed the *prisoners' dilemma*, from game theory, that has some useful lessons on how selfish agents might cooperate. This problem is an example of a non-zero-sum-game.

In the prisoners' dilemma, two suspects, A and B, are arrested by the police. There is not enough evidence for a conviction. The police separate the two prisoners, and, separately, offer each the same deal: if the prisoner testifies against (betrays) the other prisoner and the other prisoner remains silent, the betrayer gets freed and the silent accomplice gets a 10-year sentence. If both testify against the other (betray), they each receive a 2-year sentence. If both remain silent, the police can only sentence both to a small 6-month term on a minor offence.

Rational selfish behavior dictates that both A and B would betray the other. This is not a Pareto-optimal solution, where a Pareto-optimal solution is one in which the overall good of all the participants is maximized. In the above example, both A and B staying silent results in a Pareto-optimal solution. The dilemma is that this is not considered the rational behavior of choice.

In the iterative prisoners' dilemma, the game is played multiple times, until an "equilibrium" is reached. Each player retains memory of the last move of both players (in more general versions, the memory extends to several past moves). After trying out various strategies, both players should converge to the ideal optimal solution of staying silent. This is Pareto-optimal.

The commonly accepted view is that the *tit-for-tat* strategy, described next, is the best for winning such a game. In the first step, a prisoner cooperates, and in each subsequent step, he reciprocates the action taken by the other party in the immediately preceding step.

The BitTorrent P2P system [11] has adopted the tit-for-tat strategy in deciding whether to allow a download of a file in solving the leaching problem. Here, cooperation is analogous to allowing others to upload local files, and betrayal is analogous to not allowing others to upload. The term *choking* refers to the refusal to allow uploads. As the interactions in a P2P system are long-lived, as opposed to a one-time decision to cooperate or not, *optimistic unchoking* is periodically done to unchoke peers that have been choked. This optimistic action roughly corresponds to the re-initiation of the game with the previously choked peer after some time epoch has elapsed.

18.7.2 Trust or reputation management

Various incentive-based economic mechanisms to ensure maximum cooperation among the selfish peers inherently depend on the notion of trust. In a P2P environment where the peer population is highly transient, there is also a need to have trust in the quality of data being downloaded. These requirements have lead to the area of trust and trust management in P2P systems [1,18,19]. As no node has a complete view of the other downloads in the P2P system, it may have to contact other nodes to evaluate the trust in particular offerers from which it could download some file. These communication protocol messages for trust management may be susceptible to various forms of malicious attack (such as man-in-the-middle attacks and Sybil attacks), thereby requiring strong security guarantees. The many challenges to tracking trust in a distributed setting include: quantifying trust and using different metrics for trust, how to maintain trust about other peers in the face of collusion, and how to minimize the cost of the trust management protocols.

18.8 Tradeoffs between table storage and route lengths

18.8.1 Unifying DHT protocols

Chord, CAN, and Tapestry are three well-known representative protocols for managing structured P2P overlays. Despite their seeming differences, Xu *et al.* [34] showed that the routing function they perform can be expressed in a uniform way by generalizing the function of classless interdomain domain routing (CIDR) used by the IP protocol. We assume that all identifiers are in the common address space. We also assume modulo arithmetic.

Routing rule

The next-hop routing to node with identifier *dest* from the current node with identifier *id* is as follows.

Let the k entries in a routing table at a node with identifier *id* be the tuples $\langle S_{id,i}, J_{id,i} \rangle$, for $1 \leq i \leq k$. If $|dest - id| \in$ the range $S_{id,i}$ then route to $R(id + J_{id,i})$, where $R(x)$ is the node responsible for key $R(x)$.

Clearly, we must have that for distinct i and j, $S_{id,i} \cap S_{id,j} = \emptyset$ and $J_{id,i} \neq J_{id,j}$. Further, $\cup_{1 \leq i \leq s} S_{id,i}$ contains all the keys not stored by node *id*. When $S_{id,i}$ and $J_{id,i}$ are independent of *id*, as is the case for CAN, Chord, and Tapestry, the subscript *id* can be deleted.

- **Chord** if $dest - id \in S_i = [2^{i-1}, 2^i)$ then node *id* routes to node $id + J_i$, where $J_i = 2^{i-1}$.

  This corresponds to looking up the ith entry in the finger table, as described in Section 18.4.3.
- **CAN** The greedy routing function for CAN was given in Section 18.5.3. Here we assume a simple uniform distribution of nodes in the address space, $x^d = n$, and that nodes are numbered by an integer in base x, where x is the number of nodes in each dimension. Routing is assumed to be done dimension by dimension (rather than using greedy routing). Wraparound routing is assumed in each dimension. Then, for each dimension i, the following holds: if *dest* and *id* differ in dimension i, route to i's neighbor in that dimension. Formally,

  If $dest - id \in (S_i =)[x^{i-1}, x^i)$ then route to $id + J_i$, where $J_i + id$ is a neighbor node in dimension in $i - 1$ and $J_i = kx^{i-1}$ for some $k \leq x$.
- **Tapestry** Let $x = log_b n$, $lvl = 1, \ldots, x$ and $j \in 0, \ldots, b - 1$. After deleting the longest common prefix between *id* and *dest*, $prefix(dest, lvl - 1)$, from *dest*, we have $suffix(dest, x - lvl + 1)$. The routing function was described in Section 18.6.2.

  If $suffix(dest, x - lvl + 1) \in S_{(lvl-1) \cdot b + j} = [j \cdot b^{x-lvl+1}, (j+1) \cdot b^{x-lvl+1})$ then node *id* routes to node $prefix(id, lvl - 1) \circ suffix(J_{(lvl-1) \cdot b + j}, x - lvl + 1)$, where $J_{(lvl-1) \cdot b + j} \in [j \cdot b^{x-lvl+1}, (j+1) \cdot b^{x-lvl+1})$.

These routing relationships are summarized in Table 18.4.

Table 18.4 Comparison of representative P2P overlays. d is the number of dimensions in CAN. b is the base in Tapestry [34].

Protocol	Chord	CAN	Tapestry
Routing table size	$k = O(log_2 n)$	$k = O(d)$	$k = O(log_b n)$
Worst case distance	$O(log_2 n)$	$O(n^{1/d})$	$O((b-1) \cdot log_b n)$
n, common name space	2^k	x^d	b^x
S_i	$[2^{i-1}, 2^i)$	$[x^{i-1}, x^i)$	$[j \cdot b^{x-lvl+1}, (j+1) \cdot b^{x-lvl+1})$
J_i	2^{i-1}	kx^{i-1}	$suffix(J_{(lvl-1) \cdot b+j}, x-lvl+1)$

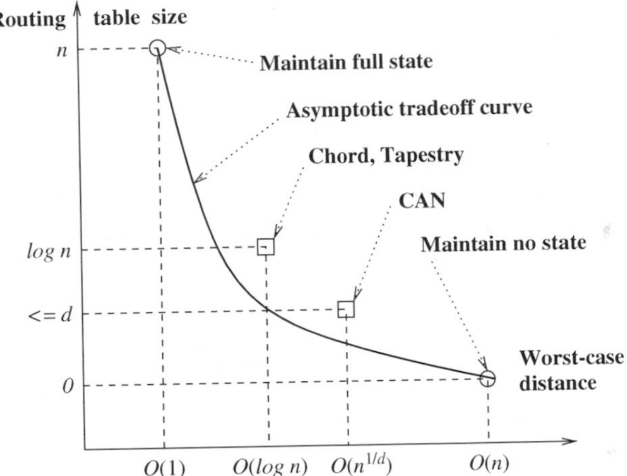

Figure 18.10 Fundamental asymptotic tradeoffs between router table size and network diameter [34].

18.8.2 Bounds on DHT storage and routing distance

Based on Table 18.4, the router table size and network diameter are represented in Figure 18.10. A fundamental question is whether the asymptotic bounds on (routing table size, network diameter as determined by the maximum number of hops) are $(log_2 n, \Omega(log_2 n))$ as for Chord and Tapestry, and $(d, \Omega(n^{1/d}))$ as for CAN. Xu *et al.* [34] used the following definitions to answer this:

- A routing algorithm is *weakly uniform* if for any nodes id and id', the jump sizes $J_{id,i} = J_{id',i}$. Thus, a weakly uniform algorithm requires the corresponding "jump sizes" for any index i to be the same for all nodes, irrespective of the node identifier.
- A routing algorithm is *strongly uniform* if it is weakly uniform and if for any nodes id and id', $S_{id,i} = S_{id',i}$. A strongly uniform algorithm requires all routing tables to also have the same corresponding sizes of the index ranges.
- A network is node-congestion-free (resp., edge-congestion-free) if all nodes (resp., edges) are handling the same average traffic. A network is congestion-free it it is node-congestion-free and edge-congestion-free.

Chord, CAN, and Tapestry are all congestion-free algorithms. A strongly uniform algorithm is node-congestion-free.

The following result has been shown by Xu *et al.* [34]:

- When the routing algorithms are weakly uniform, $\Omega(log_2 n)$ and $\Omega(n^{1/d})$ are the lower bounds on the diameter in networks with routing tables of sizes $O(log\, n)$ and d, respectively. As Chord, CAN, and Tapestry are strongly uniform, they achieve the asymptotic lower bounds in the tradeoff.

18.9 Graph structures of complex networks

P2P overlay graphs can have different structures. An intriguing question is to characterize the structure of overlay graphs. This question is a small part of a much wider challenge of how to characterize large networks that grow in a distributed manner without any coordination [4]. Such networks exist in the following:

- Computer science: the WWW graph (WWW), the Internet graph that models individual routers and interconnecting links (INTNET), and the autonomous systems (AS) graph in the Internet.
- Social networks (SOC), the phonecall graph (PHON), the movie actor collaboration graph (ACT), the author collaboration graph (AUTH), and citation networks (CITE).
- Linguistics: the word co-occurrence graph (WORDOCC), and the word synonym graph (WORDSYN).
- The power distribution grid (POWER).
- Nature: in protein folding (PROT), where nodes are proteins and an edge represents that the two proteins bind together, and in substrate graphs for various bacteria and micro-organisms (SUBSTRATE), where nodes are substrates and edges are chemical reactions in which substrates participate.

It is widely intuited that such complex graphs must display some organizational principles that are encoded in their topology in some subtle ways. This has driven research on a unification theory to determine a suitable model in which all such uncontrolled graphs are instantiations.

The first logical attempt to model large networks without any known design principles is to use random graphs. The random graph model, also known as the Erdos–Renyi (ER) model [14], assumes n nodes and a link between each pair of nodes with probability p, leading to $n(n-1)p/2$ edges. Many interesting mathematical properties have been shown for random graphs. However, the complex networks encountered in practice are not entirely random, and show some, somewhat intangible, organizational principles.

Three ideas have received much investigative attention in recent times [4]:

- **Small world networks** Even in very large networks, the path length between any pair of nodes is relatively small. This principle of a "small world" was popularized by sociologist Stanley Milgram by the "six degrees of separation" uncovered between any two people [24].

  As the average distance between any pair of nodes in the ER model grows logarithmically with n, the ER graphs are small worlds.
- **Clustering** Social networks are characterized by cliques. The degree of cliques in a graph can be measured by various clustering coefficients, such as the following. Consider a node i having k_i out-edges. Let l_i be the actual number of edges among the k_i nearest neighbors of i. If these k_i nearest neighbors were in a clique, they would have $k_i.(k_i - 1)/2$ edges among them. The clustering coefficient for node i is $C_i = 2l_i/(k_i(k_i - 1))$. The network-wide clustering coefficient is the average of all C_is, for all nodes i in the network.

  The random graph model has a clustering coefficient of exactly p. As most real networks have a much larger clustering coefficient, this random graph model (ER) is unsatisfactory.
- **Degree distributions** Let $P(k)$ be the probability that a randomly selected node has k incident edges. In many networks – such as INTER, AS, WWW, SUBST – $P(k) \sim k^{-\gamma}$, i.e., $P(k)$ is distributed with a power-law tail. Such networks that are free of any characteristic scale, i.e., whose degree characterization is independent of n, are called **scale-free networks**.

  In a random graph, the degree distribution is Poisson-distributed with a peak of $P(\langle k \rangle)$, where $\langle k \rangle$, which is a function of n, is the average degree in the graph. Thus, random graphs are not scale-free. While some real networks have an exponential tail, the actual form of $P(k)$ is still very different from that for a Poisson distribution.

Current empirical measurements show the following properties of some commonly occuring graphs:

WWW In-degree and out-degree distributions both follow power laws; it is a small world; and is a directed graph, but does show a high clustering coefficient.

INTNET Degree distributions follow power law; small world; shows clustering.

AS Degree distributions follow power law; small world; shows clustering.

ACT Degree distributions follow power law tail; small world (similar path length as ER); shows high clustering.

AUTH Degree distributions follow power law; small world; shows high clustering.

SUBSTRATE In-degree and out-degree distributions both follow power laws; small world; large clustering coefficient.

PROT Degree distribution has a power law with exponential cutoff.
PHON In-degree and out-degree distributions both follow power laws.
CITE In-degree follows power law, out-degree has an exponential tail.
WORDOCC Two-regime power-law degree distribution; small world; high clustering coefficient.
WORDSYN Power-law degree distribution; small world; high clustering coefficient.
POWER Degree distribution is exponential.

Efforts on developing models focus on random graphs to model random phenomena, small worlds to interpolate between random graphs and structured clustered lattices, and scale-free graphs to study network dynamics and network evolutions.

18.10 Internet graphs

18.10.1 Basic laws and their definitions

In this section, we consider some properties of the Internet, that demonstrate a power-law behavior as measured empirically. The power law informally implies that large occurrences are very rare, and the frequency of the occurrence increases as the size decreases. Examples pertaining to the Web are: the number of links to a page, the number of pages within a Web location, and the number of accesses to a Web page. We begin by taking the example of the popularity of Websites to illustrate the definitions of three related observed laws [2]: Zipf's law, the Pareto law, and the Power law:

- **Power law** $P[X = x] \sim x^{-a}$
  This law is stated as a probability distribution function (PDF). It says that the number of occurrences of events that equal x is an inverse power of x. Figure 18.11(a) and (b) show the typical Power law PDF plots on both

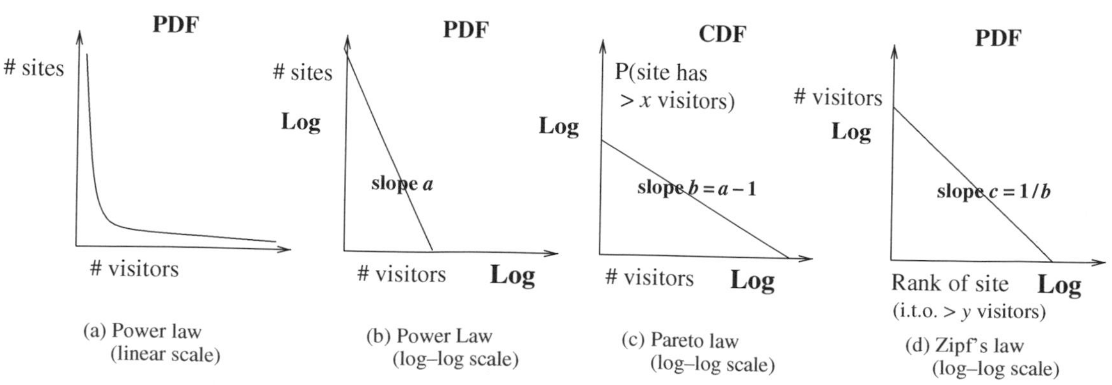

Figure 18.11 The popularity of Websites. (a) Power law showing the PDF using a linear scale. (b) Power law showing the PDF using a log–log scale. (c) Pareto law showing the CDF using a log–log scale. (d) Zipf's law using a log–log scale [2].

linear and log–log scales, respectively. In the log–log plot, the slope is a.

In our example, this corresponds to the number of sites that have exactly x visitors.

- **Pareto law** $P[X \geq x] \sim x^{-b} = x^{-(a-1)}$
This law is stated as a cumulative distribution function (CDF). The number of occurrences larger than x is an inverse power of x. The CDF can be obtained by integrating the PDF. The exponents a and b of the Pareto (CDF) and Power laws (PDF) are related as $b+1=a$. Figure 18.11(c) shows the Pareto law CDF plot on a log–log scale. In the log–log plot, the slope is $b=a-1$.
In our example, this corresponds to the number of sites that have at least x visitors.
- **Zipf's law** $n \sim r^{-c}$
This law states the count n (i.e., the number) of the occurrences of an event, as a function of the event's rank r. It says that the count of the rth largest occurrence is an inverse power of the rank r. Figure 18.11(d) shows the Zipf plot on a log–log scale. In the log–log plot, the slope is c, which, as we see below, is $1/b = 1/(a-1)$.
The context initally used by Zipf was the frequency of occurrence of words in English, where the most frequently occurring word had rank 1. The Zipf law is widely occurring, e.g., both the magnitude of earthquakes and the populations of cities also follow this law. In our example, this corresponds to the number of visits to the rth most popular site.

Clearly, the Pareto law (CDF) and Power law (PDF) are related. Zipf's law $n \sim r^{-c}$, states that "the r-ranked object has $n = r^{-c}$ occurrences," and can be equivalently expressed as: "r objects (x-axis) have $n = r^{-c}$ (y-axis) or more occurrences." This becomes the same as the Pareto law's CDF after transposing the x and y axes, i.e., by restating as: "the number of occurrences larger than $n = r^{-c}$ (x-axis) happens for r objects (y-axis)."

From Zipf's law, $n = r^{-c}$, hence, $r = n^{-1/c}$. Hence, the Pareto exponent b is $1/c$. As $b = (a-1)$, where a is the Power law exponent, we see that $a = 1 + (1/c)$. Hence, the Zipf's law distribution also satisfies a Power law PDF.

18.10.2 Properties of the Internet

The Internet is a prime example of a complex entity that exhibits power-law behavior. Based on extensive empirical measurements, Siganos *et al.* [31] showed the following results:

- **Rank exponent/Zipf law** The nodes in the Internet graph are ranked in decreasing order of their degree. When the degree d_i is plotted as a function of the rank r_i on a log–log scale, the graph is like Figure 18.11(d). The slope is termed the rank exponent $\mathcal{R}$, and $d_i \propto r_i^{\mathcal{R}}$. If the minimum degree

$d_n = m$ is known, then $m = d_n = Cn^{\mathcal{R}}$, implying that the proportionality constant C is $m/n^{\mathcal{R}}$. Exercise 18.6 asks you to estimate the number of edges as a function of the rank exponent and the number of nodes.

- **Degree exponent/ PDF and CDF** Let the CDF f_d of the node degree d be the fraction of nodes with degree greater than d. Then $f_d \propto d^{\mathcal{D}}$, where $\mathcal{D}$ is the degree exponent that is the slope of the log–log plot of f_d as a function of d.

  Analogously, let the PDF be g_d. Then $g_d \propto d^{\mathcal{D}'}$, where $\mathcal{D}'$ is the degree exponent that is the slope of the log-log plot of g_d as a function of d.

  Empirically, $D' \sim D+1$, as theoretically predicted. Further, $\mathcal{R} \sim (1/\mathcal{D})$, also as theoretically predicted. The imperfect match is attributed to imperfect measurements and approximations in curve-fitting. In practice, the CDF is preferred as it can be estimated with greater accuracy.
- **Eigen exponent $\mathcal{E}$** For the adjacency matrix A of a graph, its eigenvalue λ is the solution to $AX = \lambda X$, where X is a vector of real numbers. The eigenvalues are related to the graph's number of edges, number of connected components, the number of spanning trees, the diameter, and other important topological properties. Let the various eigenvalues be λ_i, where i is the order and between 1 and n. Then the graph of λ_i as a function of i is a straight line, with a slope of $\mathcal{E}$, the eigen-exponent. Thus, $\lambda_i \propto i^{\mathcal{E}}$. More intriguingly, when the eigenvalues and the degree are sorted in descending order, it is found that $\lambda_i = \sqrt{d_i}$, implying that $\mathcal{E} = \mathcal{D}/2$.

The following additional hypotheses have not been very vigorously tested and verified. Nevertheless, they offer insightful looks into the prevalance and use of power laws in complex uncontrolled entities such as the Internet. Two definitions are useful at this stage:

- *PN*(h) is the number of pairs of nodes within h hops, counting self-pairs, and counting all other pairs twice due to the dual edge incidence.
- *NN*(h), the neighborhood, is the expected number of nodes within h hops.

- **Hop-plot exponent**, $\mathcal{H}$ Experimental measurements have shown that *PN*(h) follows a power law regime more closely, rather than the exponential regime as previously estimated. Thus, $PN(h) \propto h^{\mathcal{H}}$, where $\mathcal{H}$ is the slope of the log-log plot of *PN*(h) as a function of h for $h \ll dia$. From the definition of *PN*(h), observe that $PN(1) = n+2l$, where l is the number of edges. Hence,

$$PN(h) = \begin{cases} (n+2l)h^{\mathcal{H}}, & \text{if } h \ll dia, \\ n^2, & \text{if } h \geq dia. \end{cases} \tag{18.9}$$

The hop-plot exponent is useful to estimate the effective diameter dia_{eff} of the network. Informally, any two nodes in the network are within dia_{eff}

hops of each other, with "high probability." When some destination node whose location is unknown needs to be reached, the use of hop-constrained broadcast is the standard solution. A large hop count takes too long, whereas a small hop count may not reach the entire network. If the hop count is set to dia_{eff}, then with high probability, the destination can be reached with just the right amount of overhead. Using n, $\mathcal{H}$, and the number of edges l, the effective diameter is defined as:

$$dia_{eff} = \left(\frac{n^2}{n+2l}\right)^{1/\mathcal{H}}.$$

This effective diameter is estimated as the abscissa of the intersection of the log–log hop-plot with slope $\mathcal{H}$ and the n^2 coverage that is expected within diameter hops.

Observe that the average size of the neighbourhood $NN(h) = (PN(h)/n) - 1$. Hence $NN(h) = ((n+2l)h^{\mathcal{H}}/n) - 1$. The $NN(h)$ is seen to be a more accurate estimate of the neighborhood than the traditional *average-degree estimate*, $NN'_d(h) = \overline{d}(\overline{d}-1)^{h-1}$. The $NN'_d(h)$ estimate assumes that the degree distribution is more uniform, and that each hop adds $\overline{d}-1$ new nodes per node at the boundary of the examined neighborhood. As the degree distribution is highly skewed, the traditional $NN'(h)$ metric is not accurate.

For all the cases above, the power law regime has so far been empirically validated. The exponent itself has been observed to change gradually over time as the networks evolve. The power law regime provides a good handle on predicting the future growth of the Internet, and building accurate graphs for simulations.

Classification of scale-free networks

Scale-free networks of different types – WWW, INTNET, AS, ACT, AUTH, SUBSTRATE, PROT, PHON, in-degree for CITE, and WORDSYN – have different degree exponents, typically ranging from 2 to 3. The quest to seek a more universal and common factor resulted in the analysis of another metric, called the "betweenness centrality" [15]. For any graph, let its geodesics, i.e., set of shortest paths, between any pair of nodes i and j, be denoted $S(i, j)$. Let $S_k(i, j)$ be a subset of $S(i, j)$ such that all the geodesics in $S_k(i, j)$ pass through node k. The betweeness centrality BC of node k, b_k, is $\sum_{i\neq j} g_k(i, j) = \sum_{i\neq j} |S_k(i, j)|/|S(i, j)|$. The b_k denotes the importance of node k in shortest-path connections between all pairs of nodes in the network.

The metric BC follows the power law $P_{BC}(g) \sim g^{-\beta}$, where β is the BC-exponent. Unlike the degree exponent which varies across different network types, the BC-exponent has been empirically found to take on values of only 2 or 2.2 for these varied network types. This interesting observation is under further study.

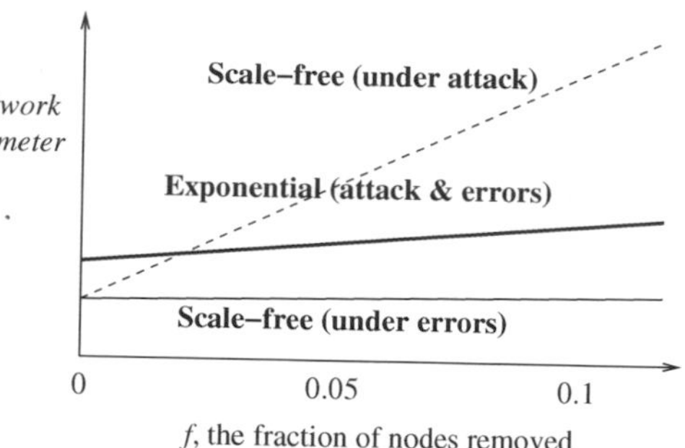

Figure 18.12 Impact of attacks and failures on the diameter of exponential networks and scale-free networks, from Albert *et al.* [5].

18.10.3 Error and attack tolerance of complex networks

Based on the node degree distribution $P(k)$, two broad classes of small world networks are the exponential networks and the scale-free networks. In exponential networks, such as the ER random graph model and the Watts–Strogatz small world model [33], $P(k)$ reaches a maximum at a $\overline{k}$ value and then $P(k)$ decreases exponentially per a Poisson distribution as k increases. In scale-free networks, such as the Web and the Internet, $P(k)$ decreases as per a power law, $P(k) \sim k^{-\gamma}$.

The following are two key differences that leads to different behavior of exponential networks and of scale-free networks, under errors and attacks: (i) nodes with a very high degree are statistically significant in scale-free networks, whereas they are close to an impossibility in exponential networks; (ii) in an exponential network, all nodes have about the same number of links, whereas in a scale-free network, some nodes have many links and the majority of the nodes have a small number of links.

Errors are simulated by removing nodes at random. Attacks are simulated by removing the nodes with highest degree. Their impact is measured on network diameter and network partitioning [5].

Impact on network diameter

Figure 18.12 is used to descibe the impact on the diameter. The graph shows only the relative trends, as empirically verified by simulations for many large networks, including the Web and Internet. Any numbers simply in the graph convey an approximate order of magnitude for the particular networks studied by Albert *et al.* [5].

- **Errors** In an exponential network, as all nodes have about the same degree, the removal of any node has approximately the same amount of small impact in terms of decrease in connectivity. The network diameter increases gradually. The diameter of scale-free networks remains almost same under errors, as nodes that are removed have small degree with very high probability and are very unlikely to alter the lengths of the paths among other nodes.

- **Attacks** As nodes in an exponential network have about the same degree, the network behaves similarly under attack as under errors. Under attack, the diameter of scale-free networks increases dramatically, as the few nodes with highest connectivity are removed, thereby greatly reducing the connectivity of the entire network.

Impact on network partitioning

The impact of removal of nodes on partitioning is measured using two metrics: S_{max}, the ratio of the size of the largest cluster to the system size, and S_{others}, the average size of all clusters except the largest.

- **Exponential networks** In Figure 18.13, as f, the fraction of nodes removed is increased, S_{others} increases from 1 to around 2 for some threshold fraction $f_{threshold}$. This implies that for very small f, where $S_{others} \sim 1$, single nodes break off. As f increases, several small but larger partitions set in, leading to a peak of S_{others} at $f_{threshold}$. For $f > f_{threshold}$, S_{others} reduces back to 1, as the isolated clusters (fragments) in the network further disintegrate. In terms of S_{max}, as f is varied from 0 to $f_{threshold}$, S_{max} decreases from 1 to a low value as small (mostly single-node) partitions break off. As $f_{threshold}$ is approached, the main cluster disintegrates, leading to S_{max} tending to 0. As f is increased beyond $f_{threshold}$, S_{max} remains near 0.

  The impact of attacks on network partitioning is the same as the impact of errors, for the same reasoning given for the analysis on the diameter.
- **Scale-free networks** In Figure 18.14, when nodes are randomly removed, S_{max} decreases from 1 very gradually. Also, S_{others} remains steady at 1, indicating that singleton nodes get removed from the main network. There is no threshold $f_{threshold}$ observed, even for high values of f, such as 0.5 error rate.

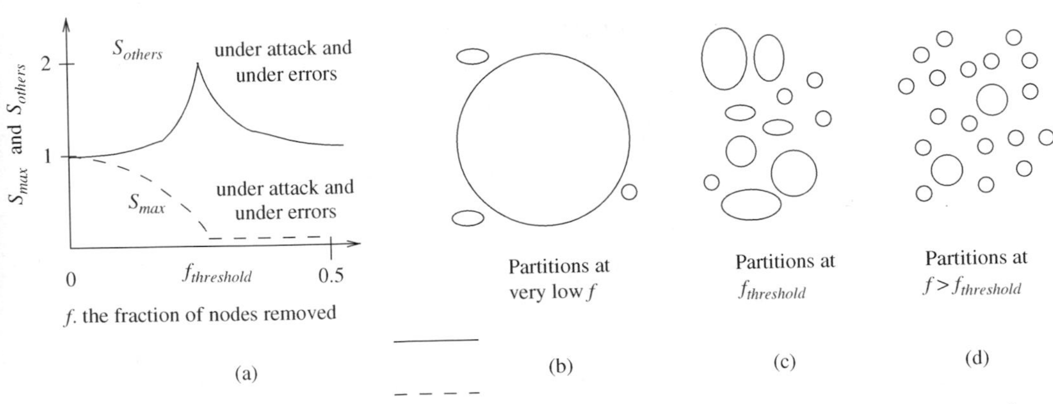

Figure 18.13 Impact of errors and attacks on cluster size of exponential networks, from Albert *et al.* [5]. (a) Graphical trend. (b) Pictoral cluster sizes for low f, i.e., $f \ll f_{threshold}$. (c) Pictoral cluster sizes for $f \sim f_{threshold}$. (d) Pictoral cluster sizes for $f > f_{threshold}$. The pictoral trend in (b)–(d) is also exhibited by scale-free networks under attack, but for a lower value of $f_{threshold}$.

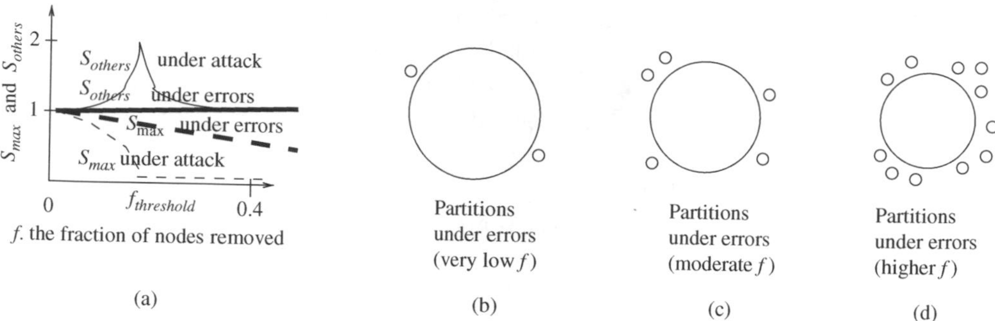

Figure 18.14 Impact of errors on cluster size of scale-free networks, from Albert *et al.* [5]. The pictoral impact of attacks on cluster sizes are similar to those in Figure 18.13. (a) Graphical trend. (b) Pictoral cluster sizes for low f under failure. (c) Pictoral cluster sizes for moderate f under failure. (d) Pictoral cluster sizes for high f under failure.

However, under attack, when the most connected nodes are removed, the behavior is similar to (but more acute than) that of the exponential network; see Figure 18.13. Thus, the threshold $f_{threshold}$ sets in at a lower value. This is because the impact of removing the highly connected nodes first causes disintegration to set in quickly.

18.11 Generalized random graph networks

Random graphs cannot capture the scale-free nature of real networks, which states that the node degree distribution follows a power law. The *generalized random graph model* uses the degree distribution as an input, but is random in all other respects. Thus, the constraint that the degree distribution must obey a power law is superimposed on an otherwise random selection of nodes to be connected by edges. These semi-random graphs can be analyzed for various properties of interest. Although a simple formal model for the clustering coefficient is not known, it has been observed that generalized random graphs have a random distribution of edges similar to the ER model, and hence the clustering coefficient will likely tend to zero as N increases.

18.12 Small-world networks

Real-world networks are small worlds, having small diameter, like random graphs, but they have relatively large clustering coefficients that tend to be independent of the network size.

Ordered lattices tend to satisfy this property that clustering coefficients are independent of the network size. Figure 18.15(a) shows a one-dimensional

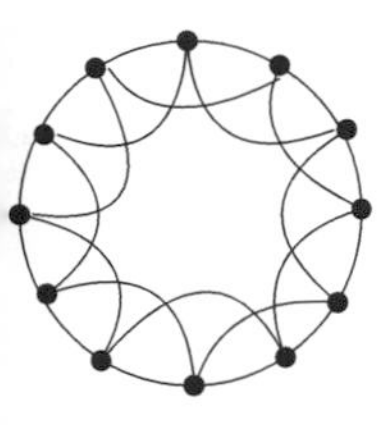

(a)

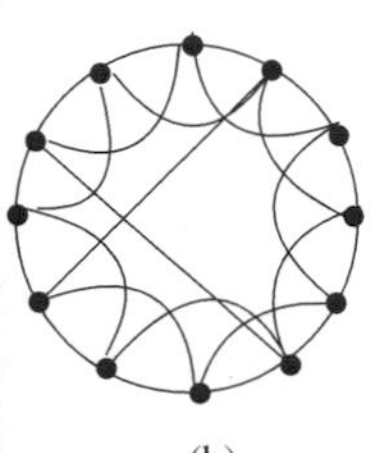

(b)

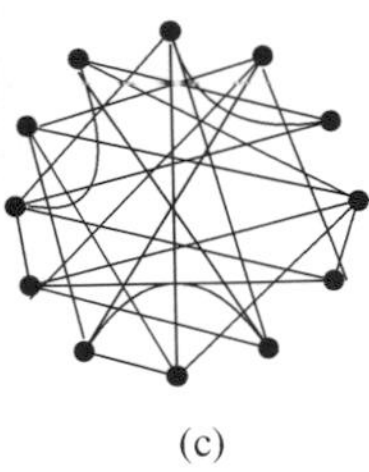

(c)

Figure 18.15. The Watts–Strogatz random rewiring procedure [4,33]. (a) Regular. (b) Small-world. (c) Random. The rewiring shown maintains the degree of each node.

lattice in which each node is connected to $k = 4$ closest nodes. The clustering coefficient is $C = \frac{3(k-2)}{4(k-1)}$.

The first model for small world graphs with high clustering coefficients and low path length is the Watts–Strogatz (WS) model [33]:

1. Define a ring lattice with n nodes and each node connected to k closest neighbors ($k/2$ on either side). Let $n \gg k \gg ln(n) \gg 1$.
2. Rewire each edge randomly with probability p. When $p = 0$, there is a perfect structure, as in Figure 18.15(a). When $p = 1$, complete randomness, as in Figure 18.15(c).

A characteristic of small-world graphs is the small average path length. When p is small, len scales linearly with n, but when p is large, len scales logarithmically. Through analytical arguments and simulations, it is now believed that the characteristic path length varies as:

$$len(n, p) \sim \frac{n^{1/d}}{k} f(pkn), \tag{18.10}$$

where the function f behaves as follows:

$$f(u) = \begin{cases} \text{constant}, & \text{if } u \ll 1, \\ ln(u)/u, & \text{if } u \gg 1. \end{cases} \tag{18.11}$$

The variable u has the intuitive interpretation that it depends on the average number of random links that provide "jumps" across the graph, and $f(u)$ is the average factor by which the distance between a pair of nodes gets reduced by the "jumps."

18.13 Scale-free networks

Many real networks are scale-free, and even for those that are not scale-free, the degree distribution follows an exponential tail that is significantly different from that of the Poisson distribution. Semi-random graphs that are constrained to obey a power law for the degree distributions and constrained to have large clustering coefficients yield scale-free networks, but do not shed any insight into the mechanisms that give birth to scale-free networks. Rather than modeling the network topology, it is better to model the network assembly and evolution process. Specifically:

Figure 18.16 The simple Barabasi–Albert model [7].

Initially, there are m_0 isolated nodes. At each sequential step, perform one of the following operations:

Growth Add a new node with m edges, (where $m \leq m_0$), that link the new node to m different nodes already in the system.

Preferential attachment The probability $\prod$ that the new node will be connected to node i depends on the degree k_i, such that:

$$\prod(k_i) = \frac{k_i}{\sum_j(k_j)}. \tag{18.12}$$

- Rather than begin with a constant number of nodes n that are then randomly connected or rewired, real networks (e.g., WWW, INTERNET) exhibit *growth* by the addition of nodes and edges.
- Rather than assume that the probability of adding (or rewiring) an edge between two nodes is a constant, real networks exhibit the property of *preferential attachment*, where the probability of connecting to a node depends on the node degree.

The simple Barabasi–Albert model [7], which captures growth and preferential attachment, is described in Figure 18.16. After t time steps, there are $t+m_0$ nodes and mt edges. Numerically, it is verified that the degree distribution follows a power law with degree $=3$, that is independent of the parameter m.

Two techniques to analyze the degree distribution of models are now described in the context of the BA model. The master-equation approach was introduced by Dogorotsev *et al.* [13] and the rate-equation approach was introduced by Krapivsky *et al.* [22].

18.13.1 Master-equation approach

Let $p(k, t_i, t)$ denote the probability that, at time t, a node i that was added at time t_i has degree k. When a new node with m edges is added to the graph, the degree of node i increases by one with probability $m \cdot \prod(k) = k/2t$. Hence, we have [4,13]:

$$p(k, t_i, t+1) = \frac{k-1}{2t} \cdot p(k-1, t_i, t) - \left[1 - \frac{k}{2t}\right] \cdot p(k, t_i, t). \tag{18.13}$$

The first term is the probability that a node with $k-1$ degree gets a new edge; the second term is the probability that a node with degree k does not get a new edge. Based on this formulation, the degree distribution can be expressed as:

$$P(k) = limit_{t\to\infty} \sum_{t_i} p(k, t_i, t)/t. \tag{18.14}$$

From Eq. (18.13), it can be shown that:

$$P(k) = \begin{cases} \frac{k-1}{k+2} P(k-1), & \text{if } k \geq m+1, \\ \frac{2}{m+2}, & \text{if } k = m. \end{cases} \tag{18.15}$$

This solves as:

$$P(k) = \frac{2m(m+1)}{k(k+1)(k+2)}. \tag{18.16}$$

18.13.2 Rate-equation approach

Let $n_k(t)$ be the average number of nodes having k edges at time t. When a new node is added, $n_k(t)$ changes as follows. New edges are added to some nodes with degree $k-1$, new edges are added to some nodes with degree k, and new nodes with m edges are added. These three changes affect $n_k(t)$ in the following manner:

$$\frac{dn_k}{dt} = m \cdot \left[\frac{(k-1) \cdot n_{k-1}(t)}{\sum_k kn_k(t)} - \frac{k \cdot n_k(t)}{\sum_k kn_k(t)} \right] + \delta_{k,m}. \tag{18.17}$$

By taking the asymptotic limit, $n_k(t) = t \cdot P(k)$, and $\sum_k kn_k(t) = 2mt$. This yields the same recursive Eq. (18.15) obtained using the master-equation approach.

18.14 Evolving networks

The BA algorithm in Figure 18.16 represents a basic model that cannot fully capture real network properties. For example, the BA model has a fixed exponent of 3 for the power law, independent of the parameter m. Real networks have an exponent that varies, typically between 1 and 3. Some real networks sometimes have exponential cutoffs that are not within the power law regime. The study of more general and flexible models that can accurately capture real networks has lead to several notable directions of investigation:

- **Preferential attachment** The BA model assumed that the probability $\prod(k)$ that a new node connects to a node i is proportional to the degree k_i. This implied that $\prod(k)$ is linearly proportional to k.

  It has been shown analytically that for *sublinear preferential attachment* as well as for *superlinear preferential attachment*, the scale-free nature of the network cannot be preserved.

  In real networks, there is a finite probability that a new node attaches to an isolated node, i.e., $\prod(0) \neq 0$ and $\prod(k) = C + k^{\alpha}$, where C denotes the *intial attractiveness*. It can be seen that initial attractiveness changes the degree exponent but preserves the scale-free nature of the degree distribution.

- **Growth** The BA model assumed that the rate of addition of nodes and edges was uniform. Many real networks, such as INTNET, AS, WEB, SUBSTRATE, and WORDOCC, have the property that the number of edges increases faster than the number of nodes, implying an increase in the average degree as the number of nodes increases. It has been shown analytically that accelerated growth does not affect the power law nature although the exponent degree is altered.
- **Local events** Real networks undergo local (microscopic) changes to the topology, such as node addition and node deletion, edge addition and edge deletion. A popular model that explores the properties of such local events is the extended Barabasi–Albert model [3], shown in Figure 18.17.
- **Growth constraints** Real networks often have bounded capacity for the number of edges (e.g., connections at a router) or a finite lifetime for the nodes (as in social networks). In the electrical power distribution network which exhibits an exponential distribution, there are practical reasons why the node degree is bounded. In the actors network, which exhibits a power law with an exponential cutoff for large k, ageing limits the accrual of new edges. Thus, ageing and finite capacity need to explicitly captured in a good model for such networks.
- **Competition** Real-world networks exhibit competition, wherein some nodes can attract more edges (e.g., via advertising) at the cost of other nodes. This feature can be modeled by a fitness parameter. Similarly, a new node may inherit edges belonging to some other node or nodes (e.g., modifying a replica of a Web page). This needs to be explicitly modeled.
- **Induced preferential attachment** Various local-level mechanisms, such as the copying mechanism (copy edges of another node as in Web

Figure 18.17 The extended Barabasi–Albert model [3].

Initially, there are m_0 isolated nodes. At each sequential step, perform one of the following operations:

With probability p, add m, where $m \leq m_0$, new edges For each new edge, one end is randomly selected, the other end with probability

$$\prod(k_i) = \frac{k_i + 1}{\sum_j (k_j + 1)}. \quad (18.18)$$

With probability q, rewire m edges To rewire an edge, randomly select node i, delete some edge (i, w), add edge (i, x) to node x that is chosen with probability $\prod(k_x)$ as per Eq. (18.18).

With probability $1 - p - q$, insert a new node Add m new edges to the new node, such that with probability $\prod(k_i)$, an edge connects to a node i already present before this step.

pages), and tracing selected walks (as in recursively following the citation trail in a citation network), need to be modeled because they implicitly introduce preferential attachment.

18.14.1 Extended Barabasi–Albert model

The extended BA model [3] is an example model for evolving networks.

Continuum theory analysis

In continuum theory, it is assumed that k_i changes continuously and the probability $\prod(k_i)$ then represents the rate at which k_i changes. Each of the three possible events in a sequential step can affect the rate at which k_i changes as follows [3]:

1. With probability p, m new links are added. For each link, one end is randomly chosen, leading to a change in k_i of pm/n. For each link, the second end attaches preferentially, leading to a change in k_i of $pm \cdot \frac{(k_i+1)}{\sum_j (k_j+1)}$. Hence,

$$\frac{dk_i}{dt} = pm\frac{1}{n} + pm\frac{k_i+1}{\sum_j (k_j+1)}. \tag{18.19}$$

2. With probability q, m existing links are rewired. For each rewired link, a randomly chosen node loses one incident edge, which then attaches preferentially. Thus, the impact on k_i is:

$$\frac{dk_i}{dt} = -qm\frac{1}{n} + qm\frac{k_i+1}{\sum_j (k_j+1)}. \tag{18.20}$$

3. With probability $(1-p-q)$, a new node is added with m links. Each of the m links connects preferentially, thus:

$$\frac{dk_i}{dt} = (1-p-q)m\frac{k_i+1}{\sum_j (k_j+1)}. \tag{18.21}$$

Summing the three effects, we have:

$$\frac{dk_i}{dt} = (p-q)m\frac{1}{n} + m\frac{k_i+1}{\sum_j (k_j+1)}. \tag{18.22}$$

As the system size and topology varies with time, we have:

$$n(t) = m_0 + (1-p-q)t; \qquad \sum_j k_j = 2mt(1-q) - m. \tag{18.23}$$

As t increases, the constants m and m_0 can be deleted. Further, for a node added at t_i, we have that $k_i(t_i) = m$ (the initialization step). Exercise 18.8 asks you to show that the solution to Eq. (18.22) has the form

$$k_i(t) = [A(p,q,m) + m + 1]\left(\frac{t}{t_i}\right)^{1/B(p,q,m)} - A(p,q,m) - 1, \qquad (18.24)$$

$$A(p,q,m) = (p-q)\left(\frac{2m(1-q)}{1-p-q} + 1\right), \qquad B(p,q,m) = \frac{2m(1-q)+1-p-}{m} \qquad (18.25)$$

Based on further algebraic derivations, Albert and Barabasi [3] showed that:

$$P(k)\ \alpha\ [k + \kappa(p,q,m)]^{-\gamma(p,q,m)}, \text{ where } \kappa(p,q,m) = A(p,q,m) + 1 \text{ and}$$
$$\gamma(p,q,m) = B(p,q,m) + 1, \qquad (18.26)$$

Equation (18.26) is valid if, for a fixed p and m,

$$q < q_{max} = min(1-p, (1-p+m)/(1+2m)).$$

There are now two cases:

$q < q_{max}$: Eq. (18.26) is valid and the degree distribution is a power law and is scale-free.

$q > q_{max}$: Eq. (18.26) is invalid, and $P(k)$ can be shown to behave like an exponential distribution. The model now behaves like the ER and WS models.

This is similar to the behavior seen in real networks – some networks show a power law while others show an exponential tail – and a single model can capture both behaviors by tuning the parameter q. The scale-free regime

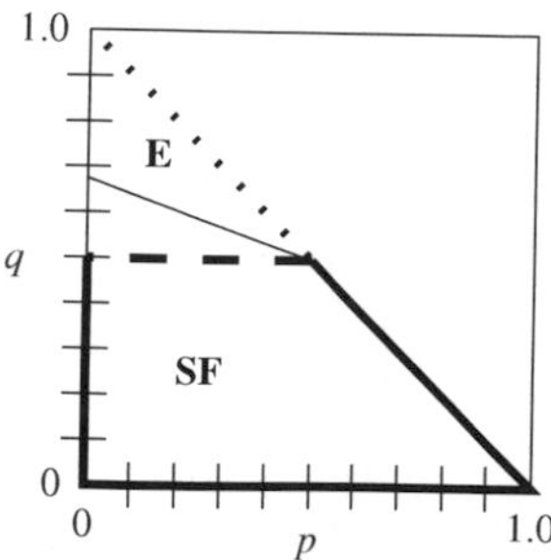

Figure 18.18 Phase diagram for the extended Barabasi–Albert model [3]. **SF** denotes the scale-free regime, which is enclosed by the thick border. **E** denotes the exponential regime that exists in the remainder of the lower diagonal region of the graph. The plain line shows the boundary for $m = 1$, having a y-axis intercept at 0.67.

and the exponential regime are marked in the graph in Figure 18.18. The boundary between the two regimes depends on the value of m and has slope $-m/(1+2m)$. The area enclosed by thick lines shows the scale-free regime; the dashed line is its boundary when $m \to \infty$ and the dotted line is its boundary when $m \to 0$.

18.15 Chapter summary

Peer-to-peer (P2P) networks allow equal participation and resource sharing among the users. This chapter first analyzed the different types of P2P networks. Unstructured P2P networks are like Gnutella and BitTorrent. We studied different search mechanisms – flooding, constrained flooding, and blind search – for such unstructured networks. We also examined some data replication strategies, and their impact on the search performance. The chapter then studied three classical structured P2P networks – Chord, CAN, Tapestry – all of which use the distributed hash table concept in their implementations. Although all the three mechanisms differ, they are similar in that they represent different tradeoffs in search efficiency, i.e., path length, and the amount of local storage for implementing the hash tables. The spectrum of P2P networks from unstructured to structured offer a wide range of tradeoffs for user requirements. The chapter also examined issues such as fairness and trust management. These issues are important because, in the P2P environment where there is no control authority, the system must be able to autonomously alllow for fairness.

The Internet, AS-AS level internets, and Web (WWW) overlays exhibit some interesting properties about how they grow and evolve. Many network overlays outside of computer science also exhibit the same properties. The chapter studied several properties of the Internet and Web graphs. Then, in a more general setting, the chapter examined random networks, small-world networks, node degree distributions, scale-free networks, and the impact of error and attack tolerance on such networks. Networks grow in an uncontrolled fashion, yet there seems to be some underlying basis for such growth. Of the several proposals to model the growth of networks, we studied the Barabasi–Albert model, which appears to be promising in its applicability to not just computer science networks, but also to networks in other disciplines and natural phenomena.

18.16 Exercises

Exercise 18.1 (Replication) Derive the values of average search size A, A_i, and utilization u_i for square-root replication. The derived answers should match the entries in Table 18.3.

Exercise 18.2 (Fault-tolerance in Chord) Adapt the code in Algorithm 18.3 so that the nodes manage a successor list of α successors, rather than a single successor.

Exercise 18.3 (Chord) In the Chord protocol, assume that the successor list at each node has $\alpha = \Omega(log\, n)$ nodes. Show the following:

1. If a Chord ring is initially stable, and if the probability of subsequent failure of each node is 0.5, then *Locate_Successor* returns the closest functional successor node to the key being searched with high probability.
2. If a Chord ring is initially stable, and if the probability of subsequent failure of each node is 0.5, it takes $O(log\, n)$ average-case time for *Locate_Successor* to complete.

Exercise 18.4 (CAN) Compute the time and message complexity of the distributed region reassignment protocol that is run periodically by the CAN protocol.

Exercise 18.5 (CAN) Identify all the changes to the base CAN protocol to accommodate the optimization of overloading coordinate regions, discussed in Section 18.5.5.

Exercise 18.6 (Power law in the Internet [31]) Show that the number of edges l in the Internet graph that obeys the power law for the rank exponent is given as follows. Let the graph have n nodes and rank exponent $\mathcal{R}$. Then:

$$l \sim \frac{1}{2(\mathcal{R}+1)}(1 - \frac{1}{n^{\mathcal{R}+1}})n.$$

Exercise 18.7 Show that Eq. (18.15) using the master-equation approach for the degree distribution in the extended BA model can be solved as Eq. (18.16).

Exercise 18.8 Show that the solution to Eq. (18.22) for the degree distribution in the extended BA model using continuum theory analysis is given by Eq. (18.25).

18.17 Notes on references

The introduction is based on the survey by Risson and Moors [29] and Androutsellis-Theotokis and Spinellis [6]. The discussion on replication and search in unstructured networks is based on Cohen and Shenker [12], and on Lv *et al.* [23], respectively. Gnutella [16,17], Napster [25], and Freenet [10] are widely implemented commercial P2P protocols. The Chord protocol was proposed by Stoica *et al.* [32]. The content addressible network (CAN) was proposed by Ratnasamy *et al.* [27]. The design of Tapestry [20,21,35,36] and the related Pastry [30] overlay was based on the ideas of Plaxton trees proposed by Plaxton *et al.* [26]. Tapestry built on the Plaxton trees by providing better fault-tolerance and resilience in the face of node joins and departures. The discussion on fundamental tradeoffs between routing table size and network diameter is based on Xu *et al.* [34] and Ratnasamy *et al.* [28]. The BitTorrent system was initially proposed by Cohen [11]. The discussion of trust management is based on Gupta *et al.* [18,19] and Aberer and Despotovic [1].

The discussion on the graph structures of complex networks is structured and based on the excellent survey by Albert and Barabasi [4]. The discussion on power laws and Zipf's law is taken from the tutorial by Adamic [2]. The power laws for the Internet were discovered by Siganos and the Faloutsos brothers [31]. The discussion on the betweenness centrality metric for graphs is based on the work by Goh *et al.* [15]. The random graphs model was proposed and analyzed by Erdos and Renyi [14]. Further results on the properties on random graphs were given by Bollobas [8,9]. The small worlds model was proposed by Watts and Strogatz [33]. The extended Barabasi–Albert model for graph evolution was given by Albert and Barabasi [3]. The analysis of error and attack tolerance on exponential networks and on scale-free networks was done by Albert *et al.* [5].

References

[1] K. Aberer and Z. Despotovic, Managing trust in a peer-to-peer information system, *Proceedings of the 10th International Conference on Information and Knowledge Management*, Atlanta, Georgia, USA, November 2001, 310–317.

[2] L. Adamic, *Zipf, Power-Laws, and Pareto – A Ranking Tutorial*, available online at: www.hpl.hp.com/research/idl/papers/ranking/ranking.html.

[3] R. Albert and A.-L. Barabasi, Topology of evolving networks: local events and universality, *Physical Review Letters*, **85**(24), 2000, 5234–5237.

[4] R. Albert and A.-L. Barabasi, Statistical mechanics of complex networks, *Review of Modern Physics*, **74**(1), 2002, 47–97.

[5] R. Albert, H. Jeong, and A. Barabasi, Error and attack tolerance of complex networks, *Nature*, **406**, 2000, 378–381.

[6] S. Androutsellis-Theotokis and D. Spinellis, A survey of peer-to-peer content distribution technologies, *ACM Computing Surveys*, **36**(4), 2004, 335–371.

[7] A.-L. Barabasi and R. Albert, Emergence of scaling in random networks, *Science*, 286, 1999, 509–512.

[8] B. Bollobas, Degree sequences of random graphs, *Discrete Math*, **33**, 1981, 1–9.

[9] B. Bollobas, *Random Graphs*, London, Academic Press, 1985.

[10] I. Clarke, O. Sandberg, B. Wiley, and T. W. Hong, Freenet: a distributed anonymous information storage and retrieval system, *Workshop on Design Issues in Anonymity and Unobservability*, Berkeley, CA, July 2000, 46–66.

[11] B. Cohen, *Incentives Build Robustness in BitTorrent*, available online at: www.bittorrent.com/bittorrentecon.pdf.

[12] E. Cohen and S. Shenker, Replication strategies in unstructured peer-to-peer networks, *ACM SIGCOMM*, 2002, 177–190.

[13] S. Dogorotsev, J. Mendes, and A. Samukhin, Structure of growing networks: exact solution of the Barabasi–Albert model, *Physical Review Letters*, **85**, 2000, 4633–4636.

[14] P. Erdos and A. Renyi, Random graphs. **6**, 1959, 290–.

[15] K. Goh, E. Oh, H. Jeong, B. Kahng, and D. Kim, Classification of scale-free networks, *Proceedings of the National Academy of Sciences*, 2002.

[16] Gnutella, `www.gnutella.com/`.

[17] The Gnutella protocol specification, available online at: `www9.limewire.com/developer/gnutella_protocol_0.4.pdf`.

[18] M. Gupta, P. Judge, and M. Ammar, A reputation system for peer-to-peer networks, *Proceedings of the 13th International Workshop on Network and Operating Systems Support for Digital Audio and Video*, Monterey, CA, June 2003, 144–152.

[19] M. Gupta, M.H. Ammar, and M. Ahamad, Trade-offs between reliability and overheads in peer-to-peer reputation tracking, *Computer Networks*, **50**(4), 2006, 501–522.
[20] K. Hildrum, J. Kubiatowicz, S. Rao, and B. Y. Zhao, Distributed object location in a dynamic network, *Proceedings of ACM SPAA 2002*, 41–52.
[21] K. Hildrum, J. Kubiatowicz, S. Rao, and B. Y. Zhao, Distributed object location in a dynamic network, *Theory of Computing Systems*, **37**, 2004, 405–440.
[22] P. Krapivsky, S. Redner, and F. Leyvraz, Connectivity of growing random networks, *Physical Review Letters*, **85**, 2000, 4629–4632.
[23] Q. Lv, P. Cao, E. Cohen, K. Li, and S. Shenker, Search and replication in unstructured peer-to-peer networks, *International Conference on Supercomputing*, 2002, 84–95.
[24] S. Milgram, The small world problem, *Psychology Today*, **1**(2), 1967, 60–67.
[25] Napster, `www.napster.com/`.
[26] C.G. Plaxton, R. Rajaraman, and A.W. Richa, Accessing nearby copies of replicated objects in a distributed environment, *Proceedings of ACM SPAA 1997*, 311–320.
[27] S. Ratnasamy, P. Francis, M. Handley, R.M. Karp, and S. Shenker, A scalable content-addressable network, *Proceedings of ACM SIGCOMM 2001*, 161–172.
[28] S. Ratnasamy, I. Stoica, and S. Shenker, Routing algorithms for DHTs: some open questions, *Proceedings of IPTPS 2002*, 45–52.
[29] J. Risson and T. Moors, Survey of research towards robust peer-to-peer networks: search methods, *Computer Networks*, **50**(17), 2006, 3485–3521.
[30] A. Rowstron and P. Druschel, Pastry: scalable, distributed object location and routing for large-scale peer-to-peer systems, *Proceedings of the IFIP/ACM Middleware 2001*, Heidelberg, Germany, November 2001, 329–350.
[31] G. Siganos, M. Faloutsos, P. Faloutsos, and C. Faloutsos, Power laws and the AS-level internet topology, *IEEE/ACM Transactions on Networking*, **11**(4), 2003, 514–524.
[32] I. Stoica, R. Morris, D. Liben-Nowell, D. Karger, M.F. Kaashoek, F. Dabek, and H. Balakrishnan, Chord: a scalable peer-to-peer lookup service for internet applications, *IEEE Transactions on Networking*, **11**(1), 2003, 17–31.
[33] D.J. Watts and S.H. Strogatz, Collective dynamics of "Small World" networks, *Nature*, No. 393, 1998, 440–442.
[34] J. Xu, A. Kumar, and X. Yu, On the fundamental tradeoffs between routing table size and network diameter in peer-to-peer networks, *IEEE Journal on Selected Areas in Communications*, **22**(1), 2004, 151–163.
[35] B.Y. Zhao, L. Huang, J. Stribling, S. Rhea, A. Joseph, and J. Kubiatowicz, Tapestry: a resilient global-scale overlay for service deployment, *IEEE Journal on Selected Areas in Communications*, **22**(1), 2004, 41–53.
[36] B.Y. Zhao, J.D. Kubiatowicz, and A.D. Joseph, *Tapestry: An Infrastructure for Fault-Resilient Wide-Area Location and Routing*, Technical Report UC Berkeley, CSD-01-1141, University of California at Berkeley, Berkeley, CA, 2001.

Index